Macromolecular Chemistry

Volume 3

A Specialist Periodical Report

Macromolecular Chemistry
Volume 3

A Review of the Literature Published during 1981 and 1982

Senior Reporters

A. D. Jenkins, *School of Molecular Sciences, University of Sussex*

J. F. Kennedy, *Department of Chemistry, University of Birmingham*

Reporters

G. Akay *Cranfield Institute of Technology*
J. C. Bevington *University of Lancaster*
D. C. Blackley *Polytechnic of North London*
C. B. Bucknall *Cranfield Institute of Technology*
J. M. G. Cowie *University of Stirling*
D. A. Crombie *The Polytechnic, Huddersfield*
J. R. Ebdon *University of Lancaster*
A. Fawcett *Queen's University, Belfast*
J. Ferguson *University of Strathclyde*
J. P. Goddard *University of Glasgow*
J. N. Hay *University of Birmingham*
J. S. Higgins *Imperial College, London*
J. W. Kennedy *K.-M. Research Group, Dedham, and Pace University, New York*
J. T. Knowler *University of Glasgow*
J. R. McCallum *University of St. Andrews*
D. G. Older *QMC Industrial Research Ltd., London*
M. I. Page *The Polytechnic, Huddersfield*
K. L. Petrak *Kodak Ltd., Harrow*

M. D. Purbrick *Kodak Ltd., Harrow*
S. M. Richardson *Imperial College, London*
K. M. Roch *Polytechnic of the South Bank, London*
J. M. Rooney *Loctite (Ireland) Ltd., Tallaght, Ireland*
S. B. Ross-Murphy *Unilever Research, Sharnbrook*
D. C. Sherrington *University of Strathclyde*
D. J. Sparrow *ICI Ltd., Organics Division, Blackley*
R. J. Sturgeon *Heriot-Watt University, Edinburgh*
P. J. T. Tait *UMIST, Manchester*
B. M. Tidswell *University of Bradford*
B. J. Tighe *University of Aston in Birmingham*
I. G. Walton *ICI Ltd., Pharmaceuticals Division, Macclesfield*
W. W. Wright *Royal Aircraft Establishment, Farnborough*
R. N. Young *University of Sheffield*

The Royal Society of Chemistry
Burlington House, London W1V 0BN

ISBN 0-85186-876-2
ISSN 0144-2988

Copyright © 1984
The Royal Society of Chemistry

All Rights Reserved
No part of this book may be reproduced or transmitted
in any form or by any means – graphic, electronic,
including photocopying, recording, taping or
information storage and retrieval systems – without
written permission from The Royal Society of Chemistry

Printed in Great Britain by J. W. Arrowsmith Ltd., Bristol

Preface

This Report, the third in the Series, is the logical successor to the previous volumes and has again been planned to give those interested in macromolecular chemistry a useful and easily-read entry to current macromolecular chemistry research. Whereas the vast amount of original research literature available prevents this work from being completely comprehensive, the degree of selection achieved ensures that as many areas as possible are adequately represented. The Report covers the new Literature available to us from January 1981 to December 1982.

The initiation of this series came principally from the Committee of Macro Group U.K., a joint subject group of the Royal Society of Chemistry and the Society of Chemical Industry, under the Chairmanship of Professor C. H. Bamford, F.R.S. Once again we are indebted to the Committee, now under the Chairmanship of Dr. D. H. Richards, for their encouragement and advice. We are also grateful to the Reporters, many of them operating for the third time, for their cooperation and dedication. It is our pleasure, therefore, to present to you the team of Reporters, drawn from a wide cross-section of research and application in macromolecular chemistry and technology, all of whom are known experts in their particular fields.

Finally, it is a pleasure to acknowledge the invaluable services of Dr. P. G. Gardam (Manager, Books) and his staff, particularly Mr. P. W. Shallis, at the Royal Society of Chemistry, in the production of this Report.

June 1984   A.D.J.
                                                                                                        J.F.K.

Contents

Chapter 1 Introduction   1
        By A. D. Jenkins and J. F. Kennedy

Chapter 2 Chain Reaction Polymerization

Part I: Co-ordination Complex Polymerization   3
By P. J. T. Tait

    1 Introduction   3

    2 Catalyst Systems   3
        Heterogeneous Ziegler–Natta Catalysts   3
            Magnesium Chloride Supported Systems   3
            Catalysts Derived from the Reduction of Titanium Tetrachloride by Alkyl Magnesium Compounds   18
            Metal Oxide Supported Catalysts   19
            One-component Supported Catalyst Systems   21
        Soluble Catalysts   22
            'Living' Co-ordination Polymerization Systems Derived from Vanadium Triacetylacetonate   22
            Soluble Vanadium Tetrachloride–Alkylaluminium Halide Catalyst Systems   24
            Soluble Dichlorobis(γ-cyclopentadienyl)titanium–Alkylaluminium Halide Catalysts   25
            Soluble Titanium and Zirconium–Aluminoxane Catalysts   26

    3 Copolymerization Studies   28

Part II: Cationic Polymerization   30
By J. M. Rooney

    1 Introduction   30

    2 Initiation Systems   30
        Metal Alkyl and Metal Halide Compounds   30
        Organic Cation Salts   30
        Photochemical and Radiation Initiation   31

    3 Propagating Species and Their Reactivities   33
        Styrene and Styrene Derivatives   33
        Hydrocarbon Monomers   34

Vinyl Ethers and Vinyl Carbazole		35
Heterocyclic Monomers		36
4 Synthetic Studies		37
Polymers from Novel Monomers		37
Block Copolymerization		38
Graft Polymerizations		39

Part III: Anionic Polymerization 40
By R. N. Young

Part IV: Radical Polymerization 49
By J. C. Bevington

1 Introduction 49

2 Initiation 50

3 Growth Reactions 53

4 Transfer Processes 55

5 Termination 56

6 Retardation and Inhibition 57

Part V: Emulsion Polymerization 59
By D. C. Blackley

1 Books and Reviews 59

2 Particle Nucleation 60

3 Kinetics and Mechanism 61

4 Initiators 63

5 Surfactants 64

6 Low-surfactant and Surfactant-free Emulsion Polymerizations 66

7 Emulsion Polymerization Reactions involving Functional-group Co-monomers 68

8 Emulsion Copolymerization 68

9 Procedural and Chemical Engineering Aspects 69

10 Miscellaneous Other Aspects 70

Part VI: Electrochemical Polymerization 72
By B. M. Tidswell

Chapter 3 Step Growth Polymerization

Part I: Polyesters, Polycarbonates, Polyamides, and Polyimides 76
By J. Ferguson

    1 Introduction 76

    2 Polyesters 76
        Synthesis 76
        Properties 79
        Morphology 80
        Blends 82

    3 Polycarbonates 82
        Synthesis 82
        Properties 83
        Morphology 84
        Blends 84

    4 Polyamides 85
        Synthesis 85
        Properties 87
        Morphology 88

    5 Polyimides 89
        Synthesis 89
        Properties 90
        Morphology 91

Part II: Developments in Polyurethanes 93
By D. J. Sparrow and I. G. Walton

    1 Introduction 93

    2 Isocyanate Products and Processes 93

    3 Polyurethane Polyols 94

    4 Catalysis and Mechanism of Isocyanate Reactions 95

    5 Rigid Foams 95

    6 Reaction Injection Moulding 96

    7 Flexible Foams 97

    8 Miscellaneous 97

Chapter 4 Natural Polymers: Polysaccharides and Glycoproteins 98
By R. J. Sturgeon

1 Introduction 98

2 Plant and Algal Polysaccharides 99
  Starch 99
  Cellulose 102
  Hemicelluloses 103
  Pectins 105
  Algal Polysaccharides 106

3 Lectins 109

4 Microbial Polysaccharides 116

5 Animal Glycoproteins and Glycosaminoglycans 121

Chapter 5 Natural Polymers: Nucleic Acids 133
By J. T. Knowler and J. P. Goddard

1 Advances in Methodology 133
  Chemical Synthesis of Oligonucleotides and its Applications 133
  Site-directed Mutagenesis 135

2 Oncogenes 136

3 The Control of Gene Expression in Eukaryotes 139
  The Influence of Chromatin Structure on Gene Expressions 139
  Base Modification of DNA and the Control of Gene Expression 141
  The Role of Specific Nucleotide Sequences in the Control of Gene Expression 142
  The Processing of Gene Transcripts in Eukaryotes 143

Chapter 6 Inorganic Polymers 147
By K. M. Roch

1 Introduction 147

2 Poly(organosiloxane)s 147
  Polymerization 147
  Molecular Structure 148
  Thermal/Oxidative Stability 150
  Applications 151

3 Organosilicon Backbone Copolymers 152
  Carbonate–Siloxane Copolymers 153

	Styrene–Siloxane Copolymers	153
	Silphenylene–Siloxane Copolymers	153
	Other Organic–Siloxane Copolymers	154
4	Poly(organophosphazene)s	154
	Synthesis	154
	Molecular Structure	156
	Properties	157

Chapter 7 Configurations 159
By S. B. Ross-Murphy

1	Introduction	159
2	Statistics of Model Polymers	159
	Unperturbed Dimensions	159
	Excluded Volume Effects	161
	Rods and Helices	166
3	Dynamics of Dilute Solutions	168
4	Effect of Increasing Concentration	172
	Dilute and Semi-dilute Solutions	172
	Molecular Rheology of More Concentrated Solutions	173

Chapter 8 The Chemical Microstructure of Synthetic Polymers Investigated by High Resolution Nuclear Magnetic Resonance 175
By J. R. Ebdon

1	Introduction	175
2	Books and Reviews	175
3	Homopolymers: Tacticity	176
4	Homopolymers: Geometrical Isomerism	178
5	Mechanisms of Polymerization	180
6	End Groups	181
7	Branching	182
8	Copolymers	182
9	Polycondensates and Uncured Resins	184
10	Chemical Modification of Polymers	186
11	Solid Polymers	186

Chapter 9 Neutron Scattering Studies   189
   By J. S. Higgins

   1 Introduction   189
   2 Amorphous Polymers   191
   3 Crystalline Polymers   193
   4 Solutions and Networks   195
   5 Charged Molecules   197
   6 Multiphase Systems   198
   7 Studies of Polymer Dynamics   201

Chapter 10 Polymer Crystallization   204
   By J. N. Hay

   1 Introduction   204
   2 The Degree of Crystallinity   204
   3 Crystallographic Analysis   206
   4 Morphology   209
      Solution Crystallization   209
      Melt Crystallization   211
      Oriented Crystallization   213
   5 Rate Measurements   214
      Spherulitic and Single Crystals Growth Rate   214
      Nucleation   215
   6 Bulk Crystallization Kinetics   216
      Isothermal   216
      Non-isothermal   219
      Non-isotropic   219
   7 Melting   220
   8 Mechanical Properties   221
   9 Conclusions   223

Chapter 11 Characterization of Synthetic Polymers   224
   By J. M. G. Cowie

   1 Introduction   224
   2 Molar Mass Measurements   224

3 Dilute Solutions	226
General Characterization	226
Aqueous Systems	227
Conformational Studies	227
Unperturbed Dimensions	233
4 Ultracentrifugation	234
5 Diffusion	235
6 Light Scattering	236
7 Osmotic Pressure	237
8 Viscosity	237
9 Prediction of Hydrodynamic Parameters	238
10 Polymers with Rigid Chains	239
11 Cyclic Structures	240
12 Chromatographic Methods	241
Calibration Methods	241
Columns and Techniques	242
Data Analysis and Detectors	244
Non-aqueous Systems	245
Aqueous Systems	246
Branching	246
Miscellaneous	247

Chapter 12 Thermodynamics of Solutions and Mixtures 248
By J. W. Kennedy

1 Introduction	248
Scope of Report	248
Importance and General Overview	248
2 Theoretical Aspects	249
Introduction	249
Phase Behaviour	249
Multiphase Separation	250
Phase Interfaces	250
Phase Equilibria in Gel Polymer Systems	250
Flory–Huggins (Lattice-Graph) Models	251
Equation-of-state Models	252
Perturbed Hard Chain Theory	252
Huggins Molecular Models	253
Renormalization Group and Scaling Models	253
Models for Stiff-chain Polymers	255

Dilute Solution Models (and Intermediate Concentrations)	256
Bridging Models (and Intermediate Concentrations)	256
Pressure Dependence	257
Molecular Weight/Composition Dependence and Fractionation	257
Solubility Parameter Theory	258

3 Methods for Obtaining Thermodynamic Data — 258

Introduction	258
Calorimetric Methods	259
Centrifugal Methods	259
Gas–Liquid Chromatography	259
Scattering Methods	259
Turbidimetry	259
Light Scattering	260
Pulse-induced Critical Scattering	261
X-Ray Scattering	261
Excimer Fluorescence Spectroscopy	261
Viscometric Methods	262
Dilute Solution Viscosity	262
Rotary Viscometers	262
Glass Transition and Related Relaxation Measurements	263
Fractionation Methods	263
Batch Fractionation	263
Column Fractionation	263
Gel Permeation Chromatography	264
Thin Layer Chromatography	264
Miscellaneous Other Methods	264
Vapour Pressure	264
Gel Swelling	264
Potentiometric Titrations	264

4 Polymer Solutions and Mixtures — 264

Introduction	264
Polymer Solutions	265
Ionic Polymer Solutions	266
Copolymers and Their Solutions	266
Random Copolymers and Their Solutions	266
Block Copolymers and Their Solutions	266
Polymer Mixtures and Their Solutions	267
Recent Data for Specific Systems	268

Chapter 13 Engineering and Technology

Part I: Rheology *By S. M. Richardson*	279
1 Introduction	279
2 Constitutive Equations	279
3 Elongational Flows	280
4 Shear Flows	281
5 Converging Flows	281
6 Numerical Solution of Viscoelastic Flow Problems	282
7 Polymer Processing	283
Part II: Engineering and Technology *By G. Akay and C. B. Bucknall*	286
1 Introduction	286
2 Mechanical Properties	286
Crazing	286
Deformation	287
Fracture	288
Fracture Mechanics	288
Composites	289
Fracture-resistant Materials	290
3 Rheology	290
Constitutive Equations	290
Fluids with Orientable Microstructure	291
Flow-induced Phenomena	292
Microstructure Redistribution	292
Microstructure Orientation	293
Rheology of Two-phase Materials	293
Structure Development during Processing	294
Mathematical Modelling	295
4 Electronic Properties	296
Ferroelectrics	296
Conductive Polymers	296

Part III: Electrical Properties 297
By D. G. Older

   1 Introduction 297

   2 Electrical Insulation 297

   3 Electrically Conducting Fillers in Polymers 298

   4 Electrically Conducting Polymers 299

   5 Solid State Cells 301

   6 Piezoelectricity in Polymers 302

Chapter 14 Reactions on Polymers 303
By D. C. Sherrington

   1 Introduction 303

   2 Chemical Modification of Polymers using Phase Transfer Catalysts 304
      Nucleophilic Displacements on Polymers 304
      Nucleophilic Polymers 308
      Use of Wittig Reactions 309
      Other PTC Modifications 309

   3 Phase Transfer Catalysed Polycondensations 310

   4 Polymer-supported Phase Transfer Catalysts 312

   5 Polymeric Acids and Bases 315

   6 Polymeric Complexes of Transition Metals 318

   7 Stoicheiometric Reactions on Polymers 320
      Systems Designed for Recycling 320
      Chiral Systems 324

   8 Polymeric Species in Photochemical Processes 326

   9 Polymers as Selective Sorbents 327

  10 Miscellaneous 329

Chapter 15 Polymer Degradation

Part I: Photo and Photo-oxidative Degradation 331
By J. R. McCallum

   1 Introduction 331

   2 General 331

3	Polyethylene	332
4	Polypropylene	332
5	Polybutadiene	333
6	Polystyrene	334
7	Poly(vinyl chloride)	335
8	Polymethacrylates and Polyketones	336
9	Polyamides	336
10	Polyesters	337
11	Polyurethanes	337
12	Poly(2,6-dimethyl-1,4-phenylene oxide)	337
13	Miscellaneous	337

Part II: Thermal and Thermo-oxidative Degradation 339
By W. W. Wright

1	Introduction	339
2	General	339
3	Polyolefins	340
4	Polydienes	341
5	Polystyrene	341
6	Poly(vinyl chloride)	342
7	Fluorine-containing Polymers	344
8	Polyacrylates	344
9	Polyacrylonitrile	345
10	Other Addition Polymers	345
11	Polyamides	346
12	Polyesters	347
13	Polyurethanes	347
14	Polyphenylene-type Polymers	347
15	Polyimides	348
16	Other Heterocyclic Polymers	348
17	Other Condensation Polymers	349

18 Silicon-containing Polymers	349
19 Phosphorus-containing Polymers	350
20 Cellulose	350

Chapter 16 Reactions in Macromolecular Systems — 351
By M. I. Page and D. A. Crombie

1 Introduction	351
2 Cyclomalto-oligosaccharides (Cyclodextrins)	351
3 Crown Ethers and Cryptands	355
Cyclophanes	359
4 Synthetic Polymers	360
Poly(ethylenimine)s	360
Poly(4-vinylpyridine) Derivatives	362
Poly(vinylimidazole)s	362
Poly(amino acids)	363
Other Polymers	363
Template Polymerization	364
Polymers as Supports and Protecting Groups	364
5 Micelles	366
6 Phase-transfer Catalysis	372
7 Ionophores	374

Chapter 17 Biomedical Applications of Polymers — 375
By B. J. Tighe

1 Introduction	375
2 Biocompatibility Studies	375
3 Applications	378
Soft Tissue Prosthesis	378
Joint Prostheses	379
Ducts and Canals	379
Vascular Prosthesis	380
Tendons and Ligaments	381
Sutures and Adhesives	381
Ophthalmic Applications	382
Artificial Skin and Wound Dressings	382
Artificial Organs	383

Polymeric Drugs and Drug Delivery Systems	384
Enzyme-related Applications	385
4 Synthetic Work and the Development of New Materials	386

Chapter 18 Computer Applications 387
By A. H. Fawcett

1 Polymer Kinetics	387
2 Polymer Characterization and Spectroscopy	389
3 Intermolecular Potentials and Force Fields	391
4 Rotational Isomeric State Calculations	394
5 Monte Carlo and Molecular Dynamic Simulations	396

Chapter 19 Selected Topics in the Photochemistry of Polymers 399
By K. L. Petrak and M. D. Purbrick

1 Introduction	399
2 Energy Transfer and Related Topics	399
3 Photoconductivity	403
4 Polymeric Resist Materials	405
Photoresists	406
Electron-beam Resists	409
Ion-beam Resists	411
X-Ray Resists	411
Plasma-developed Resists	412

Author Index 414

1
Introduction

BY A. D. JENKINS AND J. F. KENNEDY

It is in the nature of macromolecules that it is hard to define any of their properties in precise terms, indeed it is no easy matter to define 'macromolecule' in a way that would be universally accepted even by those intimately concerned with such materials. Consequently, in setting out to delineate the areas to be covered in a report of recent research in the field of macromolecules, many demarcation issues have again had to be faced and resolved.

In aiming at a list of contents for this volume, the Senior Reporters have had to pay attention not only to the subject matter itself but also to the existence of other series of Specialist Periodical Reports dealing with contiguous areas of chemistry. Without the least desire to poach on other people's preserves, some small degree of overlap seems to be the only reasonable solution. This occurs, particularly, in dealing with fields such as colloids, carbohydrates, proteins, and nucleic acids. It has seemed reasonable in the interests of providing a comprehensive treatment of the range of macromolecules to feel justified in including modest discussion in the context of 'macromolecules' rather than simply referring the reader to the individual volumes in which the information he requires may be buried within a large bulk of (to him or her) irrelevant material.

Another problem concerns the frequency with which individual topics should be examined. The more global subjects, like polymerization chemistry, will no doubt be treated in each issue, but smaller topics, for example, specific techniques for characterization, may adequately be dealt with if they are reviewed at intervals. Of course, much depends on whether a particular topic is advancing rapidly, in which case we recognize that there is an obligation to bring the reader as up to date as possible. In the production of this volume and Series, the satisfactory resolutions of other factors have also to be superimposed, in particular, the trend of polymer chemistry, since this is now the third volume. In many instances, Reporters have had to sift through thousands of references in order to identify the major works of progress under the heading of any area of macromolecular chemistry.

It will be apparent from the foregoing paragraphs that the list of chapter headings cannot be a constant factor although, as a general principle, we maintain a watching brief over the following broad areas: Polymerization Chemistry; Particular Classes of Polymers; Natural Polymers; Degradation; Polymers as Catalysts and Reagents; Properties of Solid Polymers; Crystalline and Amorphous Polymers; Properties of Polymer Solutions; Characterization Techniques; Theoretical Treatment of Polymers; Applications of Polymers; Polymer Engineering.

Inclusion of the last two areas states our intention to embrace technology as well as pure science. Macromolecules, in the shape of synthetic plastics, fibres, films, paints, adhesives and the like, make an enormous contribution to everyday life; they occupy a large slice of the chemical industry in preparation and processing operations and the borderline between polymer science and plastics technology is very diffuse. It is fully in accordance with the attitude of many people in the field, and certainly of the Science and Engineering Research Council, that one should as far as possible integrate the more academic and the more practical aspects of research on polymers, and that is the stand adopted here. However, with increasing industrial interest in and application of polymer technology and biotechnology, the whole field of reporting on polymers, particularly biopolymers, becomes more and more open-ended.

In keeping with the original plan of biennial frequency of publication, the literature survey represented in this volume is principally concerned with the years 1981 and 1982. Work earlier than 1977, the initial year of coverage of the Series, was cited only in Volume 1 where it provided an important basis for current papers and an introduction of certain phenomena into the Series.

Each chapter opens with an introduction which is specialized with respect to the contents of the chapter and others outline the context of the chapter particularly for those not completely familiar with the subject treated. Reference to the corresponding chapter in Volume 1 will also be helpful in this respect.

Our Reporters have been asked to collate rather than to criticize but they have not been debarred from offering a personal opinion on points of particular interest. It is our hope that the reader will find this book a useful guide to the most important recent literature on the chemistry of macromolecules.

2
Chain Reaction Polymerization

BY P. J. T. TAIT, J. M. ROONEY, R. N. YOUNG,
J. C. BEVINGTON, D. C. BLACKLEY, AND B. M. TIDSWELL

PART I Co-ordination Complex Polymerization
by P. J. T. Tait

1 Introduction

The excitement associated with new discoveries has been fully justified in the field of co-ordination polymerization during recent years, and as a result the increase in the number of publications and patent citations forbids any attempt to present a comprehensive survey of the field. Nevertheless these last four or five years have been very important in the history of co-ordination polymerization, not only because of the discovery of new catalyst systems, but because of the valuable insights gained into some of the older problems associated with this field. In particular, many of the basic chemical and physical principles involved in catalyst preparation and chain initiation have become more apparent. The presence of a new type of polyethylene prepared by copolymerization of ethylene and a suitable higher α-olefin, and simulating some of the properties of low-density polyethylene, has become recognized, and has highlighted the importance of catalysis to a commercial world that is becoming increasingly aware of energy costs.

2 Catalyst Systems

Heterogeneous Ziegler–Natta Catalysts.—*Magnesium Chloride Supported Systems.* In the field of propylene polymerization considerable activity has resulted from discoveries by Montecatini Edison Co.[1] and Mitsui Petrochemicals Ind.[2] that catalysts prepared from magnesium chloride, titanium tetrachloride, and electron donors, and activated by a mixture of trialkylaluminium and an electron donor, could polymerize propylene with a high yield (> 50 kg polypropylene per g Ti per h) and with a good stereospecificity (isotactic index > 90%). The relevant patents describe two basic routes for the preparation of highly active catalysts. Typically, a 30-fold molar excess of dried anhydrous $MgCl_2$ is ball-milled with a $TiCl_4$–electron donor complex, such as $TiCl_4$-ethyl benzoate, washed with n-heptane, and dried. The catalyst is then activated by a mixture of trialkylaluminium (TAA) and an electron donor such that TAA:EB:Ti ≈ 300:100:1. Alternatively, dried anhydrous $MgCl_2$ is ball-milled for 20 h at 0—5 °C with ethyl benzoate (EB) ($MgCl_2$:EB ≈ 1:0.15) and then treated with neat $TiCl_4$ at 80—130 °C for 2 h, washed with n-heptane, and then dried to yield a pale yellow solid catalyst containing typically

---

[1] Montecatini Edison Co., Br. P. 1 286 867, 1968.
[2] Mitsui Petrochemicals Ind., Ital. P. 912 345, 1968.

1—5% Ti and 5—20% EB. This catalyst is then activated by treatment with a mixture of trialkylaluminium and an electron donor (*e.g.*, ethyl benzoate, *p*-ethyl anisate, *p*-methyl toluate, *etc.*).

(a) *Effects of ball-milling magnesium chloride.* 'Activated' magnesium chloride was first described by Bryce-Smith[3] and later detailed by Kamienski.[4] 'Activated' magnesium chloride may be prepared by (*i*) treating $MgCl_2$ with activating agents such as electron donors, (*ii*) ball-milling anhydrous $MgCl_2$, and (*iii*) reaction of Grignard reagents with chlorinating agents.

Crystalline $MgCl_2$ exists in the form of agglomerates of small primary crystallites and when ball-milled in the presence of a donor such as ethyl benzoate these agglomerates are broken down. The small $MgCl_2$ crystallites produced are believed to be stabilized by the ethyl benzoate which is adsorbed onto the freshly cleaved surfaces, thus largely preventing reaggregation of the primary crystallites (*cf.*, protected colloids).

The structure of the usual crystalline form of $MgCl_2$ is quite similar to that of γ-$TiCl_3$ and may be represented in terms of a cubic close-packed structure of double chlorine layers with interstitial Mg^{2+} ions in 6-fold co-ordination.[5] This layer structure is of the $CdCl_2$ type and leads to a characteristic *X*-ray diffraction spectrum with a strong (104) reflection at $d = 2.56$ Å.[6] Another less stable crystalline form of $MgCl_2$ is also known,[7] and has a hexagonal close-packed structure similar to that of α-$TiCl_3$, and gives a strong (104) *X*-ray reflection at $d = 2.78$ Å.

Youchang *et al.*,[8] from their *X*-ray diffraction studies on samples of milled $MgCl_2$, conclude that during milling the shear of the steel (or porcelain) balls causes the Cl–Mg–Cl double layers to slide over each other, leading to cleavage along the double layers, producing hexagonal $MgCl_2$ microcrystallites of only a few layers in thickness. Figure 1 shows representations of both commercial and ball-milled $MgCl_2$ crystallites.[9]

During ball-milling there is a gradual disappearance of the (104) reflection with the appearance of a broad halo indicating stacking faults.[6] Thus the stacking sequence no longer corresponds to that of a cubic close-packed structure. The spacing of the broadened halo ($d = 2.65$ Å) lies between those observed for the cubic and hexagonal structures. Galli *et al.*[9] have concluded that the model describing the structural disorder of activated $MgCl_2$ (in the presence of $TiCl_4$) is similar to that proposed by Allegra and Bassi[10] for δ-$TiCl_3$, although differing in mathematical details.[11,12] They report that some features of the *X*-ray powder pattern of activated $MgCl_2$ such as the shifting and broadening of the (104) peak, in comparison with that of the α-form, denotes the presence of rotational disorder in the stacking of the

---

[3] G.D.R. P. 955 807, 1959.
[4] C. W. Kamienski, 'Synthesis and properties of diorganomagnesium compounds', Ph.D. thesis, University of Tennessee, 1967.
[5] G. Bruni and A. Ferrari, *Rend. Accad. Naz. Lincei*, 1925, **2**, 457.
[6] U. Giannini, *Makromol. Chem. Suppl.*, 1981, **5**, 216.
[7] I. W. Bassi (unpublished results, *cf.* ref. 6).
[8] X. Youchang, G. Linlin, L. Wanqi, B. Naiyu, and T. Yougi, *Sci. Sin.*, 1979, **22**, 1045.
[9] P. Galli, P. Barbe, G. Guidetti, R. Zannetti, A. Martorana, A. Marigo, M. Bergozza, and A. Fichera, *Eur. Polym. J.*, 1983, **19**, 19.
[10] G. Allegra and I. W. Bassi, *Gazz. Chim. Ital.*, 1980, **110**, 437.
[11] G. Guidetti, R. Zannetti, D. Ajox, A. Marigo, and M. Vidali, *Eur. Polym. J.*, 1980, **16**, 1007.
[12] A. Martorana, R. Zannetti, G. Granozzi, and D. Ajo, *Inorg. Chim. Acta*, 1980, **41**, 207.

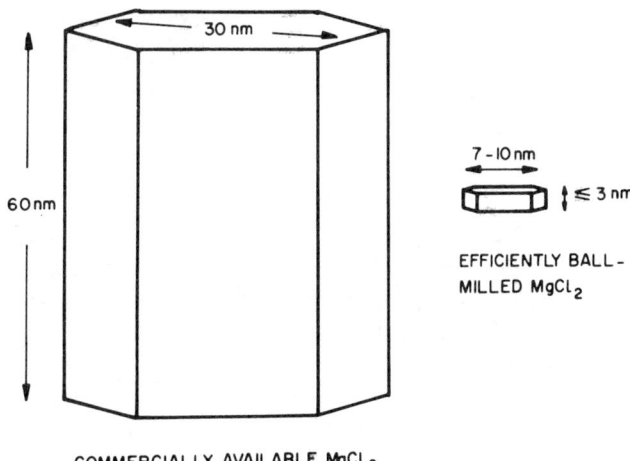

Figure 1 *Representations of commercial and ball-milled magnesium chloride crystallites*

Cl–Mg–Cl triple layers, and have proposed an appropriate model. Giannini[6] has pointed out that, unlike the atoms in the bulk phase, the cations located on the lateral surfaces and the crystal edges are co-ordinatively unsaturated, and can therefore form bonds with adsorbed molecules. Goodall[13] has distinguished three different types of surface Mg ions:

(i) *type 1* on the lateral faces; single vacancy Mg ions having an effective charge of 0 e;

(ii) *type 2* on the corners: single vacancy Mg ions having an effective charge of $-\frac{2}{3}$ e;

(iii) *type 3* on the corners: double vacancy Mg ions having an effective charge of $+\frac{1}{3}$ e.

The different types of surface Mg ions are shown in Figure 2.

X-Ray diffraction has also been used by Keszler, Bodor, and Simon[14] to study the effects of ball-milling on particle size. Crystal sizes of unground $MgCl_2$ in directions perpendicular to the (001) and (110) planes were found to be different, 111 and 65 nm, respectively. This difference was found to disappear after short grinding periods. There was also a marked reduction in the overall crystalline particle size. In a later paper, Keszler and Simon[15] have reported on catalyst activity as a function of the time of grinding for both $MgCl_2/TiCl_4$ and $MgCl_2/EB/TiCl_4$ catalysts. In the absence of ethyl benzoate the optimum results were obtained for 20 h grinding, whereas for $MgCl_2/EB$ supports the activity of the catalyst increased continuously as a function of grinding time up to 100 h. Recently, Galli et al.[9] have concluded that the activation process for grinding at 35 °C ends after milling for about 70 h and

---

[13] B. L. Goodall in 'Transition Metal Catalysed Polymerizations, Alkenes and Dienes, Part A', ed. R. P. Quirk *et al.*, Harwood Academic, New York, 1983, p. 355.

[14] B. Keszler, G. Bodor, and A. Simon, *Polymer*, 1980, **21**, 1037.

[15] B. Keszler and A. Simon, *Polymer*, 1982, **23**, 916.

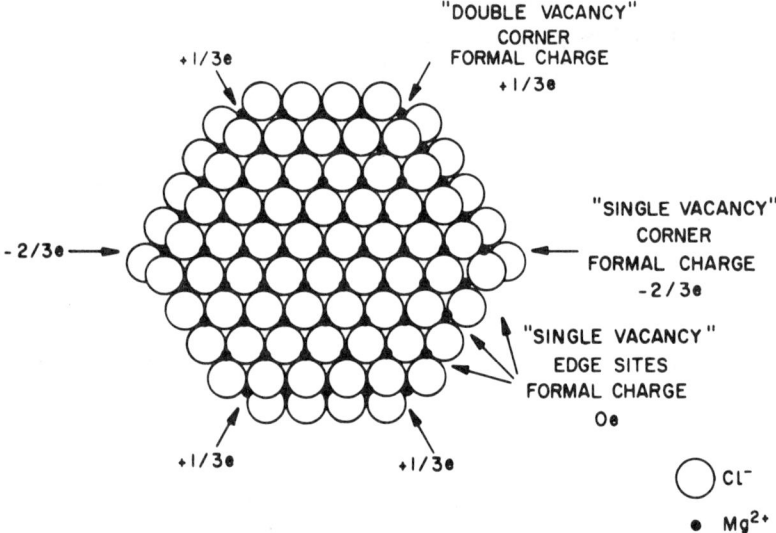

Figure 2 *Types of surface magnesium ions*

Table 1 *Results of the analyses on $MgCl_2$ + $TiCl_4$ co-milled samples*

Sample No.*	Milling time/h	D_{11+}† Å	Surface area by BET/ $m^2 g^{-1}$	Ti content after washing/ wt %	Activity (kg polyethylene/g Ti)	
					Before washing	After washing
1	20	105	114.5	2.7	170	215
2	30	79	117.5	2.85	215	260
3	40	72	119.7	2.95	230	265
4	70	57	97.5	3.15	270	290
5	100	59	92.6	3.10	255	275
6	140	54	86.1	3.10	260	280
7	180	52	80.0	3.15	230	245

* Co-milling $MgCl_2$ + $TiCl_4$, Ti content = 3.4%
† Calculated by Scherrer's formula

then decreases with further milling in contrast to reports by earlier workers.[8,16] The effects of milling on the crystallite size were also investigated and are shown in Table 1. As can be seen, the lowering of the activity (referred to unit Ti) does not correspond to any significant variation in the residual Ti or to the crystallite size, but to a reduction in the surface area.

(b) *Interaction of magnesium chloride with ethyl benzoate.* The interaction between $MgCl_2$ and ethyl benzoate has been extensively studied. Keszler *et al.*[15] from their

---

[16] N. Kashiwa, *Polym. J.*, 1980, **12**, 603.

thermal studies have concluded that the interaction involves a two-step exothermic process. Initially there is rapid adsorption of ethyl benzoate onto the surface of the $MgCl_2$. This adsorption is then followed by the much slower process of complex formation.

Infrared studies have also been used to study the interaction between $MgCl_2$ and ethyl benzoate.[13,16–18] Goodall[13] has concluded that ethyl benzoate is co-ordinated to the $MgCl_2$ crystalline surface *via* the C=O group. The C=O stretching frequency changes from 1721 cm^{-1} in the free ester to 1680 cm^{-1} in the adsorbed ester (*cf.*, the corresponding values of 1720 cm^{-1} and 1650 cm^{-1} reported by Kashiwa[18]), and is believed to arise from ester molecules bound to the basal edges and corners of the crystallites. The adsorption band observed at 1700 cm^{-1} is believed to arise from ethyl benzoate co-ordinated to lateral faces. Chien *et al.*[19] have concluded that the complexes (1) and (2) were responsible for the shifts observed in the infrared spectrum for both the C=O and C–O stretching frequencies.

$$Mg \cdots O=C \begin{array}{c} O{-}Et \\ Ph \end{array} \qquad Mg \cdots O=C \begin{array}{c} O(Et){\cdots}Mg \\ Ph \end{array}$$

(1) (2)

(*c*) *Treatment of ball-milled $MgCl_2$–ethyl benzoate with titanium tetrachloride.* The second step of the catalyst preparation involves treatment of the solid products from ball-milling with hot neat $TiCl_4$ at 80—130 °C, or by ball-milling the solid products with $TiCl_4$. The resulting solid is then thoroughly washed and dried. During treatment with $TiCl_4$, ethyl benzoate is extracted by the $TiCl_4$ so that typical catalysts contain only 5—20% ethyl benzoate after treatment with $TiCl_4$. At the same time $TiCl_4$ is adsorbed onto free $MgCl_2$ surface vacancies. Goodall[13] has reported that the ethyl benzoate is lost not only from the weakly bonding sites on the lateral faces (1700 cm^{-1} adsorption) but also from the strongly co-ordinating corner sites (1650 cm^{-1} absorption). This would seem to indicate that the $TiCl_4$ co-ordinates to these sites by replacement of the ethyl benzoate molecules.

The oxidation state of the adsorbed titanium has been studied by some workers. Baulin *et al.*[20] have established that all the titanium is in the Ti^{IV} state. Chien *et al.*[19] report 8% as Ti^{II}, 38% as Ti^{III}, and 54% as Ti^{IV}. It should be noted, however, that the catalyst studied by Chien *et al.* is more complex, having been treated with *p*-cresol and with $AlEt_3$ prior to reaction with $TiCl_4$.

In a significant series of investigations Chien and Wu[21] have examined the electron paramagnetic resonance spectra of a $MgCl_2$ supported catalyst. Treatment of the $MgCl_2$ support with HCl at 380—430 °C, with ethyl benzoate by ball-milling,

[17] B. Keszler, A. Grobler, E. Takacs, and A. Simon, *Polymer*, 1980, **22**, 818.
[18] N. Kashiwa in 'Transition Metal Catalysed Polymerizations, Alkenes and Dienes, Part A', ed. R. P. Quirk *et al.*, Harwood Academic, New York, 1983, p. 379.
[19] J. C. W. Chien, J.-C. Wu, and C.-I. Kuo, *J. Polym. Sci., Polym. Chem. Ed.*, 1982, **20**, 2019.
[20] A. A. Baulin, Y. I. Novikova, G. Y. Mal'kova, V. L. Maksimov, L. I. Vyshinskaya, and S. S. Ivanchev, *Polym. Sci. USSR*, 1980, **22**, 205.
[21] J. C. W. Chien and J.-C. Wu, *J. Polym. Sci., Polym. Chem. Ed.*, 1982, **20**, 2461.

with p-cresol at 50 °C, with AlEt$_3$, and finally with TiCl$_4$ at 100 °C, produced a catalyst containing a single e.p.r. observable Ti^{III} species which was strongly attached to the catalyst surface, being bonded to at least two, or even three, surface sites, e.g. (3). This Ti^{III} species is co-ordinatively unsaturated and the vacant co-

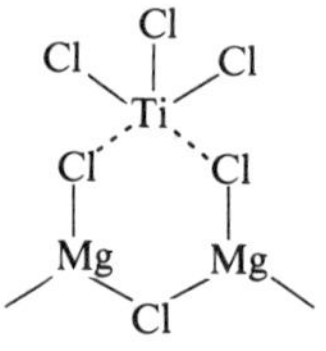

(3)

ordination position may be occupied by a weak ligand, or solvent molecule. It is unlikely that ethyl benzoate would be co-ordinated to the above species since it is known to be strongly complexed to the MgCl$_2$.

The e.p.r. observable Ti^{III} species, however, constituted only about 20% of the Ti^{III} present. It was concluded that the remainder of the Ti^{III} must have adjacent sites occupied by one or more Ti^{III} ions. Chien and Wu have concluded that the occurrence of chlorine bridge structures such as (4) and (5) are responsible for the e.p.r. silence.

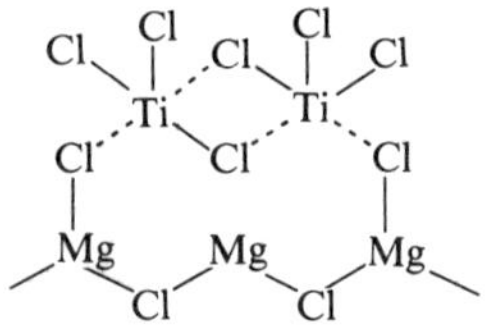

(4) 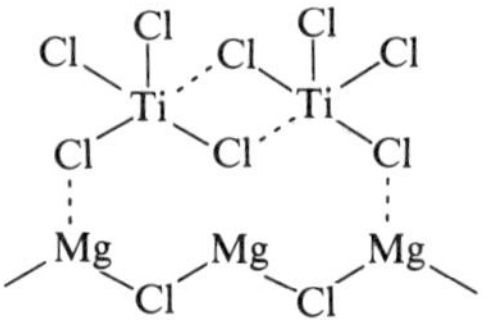   (5)

Baulin et al.[20] have also reported that the e.p.r. determination of Ti^{III} on the surface of carriers containing magnesium did not agree with the results of chemico-analytical determinations, since the values obtained did not record more than 5—12% of the total Ti^{III}. However, when an aluminosilicate carrier was used, up to 90% of the Ti^{III} detected by a chemico-analytical method could be recorded by e.p.r. Baulin et al.[20] have suggested that, in the case of magnesium containing carriers, the observed behaviour was caused either by (i) a significant proportion of the Ti^{III} existing in a dimer form and so not giving an e.p.r. signal or (ii) the e.p.r. signal being poorly recorded as a consequence of the very rapid spin–lattice relaxation of Ti^{III} ions.

(d) *The co-catalyst.* The co-catalyst employed in highly active MgCl$_2$ catalyst systems is invariably a trialkylaluminium compound, and in order to improve the stereospecificity of the polymerization an electron donor is also used.[1,2] In particular, carboxylic esters and tertiary amines have been used extensively.

Considerable interest has been focused on the role of electron donors in these polymerization systems. It is important to realize that when carboxylic esters are employed, various reactions are possible and these include[22-25] complex formation, alkylation, reduction, and elimination, leading to a multiplicity of possible products. The reactivity of esters with $AlEt_3$ is strongly dependent not only on the molar ratio of the components and temperature but also on concentrations and solvent. Thus ethyl anisate reacts with $AlEt_3$ as shown in Scheme 1.[13] Chien et al.[26]

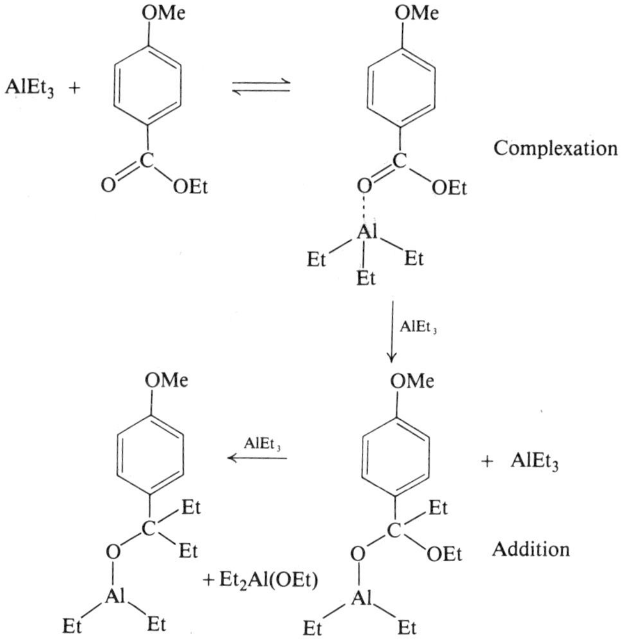

Scheme 1

have carried out a thorough investigation of the interactions between methyl p-toluate and $AlEt_3$ and ethyl benzoate and $AlEt_3$ and have formulated the reactions given below.

(i) *Complexation:*

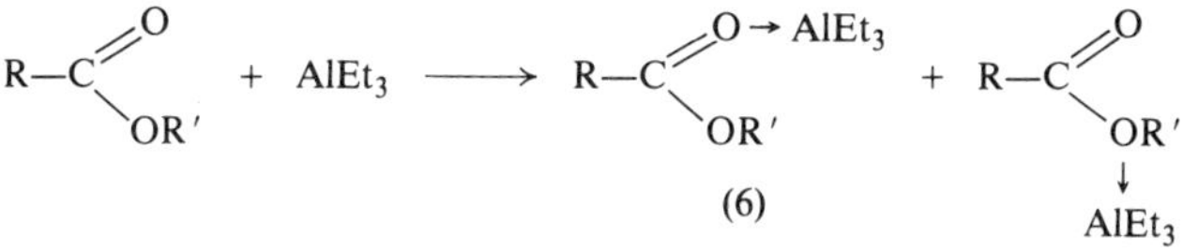

$$R-C(=O)(OR') + AlEt_3 \longrightarrow R-C(=O \to AlEt_3)(OR') \quad (6)$$

$$+ \quad R-C(=O)(OR' \downarrow AlEt_3) \quad (7)$$

---

[22] S. Pasynkiewicz and K. B. Starowieyski, *Rocz. Chem.*, 1967, **41**, 1139.
[23] S. Pasynkiewicz, L. Kozerski, and B. Grabowski, *J. Organomet. Chem.*, 1967, **8**, 233.
[24] K. B. Starowieyski, *J. Organomet. Chem.*, 1976, **117**, C1.
[25] Y. Baba, *Bull. Chem. Soc. Jpn.*, 1968, **41**, 1022.
[26] J. C. W. Chien and J.-C. Wu, *J. Polym. Sci., Polym. Chem. Ed.*, 1982, **20**, 2445.

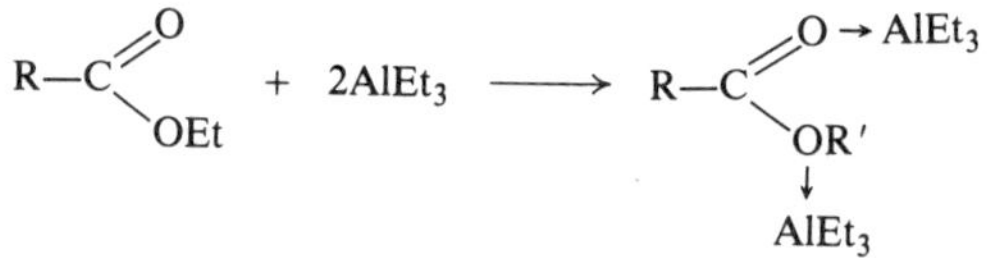

$$R-C{\overset{O}{\underset{OEt}{\diagup}}} + 2AlEt_3 \longrightarrow R-C{\overset{O \rightarrow AlEt_3}{\underset{OR'}{\diagup}}}$$
$$\downarrow$$
$$AlEt_3$$
(8)

(ii) *Alkylation to form ketones*

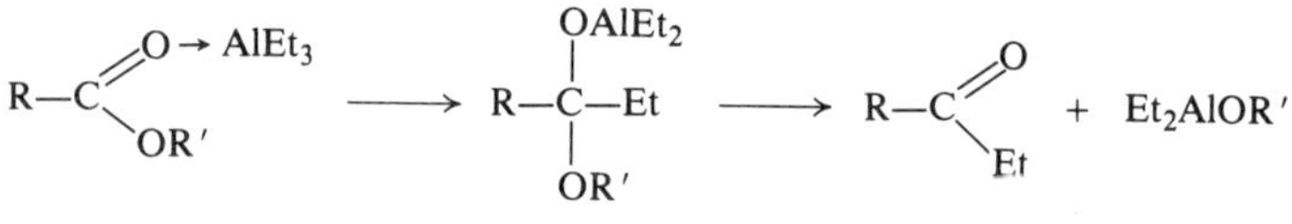

$$R-C{\overset{O \rightarrow AlEt_3}{\underset{OR'}{\diagup}}} \longrightarrow R-\underset{OR'}{\overset{OAlEt_2}{\underset{|}{C}}}-Et \longrightarrow R-C{\overset{O}{\underset{Et}{\diagup}}} + Et_2AlOR'$$

(9) (10) (11)

$$R-C{\overset{O}{\underset{Et}{\diagup}}} + AlEt_3 \longrightarrow R-C{\overset{O \rightarrow AlEt_3}{\underset{Et}{\diagup}}}$$

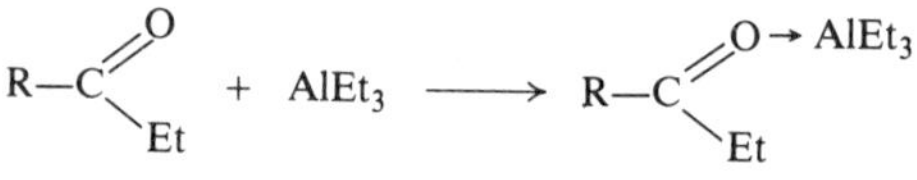

(12)

(iii) *Alkylation of the ketone*

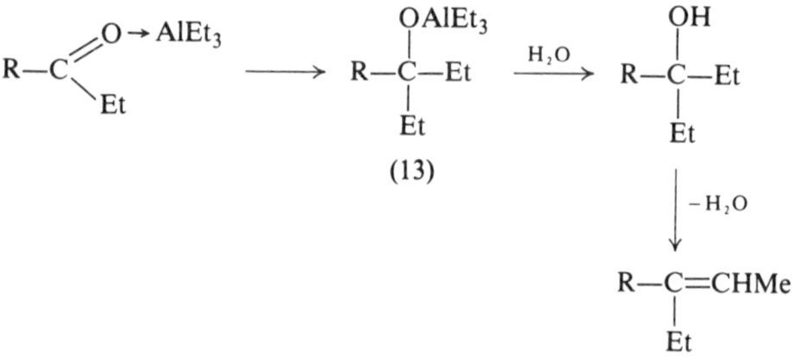

$$R-C{\overset{O \rightarrow AlEt_3}{\underset{Et}{\diagup}}} \longrightarrow R-\underset{Et}{\overset{OAlEt_3}{\underset{|}{C}}}-Et \xrightarrow{H_2O} R-\underset{Et}{\overset{OH}{\underset{|}{C}}}-Et$$

(13)

$$\downarrow -H_2O$$

$$R-\underset{Et}{\overset{}{\underset{|}{C}}}=CHMe$$

(14)

Complexation between $AlEt_3$ and an ester occurs instantaneously with formation of complexes (6)—(8). When using equimolar mixtures of $AlEt_3$ and an ester the major products are (6), (7), and (9)—(11). When there is an excess of $AlEt_3$, then reactions involving reduction of the ester to form aldehydes become possible. It should be noted that the presence of *p*-cresol in the catalyst used by Chien *et al.* causes complications in that cresol reacts quantitatively with $AlEt_3$ liberating ethane.

The role of the ester in high activity $MgCl_2$ catalysts has also been investigated by Kashiwa.[18]

Sterically hindered ketones and amines have been used to eliminate some of the side reactions mentioned in this report.[27,28] Langer *et al.*[29] have reported on the use

---

[27] A. W. Langer, U.S. P. 4 148 756, 1978.
[28] A. W. Langer, U.S. P. 4 224 182, 1979.
[29] A. W. Langer, T. J. Burkhardt, and J. J. Steger, 'Transition Metal Catalysed Polymerizations, Alkenes and Dienes, Part A', ed. R. P. Quirk *et al.*, Harwood Academic, New York, 1983, p. 421.

Chain Reaction Polymerization

of compounds of 2,2,6,6-tetramethyl piperidine with different trialkylaluminium compounds as the modifier–co-catalyst system for some supported catalysts. Complexes of $TiCl_4$ with various analogous amines, such as 1,2,4-trimethylpiperazine and 2,3,4,5-tetraethylpiperidine, have been used as starting materials, with the amine remaining as a built-in modifier in the catalyst, of catalysts consisting of titanium chloride supported on $MgCl_2$.[30,31] Karayannis and Lee[32,33] have used sterically hindered alicyclic secondary amines and their nitroxide free radicals along with conventional non-supported catalysts.

It is perhaps pertinent to point out that the reactivity of the co-catalyst, the catalyst, and the electron donor towards each other highlights the importance of the order of addition of the components of the polymerization system, and the need for authors of publications to state carefully the reaction conditions which are employed.

Activation of a $MgCl_2$ supported catalyst by a mixture of $AlEt_3$ and methyl p-toluate (3:1) leads to a reduction of 90% of the Ti^{IV} to lower valence states with the e.p.r.-observable fraction of the resulting Ti^{III} species being about 25%.[21] It was further observed by Chien et al.[21] that activation of the catalyst converts the axially symmetric e.p.r. spectra of species (3) to the rhombic spectra of a species which is believed to have a structure of the form (15). This species was found to be unstable,

(15)

and on aging is converted to a Ti^{III} species with axial symmetry. Complex (15) is co-ordinatively unsaturated and contains a Ti^{III} atom with a vacant site and an asymmetric field, and thus fulfills the requirements of a site capable of stereospecific polymerization.

Activation of the same catalyst by $AlEt_3$ alone produced at least three Ti^{III} species, including species (15), indicating that the methyl p-toluate moderates the alkylation potential of the $AlEt_3$ when in a complexed form. It is therefore to be noted that the presence of an electron donor complexed with the $AlEt_3$ is important, not only in increasing the stereospecificitity of the catalyst system, but in determining the actual types of species which are present. Obviously, the two are related.

(e) *Polymerization studies.* A number of studies on the kinetics of polymerization of both ethylene and propylene by highly active $MgCl_2$-supported catalysts have appeared in the literature over the past few years. While it should not be presupposed that all catalysts prepared from $MgCl_2$ should have exactly similar

---

[30] Montedison Sp.A., U.S. P. 4 226 963, 1980.
[31] Shell International Research, Eur. P. Appl., 19 312, 1980.
[32] N. M. Karayannis and S. S. Lee, *Makromol. Chem.*, 1982, **183**, 1171.
[33] N. M. Karayannis and S. S. Lee, *Makromol. Chem., Rapid Commun.*, 1982, **3**, 255.

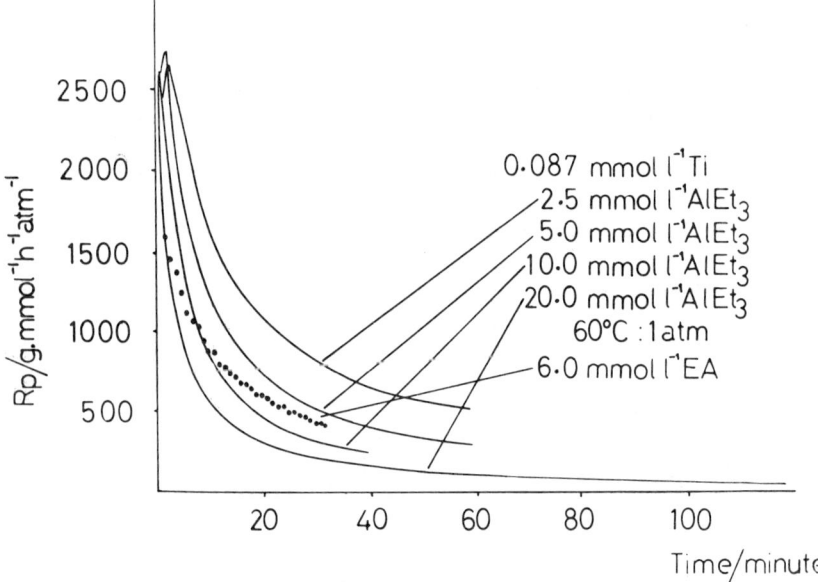

Figure 3 Plot of overall rate of polymerization against time for the polymerization of propylene using a high-activity Montedison-type catalyst. EA = ethyl anisate

characteristics, certain features appear to be common to many polymerization systems.

(i) *Variation of the overall rate of polymerization with time.* Catalysts of the type $MgCl_2/EB/TiCl_4$–$AlEt_3/EB$ when used for the polymerization of propylene are characterized by having very high activities and high stereospecificities. Many such catalysts have activities of $\times 100$—500 those of first generation δ-$TiCl_3 \cdot \frac{1}{3}AlCl_3$ –$AlEt_2Cl$-type systems. Most, although not all, are characterized by having exceedingly high initial rates of polymerization which decrease rapidly with time. A typical rate–time profile is shown in Figure 3.

This type of behaviour has been reported for the polymerization of propylene by Suzuki et al.[34] for the catalyst system $MgCl_2/EB/TiCl_4$ (containing 0.84% Ti) and activated by $AlEt_3$; Vermel et al.[35] for the catalyst system $MgCl_2/TiCl_4 \cdot EB$ (containing 3% Ti) and activated by $AlEt_3$; Giannini[6] for a catalyst of undisclosed composition; Kashiwa[18] for the catalyst system $MgCl_2/TiCl_4$ and activated by $AlEt_3$; Keii[36] for the catalyst system $MgCl_2/EB/TiCl_4$ (containing 0.41—5.4% Ti) and activated by $AlEt_3$.

The rapid decay in rate is not associated with monomer diffusion through an ever-increasing thickness of polymer layer covering the surface of the catalyst, since Keii[36] has shown that deactivation occurs when a catalyst is aged in the absence of

---

[34] E. Suzuki, M. Tamura, Y. Doi, and T. Keii, *Makromol. Chem.*, 1979, **180**, 2235.
[35] Y. Y. Vermel, V. A. Zakharov, Z. K. Bukatova, G. P. Shkurina, L. G. Yechevskaya, E. M. Moroz, and S. V. Sudakova, *Polymer Sci., USSR*, 1980, **22**, 23.
[36] T. Keii, *Makromol. Chem.*, 1982, **183**, 2285.

monomer where no polymer is being formed. Also, Giannini[6] concludes from an analysis of the kinetics of polymerization at different monomer concentrations that the decrease of the polymerization rate with time is only slightly affected by the amount of polymer produced per gram of catalyst, indicating that the decay of the catalyst activity cannot be explained in terms of a diffusion barrier of growing polymer. From a kinetic viewpoint the behaviour is very similar to that observed with many soluble systems[37,38] and may well arise from the formation of dimer-type structures, such as those proposed by Chien et al.[21]

While most of these rate-time decay-type plots can be fitted satisfactorily to second-order decay plots, the situation is more complex for some catalyst systems, and under some polymerization conditions.[39] Keii[36] has shown that at temperatures < 23 °C the second-order decay law is not applicable; nor does it hold during the initial stage (< 10 min) of the polymerization, nor during the latter stage (> 3 h) where a first-order decay seems more applicable. Nevertheless, for many catalyst systems the second-order decay law [equation (1)] is applicable throughout most of

$$-\frac{dC_t^*}{dt} = k_d C_t^{*2} \qquad (1)$$

the polymerization, where C_t denotes the number of surface active sites at polymerization time, t, and k_d is the second-order decay rate constant.

Integration between $t = t$ and $t = 0$ gives equation (2). Now if we can suppose

$$\frac{1}{C_t^*} = \frac{1}{C_0^*} + k_d t \qquad (2)$$

that rates are proportional to the numbers of active centres, and that the proportionality constant which is involved remains constant, then equation (3)

$$\frac{1}{(R_p)_t} = \frac{1}{(R_p)_0} + k_d t \qquad (3)$$

allows values of k_d to be obtained under varying experimental conditions.[40] Values of k_d have been calculated by Keii and their dependence on the concentration of $AlEt_3$ examined.[36]

(ii) *Variation of the overall rate of polymerization with alkyl concentration.* The dependence of the maximum overall rate of polymerization on the concentration of $AlEt_3$ has been shown[36] to follow a Langmuir–Hinshelwood adsorption equation [equation (4)] where K_A is the adsorption equilibrium constant[41] and [A] is the

$$R_p = \frac{k \cdot K_A[A]}{(1 + K_A[A])^2} \qquad (4)$$

equilibrium concentration of alkyl.

K_A has been found to increase slightly with time,[36] indicating either a slow

---

[37] G. Henrici-Olivé and S. Olivé, *Angew. Chem. Int. Ed. Engl.*, 1967, **6**, 790.
[38] A. G. Chesworth, R. N. Haszeldine, and P. J. T. Tait, *J. Polym. Sci., Polym. Chem. Ed.*, 1974, **12**, 1703.
[39] S. Davies and P. J. T. Tait, to be published.
[40] G. R. Shore and P. J. T. Tait, to be published.
[41] D. R. Burfield, I. D. McKenzie, and P. J. T. Tait, *Polymer*, 1972, **13**, 302.

occupation of vacant active sites by AlEt$_3$, or a fast deactivation of active sites with small K_A values in the case where sites of varying activity are present.

(iii) *Variation of the overall rate of polymerization with monomer concentration.* The overall rate of polymerization has been shown to be first order with respect to monomer concentration. Giannini[6] has presented data for propylene concentrations up to 5 mol dm^{-3} for $t = 5$ min, and Keii[36] data up to 0.5 mol dm^{-3} for various polymerization times.

The overall rate equation may now be written as equation (5), where [M] is the monomer concentration at the active sites.

$$R_p = k_p[M] \frac{K_A[A]}{(1 + K_A[A])^2} \quad (5)$$

(iv) *Variation of the overall rate of polymerization with temperature.* Keii[36] has investigated the variation of the overall rate of polymerization in the temperature range 1—65 °C, and has reported apparent activation energies of 50 kJ mol^{-1} for $T < 41$ °C and -21 kJ mol^{-1} for $T > 41$ °C. In an earlier publication Keii et al.[42] explained the break in the plot of log $(R_p/[M])$ *versus* $1/T$ as arising from an irreversible decrease in the number of polymerization centres with increasing temperature. Keii[36] has now advanced another explanation for this behaviour.

The constants k_p and K_A may be expressed as equations (6) and (7) where E_p is the

$$k_p = k_p^0 e^{-E_p/RT} \quad (6)$$

$$K_A = K_A^0 e^{-E_A/RT} \quad (7)$$

activation energy of the rate constant k_p, and E_A is the adsorption energy of AlEt$_3$ onto active sites.

At lower temperatures equation (5) reduces to (8) while at higher temperatures it reduces to (9).

$$R_p = k_p[M]/K_A[A] \quad \text{since} \quad K_A[A] \gg 1 \quad (8)$$

$$R_p = k_p[M]K_A[A] \quad \text{since} \quad K_A[A] \ll 1 \quad (9)$$

The apparent activation energy, E_{obs}, at lower temperatures is given by equation (10) and at higher temperatures by equation (11).

$$E_{obs} = E_p + E_A \quad (10)$$

$$E_{obs} = E_p - E_A \quad (11)$$

Keii reports a value of 15 kJ mol^{-1} for E_p and 36 kJ mol^{-1} for E_A.

(v) *Active centre determination and evaluation of propagation rate constant.* As a means of characterizing the behaviour of high-activity MgCl$_2$-supported catalysts there has been considerable activity during recent years in the determination of active centre determination, and in the evaluation of the rate constant for chain

---

[42] T. Keii, K. Soga, K. Go, and M. Kojima, *J. Polymer Sci., Part C*, 1968, **23**, 453.

propagation. Progress in these areas has been reviewed recently by Tait.[43] However, it should be realized that in comparing values of C^* and k_p for different catalysts, even when these all employ a ball-milled $MgCl_2$ support, different numerical values may be obtained where the conditions of catalyst preparation and polymerization are not the same. Nevertheless, some important characteristics for k_p and C^* values for supported $MgCl_2$ catalysts are evident.

(a) *Higher initial values for k_p and higher average values for k_p.* Suzuki et al.[34] have evaluated k_p values for the polymerization of propylene at 41 °C by the catalyst system $MgCl_2/EB/TiCl_3$ (containing 0.84% Ti), and activated by $AlEt_3$, by using molecular-weight data determined during the initial stages of polymerization ($t = 5$ s).

A minimum value for the overall propagation rate constant, $\bar{k}_p$, of active centres may be evaluated from the relationship[44] in equation (12), where $\bar{P}_n$ is the number

$$\bar{P}_n = \frac{\bar{M}_n}{M} = k_p[M]t \tag{12}$$

average degree of polymerization, $\bar{M}_n$ the number average molecular weight, and M the molecular weight of the monomer.

The use of equation (12) depends on the assumptions:[45] (i) formation of active centres and chain initiation are instantaneous; (ii) chain transfer with monomer and aluminium alkyl are negligible; (iii) the polymerization system is in a state of equilibrium as far as mixing and temperature are concerned. Although not all of these criteria are likely to be valid, the values obtained correspond fairly well with those derived from ^{14}CO radio-labelling.[6]

Since the polymer yield, Y, after 5 s is given by equation (13), values for both $\bar{k}_p$ and C^* may be obtained.

$$Y = \bar{P}_n[C^*] = k_p[M]C^*t \tag{13}$$

Suzuki et al. report values of $\bar{k}_p > 4.4 \times 10^2$ dm^3 mol^{-1} s^{-1} with a value of $(\bar{k}_p)_{isotactic} > 9.7 \times 10^2$ dm^3 mol^{-1} s^{-1}, and with $C^* < 2.3$ mol% Ti. Relevant comparative data are listed in Table 2. The values of k_p are much higher than those usually reported for first-generation type catalysts (1.70 dm^3 mol^{-1} s^{-1}),[4] and this led Suzuki et al. to conclude that the higher activity observed with their $MgCl_2$ supported catalyst arose from an increase in the value of k_p, rather than an increase in C^*.

Giannini[6] has studied the polymerization of propylene using a high-activity $MgCl_2$-supported catalyst and the results which he reports highlight some very important features of these polymerization systems: (a) concentrations of active centres were much higher for the high activity catalyst ($\times 20$—30), corresponding to 8—10 mol% Ti content of the catalyst; (b) the active centre concentration decreased fairly rapidly with time; (c) the active centre concentration decreased less quickly with time than did the overall rate of polymerization. Consequently, since k_p is

---

[43] P. J. T. Tait, in 'Transition Metal Catalysed Polymerizations, Alkenes and Dienes, Part A', ed. R. P. Quirk et al., Harwood Academic, New York, 1983, p. 115.
[44] L. L. Böhm, *Polymer*, 1978, **19**, 562.
[45] P. J. T. Tait in 'Developments in Polymerization', ed. R. N. Haward, Applied Science Publishers, 1979, Vol. 2, p. 81.

derived from equation (14) the catalyst showed higher values for k_p at short polymerization times. Typical values of k_p for a high activity and a conventional catalyst are listed in Table 2.

$$R_p = k_p C^*[M] \tag{14}$$

It is interesting to note that the limiting value obtained for k_p is probably very similar to that reported by Suzuki et al.,[34] illustrating the need for care when making comparisons.

Kashiwa and Yoshitake have reported on values of k_p and C^* for two $MgCl_2$ supported catalysts, $MgCl_2$/EB/$TiCl_4$–$AlEt_3$ (ref. 46) and $MgCl_2$/$TiCl_4$–$AlEt_3$ (ref. 47). The values of k_p were determined by the method of initial molecular weights and are reported in Table 2. Their results indicate that for the $MgCl_2$/EB/$TiCl_4$ system k_p is higher by a factor of 50 than for conventional catalysts,[48] whilst the $MgCl_2$/$TiCl_4$ catalyst yielded values higher by a factor of 30 for $(k_p)_{isotactic}$. They were also able to conclude that $(k_p)_{isotactic}$ was greater than $(k_p)_{atactic}$ in agreement with the earlier results reported by Tait.[49]

Shepelev et al.[50] have investigated the polymerization of propylene using the catalyst system $MgCl_2$/$TiCl_4 \cdot$ EB–$AlEt_3$ and also report high initial values for k_p (cf., Table 2).

(b) *Higher values for C^**. Giannini[6] has studied the polymerization of but-1-ene at 0 °C and propylene at 65 °C using a supported $MgCl_2$ catalyst, and compared the values of k_p and C^* with those obtained for δ-$TiCl_3 \cdot 0.33$ $AlCl_3$–$Al(Bu^i)_3$ catalysts.

In the polymerization of but-1-ene values of k_p were determined from the variation of the dependence of the molecular weight of the polymer on the polymerization time.[6]

Values of 27 and 53 dm^3 mol^{-1} s^{-1} for k_p were obtained for the δ-$TiCl_3$–$Al(Bu^i)_3$ and supported catalysts, respectively (see Table 2), while the activity of the supported catalyst was higher by about $\times 100$. Giannini, therefore, concluded that the higher activity ($\sim \times 50$) of the supported catalyst arose mainly from a higher number of active centres, in apparent contrast to the findings of Suzuki et al.[34]

Values for C^* and k_p for the polymerization of propylene by the catalyst system $MgCl_2$/$TiCl_4 \cdot$ EB–$AlEt_3$ have been reported by Bukatov et al.[51] and are listed in Table 2. ^{14}CO was used to determine C^*. In the initial stages of the polymerization k_p was again found to have a high value ($\sim 10^3$ dm^3 mol^{-1} s^{-1}) while C^* for systems activated by $Al(Bu^i)_3$ and ethyl p-methoxy benzoate are reported to have values of 7 mol% Ti for $(C^*)_{isotactic}$. Bukatov et al. conclude that the distinctive features of all titanium–magnesium catalysts in comparison with δ-$TiCl_3 \cdot 0.3 AlCl_3$ catalyst are a drastic increase in the number of active centres, and an increase of the reactivity of

---

[46] N. Kashiwa and J. Yoshitake, *Makromol. Chem., Rapid Commun.*, 1982, **3**, 211.
[47] N. Kashiwa and J. Yoshitake, *Makromol. Chem., Rapid Commun.*, 1983, **4**, 41.
[48] H. W. Coover, J. E. Guillet, R. L. Combs, and F. B. Joyner, *J. Polym. Sci., Part A-1*, 1966, **4**, 2583.
[49] P. J. T. Tait, in 'Preparation and Properties of Stereoregular Polymers', ed. R. W. Lenz, D. Reidel Publishing Co., Rehovot, 1978, p. 85.
[50] S. N. Shepelev, G. D. Bukatov, V. A. Zakharov, and Y. I. Yermakov, *Kinet. Katal.*, 1981, **22**, 258; *Chem. Abstr.*, 1981, **94**, 175653.
[51] G. D. Bukatov, S. H. Shepelev, V. A. Zakharov, S. A. Sergeev, and Y. I. Yermakov, *Makromol. Chem.*, 1982, **183**, 2657.

Table 2 Values of C^* and k_p for polymerization by various $MgCl_2$ supported catalysts

Catalyst system	Monomer	Temp./°C	Time	C^*/mol% Ti	k_p/dm^3 mol^{-1} s^{-1}	Method	Ref.
1. $MgCl_2$/EB/$TiCl_4$–$AlEt_3$ (Ti content = 0.84%)	propylene	41	5 s	2.3	440	Initial molecular weight	Suzuki et al.[34]
2. $MgCl_2$ supported high activity catalyst	propylene	65	5 min	8—10*	970*	^{14}CO radio-labelling	Giannini[6]
			15 min		500		
			30 min		340		
			60 min		300		
			180 min		210		
					140		
δ-$TiCl_3 \cdot 0.33AlCl_3$–$Al(Bu^i)_3$	propylene	65	15 min		130	^{14}CO radio-labelling	Giannini[6]
			60 min		100		
			180 min		100		
3. $MgCl_2$/EB/$TiCl_4$–$AlEt_3$ (Ti content = 2% Ti)	propylene	60	7—120 s	2.8*	2700	Initial molecular weight	Kashiwa et al.[46]
4. $MgCl_2$/$TiCl_4 \cdot$ EB–$AlEt_3$	propylene	40	5 s	0.27*	500*	Initial molecular weight	Shepelev et al.[50]
5. $MgCl_2$/$TiCl_4 \cdot$ EB–$AlEt_3$ (Ti content = 2.5%)	propylene	70		0.68*	870*	^{14}CO radio-labelling	Bukatov et al.[51]
	propylene	70			1250*	^{14}CO radio-labelling	
6. $MgCl_2$/$TiCl_4$–$AlEt_3$	propylene	60	5 s	20—60	240—730	Initial molecular weight	Kashiwa et al.[47]
				2—6*	500—1500		
7. High activity	but-1-ene	0	300 s		53	Molecular weight	Giannini[6]
δ-$TiCl_3$–$Al(Bu^i)_3$	but-1-ene	0	300 s		27	Molecular weight	Giannini[6]
8. $MgCl_2$/$TiCl_4 \cdot$ EB–$Al(Bu^i)_3$	propylene		15 min	7*	1000*	Initial molecular weight	Bukatov et al.[51]

* Isotactic; EMB = ethyl p-methoxybenzoate

the active centres in propylene polymerization (the k_p values for ethylene polymerization remain unchanged).

Catalysts Derived from the Reduction of Titanium Tetrachloride by Alkyl Magnesium Compounds. Catalysts prepared from titanium tetrachloride and alkyl magnesium compounds are characterized by high activities, and, in the polymerization of propylene, by low stereospecificities.

Licchelli et al.[52] have studied the polymerization of propylene using catalysts derived from halogen-free magnesium alkyls, *viz*, $Mg(n-C_8H_{17})_2$ and $Mg(n-C_{12}H_{25})_2$ activated by several metal alkyls, *e.g.*, $AlEt_3$, $Al(Bu^i)_3$ and $Al(n-C_8H_{17})_3$. These catalysts are somewhat similar to those prepared using Grignard reagents and showed a number of interesting features: (*i*) all catalysts were highly amorphous; (*ii*) only moderate activities (0.5—2.7 kg PP/gTi in 3 h) and low stereospecificities (~40%) are reported; (*iii*) the rate–time profile shows an initial settling period during which the overall rate of polymerization increases to a maximum value – the rate then decreases quite sharply as has been observed for many $MgCl_2$-supported catalysts. This deactivation is more marked with $AlEt_3$ than with either $Al(Bu^i)_3$ or $Al(n-C_8H_{17})_3$. The use of long-chain aluminium alkyls to prevent over-reduction of the titanium entity is a feature of many catalyst preparations in this area, although this can be achieved by the use of appropriate donors;[1,2] (*iv*) the decrease in rate does not appear to be due to monomer diffusion control; (*v*) the apparent activity of the catalysts increased as their titanium content decreased, as is observed for many $MgCl_2$-supported catalysts; (*vi*) catalysts obtained using a slurry of $Mg(n-C_8H_{17})_2$ showed higher activities at 40 °C when using $Al(Bu^i)_3$ as activator than RMgX-reduced catalysts, but little improvement in stereospecificity.

Bukatov et al.[51] have used one-component catalysts, $TiCl_m(Mg)$, prepared from the reduction of $TiCl_4$ in hexane with $MgBu_2$ and activated by $AlEt_3$, to polymerize both ethylene and propylene. Polymerizations were performed in an autoclave at constant monomer pressure (3—5 atm) and temperature. Stereospecificities, active centre concentrations, and propagation rate constants are reported and compared with those of catalysts, $TiCl_n(Al)$, prepared by the reduction of $TiCl_4$ by $AlEt_3$. Typical values for C^*, determined by ^{14}CO radio-labelling, are listed in Table 3. From a study of Table 3 the following points emerge.

(*i*) For ethylene polymerization the higher activity of the $TiCl_m(Mg)$ catalyst compared with that of the $TiCl_n(Al)$ catalyst is due to a higher number of active centres.

(*ii*) For ethylene polymerization the values of k_p obtained for both catalyst systems correspond to the value of k_p reported for $TiCl_2$ (ref. 53) and $TiCl_4/R_xMgCl_y$ catalysts.[54]

(*iii*) For propylene polymerization the higher activity of the $TiCl_m(Mg)$ catalyst compared with that of $TiCl_n(Al)$ is due to both higher numbers of active centres and higher values of k_p.

---

[52] J. A. Licchelli, R. N. Haward, and I. W. Parsons, *Polymer*, 1981, **22**, 667.
[53] V. A. Zakharov, G. D. Bukatov, V. K. Dudchenko, and Y. I. Yermakov, *Kinet. Katal.*, 1975, **16**, 417.
[54] V. A. Zakharov, N. B. Chumaevskii, S. I. Makhtarulin, G. D. Bukatov, and Y. I. Yermakov, *React. Kinet. Katal. Lett.*, 1975, **2**, 329.

Table 3 *One-component magnesium alkyl and aluminium alkyl reduced catalysts: some typical values for C^* and k_p*

Catalyst	Ethylene		Propylene*	
	C^* mol% Ti	k_p dm^3 mol^{-1} s^{-1}	C^* mol% Ti	k_p dm^3 mol^{-1} s^{-1}
TiCl$_m$(Mg)	0.15	14 500	0.014	470
TiAl$_n$(Al)	0.012	13 000	0.004	100

* Data for propylene refer to isotactic fractions

(*iv*) For propylene polymerization the value of k_p for TiCl$_n$(Al) is very similar to the value obtained for TiCl$_2$,[55] whilst the value of k_p for TiCl$_m$(Mg) is higher.

These observations are important in that they highlight the differences in both steric and electronic effects in the case of propylene polymerization when compared with that of ethylene.

(*v*) The amount of Ti involved in active centre formation for TiCl$_m$(Mg) catalysts is much less than for MgCl$_2$/EB/TiCl$_4$ catalysts.[56—59]

Metal Oxide Supported Catalysts. Catalysts derived from the reaction of transition metal halides with metal oxides continue to be of interest.

(*a*) *Catalyst supports and preparation.* A molecular deposition method has been used by Damyanov and Velikova[60] to prepare vanadium supported SiO$_2$ catalysts. Samples of amorphous SiO$_2$ of varying silanol content, prepared by calcination at temperatures between 200—800 °C, were treated with VOCl$_3$ vapour at 150 °C for 2 h. The reactions (15) and (16) between surface hydroxyl groups and the VOCl$_3$

$$\equiv\text{Si-OH} + \text{VOCl}_3 \longrightarrow \equiv\text{Si-OVOCl}_2 + \text{HCl} \qquad (15)$$

$$\begin{array}{c} \equiv\text{Si OH} \\ + \text{VOCl}_3 \longrightarrow \\ \equiv\text{Si-OH} \end{array} \quad \begin{array}{c} \equiv\text{Si-O} \\ \diagdown \\ \text{VOCl} + 2\text{HCl} \\ \diagup \\ \equiv\text{Si-O} \end{array} \qquad (16)$$

vapour are considered to be consistent with the absence of the 3750 cm^{-1} band in the i.r. spectra of these catalysts. Catalysts containing 1.43—3.36% V were produced and used with AlEt$_2$Cl as activator in the polymerization of ethylene at 60 °C. The catalyst activity depended strongly on the amount of V on the support surface, and productivities of up to 22 kg PE/g V are reported.

The influence of the nature of the support on the catalytic activity of silica gel and silica alumina supports has been investigated by Muñoz-Escalona *et al.*[61] Supports

---

[55] V. I. Yermakov, V. A. Zakharov, and G. D. Bukatov, in 'Catalysis', Proceedings of the Fifth Int. Congr. on Catalysis, Amsterdam–New York, 1973, Vol. 1, p. 399.
[56] D. G. Boucher, I. W. Parsons, and R. N. Haward, *Makromol. Chem.*, 1974, **175**, 3461.
[57] H. Meyer and K. H. Reichert, *Angew. Makromol. Chem.*, 1977, **57**, 211.
[58] L. L. Böhm, *Polymer*, 1978, **19**, 553.
[59] A. A. Baulin, V. N. Sokolov and A. A. Semenova, *Vysokomol. Soedin.*, Ser. A, 1976, **17**, 46.
[60] D. Damyanov and M. Velikova, *Eur. Polym. J.*, 1979, **15**, 1075.
[61] A. Muñoz-Escalona, A. Martin, and J. Hidalgo, *Eur. Polym. J.*, 1981, **17**, 367.

containing varying amounts of Al_2O_3, viz, 5, 13, and 25%, were prepared, and also modified by impregnation with $ZnCl_2$. Treatment of these supports with $TiCl_4$ at 50 °C for 15 h yielded catalysts containing from 5.3 to 12.9% Ti. Ethylene polymerizations were carried out with $AlEt_2Cl$ as activator using high Al:Ti ratios (20:1). The catalyst activity was found to increase with increase in Al_2O_3 content up to 8—12%, and then to decrease. This behaviour was shown by catalysts with and without supported Zn, and is believed to arise from the surface acidity of the support which shows a similar relationship with Al_2O_3 content. A higher activity was found for catalysts containing Zn metal, but these catalysts showed the same type of relationships.

The carrier thus behaves as more than an inert surface, and in particular regulates the microenvironment of the supported Ti atoms affecting the reduction of Ti^{IV} to Ti^{III}. Muñoz-Escalona et al.[61] conclude that the activity of metal oxide-supported catalysts depends primarily on the chemical composition of the carrier surface rather than on the surface area of the catalyst.

(b) *Kinetic studies.* The kinetics of the polymerization of ethylene by supported $SiO_2/TiCl_4$ catalysts modified by metal chlorides, e.g., $CaCl_2$, $BaCl_2$, $SrCl_2$, $ZnCl_2$, have been studied by Muñoz-Escalona et al.[62] Owing to the high catalyst efficiency (overall catalyst yield ~ 12 kg PE/g Ti) the initial polymerization rate was difficult to measure accurately, and consequently rate studies were carried out using low catalyst concentrations (5×10^{-3} g Ti dm^{-3}). After an initial settling period the polymerization rate decays to a constant value which is maintained for several days. The stationary rate of polymerization, R_∞, was shown to be first order with respect to monomer concentration and to be dependent on the Al:Ti ratio, showing a maximum value for Al:Ti $\approx$ 30:1.

The applicability of both Rideal and Langmuir–Hinshelwood models for the dependence of R_∞ on the $AlEt_2Cl$ concentration was tested, better agreement being obtained when the Langmuir–Hinshelwood model was used.

(c) *Active centre determination – effect of nature of support on C^* and k_p.* Magnesium oxide and aluminosilicates have been employed by Baulin et al.[63] as supports for $TiCl_4$ in the polymerization of ethylene, propylene, but-1-ene, and hex-1-ene. Catalyst activation was carried out by $AlEt_3$ and active centre concentrations were determined by the use of tritiated methanol, correction being made for the kinetic isotope effect and chain transfer with $AlEt_3$. Some typical results are shown in Table 4.

From a consideration of Table 4 the following conclusions were reached.

(i) The percentage of the supported Ti involved in chain propagation is quite high, and higher for $MgO/TiCl_4$ than for $SiO_2 \cdot Al_2O_3/TiCl_4$ catalysts systems. Comparison for the values of C^* with those in Tables 2 and 3 show that a much higher proportion of the supported Ti for $MgO/TiCl_4$ catalyst systems is active in polymerization than for $MgCl_2/EB/TiCl_4$ systems.

---

[62] A. Muñoz-Escalona, C. Martinez, and J. Hidalgo, *Polymer*, 1981, **22**, 1118.
[63] A. A. Baulin, A. G. Rodionov, S. S. Ivanchev, and N. M. Domareva, *Eur. Polym. J.*, 1980, **16**, 937.

Chain Reaction Polymerization

Table 4 *Values of C* and k_p for various olefins for polymerizations (70 °C) using metal oxide supported catalysts*

Catalyst system	Monomer	C_p^* mol% Ti	k_p dm^3 mol^{-1} s^{-1}
MgO/TiCl$_4$–AlEt$_3$ (0.2% Ti)	C$_2$H$_4$	39 ± 2	2400 ± 240
	C$_3$H$_6$	33 ± 2	4.8 ± 0.5
	C$_4$H$_8$	34 ± 2	4.6 ± 0.5
	C$_6$H$_{12}$	36 ± 2	2.5 ± 0.3
SiO$_2$·Al$_2$O$_3$/TiCl$_4$–AlEt$_3$	C$_2$H$_4$	23 ± 1	111 ± 11
	C$_3$H$_6$	22 ± 1	1.0 ± 0.1
	C$_4$H$_8$	20 ± 1	0.13 ± 0.01

(*ii*) Values for C^* depend only on the catalyst support used and not on the monomer used. Hence the different geometrical dimensions of the olefins do not appear to have any influence on values for C^*.

(*iii*) Very different values of k_p were obtained for different supports when using the same monomer. The k_p values for ethylene were significantly higher than those for the α-olefins studied.

(*iv*) Values of k_p for MgO/TiCl$_4$ and SiO·Al$_2$O$_3$/TiCl$_4$ catalysts were significantly lower than for either MgCl$_2$·EB/TiCl$_4$, or for TiCl$_m$(Mg) and TiCl$_n$(Al) catalysts (*cf.* Tables 2 and 3).

In conclusion, it is becoming very apparent that the nature of the support can bring about significant changes in not only C^* but also k_p.

One-component Supported Catalyst Systems. One-component supported catalyst systems have attracted considerable interest since complications due to the many possible reactions involving the organometallic co-catalyst are eliminated. These systems, however, have proved themselves to be very susceptible to the presence of impurities.

The following types of one-component catalyst systems can be distinguished in the published literature:

(*i*) metal oxide supported catalyst systems, *e.g.*, M$_x$O$_y$/TiCl$_4$;
(*ii*) supported organometallic compounds of transition metals, *e.g.*, Al$_2$O$_3$/ZrBz$_4$;
(*iii*) halides of transition metals, *e.g.*, TiCl$_2$;
(*iv*) metal alkyl reduced transition metal halide catalysts, *e.g.*, TiCl$_m$(Mg).[51] Since these catalysts are produced in the presence of organo-magnesium or organo-aluminium compounds, great care must be taken to remove all traces of such compounds when one-component catalysts are being prepared.

One-component catalyst systems have been prepared by Soga *et al.*[64] by reacting TiCl$_4$ with γ-Al$_2$O$_3$ or SiO$_2$. Before use, these catalysts were heated to 100—800 °C in vacuum for 6 h. The catalyst system γ-Al$_2$O$_3$/TiCl$_4$ polymerized propylene

---

[64] K. Soga, K. Izumi, M. Terano, and S. Ikdea, *Makromol. Chem.*, 1980, **181**, 657.

without the use of an activator, while the $SiO_2/TiCl_4$ system was inactive. Addition of $AlEt_2Cl$ to the $SiO_2/TiCl_4$ system produced an active polymerization system.

E.s.r. spectral studies at -196 °C in the case of γ-$Al_2O_3/TiCl_4$–$AlEt_2Cl$ and $SiO_2/TiCl_4$–$AlEt_2Cl$ catalysts showed broad spectral lines with g-values 1.942 and 1.952, respectively, which were assigned to a Ti^{III} species. The e.s.r. spectra of the γ-$Al_2O_3/TiCl_4$ catalyst in the absence of any activator showed the same signal. Additionally, the catalyst activity was found to be roughly proportional to the concentration of the Ti^{III} species, and Soga et al. concluded that Ti^{III} is the active form of the titanium in these polymerizations. The heat treatment evidently leads to a reduction of Ti^{IV} to Ti^{III}. It would obviously be of interest to see whether these catalysts contain e.s.r. silent Ti^{III} as has been found for $MgCl_2/EB/TiCl_4$ catalysts.[21]

In an earlier publication Soga et al.[65] from their studies on Y-zeolite supported catalysts concluded that both Ti^{III} and Ti^{IV} species were active. Polymerization rates, however, were very low.

One-component catalysts prepared from SiO_2 and $TiCl_4$ are believed[66] to be inactive because of the difficulty of co-ordination of propylene with the Ti^{IV} atom surrounded by three Cl ligands. Treatment with $AlEt_2Cl$ leads to a reduction of Ti^{IV} to Ti^{III}: the effective extraction of a Cl ligand allowing a site for the co-ordination of monomer.

Soga et al.[64] have drawn attention to the resemblance of one-component catalyst systems to the Phillips type of supported oxide catalysts as was done earlier by Tait.[45]

Bukatov et al.[51] have carried out extensive research using $TiCl_m(Mg)$ and $TiCl_n(Al)$-type catalysts and the results obtained have already been analysed (see p. 18).

Soluble Catalysts.—'*Living' Co-ordination Polymerization Systems Derived from Vanadium Triacetylacetonate.* Although the development of 'living' polymerization systems for the anionic polymerization of vinyl monomers was pioneered by Szwarc[67] some 27 years ago, it was not until recently that Doi et al.[68,69] reported that vanadium triacetylacetonate, $V(acac)_3$, combined with $AlEt_2Cl$ polymerized propylene at -78 °C to give a 'living'-type polymer with a narrow molecular weight distribution ($\bar{M}_w/\bar{M}_n = 1.19 \pm 0.05$). The use of the $V(acac)_3/AlEt_2Cl$ for the polymerization of propylene was first discovered by Natta et al.[70]

'Living' polymerization can only take place when the following criteria are fulfilled:

(i) chain transfer processes are absent, and
(ii) irreversible chain termination does not take place.

Further, if narrow molecular weight distributions are to be obtained then chain initiation should be more or less instantaneous.

---

[65] K. Soga, T. Sano, and S. Ikeda, *Polym. Bull.*, 1979, **1**, 665.
[66] K. Soga, T. Sano, and S. Ikeda, *Polym. Bull.*, 1980, **2**, 817.
[67] M. Szwarc, *Nature (London)*, 1956, **178**, 1168.
[68] Y. Doi, S. Ueki, and T. Keii, *Makromol. Chem.*, 1979, **180**, 1359.
[69] Y. Doi, S. Ueki, and T. Keii, *Macromolecules*, 1979, **12**, 814.
[70] G. Natta, I. Pasquon, and A. Zambelli, *J. Am. Chem. Soc.*, 1962, **84**, 1488.

Doi et al.[68,69] have been able to demonstrate that for polymerizations below −65 °C the polymerization proceeded without catalyst deactivation, and without any chain transfer reactions, since the number of polymer molecules produced per vanadium atom remained almost constant during the course of the polymerization. The polymerization reaction showed no settling period, indicating that the formation of propagation centres was completed almost immediately the polymerization reaction began, and as a consequence the polypropylenes produced showed a higher degree of monodispersity than had so far been obtained using Ziegler–Natta-type catalysts. ^{13}C n.m.r. spectroscopy showed that the polypropylene produced was predominately syndiotactic, as reported by Natta et al.[70]

Substitution of $AlEt_2Br$ for $AlEt_2Cl$ still yielded a living polymerization system at −78 °C. Use of either $Al_2Et_3Cl_3$ or $AlEtCl_2$ yielded polymerization systems in which the number of polymer molecules produced per vanadium atom increased with the polymerization time, indicating the presence of some chain transfer reactions.[69]

A detailed kinetic study on the 'living' polymerization of propylene by the soluble catalyst system $V(acac)_3$–$Al(Bu^i)_2Cl$ has been reported by Doi et al.[71] Polymerizations were effected at −78 °C producing monodisperse polymers with $\bar{M}_w/\bar{M}_n = 1.15 \pm 0.10$. The following features for the polymerization were established.

(i) The polymer yield was strictly proportional to the polymerization time, indicating that the formation of the propagating centres took place at the start of the polymerization, and that subsequent polymerization proceeded without catalyst deactivation.

(ii) The number of polymer chains produced per vanadium atom remained constant during the polymerization.

(iii) The polymer yield at a given polymerization time was first roder with respect to the concentration of $V(acac)_3$ at a constant concentration of $Al(Bu^i)_2Cl$.

(iv) The effect of the concentration of $Al(Bu^i)_2Cl$ on the yield was complex, and the following relationship was established

$$[N] = \alpha \frac{K_A[A]^2}{1 + K_A[A]^2} \qquad (17)$$

where [N] is the number of polymer chains produced per vanadium atom.

(v) The effect of the concentration on the yield was given by the expression

$$Y = k_p \left(\frac{K_M[M]}{1 + K_M[M]} \right)[N]_t \qquad (18)$$

where Y is the yield of polymer at time t.

These results were interpreted in terms of a two-step propagation reaction, as proposed by Tait et al.[41] for Ziegler–Natta polymerization, viz:

$$\overset{*}{V}-P_n + C_3H_6 \underset{}{\overset{K_M}{\rightleftharpoons}} \overset{C_3H_6 \downarrow}{V-P_n} \overset{k_p}{\longrightarrow} \overset{*}{V}-P_{n+1}$$

[71] Y. Doi, S. Ueki, S. Tamura, S. Nagahara, and T. Keii, in 'Transition Metal Catalysed Polymerizations, Alkenes and Dienes, Part A', ed. R. P. Quirk et al., Harwood Academic, New York, 1983.

Soluble Vanadium Tetrachloride–Alkylaluminium Halide Catalyst Systems. A study has been carried out of the molecular weight distribution and the kinetics of the low-temperature polymerization of propylene using soluble vanadium-based catalysts.[72] The catalysts employed were $VCl_4/AlEt_2Cl$, $VCl_4/AlEt_2Br$, and $VCl_4/AlEt_3$ and polymerizations were carried out at $-78\ °C$.

Analysis of kinetic results was carried out using the scheme:

Co-ordination of monomer

$$R_n + M \underset{}{\overset{K_M}{\rightleftharpoons}} C_n$$

Propagation

$$C_n \rightarrow R_{n+1}$$

Chain transfer with monomer

$$C_n \overset{k_{tm}}{\rightarrow} R_1 + Y_n$$

Chain transfer with alkylaluminium

$$R_n + A \overset{k_{ta}}{\rightarrow} R_1 + Z_n$$

Consequently,

$$R_p = k_p \sum_1^\infty [C_n] = k_p K_M \sum_1^\infty [R_n][M] \tag{19}$$

$$\therefore \quad R_p = k_p \left(\frac{K_M[M]}{1 + K_M[M]} \right) \sum_1^\infty ([C_n] + [R_n]) \tag{20}$$

and

$$R_t = (k_{tm} K_M [M] + k_{ta}[A]) \sum_1^\infty [R_n] \tag{21}$$

The number average degree of polymerization, $\bar{P}_n$, at time, t, is given by

$$\bar{P}_n = \frac{\int_0^t R_p\, dt}{\sum_1^\infty ([C_n] + R_n) + \int_0^t R_t\, dt} \tag{22}$$

and consequently:

$$\frac{1}{\bar{P}_n} = \left(\frac{1 + K_M[M]}{k_p K_M[M]} \right) \cdot \frac{1}{t} + \frac{R_t}{R_p} \tag{23}$$

and when $t \gg 0$, equation (23) yields

$$\frac{1}{\bar{P}_n} \approx \frac{k_{tm}}{k_p} + \frac{k_{ta}}{k_p K_M} \frac{[A]}{[M]} \tag{24}$$

[72] Y. Doi, M. Takada, and T. Keii, *Bull. Chem. Soc. Jpn.*, 1979, **52**, 1802.

Table 5 *Rate coefficients for the elementary reactions in the polymerization of propylene with $VCl_4/AlEt_2X$ (X = Cl, Br, Et) at $-78\,°C$*

	$AlEt_2Cl$	$AlEt_2Br$	$AlEt_3$
$K_M/dm^3\,mol^{-1}$	0.26	0.02	0.04
$K_A/dm^3\,mol^{-1}$	6.8	31	46
k_p/h^{-1}	4.3×10^3	2.5×10^2	5.3×10^4
k_{tm}/h^{-1}	5.4	1.7	2.3
$k_{ta}/dm^3\,mol^{-1}\,h^{-1}$	18	3.5	9.0

The rate coefficients of the elementary reactions for the polymerization of propylene were evaluated and are listed in Table 5.

The effect of alkylaluminium concentration on the rate of polymerization was found to be complex and to follow the equation (25). This equation would seem to

$$R_p \propto \frac{K_A[A]}{(1 + K_A[A])^2} \qquad (25)$$

suggest that the alkylaluminium is involved in some deactivation of the active centres as well as in the formation of centres through alkylation and complexation. The influence of the alkylaluminium on the values of the elementary reaction rates was interpreted in terms of a bimetallic structure for the polymerization center as proposed by Zambelli *et al.*[73]

Soluble Dichlorobis(γ-cyclopentadienyl)titanium–Alkylaluminium Halide Catalysts. A soluble catalyst has been prepared by reaction of Cp_2TiCl_2 with either $Al_2Et_3Cl_3$ or $AlEt_2Cl$ and used to polymerize ethylene using methylene dichloride as solvent.[74] Examination of the e.s.r. spectra showed that three spectrally distinguishable species were present. Reduction of Ti^{IV} to Ti^{III} occurred both in the presence and in the absence of ethylene. It was also observed that during the period of significant change in the e.s.r. spectra the rate of polymerization remained constant. The termination and reinitiation sequence of reactions was therefore summarized as being:

$$Cp_2TiCl_2 + Al_2Et_3Cl_3 \rightarrow Ti^* + Ti(1) + Ti(2) + Ti(3)$$

where Ti^* is the active species, probably in the maximum oxidation state since it was not detected by e.s.r. Ti(1), Ti(2), and Ti(3) are the species giving rise to e.s.r. spectra, and are formulated as shown below.

The CH_2Cl_2 thus maintains the activity of the catalyst system through re-oxidation of the reduced inactive Ti^{III} to Ti^{IV}.

Cihlar *et al.*[75] have published a detailed study of the kinetics of the polymerization of ethylene using the catalyst system $Cp_2TiEtCl/AlEtCl_2$ in toluene. Their

---

[73] A. Zambelli, I. Pasquon, R. Signorini, and G. Natta, *Makromol. Chem.*, 1968, **112**, 160.
[74] A. E. Yildirim, *Eur. Polym. J.*, 1981, **17**, 551.
[75] J. Cihlar, J. Mejzlik, O. Hamrik, P. Hudee, and J. Majer, *Makromol. Chem.*, 1980, **181**, 2549.

$$\text{Cp}_2\text{Ti}(\mu\text{-Cl})_2\text{AlClEt}$$
Ti(1)

$$\text{Cp}_2\text{Ti}(\mu\text{-Cl})_2\text{AlCl}_2$$
Ti(2)

$$\text{Cl}_2\text{Al}(\mu\text{-Cl})_2\text{Ti(Cp)}(\mu\text{-Cl})_2\text{AlCl}_2$$
Ti(3)

$$\text{Ti*} + n\text{C}_2\text{H}_4 \longrightarrow (\text{C}_2\text{H}_4)_n + \text{Ti}^{(\text{III})}$$

$$\text{Ti}^{(\text{III})} + \text{CH}_2\text{Cl} \longrightarrow \text{Ti}^{(\text{IV})} \xrightarrow{\text{Al}_2\text{Et}_3\text{Cl}_3} \text{Ti*}$$

experimental results are somewhat similar to those already reported by Reichert et al.[76—78] and Fink,[79,80] and may be summarized as follows.

(i) The rate–time profiles for polymerizations at Al:Ti ratios lower than about 8:1 were characterized by an initial stationary period, followed by a sharp increase in the rate and then by a fast decay in the rate. The increase in the rate of polymerization coincided roughly with the precipitation of polyethylene. The decay in the rate did not follow either a first- or a second-order dependence, nor did it appear to be monomer diffusion controlled.

(ii) A uni-modal molecular weight distribution was obtained, indicating the presence of only one type of catalyst site.

(iii) A high utilization of the Ti-component in active site formation was established.

(iv) The propagation rate constant had values of 3.9 and 7.2 $dm^3 \, mol^{-1} \, s^{-1}$ for Al:Ti ratios of 2:1 and 8:1.

In further experiments the hydrolysis of $AlEtCl_2$ in benzene was carried out, and the resulting oxyaluminium compounds were used as co-catalysts with $Cp_2TiEtCl$ as catalyst. The rate–time profiles for the polymerization changed dramatically, and much higher rates of polymerization were observed, the maximum rate being obtained after 20 s of the addition of the oxyaluminium compound. Values of k_p of 96 and 290 $dm^3 \, mol^{-1} \, s^{-1}$ for Al:Ti ratios of 6:1 and 15:1 were obtained. These values are 15—35 times higher than those found for the $Cp_2TiEtCl/AlEtCl_2$ system at the same Al:Ti ratios. It was believed that the higher activity of the sites was due to a higher Lewis acid acidity of the oxyaluminium compounds, forming a bridge complex with $Cp_2TiEtCl$.

Soluble Titanium and Zirconium–Aluminoxane Catalysts. One of the most fascinating discoveries in recent years in the field of transition-metal polymerization

---

[76] K. Meyer and K. H. Reichert, *Angew. Makromol. Chem.*, 1970, **12**, 175.
[77] K. H. Reichert, *Angew. Makromol. Chem.*, 1970, **13**, 177.
[78] K. H. Reichert and K. Meyer, *Makromol. Chem.*, 1973, **169**, 163.
[79] D. Schnell and G. Fink, *Angew. Makromol. Chem.*, 1974, **39**, 131.
[80] G. Fink, R. Rottler, D. Schnell, and W. Zoller, *J. Appl. Polym. Sci.*, 1976, **20**, 2779.

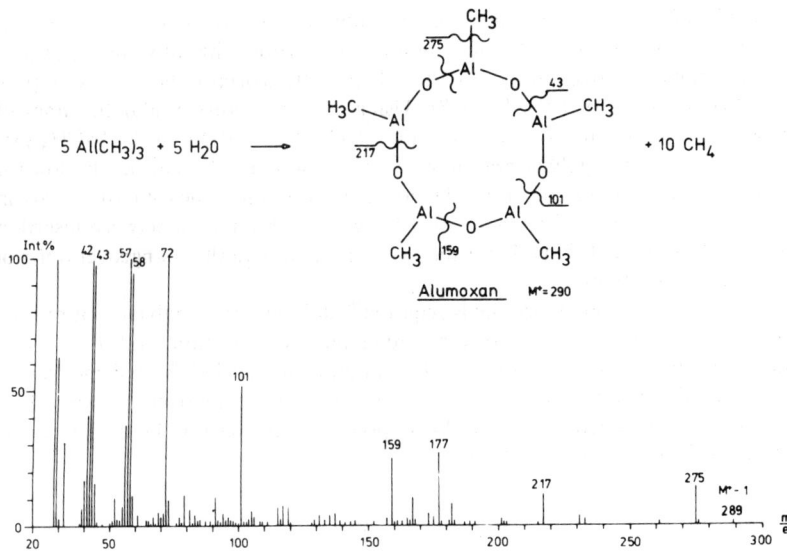

Figure 4 *Cyclic aluminoxane isolated by fractional precipitation and characterized by mass spectrometry*

has been the discovery that very active homogeneous catalysts for the polymerization of ethylene[81] could be prepared when AlEt$_3$, previously treated with water, was added to dialkylbis(cyclopentadienyl)titanium catalysts. Long and Breslow[82] had previously observed that addition of water to homogeneous chloride-containing systems based on bis(cyclopentadienyl) titanium(IV) compounds increased the activity of these catalyst systems, and subsequently this effect was studied by Cihlar et al.[83] who proposed the stabilized complexes (16) and (17).

Addition of water to AlMe$_3$ leads[84] to the formation of an oligomeric aluminoxane containing the structure $[-O-Al(Me)-]_n$. Further condensation with the elimination of AlMe$_3$ is believed to give rise to the formation of cyclic structures, and Sinn and Kaminsky[85] have proposed the structure shown in Figure 4.

---

[81] A. Andresen, H. G. Cordes, J. Herwig, W. Kaminsky, A. Merk, R. Mottweiler, J. H. Sinn, and H.-J. Vollmer, *Angew. Chem., Int. Ed. Engl.*, 1976, **15**, 630.
[82] W. P. Long and D. S. Breslow, *Liebigs' Ann. Chem.*, 1975, 463.
[83] J. Cihlar, J. Mejzlik, and O. Hamrik, *Makromol. Chem.*, 1978, **179**, 2553.
[84] E. A. Grogorjan, F. S. Dyachkovskii, and A. E. Shilov, *Vysokomol. Soedin.*, 1965, **7**, 145.
[85] H. Sinn and W. Kaminsky, *Adv. Organomet. Chem.*, 1980, **18**, 99.

Sinn et al.[86] have shown that cyclopentadienyl derivatives of zirconium, e.g., bis(cyclopentadienyl) dimethylzirconium in conjunction with alkyl aluminoxanes are exceptionally active catalysts for the polymerization of ethylene, and they report activities in excess of 10^8 g PE/g Zr. Using concentrations of aluminoxanes of between 1.5×10^{-2} and 3×10^{-3} mol dm^{-3} and concentrations of Cp$_2$Zr(CH$_3$)$_2$ < 10^{-6} mol dm^{-3}, the polymerization reaction was found to be sufficiently slow for monomer diffusion not to control the polymerization rate. The catalyst activity in this range was found to be about 10^7 g PE/g Zr, with the time between insertion steps of the order of 0.3 ms. This value is comparable with the changeover times of fast enzyme reactions.

A large excess of aluminoxane is required[86] and the rate of polymerization was found to be proportional to the zirconium concentration and to depend quadratically on the aluminoxane. After an induction period the polymerization rate reached a maximum value and then remained constant for over 20 h, indicating a 'living' polymerization system. As a speculative model for the active centre, Kaminsky[87] has proposed structure (18).

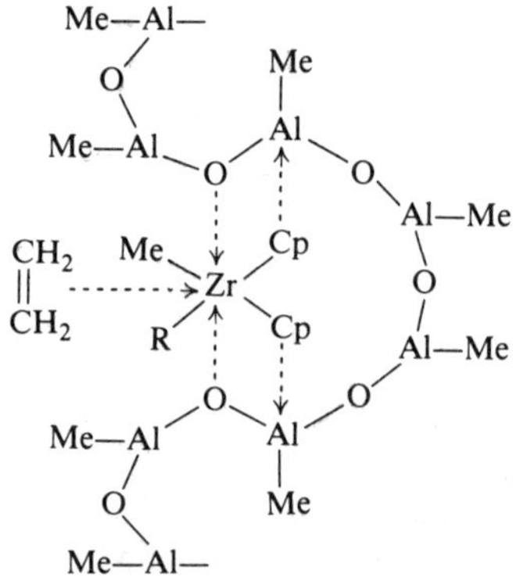

(18)

These catalysts can also be used to polymerize propylene, producing atactic polypropylene, and to copolymerize ethylene and propylene, as well as to prepare copolymers of propylene and ethylene. Kaminsky[87] has also reported that carbon hydrates such as starch grains or cellulose can be graft polymerized, with the starch grains becoming covered by polyethylene.

3 Copolymerization Studies

The field of copolymerization of ethylene and other α-olefins using co-ordination polymerization catalysts has been one of considerable activity during the past years, although nearly all of the publications have been by way of patent applications. Considerable interest stems from the fact that it is now possible by means of suitable co-ordination catalysts to copolymerize ethylene and other α-olefins, at low

---

[86] H. Sinn, W. Kaminsky, H.-J. Vollmer, and R. Woldt, *Angew. Chem., Int. Ed. Engl.*, 1980, **19**, 390.
[87] W. Kaminsky, in 'Transition Metal Catalysed Polymerizations, Alkenes and Dienes, Part A', ed. R. P. Quirk et al., Harwood Academic, New York, 1983, p. 225.

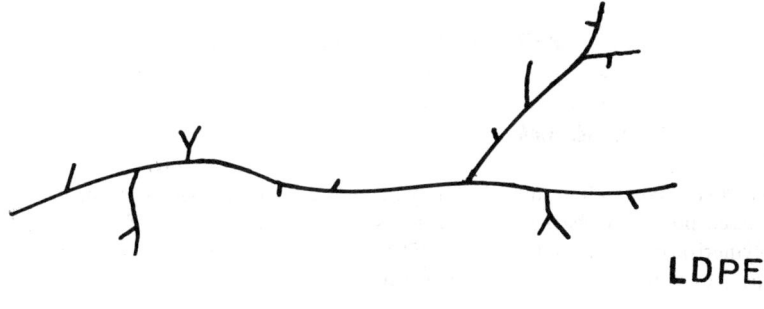

LDPE

LLDPE

Figure 5 *Schematic representations of* LDPE *and* LLDPE *molecules*

temperatures and pressures, to produce a linear low-density polyethylene, LLDPE, which simulates many of the properties of low-density polyethylene, LDPE. The discovery in itself is not new, since Du Pont of Canada has produced such resins[88,89] for several years on a commercial basis under the trade names of Sclair or Rotothene. The commercial potential of these resins has been given a considerable boost in recent years due to the publicity and claims made for the Union Carbide gas phase Unipol process.[90,91]

The molecules of LLDPE consist of long sequences of methylene units with periodic short side branches introduced *via* the incorporation of the comonomer. In contrast, the LDPE molecules contain a variety of types of short-chain as well as long-chain branches.[92] Figure 5 gives schematic representations of the molecular chains for these two types of polyethylene.

The α-olefin used depends on the catalyst and process employed. The use of but-1-ene, 4-methyl pent-1-ene, hex-1-ene, oct-1-ene, dec-1-ene, octadec-1-ene, tetradec-1-ene, *etc.*, have all been quoted in the patent literature.[88,89] For gas-phase processes but-1-ene is preferred, while for solution processes higher α-olefins, *e.g.*, hex-1-ene or oct-1-ene may be used. Short[93] has described the use of different olefins in a number of commercial operations.

---

[88] Du Pont Canada, U.S. P. 3 645 992, 1972.
[89] Du Pont, U.S. P. 4 076 698, 1978.
[90] Union Carbide, U.S. P. 4 011 382, 1977.
[91] Union Carbide, U.S. P. 4 003 712, 1977.
[92] J. P. Hogan, B. E. Nasser, and R. T. Werkman, Symposium on Olefin Copolymers, I.U.P.A.C. Meeting, Boston, MA, July 1971.
[93] J. N. Short, 'Kirk-Othmer Encyclopedia of Chemical Technology', 1981, Vol. 16, p. 385.

PART II Cationic Polymerization
by J. M. Rooney

1 Introduction

This review covers literature dealing with polymerizations which are propagated through positively charged active centres. A book on the topic of carbocationic polymerizations was published in 1981,[1] and a review dealing with initiation processes in these systems also appeared.[2]

2 Initiation Systems

Metal Alkyl and Metal Halide Compounds.—In principle, metal alkyls or metal halides can initiate the polymerization of nucleophilic monomers either through direct addition or through formation of cations by reaction with co-initiators. A competing reaction involves complexation of the metal halide by monomer.[3] Co-initiators may be adventitious or deliberately added. For example, when butan-1-ol is added to a BF_3-cyclopentene mixture, initiation occurs through protonation of the monomer.[4] In analogous systems, the order of mixing the reagents has been shown to affect product yields and distributions.[5] Since metal alkyls can also function as co-ordination catalysts, the choice of co-initiators can influence the mechanism of polymerization as studies on induced stereoregularity have confirmed.[6—8] Halogens can be used as co-initiators, generating halonium ions.[9] Metal halides are readily attached to polymeric supports and,[10,11] where direct addition does not occur, can be recycled for a limited number of cycles due to partial deactivation. One of the few instances of a direct addition initiation by a metal halide involves the formation of a zwitterion when BF_3 reacts with epichlorohydrin.[12] In general, yields of carbocations from the interaction of pure metal halides with hydrocarbon monomers are quite low.[13]

Organic Cation Salts.—By simply altering the nature of the co-initiator, metal halides may be used to generate stable or semi-stable carbocation salts.[14] In contrast

---

[1] J. P. Kennedy and E. Marechal, 'Carbocationic Polymerization', Wiley, New York, 1981.
[2] J. P. Kennedy and E. Marechal, *Macromol. Rev.*, 1981, **16**, 123.
[3] F. M. Nasirov, F. R. Khalafov, Z. M. Alieva, N. E. Mel'nikova, and T. N. Shakhtakhtinskii, *Dokl. Akad. Nauk Az. SSR*, 1981, **37**, 54.
[4] A. Onopchenko, B. L. Cupples, and A. N. Kresge, *Macromolecules*, 1982, **15**, 1201.
[5] T. Higashimura, Y. Miyoshi, and H. Hasegawa, *J. Appl. Polym. Sci.*, 1982, **27**, 2593.
[6] J. Kops and H. Spanggaard, *Macromolecules*, 1982, **15**, 1200.
[7] M. Biswas and G. C. Mishra, *J. Polym. Sci., Polym. Chem. Ed.*, 1981, **19**, 3081.
[8] T. Kawamura and K. Matsuzaki, *Makromol. Chem.*, 1981, **182**, 3003.
[9] M. DiMaina, P. Narducci, and G. Pizzirani, *Chim. Ind. (Milan)*, 1981, **63**, 253.
[10] Y. A. Sangalov, I. F. Gladkikh, and K. S. Minsker, *Vysokomol. Soedin., Ser. B*, 1982, **24**, 356.
[11] J. F. Kinstle and G. L. Quinlan, *Am. Chem. Soc., Div. Polym. Chem., Polym. Prepr.*, 1981, **22**(1), 166.
[12] T. V. Grinevich, A. N. Shupik, G. V. Korovina, and S. G. Entelis, *Eur. Polym. J.*, 1981, **17**, 1107.
[13] M. Masure, N. A. Hung, G. Sauvet, and P. Sigwalt, *Makromol. Chem.*, 1981, **182**, 2695.
[14] V. S. Byrikhin, A. S. Maloshitskii, V. B. Murachev, E. A. Ezhova, A. I. Nesmelov, and A. N. Pravednikov, *Dokl. Akad. Nauk SSR*, 1981, **261**, 657.

to the butan-1-ol/BF_3 system, tertiary alcohols yield the corresponding carbenium ions when treated with BCl_3.[15] Salts of this type, including acetyl,[16] terephthaloyl,[17] oxopropylium,[18] and particularly the triphenylmethyl (trityl)[19] salts, are being used with increased frequency in studies of cationic initiation. Two principal modes of initiation have been observed: (1) direct addition, which occurs when trityl carbenium ions react with thietanes[20] or certain substituted alkoxystyrenes[21] and (2) hydride abstraction, which can occur when the same initiators react with less nucleophilic monomers.[22,23] The relative stability of these initiators permits analysis of the effects of the nature of the counterion on the course of the polymerization.[24]

One interesting class of organic cation salts requires the presence of a co-initiator (typically water) to produce a protonic acid which initiates polymerization.[25] For example, a mixture of trisacetylacetonato-silicon(IV) hexafluoroantimonate and epoxy resin remains stable for several weeks in the absence of moisture but solidifies in hours when exposed to ambient moisture at room temperature.

Photochemical and Radiation Initiation.—The topics of photochemical initiation of cationic polymerizations,[26] its relationship to macromolecular synthesis,[27] and radiation-induced cationic polymerization together with synthetic applications have been reviewed recently.[28]

Diaryliodonium and triarylsulphonium[29–31] salts have been used widely as photoinitiators for cationic polymerizations. Ion-pair dissociation equilibria for these compounds suggest that ion clustering must be prevalent under typical reaction conditions.[32] The scope of possible applications for the salts has been increased by photosensitization which renders them susceptible to longer wavelength light.[33–35] Further extensions of their use in non-photochemical redox systems are outlined in reactions (1)—(4), where AH_2 represents ascorbic acid, which

$$AH_2 + 2\,CuY_2 \rightarrow DA + 2\,CuY + 2\,HY \qquad (1)$$

$$Ar_2I^+X^- + CuY \rightarrow CuXY + ArI + Ar^{\bullet} \qquad (2)$$

[15] H. A. Nguyen and J. P. Kennedy, *Polym. Bull. (Berlin)*, 1981, **6**, 41.
[16] L. Garrido, J. Guzman, and E. Riande, *Makromol. Chem., Rapid Commun.*, 1981, **2**, 379.
[17] R. Alamo, J. Guzman, and J. G. Fatou, *Makromol. Chem.*, 1981, **182**, 725 and 731.
[18] S. D. Pask and P. H. Plesch, *Makromol. Chem.*, 1981, **182**, 3031.
[19] J. Vohlidal, M. Pacovska, and J. Dvorak, *Collect. Czech. Chem. Commun.*, 1981, **47**, 2351.
[20] S. M. Florquin and E. J. Goethals, *Makromol. Chem.*, 1981, **182**, 3371.
[21] J. M. Rooney, *Polym. Bull. (Berlin)*, 1982, **8**, 101.
[22] I. Panayotov, N. Manolova, and R. Velichkova, *Polym. Bull. (Berlin)*, 1981, **4**, 653.
[23] R. Velichkova, I. M. Panayotov, J. Doicheva, G. Heublein, H. Schutz, P. Adler, S. Spange, and R. Wondraczek, *J. Polym. Sci., Polym. Chem. Ed.*, 1982, **20**, 2895.
[24] T. Kunitake and K. Takarabe, *Makromol. Chem.*, 1981, **182**, 817.
[25] J. A. Cella, A. W. Schwabacher, and A. Schultz, *Am. Chem. Soc., Div. Polym. Chem., Polym. Prepr.*, 1981, **22**(2), 113.
[26] S. P. Pappas, *Radiat. Curing*, 1981, **8**, 28.
[27] T. Saegusa, *Top. Curr. Chem.*, 1982, **100**, 75.
[28] V. T. Stannett, *Br. Polym. J.*, 1981, **13**, 93.
[29] R. S. Davidson and J. W. Goodin, *Eur. Polym. J.*, 1982, **18**, 589.
[30] J. V. Crivello, *Dev. Polym. Photochem.*, 1981, **2**, 1.
[31] J. V. Crivello and J. L. Lee, *Polym. Photochem.*, 1982, **2**, 219.
[32] A. Ledwith, S. Al-Kass, D. C. Sherrington, and P. Bonner, *Polymer*, 1981, **22**, 143.
[33] A. Ledwith, *Am. Chem. Soc., Div. Polym. Chem., Polym. Prepr.*, 1982, **23**(1), 323.
[34] L. R. Gatechair and S. P. Pappas, *Am. Chem. Soc., Org. Coat. and Appl. Polym. Sci. Proc.*, 1981, **46**, 701.
[35] J. V. Crivello and J. L. Lee, *Macromolecules*, 1981, **14**, 1141.

$$AH_2 + 2\,CuXY \rightarrow DA + 2\,CuY + 2\,HX \tag{3}$$

$$nM + HX \rightarrow H(M)_{n-1}M^+X^- \tag{4}$$

reduces the copper(II) salt and yields dehydroascorbic acid, DA. The aryliodonium salt is in turn reduced by the copper(I) product, forming a mixed copper salt which reacts with ascorbic acid to generate protonic acid initiators.[36,37] Iodonium and sulphonium compounds have been found to enhance the rates of γ-radiation-induced cationic polymerizations of styrene, apparently functioning as sources of counterions which can stabilize the propagating styryl carbenium ions.[38] Diaryliodonium hexafluorophosphate complexes also serve as oxidizing agents, according to reactions (5)—(7), promoting formation of dimer radical cations, D^+ from radicals, $R^\bullet$, and monomer, M.[39]

$$R^\bullet + Ar_2I^+PF_6^- \rightarrow R^+ + PF_6^- + ArI + Ar^\bullet \tag{5}$$

$$R^+ + M \rightarrow M^{\ddagger} + R^\bullet \tag{6}$$

$$M^{\ddagger} + M \rightarrow D^{\ddagger} \tag{7}$$

The concept of using metal ions as electron acceptors in photoinitiated ionic polymerizations has been applied recently to tetrahydrofuran–metal triflate systems.[40]

Pulse radiolytic techniques have been applied to the study of the fundamental processes of radiation-induced cationic polymerizations of styrene[41] and analogous monomers.[42] Reaction temperature exercises a significant influence on the product distribution in these systems.[43] The degree of monomer dryness is equally important, since water acts as a degradative chain transfer agent.[44] A wide range of radiation dose rates, obtained with an electron beam source,[45] produced polybutadienes of similar characteristics,[46,47] although discrepancies in the behaviour of bulk and solvent systems were noted. Where the solvent concerned has a relatively high dielectric constant, specific solvation of propagating ions is thought to effect a reduction in propagation rates[48,49] with implications for synthetic reactions as well.[50,51]

---

[36] J. V. Crivello and J. L. Lee, *Am. Chem. Soc., Div. Polym. Chem., Polym. Prepr.*, 1981, **22**(2), 114.
[37] J. V. Crivello and J. H. W. Lam, *J. Polym. Sci., Polym. Chem. Ed.*, 1981, **19**, 539.
[38] S. Mah, Y. Yamamoto, and K. Hayashi, *J. Polym. Sci., Polym. Chem. Ed.*, 1982, **20**, 1709.
[39] S. Mah, Y. Yamamoto, and K. Hayashi, *J. Polym. Sci., Polym. Chem. Ed.*, 1982, **20**, 2151.
[40] M. E. Woodhouse, F. D. Lewis, and T. J. Marks, *J. Am. Chem. Soc.*, 1982, **104**, 5586.
[41] K. Hayashi, *Mem. Inst. Sci. Ind. Res., Osaka Univ.*, 1982, **39**, 77.
[42] O. Brede, J. Boes, W. Helmstreit, and R. Mehnert, *Radiat. Phys. Chem.*, 1982, **19**, 1.
[43] Y. Yamamoto, M. Miki, and K. Hayashi, *Macromolecules*, 1981, **14**, 208.
[44] A. M. Adur and F. Williams, *J. Polym. Sci., Polym. Chem. Ed.*, 1981, **19**, 669.
[45] K. Hayashi and S. Okamura, *Radiat. Phys. Chem.*, 1981, **18**, 1133.
[46] K. Hayashi, Y. Tanaka, and S. Okamura, *J. Polym. Sci., Polym. Chem. Ed.*, 1981, **19**, 1435.
[47] K. Hayashi, K. Kagawa, and S. Okamura, *J. Polym. Sci., Polym. Chem. Ed.*, 1981, **19**, 1977.
[48] A. Deffieux, W. C. Hsieh, D. R. Squire, and V. T. Stannett, *Polymer*, 1982, **23**, 65.
[49] W. C. Hsieh, A. Deffieux, D. R. Squire, and V. T. Stannett, *Polymer*, 1982, **23**, 427.
[50] V. Ya. Kabanov, L. P. Sidorova, and R. D. Aliev, *ZFI-Mitt*, 1982, **43B**, 475.
[51] V. N. Kudryavtsev and V. Ya. Kabanov, *Vysokomol Soedin., Ser. A*, 1982, **24**, 401.

3 Propagating Species and Their Reactivities

Molecular orbital calculations provide a basis for attempting to predict the mechanisms and relative rates of initiation[52] and propagation reactions. Models of propagating carbenium ions and their complexes with monomer and counterion have been devised[53] and compared with experimental data.[54] However, the assumptions implicit in standard theoretical treatments of monomer reactivity are open to question and a new approach has been suggested.[55]

Styrene and Styrene Derivatives.—Although the kinetics of the fundamental processes involved in the polymerization of styrene monomers continue to receive attention,[56—58] there is a discernible shift towards the use of product analysis (facilitated by advances in instrumentation) as an investigative method. For example, studies of the oligomers and polymers obtained by cationically polymerizing α-methyl styrene in the presence of sterically hindered bases have led to the postulation of mechanistic schemes.[59,60] Ring substituents on α-methyl styrene can be used to adjust the nucleophilicity of the monomer providing information on the effect of this parameter on chain end equilibria.[61] The propensity of carbenium ions to attack aromatic substrates is demonstrated by the ready copolymerization of difunctional styrenes with dimethoxybenzene.[62] Indeed, with certain alkoxystyrenes indanyl dimer is formed exclusively as a result of internal ring attack.[63] Despite the difficulties posed by such side reactions, accurate measurements of the propagation rate constant for anethole have been made by spectroscopic techniques,[64] and selective preparations of linear dimers and trimers have been achieved.[65] The use of similar dimers as macromers in further polymerization was explored,[66] as was the polymerization of styryl monomers based on 1-phenyl-1,3-butadiene.[67] The structure and stabilization of poly(α-fluoro styrene) which bears a steric resemblance to poly(α-methylstyrene) were the subjects of a brief report.[68] Stereoregularity was shown to be influenced by the mode of initiation as is the case with other styrene derivatives.[69,70]

[52] F. L. Tobin, P. C. Hariharan, J. J. Kaufman, and R. S. Miller, *Int. J. Quantum Chem. Quantum Chem. Symp.*, 1981, **15**, 203.
[53] P. Hallpap, G. Heublein, C. Gruentzig, H. Zwanziger, and J. Reinhold, *Acta Polym.*, 1982, **33**, 358 and 362.
[54] G. Heublein and S. Spange, *J. Prakt. Chem.*, 1982, **324**, 187.
[55] C. Bunel and E. Marechal, *J. Polym. Sci., Polym. Phys. Ed.*, 1982, **20**, 131.
[56] W. Obrecht and P. H. Plesch, *Makromol. Chem.*, 1981, **182**, 1459.
[57] K. Takarabe and T. Kunitake, *Makromol. Chem.*, 1981, **182**, 1587.
[58] T. B. Bogomolova and A. R. Gantmakher, *Vysokomol. Soedin., Ser. B*, 1982, **24**, 214.
[59] Y. Kawakami, N. Toyoshima, T. Ando, and Y. Yamashita, *Polym. J.*, 1981, **13**, 947.
[60] J. P. Kennedy and T. Kelen, *J. Macromol. Sci. Chem.*, 1982, **18**, 129.
[61] R. W. Lenz, J. M. Jonte, and J. G. Faullimmel, *Am. Chem. Soc., Org. Coat. and Appl. Polym. Sci. Proc.*, 1981, **46**, 672.
[62] R. A. Smith, D. B. Patterson, and H. A. Colvin, *Am. Chem. Soc., Div. Polym. Chem., Polym. Prepr.*, 1982, **23**(2), 97.
[63] R. Alexander, A. Jefferson, and P. D. Lester, *J. Polym. Sci., Polym. Chem. Ed.*, 1981, **19**, 695.
[64] P. Cerrai, M. Tricoli, A. Lucchesi, and P. Giusti, *Conv. Ital. Sci. Macromol. (ATTI) 5th*, 1981, 213.
[65] T. Higashimura and M. Hiza, *J. Polym. Sci., Polym. Chem. Ed.*, 1981, **19**, 1957.
[66] M. Sawamoto and T. Higashimura, *Macromolecules*, 1981, **14**, 467.
[67] M. Wanatabe, H. Ishida, and I. Yamaji, *Kobunshi Ronbunshu*, 1981, **38**, 607.
[68] R. N. Majumdar, M. K. Niknam, H. A. Nguyen, and H. J. Harwood, *Makromol. Chem. Rapid Commun.*, 1982, **3**, 421.
[69] T. Kawamura, T. Uryu, and K. Matsuzaki, *Makromol. Chem.*, 1982, **183**, 125.
[70] J. C. Favier, M. Moreau, J. P. Vairon, and J. Leonard, *Polymer*, 1982, **23**, 1501.

Analysis of copolymerization products often provides a guide to the relative reactivities of propagating carbenium ions. Styrene has been used as the reference monomer in a survey of the reactivities of isopropenylnaphthalene[71] and substituted indenes.[72] The influence of reaction conditions on the copolymerization of α-methyl styrene with isobutyl vinyl ether,[73] isobutylene,[74] and indene[75] were analysed. This styrene derivative was also copolymerized cationically with cyclopentadiene and the oxidation resistance of the product examined.[76] Improved thermal resistance was sought in the copolymerization product of divinylbenzene and N-vinyl carbazole.[77]

Styrene derivatives can be designed so as to favour cyclopolymerization over linear propagation. For example, reaction (8) proceeds readily when the monomer is

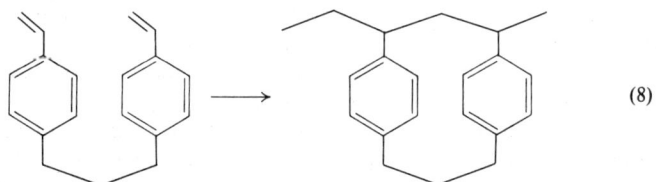

(8)

present in relatively low concentrations.[78] For monomers with fewer than three carbon atoms between the rings, the cyclic conformation is too strained to be achieved.

Hydrocarbon Monomers.—Control over the molecular weight of polyisobutylenes obtained from cationic systems can be exercised by the use of chain transfer agents such as alcohols and alkyl halides,[79] which can also function as co-initiators.[80] Low-molecular-weight isobutylene oligomers were synthesized by telomerization of the monomer[81] and by further polymerization of the dimer.[82] In a series of publications the copolymerization of isobutylene with various conjugated dienes and trienes was studied.[83—85]

Homopolymerizations of monomers containing more than one carbon–carbon multiple bond are conveniently studied by ^{13}C n.m.r. spectroscopy.[86] Cyclo-

---

[71] C. Bunel and E. Marechal, *Polymer*, 1981, **22**, 844.
[72] H. Garreau and E. Marechal, *Polym. Bull. (Berlin)*, 1981, **4**, 669.
[73] P. D. Trivedi, H. K. Acharya, and I. S. Bhardwaj, *Polym. Bull. (Berlin)*, 1981, **5**, 393.
[74] R. H. Wondraczek, W. Mueller, H. Scheutz, and G. Heublein, *J. Polym. Sci., Polym. Chem. Ed.*, 1982, **20**, 1517.
[75] H. E. Hotzel, R. H. Wondraczek, and G. Heublein, *Polym. Bull. (Berlin)*, 1982, **6**, 521.
[76] G. Heublein and G. Albrecht, *Acta Polym.*, 1982, **33**, 505.
[77] M. Biswas and G. V. Mishra, *Makromol. Chem. Phys.*, 1981, **182**, 261.
[78] J. Nishimura and S. Yamashita, *Am. Chem. Soc., Symp. Ser.*, 1982, **195**, 177.
[79] Y. N. Prokof'ev, A. P. Orlova, E. P. Kopylov, and P. G. Pautov, *Prom.-St. Sint. Kauch.*, 1982, 12.
[80] V. B. Murachev, E. A. Ezhova, S. I. Guzenko, V. S. Byrikhin, N. S. Kotsareva, L. K. Kurnosova, V. N. Zaboristov, V. I. Anosov, N. A. Knonvalenko, and A. N. Pravednikov, *Vysokomol. Soedin., Ser. B*, 1981, **23**, 308.
[81] M. Taha, G. Rigal, Y. Pietrasanta, N. Platzer, P. Sudres, and S. Raynal, *Makromol. Chem.*, 1981, **182**, 24.
[82] H. Hasegawa and T. Higashimura, *J. Appl. Polym. Sci.*, 1982, **27**, 171.
[83] A. Priola, C. Corno, M. Bruzzone, and S. Cesca, *Polym. Bull. (Berlin)*, 1981, **4**, 743.
[84] C. Corno, A. Priola, and S. Cesca, *Macromolecules*, 1982, **15**, 840.
[85] A. Priola, C. Corno, and S. Cesca, *Polym. Bull. (Berlin)*, 1982, **7**, 599.
[86] Z. Sharaby, N. Martan, and J. Jagur-Grodzinski, *Macromolecules*, 1982, **15**, 1167.

pentadiene remains reactive in the presence of metal halides even at extremely low temperatures,[87] and diacetylene must be handled with caution under cationic polymerization conditions.[88]

Because of the unusual ring-opening mechanism of the cationic polymerization of β-pinene, the propagating carbenium ion of this monomer resembles that of isobutylene as shown in reaction (9). The homopolymerization and copolymeriza-

$$CH_2 = \text{[bicyclic structure]} \xrightarrow{H+} Me-\text{[cyclohexene]}-\underset{Me}{\overset{Me}{\underset{|}{C}}}+ \qquad (9)$$

tion of β-pinene in the presence of metal halide catalysts have been examined recently.[89] Reactivity ratios were estimated for copolymerizations with styrene oxide, N-vinyl pyrrolidone,[90] and epichlorohydrin.[91] The unconjugated diene 1,4-dimethylene cyclohexane has also been homopolymerized cationically.[92]

Rigorously purified acenaphthylene was used in kinetic measurements which yielded a rate constant of propagation by unpaired carbenium ions.[93] A mechanism was postulated in which chromophore development was attributed to the formation of allylic cations from unsaturated chain ends. Chain-breaking was presumed to occur principally through transfer to monomer. Relative rate constants for chain-breaking reactions in the cationic polymerizations of vinyl naphthalenes have been determined by means of modified Schulz-Harborth plots.[94] The structure of cationically polymerized vinyl anthracenes was elucidated with the aid of spectroscopic studies on deuteriated model compounds.[95]

Vinyl Ethers and Vinyl Carbazole.—Deuteriated ethyl vinyl ether was employed in acetal addition experiments designed to model the propagation reaction in cationic polymerizations of the monomer,[96] and similar models for propenyl ether polymerizations were devised.[97] Although cationic polymerizations of vinyl ethers are prone to chain-breaking reactions, conditions can be adjusted to foster a 'quasi-living' polymerization. For example, at temperatures below −70 °C in methylene dichloride with carbenium salt initiators formed *in situ* and stable counterions, isobutyl vinyl ether polymerizations show a linear relationship between the monomer/initiator ratio and degree of polymerization.[98] The copolymerization of vinyl ethers and vinyl sulphides was investigated.[99]

The most reactive of monomers towards carbenium ions, N-vinyl carbazole, has attracted attention because the relative ease with which it undergoes cationic

[87] G. B. Sergeev and V. S. Komarov, *Vysokomol. Soedin., Ser. B*, 1982, **24**, 313.
[88] P. J. Russo and M. M. Labes, *J. Chem. Soc., Chem. Commun.*, 1982, 53.
[89] M. I. Valenzuela, J. Retuert, and F. O. Martinez, *Bol. Soc. Chil. Quim.*, 1982, **27**, 177.
[90] F. Martinez, R. Florshiem, and H. Martinez, *J. Polym. Sci., Polym. Chem. Ed.*, 1982, **20**, 1279.
[91] F. Martinez and I. Gajardo, *J. Polym. Sci., Polym. Chem. Ed.*, 1981, **19**, 1533.
[92] L. E. Ball, A. Sebenik, and H. J. Harwood, *Am. Chem. Soc., Symp. Ser.*, 1982, **195**, 207.
[93] S. D. Pask, P. H. Plesch, and S. B. Kingston, *Makromol. Chem.*, 1981, **182**, 3031.
[94] P. Blin, C. Bunel, and E. Marechal, *J. Polym. Sci., Polym. Chem. Ed.*, 1982, **19**, 891.
[95] J. Coudane, *Bull. Soc. Chim. Fr. I*, 1982, 114.
[96] H. Morii, S. Fujishige, K. Matsuzaki, and T. Uryu, *Makromol. Chem.*, 1982, **183**, 1445.
[97] K. Matsuzaki, H. Morii, N. Inoue, T. Kanai, and T. Higashimura, *Makromol. Chem.*, 1981, **182**, 2421.
[98] M. Sawamoto and J. P. Kennedy, *Am. Chem. Soc., Symp. Ser.*, 1982, **193**, 213.
[99] B. A. Trofimov, L. V. Morozova, and S. V. Amosova, *Zh. Prikl. Khim. (Leningrad)*, 1982, **55**, 2117.

polymerization in a wide range of solvents makes possible the study of the influence of solvent polarity on the stereochemistry of cationic polymerizations.[100] However, the high reactivity also renders attainment of truly homogeneous reagent concentrations difficult, possibly explaining the observed anomalies in polymer molecular weight distributions.[101] Indeed, vinyl carbazole derivatives are susceptible to photochemically and mechanochemically induced polymerizations in the solid state.[102]

Heterocyclic Monomers.—Tetrahydrofuran (THF) remains the most widely studied cationically polymerizable heterocycle. Elegant kinetic isotope experiments confirm the oxonium nature of the propagating ions and indicate the existence of a monomer complexation step preceding the propagation reaction.[103] Active centres have also been analyzed by trapping reactions with various phosphines.[104] When THF is dehydrogenated at the 4- and 5-positions, a highly reactive vinyl ether is produced[105] which undergoes metathesis polymerization in the presence of carbenes. Methyl-substituted THF derivatives are homopolymerized with great difficulty[106] but can be copolymerized readily with other cyclic ethers.[107] Copolymerization of THF with oxiranes can be controlled to yield closed ring structures such as crown ethers.[108,109]

The cationic polymerization of 2,2-dimethyloxactclobutane appears to propagate through both carbenium and oxonium ion species.[110] Extensive chain transfer reduces the molecular weight of the polymer product to about 1000. A detailed kinetic study of the sulphuric acid-induced polymerization of epichlorohydrin was undertaken,[111] as was a feasibility study on the polymerizability of 4,5-epoxy-2-pentenal isomers.[112] Copolymerizations of ethylene oxide and 7-oxabicyclo[2.2.1]heptane showed an alternating tendency.[113]

Under certain conditions cationic polymerizations of cyclic acetals can be considered as 'living' systems.[114] Stereoregularity can be induced in polymeric 4-bromo-6,8-dioxabicyclo[3.2.1]octane by use of the pure equatorially substituted stereoisomer.[115–117] Methylene-substituted dioxolanes and dioxepins polymerize in the presence of both cationic and free-radical initiators.[118,119] Selective syntheses of

---

[100] T. Kawamura, M. Sakuma, and K. Matsuzaki, *Makromol. Chem. Rapid Commun.*, 1982, **3**, 475.
[101] D. R. Terrell and F. Evers, *J. Polym. Sci., Polym. Chem. Ed.*, 1982, **20**, 2529.
[102] S. Tazuke, O. Supakorn, and T. Inoue, *J. Polym. Sci., Polym. Chem. Ed.*, 1982, **20**, 2239.
[103] V. A. Ponomarenko, E. L. Berman, A. M. Sakharov, and Z. N. Nysenko, *Eur. Polym. J.*, 1981, **17**, 1111.
[104] K. Matyjaszewski and S. Penczek, *Makromol. Chem.*, 1981, **182**, 1735.
[105] C. T. Thu, T. Bastelberger, and H. Hoecker, *Makromol. Chem., Rapid Commun.*, 1981, **2**, 383.
[106] L. Garrido, J. Guzman, and E. Riande, *Macromolecules*, 1981, **14**, 1132.
[107] M. Malanga and O. Vogl, *J. Polym. Sci., Polym. Chem. Ed.*, 1982, **20**, 2033.
[108] I. M. Robinson, E. Pechhold, and G. Pruckmayr, *Am. Chem. Soc., Symp. Ser.*, 1981, **172**, 197.
[109] M. Hu, J. Zhang, Y. Xu, and Y. Yang, *J. Liq. Chromatogr.*, 1982, **5**, 1423.
[110] J. Kops and H. Spanggaard, *Macromolecules*, 1981, **15**, 1225.
[111] G. A. Kazaryan, V. A. Sarkisyan, R. S. Arutyunyan, R. G. Grigoryan, and A. I. Kuzaev, *Vysokomol. Soedin., Ser. A*, 1981, **23**, 925.
[112] Y. Yokoyama and H. K. Hall, *J. Polym. Sci., Polym. Chem. Ed.*, 1982, **20**, 2195.
[113] M. Paci, F. Andruzzi, and G. Ceccarelli, *Macromolecules*, 1982, **15**, 835.
[114] W. Chwialkowska, P. Kubisa, and S. Penczek, *Makromol. Chem.*, 1982, **183**, 753.
[115] M. Okada, H. Sumitomo, and A. Sumi, *Macromolecules*, 1982, **15**, 1238.
[116] M. Okada, H. Sumitomo, and A. Sumi, *Polym. J.*, 1982, **14**, 59.
[118] I. Cho and M. S. Gong, *J. Polym. Sci., Polym. Lett.*, 1982, **20**, 361.
[119] Y. Yokoyama and H. K. Hall, *Macromolecules*, 1981, **14**, 471.

Chain Reaction Polymerization

macrocyclic esters are possible with controlled oligomerization of 6,8-dioxabicyclo-[3.2.1]octan-7-one.[120]

Influences of reaction conditions on the homopolymerization[121] and copolymerization[122] of trioxane have been examined. Ionic aggregation[123] and alcohol solvation[124] produced discernible effects on the kinetics of lactone polymerizations. Nitrogen-based heterocycles studied recently include lactams[125,126] aziridines,[127,128] and oxazolines.[129,130] Electron-withdrawing ring substituents were found to enhance the polymerizability of 2-phenyl-2-oxazolines. Polymeric 2-methyl-2-oxazoline formed in the presence of sulphobenzoic anhydride possess a zwitterionic nature and propagates without either transfer or termination.

Ring-opening polymerizations of 1-oxa-3-thiacyclopentane appear to propagate by expansion of the cyclic sulphonium ion in a manner analogous to that proposed for polymerizations of 1,3-dioxolane.[131] The resultant polymer resembles an ideally alternating copolymer of 1,3-dioxolane and 1,3-dithiolane. Polymeric 3-chlorothietane prepared in the presence of acid catalysts undergoes rearrangements assisted by neighbouring sulphide linkages in the polymer backbone.[132] Analogies between the cationic polymerizations of dioxolanes and cyclosiloxanes have also been drawn.[133]

4 Synthetic Studies

Polymers from Novel Monomers.—In general, the design of novel cationically polymerizable monomers incorporates reactive groups which are reminiscent of vinyl ethers, vinyl carbazole, and previously studied oxygen heterocycles.

Various synthetic routes to vinyl ether macromers have been explored. Chloroethyl vinyl ether, treated with metal alkoxides, reacts with catechols to yield difunctional products suitable for use in the formation of crown ether analogs.[134] Transvinylation involving vinyl acetate and fluorinated alcohols provided a pathway to fluoroalkyl vinyl ethers,[135] and vinylation of alcohols with acetylene was also optimized.[136] Interest in photopolymerizable monomers activated by cationic initiators led to kinetic and mechanistic studies on divinyl ethers.[137,138]

---

[120] I. Tajima, M. Okada, and H. Sumitomo, *Macromolecules*, 1981, **14**, 1180.
[121] L. Terlemezyan and M. Mihailov, *Eur. Polym. J.*, 1981, **17**, 1115.
[122] J. Fejgin, *Angew. Makromol. Chem.*, 1981, **97**, 189.
[123] E. B. Lyudvig, B. G. Belen'kaya, and A. K. Khomyakov, *Eur. Polym. J.*, 1981, **17**, 1097.
[124] B. G. Belen'kaya, E. B. Lyudvig, A. L. Izyuminkov, and Y. I. Kul'velis, *Vysokomol. Soedin., Ser. A*, 1982, **24**, 288.
[125] R. Alijev, M. Budesinsky, J. Kondelikova, and J. Kralicek, *Angew. Makromol. Chem.*, 1982, **105**, 107.
[126] L. N. Mizerovskii, *Vysokomol. Soedin., Ser. B*, 1982, **24**, 582.
[127] E. J. Goethals, A. Munir, R. Deveux, and L. Vandenberghe, *Makromol. Chem., Rapid Commun.*, 1982, **3**, 515.
[128] M. J. Han, J. Y. Chang, and Y. Y. Lee, *Macromolecules*, 1982, **15**, 255.
[129] S. Hashimoto and T. Yamashita, *J. Macromol. Sci., Chem.*, 1982, **17**, 559.
[130] S. Kobayashi, T. Tokuzawa, and T. Saegusa, *Macromolecules*, 1982, **15**, 707.
[131] E. Riande and J. Guzman, *Macromolecules*, 1981, **14**, 1511.
[132] M. P. Zussman and D. A. Tirrell, *Polym. Bull. (Berlin)*, 1982, **7**, 439.
[133] L. Wilczek and J. Chojnowski, *Macromolecules*, 1981, **14**, 9.
[134] G. B. Butler and Q. S. Lien, *Am. Chem. Soc., Symp. Ser.*, 1982, **195**, 149.
[135] V. S. Sukhinin and S. I. Mineev, *Zh. Vses. Khim. O-Va.*, 1981, **26**, 344.
[136] L. J. Mathias, J. B. Canterberry, and M. South, *J. Polym. Sci., Polym. Lett.*, 1982, **20**, 473.
[137] T. Nishikubo, T. Iizawa, A. Yoshinaga, and M. Nitta, *Makromol. Chem.*, 1982, **183**, 789.
[138] J. V. Crivello, J. L. Lee, and D. A. Conlon, *Proc. Radiat. Curing VI*, 1982, pp. 4—28.

Potential applications of vinyl carbazole derivatives as photo-conductive polymers prompted the synthesis and polymerization of trans-1-(3-vinyl-9-carbazolyl)-2-(9-carbazolyl) cyclobutane.[139]

With the desire to develop a non-migrating photo-stabilizer for polypropylene in mind, researchers synthesized 2-hydroxy-4-(2,3-epoxypropoxy) benzophenone and studied the ring-opening cationic polymerization by a calorimetric technique.[140] Polymerizations which result in little or no shrinkage can be achieved with spirocyclic orthoesters.[141] Cationic polymerizations of anhydro sugar derivatives through ring-opening reactions are often characterized by transfer and termination involving free hydroxyl groups. The effects of these chain-breaking reactions were analysed systematically.[142] A novel spirophosphorane, 5-phenyl-1,4,6,9-tetraoxa-5-phosphaspiro[4.4]nonane, was polymerized by a ring-opening mechanism in the presence of BF_3 etherate, trifluoromethanesulphonic acid and triethyloxonium tetrafluoroborate.[143]

Block Copolymerization.—The strategy of cationic block copolymerization involves either the direct sequential polymerization of two monomers or the preparation of a macro-initiator, which is then used to induce the polymerization of a second monomer. In the latter case either or both polymerizations may be cationic in nature.

Due to the relative instability of most cationic propagating species, the macro-initiator technique is far more common in practice. For example, the oxonium ion polymerization of THF may be terminated by the addition of sodium methacrylate[144] or potassium isopropenylbenzyloxide,[145] yielding acrylic or styryl macromers capable of participating in anionic or free-radical polymerizations. In a reversal of roles, THF is polymerized cationically by a polystyryl macromer prepared by deactivating an anionic polymerization.[146] Sequential cationic polymerization was observed when t-butyl aziridine was added to a 'living' poly-THF.[147] This aziridine also displays 'living' characteristics under certain conditions[148,149] and acrylic macromer derivatives have been prepared.[150]

Macro-initiators can be synthesized by the use of initiators which also function as transfer agents. With suitably chosen silanes and alkyl aluminium co-initiators, polyisobutylenes with Si–H and Si–Cl end groups were isolated.[151] Similar preparations of α,ω-diphenyl,[152] -dichloro,[153] -diacrylate,[154] and -diepoxide[155]

---

[139] T. Inoue and S. Tazuke, J. Polym. Sci., Polym. Chem. Ed., 1981, 19, 2861.
[140] Z. Manasek, J. Luston, and F. Vass, J. Macromol. Sci., Chem., 1982, 17, 653.
[141] E. Klemm, L. Haase, D. Gorski, and H. H. Hoerhold, Acta Polym., 1982, 33, 429.
[142] T. Uryu, K. Kitano, and K. Matsuzaki, J. Polym. Sci., Polym. Chem. Ed., 1982, 20, 2181.
[143] S. Kobayashi, M. Kaku, and T. Saegusa, Polym. Bull. (Berlin), 1981, 5, 325.
[144] M. Takaki, R. Asami, and T. Kuwabara, Polym. Bull. (Berlin), 1982, 7, 521.
[145] J. Sierra-Vargas, P. Masson, G. Beinert, P. Rempp, and E. Franta, Polym. Bull. (Berlin), 1982, 7, 177.
[146] M. Kucera, F. Bozek, and K. Majerova, Polymer, 1982, 23, 207.
[147] Y. Tezuka and E. J. Goethals, Am. Chem. Soc., Div. Polym. Chem., Polym. Prepr., 1981, 22(2), 313.
[148] E. J. Goethals, A. Munir, and P. Bossaer, Pure Appl. Chem., 1981, 53, 1753.
[149] A. Munir and E. J. Goethals, J. Polym. Sci., Polym. Chem. Ed., 1981, 19, 1985.
[150] E. J. Goethals and M. A. Vlegels, Polym. Bull. (Berlin), 1981, 4, 521.
[151] J. P. Kennedy, V. S. C. Chang, and A. Guyot, Adv. Polym. Sci., 1982, 43, 1.
[152] J. P. Kennedy, D. Y. L. Chung, and A. Guyot, J. Polym. Sci., Polym. Chem. Ed., 1981, 19, 2737.
[153] R. H. Wondraczek, J. P. Kennedy, and R. F. Storey, J. Polym. Sci., Polym. Chem. Ed., 1982, 20, 43.
[154] T. Liao and J. P. Kennedy, Polym. Bull. (Berlin), 1981, 6, 135.
[155] J. P. Kennedy, V. S. C. Chang, and W. P. Francik, J. Polym. Sci., Polym. Chem. Ed., 1982, 20, 2809.

Chain Reaction Polymerization

polyisobutylenes were achieved. An interesting extension of this work involves the preparation of three-arm star chlorine-terminated polyisobutylene and subsequent block copolymerization of α-methyl styrene from the chain ends.[156] More recently, these synthetic approaches have been applied to β-pinene polymerizations.[157]

The discovery that divinylbenzene polymerizes in the presence of acetyl perchlorate to yield linear polymers without cross-linking opened the way to the development of α,ω-bifunctional linear poly(divinyl benzene).[158]

Graft Polymerizations.—The techniques of cationic initiation involving alkyl halide co-initiators lend themselves readily to grafting reactions from polymeric substrates containing labile halides. A typical example of this method of grafting is the modification of polychloroprene with BCl_3 and isobutylene resulting in a polymer backbone with chlorine-terminated polyisobutylene side chains[159] which could then be used in further block copolymerizations.[160]

The possibility of rendering synthetic polymers biocompatible by cationic grafting of polysaccharide side chains has been raised.[161] In one series of experiments an anhydro sugar derivative monomer was grafted by a ring-opening mechanism from chlorosulphonated polyethylene in the presence of metal halide co-initiators. Model experiments demonstrated that grafting could occur at secondary and tertiary chlorine sites as well as at sulphonyl chloride sites.

Radiation-induced graft copolymerizations tend to be less selective in the site of attack on the substrate[162] but are profoundly affected by the physical state of the substrate[163] and reaction conditions.[164]

---

[156] J. P. Kennedy, S. C. Guhaniyogi, and L. R. Ross, *Am. Chem. Soc., Org. Coat. and Appl. Polym. Sci. Proc.*, 1981, **46**, 178.
[157] J. P. Kennedy, T. Liao, S. Guhaniyogi, and V. S. C. Chang, *J. Polym. Sci., Polym. Chem. Ed.*, 1982, **20**, 3219 and 3229.
[158] T. Higashimura, S. Aoshima, and H. Hasegawa, *Macromolecules*, 1982, **15**, 1221.
[159] J. P. Kennedy, S. S. Plamthottam, and B. Ivan, *J. Macromol. Sci., Chem.*, 1982, **17**, 637.
[160] J. P. Kennedy and S. S. Plamthottam, *Polym. Bull. (Berlin)*, 1982, **7**, 337.
[161] T. Uryu, A. Hagino, K. Terui, and K. Matsuzaki, *J. Polym. Sci., Polym. Chem. Ed.*, 1981, **19**, 2313.
[162] V. T. Stannett, *Br. Polym. J.*, 1981, **13**, 93.
[163] V. N. Kudryavtsev, V. Y. Kabanov, A. E. Chalykh, and V. I. Spitsyn, *Dokl. Akad. Nauk SSSR*, 1981, **261**, 418.
[164] V. Y. Kabanov, R. E. Aliev, V. N. Kudryavtsev, L. P. Sidorova, and V. I. Spitsyn, *Izv. Akad. Nauk SSSR, Ser. Khim.*, 1982, 875.

PART III Anionic Polymerization
by R. N. Young

The full texts of the papers presented at the 1980 ACS meeting on anionic polymerization have been published in book form[1] and are recommended as a source for valuable reviews of some thirty-five areas in anionic polymerization as well as much original work. Sigwalt has written a review[2] of ring-opening polymerizations of heterocycles.

In general, over the last two years there has been a reduction in the amount of fundamental work on anionic polymerization and a growth in the application of the anionic technique to the synthesis of block and other polymers having well-defined architectures. In the light of the many unanswered basic questions concerning mechanism and kinetics, this shifting of emphasis must be regarded with a little regret – without in any way denying the enormous importance of practical exploitation.

Shamanin, Melenevskaya, and Zgonnik believe[3] that the associated forms of polybutadienyl-lithium participate in propagation and in particular that the dimeric associate is responsible for 1,2-enchainment. The extent of aggregation of living polymers can be conveniently assessed by measurement of the viscosity of concentrated solutions, although Worsfold has expressed the view[4] that this approach may break down if the mean lifetime of an associate becomes sufficiently short compared with the lifetime of an entanglement. Published data of Wang and Szwarc[5] implied their measurement of solutions having viscosities in excess of 10^5 poise in an evacuated Ubbelohde viscometer. Fetters and Young noted[6] that 10^3 P is generally regarded as the upper limit for this technique. In their reply, Wang and Szwarc report[7] a typographical error in their original publication which, on correction, lowers their maximum viscosity to 2×10^4 P. Szwarc asserts[8] that 'there is no inherent limit for viscosity measurements, even in conventional types of viscometers, provided that the tubes are sufficiently wide'. His viscometer 'capillary' was 2 cm in length and 0.8 cm in bore.[7]

It is often assumed without experimental verification that the carbanionic end groups of hydrocarbon polymers have great longevity provided that suitably purified materials are employed under high vacuum conditions. Several recent studies have demonstrated that such pious optimism is frequently ill-founded. Vinogradova *et al.*[9] report that the molar conductance of polybutadienyl-lithium and polyisoprenyl-lithium increases dramatically in THF and in DME over the

---

[1] 'Anionic Polymerization, Kinetics, Mechanisms and Syntheses', ed. J. E. McGrath, ACS Symposium Series, 1981.
[2] P. Sigwalt, *Angew. Makromol. Chem.*, 1981, **94**, 161.
[3] V. V. Shamanim, E. Yu. Melenevskaya, and V. N. Zgonnik, *Acta Polym.*, 1982, **33**, 175.
[4] D. J. Worfsold, *J. Polym. Sci., Polym. Phys. Ed.*, 1982, **20**, 99.
[5] H. C. Wang and M. Szwarc, *Macromolecules*, 1980, **13**, 452.
[6] L. J. Fetters and R. N. Young, *Macromolecules*, 1982, **15**, 206.
[7] M. Szwarc and H. C. Wang, *Macromolecules*, 1982, **15**, 208.
[8] M. Szwarc, in ref. 1, p. 1.
[9] L. V. Vinogradova, N. I. Nikolaev, V. N. Sgonnik, B. L. Erussalimsky, G. V. Sinitsina, Ch. B. Tsvetanov, and I. M. Panayotov, *Eur. Polym. J.*, 1981, **17**, 517.

course of just a few hours. Simultaneously, the absorbances of their electronic spectra (which were exceedingly broad) decreased. The cause of these changes was not investigated. Podolskii et al. have reported[10] that the stability of the disodium salt of α-methylstyrene tetramer in THF is enhanced by the presence of monomer as a consequence of the formation of a donor–acceptor complex between these species. Ades, Fontanille, and Leonard[11] examined the bulk polymerization of α-methylstyrene with t-butyl-lithium and found significant decay of the absorption due to the anion within a matter of a few minutes at 25 °C; the final absorbance was non-zero. They attributed these changes to the loss of lithium hydride from the carbanionic chain end, followed by the addition of the hydride to residual monomer. The stability of the active chains was enhanced by the presence of an equimolar concentration of $NNN'N'$-tetramethylethylenediamine (TMEDA).

Brosse, Maidunny, and Soutif[12] have examined the metallation of polyisoprene and squalane (as a model for polyisoprene) by s-butyl-lithium complexed with TMEDA. The resulting products were treated with a variety of reagents including $Me_3SiCl, Me_2SO_4, C_6H_5CH_2Cl, (C_6H_5)_2CO$, and $(C_6H_5)_2CCH_2$. With the exception of Me_3SiCl, it was found that all these reagents led to products containing much larger amounts of *exo*-methylene structures than were present in the parent hydrocarbon as a result of the sequence:

$$\sim\sim\underset{|}{\overset{Me}{C}}=CH\sim\sim \xrightarrow{-H+} \sim\sim\underset{|}{\overset{CH_2}{C}}=CH\sim\sim \xrightarrow{X} \sim\sim\underset{\parallel}{\overset{CH_2}{C}}-CHX\sim\sim$$

Conditions for the optimization of the silylation were explored. The best conditions for the metallation of methyl-4 octene-4 were found[13] to employ TMEDA:s-BuLi in the ratio 1/2.

The increasing difficulty of replacing successive chlorine atoms in $MeSiCl_3$ was successfully exploited by Roovers and Toporowski[14] to prepare $(polystyrene)_2Si(Me)Cl$ from a slight excess of polystyryl-lithium over that demanded stoicheiometrically. This material was the intermediate in the synthesis of H-shaped polystyrene chains.

Teixeira-Barreira et al.[15,16] have made a systematic investigation of the reaction of alkylhalides with very low molecular weight oligomeric polyisoprenyl-lithium. In addition to the normal reaction product resulting from direct coupling, considerable amounts of products arising as a consequence of halogen–lithium exchange were identified:

$$RM_nLi + R'I \rightleftharpoons RM_nI + R'Li$$

$$RM_nI + RLi \rightarrow RM_nR + LiI$$

$$RM_nI + RM_mLi \rightarrow RM_nM_mR + LiI$$

---

[10] A. F. Podolski, A. A. Taran, V. V. Shamanin, and K. Kalnins, *Vysokomol. Soedin., Ser. A*, 1981, **23**, 2792.
[11] D. Ades, M. Fontanille, and J. Leonard, *Can. J. Chem.*, 1982, **60**, 564.
[12] J.-C. Brosse, Z. A. B. Maidunny, and J.-C. Soutif, *Makromol. Chem.*, 1982, **183**, 1595.
[13] J.-C. Brosse, Z. A. B. Maidunny, and J.-C. Soutif, *Makromol. Chem., Rapid Commun.*, 1982, **3**, 1.
[14] J. Roovers and P. M. Toporowski, *Macromolecules*, 1981, **14**, 1174.
[15] S. R. Teixeira-Barreira, J. Chaineaux, R. Mechin, and C. Tanielian, *J. Organomet. Chem.*, 1981, **212**, 11.
[16] S. R. Teixeira-Barreira, R. Mechin, and C. Tanielian, *Macromolecules*, 1982, **15**, 450.

In benzene solution, the extent of the formation of exchange products was much greater when starting with high ratios of monomer to t-butyl-lithium (1:2) than at lower ratios (1:1 to 3:1). The propensity for exchange was greater with ethyl iodide than with ethyl bromide or with methyl iodide. Extension of the study to pentadiene and 2,3-dimethylbutadiene showed that the living oligomers of the latter were particularly susceptible to exchange reactions. In general, as a consequence of exchange prior to coupling, different molecular weight distributions were obtained on terminating dienyl-lithiums of extremely low DP with alkylhalides than were found in the products of protonation.

The reaction of t-butylchloride with THF solutions of dianionic oligomers (DP $\simeq$ 4) of α-methylstyrene has been examined by East and Ellis[17] using spectrophotometry. The principal process led to the formation of isobutene (*ca.* 70% with the potassium salt); some introduction of the t-butyl residue into the polymer occurred but no quantitative estimate was made. The rate was slightly greater for the sodium salt than for the potassium.

Majid, George, and Barrie[18] have prepared poly(epichlorohydrin-g-styrene) by reaction of polystyryl anions under argon with polyepichlorohydrin. In benzene or toluene, grafting was accompanied by cross-linking resulting from lithium–chlorine exchange; such cross-linking was minimized when THF was employed as solvent. Hirao *et al.*[19] have developed synthetic procedures for the introduction of a primary amine group to the chain end of polystyrene. Reaction of $BrCH_2CH_2N(SiMe_3)_2$ with polystyryl-lithium was rapid even at -78 °C and was not accompanied by any coupling. An alternative and more efficient route was by reaction of polystyryl-lithium with $PhC(Et)NSiMe_3$.

A detailed investigation of the carbonation of polymeric organolithium compounds has been made by Quirk and Chen.[20] When high vacuum techniques and high-purity gaseous carbon dioxide were used, polystyryl-lithium and polyisoprenyl-lithium in benzene were found to yield mixtures of the corresponding carboxylic acids and ketones but no tertiary alcohols. It was concluded that the formation of alcohol reported by some workers is the consequence of hydrolysis by traces of water.

$$PLi \longrightarrow PCO_2Li \longrightarrow P-\underset{\underset{OLi}{|}}{\overset{\overset{OLi}{|}}{C}}-P \xrightarrow{H_2O} P_2CO \xrightarrow{PLi} P_3COLi$$

Catala *et al.* have reported[21] that reaction of oligomeric organolithiums with elemental sulphur in benzene yields predominantly coupled products having three or four sulphur atoms in the link. Duda and Penczek[22] have prepared copolymers of sulphur with propylene sulphide in benzene using sodium thiophenoxide complexed with dibenzo-18-crown-6; the products have the repeat unit

---

[17] G. C. East and H. A. Ellis, *J. Polym. Sci., Polym. Chem. Ed.*, 1981, **19**, 869.
[18] M. A. Majid, M. H. George, and J. A. Barrie, *Polymer*, 1981, **22**, 1104.
[19] A. Hirao, I. Hattori, T. Sasagawa, K. Yamaguchi, S. Nakahama, and N. Yamazaki, *Makromol. Chem., Rapid Commun.*, 1982, **3**, 59.
[20] R. P. Quirk and W. C. Chen, *Makromol. Chem.*, 1982, **183**, 2071.
[21] J. M. Catala, J. F. Boscato, E. Franta, and J. Brossas, in ref. 1, p. 483.
[22] A. Duda and S. Penczek, *Macromolecules*, 1982, **15**, 36.

{CH$_2$CH(Me$_3$)S$_x$} where x can be as large as eight. The same authors have written a review[23] of anionic copolymerization involving sulphur. Sulphinate and sulphone groups have been introduced to polystyrene by ring-metallation by means of the complex [t-butyl-lithium:TMEDA] followed by reaction with sulphur dioxide.[24] Some 30% of the rings were thus derivatized; the orientation of substitution was para:meta $\simeq$ 1:2.

$$-CH_2-CH- \qquad -CH_2-CH-$$
$$\quad\; | \qquad\qquad\qquad\quad\; |$$
$$\;\; C_6H_5 \longrightarrow \quad C_6H_4Li \longrightarrow$$

$$\qquad\qquad -CH_2-CH- \qquad\qquad -CH_2-CH-$$
$$\qquad\qquad\qquad\quad\; | \qquad\qquad\qquad\qquad\qquad\; |$$
$$\qquad\qquad\quad C_6H_4SO_2Li \longrightarrow \quad C_6H_4SO_2Me$$

Masson, Franta, and Rempp[25] have developed a route for the introduction of a terminal methacryloyl residue into polystyrene by reaction of polystyryl-lithium first with ethylene oxide and thereafter with methacryloyl chloride. The product of $\bar{M}_n$ of 1—2×10^3 was successfully chain-extended anionically to high polymer. Kucera, Bozek, and Majerova have shown[26] that difunctional polystyrylpotassium in THF can react with a dicationic silenium perchlorate in the proportion of one molecule of the former to two of the latter. With this stoicheiometry the original polymer is transformed with better than 95% efficiency into one bearing a cation at both ends, capable of initiating the polymerization of the THF on lowering the temperature.

Quirk and McFay[27] employed calorimetry to investigate the interaction of polystyryl-lithium and polyisoprenyl-lithium with TMEDA. They concluded from the dependence of the enthalpy change upon the [base]:[Li] ratio that the stoicheiometry involving polystyryl-lithium (PSLi) is

$$(PSLi)_2 + 2\,TMEDA \rightarrow 2\,[PSLi \cdot TMEDA]$$

The behaviour of polyisoprenyl-lithium (PILi) was much less amenable to unambiguous interpretation. Assuming that PILi is dimerically aggregated (as most authors agree) they tentatively proposed the scheme

$$(PILi)_2 + TMEDA \rightarrow (PILi)_2 \cdot TMEDA$$

$$(PILi)_2 TMEDA + TMEDA \rightarrow 2\,[PILi \cdot TMEDA]$$

Kminek, Kaspar, and Trekoval[28] made calorimetric measurements on the interaction of n-butyl-lithium with several Lewis bases. The largest enthalpy changes were found with TMEDA and dimethoxyethane and the smallest with diethyl ether and anisole. In a similar study of polystyryl-lithium they found[29] the enthalpy decreased in the sequence THF > TMEDA > DME whereas with polyisoprenyl-lithium the sequence was TMEDA > DME > THF (a base to lithium ratio of unity being used throughout). The same authors measured[29] the

---

[23] S. Penczek and A. Duda, *Pure Appl. Chem.*, 1981, **53**, 1679.
[24] A. J. Hagen, M. J. Farrall, and J. M. J. Frechet, *Polym. Bull.*, 1981, **5**, 111.
[25] P. Masson, E. Franta, and P. Rempp, *Makromol. Chem., Rapid Commun.*, 1982, **3**, 499.
[26] M. Kucera, F. Bozek, and F. Majerova, *Polymer*, 1982, **23**, 207.
[27] R. P. Quirk and D. McFay, *J. Polym. Sci., Polym. Chem. Ed.*, 1981, **19**, 1445.
[28] J. Kminek, M. Kaspar, and J. Trekoval, *Collect. Czech. Chem. Commun.*, 1981, **46**, 1124.
[29] I. Kminek, M. Kaspar, and J. Trekoval, *Collect. Czech. Chem. Commun.*, 1981, **46**, 2371.

enthalpy change arising from the cross-association of butyl-lithium with polyisoprenyl-lithium. They were unable to distinguish between the possibilities:

$$(BuLi)_6 + (PLi)_4 \rightleftharpoons 2\,(BuLi)_3(PLi)_2$$

and

$$(BuLi)_6 + (PLi)_2 \rightleftharpoons 2\,(BuLi)_3 PLi$$

However, the formation of a species containing five lithium atoms seems improbable.

Davidjan et al. have reported[30] on the effect of DME upon the molecular weight distribution, and upon the variation of the stereochemistry with molecular weight, when isoprenyl-lithium propagates at $-30\,°C$ at a [DME]:[Li] ratio of 0.01. The fraction having the lowest molecular weight (at 10% conversion) had the highest 3,4-content (76%) and the highest molecular weight fraction had the lowest 3,4-content (30%). As conversion increased these figures decreased – respectively to 49% and 24% (at 70% conversion). It was concluded that complexation of DME with the monomeric form of the chain end reduces the reactivity of the latter. It must be said, however, that the shape of their molecular weight distribution curve is very strange – not least in that it extends to a molecular weight of 10^6 at 10% conversion for a sample whose calculated $\bar{M}_n$ would seem to be ca. 8×10^3.

It is well known that the extent to which butadiene is polymerized by organolithium initiators in a 1,2-sense is greatly increased by the presence of Lewis bases. Halasa, Lohr, and Hall have reported[31] that particularly striking effects are induced by 1,2-dipiperidinoethane (DPE) and by the corresponding bismorpholino, pyrrolidino, and N-methylpiperazino-ethanes. At 5 °C and with a [DPE]:[Li] ratio of unity, the concatenation was 99% in a 1,2-sense. Raising the temperature, or decreasing the amount of base, decreased the 1,2-content. They proposed that propagation is through ionic allylic centres complexed with both DPE and butadiene.

Luxton et al. have shown[32] that, in the presence of TMEDA, the propagating chain end of polybutadienyl-lithium can cyclize by attacking a penultimate vinyl substituent to yield vinylcyclopentane structures amounting to as much as 45% by weight of the polymer.

Collet-Marti et al.[33] have extended their earlier study of the electronic and ^1H n.m.r. spectra of polyisoprenyl-lithium in the presence of TMEDA and of pentamethyldiethylenetriamine (PMDT). The addition of both bases caused the conformational equilibrium involving the allylic chain end to shift in favour of the cis isomer; at a base:lithium ratio (r) of ca. 1/2 the conformation became exclusively cis. The addition of base caused an upfield shift of the protons which reached a limit at $r \simeq 0.5$ with PMDT and a larger upfield shift with TMEDA at $r > 1$. It was concluded that complexes are formed having the compositions $(IsLi)_2 \cdot PMDT$, $(IsLi)_2 \cdot TMEDA$, and $IsLi \cdot TMEDA$.

Worsfold et al.[34] have employed ^{13}C n.m.r. spectroscopy to establish the influence

---

[30] A. Davidjan, N. Nikolaev, V. Sgonnik, V. Krasikov, B. Belenkii, and B. Erussalimsky, *Makromol. Chem.*, 1981, **182**, 917.

[31] A. F. Halasa, D. F. Lohr, and J. E. Hall, *J. Polym. Sci., Polym. Chem. Ed.*, 1981, **19**, 1357.

[32] A. R. Luxton, M. E. Burrage, G. Quack, and L. J. Fetters, *Polymer*, 1981, **22**, 382.

[33] V. Collet-Marti, S. Dumas, J. Sledz, and F. Schue, *Macromolecules*, 1982, **15**, 251.

[34] D. J. Worsfold, S. Bywater, F. Schue, J. Sledz, and V. Collet-Marti, *Makromol. Chem., Rapid Commun.*, 1982, **3**, 239.

of DPE upon the 1:1 adducts of t-butyl-lithium with isoprene and butadiene in cyclopentane solution. The introduction of DPE causes the γ protons of both adducts to move upfield, the movement being greater for the *cis* than for the *trans* protons. In the case of isoprene, the DPE caused the conformational equilibrium of the active centres to decrease from 66% *trans* in pure cyclopentane to 15% *trans* at a base:lithium ratio (r) of 0.95; larger amounts of base increased the *trans* content to 32% at $r = 3.65$. In the case of butadiene, the influence of DPE upon the active centre conformation was much smaller; the *trans* content changed little from its value of 77% in the absence of base to $r = 0.61$. With larger amounts of DPE there was a gradual increase in *trans* content to 93% at $r = 2.1$.

Milner and Young reported[35] on the ^{1}H n.m.r. spectroscopy of the complexation of polybutadienyl-lithium by TMEDA. The introduction of this base caused a large upfield shift of the γ protons of the active centres and a much smaller downfield shift of the β protons. These changes were attributed to an increase in ionic character consequent upon chelation of the lithium. With values of $r < 1$ there was a marked preference for the solvation of one conformer.

Matsuzaki *et al.* have determined[36] the ^{1}H and ^{13}C n.m.r. spectra of THF solutions of model compounds for lithium, potassium, and caesium poly(α-methylstyrene). The lithium compounds had a higher charge density at the α-carbon than did the potassium and caesium salts and also was the only one to exhibit phenyl rotation at elevated temperatures. Quantum calculations suggested that the larger cations interact with the phenyl ring.

Organodilithium initiators, free from the presence of polar additives, offer the prospect of the synthesis of novel block copolymers and of model networks having a narrow range of molecular weight between crosslinks. Consequently, the quest for such species has attracted a number of groups. Guyot *et al.* have reported[37] the generation of 'perfectly difunctional' initiators by the addition of t- or s-butyl-lithium to α,ω-bis(phenyl vinylidenyl) alkanes $CH_2=C(C_6H_5)(CH_2)_nC(C_6H_5)=CH_2$ or to α,ω-diisopropenyl diphenylalkanes $CH_2=C(Me)C_6H_4(CH_2)_nC_6H_4C(Me)CH_2$ in hexane solution. An excess of butyl-lithium was employed to hasten the reaction; the di-adduct precipitated from solution and could readily be freed from monofunctional species which are soluble. Reaction of the initiators with dienes results in dissolution and good-quality styrene–butadiene triblock copolymers were prepared and characterized.[38]

Lutz, Beinert, and Rempp report[39] that butyl-lithium in THF at -30 °C causes the reversible polymerization of 1,3-diisopropenyl-benzene to yield, at low conversions, polymer containing approximately one residual double bond per monomer unit. At higher conversions, gelation sets in. The isomeric 1,4-diisopropenylbenzene gels less rapidly. The same group has extended[40] its earlier work on the products of the reaction of one mole 1,3-diisopropenylbenzene with two moles of s-BuLi in benzene solution. The addition reaction is slow and rather complicated; all reaction mixtures contained oligomers as well as the diadduct, and

---

[35] R. Milner and R. N. Young, *Polymer*, 1982, **23**, 1636.
[36] K. Matsuzaki, Y. Shinohara, and T. Kanai, *Makromol. Chem.*, 1981, **182**, 1533.
[37] P. Guyot, J. C. Favier, H. Uytterhoeven, M. Fontanille, and P. Sigwalt, *Polymer*, 1981, **22**, 1724.
[38] P. Guyot, J. C. Favier, M. Fontanille, and P. Sigwalt, *Polymer*, 1982, **23**, 73.
[39] P. Lutz, G. Beinert, and P. Rempp, *Makromol. Chem.*, 1982, **183**, 2787.
[40] P. Lutz, E. Franta, and P. Rempp, *Polymer*, 1982, **23**, 1953.

ultimately precipitation occurred. Polystyrenes were prepared by using 'diadduct' solutions having varying proportions of residual unreacted isopropenyl groups, values of $\bar{M}_w/\bar{M}_n$ (from gpc) ranged from 1.24 at 75% residual unsaturation to 1.15 at ca. 0%. The authors noted, but were unable to account for, the complete conversion of the oligomeric initiator to difunctional initiator once styrene was introduced.

Methyl methacrylate, which was long regarded as something of an anionic Cinderella, has become a popular monomer in recent years, along with the other methacrylates. In large measure, this has been the consequence of the discovery of conditions under which side-reactions to propagation became negligible – the fruit largely of German and Japanese workers. Muller has written a useful review.[41] He has shown that the polymerization of t-butyl methacrylate with Na^+ or Cs^+ as counterions proceeds, even at room temperature, to yield a near monodisperse product ($\bar{M}_n/\bar{M}_n \simeq 1.01$). Linear Arrhenius plots gave for Na^+ the values $E = 30$ kJ mol^{-1} and $\log A = 8.5$, while for Cs^+ the corresponding parameters are 23 kJ mol^{-1} and 9.5. This difference between Na^+ and Cs^+ contrasts[42] with the situation where the monomer is methyl methacrylate; that system exhibits little difference between cations ($E = 19.5$ kJ mol^{-1}, $\log A = 7.3$). To judge from the A values, the t-butyl substituents give rise to lower steric requirement than do methyl groups; the E values suggest that the cation is more intimately in contact with the carbanion in the butyl ester, probably as a consequence of the inductive effect. Jeuck and Muller note[43] a linear dependence of $\log k_p$ and $(r_+ + 1.5)^{-1}$, where r_+ is the cation crystal radius (in Å), for methyl methacrylate in THF at -98 °C for the Li^+, Cs^+, 222-cryptated[44] Na^+ and the free carbanion. The points for K^+ and Na^+ (having almost exactly the same k_p as Cs^+) do not lie on the line. An unexpected observation[43] was that, whereas diphenylmethyl-lithium behaves straightforwardly, 1,1-diphenylhexyl-lithium gives rise to a product having a strongly bimodal distribution up to about 30% conversion. The tacticities of the high and low molecular weight fractions are markedly different. The cause of this behaviour was not established.

Hatada et al.[45] initiated the polymerization of d-8 methyl methacrylate using n-butyl-lithium in toluene at -78 °C and the resulting polymer and oligomer were analysed to establish the fate of the initiator. It was found that the polymer contained an average of 1.9 butyl groups per chain and the oligomer 1.7 groups. It was proposed that reaction of BuLi with monomer can generate butylisopropenyl ketone which can add to a propagating chain:

$$CH_2=\underset{\underset{}{Me}}{C}-CO_2Me \longrightarrow CH_2=\underset{\underset{OLi}{|}}{\overset{Me}{C}}-\underset{\underset{}{Bu}}{C}-OMe \longrightarrow CH_2=\underset{\underset{}{Me}}{C}-\underset{\underset{}{Bu}}{C}=O$$

$$\longrightarrow Bu(CH_2-\underset{\underset{CO_2Me}{|}}{\overset{Me}{C}})_n CH_2-\underset{\underset{COBu}{|}}{\overset{Me}{C}}Li$$

[41] A. H. E. Muller, in ref. 1, p. 441.
[42] A. H. E. Muller, *Makromol. Chem.*, 1981, **182**, 2863.
[43] H. Jeuck and A. H. E. Muller, *Makromol. Chem., Rapid Commun.*, 1982, **3**, 121.
[44] C. Johann and A. H. E. Muller, *Makromol. Chem., Rapid Commun.*, 1981, **2**, 687.
[45] K. Hatada, T. Kitayama, K. Fumikawa, K. Ohta, and H. Yuki, in ref. 1, p. 327.

This last species is of low reactivity and only a fraction of such structures actually succeed in adding further methyl methacrylate.

Piejko and Hoecker[46] have found that the adduct of s-butyl-lithium and 1,1-diphenylethylene in toluene gives rise to the formation of ketone-containing oligomers (DP ≤ 3) which can participate in the subsequent polymerization. Okamoto et al. have examined[47] the asymmetric selective polymerization of racemic methacrylates using cyclohexylmagnesium bromide complexed by (−)-sparteine. Matsuzaki, Tanaka, and Kanai conclude[48] that the polymerization of methyl methacrylate by phenylmagnesium bromide in toluene/THF proceeds through the intermediacy of three propagating species which propagate at different rates to yield products of different tacticities. Allen et al.[49] propose a similar kind of interpretation for initiators containing t-butyl and phenylmagnesium compounds.

Tritylcalcium chloride and the corresponding strontium and barium compounds have been employed as initiators for the polymerization of methyl methacrylate.[50] The resulting polymers had very broad molecular weight distributions which were regarded as due to the participation of several propagating species.

Poly(methyl methacrylate)s having a terminal hydroxyl group have been synthesized by Anderson et al.[51] Ethyl 3-lithiopropyl acetaldehyde acetal was added to 1,1-diphenylethylene and the product was used to initiate polymerization with 100% efficiency at −78 °C. After acid hydrolysis, completely functionalized polymer of fairly narrow molecular weight distribution ($\bar{M}_w/\bar{M}_n < 1.2$) was obtained.

$$\underset{\underset{\text{OEt}}{|}}{\text{MeCHO(CH}_2)_4\text{CPh}_2\text{Li}} \longrightarrow \text{HO(CH}_2)_4\text{CPh}_2(\text{MMA})_n\text{H}$$

The same authors found that essentially monodisperse poly(methyl methacrylate) could be obtained using n-butyl-lithium in the range −78 to −20 °C in a 30:70 pyridine–toluene mixture. Analysis of the polymer showed that the actual initiator was the adduct:

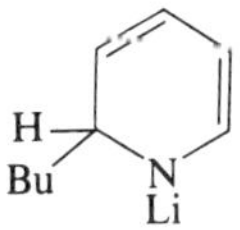

Similar results were obtained in THF using equimolar quantities of s- or t-butyl-lithium, but only in the temperature range −78 to −65 °C. It would be interesting to know how significant such ring additions are with vinyl pyridines – processes known to occur, but often ignored.

Soum and Fontanille[52] have interpreted the ^{13}C n.m.r. spectrum of poly(2-vinylpyridine) initiated by organomagnesium compounds in non-polar solvents to conclude that the propagation is first-order Markov in character. The spectrum of

---

[46] K. E. Piejko and H. Hoecker, Makromol. Chem., Rapid Commun., 1982, 3, 243.
[47] Y. Okamoto, K. Urakawa, and H. Yuki, J. Polym. Sci., Polym. Chem. Ed., 1981, 19, 1385.
[48] K. Matsuzaki, H. Tanaka, and T. Kanai, Makromol. Chem., 1981, 182, 2905.
[49] P. E. M. Allen, M. C. Fisher, C. Mair, and E. H. Williams, in ref. 1, p. 185.
[50] W. E. Lindsell, F. C. Robertson, I. Soutar, and D. H. Richards, Eur. Polym. J., 1981, 17, 107.
[51] B. C. Anderson, G. D. Andrews, P. Arthur, H. W. Jacobson, L. R. Melby, A. J. Playtis, and W. H. Sharkey, Macromolecules, 1981, 14, 1599.
[52] A. Soum and M. Fontanille, Makromol. Chem., 1982, 183, 1145.

benzyl-α-picolylmagnesium – a model for the active chain end – was in accord with sp^3 hybridization. A mechanism was proposed to account for the highly isotactic propagation in which the magnesium is co-ordinated by the terminal and penultimate pyridine rings.

The polymerization of 1,3-cyclohexadiene by lithium naphthalene in THF at −20 °C yields polymers having fairly narrow molecular weight distributions. The initiation mechanism is complex.[53]

The reactivity ratios for the copolymerization of styrene and p-t-butylstyrene initiated by s-butyl-lithium in benzene have been determined.[54] Within experimental error the values are unchanged when THF is added at a level of 10 molecules per lithium atom. Copolymerization of styrene and butadiene in hexane by lithium morpholinide results in the formation of a tapered block copolymer, the butadiene being the more reactive monomer.[55] The addition of small amounts of ethers or tertiary amines randomizes the distribution of the monomers and the vinyl content could be varied from 11 to 95% depending upon the choice of modifier. Sequential polymerization of styrene and 1-phenyl-1,3-butadiene (or the 2-phenyl isomer) to generate AB diblocks proceeds with 100% efficiency in toluene regardless of which monomer is chosen as the first.[56] However, when the solvent is THF, efficient synthesis of diblock is only possible when styrene is the first monomer to be polymerized.

---

[53] Z. Sharaby, J. Jagur-Grodzinski, M. Martan, and D. Vofsi, *J. Polym. Sci., Polym. Chem. Ed.*, 1982, **20**, 901.
[54] J. Chen and L. J. Fetters, *Polym. Bull.*, 1981, **4**, 275.
[55] T. C. Cheng, *Macromolecules*, 1981, **14**, 664.
[56] Y. Tsuji, T. Suzuki, Y. Watanabe, and Y. Takegami, *Macromolecules*, 1981, **14**, 1194.

PART IV Radical Polymerization
by J. C. Bevington

1 Introduction

Much of the recent work on radical polymerization can be regarded as extending or improving earlier investigations, but there have been some developments which may prove to be very significant. They include studies of the initiation process in detail not previously achieved, the preparation of polyesters by radical polymerization, the discovery of methods for the controlled preparation of 'pure' block and graft copolymers by procedures involving radicals, and the first reports of a new type of transfer process which may be valuable for the production of useful oligomers. These topics are discussed in later sections of this article.

Recent general papers include a review[1] of abnormal groups in polymers, their formation and their significance. End-groups have been considered in the special case of poly(vinyl chloride),[2] a material for which structural irregularities are probably of great importance as sites for instability.

Attention continues to be paid to deviations from 'ideal' kinetics for certain radical polymerizations even in homogeneous systems at low conversions. In this connection, primary radical termination and degradative transfer, including reaction with initiator, have been discussed again. The 'hot radical' theory has been applied to systems of several types including polymerizations and copolymerizations involving methyl and ethyl acrylates[3—5] and transfer to carbon tetrachloride.[6] Complexing between growing radicals and monomer or solvent has been considered again to explain the kinetics of the polymerization of butyl acrylate in aromatic solvents.[7] A critical comparison[8] has been made between the hot radical and complexing theories. There has been an attempt[9] to put on a quantitative basis the view that kinetic deviations might arise from preferential solvation of growing radicals by one of the components of the reaction mixture; the concentration of a reactant in the immediate vicinity of a reactive site is regarded as the effective concentration which may be distinctly different from the overall concentration normally used in kinetic treatments.

There have been further accounts of the large effects of additives, such as zinc chloride and ethyl aluminium sesquichloride, upon some radical polymeriza-

---

[1] D. H. Solomon, *J. Macromol. Sci., Chem.*, 1982, **17**, 337.
[2] T. Hjertberg and E. M. Sörvik, *J. Macromol. Sci., Chem.*, 1982, **17**, 983.
[3] I. Czajlik, T. Földes-Bereznich, F. Tüdös, and E. Vértes, *Eur. Polym. J.*, 1981, **17**, 131.
[4] A. Fehérvári, T. Földes-Bereznich, and F. Tüdös, *J. Macromol. Sci., Chem.*, 1982, **18**, 337.
[5] A. Fehérvári, E. Boros-Gyevi, T. Földes-Bereznich, and F. Tüdös, *J. Macromol. Sci., Chem.*, 1982, **18**, 431.
[6] T. Földes-Bereznich, M. Szesztay, E. Boros-Gyevi, and F. Tüdös, *J. Macromol. Sci., Chem.*, 1981, **16**, 977.
[7] M. Kamachi, M. Fujii, S.-I. Ninomiya, S. Katsuki, and S.-I. Nozakura, *J. Polym. Sci., Polym. Chem. Ed.*, 1982, **20**, 1489.
[8] J. K. Fink, *Polym. Bull. (Berlin)*, 1982, **7**, 159.
[9] J. Pavlinec, J. Jergušová, and Š. Florián, *Eur. Polym. J.*, 1982, **18**, 279.

tions.[10-12] Production of alternating copolymers under the influence of these additives can satisfactorily be explained by dominance of cross-propagations resulting from modification by complexing of the reactivities of monomers.[13-15]

2 Initiation

New information has been obtained on the reactions with monomers of several common initiating radicals.[16-18] In some cases, interaction is not solely by tail-addition of the radical to the unsubstituted end of the double bond; there are significant contributions from head-addition, hydrogen-abstraction, and aromatic substitution for appropriate monomers.[19-21] Head-addition of $C_6H_5.CO.O.$ to vinyl acetate is, as expected, more pronounced than to styrene[22] but its importance is considerably greater than that of head-to-head propagation. The overall reactivities of monomers towards $C_6H_5.CO.O.$ closely resemble those found earlier from end-group analyses of polymers.[23]

Appreciable initiation by hydrogen-abstraction (or any other process not leading to incorporation in polymer of an initiator fragment) would have serious implications for derivation of quantities such as kinetic chain length from analyses of polymers for initiator fragments.[24] The new results of the preceding paragraph were obtained by using radical traps so that growth of polymer radicals was stopped at a very early stage. It is conceivable that certain small radicals, *e.g.*, those formed by hydrogen-abstraction from monomer, might engage in side-reactions and never grow into macromolecules even in the absence of radical traps. It is desirable therefore that tests for 'abnormal' initiation should be performed also on high polymers, presumably by special end-group analyses. Relevant information has been obtained from n.m.r. examination of polystyrene prepared using $(C_6H_5.^{13}CO.O)_2$.[25] Both $C_6H_5.CO.O.CH_2.CH(C_6H_5)-$ and $C_6H_5.CO.O.CH(C_6H_5).CH_2-$ groups were detected; some at least of the latter type resulted from head-addition of the primary radical.

The most notable feature of the n.m.r. method for examining the initiation process is the opportunity provided for obtaining information about the precise nature of the resulting end-groups and their adjacent monomer units. This point can be

---

[10] A. K. Srivastava and G. N. Mathur, *Polymer*, 1981, **22**, 391.
[11] E. L. Madruga and J. S. Román, *J. Polym. Sci., Polym. Chem. Ed.*, 1981, **19**, 1101.
[12] A. Merlin, D.-J. Lougnot, and J. P. Fouassier, *Eur. Polym. J.*, 1981, **17**, 755.
[13] C. H. Bamford and P. J. Malley, *J. Polym. Sci., Polym. Lett.*, 1981, **19**, 239.
[14] C. H. Bamford and P. J. Malley, *J. Chem. Soc., Faraday Trans. 1*, 1982, **78**, 2497.
[15] C. H. Bamford and X.-Z. Han, *J. Chem. Soc., Faraday Trans. 1*, 1982, **78**, 855 and 869.
[16] M. Kamachi, Y. Kuwae, and S. Nozakura, *Polym. Bull. (Berlin)*, 1982, **6**, 143.
[17] M. J. Cuthbertson, G. Moad, E. Rizzardo, and D. H. Solomon, *Polym. Bull. (Berlin)*, 1982, **6**, 647.
[18] E. Rizzardo, A. K. Serelis, and D. H. Solomon, *Aust. J. Chem.*, 1982, **35**, 2013.
[19] P. G. Griffiths, E. Rizzardo, and D. H. Solomon, *J. Macromol. Sci., Chem.*, 1982, **17**, 45.
[20] G. Moad, E. Rizzardo, and D. H. Solomon, *J. Macromol. Sci., Chem.*, 1982, **17**, 51.
[21] G. Moad, E. Rizzardo, and D. H. Solomon, *Macromolecules*, 1982, **15**, 909.
[22] G. Moad, E. Rizzardo, and D. H. Solomon, *Makromol. Chem., Rapid Commun.*, 1982, **3**, 533.
[23] J. C. Bevington and J. R. Ebdon, in 'Developments in Polymerization – 2', ed. R. N. Haward, Applied Science, London, 1979, p. 11.
[24] J. C. Bevington, in 'Macromolecular Chemistry' (Specialist Periodical Reports), ed. A. D. Jenkins and J. F. Kennedy, Royal Society of Chemistry, London, 1980, Vol. 1, p. 47.
[25] G. Moad, D. H. Solomon, S. R. Johns, and R. I. Willing, *Macromolecules*, 1982, **15**, 1188.

further illustrated by the results of studies on polymers and copolymers made using ^{13}C-azoisobutyronitrile.[26,27] From the n.m.r. spectra of the products, it was possible to compare the numbers of initiator fragments attached to the various types of monomer units in copolymers and then to derive quantitative information on the reactivities of monomers towards the $(Me)_2C(CN)$. radical. Examination by n.m.r. of polymers made using the azonitrile labelled with deuterium, carbon-13, or nitrogen-15[28] showed the absence of $(Me)_2C:C:N-$ end-groups. Spin-trapping studies[29] also gave no indication that the radical $(Me)_2C:C:N$. arises during thermolysis of the azonitrile. Clearly further consideration needs to be given to the formation of appreciable quantities of $(Me)_2C(CN).N:C:C(Me)_2$ when the azonitrile is decomposed in unreactive solvents.

For the polymerization of butadiene, initiation by .OH or certain aroyloxy radicals leads to polymer unexpectedly rich in 1,2-units near the ends of the chains.[30,31] End-group studies by n.m.r. methods have also been made for other polymers prepared using initiation by .OH radicals.[32,33]

Several new azo initiators have been described.[34—37] The use of peroxides in high-pressure polymerizations has been reviewed[38] and an account has been given of polymeric peroxides of various types.[39] Alkyl boranes with oxidizing agents[40] and bistetraphenylethylene orthosilicate[41] have been investigated as sources of radicals. Telomerizations brought about by redox initiation have been described.[42—45] Polymerizations involving perphosphates[46—48] and monopersulphate[49—51] have been reported. Quaternary ammonium salts have been used with potassium persulphate in non-aqueous systems.[52]

---

[26] J. C. Bevington, J. R. Ebdon, T. N. Huckerby, and N. W. E. Hutton, *Polymer*, 1982, **23**, 163.
[27] J. C. Bevington, T. N. Huckerby, and N. W. E. Hutton, *J. Polym. Sci., Polym. Chem. Ed.*, 1982, **20**, 2655.
[28] J. C. Bevington, T. N. Huckerby, and N. W. E. Hutton, *Eur. Polym. J.*, 1982, **18**, 963.
[29] J. C. Bevington, P. F. Fridd, and B. J. Tabner, *J. Chem. Soc., Perkin Trans. 2*, 1982, 1389.
[30] J. C. Brosse, M. Bonnier, and G. Legeay, *Makromol. Chem.*, 1982, **183**, 303.
[31] L. S. Bresler, E. N. Barantsevich, V. I. Polyansky, and S. S. Ivantchev, *Makromol. Chem.*, 1982, **183**, 2479.
[32] J.-C. Lenain and J.-C. Brosse, *Makromol. Chem., Rapid Commun.*, 1982, **3**, 609.
[33] J.-C. Brosse, J.-C. Lenain, and A. Sabet, *Makromol. Chem., Rapid Commun.*, 1982, **3**, 765.
[34] C. Oppenheimer and W. Heitz, *Angew. Makromol. Chem.*, 1981, **98**, 167.
[35] Y. Ouishi, K. Kodaira, and K. Ito, *Polymer*, 1982, **23**, 630.
[36] D. Braun and R. Jakobi, *Angew. Makromol. Chem.*, 1982, **105**, 217.
[37] O. Nuyken, H. Schuster, and R. Kerber, *Makromol. Chem.*, 1982, **183**, 1733.
[38] H. Seidl and G. Luft, *J. Macromol. Sci., Chem.*, 1981, **15**, 1.
[39] K. Kishore, V. Gayathri, and K. Ravindran, *J. Macromol. Sci., Chem.*, 1981, **16**, 1359.
[40] S. S. Ivanchev, L. V. Shumnyi, and V. V. Konovalenka, *Polym. Sci. USSR, Engl. Transl.*, 1980, **22**, 3000.
[41] D. Braun and R. Rengl, *Angew. Makrol. Chem.*, 1981, **98**, 265.
[42] B. Boutevin, C. Maubert, A. Mebkhout, and Y. Pietrasanta, *J. Polym. Sci., Polym. Chem. Ed.*, 1981, **19**, 499.
[43] B. Boutevin, C. Maubert, Y. Pietrasanta, and P. Sierra, *J. Polym. Sci., Polym. Chem. Ed.*, 1981, **19**, 511.
[44] S. Rubio, B. Serre, J. Sledz, F. Schué, and G. Chapelet-Letourneux, *Polymer*, 1981, **22**, 519.
[45] B. Boutevin, Y. Piétrasanta, and M. Taha, *Makromol. Chem.*, 1982, **183**, 2977, 2985, and 2995.
[46] P. L. Nayak, S. Lenka, and M. K. Mishra, *J. Polym. Sci., Polym. Chem. Ed.*, 1981, **19**, 839.
[47] S. Sarasvathy and K. Venkatarao, *Makromol. Chem., Rapid Commun.*, 1981, **2**, 219.
[48] S. Lenka and A. K. Dhal, *Eur. Polym. J.*, 1982, **18**, 347.
[49] R. K. Samal, P. R. Das, D. P. Das, M. C. Nayak, G. Panda, and G. V. Suryanarayan, *J. Polym. Sci., Polym. Chem. Ed.*, 1981, **19**, 2751.
[50] R. K. Samal, M. C. Nayak, and D. P. Das, *Eur. Polym J.*, 1982, **18**, 313.
[51] R. K. Samal, S. C. Satrusallya, and B. L. Nayak, *J. Polym. Sci., Polym. Chem. Ed.*, 1982, **20**, 2409.
[52] J. K. Rasmussen and H. K. Smith II, *Makromol. Chem.*, 1981, **182**, 701.

Symmetrical trifunctional initiators, one having a central azo group with flanking peroxide groups[53] and the other having three peroxide groups,[54] have been described. The latter has been used in kinetic and g.p.c. studies of the polymerization of styrene. When using an initiator of this type, polymer formed early in a reaction possesses reactive end-groups which later give radicals leading to renewed growth; to achieve the best results, the central group of the initiator should be the most labile. In some polyfunctional initiators, the unstable groups may be involved in facile concerted homolysis; the compound $(Me)_2C(O.O.CO.C_6H_5).N:N.C_6H_5$ very readily decomposes to acetone and nitrogen with benzoyloxy and phenyl radicals even at 10 °C.[55]

The perester $C_6H_5.CO.p-C_6H_4.CO.O.O.C(Me)_3$ exhibits interesting behaviour in photolysis; it dissociates thermally at a convenient rate at about 100 °C.[56] Substituted benzoyloxy radicals result from cleavage of the O—O bond and substituted phenyl radicals may be formed from them but the decarboxylation is comparatively unimportant when monomer is present. Both types of radical possess carbonyl groups which can be utilized for spectroscopic analyses for end-groups of polymer molecules and for chemical modification.

Transformation of anionic to radical centres can be achieved through the agency of organolead compounds.[57] The termination $-M^-Li^+ + R_3PbCl \rightarrow -M.PbR_3 +$ LiCl leads to end-groups which can furnish radicals by direct thermolysis or through reaction with a transition-metal salt. Such a system may be useful for production of block copolymers or of stereoblock homopolymers.

Recent papers on photo-initiation include a re-examination of the radical $C_6H_5.\dot{C}H.OMe$ formed from benzoin methyl ether or from benzophenone and benzyl methyl ether under irradiation. It has been shown[58] that the radical readily initiates polymerization and that its incorporation as an end-group is not due to primary radical termination. Tetraphenylphosphonium salts have been reported[59] as photo-initiators of polymerization through the agency of phenyl radicals.

The promotion in photo-systems of simultaneous radical and cationic polymerizations of N-vinyl carbazole has been considered again.[60-62] This explanation of unusual features of the polymerization initiated thermally by benzoyl peroxide has been challenged[63] but the occurrence in parallel of two types of polymerization has been confirmed[64] for systems containing the carbazole monomer, maleic anhydride, and azoisobutyronitrile.

---

[53] A. S. Shaikh, E. Comănitá, S. Dumitriu, and C. Simionescu, *Angew. Makromol. Chem.*, 1981, **100**, 147.
[54] S. S. Ivanchev, N. G. Podosenova, V. V. Konovalenko, T. A. Kuznetsova, E. G. Zotikov, and V. P. Budtov, *Polym. Sci. USSR, Engl. Transl.*, 1982, **24**, 98.
[55] A. S. Nazran and J. Warkentin, *J. Am. Chem. Soc.*, 1982, **104**, 6405.
[56] S. N. Gupta, I. Gupta, and D. C. Neckers, *J. Polym. Sci., Polym. Chem. Ed.*, 1981, **19**, 103.
[57] M. J. M. Abadie, F. Schué, T. Souel, and D. H. Richards, *Polymer*, 1981, **22**, 1076.
[58] S. P. Pappas and R. A. Asmus, *J. Polym. Sci., Polym. Chem. Ed.*, 1982, **20**, 2643.
[59] D. Asai, A. Okada, S. Kondo, and K. Tsuda, *J. Macromol. Sci., Chem.*, 1982, **18**, 1011.
[60] R. G. Jones and N. Khalid, *Eur. Polym. J.*, 1982, **18**, 285.
[61] K. M. Z. Al-Abidin and R. G. Jones, *J. Chem. Soc., Faraday Trans. 1*, 1982, **78**, 513.
[62] D. R. Terrell, *Polymer*, 1982, **23**, 1045.
[63] P. K. Sengupta and G. Mukhopadhyay, *Makromol. Chem.*, 1982, **183**, 1093.
[64] Y. Shirota, K. Takemura, H. Mikawa, T. Kawamura, and K. Matsuzaki, *Makromol. Chem., Rapid Commun.*, 1982, **3**, 913.

3 Growth Reactions

Materials referred to as 'macromonomers' or 'macromers' have been described.[65,66] They can be prepared by polymerization of monomers such as styrene in a living monofunctional anionic system, followed by termination in a special manner to produce end-groups which are unsaturated and polymerizable. The products can be used as comonomers in radical polymerizations to give what are essentially graft copolymers free from contaminating homopolymer. Appropriate choice of conditions leads to products resembling triblock copolymers in their physical properties. The wide applicability of radical polymerization means that the monomer copolymerized with a macromer can be of a type not polymerizable anionically and so not usable in the controlled production of block copolymers in living anionic systems.

Polymers and copolymers with reactive side-groups can be prepared using $CH_2:CH.p-C_6H_4.CO.p-C_6H_4.CO.O.OBu^t$;[67] they can be used as photo-initiators. Products with photo-active pendant groups have also been made from $CH_2:C(Me).CO.O.CH_2.CH_2.O.p-C_6H_4.CO.R$.[68] Complex kinetic behaviour is exhibited by $CH_2:CH.C:C.C(Me)_2.O.O.C(Me)_3$;[69] its copolymer with acrylic acid forms salts which are water soluble polymeric peroxides.[70] Acrylate and methacrylate polymers with pendant azo groups have been made;[71] in some cases the side-groups are mesogenic.[72]

Various little-known monomers have been used in radical polymerizations to make polymers and copolymers, some with rather interesting properties. They include some heterocyclic monomers,[73–75] an organo-tungsten compound,[76,77] pentachlorophenyl acrylate,[78] methacryloyl fluoride,[79] perfluoroalkyl derivatives of styrene,[80] 6-vinyl chrysene,[81] d-limonene,[82] vinyl chloroformate,[83–85] monomers

---

[65] T. Nishimura, M. Maeda, Y. Nitadori, and T. Tsuruta, *Makromol. Chem., Rapid Commun.*, 1980, **1**, 573.
[66] G. O. Schulz and R. Milkovich, *J. Appl. Polym. Sci.*, 1982, **27**, 4773.
[67] S. N. Gupta, L. Thijs, and D. C. Neckers, *J. Polym. Sci., Polym. Chem. Ed.*, 1981, **19**, 855.
[68] P. Hrdlovič, I. Lukáč, and I. Zvara, *Eur. Polym. J.*, 1981, **17**, 1121
[69] S. A. Voronov, V. A. Puchin, V. S. Tokarev, and Yu. A. Lastukhin, *Polym. Sci. USSR, Engl. Transl.*, 1980, **22**, 635.
[70] S. A. Voronov, V. A. Puchin, V. S. Tokarev, and V. S. Kurganskii, *Polym. Sci. USSR, Engl. Transl.*, 1980, **22**, 973.
[71] J. Lokaj, H. Pivcová, and F. Hrabák, *Makromol. Chem.*, 1981, **182**, 1929.
[72] V. V. Tsukruk, V. V. Shilov, I. I. Konstantinov, Yu. S. Lipatov, and Yu. B. Amerik, *Eur. Polym. J.*, 1982, **18**, 1015.
[73] V. Bertini, M. Pocci, F. Provenzano, and A. de Munno, *Macromolecules*, 1981, **14**, 1833.
[74] G. Wouters and G. Smets, *Makromol. Chem.*, 1982, **183**, 1861.
[75] M. G. Joshi and F. Rodriguez, *J. Appl. Polym. Sci.*, 1982, **27**, 3151.
[76] C. U. Pittman, jun., T. V. Jayaraman, R. D. Priester, jun., S. Spencer, M. D. Rausch, and D. Macomber, *Macromolecules*, 1981, **14**, 237.
[77] C. U. Pittman, jun., R. D. Priester, jun., and T. V. Jayaraman, *J. Polym. Sci., Polym. Chem. Ed.*, 1981, **19**, 3351.
[78] C. U. Pittman, jun., and G. A. Stahl, *J. Appl. Polym. Sci.*, 1981, **26**, 2403.
[79] M. Ueda, T. Kumakura, Y. Imai, and C. U. Pittman, jun., *J. Polym. Sci., Polym. Chem. Ed.*, 1982, **20**, 2829.
[80] B. Bömer and H. Hagemann, *Angew. Makromol. Chem.*, 1982, **109/110**, 285.
[81] E. Chiellini, R. Solaro, and F. Ciardelli, *Makromol. Chem.*, 1982, **183**, 103.
[82] T. Doiuchi, H. Yamaguchi, and Y. Minoura, *Eur. Polym. J.*, 1981, **17**, 961.
[83] G. Meunier, P. Hémery, S. Boileau, J.-P. Senet, and H. Chéradame, *Polymer*, 1982, **23**, 849.
[84] G. Meunier, P. Hémery, J.-P. Senet, and S. Boileau, *Polym. Bull. (Berlin)*, 1981, **4**, 699 and 705.
[85] C. Gamichon, P. Hémery, B. Raynal, and S. Raynal, *J. Polym. Sci., Polym. Chem. Ed.*, 1982, **20**, 3255.

containing sulphoxide,[86] thioester,[87] or sulphonium ylide[88] groups, N-vinyl succinimide,[89] N-aryl itaconimides,[90] hydroxyalkyl methacrylates,[91] chloromethyl methacrylate,[92] carbomethoxy[93] and bis(carbomethoxy) derivatives[94] of maleic anhydride, and methacrylate and fumarate esters containing naphthalene residues;[95] in certain cases, the polymerizations have received fairly detailed attention and values of Q and e have been recorded for some of the monomers. For 1-vinyl imidazole,[96] a degradative reaction between growing radical and monomer occurs not by transfer but by an abnormal addition. Diethyl maleate can apparently be polymerized radically if a suitable amine is present but reaction is preceded by isomerization to the fumarate.[97—99]

Interest has been renewed in the polymerizations of acrylic and methacrylic anhydrides.[100—102] Intramolecular reaction of the radical $-CH_2 . \dot{C}R . CO . O . CO . CR:CH_2$ (R = H or Me), leading to 5- or 6-membered rings by head-to-head or head-to-tail processes, competes with intermolecular reaction with monomer. Radical cyclopolymerizations have been described for other monomers with two olefinic bonds suitably placed.[103—105] Reviews of these polymerizations[106,107] include discussion of control by kinetic and thermodynamic factors.

During radical polymerizations involving carborane exocyclic vinyl silanes, some ring-opening occurs.[108] The process has been reported again for *spiro ortho* esters[109] and for cyclic ketene acetals[110—113] to give polyesters according to equation (1). In

$$P. + CH_2{:}C\begin{array}{c}O-CH.C_6H_5\\|\\O-CH_2\end{array} \longrightarrow P.CH_2.CO.O.\dot{C}H_2.CH(C_6H_5) \quad (1)$$

[86] K. Ogura, K. Itoh, S.-I. Isogai, S. Kondo, and K. Tsuda, *J. Macromol. Sci., Chem.*, 1982, **17**, 1371.
[87] D. Kokkiaris, C. Touloupis, and N. Hadjichristidis, *Polymer*, 1981, **22**, 63.
[88] M. Senga, S. Kondo, and K. Tsuda, *J. Polym. Sci., Polym. Lett.*, 1982, **20**, 657.
[89] S. G. Bondarenko, A. F. Nikolayev, S. A. Baranova, N. I. Plyashechnik, G. A. Smirnova, S. V. Obukhova, I. V. Baidenok, Ye. M. Stepanov, I. N. Glushchenok, and Ye. D. Andreyeva, *Polym. Sci. USSR, Engl. Transl.*, 1981, **23**, 2860.
[90] T. Pyriadi and M. Fraih, *J. Macromol. Sci., Chem.*, 1982, **18**, 159.
[91] M. Macret and G. Hild, *Polymer*, 1982, **23**, 81.
[92] M. Ueda, K. Iri, Y. Imai, and C. U. Pittman, jun., *Macromolecules*, 1981, **14**, 1046.
[93] H. K. Hall, jun., J. W. Rhoades, P. Nogues, and G. K. C. Wai, *Polym. Bull. (Berlin)*, 1981, **4**, 629.
[94] H. K. Hall, jun., R. C. Sentman, and P. Nogues, *J. Org. Chem.*, 1982, **47**, 3647.
[95] D. A. Holden and J. E. Guillet, *Macromolecules*, 1982, **15**, 1475.
[96] C. H. Bamford and E. Schofield, *Polymer*, 1981, **22**, 1227.
[97] T. Otsu and N. Toyoda, *Makromol. Chem., Rapid Commun.*, 1981, **2**, 79.
[98] T. Otsu, O. Ito, N. Toyoda, and S. Mori, *Makromol. Chem., Rapid Commun.*, 1981, **2**, 725.
[99] T. Otsu, O. Ito, and N. Toyoda, *Makromol. Chem., Rapid Commun.*, 1981, **2**, 729.
[100] G. B. Butler and A. Matsumoto, *J. Polym. Sci., Polym. Lett.*, 1981, **19**, 167.
[101] A. Matsumoto, T. Kitamura, M. Oiwa, and G. B. Butler, *J. Polym. Sci., Polym. Chem. Ed.*, 1981, **19**, 2531.
[102] A. Matsumoto, T. Kitamura, M. Oiwa, and G. B. Butler, *Makromol. Chem., Rapid Commun.*, 1981, **2**, 683.
[103] A. Matsumoto, K. Iwanami, and M. Oiwa, *J. Polym. Sci., Polym. Chem. Ed.*, 1981, **19**, 213.
[104] S. A. Stone-Elander, G. B. Butler, J. H. Davis, and G. J. Palenik, *Macromolecules*, 1982, **15**, 45.
[105] A. Matsumoto, H. Ishido, M. Oiwa, and K. Urushido, *J. Polym. Sci., Poly. Chem. Ed.*, 1982, **20**, 3207.
[106] G. Butler, ed., *Am. Chem. Soc., Symp. Ser.*, 1982, **195**.
[107] G. C. Corfield and G. B. Butler, in 'Developments in Polymerisation – 3', ed. R. N. Haward, Applied Science, London, 1982, p. 1.
[108] T. M. Frunze, A. A. Sakharova, O. A. Mel'nik, B. A. Izmailov, and V. N. Kalinin, *Polym. Sci. USSR, Engl. Transl.*, 1981, **23**, 2261.
[109] T. Endo, M. Okawara, N. Yamazaki, and W. J. Bailey, *J. Polym. Sci., Polym. Chem. Ed.*, 1981, **19**, 1283.
[110] I. Cho and M. S. Gong, *J. Polym. Sci., Polym. Lett.*, 1982, **20**, 361.
[111] W. J. Bailey, Z. Ni, and S.-R. Wu, *Macromolecules*, 1982, **15**, 711.
[112] W. J. Bailey, Z. Ni, and S.-R. Wu, *J. Polym. Sci., Polym. Chem. Ed.*, 1982, **20**, 3021.
[113] W. J. Bailey, S.-R. Wu, and Z. Ni, *Makromol. Chem.*, 1982, **183**, 1913.

Chain Reaction Polymerization

this example, polymerization proceeds almost exclusively by ring-opening with complete regioselectivity because of the stability of the product radical. If the phenyl group is replaced by an alkyl group such as n-decyl, polymerization without ring-opening and growth reactions leading to the radical $P.CH_2.CO.O.CH_2.CHR.$ and $P.CH_2.CO.O.CHR.CH_2.$ are of comparable importance.[114] The cyclic ketene acetals may be useful as comonomers for introduction of in-chain ester groups and certain pendant groups into polymers; these structural features may produce desirable changes in chemical and physical properties. Another exo-methylene cyclic monomer with a sizable ring, viz, α-methylene-γ-butyrolactone, readily undergoes radical polymerization without ring-opening;[115] it is evident therefore that steric effects need not be of great importance for monomers of this type.

4 Transfer Processes

A new effect in transfer reactions has been described.[116,117] A cobalt complex of a porphyrin derivative appears to be an especially powerful transfer agent in the radical polymerization of methyl methacrylate, without being consumed so that its effect persists throughout the reaction. The complex is regarded as catalyzing the transfer of hydrogen from growing radical to monomer. Comparatively narrow distributions of molecular weight in the resulting polymers have been attributed[118] to the additive being less effective for small polymer radicals than for large. The cobalt complex behaves similarly with styrene but its action is less marked than with methyl methacrylate and there are side effects.[119] Use of the cobalt complex is thought to be impracticable for production of oligomers but suitable alternatives may be found. Transfer to monomer has been considered generally and the resulting unsaturated end-groups have been examined by chemical and spectroscopic methods.[120] Transfer during polymerization of tetrafluoroethylene in organic media has been studied[121] with emphasis on the production of comparatively low polymers.

Reagents functioning both as initiators and as reactive transfer agents are known as inifers; the descriptive name iniferter has been used when the transfer is markedly degradative and is accompanied by appreciable retardation. In radical systems, particular attention has been paid to sulphur-containing substances, e.g., tetraethyl-

---

[114] W. J. Bailey, S.-R. Wu, and Z. Ni, J. Macromol Sci., Chem., 1982, **18**, 973.
[115] M. Ueda, M. Takahashi, Y. Imai, and C. U. Pittman, jun., J. Polym. Sci., Polym. Chem. Ed., 1982, **20**, 2819.
[116] N. S. Enikolopyan, B. R. Smirnov, G. V. Ponomarev, and I. M. Belgovskii, J. Polym. Sci., Polym. Chem. Ed., 1981, **19**, 879.
[117] B. R. Smirnov, A. P. Marchenko, G. V. Korolev, I. M. Bel'govskii, and N. S. Yenikolopyan, Polym. Sci. USSR, Engl. Transl., 1981, **23**, 1158.
[118] B. R. Smirnov, A. P. Marchenko, V. D. Plotnikov, A. I. Kuzayev, and N. S. Yenikolopyan, Polym. Sci. USSR, Engl. Transl., 1981, **23**, 1169.
[119] B. R. Smirnov, V. D. Plotnikov, B. V. Ozerkovskii, V. P. Roshchupkin, and N. S. Yenikolopyan, Polym. Sci. USSR, Engl. Transl., 1981, **23**, 2807.
[120] F. Hrabák, J. Lokaj, J. Bilá, K. Bouchal, and J. Štokr, Makromol. Chem., Rapid Commun., 1982, **3**, 891.
[121] N. Yu. Andreyeva, Yu. A. Panshin, M. A. Andreyeva, L. N. Pirozhnaya, N. Ye. Shadrina, and L. I. Tarutina, Polym. Sci. USSR, Engl. Transl., 1981, **23**, 2776.

thiuram disulphide.[122—124] The resulting $(Et)_2N.CS.S-$ end-groups are themselves labile so that a seemingly 'dead' polymer molecule can give rise to radicals and a system resembling a living radical polymerization can be achieved.

Further investigation of transfer to polymer, leading to branching, is needed; it has been established that the process may be studied by monitoring changes in molecular weight distribution during polymerization.[125] A theoretical treatment of branching has been developed for the polymerization of ethylene at high pressure;[126] in that system, the presence of water apparently increases the number of long branches.[127] Formation of short branches in poly(vinyl chloride) has been considered again.[128] Branching in oligomers made radically has been associated with transfer involving primary radicals.[129] Copolymers of vinyl benzyl thiol with styrene contain units susceptible to transfer and so acting as sites for long branches.[130,131]

5 Termination

There is now general acceptance that k_t depends upon radical size; several authors have commented upon or made use of the dependence.[132—136] Primary radicals termination still attracts attention, often being postulated to explain abnormalities in the kinetics of the overall polymerization.[137—141] A new and direct method of study involves search for end-groups such as $X.CH(C_6H_5).CH_2-$, where X is an initiating radical; it is necessary to distinguish between groups of this type formed by the combination of a primary radical with a growing polymer radical and those resulting from 'abnormal' initiation (see Section 2) or from transfer to initiator.[25]

New treatments of high-conversion polymerizations have been developed;[142–146] the case of vinyl acetate has been considered.[147] Further attention has been given to

[122] T. Otsu, M. Yoshida, and A. Kuriyama, *Polym. Bull. (Berlin)*, 1982, **7**, 45.
[123] T. Otsu and M. Yoshida, *Makromol. Chem., Rapid Commun.*, 1982, **3**, 127.
[124] T. Otsu, M. Yoshida, and T. Tazaki, *Makromol. Chem., Rapid Commun.*, 1982, **3**, 133.
[125] N. G. Taganov, *Polym. Sci. USSR, Engl. Transl.*, 1981, **23**, 3009.
[126] W. L. Mattice and F. C. Stehling, *Macromolecules*, 1981, **14**, 1479.
[127] D. Constantin, M. Hert, and J. P. Machon, *Eur. Polym. J.*, 1981, **17**, 115.
[128] T. Hjertberg and E. Sörvik, *J. Polym. Sci., Polym. Lett.*, 1981, **19**, 363.
[129] Ye. N. Barantsevich, V. P. Kartavykh, V. A. Drach, S. S. Ivanchev, T. P. Nasonova, and E. B. Rotenberg, *Polym. Sci. USSR, Engl. Transl.*, 1981, **23**, 2460.
[130] L. H. Tung, A. T. Hu, S. V. McKinley, and A. M. Paul, *J. Polym. Sci., Polym. Chem. Ed.*, 1981, **19**, 2027.
[131] L. H. Tung, *J. Polym. Sci., Polym. Chem. Ed.*, 1981, **19**, 3209.
[132] T. Yasukawa and K. Murakami, *Macromolecules*, 1981, **14**, 227.
[133] S. K. Soh and D. C. Sunberg, *J. Polym. Sci., Polym. Chem. Ed.*, 1982, **20**, 1299, 1315, 1331, and 1345.
[134] O. F. Olaj and G. Zifferev, *Makromol. Chem., Rapid Commun.*, 1982, **3**, 549.
[135] H. M. J. Boots, *J. Polym. Sci., Polym. Phys. Ed.*, 1982, **20**, 1695.
[136] P. G. de Gennes, *J. Chem. Phys.*, 1982, **76**, 3316 and 3322.
[137] E. F. Okieimen, *Eur. Polym. J.*, 1981, **17**, 641.
[138] E. F. Okieimen, *Polymer*, 1981, **22**, 1737.
[139] P. C. Deb, *Eur. Polym. J.*, 1982, **18**, 769.
[140] P. C. Deb and A. B. Samui, *Angew. Makromol. Chem.*, 1982, **103**, 77.
[141] S. Sarasvathy and K. Venkatarao, *Polymer*, 1982, **23**, 1999.
[142] M. Stickler and G. Meyerhoff, *Polymer*, 1981, **22**, 928.
[143] V. P. Budtov, E. G. Zotikov, and N. G. Podosenova, *Polym. Sci. USSR, Engl. Transl.*, 1981, **23**, 1603.
[141] T. J. Tulig and M. Tirrell, *Macromolecules*, 1981, **14**, 1501.
[145] D. J. T. Hill and J. H. O'Donnell, *J. Polym. Sci., Polym. Chem. Ed.*, 1982, **20**, 241.
[146] F. L. Marten and A. E. Hamielec, *J. Appl. Polym. Sci.*, 1982, **27**, 489.
[147] W. Baade, H. U. Moritz, and K. H. Reichert, *J. Appl. Polym. Sci.*, 1982, **27**, 2249.

Chain Reaction Polymerization

relationships between the onset of the gel effect and the characteristics of material produced early in the reaction or added to the system as preformed polymer.[148,149] An early gel effect has been confirmed for methyl methacrylate under circumstances leading to polymer of particularly high molecular weight.[150]

Calorimetric methods have been used again for polymerizations at high conversions and in systems of high viscosity.[151,152] If bulk polymerizations are conducted in the presence of fluorescent probes, fluorescence increases abruptly and substantially as the system approaches a glassy state;[153,154] it has been suggested that the phenomenon may be used to obtain information about molecular motions in these systems.

Auto-acceleration in the bulk polymerization of acrylonitrile is usually explained in terms of occlusion of growing radicals in the precipitated polymer and consequent reduction in the rate of termination. This explanation has been challenged[155] and the acceleration has instead been associated with an increase in the rate of propagation caused by a matrix effect resulting from interaction of monomer and polymer. Absence of auto-acceleration for polymerization in a solvent for the polymer has been explained by breakdown of the polymer/monomer interaction rather than by removal of the restrictions on the mobility of the polymer radicals.

6 Retardation and Inhibition

Inhibitors are commonly used for measurements of rates of initiation in radical polymerizations; there is always interest in new inhibitors for which the stoicheiometry of the reaction with primary radicals may be clear. Radical traps are used also for identification of the radicals formed in initiating systems. In the latter connection, 2,2,6,6-tetramethylpiperidinoxyl has been applied to the photodecomposition of some benzoin derivatives.[156] This inhibitor has been used in studies of special aspects of the polymerization of a dimethacrylate.[157,158] The trapping technique has been extended to certain nitroxides, *e.g.*, 1,1,3,3-tetramethylisoindolinyl-2-oxy, which do not react irreversibly with the oxygen-centred radicals generated from some widely used initiators. On the other hand, with the carbon-centred radicals produced by addition of the primary radicals to monomers, the nitroxides give stable products for which reliable analyses can be performed;[17] by this procedure, important information on initiation processes has been obtained (see Section 2). The nitroxides may, however, engage in side-reactions including

---

[148] N. M. Bityurin, V. N. Genkin, V. P. Zubov, and M. B. Lachinov, *Polym. Sci. USSR, Engl. Transl.*, 1981, **23**, 1873.
[149] T. J. Tulig and M. Tirrell, *Macromolecules*, 1982, **15**, 459.
[150] D. R. Johnson, Y. Osada, A. T. Bell, and M. Shen, *Macromolecules*, 1981, **14**, 118.
[151] E. Takács and J. Dobó, *Polym. Bull. (Berlin)*, 1981, **5**, 551.
[152] I. M. Barkalov and D. P. Kiryukhin, *Polym. Sci. USSR, Engl. Transl.*, 1980, **22**, 797.
[153] R. O. Loutfy, *Macromolecules*, 1981, **14**, 270.
[154] R. O. Loutfy, *J. Polym. Sci., Polym. Phys. Ed.*, 1982, **20**, 825.
[155] G. Burillo, A. Chapiro, and Z. Mankowski, *Eur. Polym. J.*, 1982, **18**, 367.
[156] H. J. Hageman and T. Overeem, *Makromol. Chem., Rapid Commun.*, 1981, **2**, 717.
[157] V. M. Lagunov, M. P. Berezin, I. V. Golikov, and G. V. Korolev, *Polym. Sci. USSR, Engl. Transl.*, 1981, **23**, 2980.
[158] V. M. Lagunov, I. V. Golikov, and G. V. Korolev, *Polym. Sci. USSR, Engl. Transl.*, 1982, **24**, 149.

induced decomposition of some initiators,[159] abstraction of hydrogen from a thermally generated dimer of styrene, and, to a minor extent, addition to styrene.[160] The hydrogen abstraction accounts for the excessively high apparent rates of initiation for the thermal polymerization of styrene, as deduced from the consumption of the nitroxide.

Polymerizations are retarded by 1,1,4,4-tetraphenyl-2-tetrazene[161,162] because of its dissociation to stabilized radicals able to react with polymer radicals but not with monomer molecules. On the other hand, 1,4-dimethyl-1,4-diphenyl-2-tetrazene acts as an initiator for styrene at 80—95 °C because it yields radicals of moderately high reactivity.[163] Molecule inhibitors have been studied[164,165] with results interpreted in terms of the hot-radical theory. The rate constant for the very rapid reaction of the polystyrene radical with ferric chloride has been determined[166] using the procedure of moderated copolymerization with methyl methacrylate as the vehicle monomer.

---

[159] G. Moad, E. Rizzardo, and D. H. Solomon, *Tetrahedron Lett.*, 1981, **22**, 1165.
[160] G. Moad, E. Rizzardo, and D. H. Solomon, *Polym. Bull. (Berlin)*, 1982, **6**, 589.
[161] K. Sugiyama, T. Oda, and T. Maeshima, *J. Macromol. Sci., Chem.*, 1981, **15**, 107.
[162] K. Sugiyama, T. Oda, and T. Maeshima, *Makromol. Chem.*, 1982, **183**, 2445.
[163] K. Sugiyama, T. Oda, and T. Maeshima, *Makromol. Chem.*, 1982, **183**, 1.
[164] I. Tánczos, F. Tüdös, and T. Földes-Berezsnich, *Eur. Polym. J.*, 1982, **18**, 295.
[165] I. Tánczos, Á. Rehák, and F. Tüdös, *Eur. Polym. J.*, 1982, **18**, 487.
[166] A. D. Jenkins and B. H. Mustafa, *J. Polym. Sci., Polym. Lett.*, 1981, **19**, 1.

PART V Emulsion Polymerization
by D. C. Blackley

1 Books and Reviews

Several important books and reviews on emulsion polymerization and related topics have appeared during the period covered by this Report. An important book edited by Piirma[1] contains chapters on the stability and instability of polymer latices (Ottewill), particle formation mechanisms (Hansen and Ugelstad), theoretical predictions of particle-size and molecular-weight distributions produced by emulsion polymerization reactions (Lichti, Gilbert, and Napper), theory of the kinetics of compartmentalized free-radical polymerization reactions (Blackley), desorption and re-absorption of free radicals in emulsion polymerization (Nomura), effects of choice of emulsifier in emulsion polymerization (Dunn), polymerization of polar monomers (Yeliseyeva), recent developments and trends in the industrial use of latices (Force), principles of the design, operation, and control of latex reactors (Hamielec and MacGregor), emulsion polymerization in continuous reactors (Poehlein), effect of additives upon the formation of monomer emulsions and polymer dispersions (Ugelstad, Mørk, Berge, Ellingsen, and Khan), and radiation-induced emulsion polymerization (Stannett).

A welcome translation into English of an important Russian book by Eliseeva, Ivanchev, Kuchanov, Lebedev, and others[2] has now appeared. This book contains chapters upon various types of latices of industrial importance, as well as chapters upon various aspects of emulsion polymerization. Although rather out-of-date because of the time it has taken for the translation to appear, this book is particularly important because it provides a convenient summary in English of emulsion polymerization from the Russian point of view.

Final revisions of many of the papers presented at the 1980 American Chemical Society Emulsion Polymerization Symposium have now appeared in a book edited by Bassett and Hamielec.[3] Attention was drawn to most of these papers in the contribution to Volume 2 of these Specialist Periodical Reports. A book edited by Calvert[4] provides a convenient up-to-date review of the cognate subject of polymer latices viewed essentially from the industrial standpoint.

The emulsion polymerization of vinyl acetate is the subject of a specialist monograph edited by El-Aasser and Vanderhoff.[5] Ham[6] has recently reviewed the important subject of emulsion copolymerization. The emulsion copolymerization of diene and vinyl monomers is the subject of a Russian review by Askerov et al.[7]

---

[1] I. Piirma, ed., 'Emulsion Polymerization', Academic, New York, 1982.
[2] V. I. Eliseeva, S. S. Ivanchev, S. I. Kuchanov, and A. V. Lebedev, 'Emulsion Polymerization and its Applications in Industry', transl. from Russian by S. J. Teague, Consultants Bureau, New York, 1981.
[3] D. R. Bassett and A. E. Hamielec, eds., 'Emulsion Polymers and Emulsion Polymerization', ACS Symposium Series No. 165, American Chemical Society, Washington DC, 1981.
[4] K. O. Calvert, ed., 'Polymer Latices and their Applications', Applied Science Publishers, London, 1982.
[5] M. S. El-Aasser and J. W. Vanderhoff, ed., 'Emulsion Polymerisation of Vinyl Acetate', Applied Science Publishers, London, 1981.
[6] G. E. Ham, *J. Macromol. Sci., Chem.*, 1982, **17**, 369.
[7] A. K. Askerov et al., *Vopr. Razvit. Neftekhim. Khim. Prisadok. Azerb.*, 1980, 78.

Poehlein has reviewed the reaction engineering aspects of emulsion polymerization.[8] Three Japanese reviews have appeared which deal with the characteristics of emulsion polymerization[9] reactions, with emulsion polymerization initiated in emulsion droplets,[10] and with emulsion polymerization generally.[11] Bolza[12] has reviewed the role of surfactants in emulsion polymerization. Other reviews of various aspects include those of Pavlyuchenko and Ivanchev,[13] of Beileryan,[14] and of Lukas et al.[15]

Fitch[16] has given a most useful review of the mechanism of particle nucleation in emulsion polymerization reactions. A 'literature analysis' of particle formation in emulsion polymerization has been given by Schneider.[17] Hearn, Wilkinson, and Goodall[18] have given an extensive review of polymer latices at model colloids; this is also the subject of a review by Overbeek.[19]

2 Particle Nucleation

Rahman and Brown[20] have reported the results of an investigation of the effects of pH upon the rate of emulsion polymerization of styrene initiated by potassium persulphate in the presence of sodium dodecyl sulphate as surfactant. The observed reduction in the rate of polymerization as the pH was reduced is interpreted in terms of an increase in the size of the sodium dodecyl sulphate micelles as the pH was reduced.

Piirma and Chang[21] have now published a full paper on the interesting effects which they reported at the 1980 American Chemical Society Emulsion Polymerization Symposium concerning the nucleation of particles in the persulphate-initiated emulsion polymerization of styrene in the presence of a non-ionic surfactant. A second particle-nucleation process occurs at about 40% conversion, and this leads to an increase in the rate of polymerization, and to the formation of a latex which has a bimodal particle-size distribution. Possible reasons for the phenomenon are discussed.

Eliseeva[22] has discussed the mechanism of particle formation and growth in emulsion polymerization with particular reference to the solubility of the monomer in water, the polarity of the organic phase, and adsorption phenomena. The potential effects of hydrodynamic action generated by the rotation of the apparatus upon the mechanism of particle formation and polymerization kinetics are

---

[8] G. W. Poehlein, *NATO Adv. Study Inst. Ser., Ser. E*, 1981, **51**, 469.
[9] K. Tadashi, *Kobunshi Kako*, 1982, **31**, 266.
[10] T. Kato, *Kobunshi Kako*, 1981, **30**, 410.
[11] K. Toshio, *Kobunshi Kako*, 1982, **31**, 255.
[12] F. Bolza, *Aust. OCCA Proc. News*, 1981, **18**(1—2), 5.
[13] V. N. Pavlyuchenko and S. S. Ivanchev, *Usp. Khim.*, 1981, **50**, 715; *Russ. Chem. Rev.*, 1981, **50**, 380.
[14] N. M. Beileryan, *Acta Polym.*, 1982, **33**, 339.
[15] R. Lukas, et al., *Chem. Prum.*, 1982, **32**, 193.
[16] R. M. Fitch, Main Lecture, 27th International Symposium on Macromolecules, 1981, ed. H. Benoit and P. Rempp, Pergamon, Oxford, 1982.
[17] H. J. Schneider, *Acta Polym.*, 1981, **32**, 667.
[18] J. Hearn, M. C. Wilkinson, and A. R. Goodall, *Adv. Colloid Interface Sci.*, 1981, **14**, 173.
[19] J. T. G. Overbeek, *Adv. Colloid Interface Sci.*, 1982, **15**, 251.
[20] A. Rahman and C. W. Brown, *J. Appl. Polym. Sci.*, 1982, **27**, 2563.
[21] I. Piirma and M. Chang, *J. Polym. Sci., Polym. Chem. Ed.*, 1982, **20**, 489.
[22] V. I. Eliseeva, *Acta Polym.*, 1981, **32**, 355.

discussed. Nalchadzhyan et al.[23] have studied the mechanism of particle formation in the emulsion polymerization of methyl methacrylate. Dunn and Hassan[24] have discussed the conditions which must be fulfilled if new particles are to be nucleated in the seeded emulsion polymerization of styrene. It is concluded that further nucleation does not occur if the particle concentration and size have increased sufficiently to ensure that nascent particles coalesce with existing particles, taking into account such factors as the magnitude of the prevailing potential energy barriers to coalescence. Bataille, Van, and Pham[25] have provided evidence for the continued nucleation of particles even at high conversion in the conventional emulsion polymerization of styrene.

3 Kinetics and Mechanism

Interest continues to be shown in methods for solving the Smith–Ewart differential difference equations, both for steady-state systems[26,27] and for non-steady-state systems.[28] Soh[29] has shown how the basic rate equation for emulsion polymerization can be rearranged to allow the rate coefficient for propagation to be determined from measurable quantities such as conversion, volume change, and average particle volume. However, the method requires the assumption that the average number of propagating radicals per locus is 0.5. Rate coefficients for radical desorption during emulsion polymerization is the subject of papers by Brooks and Makanjuola[30] (for styrene and methyl methacrylate), and by Nomura and Harada[31] (for vinyl acetate and vinyl chloride). Brooks and Makanjuola have shown that the values of the rate coefficients for radical desorption for styrene and methyl methacrylate are not greatly affected by monomer concentration or by degree of surface coverage by sodium dodecyl sulphate. When cetyl pyridinium chloride is used as surfactant, the rate coefficient depends upon surface coverage in the case of methyl methacrylate, but not in the case of styrene. Hawkett, Napper, and Gilbert[32] have reported results for radial-capture efficiencies in emulsion polymerization.

Interest continues to be shown in the seeded emulsion polymerization of styrene,[33–35] principally because studies of such reactions provide detailed information concerning the mechanism of emulsion polymerizations during Interval II. Of particular interest are the results of Lichti et al.[35] for the kinetics of seeded emulsion polymerization of styrene in the presence of transfer agents such as carbon tetrachloride and carbon tetrabromide. In a further paper, Lichti et al.[36]

[23] S. O. Nalchadzhyan, et al., Arm. Khim. Zh., 1982, **35**, 561.
[24] A. S. Dunn and S. A. Hassan, Preprints for PRI Emulsion Polymers Conference, London, 1982, Paper No. 1.
[25] P. Bataille, B. T. Van, and Q. B. Pham, J. Polym. Sci., Polym. Chem. Ed., 1982, **20**, 795.
[26] M. J. Ballard, R. G. Gilbert, and D. H. Napper, J. Polym. Sci., Polym. Lett. Ed., 1981, **19**, 533.
[27] B. W. Brooks, J. Chem. Soc., Faraday Trans. 1, 1982, **78**, 3137.
[28] D. T. Birtwistle and D. C. Blackley, J. Chem. Soc., Faraday Trans. 1, 1981, **77**, 1351.
[29] S. K. Soh, J. Appl. Polym. Sci., 1980, **25**, 2993.
[30] B. W. Brooks and B. O. Makanjuola, J. Chem. Soc., Faraday Trans. 1, 1981, **77**, 2659.
[31] M. Nomura and M. Harada, J. Appl. Polym. Sci., 1981, **26**, 17.
[32] B. S. Hawkett, D. H. Napper, and R. G. Gilbert, J. Polym. Sci., Polym. Chem. Ed., 1981, **19**, 3173.
[33] B. C. Y. Whang, et al., J. Chem. Soc., Faraday Trans. 1, 1982, **78**, 1117.
[34] S. M. Hasan, J. Polym. Sci., Polym. Chem. Ed., 1982, **20**, 3031.
[35] G. Lichti, et al., J. Chem. Soc., Faraday Trans. 1, 1982, **78**, 2129.
[36] G. Lichti, et al., J. Polym. Sci., Polym. Chem. Ed., 1981, **19**, 925.

have reported upon the time evolution of particle-size distribution in seeded and *ab initio* styrene emulsion polymerizations. This is also the subject of a paper by Cauley and Thompson,[37] who have been particularly concerned with changes in the breadth of the particle-size distribution. Cauley and Thompson have developed a model which utilizes a surface-area-dependent volumetric growth rate of a single particle. This model predicts a time-invariant standard deviation of the size distribution during periods of particle growth only. This behaviour is reconciled with some experimental observations by considering the occurrence of particle nucleation during some part of the growth interval. It is concluded that narrow particle-size distribution is favoured by high initiator level and low surfactant level. Chamberlain, Napper, and Gilbert[38] have reported results for the polymerization of styrene in emulsified droplets. Hawkett, Napper, and Gilbert[39] have presented an analysis of data for the kinetics of emulsion polymerization in Interval III, and have shown how estimates of the rate coefficients for radical entry and radical desorption can be obtained from such data. Fitch[40] has discussed the control of particle-size and particle-size distribution during emulsion polymerization. Dunn and Said[41] have reported results for the effects of electrolyte (potassium chloride) upon the emulsion polymerization of styrene using potassium decanoate and potassium octadecanoate as emulsifiers.

Other papers on the subject of seeded emulsion polymerization include those of Brooks and Makanjuola[42] for vinyl acetate and of Hagiopol et al.[43] for various vinyl monomers, both water-soluble and water-insoluble. Brooks and Makanjuola have been particularly concerned with the order of reaction with respect to initiator concentration; this they find to be 0.7 in all cases investigated. Sathpathy et al. have reported on the kinetics and mechanism of the emulsion polymerization of methyl acrylate,[44] and also upon the emulsion polymerization of ethyl methacrylate.[45] Cao et al.[46,47] have presented a mathematical model for emulsion polymerization reactions. Padwa and Paster[48] have presented results for the kinetics of high-temperature redox-initiated emulsion polymerization of butadiene; the initiator used was of the iron–hydroperoxide–sulphoxylate type.

Two papers by Kast and Funke[49,50] are concerned with the thermally initiated emulsion polymerizations of styrene and 1,4-divinylbenzene respectively. In the case of styrene, the order of reaction with respect to surfactant concentration is reported as 0.7 up to 0.2M, and thereafter as zero. Two papers by Goebel et al.[51,52] report results for the kinetics of the emulsion polymerization of vinyl chloride.

[37] D. A. Cauley and R. W. Thompson, *J. Appl. Polym. Sci.*, 1982, **27**, 363.
[38] B. J. Chamberlain, D. H. Napper, and R. G. Gilbert, *J. Chem. Soc., Faraday Trans. 1*, 1982, **78**, 591.
[39] B. S. Hawkett, D. H. Napper, and R. G. Gilbert, *J. Chem. Soc., Faraday Trans. 1*, 1981, **77**, 2395.
[40] R. M. Fitch, *Proc. Water-borne Higher-Solids Coat. Symp.*, 1981, **8**, 25.
[41] A. S. Dunn and Z. F. M. Said, *Polymer*, 1982, **23**, 1172.
[42] B. W. Brooks and B. O. Makanjuola, *Polymer*, 1982, **23**, 77.
[43] C. Hagiopol, et al., *Acta Polym.*, 1981, **32**, 390.
[44] M. Banerjee, U. Sathpathy, T. K. Paul, and R. S. Konar, *Polymer*, 1981, **22**, 1729.
[45] U. Sathpathy, T. K. Paul, M. Banerjee, and R. S. Konar, *J. Macromol. Sci., Chem.*, 1981, **15**, 1495.
[46] T. Cao, et al., *Huagong Xuebao*, 1982, 14.
[47] T. Cao, et al., *Tianjin Daxue Xuebao*, 1981, 44.
[48] A. R. Padwa and M. D. Paster, *J. Appl. Polym. Sci.*, 1982, **27**, 1385.
[49] H. Kast and W. Funke, *Makromol. Chem.*, 1981, **182**, 1553.
[50] H. Kast and W. Funke, *Makromol. Chem.*, 1981, **182**, 1567.
[51] K. H. Goebel, et al., *Acta Polym.*, 1981, **32**, 117.
[52] K. H. Goebel, et al., *Acta Polym.*, 1982, **33**, 49.

Chain Reaction Polymerization

Other papers on the kinetics and mechanism of emulsion polymerization include those by Anzur et al.[53] on correlations between particle size, conversion, heat of polymerization and molecular weight using mixed surfactants, by Nomura et al.[54] on the role of chain transfer agents in the emulsion polymerization of styrene, by Akopyan and Beileryan[55] on the emulsion polymerization of styrene in the presence of a non-ionic surfactant, and by Sundardi et al.[56] on radiation-initiated emulsion polymerization.

4 Initiators

Interest continues to be shown in the kinetics of the decomposition of persulphate initiators. Brooks and Makanjuola[57] have reported results for the rate of decomposition of ammonium persulphate in the presence of polymer latices. The rate of decomposition in water containing sodium dodecyl sulphate at 60 °C was found to be unaffected by the presence of polymethyl methacrylate particles, but was found to be increased 5-fold by polystyrene particles and 3-fold by polyvinyl acetate particles. Pirumyan[58] has published the results of a colorimetric study of the decomposition of potassium persulphate in aqueous solutions of polyoxyethylated hexadecanols. The effect of ethylene glycol and alkylethoxylates upon the decomposition of persulphate salts has been reported by Volkov et al.[59] Ryabova et al.[60] have investigated the rate of decomposition of potassium persulphate in aqueous emulsions of various hydrocarbons (ethylbenzene, styrene, and α-methylstyrene) in the presence of various surfactants. The rates were found to be 20—70% higher than in pure water. Bataille, Van, and Pham[61] have shown that silver(I) ions accelerate the emulsion polymerization of styrene initiated by potassium persulphate, whereas iron(II) ions reduce the rate of polymerization.

Some interest continues to be shown in the use of chelate metal complexes as initiators for emulsion polymerization. Karamyan et al.[62] have described the emulsion copolymerization of vinyl acetate, ethylene, and vinyl chloride using manganese triacetylacetonate as initiator. Kukushkina and Belogovodskaya[63] have investigated the effect of the nature of a buffer additive upon the kinetics of the emulsion polymerization of vinyl acetate using this initiator.

Several papers which have appeared during the period covered by this Report are concerned with the interesting topic of emulsion polymerization initiators which also function as surfactants. Dicke and Heitz[64] have described the preparation and characteristics of azo initiators of this type which are prepared by polymerizing

---

[53] I. Anzur, et al., *Vestn. Slov. Kem. Drus.*, 1982, **29**, 91.
[54] M. Nomura, Y. Minamino, K. Fujita, and M. Harada, *J. Polym. Sci., Polym. Chem. Ed.*, 1982, **20**, 1261.
[55] G. D. Akopyan and N. M. Beileryan, *Arm. Khim. Zh.*, 1981, **34**, 717.
[56] F. Sundardi, et al., *Radiat. Phys. Chem.*, 1981, **18**, 1109.
[57] B. W. Brooks and B. O. Makanjuola, *Makromol. Chem., Rapid Commun.*, 1981, **2**, 69.
[58] G. P. Pirumyan, *Arm. Khim. Zh.*, 1982, **35**, 360.
[59] V. A. Volkov, et al., *Zh. Prikl. Khim.*, 1981, **54**, 2297.
[60] M. S. Ryabova, et al., *Zh. Prikl. Khim.*, 1982, **55**, 632.
[61] P. Bataille, B. T. Van, and Q. B. Pham, *J. Polym. Sci., Polym. Chem. Ed.*, 1982, **20**, 811.
[62] D. R. Karamyan, et al., *Zh. Prikl. Khim.*, 1981, **54**, 2730.
[63] N. P. Kukushkina and K. V. Belogovodskaya, Deposited Doc., 1980, SPSTL 605 Khp-D80, 27.
[64] H. R. Dicke and W. Heitz, *Colloid Polym. Sci.*, 1982, **260**, 3.

acrylamide using as initiator the reaction product of azobisisobutyronitrile and tetraethylene glycol or 1,6-hexanediol. Initiators of this type are also described in a recent patent.[65] Ershov et al.[66] have described a novel type of polymeric emulsifier-initiator which has the general structure

$$(CH_2CRCO_2M)_m(CH_2CRCN)_n(CH_2CHC\equiv CCMe_2OOCMe_3)_p$$

where R = H or Me, M = K, Na, or NH_4, m = 50—80 mol %, n = 0—43.5 mol %, and p = 2—20 mol %. Pavlyuchenko et al.[67] have reported on the use of surface-active hydroperoxide initiators in emulsion polymerization reactions. Rozhkova et al.[68] have described the use of 1-(α-alkylacryloyloxy)-1-(t-butylperoxy)ethane as a surface-active initiator for the emulsion polymerization of vinyl monomers. Other publications which deal with hydroperoxidic surface-active initiators include those of Voronov et al.[69] and of Ivanchev et al.[70,71]

Bonamy, Fouassier, and Lougnot[72] have reported on the use of the sodium salt of (4-sulphomethylphenyl) phenylethanedione as a very efficient photo-initiator for emulsion polymerization reactions. Soltes et al.[73] have described the production of polystyrene of very high molecular weight by emulsion polymerization using a macromolecular initiator.

5 Surfactants

An important paper by Karsa[74] summarizes the results of an extensive investigation into the influence of the structure of surfactants of the alkylbenzenesulphonate type on their performance as primary surfactants in emulsion polymerization reactions. The emulsion polymerizations investigated were those of styrene and styrene–butadiene. Surfactants of the alkylbenzenesulphonate type are extensively used industrially for the production of, inter alia, carboxylated styrene–butadiene copolymer latices. It is well known that, in recipes of industrial importance, the nature of the alkylbenzenesulphonate surfactant can be of great importance in determining whether or not the reaction goes smoothly; the paper by Karsa is probably the first which seeks to establish correlations between structure and performance for this class of surfactant. At least two other publications[75,76] have appeared which are also concerned with the use of surface-active alkylbenzenesulphonates in emulsion polymerization reactions. Three publications[77-79] have appeared which are concerned with the use of polyalkenoxylsulphonates. Nakae,

---

[65] W. Heitz, H. G. Stahl, and R. Dicke, Ger. P., 3 005 889, 1981.
[66] A. Ershov, et al., USSR P., 937 467, 1982.
[67] V. N. Pavlyuchenko, et al., Dokl. Akad. Nauk SSSR, 1981, **259**, 641.
[68] D. A. Rozhkova, et al., USSR P., 852 879, 1981.
[69] S. A. Voronov, et al., Colloid J. USSR, 1980, **42**, 373; Kolloidn. Zh., 1980, **42**, 452.
[70] S. S. Ivanchev, et al., Neth. P., 79 07 193, 1981.
[71] S. S. Ivanchev, et al., Ger. P., 2 938 745, 1981.
[72] A. Bonamy, J. P. Fouassier, and D. J. Lougnot, J. Polym. Sci., Polym. Lett. Ed., 1982, **20**, 315.
[73] L. Soltes, et al., Chem. Zvesti, 1981, **35**, 533.
[74] D. R. Karsa, Preprints for PRI Emulsion Polymers Conference, London, 1982, Paper No. 2.
[75] Diamond Shamrock Corp., Jpn. P., 80 142 019, 1980.
[76] Lion Corp., Jpn. P., 82 53 504, 1982.
[77] P. M. Chakrabarti and D. G. Kirchner, US P., 4 329 268, 1982.
[78] Kao Soap Co. Ltd., Jpn. P., 82 78 936, 1982.
[79] Kao Soap Co. Ltd., Jpn. P., 82 78 937, 1982.

Chain Reaction Polymerization

Tsuji, and Yamanaka[80] have published details of a method of determining the distribution of alkyl chain lengths of alkylbenzenesulphonates by liquid chromatography. Asbeck et al.[81] have described the use of novel sulphosuccinate surfactants in emulsion polymerization reactions; this is also the subject of two other publications.[82,83]

Various publications have appeared which are concerned with the use of surface-active carboxylates as surfactants in emulsion polymerization reactions.[84-90] Of particular interest are those by Shapiro et al.,[84] which is concerned with the effect of the nature of the surfactant upon the microstructure of butadiene–methyl methacrylate copolymers produced by emulsion polymerization; of Boiesan, Miron, and Cosaveanu,[85] which is concerned with the influence of the structure of carboxylate surfactants upon performance in the emulsion polymerization of butadiene; and of Force,[86] which describes modified carboxylate surfactants which can be used for the production of synthetic rubber latices having increased particle size.

Two patents granted to the Kao Soap Co.[91,92] disclose the use of alkoxylated phenols as surfactants for the emulsion polymerization of vinyl and diene monomers. A patent to Aika Industry Co.[93] describes the use of non-ionic surfactants such as alkylphenyl ethers of polyethylene glycols as surfactants for the emulsion polymerization of vinyl acetate. Several other publications[94-98] have appeared which describe novel non-polymerizable surfactants. Publications have appeared which describe the use of copolymerizable surfactants[99-100] and polymeric surfactants[101-122] in emulsion polymerization reactions; some of the latter surfactants may also be copolymerizable.

[80] A. Nakae, K. Tsuji, and M. Yamanaka, *Anal. Chem.*, 1981, **53**, 1818.
[81] A. Asbeck, et al., US P., 4 299 975, 1981.
[82] L. I. Kovtunenko, et al., *Prom-St. Sint. Kauch.*, 1980 (10), 10.
[83] L. I. Kovtunenko, et al., *Prom-St. Sint. Kauch.*, 1982 (4), 11.
[84] Yu. Ye. Shapiro, et al., *Polym. Sci. USSR*, 1981, **23**, 1522; *Vysokomol. Soedin., Ser. A*, 1981, **23**, 1374.
[85] V. Boiesan, M. Miron, and A. Cosaveanu, *Mater. Plast.*, 1981, **18**, 88.
[86] C. G. Force, US P., 4 259 459, 1981.
[87] E. Lashkina, et al., *Prom-St. Sint. Kauch.*, 1980 (11), 12.
[88] A. P. Titov, et al., USSR P., 765 286, 1980.
[89] T. G. Balyberdina and N. M. Mironova, Deposited Doc., 1981, SPSTL 226 Khp D81.
[90] K. Sugiyama, *Kinki Daigaku Kogakubu Kenkyu Hokoku*, 1981, **15**, 1.
[91] Kao Soap Co. Ltd., Jpn. P., 81 139 502, 1981.
[92] Kao Soap Co. Ltd., Jpn. P., 81 139 503, 1981.
[93] Aika Industry Co. Ltd., Jpn. P., 82 00 105, 1982.
[94] H. Perrey, et al., Can. P., 1 112 976, 1981.
[95] Arakawa Chem. Ind. Ltd., Jpn. P., 82 100 101, 1982.
[96] Lion Corp., Jpn. P., 82 53 503, 1982.
[97] H. Inagaki, *Kobe Joshi Daigaku Kiyo*, 1980 (10), 57.
[98] Kuraray Co. Ltd., Jpn. P., 82 55 913, 1982.
[99] J. H. Deutsch, US P., 4 246 387, 1981.
[100] E. B. Malyukova, et al., *Dokl. Akad. Nauk SSSR*, 1982, **265**, 375.
[101] J. Hambrecht, et al., Ger. P., 3 047 688, 1982.
[102] R. A. Wessling and D. M. Pickelman, US P., 4 337 185, 1982.
[103] Nippon Synthetic Chemical Industry Co. Ltd., Jpn. P., 81 141 825, 1981.
[104] Kao Soap Co. Ltd., Jpn. P., 82 18 709, 1982.
[105] W. D. Del Vecchio, I. E. Isgur, and J. L. Ohlson, Fr. P., 2 474 511, 1981.
[106] Agency of Industrial Science and Technology, Jpn. P., 81 43 303, 1981.
[107] Nippon Synthetic Chemical Industry Co. Ltd., Jpn. P., 82 94 002, 1982.
[108] Nippon Synthetic Chemical Industry Co. Ltd., Jpn. P., 81 161 828, 1981.
[109] Sekisui Chemical Co. Ltd., Jpn. P., 81 135 503, 1981.
[110] Dainippon Ink and Chemical Co. Ltd., Jpn. P., 82 108 113, 1982.

Shevchuk et al.[123] have reported on the effect of the nature and concentration of ionogenic surfactants on graft copolymerization during the production of methyl methacrylate–butadiene–styrene copolymers by emulsion polymerization. Other papers which contain information relating to the behaviour of surfactants in emulsion polymerization include those of Haq and Thompson[124] on the significance of the polymer glass-transition temperature for the stabilization of latices by non-ionic surfactants, of Vijayendran, Bone, and Gajria[125] on surfactant interactions in poly(vinyl acetate) and vinyl acetate–butyl acrylate copolymer latices, of Kronberg et al.[126] on the adsorption of non-ionic surfactants on polystyrene and poly(vinyl chloride) latices, and of Yablonskii et al.[127] on the mechanism of the solubilization of non-polar monomers in the micelles of a sodium alkylbenzenesulphonate surfactant. Al-Shahib and Dunn[128] have proposed emulsion polymerization as a method for determining the critical micelle concentration of surfactants of low solubility in water; the method depends upon the observation that, as the surfactant concentration is increased, so the number of latex particles formed and the rate of polymerization increase sharply over the concentration range where micelles first become present in significant numbers.

6 Low-surfactant and Surfactant-free Emulsion Polymerizations

Interest continues to be shown in emulsion polymerizations in low-surfactant and surfactant-free reaction systems. Arai, Arai, and Saito[129] have studied the surfactant-free emulsion polymerization of methyl methacrylate in water in the presence of calcium sulphite. It is claimed that solid calcium sulphite particles can function as loci for polymerization. Arai et al.[130] have also studied the effect of agitation upon the surfactant-free emulsion polymerization of methyl methacrylate in water. The results indicate that the rate of transfer of monomer from the droplets to the aqueous phase is an important factor in determining the rate of polymerization. Konno et al.[131] have also investigated the matter of the distribution of

---

[111] Showa Highpolymer Co. Ltd., Jpn. P., 80 145 703, 1980.
[112] Sekisui Chemical Co. Ltd., Jpn. P., 80 144 002, 1980.
[113] Sekisui Chemical Co. Ltd., Jpn. P., 81 28 204, 1981.
[114] Sekisui Chemical Co. Ltd., Jpn. P., 82 87 401, 1981.
[115] Lion Corp., Jpn. P., 82 53 502, 1982.
[116] Aika Industry Co. Ltd., Jpn. P., 82 67 608, 1982.
[117] Nichiden Kagaku Co. Ltd., Jpn. P., 80 161 806, 1980.
[118] Nippon Synthetic Chemical Industry Co. Ltd., Jpn. P., 80 152 707, 1980.
[119] Nippon Synthetic Chemical Industry Co. Ltd., Jpn. P., 81 93 702, 1981.
[120] Kuraray Co. Ltd., Jpn. P., 81 99 204, 1981.
[121] M. Buday, et al., Czech. P., 209 105, 1981.
[122] Kuraray Co. Ltd., Jpn. P., 80 98 201, 1980.
[123] L. M. Shevchuk, et al., Vysokomol. Soedin., Ser. A, 1981, 23, 913.
[124] Z. Haq and L. Thompson, Colloid Polym. Sci., 1982, 260, 212.
[125] B. R. Vijayendran, T. Bone, and C. Gajria, J. Appl. Polym. Sci., 1981, 26, 1351.
[126] B. Kronberg, et al., J. Dispersion Sci. Technol., 1981, 2, 215.
[127] O. P. Yablonskii, et al., Kolloid. Zh., 1982, 44, 391.
[128] W. A. Al-Shahib, and A. S. Dunn, J. Dispersion Sci. Technol., 1981, 2, 175.
[129] M. Arai, K. Arai, and S. Saito, J. Polym. Sci., Polym. Chem. Ed., 1982, 20, 1021.
[130] K. Arai, et al., J. Polym. Sci., Polym. Chem. Ed., 1981, 19, 1203.
[131] M. Konno, et al., J. Polym. Sci., Polym. Chem. Ed., 1982, 20, 3251.

monomer in reaction systems for the emulsifier-free emulsion polymerization of methyl methacrylate. Okuba et al.[132] have investigated the surfactant-free emulsion polymerization of styrene in mixtures of acetone and water using potassium persulphate as initiator. Below an acetone concentration of about 40% by volume, stable latices can be prepared at rates of polymerization which are much faster than those in pure water. The particle size of the latices decreases with increasing acetone content; the particle-size distributions are very sharp. Laaksonen and Stenius[133] have reported on the surfactant-free emulsion polymerization of vinyl chloride. Ohtsuka, Kawaguchi, and Hayashi[134] have reported on the copolymerization of styrene and 4-vinylpyridine in a surfactant-free aqueous medium to give a cationic copolymer latex. As might be expected, the latex obtained by polymerization under acid conditions differs somewhat from that obtained by polymerization under basic conditions. Kawaguchi, Sugi, and Ohtsuka[135,136] have described surfactant-free emulsion copolymerizations of styrene and various acrylamide derivatives in water. Kawaguchi, Hoshino, and Ohtsuka[137] have described a method for preparing amphoteric latices by modifying styrene–acrylamide copolymer latices which have been prepared by surfactant-free emulsion polymerization in water. The latices are treated with mixtures of sodium hypochlorite and sodium hydroxide to give latices which exhibit amphoteric properties because of the presence of amino and carboxyl groups at the surface of the particles. Latices having the same particle size but different surface-charge densities and isoelectric points could be prepared by varying the reaction conditions.

Several other publications on the subject of low-surfactant and surfactant-free emulsion polymerizations have appeared.[138–145] Of particular interest is a paper by Furusawa,[138] which describes the preparation of polystyrene latices having a wide range of surface-charge densities by surfactant-free emulsion polymerization. Conductometric titrations show that the charge associated with the Stern layer increases at approximately the same rate as the surface charge, leading to a diffuse layer whose nature is virtually independent of the surface-charge density. Steffers, Rothenhaeusser, and Voelzke[139] have disclosed a method for producing mechanically stable butadiene copolymer latices having low surfactant contents; the colloid stabilizer comprises a combination of an anionic surfactant and a fatty alcohol. Batz et al.[145] have described the production of latices of hydrophilic polymers (such as polyglycidyl methacrylate) by surfactant-free emulsion polymerization.

[132] M. Okuba, et al., J. Appl. Polym. Sci., 1981, **26**, 1675.
[133] J. Laaksonen and P. Stenius, Plast. Rubber: Mater. Appl., 1980, **5**, 21.
[134] Y. Ohtsuka, H. Kawaguchi, and S. Hayashi, Polymer, 1981, **22**, 658.
[135] Y. Ohtsuka, H. Kawaguchi, and Y. Sugi, J. Appl. Polym. Sci., 1981, **26**, 1637.
[136] H. Kawaguchi, Y. Sugi, and Y. Ohtsuka, J. Appl. Polym. Sci., 1981, **26**, 1649.
[137] H. Kawaguchi, H. Hoshino, and Y. Ohtsuka, J. Appl. Polym. Sci., 1981, **26**, 2015.
[138] K. Furusawa, Bull. Chem. Soc. Jpn., 1982, **55**, 48.
[139] F. Steffers, B. Rothenhaeusser, and W. Voelzke, East Ger. P., 149 226, 1981.
[140] S. Besecke, et al., Eur. P., 48 320, 1982.
[141] S. Yamazaki, Tokyo Kogyo Shikensho Hokoku, 1980, **75**, 341.
[142] C. N. Bush, PCI Int. P., 81 02 158, 1981.
[143] J. C. Salamone, et al., Polym. Amines Ammonium Salts, Invited Lect. Contrib. Pap. Int. Symp. 1979 (published 1980), p. 105.
[144] T. Suwa, Report 1979, JAERI-M-8529.
[145] H. G. Batz, et al., Eur. P., 54 685, 1982.

7 Emulsion Polymerization Reactions involving Functional-group Co-monomers

Publications which deal with emulsion polymerization reaction systems which contain functional-group co-monomers inevitably overlap to some extent with those concerned with low-surfactant and surfactant-free emulsion polymerizations. This is because functional-group co-monomers can be used to provide additional stabilizing moieties in low-surfactant and surfactant-free emulsion polymerizations.

Several publications[146—156] are concerned with emulsion polymerization reaction systems which contain polymerizable carboxylic acids. Egusa and Makuuchi[146,147] have produced carboxylated acrylic latices by radiation-initiated emulsion polymerization, and compared the products with similar latices produced by conventional chemically initiated emulsion polymerization. Three publications[154—156] are concerned with the interesting and industrially important subject of the production of alkali-thickenable carboxylated latices; much of the information contained in the third of these publications is also available in a paper which appears in ref. 3.

Liu and Krieger[157] have described the use of methylvinylpyridinium salts as cationic co-monomers. Chonde and Krieger[158] have published results for the emulsion polymerization of styrene with ionic co-monomers in the presence of methanol. Other publications in this area include those of the Kuraray Co.,[159] and of Kurenkov et al.,[160,161] all of which are concerned with emulsion polymerizations in which ionic co-monomers such as potassium styrenesulphonate are present.

8 Emulsion Copolymerization

Again, some overlap between some of the publications considered under this heading and some of those noted elsewhere in this report is inevitable. Theoretical treatments of the kinetics of emulsion copolymerization have been given by Ballard et al.[162,163] and by Nomura, Kubo, and Fujita.[164] Lin, Ku, and Chin[165] have presented a simulation model for the emulsion copolymerization of acrylonitrile

---

[146] S. Egusa and K. Makuuchi, *J. Polym. Sci., Polym. Chem. Ed.*, 1982, **20**, 863.
[147] S. Egusa and K. Makuuchi, *J. Colloid Interface Sci.*, 1981, **79**, 350.
[148] P. Kofinov, et al., *Kolloidn. Zh.*, 1982, **44**, 582.
[149] F. Steffers, et al., East Ger. P., 143 621, 1980.
[150] F. Steffers, et al., East Ger. P., 142 346, 1980.
[151] F. Steffers and B. Rothenhaeusser, East Ger. P., 144 925, 1980.
[152] F. Steffers and B. Rothenhaeusser, East Ger. P., 146 612, 1981.
[153] F. Heins, et al., Eur. P., 54 766, 1982.
[154] M. Okubo, et al., *Kobunshi Ronbunshu*, 1982, **39**, 157.
[155] M. Okubo, et al., *Kobunshi Ronbunshu*, 1982, **39**, 345.
[156] S. Nishida, Ph.D. Thesis, Lehigh University, 1981.
[157] L.-J. Liu and I. M. Krieger, *J. Polym. Sci., Polym. Chem. Ed.*, 1981, **19**, 3013.
[158] Y. Chonde and I. Krieger, *J. Appl. Polym. Sci.*, 1981, **26**, 1819.
[159] Kuraray Co. Ltd., Jpn. P., 81 62 811, 1981.
[160] V. F. Kurenkov, et al., *Izv. Vyssh. Uchebn. Zaved. Khim. Khim. Tekhnol.*, 1982, **25**, 221.
[161] V. F. Kurenkov, et al., Deposited Doc., 1981, SPSTL 13 Khp-D81, 116.
[162] M. J. Ballard, D. H. Napper, and R. G. Gilbert, *J. Polym. Sci., Polym. Chem. Ed.*, 1981, **19**, 939.
[163] M. J. Ballard, R. G. Gilbert, and D. H. Napper, *J. Dispersion Sci. Technol.*, 1981, **2**, 163.
[164] M. Nomura, M. Kubo, and K. Fujita, *Fukui Daigaku Kogakubu Kenkyu Hokoku*, 1981, **29**, 167.
[165] C.-C. Lin, H.-C. Ku, and W.-Y. Chin, *J. Appl. Polym. Sci.*, 1981, **26**, 1327.

Chain Reaction Polymerization

and styrene under azeotropic conditions. Guillot[166] has reported on kinetic and thermodynamic aspects of the emulsion copolymerization of acrylonitrile and styrene. Nomura et al.[167,168] have published two papers on the emulsion copolymerization of styrene and methyl methacrylate; the second of these is concerned with the effect of radical desorption upon the rate of polymerization. Goldwasser and Rudin[169] have also reported on the emulsion copolymerization of styrene and methyl methacrylate. They report that the monomer composition in the loci is the same as that in the monomer droplets, notwithstanding the higher water solubility of methyl methacrylate, and that monomer reactivity ratios for the emulsion reaction are not significantly different from those for the bulk reaction.

Amongst the many other publications[170—187] which have appeared on the subject of emulsion copolymerization, those of Rios Guerrero and Guillot[170,171] on the emulsion copolymerization of acrylonitrile, styrene, and methyl acrylate, of Siadat[172] on the emulsion copolymerization of isoprene and sodium styrenesulphonate, and of Kovarik and Horak[173] on the use of gas chromatography to follow the course of emulsion copolymerization reactions may be especially noted. Closely related to the subject of emulsion copolymerization is a paper by Hourston and Satgurunathan[188] which describes the preparation of interpenetrating polymer networks by emulsion polymerization.

9 Procedural and Chemical Engineering Aspects

Interest continues to be shown in ways of producing structured latex particles by emulsion polymerization.[189—195] Thus Chainey, Hearn, and Wilkinson[189] have

---

[166] J. Guillot, *Acta Polym.*, 1981, **32**, 593.
[167] M. Nomura, et al., *Kobunshi Kako*, 1981, **30**, 459.
[168] M. Nomura, et al., *J. Appl. Polym. Sci.*, 1982, **27**, 2483.
[169] J. M. Goldwasser and A. Rudin, *J. Polym. Sci., Polym. Chem. Ed.*, 1982, **20**, 1993.
[170] L. Rios and J. Guillot, *Makromol. Chem.*, 1982, **183**, 531.
[171] J. Guillot and L. Rios Guerrero, *Makromol. Chem.*, 1982, **183**, 1979.
[172] B. Siadat, *J. Dispersion Sci. Technol.*, 1982, **2**, 147.
[173] J. Kovarik and J. Horak, *Makrotest*, 1980, 190.
[174] K.-Y. Moon, et al., *Pollimo*, 1981, **5**(2), 122.
[175] G. D. Akopyan and N. M. Beileryan, *Arm. Khim. Zh.*, 1981, **34**, 801.
[176] V. D. Malkov, et al., Deposited Doc., 1980, SPSTL 977 Khp-D80.
[177] T. Makawinata, M. S. El-Aasser, J. W. Vanderhoff, and C. Pichot, *Acta Polym.*, 1981, **32**, 583.
[178] A. Hu and M. M.-P. Lee, *J. Chin. Inst. Chem. Eng.*, 1981, **12**(2), 53.
[179] O. K. Shvetsov, et al., *Acta Polym.*, 1981, **32**, 403.
[180] V. Jarm and M. Kovac-Filipovic, *Polimeri*, 1982, **3**, 29.
[181] G. Kojima and M. Hisasue, *Makromol. Chem.*, 1981, **182**, 1429.
[182] V. N. Pavlyuchenko, et al., *Vysokomol. Soedin., Ser. A*, 1981, **23**, 2204.
[183] M. Georgescu, et al., *Rev. Chim.*, 1981, **32**, 845.
[184] A. K. Askerov, et al., Mater. Sumgaitskoi Gor. Nauchno-Tekh. Konf. Probl. 'Org. Khlororg. Sint', 1980, p. 54.
[185] L. M. Shevchuk, et al., *Polym. Sci. USSR*, 1981, **23**, 1022; *Vysokomol. Soedin., Ser. A*, 1981, **23**, 913.
[186] I. Ismailov, et al., *Dokl. Akad. Nauk. Uzb. SSR*, 1981, 41.
[187] A. Klein and E. S. Barabas, *Polym. Prepr., Am. Chem. Soc., Div. Polym. Chem.*, 1979, **20**(1), 199.
[188] D. J. Hourston and R. Satgurunathan, Preprints for PRI Emulsion Polymers Conference, London, 1982, Paper No. 7.
[189] M. Chainey, J. Hearn, and M. C. Wilkinson, *Br. Polym. J.*, 1981, **13**, 132.
[190] A. Kowalksi, M. Vogel, and R. M. Blankenship, Eur. P., 22 633, 1981.
[191] T. Ishikawa and D. I. Lee, Eur. P., 31 964, 1981.
[192] M. Baer, Eur. P., 55 890, 1982.
[193] J. R. Erickson and R. J. Seidewand, US P., 4 226 752, 1980.
[194] M. Suzuki and M. Kobori, Ger. P., 3 035 911, 1981.
[195] M. Okubo, Y. Katsuta, and T. Matsumoto, *J. Polym. Sci., Polym. Lett. Ed.*, 1982, **20**, 45.

described the preparation of overcoated polymer latex particles by a 'shot-growth' technique. Three recent patents[190–192] disclose novel methods for preparing structured latex particles. The first of these [190] is particularly interesting in that it describes the preparation of latex particles which have acidic cores that are at least partially encased in sheath polymers which are permeable to volatile bases; such particles do not swell appreciably in an alkali such as potassium hydroxide, but do swell in an alkali such as ammonia.

Several publications have appeared on the subject of continuous and semi-continuous emulsion polymerization.[196–207] Poehlein[196] has published a useful review of emulsion polymerization in continuous reactor systems, which includes a review of the differences between batch and continuous emulsion polymerizations. In a second paper, Poehlein et al.[205] have presented a steady-state model for emulsion polymerization in a continuous stirred-tank reactor which is being fed with a seed latex of narrow particle-size distribution. Several publications concerned with the control of emulsion polymerization reactions have appeared.[208–219] A patent to Abbey[218] discloses a process for preparing latices having a bimodal particle-size distribution. A patent to Mitsubishi Electric Corp.[219] discloses a procedure for detecting the starting-point of emulsion polymerization reactions in which a light beam is passed by means of an optical fibre through the reaction system, and the light scattered by the polymer particles is measured.

10 Miscellaneous Other Aspects

Note can be taken here of but a selection of the many publications which have appeared which fall under this broad heading. Lukhovitskii[220] has reported on the emulsion polymerization of monomers which function as chain-transfer agents. Zuikov and Solov'ev[221] have studied the influence of monomer polarity upon the particle-size distribution of the latex produced. Various publications have appeared which deal with aspects of the emulsion polymerization of monomers such as vinyl

[196] G. W. Poehlein, *Br. Polym. J.*, 1982, **14**, 153.
[197] M.-K. Lu, Y.-W. Huang, and C.-Z. Hsu, *Ying Yung Chieh Mien Hua Hsueh*, 1981, **11**, 11.
[198] T. V. Kreitser, *et al.*, *Zh. Prikl. Khim.*, 1982, **55**, 1647.
[199] M. J. Heckel, *et al.*, East Ger. P., 145 171, 1980.
[200] J. Stoeckel, *et al.*, East Ger. P., 148 225, 1981.
[201] J. Snuparek and K. Kaspar, *J. Appl. Polym. Sci.*, 1981, **26**, 4081.
[202] J. Snuparek, *Acta Polym.*, 1981, **32**, 368.
[203] T. Makawinata, Thesis, Lehigh University, 1982.
[204] F. J. Schork, Thesis, University of Wisconsin, 1981.
[205] G. W. Poehlein, *et al.*, *Br. Polym. J.*, 1982, **14**, 143.
[206] C.-C. Lin and W.-Y. Chiu, *J. Chin. Inst. Chem. Eng.*, 1980, **11**(4), 203.
[207] C.-C. Lin and W.-Y. Chiu, *J. Chin. Inst. Chem. Eng.*, 1981, **12**(2), 79.
[208] K. W. Leffew, Thesis, University of Louisville, 1981.
[209] C. Kiparissides, J. F. MacGregor, and A. E. Hamielec, *AIChE J.*, 1981, **27**, 13.
[210] A. A. Abdullaev, *et al.*, USSR P., 804 641, 1981.
[211] A. A. Abdullaev, *et al.*, USSR P., 852 878, 1981.
[212] A. A. Abdullaev, *et al.*, USSR P., 937 465, 1982.
[213] Z. M. Guseinov, *et al.*, USSR P., 956 490, 1982.
[214] R. Hanna, US P., 4 331 577, 1982.
[215] T. I. Ismailov, USSR P., 761 482, 1980.
[216] T. I. Ismailov, USSR P., 761 486, 1980.
[217] T. I. Ismailov, USSR P., 943 248, 1982.
[218] K. J. Abbey, US P., 4 254 004, 1981.
[219] Mitsubishi Electric Corp., Jpn. P., 81 20 001, 1981.
[220] V. I. Lukhovitskii, *Vysokomol. Soedin., Ser. A*, 1981, **23**, 2039.
[221] A. V. Zuikov and Yu. V. Solov'ev, *Colloid J. USSR*, 1980, **42**, 454; *Kolloidn. Zh.*, 1980, **42**, 549.

acetate,[222,223] vinyl benzoate,[224] butyl acrylate,[225] fluorine-containing monomers,[226] and cyclosiloxanes.[227,228] Hasan[229] has reported on the effects of purity of ingredients, oxygen, and the presence of inorganic particles upon the emulsion polymerization of styrene. Merlin and Fouassier[230] have studied photo-initiated emulsion polymerization of methyl methacrylate in the presence of a saccharide. Yu and Funke[231] have prepared so-called 'reactive microgels' by the emulsion polymerization of unsaturated polyester resins. Some interest has been shown in the preparation of coloured latices by emulsion polymerization,[232,233] and a method has been disclosed for preparing latices which contain magnetic particles.[234] Uranek and Clark[235] have shown that the reactivity of mercaptan modifiers in styrene–butadiene emulsion copolymerizations can be greatly enhanced by adding them as microemulsion prepared using a surfactant/alcohol combination. Blackley and Jarm[236] have reported further observations on the effect of organic diluents upon emulsion polymerization reactions. Dunn and Said[237] have reported on the effects of ionic strength upon the emulsion polymerization of styrene. Ivanchev and Pavlyuchenko[238] have studied the emulsion polymerization of styrene under conditions such that radical formation is believed to be localized within the adsorption layer of the surfactant. Wilkinson et al.[239] have described a microfiltration technique for the cleaning of polymer latices. Corner[240] has described the preparation of polyelectrolyte-stabilized latices. Several publications have appeared on the subject of inverse emulsion polymerization in water-in-oil emulsions;[241–246] most of these publications are concerned with inverse emulsion polymerizations and copolymerizations of acrylamide, but one[246] is concerned with the inverse emulsion polymerization of N-vinylpyrrolidone.

Acknowledgement. The assistance of Dr. A. C. Haynes in preparing this report is gratefully acknowledged.

[222] D. Donescu, et al., *Mater. Plast.*, 1980, **17**, 146.
[223] W. D. Hergeth, et al., *Acta Polym.*, 1980, **31**, 704.
[224] B. Plavljonić and Z. Janović, *J. Polym. Sci., Polym. Chem. Ed.*, 1981, **19**, 1795.
[225] I. V Malumyan, et al., *Plast. Massy*, 1981 (9), 8.
[226] S. Machi, *Polym. Prepr., Am. Chem. Soc., Div Polym. Chem.*, 1979, **20**(1), 419.
[227] X. Zhang, et al., *Gaofenzi Tongxun*, 1982, 154.
[228] S. Liu and P. Li, *Gaofenzi Tongxun*, 1982, 257.
[229] S. M. Hasan, *J. Polym. Sci., Polym. Chem. Ed.*, 1982, **20**, 2969.
[230] A. Merlin and J.-P. Fouassier, *J. Polym. Sci., Polym. Chem. Ed.*, 1981, **19**, 2357.
[231] Y. C. Yu and W. Funke, *Angew. Makromol. Chem.*, 1982, **103**, 187.
[232] S. A. Voronov, et al., USSR P., 887 572, 1981.
[233] Y. J. Shih, Thesis, Lehigh University, 1981.
[234] J. C. Daniel, J. L. Schuppiser, and M. Tricot, Eur. P., 38 730, 1981.
[235] C. A. Uranek and E. Clark, *J. Appl. Polym. Sci.*, 1981, **26**, 107.
[236] D. C. Blackley and V. Jarm, Preprints for 27th International Symposium on Macromolecules, Strasbourg, 1981, p. 355.
[237] A. S. Dunn and Z. F. M. Said, Preprints for 27th International Symposium on Macromolecules, Strasbourg, 1981, p. 333.
[238] S. S. Ivanchev and V. N. Pavlyuchenko, *Acta Polym.*, 1981, **32**, 407.
[239] M. C. Wilkinson, et al., *Br. Polym. J.*, 1981, **13**, 82.
[240] T. Corner, *Colloids Surf.*, 1981, **3**, 119.
[241] Y. S. Leong and F. Candau, *J. Phys. Chem.*, 1982, **86**, 2269.
[242] F. Candau and Y. S. Leong, Preprints for 27th International Symposium on Macromolecules, Strasbourg, 1981, p. 326.
[243] M. T. McKechnie, Preprints for PRI Emulsion Polymers Conference, London, 1982, Paper No. 3.
[244] K. U. Kim, et al., *Pollimo*, 1982, **6**, 188.
[245] K. U. Kim, et al., *Pollimo*, 1982, **6**, 197.
[246] K. U. Kim, et al., *Pollimo*, 1982, **6**, 305.

PART VI Electrochemical Polymerization
by B. M. Tidswell

When an electric current is passed through a conducting solution of a suitable monomer, polymer may be produced. In many systems the polymer forms a coherent adhering coating on the electrode. This may be used to advantage and has been the subject of a short review by Dhyaneswari et al.[1] A more comprehensive review of the whole field of electropolymerization is that by Simionescu.[2] The majority of papers in the field of electropolymerization which have appeared over the past two years are concerned with this direct, *in situ*, film formation. Tidswell and Mortimer[3,4] studied the kinetics and locus of polymerization of acrylonitrile and methacrylonitrile on tin-plated steel using several electroanalytical techniques. Lecayon *et al.* have studied the use of acrylonitrile as a suitable monomer for the production of thin organic films[5] and have published a patent on the subject.[6] Porta, Wahl, and Gillot[7] have also patented a method for the direct formation of polymer films using diacetone acrylamide and NN'-methylene-bis(acrylamide) in the presence of anionic and non-ionic surfactants in aqueous H_2SO_4. Polyacrylamide has been deposited electrochemically on metal cathodes from aqueous solutions of the monomer containing $ZnCl_2$ to give a hydrogel coating.[8,9] The process was investigated using a rotating disc Ni electrode and cyclic chronovoltammetry.[10] In the presence of $\geqslant 4\%$ V/V HCHO Mironov *et al.*[11] produced polyacrylamide-based antifriction coatings. The presence of HCHO greatly improves the rate of formation and properties of the films, whereas inclusion of dispersed graphite improves the frictional properties. All other properties and the preparation suffer.

Koval'chuk *et al.*[12] have produced coatings on steel using vinyl co-monomer mixtures electrolysed in aqueous methanol using $K_2S_2O_8$ as electrolyte to produce even, smooth, anticorrosive films superior to those of homopolymers. Karpinets and Bezuglyi[13] have copolymerized methyl methacrylate and styrene, as have Zytner and Makarov,[14] who have also patented a process for the direct coating of metal surfaces.[15] Other Russian workers have also been involved in the electro-

---

[1] E. S. Dhyaneswari, S. Guruvaih and K. S. Rajogapalan, *Paintindia*, 1981, **31**, 3.
[2] C. Simionescu and M. Grovu, *Mater. Plast. (Bucharest)*, 1981, **18**, 73.
[3] B. M. Tidswell and D. A. Mortimer, *Eur. Polym. J.*, 1981, **17**, 735.
[4] B. M. Tidswell and D. A. Mortimer, *Eur. Polym. J.*, 1981, **17**, 745.
[5] G. Lecayon, C. LeGressus, C. Borziau, Y. Bourzem, C. Reynard, and C. Juret, *Chem. Phys. Lett.*, 1982, **91**, 506.
[6] G. Lecayon, C. LeGressus, and A. LeMoel, Eur. P., 38 244, 1981.
[7] A. Porta, B. Wahl, and M. A. Gillot, Swiss P., 623 608, 1981.
[8] Ya. D. Zytner, K. A. Makarov, and O. K. Lebedkina, *Isv. Vyssh. Uchebn. Zaved. Khim. Khim. Technol.*, 1980, **23**, 1532.
[9] Ya. D. Zytner and K. A. Makarov, *Vysokomol. Soedin.*, Ser. A, 1980, **22**, 2612.
[10] L. S. Tikhonova, K. A. Makarov, Ya. D. Zytner, V. Yu. Kostyushkina, and K. I. Tikhonov, *Zh. Prikl. Khim. (Leningrad)*, 1981, **54**, 1867.
[11] V. S. Mironov, Yu. M. Pleskachevskii, and A. F. Klimovich, *Zh. Prikl. Khim. (Leningrad)*, 1981, **51**, 87.
[12] E. P. Koval'chuk, L. A. Morkind, and E. I. Aksiment'eva, *Lakokras. Mater. Ikh. Primen.*, 1982, 17.
[13] A. P. Karpinets and V. D. Bezuglyi, *Zh. Prikl. Khim. (Leningrad)*, 1982, **55**, 1436.
[14] YA. D. Zytner and K. A. Makarov, *Vysokomol. Soedin.*, Ser. A, 1980, **22**, 2035.
[15] Ya. D. Zytner and K. A. Makarov, USSR P., 943 333, 1982.

chemical study of polymer coatings.[16] Bezuglyi et al.[17] have studied the effect of the presence of benzene on the formation of poly(methyl methacrylate) on metal cathodes; at $\geqslant 10\%$ benzene the rate of formation increases but the molecular weight of the resulting polymer decreases.

The protection of metals other than steel has also been investigated. Mengoli, Munari, and Folonari have reported the formation of polynitroanilide films on copper, possibly by an oxidative coupling process to give oligomeric structures.[18] Other Italian workers of the Pisa School have produced protective coatings with improved salt spray resistance on both copper and iron by the anodic oxidation of 2-methylphenol in aqueous solution.[19]

Subramanian and Jakubowski have improved the properties of high-temperature resistant polyimide and polyquinoxaline composites by electrolytically coating graphite films with polyimide resins;[20] the process has been patented.[21] MacCallum and MacKerron have recently reported their work on the electropolymerization of methyl methacrylate[22] and acrylamide[23] on carbon fibres. They propose that initiation is brought about by active hydrogen atoms and that post-polymerization effects, in the case of methyl methacrylate, indicate the presence of living polymer radicals.

Zytner et al. have studied the production of organosilicon polymer coatings derived from the electrochemically initiated polymerization of (diethyloxymethylsilyl)methyl methacrylate in NN'-dimethyl formamide or acetone using sodium perchlorate as electrolyte.[24-26]

Several studies have been reported on the production and properties of electroactive polymers. Funt and Hsu[27] have shown that for benzoylated polystyrene samples with constant spacing of the PhCO-groups along the chain, the electroactive centres behave independently of one another. Shaw, Haight, and Faulkner[28] have extended the approach of Funt and Hsu to produce electrochemically, electrodes coated with crosslinked poly(styrene sulphonate), while Kaufman et al. have stabilized the morphology of the polymer on polymer-coated electrodes by an electrochemical crosslinking reaction.[29]

The production of homogeneous and very adherent thin films on a variety of metal surfaces of amino-substituted poly(phenylene oxide) has been reported by

---

[16] L. M. Apraksina and D. M. Poddubnaya, Ref. Zh. Khim., 1981, Abs. No. 14T523.
[17] V. D. Bezuglyi, V. F. Mikhailik, and T. A. Alekseeva, Vysokomol. Soedin., Ser. B, 1980, 22, 681.
[18] G. Mengoli, M. T. Munari, and C. Folonari, J. Electroanal. Chem., Interfacial Electrochem., 1981, 124, 237.
[19] A. Lucchesi, G. Maschio, P. Cerrai, and G. Angeli, Conv. Ital. Sci. Macromol. (Atti)5th, 1981, p. 390.
[20] R. V. Subramanian and J. J. Jakubowski, Org. Coat. Plast. Chem., 1979, 40, 688.
[21] J. J. Jakubowski and R. V. Subramanian, US P., 4 272 346, 1979.
[22] J. R. MacCallum and D. H. MacKerron, Eur. Polym. J., 1982, 18, 717.
[23] J. R. MacCallum and D. H. MacKerron, Br. Polym. J., 1982, 14, 14.
[24] Ya. D. Zytner, O. K. Lebedkina, and K. A. Makarov, Izv. Vyssh. Uchebn. Zaved. Khim. Khim. Technol., 1981, 24, 1147.
[25] Ya. D. Zytner, K. A. Makarov, and O. K. Lebedkina, Vysokomol. Soedin., Ser. A, 1982, 24, 944.
[26] Ya. D. Zytner, O. K. Lebedkina, G. U. Ostrovidova, and K. A. Makarov, Vysokomol. Soedin., Ser. B, 1982, 24, 706.
[27] B. L. Funt, L.-C. Hsu, and P. M. Hoang, J. Polym. Sci., Polym. Chem. Ed., 1981, 19, 203.
[28] B. R. Shaw, G. P. Haight, and L. R. Faulkner, J. Electroanal. Chem., Interfacial Electrochem., 1982, 140, 147.
[29] F. B. Kaufman, A. H. Schroeder, V. V. Patel, and K. H. Nichols, J. Electroanal. Chem., Interfacial Electrochem., 1982, 132, 151.

Dubois, Lacaze, and Pham.[30] In the case of N-(o-hydroxybenzyl)aniline a further grafting of ferrocene entities through condensation of the amine groups of the polymer with the aldehyde group of ferrocenecarboxaldehyde produced an interesting electroactive polymer.

The electrochemical preparation and properties of polypyrrole and poly(N-alkyl pyrrole) polymer films has been studied extensively by Diaz and co-workers.[31–37] Polypyrrole may be produced as thin films on Pt anodes by electrolysis in a wide variety of solvents and using a wide variety of electrolyte salts. The films are electroactive and when in the oxidized form display metal-like conductivity, they are stable in air and in most solutions. They can be driven repeatedly between the oxidized and the neutral (non-conducting) states. Piejza et al. have also studied the electropolymerization of pyrrole.[38] The electrochemical behaviour of polypyrrole films on Pt and Ta has also been described by Bull, Fan, and Bond,[39] while Noufi, Tench, and Warren[40] have described the protection of semiconductor photoanodes with photoelectrochemically generated polypyrrole films. Tourillon and Garnier[41] have produced electrochemically conducting polymers from pyrrole, indole, thiophene, furan, and azulene in acetonitrile using either tetrabutylammonium perchlorate or fluoroborate as electrolyte.

Viologen polymers based on bipyridines and alkylated bipyridines have been produced electrochemically. Vinyl groups have been introduced into $\alpha\alpha$-bipyridine and o-phenanthroline. These groups were used to attach ruthenium bipyridyl ions, $Ru(bipy)_3^{2+}$, and ruthenium phenanthrolyl ions, $Ru(phen)_3^{2+}$, to Pt electrodes by cycling the electrode to negative potentials to induce anionic polymerization.[42] The polymer films are electroactive; scanning to potentials sufficiently negative to reduce Ru^{2+} causes an immediate loss in electroactivity. Abruhna and Bard[43] have reported electrogenerated chemiluminescence in a polymer based on the tris-(4-vinyl-4'-methyl-2,2'bipyridyl) ruthenium(II) system. The electrochemical production of viologen polymers has also been studied by Calvert et al.[44] and by Murray and co-workers.[45,46]

[30] J. E. Dubois, P. C. Lacaze, and M. C. Pham, J. Electroanal. Chem., Interfacial Electrochem., 1981, 117, 233.
[31] A. F. Diaz, J. M. Vasquez Vallejo, and A. Martinez Duran, IBM J. Res. Dev., 1981, 25, 42.
[32] A. F. Diaz, Chem. Scr., 1981, 17, 145.
[33] A. F. Diaz, K. K. Kanazawa, J. L. Castillo, and J. A. Logan, Polym. Sci. Technol., 1981, 15, 149.
[34] K. K. Kanazawa, A. F. Diaz, W. D. Gill, P. M. Grant, G. B. Street, G. P. Gardini, and J. E. Kwak, Synth. Met., 1980, 1, 329.
[35] A. F. Diaz, J. I. Castillo, J. A. Logan, and W. Y. Lee, J. Electroanal. Chem., Interfacial Electrochem., 1981, 129, 115.
[36] A. F. Diaz, A. Martinez, K. K. Kanazawa, and M. Salmon, J. Electroanal. Chem., Interfacial Electrochem., 1981, 130, 181.
[37] A. F. Diaz, J. I. Castillo, K. K. Kanazawa, J. A. Logan, M. Salmon, and O. Fajardo, J. Electroanal. Chem., Interfacial Electrochem., 1982, 133, 233.
[38] J. Piejza, I. Lundstrom, and T. Skotheim, J. Electrochem. Soc., 1982, 129, 1685.
[39] R. A. Bull, F. R. F. Fan, and A. J. Bond, J. Electrochem. Soc., 1982, 129, 1009.
[40] R. Noufi, D. Tench, and L. F. Warren, J. Electrochem. Soc., 1981, 128, 2596.
[41] G. Tourillon and F. Garnier, J. Electroanal. Chem., Interfacial Electrochem., 1982, 135, 173.
[42] P. K. Ghosh and T. C. Spiro, J. Electrochem. Soc., 1981, 128, 1281.
[43] H. D. Abruhna and A. J. Bard, J. Am. Chem. Soc., 1982, 104, 2641.
[44] J. M. Calvert, B. P. Sullivan and T. J. Meyer, ACS Symp. Ser., 1982, 192, 159.
[45] P. Denisevich, H. D. Abruhna, C. R. Leidner, T. J. Meyer, and R. W. Murray, Inorg. Chem., 1982, 21, 2153.
[46] K. W. William and R. W. Murray, J. Electroanal. Chem., Interfacial Electrochem., 1982, 133, 211.

Simionescu and Vladimir[47] have patented an electrochemical process for the production of low-molecular-weight, heat-resistant polymers based on acetylene derivatives, while Przyluski et al.[48] have electrochemically oxidized polyacetylene in the presence of $LiCl/FeCl_3$ in acetonitrile to give a highly conducting polymer with p-type conductivity.

Alekseeva and co-workers[49—51] have studied the cationic electropolymerization of acenaphthalene and its copolymerization with styrene[52] and with methyl methacrylate.[53] Mano and Calafate have also studied the electropolymerization of acenaphthalene and indene.[54,55] Andruzzi et al.[56] have continued their studies into the cationic electropolymerization of oxyheterocyclic compounds.

Shapoval has reported on the nature of the reduction products initiating the anionic polymerization of methyl methacrylate[57,58] and Koval'chuk et al.[59] have studied the electrochemical polymerization of acrylonitrile and co-monomers in aqueous alkali and alkaline-earth metal perchlorates plus inorganic peroxides at iron electrodes. Koval'chuk has also described and measured the electrochemical luminescence observed during the electrolysis of diazonium salts in the presence of free radical acceptors, including vinyl monomers which undergo free-radical polymerization during the electrolysis.[60]

Finally, the author must report, as in the previous edition of this review, the work of Salyer and Usmani,[61] who have now patented their process for the production, at the anode, of porous particles (< 10 μm diameter) made by the electrolysis of aqueous solutions of either urea/formaldehyde or melamine/formaldehyde prepolymers in aqueous sulphuric acid.

---

[47] C. F. Simionescu and G. M. Vladimir, Rom. P., 69 490, 1980.
[48] J. Przyluski, M. Zag'orska, K. Conder, and A. Pron, Polymer, 1982, 23, 1872.
[49] T. A. Alekseeva, E. G. Ol'khovskaya, and V. D. Bezuglyi, J. Appl. Chem. USSR, 1981, 54, 1845.
[50] T. A. Alekseeva, E. G. Ol'khovskaya, and V. D. Bezuglyi, Zh. Prikl. Khim., 1981, 54, 2107.
[51] M. M. Gerner, E. G. Ol'khovskaya, and T. A. Alekseeva, Electrokhimiya, 1981, 17, 931.
[52] T. A. Alekseeva, E. G. Ol'khovskaya, and V. D. Bezuglyi, Vysokomol. Soedin., Ser. B, 1981, 23, 535.
[53] T. A. Alekseeva, E. G. Ol'khovskaya, and V. D. Bezuglyi, USSR P., 883 068, 1982.
[54] E. B. Mano and B. A. L. Calafate, Rev. Quim. Ind. (Rio de Janeiro), 1981, 50, 10.
[55] E. B. Mano and B. A. L. Calafate, J. Polym. Sci., Polym. Chem. Ed., 1981, 19, 3325.
[56] F. Andruzzi, P. Cerrai, G. Guerra, L. Nucci, A. Pescia, and M. Tricoli, Eur. Polym. J., 1982, 18, 685.
[57] G. S. Shapoval, E. S. Shevchuk, and L. V. Marchuk, Ukr. Khim. Zh., 1981, 47, 408.
[58] G. S. Shapoval, J. Macromol. Sci., Chem., 1982, 17, 453.
[59] E. P. Koval'chuk, N. S. Tsvetkov, and E. I. Aksiment'eva, USSR P., 767 128, 1980.
[60] E. P. Koval'chuk, N. I. Ganushchak, V. M. Prisyazhnyi, and N. D. Obuschak, Ukr. Khim. Zh., 1982, 48, 491.
[61] I. O. Salyer and A. M. Usmani, US P., 4 230 551, 1980.

3
Step Growth Polymerization

BY J. FERGUSON, D. J. SPARROW, AND I. G. WALTON

PART I Polyesters, Polycarbonates, Polyamides, and Polyimides
by J. Ferguson

1 Introduction

Since the last review two years ago in Volume 2 of this Series, notable changes have occurred in the literature on this group of polymers. As might be expected, the number of papers published on polyesters and polyamides continues to be very much larger than those on polycarbonates and polyimides. Nevertheless within each group the distribution of the papers by topic is quite different. Aromatic and liquid crystalline polyesters are now major research topics, while morphology of polycarbonates continues to be of great interest. There is also a large amount of work being published on the synthesis of novel polyamides, while the main preoccupation shown in papers on polyimides is with synthesis and structure/property relationships. Surprisingly, not nearly as many papers have appeared on polymer blends as might have been expected.

2 Polyesters

Synthesis.—The industrial synthesis of poly(ethylene terephthalate) has been modelled for a semi-batch ester interchange reactor in order to investigate the effect of process variables on conversion rate.[1] This work has been developed by means of mathematical modelling of the transesterification process,[2] and of the pre-polymerization reaction.[3] A similar technique has been employed for modelling reversible poly(ethylene terephthalate) reactors.[4] The equations for the calculation of the rate constant and the molecular weight distribution (MWD) have been solved numerically; where ethylene glycol was not removed from the reaction medium, MWD was found to be unaffected by choice of the redistribution rate constant. This work compliments a study on the kinetics of polyesterification of reactions carried out under isothermal and non-isothermal heating.[5] Two fundamental equations, one empirical and one theoretical containing the kinetic constants for the reaction, enabled the process to be followed continuously.

---

[1] K. Ravindranath and R. A. Mashelkar, *J. Appl. Polym. Sci.*, 1981, **26**, 3179.
[2] K. Ravindranath and R. A. Mashelkar, *J. Appl. Polym. Sci.*, 1982, **27**, 771.
[3] K. Ravindranath and R. A. Mashelkar, *J. Appl. Polym. Sci.*, 1982, **27**, 2625.
[4] A. Kumar, S. K. Gupta, B. Gupta, and D. Kunzru, *J. Appl. Polym. Sci.*, 1982, **27**, 4421.
[5] T. Pomakis and I. Simitzis, *Angew. Makromol. Chem.*, 1981, **99**, 145.

Phase-transfer catalysis has been used to induce polymerization between bifunctional nucleophiles and electrophiles.[6] Dipotassium sebacate, for example, forms with m-xylylene dibromide. Acetonitrile and acetonitrile/benzene were the only solvents which yielded polymers in these systems. Related work had also been reported on the polymerization[7] of salts of 4-(p-bromoacetylphenyl) butanoic acid.

One of the interesting features of the literature has been the growing interest in the synthesis of poly(butylene terephthalate).[8,9] The kinetics of the polymerization in the presence of titanium tetrabutylate both by model compounds and in terms of the secondary reaction leading to thermal degradation has enabled a mechanism for the formation of tetrahydrofuran to be proposed. Cyclic oligomers have been extracted from this polymer and identified by HPLC.[10] The solid state polycondensation of poly(butylene terephthalate) has also been studied in vacuum at 200 °C. The results suggest that the increase in $\bar{M}_n$ can be explained by diffusion together of hydroxyl end groups. The reaction ceases when a limiting value of hydroxyl group concentration is attained.[11]

The structure and properties of poly(hexamethylene terephthalate) have also been described. The polymers prepared were found to crystallize in one of three possible allomorphs, one of which is favoured when crystallization occurs under stress giving a fibrillar material, while a more spherulitic structure is obtained from the high-temperature relaxed and annealed polymer.[12]

Poly(methylene terephthalate) has been formed from methylene bromide and caesium terephthalate. The procedure could not be used with aliphatic dicarboxylic acid.[13]

Some new polyesters have been described, for example, macrocyclic polyesters formed by catalytic cyclization,[14] polyesters by thionyl chloride activated polycondensation,[15] from β-lactones,[16] and valerolactone.[17] The synthesis characterization and electrical conductivity of polyesterification products with tetrathiofulvalene units show interesting electrical conductivity.[18] As might be expected, the possible production of polymers for implant surgery is also of some interest and a stereo-regular bioresorbable polyester from alphahydroxy acid has been found to be bio-compatible.[19]

A number of papers report new copolyesters. The kinetics of the reaction of dimethyl terephthalate ethylene glycol and 1,4-bis-β-hydroxyethoxy-2,3,5,6-tetrabromobenzene has been studied and mathematically modelled.[20] Poly(ethylenetetramethylene terephthalate)[21] and poly caprolactone-based block copolymers

---

[6] G. G. Cameron and K. S. Law, *Polymer*, 1981, **22**, 272.
[7] G. G. Cameron, G. M. Buchan, and K.-S. Law, *Polymer*, 1981, **22**, 558.
[8] F. Piloti, P. Manaresi, B. Fortunoto, A. Munari, and V. Passalacqua, *Polymer*, 1981, **22**, 799.
[9] F. Piloti, P. Manaresi, B. Fortunoto, A. Murari, and V. Passalacqua, *Polymer*, 1981, **22**, 1566.
[10] G. C. East and A. M. Girshab, *Polymer*, 1981, **23**, 323.
[11] B. Fortunoto, F. Piloti, and P. Manaresi, *Polymer*, 1981, **22**, 655.
[12] I. Hall and B. A. Ibrahim, *Polymer*, 1982, **23**, 805.
[13] G. C. East and M. Morshed, *Polymer*, 1982, **23**, 1555.
[14] A. Bachrach and A. Zilkha, *Eur. Polym. J.*, 1982, **18**, 421.
[15] H. G. Elias and R. J. Warner, *J. Macromol. Chem., Phys.*, 1981, **182**, 681.
[16] R. W. Lenz, *Pure Appl. Chem.*, 1981, **53**, 1729.
[17] M. Aubin and R. E. Prud'homme, *Polymer*, 1981, **22**, 1223.
[18] G. Kobmehl and M. Rohde, *Makromol. Chem.*, 1982, **183**, 2077.
[19] M. Vert, F. Chabot, J. Leray, and P. Christel, *Makromol. Chem. Suppl.*, 1981, **5**, 30.
[20] K. Dimov, J. Alba, and R. Lasarova, *Angew. Makromol. Chem.*, 1982, **102**, 87.
[21] E. Ponnusamy, C. T. Vijayakumar, T. Balakrishnan, and H. Kothandaraman, *Polymer*, 1982, **23**, 1391.

synthesized using anionic co-ordination catalysts have also been described.[22] However, the main interest is clearly in producing copolymers containing aromatic groups, and these have been dealt with in a series of papers.[23-26]

Aromatic polyesters, their synthesis, and properties, represent now a very important and growing area of research, particularly those polymers showing liquid crystalline behaviour. A new homologous series of thermotropic mesomorphic polyesters in which the mesogenic element, aromatic rings, are separated by flexible spacers[27] have been produced with high inherent viscosity. Flexible spacers in the form of aliphatic groups[28] and disiloxane units[29] have also been described. In the latter case all the polymers formed nematic phases and the thermodynamic characteristics of their nematic/isotropic phase transitions have been explained on the basis of their structural features. Nematic and cholesteric thermotropic polyesters with azoxybenzene mesogenic units,[30] and with aromatic ester mesogenic units, were synthesized. Flexible spacers of poly(alkylene oxide)s had a profound effect on liquid crystalline behaviour.[31] Spacers having more than ten units were not liquid crystalline, whereas those of shorter length exhibited mesophase properties. The polymers containing two, three, or four oxyethylene units showed two mesophases with textures suggesting that both smectic and nematic phases were present. Polyesters with bisphenol spacers have also been described.[32] The thermal properties of a series of aromatic–cycloaliphatic polyesters[33] fall between those of fully aromatic and aliphatic–aromatic polymers. Other polymers have been prepared from α-ethyl, α-phenyl, β-propiolactone.[34] 3-Hydroxy benzoates[35] and liquid crystalline 4-acetoxypropyloxybenzenoic acid-4-carboxyphenyl ester[36] were used as reagents. It has been claimed that thermotropic polyesters containing azoxy groups have moderately low transition temperatures.[37] Other thermotropic polyesters with rigid backbones have revealed that unsymmetrically substituted phenylene rings in the mesogenic units reduced the thermal stability of the mesophase.[38,39] The permeability of aromatic polyesters with polynaphthalene systems in the main chain has been reported.[40] Two interesting papers have reported the synthesis of aromatic polyesters containing peroxide linkages,[41] and dibenzo-

---

[22] J. Heuschen, R. Jérôme, and Ph. Teyssié, *Macromolecules*, 1981, **14**, 242.
[23] B. P. Bajaj and D. N. Khanna, *Eur. Polym. J.*, 1981, **17**, 275.
[24] B. P. Bajaj and D. N. Khanna, *Polymer*, 1982, **22**, 1522.
[25] A. S. Cěgolja, G. P. Michailov, V. V. Sévcěnko, and S. N. Lavrova, *Acta Polym.*, 1981, **32**, 532.
[26] A. Wojcik and T. Matynia, *Angew. Makromol. Chem.*, 1981, **93**, 211.
[27] L. Strzelecki and L. Liebert, *Eur. Polym. J.*, 1981, **17**, 1271.
[28] A. Blumstein, K. N. Sivanamakrishnan, R. B. Blumstein, and S. B. Clough, *Polymer*, 1982, **23**, 77.
[29] B.-W. Jo, J.-I. Jin, and R. N. Lenz, *Eur. Polym. J.*, 1982, **18**, 233.
[30] A. Blumstein, S. Vilasagar, S. Ponratham, S. B. Clough, and R. B. Blumstein, *J. Polym. Sci., Polym. Phys. Ed.*, 1982, **20**, 877.
[31] G. Galli and E. Chiellini, *Makromol. Chem.*, 1982, **183**, 2893.
[32] R. W. Lenz and J.-I. Jin, *Macromolecules*, 1981, **14**, 1405.
[3] V. Bulacovschi, M. Darângă, E. Botan, and C. Simionescu, *Acta Polym.*, 1982, **33**, 527.
[34] F. J. Carriere and C. D. Eisenbach, *J. Macromol. Chem. Phys.*, 1981, **182**, 325.
[35] H. R. Kricheldorf, Q. Z. Zang, and C. Schwarz, *Polymer*, 1982, **23**, 1821.
[36] D. Van Luyen, *Acta Polym.*, 1981, **32**, 284.
[37] A. Blumstein and S. Vilasagar, *Mol. Cryst. Liquid Cryst.*, 1981, **72**, 1.
[38] B. Wook Jo, R. W. Lenz, and J. I. Jin, *Makromol. Chem., Rapid Commun.*, 1982, **3**, 23.
[39] A. C. Griffin and S. J. Hovens, *J. Polym. Sci., Polym. Phys. Ed.*, 1981, **19**, 951.
[40] J. Popkowka, D. Sek, and A. Puchalik, *Eur. Polym. J.*, 1981, **17**, 377.
[41] W. Kuran and J. Pelrus, *Macromol. Sci., Chem.*, 1981, **15**, 381.

Step Growth Polymerization

phosphole units.[42] These last-named polymers have initial decomposition temperatures comparable to those of phosphorus-free polyarylates.

Graft polymerization continues to be used extensively, particularly to modify the properties of poly(ethylene terephthalate) (PET) fibres. Acrylic acid,[43,44] acrylamide,[45] methyl methacrylate,[46–48] and vinylacetate[49] have all been polymerized on to PET by grafting, as have acrylated azo dyes.[50]

Properties.—Studies on the conformation of PET have involved effect of composition,[51] far infrared absorption,[52] and Raman and n.m.r. spectra.[53] Molecular weight distribution of non-linear[54] and unsaturated[55] polyesters have shown good agreement with that expected from the theory of polycondensation kinetics. The thermodynamics and hydrodynamic properties of poly(β-propiolactone) show, from the relationship between intrinsic viscosity and molecular weight, that this polymer has a linear flexible chain.[56] As might be expected, the conformation of aromatic polyesters is clearly of importance. It has been found, for example, that orientation of the nematic phase by a magnetic field is preserved during solidification of the polymer.[57] The temperature coefficient of the unperturbed dimensions of poly(diethylene glycol terephthalate) end-linked onto a non-crystallizable trifunctional network using an aromatic triisocyanate was found, and the mean square end-to-end distance calculated.[58] Transition temperatures of polyitaconic acid esters containing pendant cyclo-aliphatic rings have been shown to be controlled by ring flexibility,[59] while sub-glass transition events are similarly controlled.[60] Eximer formation by polyesters with pyrenylmethyl groups and their dimer model compounds showed that the peak wavelength of eximer emission was 475 nm.[61] Other structural studies involving conformation measurements have been made on poly(α,ω-alkanedioyl)(1′,4′-dioxyphenyl)-oxy-4-benzoates,[62] on segmented copolymers,[63] and on flowing PET melts.[64]

---

[42] H. Kondo, M. Sato, and M. Yokoyama, *Eur. Polym. J.*, 1981, **17**, 583.
[43] A. Hebeish, S. E. Sholsby, and A. M. Bayazeed, *J. Appl. Polym. Sci.*, 1981, **26**, 3245.
[44] A. Hebeish, S. Shaloby, and A. M. Bayazeed, *J. Appl. Polym. Sci.*, 1982, **27**, 197.
[45] M. K. Mishra and A. K. Tripathy, *J. Appl. Polym. Sci.*, 1982, **27**, 1845.
[46] A. K. Praahan, N. C. Pati, and P. L. Nayak, *J. Appl. Polym. Sci.*, 1982, **27**, 2131.
[47] S. Mirsa, P. L. Nayak, and C. Sahu, *J. Appl. Polym. Sci.*, 1982, **27**, 3867.
[48] A. Hebeish, S. E. Shaloby, and A. M. Bayazeed, *J. Appl. Polym. Sci.*, 1982, **27**, 3683.
[49] S. A. Faterpeker and S. P. Potuis, *J. Appl. Polym. Sci.*, 1982, **27**, 3349.
[50] S. Calgari, E. Selli, and I. R. Bellobono, *J. Appl. Polym. Sci.*, 1982, **27**, 527.
[51] C. S. Wang and C. S. Y. Yeh, *Polymer*, 1982, **23**, 505.
[52] W. F. X. Frank, W. Strohmeier, and M. L. Hollensleben, *Polymer*, 1981, **22**, 615.
[53] J. Štokr, B. Schneider, D. Doskočilova, and J. Lövy, *Polymer*, 1982, **23**, 717.
[54] P. Helias, D. Durand, J. P. Busnel, and C. M. Bruneau, *Eur. Polym. J.*, 1982, **18**, 647.
[55] A. Kostanck, T. Zelenka, and K. Hajek, *J. Appl. Polym. Sci.*, 1981, **26**, 4117.
[56] D. Rosenvasser, A. Sagrario Casas, and R. V. Figini, *Makromol. Chem.*, 1982, **183**, 3067.
[57] L. Liebert, L. Strzelecki, D. Van Luyen, and A. M. Levelut, *Eur. Polym. J.*, 1981, **17**, 71.
[58] E. Riande, J. Guzman, and M. A. Llorente, *Macromolecules*, 1982, **15**, 298.
[59] J. M. G. Cowie and I. J. McEwen, *Eur. Polym. J.*, 1981, **17**, 619.
[60] J. M. G. Cowie, R. Ferguson, and I. J. McEwen, *Polymer*, 1982, **23**, 605.
[61] S. Tazuke, H. Ooki, and K. Sato, *Macromolecules*, 1982, **15**, 400.
[62] C. Merienne, L. Liebert, and L. Strzelecki, *Eur. Polym. J.*, 1982, **18**, 137.
[63] L. W. Jelinski, F. C. Schilling, and F. A. Bovey, *Macromolecules*, 1981, **14**, 581.
[64] P. T. Hendra, D. B. Morris, R. D. Sang, and H. A. Willis, *Polymer*, 1982, **23**, 9.

The chemical reactions of polyester, in particular PET, have centred mainly on degradation reactions. Hydrolysis and aminolysis studies,[65–69] have shown that pitting of the surface of polymers followed by a brittleness associated with a drop in molecular weight can be expected. Aromatic and aliphatic block polyesters have also been hydrolysed by lipase.[70] The reaction has been turned to advantage in the production of bio-degradable polymers.[71,72] The other main area of degradation has been photolysis where far ultraviolet light has been found to cause etching of PET film.[73] The photolysis and energy transfer in a series of PET–co-4,4'-biphenyldicarboxylate apparently occur from excited terephthalate units to ground state biphenyldicarboxalate.[74] A spectroscopic study of the photo-oxidation of poly-(propylene-1,2-maleate), etc., has also been reported.[75]

Pyrolysis studies have centred mainly on PET and its derivatives. The role of flame retardants containing bromine and phosphorous compound[76–78] has come under scrutiny. It is claimed that phosphorous-only retardants have no effect on the solid-phase pyrolysis of PET, while bromine-containing retardants have been claimed to alter the balance of secondary competing pyrolysis reactions.[79] The pyrolysis[80] and flame retardant behaviour of unsaturated polyesters uncured and in the presence of styrene[81] involves a reverse Diels–Alder reaction in certain cases.

Because of their importance in composite formation, studies on reactivity have centred on cross-linking. The effect of substituents, particularly styrene,[82,83] and the sequence distribution of cross-links,[84] have been examined by a variety of techniques including n.m.r. This particular method required the degradation of the cross-linked polyester before examination.

Morphology.—As in previous years, a considerable number of papers have appeared on the morphology of PET, no doubt because of the commercial importance of this material in fibre and film form. The influence of crystal perfection on melting point,[85] the intrinsic birefringence,[86] and adsorption properties of glassy

---

[65] H. M. Berry and K. Hayes, *Text. Res. J.*, 1982, **52**, 286.
[66] S. M. Milnera, P. Z. Sturgeon, and B. J. Carlsson, *Colloid Polym. Sci.*, 1981, **259**, 47141.
[67] E. M. Sanders and S. H. Zeronian, *J. Appl. Polym. Sci.*, 1982, **27**, 4477.
[68] M. S. Ellison, L. D. Fisher, K. W. Alger, and S. H. Zevonian, *J. Appl. Polym. Sci.*, 1982, **27**, 277.
[69] G. Hindrickscen, H. Krebs, and H. Springer, *Colloid Polym. Sci.*, 1982, **260**, 502.
[70] Y. Tokiwa and T. Suzuki, *J. Appl. Polym. Sci.*, 1981, **26**, 441.
[71] A. M. Reed and D. K. Gilding, *Polymer*, 1981, **22**, 499.
[72] C. G. Pitt, F. I. Chosalow, Y. M. Hibionado, D. M. Klimos, and A. Schindler, *J. Appl. Polym. Sci.*, 1981, **26**, 3779.
[73] R. Srinvasan, *Polymer*, 1982, **23**, 1863.
[74] T. A. Dellinger and C. N. Roberts, *J. Appl. Polym. Sci.*, 1981, **26**, 1321.
[75] J. Lucki, J. F. Rabek, B. Ranby, and C. Ekstrom, *Eur. Polym. J.*, 1981, **17**, 919.
[76] M. Day, T. Suprunchuk, and D. M. Wiles, *J. Appl. Polym. Sci.*, 1981, **26**, 3085.
[77] M. Day, V. Parfenov, and D. M. Wiles, *J. Appl. Polym. Sci.*, 1982, **27**, 575.
[78] V. Freudenberger and F. Jakob, *Angew. Makromol. Chem.*, 1982, **105**, 203.
[79] M. E. Bednas, M. Day, K. Ho, R. Sander, and D. M. Wiles, *J. Appl. Polym. Sci.*, 1981, **26**, 277.
[80] G. L. Marshall, *Eur. Polym. J.*, 1982, **18**, 53.
[81] C. Thangavel Vijayakumar and J. K. Fink, *J. Appl. Polym. Sci.*, 1982, **27**, 1629.
[82] P. E. Froehling, *J. Appl. Polym. Sci.*, 1982, **27**, 3577.
[83] W. Douglas and G. Pritchard, *J. Polym. Sci., Polym. Phys. Ed.*, 1982, **20**, 1223.
[84] A. W. Birley, J. V. Dawkins, D. Kyriacos, and A. Bunn, *Polymer*, 1981, **22**, 812.
[85] F. Fontaine, T. Ledent, G. Groeninckx, and H. Reynaers, *Polymer*, 1982, **23**, 185.
[86] S. K. Garag, *J. Appl. Polym. Sci.*, 1982, **27**, 2857.

PET,[87,88] have been described. Orientation effects and crystallization from solvent,[89] the stress relaxation,[90] and the effect of molecular weight on orientation,[91] have all been described. In this last paper[91] it has been reported that the formation of a permanent physical network is very sensitive to molecular weight. DTA of drawn PET film,[92] and the orientational loss above T_g in orientated PET has been found using Brillouin light scattering;[93] a vibrational spectroscopic study of PET crystallized by annealing in the orientated state was followed by Raman spectra using the width of the 1730 carbonyl stretching vibration.[94] Annealing in the highly orientated state studied by dynamic mechanical properties,[95] indicates that α and β processes depend on the annealing temperature in different ways for the drawn and undrawn material. Zone annealing has also been used to produce very high modulus and strength fibres.[96] Physical aging in glassy[97] and semi-crystalline PET[98] shows that relaxation of the glass can occur and that drawing behaviour changes significantly. The surface characterization of PET has been examined by gas chromatography[99] and by external reflection Fourier transform i.r. spectroscopy.[100]

Several papers have appeared on crystallinity in poly(butylene terephthalate).[101—103] Model compounds such as butyl dibenzoate has been used to show that the crystal conformation of the aliphatic segments in the low-molecular-weight analogues closely resemble the extended all-*trans* β-form of poly(butylene terephthalate).[104] Studies on poly(1,4-*trans*-cyclohexanediyldimethylene terephthalate) indicate the existence of a triclinic unit cell.[105] Polymorphism[106] has also been found in poly(hexamethylene terephthalate) and poly(decamethylene terephthalate), while in another interesting paper the crystallization and melting behaviour of a semi-crystalline polyester and its Mg^{II} complexes has been described.[107]

Parallel with the increased interest in the synthesis of aromatic and liquid-crystal-forming polyesters a number of papers have appeared on the morphology of these compounds, almost all of which are copolymers. The aromatic copolyesters of poly(ethylene terephthalate) and 80 mol% of *p*-acetoxy benzoic acid and of poly(ethylene terephthalate)–co–*p*-oxetoxybenzoate show the existence of ordered

---

[87] A. L. Volynskii, V. S. Loginov, and N. F. Bakeyev, *Polym. Sci. (USSR)*, 1981, **23**, 1178.
[88] A. L. Volynskii, V. S. Loginov, and N. F. Bakeyev, *Polym. Sci. (USSR)*, 1981, **23**, 1350.
[89] H. Tameel, T. Waldmon, and L. Rebenfeld, *J. Appl. Polym. Sci.*, 1981, **26**, 1795.
[90] J. H. Hawthorne, *J. Appl. Polym. Sci.*, 1981, **26**, 3317.
[91] J. C. Engelaeve, J. P. Corot, and F. Rietsch, *Polymer*, 1982, **23**, 766.
[92] H. Springer, N. Brinkman, and G. Hinrichsen, *Colloid Polym. Sci.*, 1981, **259**, 38.
[93] D. B. Cavanaugh and G. H. Wang, *J. Polym. Sci., Polym. Phys. Ed.*, 1981, **19**, 1273.
[94] G. M. Venkotesh, D. T. Bose, A. H. Khan, J. P. Sibilo, and S. L. Hsu, *J. Appl. Polym. Sci.*, 1981, **26**, 223.
[95] S. Fakirov and D. Stahl, *Angew. Makromol. Chem.*, 1982, **102**, 117.
[96] T. Kunugi, A. Susuki, and M. Hooshimoto, *J. Appl. Polym. Sci.*, 1981, **26**, 1951.
[97] A. A. Azar and J. N. Hay, *Polymer*, 1982, **23**, 1129.
[98] M. R. Taut and G. L. Wilkes, *J. Appl. Polym. Sci.*, 1981, **26**, 2813.
[99] T. Anhang and D. G. Gray, *J. Appl. Polym. Sci.*, 1982, **27**, 71.
[100] G. Gillberg and D. Kemp, *J. Appl. Polym. Sci.*, 1981, **26**, 2023.
[101] W. Strohmeier and W. F. X. Frank, *Colloid Polym. Sci.*, 1982, **260**, 937.
[102] W. P. Leung and C. L. Choy, *J. Appl. Polym. Sci.*, 1982, **27**, 2693.
[103] K. Nakamae, M. Kameyama, M. Yoshikawa, and T. Matsumoto, *J. Polym. Sci., Polym. Phys. Ed.*, 1982, **20**, 319.
[104] P. C. Gillette, S. D. Dirlikov, T. L. Koenig, and T. B. Lando, *Polymer*, 1982, **23**, 1759.
[105] B. Rémillard and F. Brisse, *Polymer*, 1982, **23**, 1960.
[106] I. Daniewska and A. Wasiak, *Makromol. Chem., Rapid Commun.*, 1982, **3**, 897.
[107] Zs. Székely-Pécsi, I. Vancsó-Szmeresányi, F. Cser, J. Varga, and K. Belina, *J. Polym. Sci., Polym. Phys. Ed.*, 1981, **19**, 703.

domains which appear as large lamellar blocks[108] and isodimorphism,[109] respectively. Other papers have reported the nature of the crystallites in quenched thermotropic liquid crystal polyesters.[110] The structure of melt-spun fibres prepared from p-hydroxy benzoic acid, 2,6-dihydroxy naphthalene, and terephthalate acid[111] show the presence of disclination in mesomorphic copolyesters.[112]

Blends.—By definition, papers on polymer blends can be covered under the heading of each of the polymer sections reported in this review. Some blends, *e.g.*, with polyamides, have been described and can be found under that heading.

The miscibility of poly(neopentyl glycol adipate)/chlorinated polymer blends[113] has been attributed to specific interaction between the polymers. Aliphatic polyesters combined with the polyhydroxy ether of bisphenol-A apparently exhibit single-composition-dependent glass-transition temperatures characteristic of miscible systems.[114] Polyesters have also been blended with PVC,[115] with nitrile rubber,[116] and with chlorinated PVC.[117] DSC of polyester poly(vinyl bromide)[118] indicates a limited degree of miscibility. The influence on film-forming behaviour of PET–low-temperature polyethylene blends manifests itself mainly in crystallization and long needle-like voids can apparently occur.[119]

3 Polycarbonates

Synthesis.—A comparatively small number of papers has appeared on the synthesis of polycarbonates and, in general, these have dealt with copolymers. The polycarbonates and alternating copolycarbonates of bithional have been described.[120] High-molecular weight macrocyclic polycarbonates were synthesized by interfacial condensation of bischloroformals of bisphenol A with bisphenol A.[121] These cyclic polymers with molecular weight exceeding 100 000 showed markedly lower intrinsic viscosity relative to their molecular weight than did linear polycarbonates. Interfacial synthesis has also been used for the phase transfer catalysed preparation of polycarbonate/polysiloxane block copolymers.[122] Other block copolymers prepared from hydroxy terminated poly(bisphenol A)–iso-terephthalate copolymers reacted with phosgene.[123] As the proportion of isophthalate increased, T_g fell. The chemical fixation of carbon dioxide has been

---

[108] A. E. Zachariades, J. Ecomony, and J. A. Logan, *J. Appl. Polym. Sci.*, 1982, **27**, 2009.
[109] W. Meesiri, J. Menczel, U. Gaur, and B. Wunderlich, *J. Polym. Sci., Polym. Phys. Ed.*, 1982, **20**, 719.
[110] D. T. Blundell, *Polymer*, 1982, **23**, 359.
[111] J. Blackwell and G. Gutierrez, *Polymer*, 1982, **23**, 671.
[112] M. R. Mackley, F. Pinaud, and G. Siekmann, *Polymer*, 1981, **22**, 437.
[113] S. L. Goh, D. R. Paul, and J. W. Barlow, *J. Appl. Polym. Sci.*, 1982, **27**, 1091.
[114] T. E. Harris, S. H. Goh, D. R. Paul, and J. W. Barlow, *J. Appl. Polym. Sci.*, 1982, **27**, 839.
[115] D. F. Varnell and M. M. Coleman, *Polymer*, 1981, **22**, 1327.
[116] D. J. Hourston and I. D. Hughes, *Polymer*, 1981, **22**, 127.
[117] G. Bélorgey, M. Aubin, and R. E. Prud'homme, *Polymer*, 1982, **23**, 1051.
[118] P. Cousin and R. E. Prud'homme, *Eur. Polym. J.*, 1982, **18**, 957.
[119] W. Wenig and R. Hammel, *Colloid Polym. Sci.*, 1982, **260**, 31.
[120] W. Deits and O. Vogl, *J. Polym. Sci., Polym. Chem. Ed.*, 1981, **19**, 403.
[121] A. Horbach, H. Vernaleken, and K. Weirauch, *Makromol. Chem.*, 1980, **181**, 111.
[122] J. S. Riffle, R. G. Freelin, A. K. Banthia, and J. E. McGrath, *J. Macromol. Sci., Chem.*, 1981, **15**, 967.

used to synthesize an alternating copolymer of carbon dioxide and epoxide using a diethyl zinc catalyst to give polycarbonates with very broad molecular weight distribution.[124] Block copolymers using propylene oxide have been formed with very narrow molecular weight distribution. One particularly interesting paper has described the synthesis of functional polycarbonates of similar structure to those but which contain pendant carbonate groups. These underwent acidic and basic hydrolysis releasing the compound attached as the pendant group *via* the carbonate linkage.[125]

Properties.—The influence of end capping ratio on the properties of polycarbonate–poly(dimethylsiloxy) block copolymers examined by n.m.r. apparently deviates from the theoretical.[126] Segmental motion and spin relaxation,[127] and molecular dynamics,[128] have been examined by p.m.r. and n.m.r. The existence of two dynamical processes, one below −40 °C and the other at higher temperature, is indicated. Optical anisotropy has been calculated for random coiled polycarbonate from molecular geometry dipole moment of the carbonate group and the anisotropic polarizability tensor for the carbonate and phenylene groups. The calculated Cotton–Mouton constant is about 30% lower than the values reported by other workers, nevertheless it is claimed that additivity of group polarizabilities is valid and that optical and anisotropy can be used as an index of conformation and structure.[129,130] Two papers have dealt with the characterization of branching,[131] and polydispersity[132] on viscosity. The effects of structure[133] and cross linking have also been reported on dynamic mechanical properties,[134] as has moisture.[135]

Thermal degradation of block polycarbonates,[136] polyester carbonates,[137] and phenylphthalein polycarbonate,[138] has been related to chemical structure with silicone block copolymers;[136] char formation is a polycondensation reaction occurring between pyrolysis products from the two blocks. The char itself functions effectively as a thermal insulator in preventing flammability.

The hydrolysis of bisphenol A polycarbonate at 95% relative humidity and 85 °C has been found to produce brown surface crystals of bisphenol A within 30 days; the remaining hydrolysis products were oligomers of bisphenol A.[139] N.m.r. spectroscopy has also been used to study the hydrolysis of this polymer; the same products

---

[123] C. P. Bocnyk, J. H. Hay, I. W. Parsons, and R. N. Haward, *Polymer*, 1982, **23**, 609.
[124] T. Aida and S. Inoue, *Macromolecules*, 1982, **15**, 682.
[125] M. Takanashi, Y. Nomura, Y. Yoshida, and S. Inoue, *Makromol. Chem.*, 1982, **183**, 2085.
[126] E. A. Williams, P. E. Donahue, and J. D. Cargioli, *Macromolecules*, 1981, **14**, 1016.
[127] J. F. O'Gara, S. B. Desjardins, and A. A. Jones, *Macromolecules*, 1981, **14**, 64.
[128] P. T. Inglefield, A. A. Jones, R. P. Lubianez, and J. F. O'Gara, *Macromolecules*, 1981, **14**, 288.
[129] B. Erman, D. Wu, P. A. Irvine, D. C. Martin, and P. J. Flory, *Macromolecules*, 1982, **15**, 670.
[130] B. Erman, D. C. Marvin, P. A. Irvine, and P. J. Flory, *Macromolecules*, 1982, **15**, 664.
[131] Z. Dobkowski, *Eur. Polym. J.*, 1982, **18**, 1051.
[132] Z. Dobkowski and J. Brzezinski, *Eur. Polym. J.*, 1981, **17**, 537.
[133] A. F. Yee and S. A. Smith, *Macromolecules*, 1981, **14**, 54.
[134] L. Makaruk and H. Polanska, *Polym. Bull.*, 1981, **4**, 127.
[135] F. P. La Mantia, G. Spadaro, and D. Acierno, *Acta Polym.*, 1982, **32**, 209.
[136] R. P. Kambour, *J. Appl. Polym. Sci.*, 1981, **26**, 861.
[137] C. P. Bosnyak, G. J. Knight, and W. W. Wright, *Polym. Degrad. Stabil.*, 1981, **3**, 273.
[138] M. S. Lin, B. J. Balkin, and E. M. Pearce, *J. Polym. Sci., Polym. Chem. Ed.*, 1981, **19**, 2773.
[139] H. E. Bair, D. R. Falcone, M. Y. Hellman, G. E. Johnson, and P. G. Kellcher, *J. Appl. Polym. Sci.*, 1981, **26**, 1777.

were detected and the reaction rate measured.[140] Photo-initiated degradation of a polycarbonate substrate by cerium(III) overcoating indicates that the cerous ion acts as an effective u.v. screen.[141] Degradation effects have also been reported in polycarbonates melts.[142,143]

Morphology.—In view of the end use of polycarbonate as a fracture resistant material, it is not surprising that considerable attention has been focused on the polymer's morphology. The existence of microgels,[144] and the effect of doping on mechanical properties[145] and dielectric relaxations,[146] were all reported. Crack fatigue studies[147] have shown that an extraordinary crack-tip plastic zone can occur under certain fatigue loading conditions. Some similar studies on craze phenomena have appeared,[148–152] as have papers on the growth of fibrils[153] and diamond cavities.[154] Annealing and heat treatment and their effects on stress,[155,156] molecular relaxation,[157] and transition phenomena,[158] show in the latter case that the β relaxation reduces greatly on annealing. Exposure of polycarbonates to methylene chloride vapour shows that after an induction period of about 5 min the intensity of a γ_3 relaxation at $-78\ °C$ decreases, while the γ_1 at $-30\ °C$ is unaffected. Mechanical behaviour has also been correlated with relaxation processes.[159]

Blends.—The number of papers on polycarbonate blends has been comparatively small. Blends of bisphenol A (cumylphenyl) carbonate/lexan polycarbonate have been examined by dielectric relaxation; these form solid solutions.[160] The theme of miscible blends has been continued in a number of other papers. Degradation studies on miscible polycarbonate and a copolyester based on 1,4-cyclohexanedimethanol terephthalate and isophthalic acid has shown that residual titanium catalyst produces colour formation.[161] Mechanical properties[162] and crystallization behaviour[163] indicate that interchange reactions can take place between the

---

[140] F. C. Schilling, W. M. Ringo, jun., N. J. A. Sloane, and F. A. Bovey, *Macromolecules*, 1981, **14**, 532.
[141] A. J. Klein, H. Yu, and N. M. Yen, *J. Appl. Polym. Sci.*, 1981, **26**, 2381.
[142] L. B. Gavrilov, Yu. A. Mikheev, D. Ya. Toptygin, and M. S. Akutin, *Vysokom. Soedin. Ser. A*, 1981, **23**, 598.
[143] K. B. Abbas, *Polymer*, 1981, **22**, 836.
[144] Z. Czlonkowska-Kohutnicka and Z. Dobkowski, *Eur. Polym. J.*, 1982, **18**, 911.
[145] L. B. Gabrilov, Ye. M. Zvonkova, Yu. A. Mikheev, D. Ya. Toptygin, and M. L. Kerber, *Vysokomol. Soedin. Ser. A*, 1981, **23**, 1552.
[146] P. C. Mehendru, J. P. Agrawal, K. Jain, and L. A. V. R. Warrier, *Thin Solid Films*, 1981, **78**, 251.
[147] M. T. Takemori and D. S. Matsomoto, *J. Polym. Sci., Polym. Phys. Ed.*, 1982, **20**, 2027.
[148] M. Dettenmaier and H. H. Kausch, *Colloid Polym. Sci.*, 1981, **259**, 937.
[149] M. Narkis and J. P. Bell, *J. Appl. Polym. Sci.*, 1982, **27**, 2809.
[150] A. T. Di Benedetto, P. Bellusci, M. Iannone, and L. Nicolais, *J. Mater. Sci.*, 1981, **16**, 2310.
[151] A. M. Donald and E. J. Kramer, *J. Mater. Sci.*, 1981, **16**, 2977.
[152] A. M. Donald and E. J. Kramer, *J. Mater. Sci.*, 1981, **16**, 2967.
[153] E. Paredes and E. W. Fischer, *J. Polym. Sci., Polym. Phys. Ed.*, 1982, **20**, 929.
[154] N. Walker, R. N. Haward, and J. N. Hay, *J. Mater. Sci.*, 1981, **16**, 817.
[155] M. Yokouchi and Y. Kobayashi, *J. Appl. Polym. Sci.*, 1981, **26**, 431.
[156] C. Bauwene-Crowet and J. C. Bauwens, *Polymer*, 1982, **23**, 1599.
[157] T. Koto and N. Yanogiliova, *J. Appl. Polym. Sci.*, 1981, **26**, 2139.
[158] K. Varadarajan and R. F. Boyer, *J. Polym. Sci., Polym. Phys. Ed.*, 1982, **20**, 141.
[159] J. Hong and O. Brittain, *J. Appl. Polym. Sci.*, 1981, **26**, 2471.
[160] J. M. Pochan, H. W. Gibson, and D. L. F. Pochan, *Macromolecules*, 1982, **15**, 1368.
[161] W. A. Smith, J. W. Barlow, and D. R. Paul, *J. Appl. Polym. Sci.*, 1981, **26**, 4233.
[162] E. Joseph, M. D. Lorenz, J. W. Barlow, and D. R. Paul, *Polymer*, 1982, **23**, 112.
[163] R. S. Barnum, J. W. Barlow, and D. R. Paul, *J. Appl. Polym. Sci.*, 1982, **27**, 7065.

components. Gas sorption in the same blends shows negative deviations of permeability and diffusion coefficients from simple additivity relations.[164] The viscoelastic[165] properties of blends with polystyrene and flow behaviour of blends with poly(methylmethacrylate) were also described.[166]

4 Polyamides

Synthesis.—The most notable feature of papers on polyamides has been the very large number of studies on the synthesis of ε-caprolactam. The simulation[167] of kinetics,[168,169] of hydrolytic polymerization seem to stress the role of oligomers. Molecular weight distribution studies of the polymerization[170] stress the same feature.[171]

Anionic polymerization of ε-caprolactam has been the subject of several papers.[171-177] These cover several aspects of the reaction and it is plain that the kinetics can only be explained on the basis that chain growth involves both free anions and ion pairs. By examining changes in rheological behaviour of the reaction systems[178] it has been claimed that kinetic constants for the reaction may be determined. The use of rheological measurements to study kinetics has been used by various workers but the difficulties associated with the technique may not always be apparent, and data obtained in this way should be viewed with some reserve. The influence of polysiloxane[179]-based activators on anionic block copolymerization of ε-caprolactam and the kinetics of the copolymerization with ω-dodecolactam[180,181] was explained on the basis of proton exchange between the various reactive centres. In a series of papers on the polymerization of lactams the hydrolytic copolymerization of ε-caprolactam with ω-dodecolactam[182] and the analysis of the copolymers[183,184] has provided information about the mean length of homogeneous blocks in the chain.

[164] P. Masi, D. R. Paul, and J. W. Barlow, *J. Polym. Sci., Polym. Phys. Ed.*, 1982, **20**, 15.
[165] Yu. S. Lipatov, V. F. Shirmsky, A. N. Govbatenko, Yu. N. Panov, and L. S. Bolotnikova, *J. Appl. Polym. Sci.*, 1981, **26**, 499.
[166] M. Kasajima, K. Ito, A. Suganuma, and D. Kunii, *Kobunshi Ronbunshu*, 1981, **38**, 239.
[167] K. Tai, Y. Arai, and T. Tagawa, *J. Appl. Polym. Sci.*, 1982, **27**, 732.
[168] K. Tai and T. Tagawa, *J. Appl. Polym. Sci.*, 1982, **27**, 2791.
[169] Y. Arai, K. Tai, H. Teranishi, and T. Tagawa, *Polymer*, 1981, **22**, 273.
[170] S. K. Gupta, C. D. Naik, P. Tandon, and A. Kumar, *J. Appl. Polym. Sci.*, 1981, **26**, 2153.
[171] S. K. Gupta, A. Kumar, P. Tandon, and D. Naik, *Polymer*, 1981, **22**, 481.
[172] A. Ya. Molkin, V. P. Beglishev, and S. A. Blgow, *Polymer*, 1982, **23**, 385.
[173] T. M. Frunze, *et al.*, *Dokl. Akad. Nauk SSSR*, 1981, **257**, 162.
[174] T. M. Frunze, V. A. Kotelnikov, T. V. Volkova, and V. V. Kurashov, *Eur. Polym. J.*, 1981, **17**, 1079.
[175] T. M. Frunze, V. A. Kotelnikov, T. V. Volkova, V. V. Kurasev, S. P. Davtjan, and I. V. Stankevic, *Acta Polym.*, 1981, **32**, 31.
[176] E. Biagini, B. Pedemonte, E. Pedemonte, and S. Russo, *Makromol. Chem.*, 1982, **183**, 2131.
[177] S. Russo, *Chim e Ind.*, 1981, **63**, 412.
[178] A. Ya. Malkin, S. G. Kulichikhin, V. G. Frolov, and M. I. Demina, *Polym. Sci. (USSR)*, 1981, **23**, 1471.
[179] A. Ya. Malkin, S. L. Ivanova, V. G. Frolov, A. N. Ivanova, and Z. S. Andrianova, *Polymer*, 1982, **23**, 1791.
[180] Ye. V. Garbunova, Yu. S. Deyev, and Ye. A. Ryabov, *Polym. Sci. (USSR)*, 1981, **23**, 907.
[181] T. M. Frunze, V. A. Kotel'nikov, M. P. Ivanov, T. V. Volkova, V. V. Kurashev, and S. P. Davtyan, *Polym. Sci. (USSR)*, 1981, **23**, 2902.
[182] R. Alijev, J. Kondelikovā, A. Moucha, and J. Kralicek, *Angew. Makromol. Chem.*, 1982, **103**, 97.
[183] R. Alijev, Z. K. Tuzar, J. Kondelikovā, H. Verlovā, J. Krőliček, and M. Buděnsinský, *Angew. Makromol. Chem.*, 1982, **105**, 99.
[184] R. Alijev, M. Buděnšinsky, J. Kondelikovā, and J. Králiček, *Angew. Makromol. Chem.*, 1982, **105**, 107.

The post-condensation of nylon-6 in the solid state[185] and the preparation of ion-exchange polycaproamide fibres[186] have also been described.

The considerable interest in the production of new polyesters has been paralleled in aromatic polyamides.[187] p-Terephthalamide polymers were synthesized thermally[188] and the gelling during reaction studied.[189] The polycondensation of terephthaloyl chloride and p-phenylenediamine takes place under liquid crystalline solution conditions. In the early stages of the polycondensation reaction, inherent viscosity increased rapidly as the system gelled.[190] High-molecular-weight polyamides containing bridged biphenylene groups,[191] N-propargyl substituents,[192] and phenthridinonediyl groups[193] were reported; the last named showed liquid crystalline behaviour in high concentration sulphuric acid. Aromatic polyamides have also been synthesized with triphenylphosphine,[194] alkyl and ether links,[195] and from 6,7-methylenedioxybenz-3,1-oxazin-2,4-dione.[196] Polyamides containing benzimidazole units have been found to be capable of conversion by polydehydrocyclocondensation into heat-resistant polybenzimidazole.[197] Other intermediates used to prepare polyamides have included 3-hydroxy 1,2-benzoxazole,[198] bisbenzoyl propionic acids,[199] and perchloroterephthaloyl dichlorides.[200] A number of papers have dealt with the synthesis of polyamides containing heterogeneous units such as benzoisothiazole,[201] mercaptobenzoxyazole,[202] and azodibenzamido groups.[203] The effect of aromatic rings on the solution and thermal features of aromatic polythioamides is to increase thermal stability, particularly when the ring and the functional groups are conjugated.[204]

Among the great variety of work described in papers on the synthesis of other polyamides are included studies on the preparation of polymers containing definite numbers of oxyethylene units,[205] asymmetric cyclopropyl groups,[206] and polyelectrolytes with phenolic side groups.[207] In this last case the condensation of nylon-6 with o-cresol sulphonic acid and formaldehyde takes place by condensation of the

---

[185] R. T. Gaymons, J. Amirtaraj, and H. Kamp, *J. Appl. Polym. Sci.*, 1982, **27**, 2513.
[186] N. Gankov, K. Dimov, P. Pavlov, D. Dimtirov, N. Simeonov, E. Terlemestan, and T. Simeonska, *Acta Polym.*, 1982, **33**, 31.
[187] Y. P. Khanna, E. M. Pearce, B. D. Forman, and D. A. Bini, *J. Polym. Sci., Polym. Chem. Ed.*, 1981, **19**, 2799.
[188] P. Costa Bizzari, C. Dalla Casa, and A. Monaco, *Polymer*, 1981, **22**, 1263.
[189] L. B. Sokolov, V. M. Savinov, V. M. Ivanov, and Z. P. Titova, *Proc. Acad. Sci. (USSR)*, 1981, **256**, 606.
[190] B. Jingsheng, Y. Anji, L. Shengzing, T. Shufan, and H. Cheng, *J. Appl. Polym. Sci.*, 1981, **26**, 1211.
[191] T. Kaneda, S. Ishikawa, H. Daimon, T. Katsura, and M. Ueda, *Makromol. Chem.*, 1982, **183**, 433.
[192] T. D. Greenwood, D. M. Armistead, and J. F. Wolfe, *Polymer*, 1982, **23**, 621.
[193] T. Kaneda, S. Ishikawa, H. Daimon, T. Katsura, and M. Ueda, *Makromol. Chem.*, 1981, **183**, 417.
[194] G. Wu, H. Tanaka, K. Sanui, and N. Ogata, *J. Polym. Sci., Polym. Lett. Ed.*, 1981, **19**, 343.
[195] M. Balasubramanian, M. J. Nanjan, and M. Santappa, *J. Appl. Polym. Sci.*, 1982, **27**, 1723.
[196] A. F. Amin, N. R. Naik, B. P. Suthar, and S. R. Patel, *Eur. Polym. J.*, 1982, **18**, 741.
[197] A. Kehayoglou, G. Karayannidis, and I. Sideridou-Karayannidou, *Makromol. Chem.*, 1982, **183**, 293.
[198] M. Ueda, T. Harada, S. Aoyama, and Y. Imai, *J. Polym. Sci.*, 1981, **19**, 1061.
[199] S. Padma, V. Mahadevan, and M. Srinivasan, *Eur. Polym. J.*, 1982, **18**, 155.
[200] F. R. Diaz, R. Larrain, and L. H. Tagle, *Eur. Polym. J.*, 1981, **17**, 1069.
[201] M. Ueda, N. Kawahanasaki, and Y. Imai, *Makromol. Chem., Rapid Commun.*, 1982, **3**, 881.
[202] M. Ueda, K. Seki, and Y. Imai, *Macromolecules*, 1982, **15**, 17.
[203] M. Balasubramanian, M. J. Nanjan, and M. Santappa, *Makromol. Chem.*, 1981, **182**, 853.
[204] J. C. Gressier and G. Levesque, *Eur. Polym. J.*, 1981, **17**, 695.
[205] H. Sato, S. Iwabuchi, V. Bohmer, and W. Kern, *Makromol. Chem.*, 1981, **182**, 755.
[206] C. G. Overberger and T. Nishiyama, *J. Polym. Sci., Polym. Chem. Ed.*, 1981, **19**, 331.
[207] J. Kapko and J. Poloczek, *Polymer*, 1981, **22**, 1544.

amide group with the acid and formaldehyde followed by elimination of a number of the sulphonic groups giving a methylene phenol side chain. Degradation of the polyamide chain can also take place. Novel preparations of speciality polyamides by interfacial and solution methods,[208] of polyesteramides in solution,[209] and of polyamides containing cytosine and hypoxanthine,[210] diacyl derivatives of N-hydroxy compounds,[211] and containing α-dicetone linkages[212] emphasize the importance placed in the search for new structures. Amino terminated polyamides synthesized *via* amide interchange reactions were investigated using NN'-bis(2-aminoethyl) sebacamide and $NNN'N'$-bis(diethylamino) sebacamide.[213] The polycondensation followed second-order kinetics, the rate constants for the cross reaction and the reactivity ratios of the co-polycondensation being calculated. The synthesis of a new polyamide, nylon-18,[214] was carried out to investigate whether the large number of methylene groups in the monomer would change chain packing. In fact, this was found to be similar to that shown by other polyamides in the series.

Graft polymerization studies have mainly involved the reaction of vinyl monomers with the preformed polyamides, usually polycaprolactam. Among the monomers grafted has been methyl methacrylate.[215–220] These have involved a number of initiation systems such as thiourea redox, N-bromosuccinamide, acetylacetonate complexes of Mn^{III}, Co^{III}, and Fe^{III}, as well as the hexavalent chromium ion and pentavalent vanadium. In one of these studies,[219] the initiation of grafting with γ-picolene–bromine charge transfer complex only took place in the presence of a sensitizer. Allyl methacrylate[221] and acrylic acid[222,223] have also been grafted, as has acrylamide.[224] It has also been found possible to graft, photochemically, acrylated azo dyes on to polyamides.[225]

Properties.—A number of studies have dealt with the conformation of polyamides in solution and in the solid state. The Kuhn–Mark–Howink–Sakurada relationship has been derived for ε-caprolactam.[226] The conformation of polyamides derived from cyclopropyl diacids,[227] isomeric rearrangements in polymers with changing chain flexibility,[228] and the relationships between flow birefringence[229] and chain

---

[208] W. Deits, S. Grossman, and O. Vogl, *J. Macromol. Sci., Chem.*, 1981, **15**, 1027.
[209] A. Tsamantakis and F. Carriere, *Angew. Makromol. Chem.*, 1982, **104**, 19.
[210] M. Muraki, Y. Miura, and M. Kinoshita, *Makromol. Chem.*, 1982, **183**, 2059.
[211] C. Lu, P. Liu, and C. Hu, *J. Polym. Sci., Polym. Chem. Ed.*, 1981, **19**, 2074.
[212] K. Nagakubo, F. Akutsu, T. U. S. Sato, and M. Miura, *Polymer*, 1982, **23**, 372.
[213] M. J. Han, *Macromolecules*, 1982, **15**, 438.
[214] G. Cojazzi, A. M. Drusiani, A. Fichera, V. Malta, F. Pilati, and R. Zannetti, *Eur. Polym. J.*, 1981, **17**, 1241.
[215] A. K. Pradhou, N. C. Pati, and P. L. Nayak, *J. Appl. Polym. Sci.*, 1982, **27**, 1839.
[216] S. Lenka and P. L. Nayak, *J. Appl. Polym. Sci.*, 1982, **27**, 3625.
[217] S. Lenka and P. L. Nayak, *J. Appl. Polym. Sci.*, 1982, **27**, 1959.
[218] P. L. Nayak, S. Lenka, and M. K. Mishara, *J. Polym. Sci.*, 1981, **26**, 2437.
[219] S. Lenka, P. L. Nayak, and A. K. Tripathy, *J. Appl. Polym. Sci.*, 1982, **27**, 1853.
[220] M. I. Khalil, Sh. Sh. Aggour, M. H. El-Rafie, and A. Hebeish, *Angew. Makromol. Chem.*, 1981, **96**, 59.
[221] M. I. Khalil, F. I. Abdel-Hay, and A. Hebeish, *Angew. Makromol. Chem.*, 1982, **103**, 143.
[222] M. B. Huglin and J. Smith, *Eur. Polym. J.*, 1981, **17**, 389.
[223] M. B. Huglin and J. Smith, *Eur. Polym. J.*, 1981, **17**, 631.
[224] M. K. Mishra, S. Lenka, and A. K. Tripathy, *J. Appl. Polym. Sci.*, 1981, **26**, 2593.
[225] I. R. Bellobono, F. Tolusso, and E. Selli, *J. Appl. Polym. Sci.*, 1981, **26**, 619.
[226] H. Muller, D. Neuray, and A. Horbach, *Makromol. Chem.*, 1981, **182**, 177.
[227] C. G. Overberger and T. Nishiyama, *J. Polym. Sci., Polym. Chem. Ed.*, 1981, **19**, 349.
[228] M. V. Shablygin and P. M. Pakhanov, *Vysokomol. Soedin. Ser. B*, 1981, **23**, 448.
[229] V. N. Tsvetkov, N. V. Pogodina, and L. V. Starchenko, *Eur. Polym. J.*, 1981, **17**, 397.

flexibility of copolymers of *para* and *meta* aromatic polyamides were all examined. The dynamics of conformational changes of polyamides with photo-responsive azobenzene groups on the chain backbone shows that *cis–trans* isomerization occurs upon irradiation with visible light. *Trans–cis* isomerization is induced with u.v. light.[230] Optical rotatory dispersion measurements show that for poly(1,2-diaminopropane sebacamide) the optical activity of the polymers and its model compound is related to the solvent composition.[231] The electron structure of rigid chain aromatic polyamides can be used to explain the high thermal oxidative ability of poly(*p*-benzamide).[232]

The influence of γ-radiation on poly(ε-caprolactam) has been examined by viscometric[233] and e.s.r. techniques.[234] The same method has been used for studying the thermal degradation of Kevlar 49.[235] Other degradation studies have been carried out on polyamides in sulphuric acid[236] and on carboborane-containing polyamides;[237] the latter show that a mechanism for the degradation can be proposed which holds good both in heterocyclic and in homolytic reactions. A number of other papers have described the photo oxidation of aliphatic polyamides.[238—241] It has been reported that N-methylene groups are attacked in the first step of the oxidation reaction.[242] Other chemical reactions which have been reported on poly(ε-caprolactam) are the dechlorination of the N-chloro polymer,[243] and the interaction with anhydrous ferric chloride.[244]

One particularly interesting paper which, in its content, covers both properties and to a certain extent morphology, describes the stress-induced chemiluminescence from nylon-66 fibres.[245] The emission of visible light has been studied over a range of strain rates and it is suggested that the measured spectral distribution of this light arises from bimolecular termination from alkyl peroxy macroradicals.

Morphology.—The morphology of poly(ε-caprolactam) has been studied with the electron microscope[246] and by annealing.[247,248] It has been found that good agreement between theoretical and experimental results on compressibility of

---

[230] M. Irig and W. Schnabel, *Macromolecules*, 1981, **14**, 1246.
[231] M. Vert, C. Braun, and Y. Huguet, *Polymer*, 1981, **22**, 1683.
[232] N. M. Aref'yev and Z. Yu. Chereiskii, *Polym. Sci. (USSR)*, 1981, **8**, 2058.
[233] W. Szymanski and B. Rymian, *Angew. Makromol. Chem.*, 1981, **99**, 89.
[234] S. Takigami, I. Matsumoto, and Y. Nakomura, *J. Appl. Polym. Sci.*, 1981, **26**, 4317.
[235] J. R. Brown and D. K. C. Hodgeman, *Polymer*, 1982, **23**, 365.
[236] P. N. Lavrenko, O. V. Okatova, and A. B. Mel'nikov, *Vysokomol. Soedin. Ser. A*, 1981, **23**, 532.
[237] V. V. Korsak, S. S. A. Pavlova, P. N. Gribkova, and T. N. Balykova, *Acta Polym.*, 1981, **32**, 61.
[238] Ye. V. Vichutinskaya, A. L. Margolin, L. M. Postnikov, and V. Ya. Shlyapintokh, *Polym. Sci. (USSR)*, 1981, **23**, 3000.
[239] L. Tang, D. Sallet, and J. Lemaire, *Macromolecules*, 1982, **15**, 1437.
[240] Y. Fujiwara and H. Zeronian, *J. Appl. Polym. Sci.*, 1981, **26**, 3729.
[241] G. A. Horsfall, *Text. Res. J.*, 1982, **52**, 197.
[242] S. Ichiro Imanura, K. Nishii, and H. Heranishi, *J. Appl. Polym. Sci.*, 1982, **27**, 1713.
[243] A. A. Koutinas, *J. Polym. Sci., Polym. Chem. Ed.*, 1981, **19**, 2269.
[244] L. C. Chow and E. P. Chang, *J. Appl. Polym. Sci.*, 1981, **26**, 603.
[245] G. A. George, G. T. Egglestone, and S. Z. Riddell, *J. Appl. Polym. Sci.*, 1982, **27**, 3999.
[246] A. Schaper, E. Schulz, R. Hirte, and C. Ruscher, *Acta Polymer.*, 1982, **33**, 227.
[247] M. Hirami, *Makromol. Chem.*, 1982, **183**, 2857.
[248] S. Gogolewski, M. Crasiorek, K. Czerniawska, and A. J. Peringa, *Colloid Polym. Sci.*, 1982, **260**, 859.

Step Growth Polymerization

nylon-6 occurs.[249] Polymorphism has been reported during orientation,[250] and the deformation mechanism[251] and conformational changes[252] can be related to the dependence of melting point and glass-transition temperatures on stress. Chain rupture detected by e.s.r.[253] is claimed to be much too small to account for changes in tensile properties and molecular weight. Gross failure in the form of cracks in photodegraded poly(ε-caprolactam) occurs perpendicular to the fibre axis of drawn filaments, but not in the undrawn material.[254]

Nylon-66 has been crystallized from solution[255] and a number of papers have concentrated on various aspects of fatigue crack propagation.[256—259] It has been claimed that resistance to crack propagation with increasing molecular weight is due to the development of a molecular entanglement network.[259] The thermal mechanical properties of nylon-66 annealed in glycerol were reported.[260]

Nylon-4 crystallized from solution has shown the effect of various solvent–precipitant systems.[261] Nylon-12 has also been solution crystallized,[262] and the influence of hydrogen bonding on the mechanical anisotropic of this polymer has been demonstrated for samples with different thermal treatments.[263] Comparatively little has appeared on the morphology of aromatic polyamides but papers have been published on surface modifications[264] of Kevlar and the influence of morphology on its compressional behaviour.[265]

5 Polyimides

Synthesis.—Several papers have dealt with the mechanism of polyimide synthesis. Autocatalysis[266] and amidocatalysis[267] were both examined. The imidization during the reaction of diethyl-3,3′,4,4′-benzophenone tetracarboxylate, ethyl-5-norborane-2,3-dicarboxylate and amine was studied by HPLC and n.m.r. Both imidization reactions proceeded directly to the imide. Neither amic acid was present in significant amounts at any stage.[268] N.m.r. spectroscopy has also been used to study norbornene end-capped polyimides.[269]

---

[249] T. Ito, *Polymer*, 1982, **23**, 1712.
[250] J. Gianchandoi, J. E. Spruiell, and E. S. Clark, *J. Appl. Polym. Sci.*, 1982, **27**, 3527.
[251] M. Matsuo, Y. Seino, T. Watanabe, S. Mariguchi, F. Ozaki, and T. Ogita, *Polym. J.*, 1981, **13**, 755.
[252] S. Fakirov and I. Seganov, *Polym. Sci. (USSR)*, 1981, **23**, 857.
[253] O. Frank and J. H. Wendorff, *Colloid Polym. Sci.*, 1981, **259**, 1047.
[254] Y. Fujiwara and S. H. Zeronian, *J. Appl. Polym. Sci.*, 1982, **27**, 2773.
[255] J. H. Magill, M. Giroloamo, and A. Keller, *Polymer*, 1981, **22**, 43.
[256] P. E. Bretz, R. W. Hertzber, and J. A. Manson, *J. Mater. Sci.*, 1981, **16**, 2061.
[257] M. T. Hahn, R. W. Hertzberg, J. A. Manson, R. W. Lang, and P. E. Bretz, *Polymer*, 1982, **23**, 1675.
[258] P. E. Bretz, R. W. Hertzberg, and J. A. Manson, *Polymer*, 1981, **22**, 1272.
[259] P. E. Bretz, R. W. Hertzberg, and J. A. Manson, *J. Appl. Polym. Sci.*, 1982, **27**, 1707.
[260] H. Mitomo and I. Kuriyama, *Polymer*, 1982, **23**, 1377.
[261] L. Kudláček, M. Kaplanová, and M. Hepner, *Acta Polym.*, 1981, **32**, 681.
[262] M. Kyotani, *J. Polym. Sci., Polym. Phys. Ed.*, 1982, **20**, 345.
[263] P. Kollross and A. J. Owen, *Polymer*, 1982, **23**, 829.
[264] M. R. Wertheimer and H. P. Schreiber, *J. Appl. Polym. Sci.*, 1981, **26**, 2087.
[265] M. C. Dobb, D. J. Johnson, and B. P. Saville, *Polymer*, 1981, **22**, 950.
[266] R. L. Kass, *J. Polym. Sci., Polym. Chem. Ed.*, 1981, **19**, 2255.
[267] B. A. Zhubanov, G. I. Boiko, A. Sh. Zainullina, and S. K. Kudaikulova, *Polym. Sci. (USSR)*, 1981, **23**, 2358.
[268] S. E. Delos, R. K. Schellenberg, and J. E. Smedley, *J. Appl. Polym. Sci.*, 1982, **27**, 4295.
[269] A. C. Wong, A. N. Garroway, and N. M. Ritchey, *Macromolecules*, 1981, **14**, 832.

As might be expected, most of the studies on the synthesis of polyimides have concentrated on the production of polymers with high temperature resistance. Linear aliphatic and cycloaliphatic bisdi-imides have been prepared and polymerized thermally to give highly cross-linked laminating resins.[270] Polyimides soluble in cresylic acid were synthesized, their solubility being correlated with solubility parameter, symmetry, and tendency to hydrogen bond.[271] Other polyamides have been prepared from carbazole and pyromellitic dianhydride,[272] N-(p-carboxyphenyl)trimellitimide and pp'-di(aminocyclohexyl) methane,[273] and 3,4-dicarboxy-4'-chloroformylbiphenyl anhydride.[274] This last reaction gave poly(amide-imides) which were found to have a surprisingly high thermal stability. Other polymers which have been prepared have been novel poly(N-oxymide)s[275] and a series of copolyimides in which the one- and two-step methods were compared.[276] A new polyester imide containing C=C bonds was prepared at low temperature from an unsaturated diacid chloride containing a cyclic imide group with ethylene glycol.[277] This was thermally stable, soluble in highly polar solvents, and could be cross-linked. Other copolyimides reported have been polybenzoxazinone imides,[278] polybenzothiazole amide-imides,[279] polyimide carbonate, and polyimide urethanes,[280] as well as aromatic block copolyimides[281] and polyimides containing various heterocyclic main-chain units.[282] In an attempt to get thermal and flame-resistant characteristics, other polymers have been prepared containing phenoxaphosphine rings,[283] and which are chlorinated.[284]

Properties.—Recent theoretical and instrumental advances have made it possible to record high-resolution ^{13}C n.m.r. spectra of polyimides by examining the shifts for both polymers and model compounds. The nature of the conjugation along the polyimide chain has been postulated. The conjugation is thought to be an important factor in the high mechanical and thermal stability of these systems.[285] Geometric parameters[286] and dynamic mechanical relaxation[287] have been reported, as has the influence of chemical structure on glass-transition temperature of polyarimides.[288] The effect of increasing the distance between, and varying the isomeric positions of,

---

[270] D. A. Scola and M. P. Stevens, *J. Appl. Polym. Sci.*, 1981, **26**, 231.
[271] H. Sheffed, *J. Appl. Polym. Sci.*, 1981, **26**, 3837.
[272] M. Biswas and S. K. Das, *Eur. Polym. J.*, 1981, **17**, 1245.
[273] S. Maiti and A. Roy, *J. Appl. Polym. Sci.*, 1982, **27**, 4345.
[274] L. H. Tagle and F. R. Diaz, *Polymer*, 1982, **23**, 1057.
[275] A. M. Ibrahim, V. Mahadevan, and M. Srinivasan, *J. Polym. Sci.*, 1981, **19**, 687.
[276] Ja. S. Vygodskij, S. V. Vinogradova, Z. M. Nagiev, V. V. Koršak, Ja. G. Urman, G. Reinisch, and G. Rafler, *Acta Polym.*, 1982, **33**, 131.
[277] S. Maiti and A. Ray, *Makromol. Chem.*, 1982, **183**, 1949.
[278] G. Neamtu and M. Brumă, *Angew. Makromol. Chem.*, 1982, **103**, 19.
[279] G. Neantu, G. Mandric, and I. Zugrăvescu, *Angew. Makromol. Chem.*, 1982, **103**, 29.
[280] H. Imajo, K. Kwita, and Y. Iwakura, *J. Polym. Sci., Polym. Chem. Ed.*, 1981, **19**, 1855.
[281] G. N. Babu and S. Samant, *Eur. Polym. J.*, 1981, **17**, 421.
[282] M. M. Koton, T. M. Kiseleva, T. I. Zhukova, S. N. Nikolayeva, L. A. Laius, and Yu. N. Sazanov, *Polym. Sci. (USSR)*, 1981, **23**, 1909.
[283] M. Sato, Y. Tada, and M. Yokoyama, *J. Polym. Sci.*, 1981, **19**, 1037.
[284] P. F. Frigerio, L. H. Tagle, and F. R. Diaz, *Polymer*, 1981, **22**, 1571.
[285] J. R. Havens, H. Ishida, and J. L. Koenig, *Macromolecules*, 1981, **14**, 1327.
[286] B. M. Ginsburg, Ye. Y. Magdelev, V. N. Volosatov, and S. Ya. Frenkel, *Vysokomol. Soedin. Ser. B*, 1981, **23**, 701.
[287] J. M. Perera, *Angew. Makromol. Chem.*, 1982, **106**, 61.
[288] L. N. Korzhaven, S. V. Bromnikov, and S. Ya. Frenkel, *Polym. Sci. (USSR)*, 1981, **23**, 409.

the amine group on the glass-transition temperatures of aromatic polyimides[289] is to cause a reduction on dilution of the imide content. Penetrant interactions in Kapton polyimide investigated with NH_3 and SO_2 are well described by Fickian diffusion.[290]

Thermal stability on degradation has been specifically dealt with in many of these papers which describe both syntheses and properties. Model compounds such as phthalimide[291] have been used to explain the mechanisms for CO_2 formation, and the influence of unit inhomogeneity has also been reported upon.[292]

Possibly because of the potential use of polyimides for micro electronics application, considerable interest has been shown in the literature in the electrical behaviour of polyimide films.[293] The insulation characteristics of the polymer for multilevel interconnections[294] and the electrical properties of polyimides containing tricyclic fused rings,[295] as well as polyimides containing palladium[296] and lithium,[297] have all come under scrutiny. It has been found, for example, that the lithium chloride addition increases electrical conductivity significantly. The question of the conductivity of pyrolized polyimides has also come under discussion.[298] Dielectric polarization of polyimide by an electric field (photo electrets) is apparently of considerable significance in semi-conductor research. Kapton has been found to have good photo-conduction properties.[299] The mechanism for polarization of poly(pyromellitimide) thermally stimulated discharge currents (TSDC) were investigated with polymer films. The TSDC spectra produces peaks at 100 °C, 196 °C, and 249 °C. These were attributed to absorbed water, and dipole orientation, space charged polarization, and surface charge, respectively.[300] These interesting electrical characteristics have also permitted the electrodeposition of polyimides from non-aqueous electrophoretic emulsions to take place.[301]

Morphology.—The pattern noted in previous years in which the work on polyimides has concentrated on the synthesis and properties of the polymers has resulted once again in very many fewer papers being published on the polymers' morphology. An attempt has been made to calculate theoretically the molecular packing in the crystalline domains using atom–atom potentials combined with X-ray structural analysis.[302] Small-angle X-ray scattering has also been used to examine molecular aggregation in aromatic polyimides[303] and its affect on mechanical properties.[304] From analysis of the correlation between molecular aggregation and the

---

[289] V. L. Bell, L. Kilzer, E. M. Hett, and G. M. Stokes, *J. Appl. Polym. Sci.*, 1981, **26**, 3805.
[290] L. R. Iler, R. C. Laundon, and W. J. Koros, *J. Appl. Polym. Sci.*, 1982, **27**, 1163.
[291] J. Zurakowska Orszagh and T. Chreptowicz, *Eur. Polym. J.*, 1981, **17**, 877.
[292] V. V. Korshak, S. S. A. Pavlova, P. N. Gribkova, I. V. Vlasova, Ya. S. Vygodskii, and S. V. Vinogradova, *Vysokomol. Soedin. Ser. A*, 1981, **23**, 1586.
[293] G. D. Khune, *J. Macromol. Sci., Chem.*, 1981, **15**, 241.
[294] A. M. Wilson, *Thin Solid Films*, 1981, **83**, 145.
[295] K. Nuime, R. Hirohashi, F. Joda, M. Hasegawa, and T. Iwakuva, *Polymer*, 1981, **22**, 649.
[296] T. L. Wohlford, J. Schaff, L. T. Taylor, A. K. St. Clair, T. A. Furtsch, and E. Khor, 'Conductive Polymers', ed. E. B. Seymour, Plenum, New York, 1981, p. 7.
[297] E. Khor and L. T. Taylor, *Macromolecules*, 1982, **15**, 379.
[298] J. I. Gittleman and E. K. Sichel, *J. Electron. Mater.*, 1981, **10**, 327.
[299] J. K. Quarma, P. K. C. Pillai, and B. L. Sharma, *Acta Polym.*, 1982, **33**, 501.
[300] J. K. Quarma, P. K. C. Pillai, and B. L. Sharma, *Acta Polym.*, 1982, **33**, 205.
[301] W. M. Alvino and L. Scola, *J. Appl. Polym. Sci.*, 1982, **27**, 341.
[302] N. V. Lukasheva, I. S. Milevskaya, and A. M. El'yashevich, *Polym. Sci. (USSR)*, 1981, **23**, 2404.
[303] S. Isoda, H. Shimada, M. Kochi, and H. Kambe, *J. Polym. Sci., Polym. Phys. Ed.*, 1981, **19**, 1293.
[304] S. Isoda, M. Kochi, and H. Kambe, *J. Polym. Sci., Polym. Phys. Ed.*, 1982, **20**, 837.

mechanical properties it was concluded that the mode of molecular motion corresponding to the α-dispersion is a long-range co-operative motion of the main chain which is associated with the glass transition.

The dichroism of adsorption bands of orientated films of polyamidoimides and polyesterimides has shown the effect of *meta* bonded phenylene rings,[305] while at a more practical level the resistance of polyether-imides to environmental stress crazing and the cracking apparently reaches a minimum in solvents having solubility parameters close to that of the resin.[306]

---

[305] Yu. G. Baklagina, I. S. Milevskaya, N. V. Mikhailova, A. V. Sidorovich, and L. K. Prokharova, *Polym. Sci. (USSR)*, 1981, **23**, 376.

[306] S. A. White, S. R. Weissman, and R. P. Kumbour, *J. Appl. Polym. Sci.*, 1982, **27**, 2675.

PART II Developments in Polyurethanes
by D. J. Sparrow and I. G. Walton

1 Introduction

The literature during the last 2 years has shown a steady evolutionary progress along the lines discussed in the last Report,[1] rather than revealing new developments. A useful review of these trends was presented at a recent conference.[2] An unusual number of new books has appeared, and these will be discussed in the relevant sections below; the discussion is divided into raw materials and applications.

2 Isocyanate Products and Processes

The very large number of patents appearing in the last 2 years again contrasts with the relative lack of scientific articles. A good general review on organic isocyanates has, however, appeared.[3] The family of isocyanates based on 4,4'-di-isocyanato diphenyl methane (MDI) has continued to receive much attention. The conventional production route – phosgenation of the appropriate aminoarenes – is still the subject of research. For example, patents have appeared on optimizing the level of the 4,4'-MDI isomer[4] and on reducing the level of impurities.[5] The major research emphasis on MDI-type isocyanates is on the development of a non-phosgene route. The favoured process is carbonylation of nitrobenzene or aniline in the presence of an alcohol, followed by condensation with formaldehyde to form a mixture of di- and multi-functional carbamates, which is then thermally decomposed to the equivalent isocyanates. The first step appears to be the most difficult to achieve at a reasonable conversion, and the patented reaction conditions fall into two groups. One[6] uses sulphur or selenium as catalyst, the other noble metals such as platinum[7] or palladium.[8] There is also evidence that copper is being investigated as a catalyst.[9] Opening of the Atlantic Richfield plant based on this route has been postponed until 1984, apparently because of the environmental difficulties associated with the use of selenium.[10] The direct carbonylation of nitro-arenes to isocyanates has also been the subject of patent activity; the catalysts cited are again noble metals such as

---

[1] D. J. Sparrow and I. G. Walton, in 'Macromolecular Chemistry', ed. A. D. Jenkins and J. F. Kennedy (Specialist Periodical Reports), The Royal Society of Chemistry, London, 1982, Vol. 2, p. 69.
[2] K. C. Frisch, talk given to Urethane Group Symposium 1982, available from the Plastics and Rubber Institute, London.
[3] D. H. Chadwick and T. H. Cleveland, in 'Kirk-Othmer Encyclopedia of Chemical Technology', 3rd edition, Wiley, New York, 1981, Vol. 13, p. 789.
[4] US P., 4 297 294.
[5] US P., 4 294 666.
[6] US P., 4 267 353.
[7] Eur. P., 28 460.
[8] Br. P., 1 592 731.
[9] US P., 4 251 667.
[10] Anon., *Chemical Age*, 1981, 6 Feb., 3.

palladium[11] and nickel.[12] A paper has appeared discussing the related area of the carbonylation of phenyl azide.[13]

Several novel isocyanates have been patented, all prepared by phosgenation of an amine. One of the more interesting patents[14] provides a route to 2,3'- and 3,4'-substituted MDI, by condensing nitrobenzyl chloride with nitrobenzene, followed by hydrogenation and then phosgenation of the resulting amine. Several aliphatic isocyanates have been claimed, mostly for use in applications where colour stability is important. They include 1,8-di-isocyanato-4-isocyanato methyloctane,[15] 1,6,11-tri-isocyanato-undecane,[16] and 1,3,5-tri-isocyanatomethyl-cyclohexane.[17]

The production of storage-stable liquid di-isocyanates remains desirable; efforts to achieve this have included chemically modifying 4,4'-MDI by the introduction of urethane,[18] carbodiimide,[19] biuret,[20] and uretonimine[21] groups. In the last case, a solid inorganic catalyst is used; it can readily be removed to avoid the possibility of the reaction proceeding further than desired on storage.

3 Polyurethane Polyols

Comprehensive summaries of the polyethers[22] and tetrahydrofuran polymers[23] used industrially have been published. In recent papers, detailed considerations have been given to the side reactions occurring in potassium hydroxide catalysed polyether preparation,[24] and to the distribution of functionalities obtained.[25]

A plate column polymerization reactor for polyester polyol manufacture has been described,[26] and papers covering polyesterification process control[27] and kinetics[28] have been published. In the rigid foam area there has been a resurgence of interest in polyols based on α-methyl glucoside.[29]

The PHD polyols, which are polyurea dispersions prepared by the *in situ* reaction of amine and isocyanate in polyether, are receiving continued attention.[30] Recently, a new group of polymer polyols formed by the *in situ* reaction of an alkanolamine with isocyanate have been introduced.[31] Polyurethane solutions in polyol, prepared

---

[11] Br. P., 2 068 939.
[12] W. E. Martin, *J. Organomet. Chem.*, 1981, **206**, C393.
[13] G. La Monica and S. Cenini, *J. Organometal. Chem.*, 1981, **216**, C35.
[14] Eur. P., 46 556.
[15] Fr. P., 2 478 088.
[16] US P., 4 276 228.
[17] Eur. P., 21 067.
[18] Eur. P., 31 207.
[19] US P., 4 284 730.
[20] US P., 4 332 953.
[21] Eur. P., 54 294.
[22] R. A. Newton, in ref. 3, Vol. 18, p. 633.
[23] P. Dreyfuss and M. P. Dreyfuss, in ref. 3, Vol. 18, p. 645.
[24] A. Penati and C. Maffezzoni, *J. Appl. Polym. Sci.*, 1981, **26**, 1059.
[25] H. U. Schimpfle and H. Becker, *Acta Polym.*, 1981, **32**, 55.
[26] Eur. P., 43 502.
[27] I. Pamakis and I. Simitzis, *Angew. Makromol. Chem.*, 1981, **99**, 145.
[28] A. Fradet and E. Marechal, *Polym. Bull. (Berlin)*, 1980, **3**, 441.
[29] J. J. Cimerol and S. Fusezi, *J. Elastomers Plast.*, 1981, **13**, 224.
[30] K. G. Spitler and J. J. Lindsey, *J. Cell. Plast.*, 1981, **17**, 43.
[31] Belg. P., 887 514.

by the reaction of isocyanate and diols in the polyol, have also been described.[32] These products are claimed to be suitable for textile lamination applications.

4 Catalysis and Mechanism of Isocyanate Reactions

Patents for individual compounds or mixture of compounds have again dominated the literature, with the usual claims for improvements in processing or in final properties. A noticeable trend is the number of citations of morpholine derivatives.[33,34]

Attention has been paid to the mechanism of catalysis of the isocyanate–hydroxyl reaction by organometallic compounds[35] and by amine/organometallic synergistic mixtures.[36] Fewer references than usual have dealt with the mechanism of trimerization of isocyanates,[37] but a number have appeared on the urea-forming reaction between isocyanates and amines.[38,39] A discussion on the selection of catalysts for flexible foam systems has been published,[40] and an attempt made to study network formation in a simple polyurethane by rheological methods.[41]

5 Rigid Foams

Most commercial rigid polyurethane foams are anisotropic materials formed from highly cross-linked polymers, which makes both chemical and physical analysis difficult. Structure–property relationships have received much attention; a book[42] and several papers (*e.g.*, refs. 43 and 44) have appeared. The data contained in these publications highlights the difficulty of dealing with the physical properties of anisotropic cellular materials; an important 1981 paper[45] derived equations which allow such data to be readily interpreted. Attention has also been directed towards more accurate measurement of the rise profile of rigid foam.[46]

The major industrial preoccupation with rigid foams has been to improve the fire properties. There are three principal ways of doing this; the introduction of thermally stable cyclic structures into the foam, the use of fire-retardant species, and the reinforcement of foam with mineral fibres. The commonest cyclic structure in polyurethane foams is the isocyanurate ring, formed by trimerizing isocyanates. A

---

[32] Eur. P., 35 687.
[33] Eur. P., 54 219.
[34] US P., 4 326 042.
[35] F. W. van der Weij, *J. Polym. Sci., Polym. Chem. Ed.*, 1981, **19**, 381 and 3063.
[36] I. S. Bechara, *ACS Symp. Ser.*, 1981, **166**, 501.
[37] J. E. Kresta, C. S. Shen, K. H. Hsieh, and K. C. Frisch, *ACS Symp. Ser.*, 1981, **166**, 501.
[38] J. M. Borsus, R. Jerome, and P. Teyssie, *J. Appl. Polym. Sci.*, 1981, **26**, 3027.
[40] M. L. Bye, F. O. Baskent, and M. R. Sandner, *Proc. SPI Annual Ureth. Div. Tech. Conf.*, 1981, **26**, 109.
[41] M. D. Hartley and H. L. Williams, *Polym. Eng. Sci.*, 1981, **21**, 135.
[42] 'Mechanics of Cellular Plastics', ed. N. C. Hilyard, Applied Science Publishers, Barking, 1982.
[43] G. F. Baumann and W. Dietrich, *J. Cell. Plast.*, 1981, **17**, 144.
[44] J. R. Dawson and J. B. Shortall, *J. Mater. Sci.*, 1982, **17**, 220.
[45] A. Cunningham, *Polymer*, 1981, **22**, 882.
[46] L. H. Hanusa and R. N. Hunt, *J. Cell. Plast.*, 1982, **18**, 100.

paper on recent developments has appeared.[47] The use of furan rings in rigid foam structures has also been discussed.[48]

In fire retardants, the trend is towards hydroxyl-containing species, which can react into the polymer network. Fire retardancy is conferred by the inclusion of phosphorus[49] or halogen[50] atoms in the polyol. Inorganic fillers such as alumina trihydrate have been claimed as a means of reducing smoke generation.[51] Another use of inorganic materials is to act as a reinforcing network to hold the foam together and thus slow down the spread of flame. Glass fibre mat is typically employed.[52] These rather empirical developments have been complemented by efforts to understand the mechanism of the thermal degradation of polyurethanes, typically by studying model compounds.[53,54]

6 Reaction Injection Moulding

The increasingly encouraging prospect of using urethane reaction injection moulding (RIM) in the automotive industry[55] has led to a large volume of publications. Several general reviews have appeared.[56,57] A chapter in a recent book authoritatively discusses the general development of urethane polymers for RIM applications;[58] other publications have dealt with specific chemical systems.[59,60] The possibilities of introducing isocyanurate groups[47] and of using interpenetrating polymer networks[61] have been considered. Glass-fibre reinforcement of RIM systems is becoming important, and the use of polymer-filler coupling agents such as titanates has been discussed.[62] A series of papers on modelling the RIM process has appeared from Macosko and co-workers, including studies of mould filling,[63] heat transfer,[64] and the effects of impingement mixing.[65] In a similar vein, a simulation of cavity filling and curing has appeared.[66]

---

[47] H. Ulrich, *J. Cell. Plast.*, 1981, **17**, 31.
[48] E. K. Moss, *J. Cell. Plast.*, 1982, **18**, 240.
[49] S. Yanai, D. Vofsi, and M. Halman, *J. Cell. Plast.*, 1981, **17**, 284.
[50] S. R. Sandler, M. M. Chan, and J. D. Miano, *J. Fire Retard. Chem.*, 1981, **8**, 193.
[51] P. V. Bonsignore, *J. Cell. Plast.*, 1981, **17**, 220.
[52] P. J. Briggs and R. Laker, *Vetrotex*, 1981 (12), 12.
[53] J. Chambers, J. Jiricny, and C. B. Reese, *Fire Mater.*, 1981, **5**, 133.
[54] S. Foti, P. Maravigna, and G. Montaudo, *J. Polym. Sci., Polym. Chem. Ed.*, 1981, **19**, 1679.
[55] Anon., *Urethanes Today*, 1981, **2**, 1.
[56] G. Kleiner, *Chem. Ind. (London)*, 1981 (22), 779.
[57] S. H. Metzger and K. Seel, *J. Cell. Plast.*, 1981, **17**, 268.
[58] D. C. Allport, C. Barker, and J. F. Chapman, in 'Development in Block Copolymers – 1', ed. I. Goodman, Applied Science Publishers, Barking, 1982.
[59] Eur. P., 82 237.
[60] J. M. O'Connor, M. L. Rosin, F. J. Preston, V. B. Jenkin, R. L. Visger, and W. J. Sessions, *J. Cell. Plast.*, 1981, **17**, 35.
[61] R. Penice, K. C. Frisch, and R. Navare, *J. Cell. Plast.*, 1982, **18**, 121.
[62] S. J. Monte, G. Sugarman, A. Damusis, and R. Patel, *J. Elastomers Plast.*, 1981, **14**, 34.
[63] J. M. Castro and C. W. Macosko, *AICHE J.*, 1982, **28**, 250.
[64] L. J. Lee and C. W. Macosko, *Int. J. Heat Mass Transfer*, 1980, **23**, 1479.
[65] P. Kolodziej, C. W. Macosko, and W. E. Ranz, *Polym. Eng. Sci.*, 1982, **22**, 388.
[66] L. T. Manzione, *Polym. Eng. Sci.*, 1981, **21**, 1234.

7 Flexible Foams

The chemistry and technology of flexible foams have been comprehensively reviewed in an excellent recent book.[67] Two booklets[68] have appeared, providing useful data and putting the fire performance of flexible foams into perspective. Papers have been published on the chemical reactions involved in the formation of water-blown foam,[69] the role of silicone surfactants,[70] and the use of amine catalysts.[71] Important industrial developments include the increasing use of MDI-based flexible foam,[72] the claimed excellent fire properties of polyimide based flexible foam,[73] and the development of a novel flexible foam slabstock machine, which is particularly suitable for low production rates.[74]

8 Miscellaneous

Elastomers have continued to attract considerable academic interest; a book has appeared.[75] The morphology and properties of polyurethane block copolymers have been thoroughly discussed in a chapter of another recent book.[76] A number of papers have dealt with the morphology of the hard segment in these elastomers.[77,78]

The use of polymeric MDI as a binder for chipboard continues to spread; a notable improvement is the availability of MDI containing an internal release agent.[79] The polyurethane tyre remains a distinct possibility for the future.[80] A number of references have appeared dealing with the use of polyurethane foam to extract metals such as rhodium,[81] iridium,[81] and palladium.[82] The recycling of scrap polyurethane foam has again been discussed; it is typically done by hydrolysis[83] to extract polyols[84] and/or polyamines.[85]

---

[67] G. Woods, 'Flexible Polyurethane Foams: Chemistry and Technology', Applied Science Publishers, Barking, 1982.
[68] Anon., 'Flexible Polyurethane Foam: The Facts', and 'Flexible Polyurethane Foam: Misconceptions', British Rubber Manufacturers Association, London, 1982.
[69] F. E. Bailey and F. E. Critchfield, *J. Cell. Plast.*, 1981, **17**, 333.
[70] G. Rossmy, H. Kollmeier, W. Lidy, and H. Shator, *J. Cell. Plast.*, 1981, **17**, 319.
[71] F. P. Carroll and M. H. Ziv, *J. Cell. Plast.*, 1982, **18**, 168.
[72] Eur. P., 22 617.
[73] US P., 4 305 796.
[74] Eur. P., 58 553.
[75] C. Hepburn, 'Polyurethane Elastomers', Applied Science Publishers, Barking, 1982.
[76] P. E. Gibson, M. A. Vallance, and S. L. Cooper, in ref. 58, p. 217.
[77] J. Blackwell and M. R. Nagarajan, *Polymer*, 1981, **22**, 202 and 1534.
[78] Y. Camberlin, J. P. Pascault, J. M. Letoffe, and P. Claudy, *J. Polym. Sci., Polym. Chem. Ed.*, 1982, **20**, 1445.
[79] Br. P., 2 090 261.
[80] Y. S. Marshall, *Eur. Rubber J.*, 1982, **164**(10), 14.
[81] S. J. Al-Bazi and A. Chow, *Anal. Chem.*, 1981, **53**, 1073.
[82] S. J. Al-Bazi and A. Chow, *Talanta*, 1982, **29**, 507.
[83] R. J. Salloum and C. C. Duff, *Polym. Plast. Technol.*, 1982, **19**, 20.
[84] US P., 4 328 368.
[85] Eur. P., 47 913.

4
Natural Polymers: Polysaccharides and Glycoproteins

BY R. J. STURGEON

1 Introduction

The first two reports of advances in polysaccharides and glycoproteins have covered publications of 1977—1980. Activity in this field has been maintained: the number of publications dealing with polysaccharides has remained fairly constant, but significant increases in literature about glycoproteins are noted. The aim of this Chapter is to highlight publications on polysaccharides and glycoproteins. Most of the polysaccharides which are being studied in depth, such as starch, cellulose, alginic acid, and carrageenans, are biopolymers, occurring in nature in large amounts and being of high economic value in biotechnology. Much of the current interest in glycoproteins centres on the function of these molecules. Thus the selection of work described in this article deals with the relationship of glycoproteins to lectins, cell adhesion, and catabolism and clearance.

Several important treatises, books, and reviews have been published. The first volume of a three-volume set, entitled 'The Polysaccharides', presents state-of-the-art coverage of polysaccharide chemistry and related aspects of biochemistry.[1] This volume deals with methodology, isolation and fractionation of polysaccharides, procedures for the determination of structure by chemical degradative and spectroscopic non-degradative methods, shapes and interactions of carbohydrate chains, and immunochemical properties of polysaccharides. Two volumes in the new series of the 'Encyclopedia of Plant Physiology' contain review articles related to plant carbohydrates, and represent a comprehensive assessment of the current viewpoint in plant carbohydrates with emphasis on those aspects which impinge on physiological processes of growth and development. One volume is devoted to macromolecular carbohydrates which occur intracellularly, *e.g.*, starch and other reserve polysaccharides, glycoproteins, and glycolipids.[2] Additional sections deal with physiological processes such as secretion, storage, and mobilization of carbohydrate reserves. The other volume contains information on cell wall structure and function in algae, fungi and higher plants, export of carbohydrates across cell walls, and cell surface interactions and the role of carbohydrate–lectin interactions in plants.[3] Solution properties of polysaccharides have been the subject of a review.[4] Further volumes of the well-received treatise on the glycoconjugates

---

[1] 'The Polysaccharides', ed. G. O. Aspinall, Academic Press, New York, 1982, Vol. 1.
[2] 'Encyclopedia of Plant Physiology, Plant Carbohydrates', ed. F. A. Loewus and W. Tanner, Springer-Verlag, Berlin, 1982, Vol. 13A.
[3] 'Encyclopedia of Plant Physiology, Plant Carbohydrates', ed. F. A. Loewus and W. Tanner, Springer-Verlag, Berlin, 1981, Vol. 13B.
[4] 'Solution Properties of Polysaccharides', ed. D. A. Brant, *Am. Chem. Soc. Symp. Ser.*, 1981, **150**.

have been published. One volume deals with the glycosylation of proteins, glycosylation and development, and glycoconjugates in cellular adhesion and aggregation.[5] The other volume is devoted to the uptake of glycoconjugates, the turnover and shedding of cell surface glycoconjugates, virus glycoproteins and glycolipids, and the applications of glycosyltransferases and glycoconjugates to chemotherapy, aging, and disease.[6] The role of fibronectin, a glycoprotein that binds to a variety of macromolecules including fibrin, collagen, cell surfaces, and glycosaminoglycans, has been reviewed.[7]

2 Plant and Algal Polysaccharides

Starch.—A method has been developed for the determination of amylose:amylopectin ratios of starches.[8] The procedure offers an alternative to iodometric methods and involves solubilization of starches in DMSO followed by debranching with isoamylase. The resulting linear components are quantitated by gel permeation chromatography. A method for determining the starch content of cereals using α-amylases has been developed.[9] The starch is solubilized with a thermostable α-amylase and the products are then completely hydrolysed with amyloglucosidase. D-Glucose is then estimated by the hexokinase-D-glucose 6-phosphate procedure.

The susceptibility of native starch granules to amylolysis has been monitored chemically and microscopically.[10] Attack by salivary α-amylase results in a gradual erosion of the surface followed by granule penetration at certain locations. Attack by glucoamylase, on the other hand, was more uniform and results in pitting and depressions all over the surface. A procedure adapted for the rapid analysis of total starch and amylose has been used on a series of 37 barley genotypes.[11] The amylose content was not related to either starch content or grain yield.

An enzymatic method of determining the A and B chains in amylopectin has led to a ratio of 1:1, not 2:1 as previously suggested.[12] Partial debranching with pullulanase gave results consistent with earlier suggestions that A chains are predominantly and selectively removed by this enzyme.

The susceptibility of starch granules to degradation by amylases depends not only on the source of the starch but also on that of the amylase.[13] Three different types of α-amylase, from *Streptomyces precox*, porcine, and *Bacillus subtilis* sources, degraded starch granules of sugary-2(su_2), waxy (wx), normal and ae maize to decreasing extents.

In studies on the characterization of legume starches, the fine structures of amylose and amylopectin have been determined by the use of pullulanase, β-amylase, and glucoamylase.[14] The results indicated that there were some $(1 \rightarrow 6)$

---

[5] 'The Glycoconjugates', ed. M. I. Horowitz, Academic Press, New York, 1982, Vol. 3.
[6] 'The Glycoconjugates', ed. M. I. Horowitz, Academic Press, New York, 1982, Vol. 4.
[7] E. H. Ruoslahti, M. Pierschbacher, E. G. Hayman, and E. Engvall, *Trends Biochem. Sci.*, 1982, **7**, 188.
[8] J. G. Sargant, *Starch*, 1982, **34**, 89.
[9] I. L. Batey, *Starch*, 1982, **34**, 125.
[10] S. V. Paramathans and R. N. Tharanathan, *Starch*, 1982, **34**, 73.
[11] J. Torp, *J. Sci. Food Agric.*, 1980, **30**, 1354.
[12] D. J. Manners and N. K. Matheson, *Carbohydr. Res.*, 1981, **90**, 99.
[13] H. Fuwa, *J. Jpn. Soc. Starch Sci.*, 1982, **29**, 99.
[14] C. G. Biliaderis, D. R. Grant, and J. R. Vose, *Cereal Chem.*, 1981, **58**, 496.

substituted α-D-glucopyranosyl residues in amylose. Similar studies were carried out on acid-treated starches.[15] A fast hydrolysis of amorphous and gel-phases occurred with a slow hydrolysis of crystalline regions. The results were consistent with the structure being a cluster type.

Bacillus circulans F-2 produces an extracellular amylase when potato starch granules, but not soluble starch, is used as the carbon source.[16] The enzyme liberates maltohexaose units from the non-reducing ends of starch.

The structure of the hydrated amylose–iodine complex is an orthorhombic unit cell with dimensions of a = 13.60 Å, b = 23.42 Å, and c (the fibre repeat) = 8.17 Å, as determined by combined methods of X-ray diffraction and stereochemical packing analysis.[17] Two amylose chains pass through the unit cell and iodide ions were present as an almost linear chain in the centre of the six-fold, left-handed amylose helix. Eight molecules of hydration per unit cell were located in good hydrogen-bonding positions between the amylose helices. The crystal structure of V_h amylose has been refined by similar methods to show another orthorhombic unit cell of dimensions a = 13.65 Å, b = 23.70 Å, and c = 8.05 Å.[18] The chain conformation is a left-handed, six-fold helix with O-6 in the position *gauche* to both O-5 and C-4. The water molecules are located inside the helical channel of the amylose and in the interstitial spaces, forming an intensive hydrogen-bonded network.

Starch has been degraded by plasma and thermolysis in the ionization chamber of a mass spectrometer and the sugars and oligosaccharides formed were characterized by direct chemical ionization mass spectrometry.[19] The technique, especially with negative ions, extends mass spectrometry to polymers.

A resonance Raman spectroscopic study has been made of the blue colouring of iodine/iodide in amylose.[20] Similar spectra were obtained regardless of the KI and I_2 concentrations, degree of polymerization, and excitation wavelengths, but the relative intensities of the lines changed. It was concluded that the basic unit changed from I_6^{2-} to I_{10}^{2-} through I_8^{2-} with decreasing KI concentration. A simple straight-line relationship has been obtained between average chain length of linear amyloglucans and extent of iodine staining.[21] The wavelength of maximal optical absorbance, the maximal absorbance, and the iodine-binding capacity per chain are all linearly related to the chain length.

Resonance Raman measurements have been conducted with solutions of synthetic amyloses (dp 25—200) complexed with iodine.[22] Findings are incompatible with I_3^- units as the only bound species, but they are compatible with I_3^-/I_2 and I_5^- sub-units under certain conditions.

C.d. measurements on amylose–iodine solutions carried out at different degrees of iodine saturation show that the decrease of the Cotton effect (dp 50) is mainly due to the decreasing complex formation constants.[23] Studies with aggregating solutions on the one hand and with very dilute solutions on the other hand, indicate

---

[15] C. G. Biliaderis, D. R. Grant, and J. R. Vose, *Cereal Chem.*, 1981, **58**, 502.
[16] H. Taniguchi, M. J. Chunng, Y. Maruyama, and M. Nakamura, *J. Jpn. Soc. Starch Sci.*, 1982, **29**, 167.
[17] T. L. Bluhm and P. Zugenmaier, *Carbohydr. Res.*, 1981, **89**, 1.
[18] G. Rappenecker and P. Zugenmaier, *Carbohydr. Res.*, 1981, **89**, 11.
[19] J. Metzger, *Fresenius Z. Anal. Chem.*, 1981, **308**, 29.
[20] T. Handa, H. Yajima, and T. Kajiura, *Biopolymers*, 1980, **19**, 1723.
[21] F. W. Fales, *Biopolymers*, 1980, **19**, 1535.
[22] B. Pfannemüller and G. Ziegast, *Int. J. Biol. Macromol.*, 1982, **4**, 9.
[23] G. Ziegast and B. Pfannemüller, *Int. J. Biol. Macromol.*, 1982, **4**, 419.

that the c.d. intensity is related to a conformational state of helix ordering and not to an ordering by chain folding or regular intermolecular association.

Differential scanning calorimetry, high-resolution ^{1}H n.m.r. spectroscopy, and optical rotation measurements have been used in a study of amylose-fatty acid complexes in starch granules and in solution.[24] Addition of palmitate to amylose results in a loss of conformational mobility by the polysaccharide chain as shown by n.m.r. spectroscopy and the adoption of a novel right-handed V-amylose structure, as shown by optical rotation, rather than the expected left-handed form characterized in the solid state.

The physico-chemical properties of potato and corn starches are affected by linseed oil during γ-irradiation.[25] The presence of the lipid in starch enhances the deterioration of the polysaccharide granules under irradiation as measured by viscosity, molecular size (gel permeation chromatography), and β-amylolysis.

The sorption of water and water-soluble alcohols by starch granules in aqueous suspension has been studied and potato and corn starch shown to absorb 33 and 28% water, respectively, at pH 7.[26] The method was not suitable for cationic starches. Alcohols were also absorbed in large amounts. The kinetics of adsorption of 2,3-dialdehydostarch can be expressed by an exponential equation which is applicable for energetically uniform heterogeneous cellulose surfaces.[27] The activation energy did not depend on the initial concentration of starch in solution but increased linearly with the increased amount of starch adsorbed.

From the temperature behaviour of the proton relaxation times of different starches, it is concluded that the water molecules at the starch structure carry out an anisotropic motion.[28] Exchange process of water protons with OH protons are observed above room temperature. ^{1}H n.m.r. observations have been interpreted to suggest that the blocking of the helices of potato starch with iodine has no influence on the structure of the water.[29] The water in starch is adsorbed interhelically and not intrahelically.

Some physicochemical properties of heat–moisture treatment of starches have been determined.[30] The water-binding capacity and enzyme susceptibility increased while the swelling power decreased. The starches with the highest moisture content before heating had the lowest swelling powers. The same treatments had adverse effects on the functional properties and baking potential of the starches.[31]

The rheological properties of potato starch pastes of small, middle, and large granules have been recorded.[32] During gelatinization of granules in water, paste properties of all three types of starch changed. Dynamic moduli and viscosities decreased, and the patterns of flow curves changed from thixotropic flow to plastic flow.

---

[24] P. V. Bulpin, E. J. Welsh, and E. R. Morris, *Starch*, 1982, **34**, 335.
[25] T. Komiya, T. Yamada, S. Kawakishi, and S. Nara, *J. Jpn. Soc. Starch Sci.*, 1982, **29**, 1.
[26] J. N. BeMiller and G. W. Pratt, *Cereal Chem.*, 1981, **58**, 517.
[27] M. Nedelcheva, E. Valcheva, S. Bencheva, and V. Valchev, *Starch*, 1981, **33**, 195.
[28] H. Lechert and I. Schwier, *Starch*, 1982, **34**, 6.
[29] I. Schwier and H. Lechert, *Starch*, 1982, **34**, 11.
[30] K. Kulp and K. Lorenz, *Cereal Chem.*, 1981, **58**, 46.
[31] K. Lorenz and K. Kulp, *Cereal Chem.*, 1981, **58**, 49.
[32] K. Yamamoto, Y. Sugai, and T. Onogaki, *J. Jpn. Soc. Starch Sci.*, 1982, **29**, 277.

A comparison has been made of the degree of crystallinity in a number of varieties of wheat starch with different swelling capacities.[33] High swelling capacities are associated with relatively disordered arrangements of polymer within granules.

Normal- and waxy-type starches have been isolated from two types of *Amaranthus hypochondriacus* and found to be round granules (1 μm diameter) with X-ray diffraction diagrams similar to those of other types.[34] The amylose contents were 14 and 0%, respectively, while the amylopectins were similar to those from rice and maize starches.

The gelatinization of wheat starch as modified by xanthan, guar, and cellulose gums has been investigated.[35] Each gum hastens the onset of initial paste viscosity. The reaction seems to involve strong association with the amylose component since the β-amylolysis limit of the amylose is reduced. The starch-xanthan gum exhibited a pseudoplastic behaviour whereas the starch-guar gum resisted shearing at low shear rates. The syneresis of curdlan gel has been repressed by the addition of starch before heating but not by other sugars.[36] The role of starch in this process was discussed.

Cellulose.—The solubilization of cellulose and other plant structural carbohydrates has been achieved in a mixture of 4-methylmorpholine *N*-oxide and DMSO.[37] The separation and purification of homogeneous polysaccharide fractions could then be made. The phase behaviour of the quasiternary system *N*-methylmorpholine *N*-oxide–water–cellulose has been studied by differential scanning calorimetry, optical and electron microscopy, and X-ray scattering.[38] *N*-Methylmorpholine *N*-oxide and its hydrate are amongst the best solvents for cellulose.[39] Raman and i.r. spectroscopic studies on the two compounds indicate that for the anhydrous molecule, the dipole–dipole interaction is characterized by a strong line at 75 cm^{-1} in the Raman spectra. The molecules contain a hydrophobic part $[O(CH_2)_2]$ not active in the dissolution of cellulose, and a hydrophilic part which is able to stabilize a chain of water molecules and reveals the importance of the pyramidal environment of water molecules in the monohydrate.

^{1}H n.m.r. spectroscopy has shown that both ramie and cotton have a unique phase structure characteristic of native cellulose.[40] The non-crystalline component did not exhibit micro-Brownian segmental motion even in the swollen state. The phase structure was distinctly different from that of regenerated cellulose. An n.m.r. spectroscopic study of the interactions between cadoxen and cellulose has been carried out using signals from ^{1}H, ^{13}C, and ^{113}Cd.[41] A well-recorded ^{13}C spectrum was obtained which could be assigned by comparison with the spectra of simpler sugars. The ^{113}Cd and ^{13}C results were not consistent with the formation of chelate–alcoholate complexes with the C-2 and C-3 hydroxy groups. Hydrogen

---

[33] R. B. K. Wong and J. Lelievre, *Starch*, 1982, **34**, 159.
[34] Y. Sugimoto, K. Yamada, S. Sakamoto, and H. Fuwa, *Starch*, 1981, **33**, 112.
[35] D. D. Christianson, J. E. Hodge, D. Osborne, and R. W. Detroy, *Cereal Chem.*, 1981, **58**, 513.
[36] K. Ishida and T. Takeuchi, *Agric. Biol. Chem.*, 1981, **45**, 1409.
[37] J. P. Joseleau, G. Chambat, and B. Chumpitazi-Hermoza, *Carbohydr. Res.*, 1981, **90**, 339.
[38] H. Chanzy, S. Nawrot, A. Peguy, P. Smith, and J. Chevalier, *J. Polym. Sci., Polym. Phys. Ed.*, 1982, **20**, 1909.
[39] T. M. Pham, M. H. Herzog-Cance, A. Potier, and J. Potier, *Can. J. Chem.*, 1982, **60**, 2777.
[40] A. Hirai, R. Kitamura, F. Horii, and I. Sakurada, *Cellul. Chem. Technol.*, 1980, **14**, 611.
[41] A. D. Bain, D. R. Eaton, R. A. Hux, and J. P. K. Tong, *Carbohydr. Res.*, 1980, **84**, 1.

bonding is probably the dominant interaction. It is further suggested that the metal ion serves the dual purpose of holding two amino groups in a favourable orientation to hydrogen bond with a pair of equatorial hydroxy groups on the cellulose.

A hydrate of cellulose II has been prepared and its crystal structure determined.[42] The unit cell contains disaccharide sections of two chains and an antiparallel arrangement of adjacent chains was assumed. The chains are stacked in the same way as in cellulose II with the water molecules between the stacks, but it was not possible to position the water molecules in a regular arrangement. The treatment of cotton and rayon with 1,2-diaminoethane resulted in the transformation of cellulose IV to cellulose III as determined by X-ray analysis.[43] Thus 1,2-diaminoethane—methanol can affect lattice modifications.

Single crystals of cellulose IV_{II} have been obtained by deacetylation and crystallization of cellulose triacetate.[44] Large lamellar crystals were obtained only with unfolded short cellulose chains and the occurrence of chain-folded samples having a high degree of polymerization could not be demonstrated.

The transformation of cellulose I into cellulose III in *Valonia macrophysa*, preformed in 1,2-diaminoethane, has maintained the external appearance of the microfibrils as shown by electron microscopy.[45] This conversion must have involved a fracturing and fibrillation of the initial crystals.

Transmission electron microscopy has indicated that the cross-section of cellulose crystallites from the cell wall of *Valonia ventricosa* is almost square with an average side of 18 nm.[48] The sub-units corresponding to elementary fibrils are not detectable within the crystals. The cellulose component of native pellicles of *Acetobacter xylinum* has a three-dimensional microfibrillar interconnected brushwood structure.[47] It may be produced by a mechanism of spontaneous association and *post facto* crystallization of preformed transient $(1\rightarrow 4)$-β-D-glucans.

The Raman spectra of cellulose gels containing 55% water show a broad band with a maximum at 200 cm^{-1}.[48] This band is not observed in gels with lower water content, suggesting that the local water structure in gels at the higher water concentrations is identical to that in liquid water.

The errors associated with the determination of number average molecular weights through osmotic measurements and of intrinsic viscosities through viscometric methods for solutions of cellulose have been calculated by statistical computations.[49] More precise values than those obtained by graphical extrapolation have been obtained.

Hemicelluloses.—The crystal structure of a regenerated form of $(1\rightarrow 3)$-α-D-glucan has been determined by X-ray diffraction analysis and stereochemical model

---

[42] D. M. Lee and J. Blackwell, *Biopolymers*, 1981, **20**, 2165.
[43] P. K. Chidambareswaran, S. Sreenivasan, N. B. Patil, and H. T. Lokhande, *J. Polym. Sci., Polym. Lett. Ed.*, 1980, **18**, 603.
[44] A. Buleon, H. Chanzy, and P. Froment, *J. Polym. Sci., Polym. Phys. Ed.*, 1982, **20**, 1081.
[45] E. Roche and H. Chanzy, *Int. J. Biol. Macromol.*, 1981, **3**, 201.
[46] J. F. Revol, *Carbohydr. Polym.*, 1982, **2**, 123.
[47] J. R. Colvin, M. Takai, L. C. Sowden, and J. Hayashi, *Int. J. Biol. Macromol.*, 1982, **4**, 244.
[48] O. F. Nielsen, T. Lindstroem, and P. A. Lund, *Acta Chem. Scand., Ser. A*, 1982, **36**, 623.
[49] D. Jadraque and J. Perena, *Angew. Makromol. Chem.*, 1982, **104**, 163.

refinement as an orthorhombic unit cell.[50] The chain conformation is almost completely extended and is very close to a 2/1 helix even though the dimer residue is the crystallographic repeat unit. The vacuum c.d. spectrum of $(1 \rightarrow 6)$-β-D-glucan has a single positive band near 177 nm.[51] As gels formed, a negative band at 190 nm appeared followed by a blue shift in both bands with aging.

The specificity of the interaction of direct dyes with polysaccharides has been studied by changes in solubility and in the fluorescence and absorption spectra of the dyes.[52] The strongest interactions were shown by polysaccharides with contiguous $(1 \rightarrow 4)$-linked β-D-glucopyranosyl residues, but $(1 \rightarrow 3)$-β-D-glucans exhibited some complex formation. Dyes differed in their affinity. In addition to having an affinity for cellulose, Congo Red and Calcofluor, representative of the azo-type and stilbene-type of dyes, undergo complex formation with oat endospermic β-D-glucan.[53]

The vacuum u.v.-c.d. spectra of guar, tara, and carob D-galacto-D-mannans, which had D-galactose contents of 39, 25, and 19%, respectively, have been recorded.[54] A positive band at 169 nm and a negative band at 149 nm were observed. For both bands the relative intensities increased systematically with decreasing D-galactose content for solid films. Some of the differences were attributed to conformational restriction of D-galactose residues by chain packing with a consequent increase in the c.d. intensities from these residues and a cancellation of D-mannan backbone contributions.

High-performance size-exclusion chromatography of guar gum has indicated a molecular weight 2.0×10^6.[55] Viscosity–molecular weight relationships, intrinsic chain flexibility, and dynamic solution properties have been measured for guar D-galacto-D-mannans.[56]

Residual carbohydrates have been removed almost completely from milled wood lignin preparations of birchwood by treating with NaOH–dioxane.[57] Treatment with HCl–dioxane did not decrease the carbohydrate content. Studies in which i.r. and n.m.r. spectroscopies were used indicated that a benzyl ester type of linkage existed between lignin and the carbohydrates. The lignin–carbohydrate complex from the milled wood lignin fraction of *Pinus densiflora* contained three subfractions on gel filtration.[58] Two of the fractions were homogeneous and their compositions were determined. Methylation analysis showed that the carbohydrate moiety exhibited multiple branching, with the major backbone being composed of $(1 \rightarrow 4)$-D-mannan chains. A water-soluble lignin–carbohydrate complex has been released from jute (*Corcharus capsularis*) fibre by reduction with borohydride.[59] Analysis of the D-xylans from the original and borohydride-treated fibres indicated that 34% of the acidic side chains of the D-xylan were linked to lignin by ester bonds. Lignin–carbohydrate complexes have been released from wheat straw by prolonged

[50] K. Ogawa, K. Okamura, and A. Sarko, *Int. J. Biol. Macromol.*, 1981, **3**, 31.
[51] A. J. Stipanovic and E. S. Stevens, *Int. J. Biol. Macromol.*, 1980, **2**, 209.
[52] P. J. Wood, *Carbohydr. Res.*, 1980, **85**, 271.
[53] P. J. Wood, *Carbohydr. Res.*, 1982, **102**, 283.
[54] L. A. Buffington, E. S. Stevens, E. R. Morris, and D. A. Rees, *Int. J. Biol. Macromol.*, 1980, **2**, 199.
[55] H. G. Barth and D. A. Smith, *J. Chromatogr.*, 1981, **206**, 410.
[56] G. Robinson, S. B. Ross-Murphy, and E. R. Morris, *Carbohydr. Res.*, 1982, **107**, 17.
[57] K. Lundquist, R. Simonson, and K. Tingsvik, *Sven. Papperstidn.*, 1980, **83**, 452.
[58] J. I. Azuma, N. Takahashi, and T. Koshijima, *Carbohydr. Res.*, 1981, **93**, 91.
[59] N. N. Das, S. C. Das, A. S. Dutt, and A. Roy, *Carbohydr. Res.*, 1981, **94**, 73.

hot-water extraction.[60] The bound sugars were identified as D-glucose, L-arabinose, D-xylose, and D-galactose residues with uronic acids being absent. The higher-molecular-weight fractions were richer in D-glucose residues with the lower-molecular-weight fractions being richer in D-xylose and L-arabinose residues.

Pectins.—An investigation of oligo- and poly-galacturonic acids has been carried out by potentiometry and c.d.[61] It was shown that no Ca^{2+} ion is ever fixed in excess of stoicheiometry. The c.d. spectra, in agreement with ^{13}C n.m.r. spectroscopic data, suggested that, in dilute solutions, the polymer adopts two conformations.

Vacuum u.v.–c.d. spectra have been reported for poly(D-galacturonic acid) and sodium and calcium poly(D-galacturonate).[62] In addition to a previously observed positive c.d. band near 208 nm, a pair of high-energy bands at 170—180 nm and 145 nm have been detected. The low-energy band, assigned to a carboxyl transition, is blue-shifted upon gelation or film formation.

The gelation of pectin under conditions of low water activity has been investigated by c.d., competitive inhibition, and mechanical properties.[63] The optimum degree of esterification for pectin gelation under conditions of low water activity is 70%. The reduction in gel strength at lower degrees of esterification is not merely due to increased charge density but also results from the loss of a significant contribution which the ester group makes to the stability of the interchain junctions. Gel permeation chromatography, c.d., and viscosity measurements showed that appreciable aggregations of aqueous solutions of pectin occur at low concentrations both in pectins of low and high ester content.[64] The aggregation is independent of divalent cations. The Arrhenius slopes were greater at lower temperatures and they increase with conditions such as pH, ionic strength, concentration and degree of esterification that are known to favour interchain association of pectin. The c.d. spectra at various pH values and in the presence of additives indicated that two modes of interchain association contribute to the network formation in calcium pectate gels.[65] The first is by interchain chelation of Ca^{2+} ions and the second is by non-ionic interchain associations analogous to those in pectin gels with low water-activity. The contributions of the latter type increases on lowering the pH, as the Ca^{2+} ion binding capacity is diminished and the interchain electrostatic repulsions are minimized. The mechanical properties and thermal stability of a gel is a function of the proportion of each type.

A relationship has been sought between the ability of various ions to inhibit growth and their ability to cause gelation of isolated pectins, their binding affinity for isolated cell walls and their binding affinity for purified pectin.[66] A good correspondence was found with the second factor. Pectic gel formation was not involved in cation-induced growth inhibition. The previously determined activity

---

[60] Y. Vered, O. Milstein, H. M. Flowers, and J. Gressel, *Eur. J. Appl. Microbiol. Biotechnol.*, 1981, **12**, 183.
[61] G. Ravanat and M. Rinaudo, *Biopolymers*, 1980, **19**, 2209.
[62] J. N. Liang and E. S. Stevens, *Int. J. Biol. Macromol.*, 1982, **4**, 316.
[63] E. R. Morris, M. J. Gidley, E. J. Murray, D. A. Powell, and D. A. Rees, *Int. J. Biol. Macromol.*, 1980, **2**, 327.
[64] M. A. F. Davis, M. J. Gidley, E. R. Morris, D. A. Powell, and D. A. Rees, *Int. J. Biol. Macromol.*, 1980, **2**, 330.
[65] M. J. Gidley, E. R. Morris, E. J. Murray, D. A. Powell, and D. A. Rees, *Int. J. Biol. Macromol.*, 1980, **2**, 332.
[66] M. Tepfer and I. E. P. Taylor, *Can. J. Bot.*, 1981, **59**, 1522.

coefficients of Ca^{2+} ion in oligo-D-galacturonates were plotted and extrapolated to infinity.[67] In molecular disperse solutions the binding of Ca^{2+} ions was electrostatic, whereas in aggregated molecules a chelate intermolecular binding also took place. The non-reducing terminal uronic acid residues bind Ca^{2+} ions less firmly than do the inner units of the chain.

Interchain associations of alginate and pectins have been studied.[68] Unexpected thickening and gelling behaviour at low pH has been observed for mixtures of the two polymers. Mixed interchain associations at a molecular level, rather than non-specific polymer incompatibility or exclusion effects, have been suggested as an explanation of this phenomenon. A molecular mechanism for the association is proposed, and compared with known structures in the condensed phase.

Algal Polysaccharides.—A method has been presented for the determination of the backbone conformations of polymer chains conforming to a given helical type.[69] The method is illustrated by its application to agarose.

Based on chemical shifts and C–H coupling constants from model molecules, the structural conformation of aqueous agarose solutions and gels has been determined from the ^{13}C n.m.r. spectra.[70] Similar studies on the agaroses of *Pterocladia* species and *Gracilaria secundata* have been reported and all signals in the spectra were assigned to various D-galactose derivatives and agaro-oligosaccharides.[71]

The high-resolution ^{13}C n.m.r. spectra of slightly depolymerized alginates have been interpreted.[72] The sequence of the monomer units L-guluronate (G) and D-mannuronate (M) markedly influence the chemical shifts. Some of the individual carbon resonances were resolved into four lines, which was evidence for a dependence on the identities of the units immediately before and after them in the polymer chain. The relative intensities of the signals permitted a rapid computation of the monomeric composition (M/G ratio), the composition of end units (M/G ratio), and the monomeric sequence in terms of a complete set of four diad and eight triad frequencies. Any region with a strictly alternating M and G sequence was very short.

Interactions of alginate with univalent cations in solution have been investigated by c.d. and rheological measurements.[73] Three modes of interaction are proposed: ion-pair formation with carboxyl groups of D-mannuronate and L-guluronate residues; specific site binding of contiguous L-guluronate residues, and co-operative 'egg-box' binding, particularly of Na^+, between poly-L-guluronate chain sequences.

The cation-specific vacuum c.d. spectra of alginate solutions, gels, and solid films have been recorded.[74] A band at 185 nm has been assigned to carboxy groups while those at 169 and 149 nm were assigned to changes in the polymer backbone.

---

[67] R. Kohn and A. Malovikova, *Collect. Czech. Chem. Commun.*, 1981, **46**, 1701.
[68] D. Thom, I. C. M. Dea, E. R. Morris, and D. A. Powell, *Prog. Food Nutr. Sci.*, 1982, **6**, 97.
[69] S. A. Foord and E. D. T. Atkins, *Int. J. Biol. Macromol.*, 1980, **2**, 193.
[70] F. M. Nicolaisen, I. Meyland, and K. Schaumburg, *Acta Chem. Scand., Ser. B*, 1980, **34**, 103.
[71] D. J. Brasch, C. T. Chuah, and L. D. Melton, *Aust. J. Chem.*, 1981, **34**, 1095.
[72] H. Grasdalen, B. Larsen, and O. Smidsrød, *Carbohydr. Res.*, 1981, **89**, 179.
[73] R. Searle, E. R. Morris, and D. A. Rees, *Carbohydr. Res.*, 1982, **110**, 101.
[74] J. N. Liang, E. S. Stevens, S. A. Frangou, E. R. Morris, and D. A. Rees, *Int. J. Biol. Macromol.*, 1980, **2**, 204.

The changes in chain conformation of inter-junction sequences in calcium alginate gels has been studied using c.d. and o.r.d.[75] Changes observed in the backbone optical activity calculated for poly-D-mannuronate and heteropolymeric chain sequences are explained in terms of stretching of interconnecting sequences between calcium poly-L-guluronate junctions in alginate gels to give a more extended chain conformation than in free solution.

ι-Carrageenan has shown substantial helix formation in the presence of Li^+, Na^+, and Me_4N^+ ions as measured by optical rotation, but the polymer does not gel.[76] Under identical conditions, the polymer does gel with K^+, Rb^+, Cs^+, and NH_4^+ ions. The gel formation is in three stages. Initially the individual coils form double helices, and then the double helices form domains which finally form aggregates around the heavier group I ions.

Rheological studies show that Ca^{2+}, K^+, and Na^+, in decreasing order of effectiveness, stimulate gel formation by ι-carrageenan.[77] Discrete melting of junction zones between chains was observed at different temperatures. N.m.r. studies show that K^+ ions interact strongly and Na^+ ions interact weakly with the polysaccharide network.

Measurements of rigidity moduli and ^{39}K and ^{23}Na n.m.r. spectra have been used to investigate the roles played by Ca^{2+}, K^+, and Na^+ ions in ι-carrageenan gels.[78] Differences in binding were reflected in the rheological properties of the gels. Potassium ι-carrageenate gels have been prepared and studied by photon correlation spectroscopy at 633 nm.[79] The molecular and vibrational motions of the gels were interpreted in terms of their elastic properties. Below the sol–gel transition temperature, oscillatory correlation functions were found with the frequency remaining constant for two weeks. Measurements of the shear modulus for Ca^{2+}, K^+, and Na^+ ι-carrageenate gels as a function of biopolymer concentration and temperature showed that the Ca^{2+} salt appears to possess a random glass-like structure whereas the K^+ and Na^+ forms show a pseudocrystalline structure.[80] The differences were tentatively attributed to differences in the solubility of the salt forms which, through controlling the gelation temperature, moderated the kinetics of the gelation process.

A detailed assignment of the ^{13}C chemical shifts of κ- and ι-carrageenan in their Na^+ and K^+ forms has been given.[81] Evidence for the conformational transition induced by temperature variation in the absence of any gel formation of κ-carrageenan is also presented. The evidence is based on both n.m.r. and optical rotation experiments.

Photon correlation spectroscopy can be used for characterizing isolated macromolecules by probing the translational and rotational diffusion of dilute suspensions of biopolymers.[82] Rheological parameters for carrageenan have been obtained with this method.

---

[75] E. R. Morris, D. A. Rees, and G. Young, *Carbohydr. Res.*, 1982, **108**, 181.
[76] G. Robinson, E. R. Morris, and D. A. Rees, *J. Chem. Soc., Chem. Commun.*, 1980, 152.
[77] V. J. Morris and P. S. Belton, *Prog. Food Nutr. Sci.*, 1982, **6**, 55.
[78] V. J. Morris and P. S. Belton, *J. Chem. Soc., Chem. Commun.*, 1980, 983.
[79] V. J. Morris and K. S. Fancey, *Int. J. Biol. Macromol.*, 1981, **3**, 213.
[80] V. J. Morris and G. R. Chilvers, *J. Sci. Food Agric.*, 1981, **32**, 1235.
[81] C. Rochas, M. Rinaudo, and M. Vincendon, *Biopolymers*, 1980, **19**, 2165.
[82] V. J. Morris, *Prog. Food Nutr. Sci.*, 1982, **6**, 119.

The amplitude of ultrasonic relaxation in aqueous solution of disordered polysaccharides, λ-carrageenan, locust bean gum, and xantham gum show a marked increase with increasing degree of coil overlap and, at low concentrations, attain values comparable with those observed in polysaccharide gels.[83] On the ultrasonic time scale, dynamic networks formed by polymer entanglement in solutions are indistinguishable from true gels. The observed intense ultrasonic relaxations were attributed predominantly to the motion of solvent within the polymer network and due to the inherent stiffness of most polysaccharides. The formation of a highly entangled network structure occurred at lower concentrations than for typical synthetic polymers.

The optical rotation and the conductivity of κ-carrageenans in aqueous solutions have been investigated as functions of temperature in the presence of various electrolytes.[84] The activity coefficients of Na^+ and K^+ ions have been determined and correlated with the conformation. The K^+ activity coefficient under ordered conformation is in agreement with a mechanism involving dimerization.

Stress relaxation and dynamic measurements on κ-carrageenan have shown that the elastic modulus increases in the presence of Li^+, Na^+, K^+, and Cs^+.[85] The electrostatic repulsion of sulphate groups in the carrageenan may be prevented by a shielding action of the alkali-metal ions which stabilizes the double helical structure of the gels. De-esterification of κ-carrageenan increases the elastic modulus of the gels.[86] This has been attributed to the increase of alternation of D-galactosyl and 3,6-anhydro-D-galactosyl residues, and then the stabilization of the double helical structure in the gel.

The dynamic Young's modulus of κ-carrageenan gels passes through a maximum at (0.2–0.3 M) LiCl, NaCl, KCl, or CsCl.[87] Further addition of the salts decreases the modulus sharply. The effect of the salts is only slight in agarose.

Measurements have been made of the shear modulus as a function of biopolymer concentration for pure potassium κ-carrageenate gels.[88] The results have been discussed on the rubber elasticity theory to investigate the 'free polymer' linkages between junction zones within the gels. Pulsed electric birefringence studies have been made on segmented κ-carrageenan as a model for the junction zones of the gel.

The enthalpy changes at the dilution and mixing of solutions of anionic polysaccharides, such as κ-carrageenan, are independent of the nature of the anionic group but have marked dependence on the nature of the cation.[89] For a given cation the enthalpy changes are temperature dependent. These effects are interpreted in terms of the interaction between the cation and the bulk solvent. The rheological behaviour of the hydrocolloid solutions is influenced by the presence of other solutes, which results in a more restricted hydration of the colloid. As a consequence the colloid molecules have a more limited extension and are less able to develop intermolecular linkages.

---

[83] M. C. Pereira, E. Wyn-Jones, and E. R. Morris, *Carbohydr. Polymers*, 1982, **2**, 103.
[84] C. Rochas and M. Rinaudo, *Biopolymers*, 1980, **19**, 1675.
[85] M. Watase and K. Nishinari, *J. Texture Stud.*, 1981, **12**, 427.
[86] M. Watase and K. Nishinari, *J. Texture Stud.*, 1981, **12**, 447.
[87] M. Watase and K. Nishinari, *Rheol. Acta*, 1982, **21**, 318.
[88] V. J. Morris, *Int. J. Biol. Macromol.*, 1982, **4**, 155.
[89] P. W. Hales, M. Jeffries, and G. Pass, *Prog. Food Nutr. Sci.*, 1982, **6**, 33.

3 Lectins

Review articles dealing with lectins include a study of lectins in higher plants.[90] The endogenous biological functions of lectins have been reviewed,[91] as have the interactions of lectins with bacteria and fungi.[92] Interactions as a means of distinguishing between microbial species and the identification of a particular sugar residue have been discussed. Affinity electrophoresis has been used to study the effect of detergents on the interactions of lectins with carbohydrates.[93]

Glycopeptides and oligosaccharides derived from N-glycosyl proteins have been used to define the specificity of 12 different lectins.[94] Lectins considered identical in terms of monosaccharide specificity possess an ability to recognize fine differences in more complex structures. The observation that glycoasparagines, glycopeptides, and glycoproteins possess a higher affinity for lectins than have the related oligosaccharides has been explained by the fact that the glycan–amino acid linkage leads to structures more rigid than those of the oligosaccharides themselves. Many of the lectins used in this study seem to possess in or near their carbohydrate-binding site a hydrophobic area such as that described for concanavalin A, and this could be the cause of non-specific hydrophobic interactions between the lectins and residues of glycopeptides or glycoproteins, as described for concanavalin A.

Photoactivatable, iodinated glycopeptides bearing oligosaccharides of defined structure have been synthesized for use as lectin binding site-specific reagents.[95] Two such glycopeptides have been examined utilizing concanavalin A, RCA_I (*Ricinus communis* agglutinin), RCA_{II} (*R. communis* toxin), and the D-galactose-specific lectin from human and rat hepatocytes. Covalent incorporation upon photoactivation only occurs with a glycopeptide which is specifically bound by the lectin and is inhibited only by the haptenic monosaccharide.

Intermolecular forces in lectin–carbohydrate interaction have been analysed on the basis of their structure and chemical features.[96] The role of water, as well as that of other physico-chemical parameters influencing the formation of such complexes in aqueous media, is also discussed. A model depicting the importance of hydrogen-bonding and charge transfer interactions as the main sources of complex stability in the association between lectins and carbohydrates is proposed.

By performing electrophoresis perpendicular to a stationary pH gradient in polyacrylamide gels containing a specific ligand either covalently fixed or entrapped in the gel matrix, it is possible to measure dissociation constants of lectins and their pH dependence in a pH range.[97] The method has been used to study the lectins from *Ricinus communis* and *Lens culinaris*. The lectins listed in Table 1 have been purified and some of their physical and chemical characteristics recorded.

---

[90] H. Lis and N. Sharon, in 'Lectins in Higher Plants; Biochemistry of Plants', ed. A. Marcus, Academic Press, New York, 1981, Vol. 6.
[91] S. H. Barondes, *Ann. Rev. Biochem.*, 1981, **50**, 207.
[92] T. G. Pistole, *Ann. Rev. Microbiol.*, 1981, **35**, 85.
[93] L. Pechova, M. Ticha, and J. Kocourek, *J. Chromatogr.*, 1982, **240**, 43.
[94] M. I. Khan, N. Surolia, M. K. Mathew, P. Balaram, and A. Surolia, *Eur. J. Biochem.*, 1981, **115**, 149.
[95] J. U. Baenziger and D. Fiete, *J. Biol. Chem.*, 1982, **257**, 4421.
[96] J. L. Ochoa, *J. Chromatogr.*, 1981, **215**, 351.
[97] K. Ek, E. Gianazza, and P. G. Righetti, *Biochim. Biophys. Acta*, 1980, **626**, 356.

Table 1 Composition and properties of some purified lectins

Source	Mol. wt.	Sub-units	Carbohydrate % Neutral	Carbohydrate % Amino sugar	Human blood type	Sugar specificity	Ref.
Plants							
Datura stramonium	8.6×10^4	2	37			Chitooligosaccharides	98
Erythrina arborescens	2.8×10^4		5.7		ABO	α-D-Gal*p*; α-D-Gal*p*NAc	99
Erythrina cristagalli	5.6×10^5	2	4.5	1.4	ABO	D-Gal; β-D-Gal*p*-(1 → 4)-β-D-GlcNAc	100
Erythrina indica	6.8×10^4	2	9.0	1.5	ABO	β-D-Gal*p*; D-Gal*p*NAc	99
Erythrina lithasperma	5.4×10^4	2	4.1		ABO	α-D-Gal*p*; D-Gal*p*NAc	99
Erythrina suberosa	2.8×10^4		6.8		ABO	D-Gal*p*NAc	99
Erythrina variegata			9–12		ABO	D-Gal*p*; D-Gal*p*NAc	101
Griffonia simplicifolia	5.8×10^4	2	3.1	0.8	Le^b	Le^b-oligosaccharides	102
Momordica charantia	1.2×10^5	4	10	2.5		β-D-Gal*p*	103
Psophocarpus tetragonolobus	4.1×10^4				ABO	D-Gal*p*NAc	104
Ulex europaeus II	1.05×10^5	4			H(O)	Chitotriose	105
Vicia cracca	4.4×10^4	4			ABO	D-Man*p*, D-Glc*p*	106
Viscum album	1.15×10^5	4			ABO	D-Gal*p*	107
	6.0×10^4	2			ABO	D-Gal*p*; D-Gal*p*NAc	107
	5.0×10^4	2			ABO	D-Gal*p*NAc	107
Animal							
Electrophorus electricus	3.5×10^4	2	2.2			β-D-Gal*p*	108
Limax flavus	4.4×10^4	2				Neup5Ac; Neup5G	109
Xenopus laevis	4.8×10^5		15	4.8		α,β-D-Gal*p*	110

The screening of peanut (*Arachis*) and its wild relatives for multiple molecular forms of peanut lectin and for genotypes devoid of the lectin has been reported.[111] A peanut lectin-negative phenotype was detected in four genotypes (0.1% of those examined). Lectin preparations from lectin-containing phenotypes were resolved into a small number of defined and clearly related isolectin profiles. Association constants and thermodynamic parameters for the binding of sugars to peanut agglutinin by u.v. difference spectroscopy have been reported.[112] The binding kinetics of methyl α- and methyl β-D-galactopyranosides to the lectin have been studied by ^{13}C n.m.r. spectroscopy employing methyl D-galactopyranosides specifically enriched in ^{13}C at C-1.[113] A two-step binding model is presented. Peanut agglutinin exhibits cryoinsolubility, which is partially reversible and totally inhibited in the presence of D-galactosides.[114] The efficacy of sugars as inhibitors of cryoinsolubility is related to their affinity for the lectin. Charge–charge interactions are of little importance, but hydrogen bonds and/or van der Waals interactions are most probably responsible for the formation of cryoprecipitates. The effects of chemical modification on the conformation and biological activity of peanut agglutinin have been recorded.[115] When free amino groups were modified with succinic anhydride and 1-isothiocyanate-4-benzenesulphonic acid, the derivatives retain their sugar-binding capacity, although the agglutinating activity with neuraminidase-treated erythrocytes and various tumour cells is reduced. When L-tyrosine residues are modified with tetranitromethane and further with 4-aminophenyl α-D-glucopyranoside and with 2-(4-aminobenzyl)-α-D-neuraminic acid, the agglutinating and mitogenic activities are not markedly altered. The influence of these modifications on the conformation of the lectin was examined by c.d. spectral studies.

The binding of methyl β-D-lactoside, specifically ^{13}C-labelled at C-1 of the D-galactosyl residue, to peanut agglutinin has been reported.[116] ^{13}C n.m.r. studies together with reaction rate constants and activation parameters reveal differences for the binding of D-galactose and the disaccharide, suggesting an extended binding site on peanut agglutinin.

The distribution of the five tetrameric *Griffonia simplicifolia* I isolectins (A_4, A_3B, A_2B_2, AB_3, B_4) isolated from seeds of different trees, has been analysed.[117] The

[98] J. F. Crowley and I. J. Goldstein, *FEBS Lett.*, 1981, **130**, 149.
[99] L. Bhattacharyya, P. K. Das, and A. Sen, *Arch. Biochem. Biophys.*, 1981, **211**, 459.
[100] J. L. Iglesias, H. Lis, and N. Sharon, *Eur. J. Biochem.*, 1982, **123**, 247.
[101] T. K. Datta and P. S. Basu, *Biochem. J.*, 1981, **197**, 751.
[102] S. Shibata, I. J. Goldstein, and D. A. Baker, *J. Biol. Chem.*, 1982, **257**, 9324.
[103] T. Mazumber, N. Guar, and A. Surolia, *Eur. J. Biochem.*, 1982, **113**, 463.
[104] P. S. Appokattan and D. Basu, *Anal. Biochem.*, 1981, **113**, 253.
[105] Y. Konami, T. Tsuji, I. Matsumoto, and T. Osawa, *Hoppe Seyler's Z. Physiol. Chem.*, 1981, **362**, 983.
[106] C. M. Baumann, A. D. Strosberg, and H. Rüdiger, *Eur. J. Biochem.*, 1982, **122**, 105.
[107] H. Franz, P. Ziska, and A. Kindt, *Biochem. J.*, 1981, **195**, 481.
[108] G. Levi and V. L. Teichberg, *J. Biol. Chem.*, 1981, **256**, 5735.
[109] R. L. Miller, J. F. Collawn, and W. W. Fish, *J. Biol. Chem.*, 1982, **257**, 7574.
[110] M. M. Roberson and S. H. Barondes, *J. Biol. Chem.*, 1982, **257**, 7520.
[111] S. G. Pueppke, *Arch. Biochem. Biophys.*, 1981, **212**, 254.
[112] K. J. Neurohr, N. M. Young, and H. H. Mantsch, *J. Biol. Chem.*, 1980, **255**, 9205.
[113] K. J. Neurohr, N. M. Young, I. C. P. Smith, and H. H. Mantsch, *Biochemistry*, 1981, **20**, 3499.
[114] M. Decastel, R. Rourrillon, and J. P. Frénoy, *J. Biol. Chem.*, 1981, **256**, 9003.
[115] D. Nonnenmacher and R. Brossmer, *Biochim. Biophys. Acta*, 1981, **668**, 149.
[116] K. J. Neurohr, H. M. Mantsch, N. M. Young, and D. R. Bundle, *Biochemistry*, 1982, **21**, 498.
[117] J. E. Lamb, F. L. Bookstein, I. J. Goldstein, and L. E. Newton, *J. Biol. Chem.*, 1981, **256**, 5874.

amount of each isolectin varies from seed to seed, each seed displaying a unique distribution of the five isolectins. It is postulated that the intact *G. simplicifolia* I B_4 isolectin undergoes proteolytic cleavage by an *endo-* or *exo-*protease to form first the AB_3 isolectin, which in turn is then processed further to yield the A_2B_2 isolectin, and so on until A_4 is produced. 4-Methylumbelliferyl glycosides have been used in binding studies with the lectins BSI-A_4, BSI-B_4, and BS II from *G. simplicifolia*.[118] Equilibrium dialysis, fluorescent enhancement, quantitative precipitation, and hapten inhibition techniques have been used to investigate the binding characteristics of four of the *G. simplicifolia* I isolectins.[119] One of the most unusual features is that, although distinctly different, both the A and B sub-units have the same affinity for α-D-galactosyl residues. Although both sub-units also bind 2-acetamido-2-deoxy-D-galactosyl residues, the A sub-unit has an association constant for the amino-sugar more than 1000-fold greater than the B sub-unit.

Affinity chromatography of divalent concanavalin A on immobilized 4-aminophenyl β-D-glucose has been analysed by means of equations based on the simplest model, in which the lectin has equivalent and independent binding sites and one lectin molecule binds to only one immobilized ligand at a time.[120]

Over wide ranges of temperature and pH, concanavalin A consists of only dimers and tetramers.[121] The large fraction of hydrolysed sub-units in commercial preparations causes significant populations of dimeric species that associate only weakly or not at all. The effect of the binding of saccharide ligands on the reversible dimer–tetramer equilibrium of the lectin has been studies by high-speed sedimentation equilibrium.[122] Contrary to earlier published work, saccharide binding does not appear to affect the dimer–tetramer equilibrium of the native concanavalin A, although it does introduce irreversible behaviour in preparations containing significant proportions of hydrolysed sub-units. The dimer–tetramer equilibrium is linked to the weak binding of calcium to a site probably in the dimer–dimer interface. Studies have been reported of the interaction of concanavalin A with 4-nitrophenyl 2-*O*-α-D-glucopyranosyl-α-D-mannopyranoside, 4-nitrophenyl-2-*O*-α-D-galactopyranosyl-α-D-mannopyranoside, and 4-nitrophenyl maltoside.[123] The existence of two binding modes for the disaccharides are proposed. Some of the specificity requirements for the interaction of concanavalin A with the non-reducing glycosyl group are characterized, and are summarized in Scheme 1.

Conformational equilibrium studies on demetallized concanavalin A have been

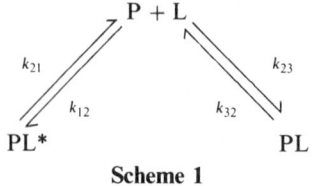

Scheme 1

[118] H. De Boeck, F. G. Loontiens, F. M. Delmotte, and C. K. De Bruyne, *FEBS Lett.*, 1981, **126**, 227.
[119] I. J. Goldstein, D. A. Blake, S. Ebisu, T. J. Williams, and L. A. Murphy, *J. Biol. Chem.*, 1981, **256**, 3890.
[120] Y. Oda, K. I. Kasai, and S. I. Ishii, *J. Biochem. (Tokyo)*, 1981, **89**, 285.
[121] D. F. Senear and D. C. Teller, *Biochemistry*, 1981, **20**, 3076.
[122] D. F. Senear and D. C. Teller, *Biochemistry*, 1981, **20**, 3083.
[123] T. J. Williams, L. D. Homer, J. A. Shafer, I. J. Goldstein, P. J. Garegg, H. Hultberg, T. Iverson, and R. Johansson, *Arch. Biochem. Biophys.*, 1981, **209**, 555.

reported.[124] By monitoring the solvent ^{1}H relaxation dispersion as equimolar concentrations of Mn^{2+} and Ca^{2+} are liberated, at 5 °C, into an apo-concanavalin A solution at 25 °C, 12.5% of the apoprotein is in the 'locked' conformation, corresponding to an energy separation of 1.2 kcal mol^{-1}.

Concanavalin A has been dissociated into monomers in the presence of propan-2-ol, while maintaining its tertiary structure.[125] Limiting studies on the solubility and stability of the lectin in the presence of propan-2-ol indicate that it is stable under dissociating conditions.

Temperature jump relaxation studies with 4-methylumbelliferyl α-D-mannopyranoside as a fluorescence indicator ligand have been used in an examination of the binding kinetics of methyl α-D-mannopyranoside to concanavalin A.[126]

A continuous titration of absorption differences is described for the binding of 4-methylumbelliferyl glycosides to concanavalin A and peanut agglutinin.[127] With both lectins the change in molecular extinction coefficient of the ligand and the association constant, valid for the entire protein saturation range, were obtained. The results compare well with those obtained by other methods, including equilibrium dialysis.

An i.r.-attenuated total reflectance spectroscopic method has been used to study the interaction of monolayers of concanavalin A with mono- and polysaccharides.[128] The effects of the film pressure, pH, urea, Mn^{2+}, and Ca^{2+} on the binding of dextran, methyl α-D-mannopyranoside and D-galactose to the lectin were studied. The thermodynamic parameters of the Mn^{2+}-binding reaction with concanavalin A in the dimer state have been determined by a calorimetric technique.[129] The results indicate that the free energy change for the process is dominated by a positive entropy change and that the two identical S_1 sites of the dimer are independent. Calorimetric studies have also been used to examine the binding of saccharides to concanavalin A.[130] The binding process is induced by both favourable enthalpic and entropic effects.

A fluorescent 1-pyreneacrylate–concanavalin A conjugate has been prepared and characterized.[131] Excitation, emission, and polarization spectra are presented as well as quantum yield and fluorescent lifetimes. Perrin plots examining the rotational behaviour of the pyrene–lectin complex show good agreement with those previously published for other fluorophore–concanavalin A conjugates. The choice of a pyrene fluorophore to label the lectin is for the selection of a probe that can be efficiently excited at 380 nm and can serve as an effective donor to fluorescein–concanavalin A derivatives in resonance energy transfer studies.

The carbohydrate-binding activities of concanavalin A containing various levels of Ca^{2+} and Mn^{2+} have been reported.[132] The precipitation activity of the lectin with glycogen as well as the binding of lectin to 4-nitrophenyl α-D-mannopyranoside is strongly affected by the number of bound Ca^{2+} ions. The kinetics of the

---

[124] R. D. Brown, S. H. Koenig, and C. F. Brewer, *Biochemistry*, 1982, **21**, 465.
[125] A. Sophianopoulos and J. A. Sophianopoulos, *Arch. Biochem. Biophys.*, 1982, **217**, 751.
[126] R. M. Clegg, F. G. Loontiens, A. van Landschoot, and T. M. Jovin, *Biochemistry*, 1981, **20**, 4687.
[127] H. De Boeck, F. G. Loontiens, and C. K. De Bruyne, *Anal. Biochem.*, 1982, **124**, 308.
[128] N. Ockman, *Biochim. Biophys. Acta*, 1981, **643**, 220.
[129] M. Toselli, E. Battistel, F. Manca, and G. Rialdi, *Biochim. Biophys. Acta*, 1981, **667**, 99.
[130] M. Dani, F. Manca, and G. Rialdi, *Biochim. Biophys. Acta*, 1981, **667**, 108.
[131] B. A. Herman and S. M. Fernandez, *Biochemistry*, 1982, **21**, 3271.
[132] F. Obata, R. Sakai, and H. Shiokawa, *J. Biochem. (Tokyo)*, 1981, **89**, 1475.

interactions of Ca^{2+} with concanavalin A in the presence of Co^{2+}, Mn^{2+}, and Zn^{2+} has been followed by measurement of the quenching of fluorescence of 4-methylumbelliferyl α-D-mannopyranoside when bound to protein.[133]

The co-operative binding properties of concanavalin A with glycoconjugates are dependent on the presence of hydrophobic binding sites.[134] The degree of co-operativity can be substantially altered by conformational changes of the ligand.

The structural requirements for the binding of oligosaccharides and glycopeptides to immobilized lentil and pea lectin columns have been studied.[135] An intact 2-acetamido-2-deoxy-β-D-glucopyranosyl residue at the reducing end of a complex oligosaccharide is essential for high affinity binding to lentil lectin but not apparently to concanavalin A. An L-asparagine residue is required for the binding of a complex-type glycopeptide to pea lectin.

Concanavalin A has been used as a non-penetrating probe to study the location of oligosaccharide lipid (D-Glcp_3-D-Manp_9-D-Glc$p$NAc$_2$-pyrophosphoryl dolichol) in microsomal vesicles prepared from cultured fibroblasts.[136] The oligosaccharide-lipid was accessible to the lectin only in leaky vesicles. It is suggested that the bulk of the mature oligosaccharide-lipid pool is on the luminal side of the membrane of the endoplasmic reticulum and that these molecules function as donors in protein glycosylation from this side of the membrane.

Hydrophobic binding to the lectin from lima beans (*Phaseolus lunatus*) has been studied by lectin-induced alterations in the fluorescence and absorption spectra of several hydrophobic ligands.[137] The carbohydrate and hydrophobic binding sites on the lectin are considered to be independent and non-interacting.

Antibodies against pure E_4- and L_4-lectins from *Phaseolus vulgaris* seeds have been made monospecific by immunoaffinity chromatography.[138] Localization of the lectins in bean seeds was then investigated by indirect immunofluorescence and electron microscopy on sections stained with colloidal gold particles coated with anti-E_4- and anti-L_4-immunoglobulin G. In parenchyma cells from the cotyledons, both E- and L-type lectins were located inside the protein bodies, but in vascular and in axis cells the two types of lectins were localized in the cytoplasm, outside the protein bodies. Different roles for the lectins are suggested. In cotyledons this may be a specific form of nitrogen storage, while in vascular and axis cells, lectins may have a more direct metabolic part to play.

Interactions of monomeric amino sugars with wheat germ agglutinin have been studied by ^1H and ^{19}F n.m.r. spectroscopy.[139] N-Acetyl-neuraminic acid and 2-acetamido-2-deoxy-D-glucose have a common binding site on the lectin. Deuterium n.m.r. resonance has been used to delineate the molecular dynamics of sugars bound to lectins.[140] ^2H relaxation times and rotational correlation times for 2-[^2H]-acetamido-2-deoxy-D-3-[^2H]-glucose in the presence and absence of wheat germ agglutinin were studied. The pyranose ring appears to have negligible motional

---

[133] P. C. Harrington, R. Moreno, and R. G. Wilkins, *Isr. J. Chem.*, 1981, **21**, 48.
[134] E. Köttgen, C. Bauer, U. Ehlerding, and W. Gerok, *Biochem. J.*, 1981, **193**, 659.
[135] K. Yamamoto, T. Truji, and T. Osawa, *Carbohydr. Res.*, 1982, **110**, 283.
[136] M. D. Snider and P. W. Robbins, *J. Biol. Chem.*, 1982, **257**, 6796.
[137] D. D. Roberts and I. J. Goldstein, *J. Biol. Chem.*, 1982, **257**, 11274.
[138] J. F. Manen and A. Pusztai, *Planta*, 1982, **155**, 328.
[139] F. Jordan, H. Bahr, J. Patrick, and P. W. K. Woo, *Arch. Biochem. Biophys.*, 1981, **207**, 81.
[140] K. J. Neurohr, N. Lacelle, H. H. Mantsch, and I. C. P. Smith, *Biophys. J.*, 1980, **32**, 931.

freedom relative to the protein on binding. The acetamido-side chain is also immobilized in the binding site, the only motion available being rotation of the C^2H_3 group about its three-fold axis.

The binding of three purified neuraminic acid-containing oligosaccharides to two isolectins of wheat germ agglutinin (WGA I and WGA II) have been quantified by measuring the broadening of a ligand resonance in the ^1H n.m.r. spectrum at 360 MHz.[141] The three ligands (1—3) used in the study represent analogues of the distal portion of L-asparagine-linked glycopeptides. The dissociation constants K_D range from 0.7 to 10 mM. Both isolectins bind the α-2,3-isomer (1) with a higher affinity than the α-2,6-form (2), and WGA I binds oligosaccharides (1, 2) more tightly than does WGA II.

Ricinus communis agglutinin has been subjected to various chemical modifications and the effect on the haemagglutinating and saccharide-binding properties have been studied.[142] At least two L-lysyl residues and an L-tyrosyl residue per average sub-unit are involved in the carbohydrate binding of the agglutinin.

$$\alpha\text{-Neu}p5\text{Ac-}(2 \rightarrow 3)\text{-}\beta\text{-D-Gal}p\text{-}(1 \rightarrow 4)\text{-D-Glc}$$

(1)

$$\alpha\text{-Neu}p5\text{Ac-}(2 \rightarrow 6)\text{-}\beta\text{-D-Gal}p\text{-}(1 \rightarrow 4)\text{-D-Glc}$$

(2)

$$\alpha\text{-Neu}p5\text{Ac-}(2 \rightarrow 6)\text{-}\beta\text{-D-Gal}p\text{-}(1 \rightarrow 4)\text{-D-GlcNAc}$$

(3)

The lectin from *Vicia graminea* binds strongly to blood-group N and Ss human sialoglycoproteins, weakly to the major sialoglycoprotein of horse erythrocyte membranes, and not at all to blood group M sialoglycoprotein.[143] The lectin appears to react only with glycopeptides having clusters of unsubstituted or sialylated 2-acetamido-2-deoxy-3-O-β-D-galactosyl-D-galactose chains located at adjacent amino-acid residues. Reaction of glycopeptides with the lectin is also dependent on other (*e.g.*, amino-acid) residues present at the lectin-binding site, enhancement by hydrophobic residues, and weakened by free amino groups.

A D-mannose/D-glucose binding lectin (M_r 4.4 × 10^4) from the seeds of *Vicia cracca* is composed of two large sub-units (M_r 1.7 × 10^4) and two small sub-units (M_r 5.7 × 10^3).[106] The amino-acid sequence of this lectin is highly homologous to the lectins from four other plants, but there is only a limited homology to the 2-acetamido-2-deoxy-D-galactosyl lectin which is also found in the seeds of *V. cracca*.

The binding of the neuraminic acid-specific lectin, carcinoscorpin, depends not only on neuraminic acid content, but also on the type of glycosidic linkage and form (branched or linear) of the oligosaccharide chains of glycoproteins.[144] The lectin has different classes of binding sites, and binding follows a phenomenon of positive co-operativity.

---

[141] K. A. Kronis and J. P. Carver, *Biochemistry*, 1982, **21**, 3050.
[142] M. I. Khan and A. Surolia, *Eur. J. Biochem.*, 1982, **126**, 495.
[143] M. Duk, E. Lisowska, M. Kordowicz, and K. Wasniowska, *Eur. J. Bioehcm.*, 1982, **123**, 105.
[144] S. Mohan, D. Thambidorai, S. Srimal, and B. K. Bachhawat, *Biochem. J.*, 1982, **203**, 253.

It has been appreciated for some time that lectins are not exclusively found in the plant kingdom. The properties and functions of lectins from animal and microbial sources are now being investigated. Bacterial adherence to surfaces of animal tissues is an important initial event in the pathogenesis of bacterial infection. This adherence has been shown in a number of cases to be mediated by adhesions on the bacterial cell surface exhibiting similar properties to lectins.[145] The adherence of most strains of *Escherichia coli* of human origin is mediated by type-1 fimbriae and is inhibited by D-mannose and methyl α-D-mannopyranoside. The combining site of the fimbrial lectin is an extended one, corresponding to the size of a trisaccharide.[146] It contains a hydrophobic region and is in the form of a pocket on the surface of the lectin. The combining site fits best structures found in short oligo-D-mannose chains present in *N*-glycosidically linked glycoproteins.

Dental caries and periodontal disease are initiated by colonization of *Streptococcus sanguis*, *S. mitis* and Gram-positive filaments onto the tooth surface. Salivary mucins are involved in the clearance of *S. sanguis*, by interaction of a bacterial lectin with the mucin.[147] The lectin, which has been isolated from *S. sanguis*, displays a specificity for an acidic trisaccharide (4) which is the major oligosaccharide of salivary mucin.[148]

$$\alpha\text{-Neu}p5\text{Ac}(2 \rightarrow 3)\text{-}\beta\text{-D-Gal}p\text{-}(1 \rightarrow 3)\text{-D-GalNAc}$$

(4)

4 Microbial Polysaccharides

Teichuronic acid isolated from the cell walls of *Micrococcus luteus* has been examined by natural abundance ^{13}C n.m.r. spectroscopy.[149] ^{1}H decoupled and ^{1}H coupled spectra were obtained for native teichuronic acid and also after the teichuronic acid had been oxidized with periodate and reduced borohydride. The spectra are consistent with structure (5). Teichuronic acid synthesized *in vitro* from suitable substrates by the particulate enzyme fraction obtained from *M. luteus* yielded a ^{13}C n.m.r. spectrum which is indistinguishable from that of the native teichuronic acid, indicating a structural identity of the teichuronic acid synthesized *in vitro* with that isolated from cell walls.

$$\rightarrow 4)\text{-}\beta\text{-D-Man}p\text{NAcA-}(1 \rightarrow 6)\text{-}\alpha\text{-D-Glc}p\text{-}(1 \rightarrow$$

(5)

^{1}H and ^{13}C n.m.r. has been shown to distinguish easily covalently and ionically associated D-alanine in lipoteichoic acids, with ^{13}C n.m.r. showing regular 1,3-linkages in the polyglycerophosphate chain and assignable resonances for those sub-units carrying D-alanine and sugar residues.[150]

---

[145] N. Sharon, Y. Eshdat, F. J. Silverblatt, and I. Ofek, *CIBA Symp. Ser.*, 1981, **80**, 119.
[146] N. Firon, I. Ofek, and N. Sharon, *Biochem. Biophys. Res. Commun.*, 1982, **105**, 1426.
[147] L. A. Tabak, M. J. Levine, I. D. Mandel, and S. A. Ellison, *J. Oral Path.*, 1982, **11**, 1.
[148] P. A. Murray, M. J. Levine, L. A. Tabak, and M. S. Reddy, *Biochem. Biophys. Res. Commun.*, 1982, **106**, 390.
[149] S. D. Johnson, K. P. Lacher, and J. S. Anderson, *Biochemistry*, 1981, **20**, 4781.
[150] M. Batley, K. Knox, J. Redmond, and A. Wicken, *Proc. Aust. Biochem. Soc.*, 1981, **14**, 110.

Two carbohydrate cell-surface antigens have been extracted from *Clostridium difficile*.[151] One, isolated by alkaline extraction of cell walls, contained D-glucose, D-mannose, 2-amino-2-deoxy-D-galactose, and phosphate (2.0:1.0:1.6:0.04). Both antigens showed partial immunological identity and both cross-reacted with *Clostridium sordellii* antiserum. They are analogues of the wall and membrane techoic acids of other Gram-positive bacteria.

The proportions of cell wall teichoic and teichuronic acids present in cultures of *Bacillus subtilis* W23, which are growing under balanced conditions, alter progressively, rather than abruptly, according to phosphate supply.[152] This alteration in composition does not reflect the presence of varied proportions of bacteria that contain either teichoic acid or teichuronic acid exclusively, but both polymers are present simultaneously in the individual bacteria.

Enzymes involved in the synthesis of teichoic acid and its linkage to the wall in *Bacillus subtilis* W23 have been measured during growth of the organism in different concentrations of inorganic phosphate.[153] All the enzymes, except teichoic acid D-glycosyltransferase, which is insensitive to changes in phosphate concentration, were undetected at 0.5 mM phosphate. At higher phosphate concentrations, the changes in the activity of the enzymes of linkage unit synthesis were sufficient to account for changes in the rate of incorporation of teichoic acid into the wall *in vivo*.

The conformational energies of complexes of alternating copolymers of 2-acetamido-2-deoxy-D-glucose and *N*-acetylmuramic acid with hen egg white lysozyme have been computed.[154] This involved a complete search of the conformational space at the active site of the enzyme available to these substrates, and minimization of the conformational energies of the non-covalent complexes. As with the homopolymer (D-Glc$p$NAc)$_6$, the hexasaccharide (D-Glc$p$NAc-MurNAc)$_2$-(D-GlcNAc)$_2$ binds preferentially on the 'left' side of the active-site cleft, involving residues such as Arg-45, Asn-46, and Thr-47 of the lysozyme molecule. The alternating copolymer (D-Glc$p$NAc-MurNAc)$_3$, however, binds with its F-site residue preferentially on the 'right' side of the active-site cleft, involving residues such as Phe-34 and Arg-114. The lactic acid side chain prevents good binding to the F site on the left side. This result can explain the higher rate of catalysis for the cell wall substrate (the alternating copolymer). The relative affinities of the disaccharide D-Glc$p$NAc-MurNAc for all sequential pairs of sites A—F (including E and F sites on both sides of the cleft) has been determined.

The solid-state conformational analysis of Ac-D-Ala-D-Ala-OH.H$_2$O, carried out by i.r. absorption and *X*-ray diffraction, has indicated that the molecules are not extended in a regular conformation, but rather that they are partially folded, the ϕ and ψ torsional angles of the carboxyl-terminal residue in particular being in the region of the left-handed helix of the Ramachandran map.[155] The acetylamino and peptide groups are found in the usual *trans* conformation, the latter exhibiting a deviation from rigid planarity. Only intermolecular hydrogen bonds occur in the crystal state. The solution conformational analysis, performed by i.r. absorption

---

[151] I. R. Poxton and T. D. I. Cartmill, *J. Gen. Microbiol.*, 1982, **128**, 1365.
[152] W. K. Lang, K. Glassey, and A. R. Archibald, *J. Bacteriol.*, 1982, **151**, 367.
[153] S. C. Cheah, H. Hussey, I. Hancock, and J. Baddiley, *J. Gen. Microbiol.*, 1982, **128**, 593.
[154] M. R. Pincus and H. A. Scheraga, *Biochemistry*, 1981, **20**, 3960.
[155] E. Benedetti, B. Di Blasio, V. Pavone, C. Pedone, C. Toniolo, and G. M. Bonora, *J. Biol. Chem.*, 1981, **256**, 9229.

and c.d. spectroscopy, has revealed that the amount of intramolecular N–H ··· O C hydrogen-bonded folded forms, if any, should be extremely small, even in deuteriochloroform at high dilution. In water, solvated unordered species largely predominate.

The potential of i.r. spectroscopy for bacterial cell wall analytical studies has been investigated.[156] All peptidoglycans investigated exhibited very similar i.r. spectra. Analysis of band half-widths revealed no high crystalline state of order of peptidoglycan, and regular conformations like α-helices or β-pleated sheets are excluded. However, several distinctive, fingerprint-like spectral features of various peptidoglycan samples have allowed a facile identification of individual peptidoglycans. Comparative analyses of amide band positions and band half-widths indicate substantial differences between the i.r. spectra of peptidoglycan and chitin, thus refuting previous models based on the assumption of a nearly crystalline chitin-like structure of the glycan chains of peptidoglycans.

When lipopolysaccharides from *Escherichia coli* B were sonicated together with pure spin-labelled phospholipids without the addition of unlabelled phospholipids, extensive line broadening was observed due to the close proximity of spin-labelled molecules to each other, a result suggesting that spin-labelled phospholipids existed in segregated domains containing few lipopolysaccharide molecules.[157] The idea that phospholipid (and most probably lipopolysaccharide) domains in mixed bilayers tend to be rather stable and persist for long periods of time has been suggested.

The mobility of membrane components, particularly lipopolysaccharide, is essential for biogenesis of the outer membrane and is a primary event in phage infection. To define the fluid dynamic properties of the outer membrane as related to function more accurately, the capability of measuring lateral diffusion coefficients *in vivo* of rhodaminated G30 lipopolysaccharide fused into *Salmonella typhimurium* G30A filamentous bacteria has been developed.[158] The method used extends the FRAP (fluorescence redistribution after photobleaching) procedure to bacteria, and the results demonstrate rapid diffusion of lipopolysaccharide ($D = 2.0 \pm 0.9 \times 10^{-10}$ cm^2 s^{-1}) over micrometre distances.

Purified lipopolysaccharide preparations from *Escherichia coli* K12 have been studied by ^{1}H–^{32}P n.m.r. double resonance methods.[159] At a number of temperatures studied, the overall rotational mobility of a sizeable fraction of the P nuclei was restricted on the time scale of the n.m.r. measurements. The studies provide a numerical estimate of the percentage of lipopolysaccharide that is motionally restricted at a given temperature.

The surface polysaccharides of four independent clinical isolates of *Escherichia coli* group O111 have been examined.[160] In three of these strains, 50% of the O-antigen is not covalently linked to the lipid A core oligosaccharide characteristic of lipopolysaccharides. This fraction of O-antigen is not a cytoplasmic component, nor is it a precursor of lipopolysaccharide as was previously believed.

[156] D. Nanmann, G. Barnickel, H. Bradaczek, H. Labischinski, and P. Giesbrecht, *Eur. J. Biochem.*, 1982, **125**, 505.
[157] Y. Takeuchi and H. Nikaido, *Biochemistry*, 1981, **20**, 523.
[158] M. Schindler, M. J. Osborn, and D. E. Koppel, *Nature (London)*, 1980, **285**, 261.
[159] W. Egan and L. Leive, *Biochim. Biophys. Acta*, 1982, **692**, 165.
[160] R. C. Goldman, D. White, D. Ørskov, I. Ørskov, P. D. Rick, M. S. Lewis, A. K. Bhattacharjee, and L. Leive, *J. Bacteriol.*, 1982, **151**, 1210.

The concentration (c) and shear rate (γ) dependence of viscosity (η) has been studied for a wide range of random coil polysaccharide solutions (including dextran).[161] The following striking generalities are observed:

(1) The transition from dilute to concentrated solution behaviour occurs at a critical concentration $c^* \approx 4/[\eta]$, when 'zero shear' specific viscosity (η_{sp}) $\approx$ 10. η_{sp} varies as $c^{1.4}$ for dilute solutions, and as $c^{3.3}$ for concentrated solutions.

(2) The shear rate dependence of viscosity, and frequency dependence of dynamic (oscillatory viscosity) are closely superimposable.

(3) Double logarithmic plots of η/η_0 against $\gamma/\gamma_{0.1}$ (where η_0 is 'zero shear' viscosity, and $\gamma_{0.1}$ is the shear rate at which $\eta = \eta_0/10$) are essentially identical for all concentrated solutions studied, and thus the two parameters η_0 and $\gamma_{0.1}$ completely define the viscosity at all shear rates of practical importance.

In a review of the theory of gel permeation chromatography for the determination of the molecular weights of polymers, suggestions have been made for the use of two systems containing TSK gel PW for the determination of the molecular weight of clinical dextran.[162]

Gel permeation chromatography of aqueous solutions of dextrans has been followed by using a three-detector system (refractometry, conductometry, and viscometry).[163] The role of the low adsorption of dextran on silica gel was noted, and the dependence of elution volume on polymer concentration is discussed in terms of the screening length of the molecules.

Frictional coefficients for aqueous solutions of dextran have been evaluated from self-diffusion coefficients by pulsed-field gradient n.m.r. spectroscopy and from sedimentation coefficients in concentrated solutions.[164] These frictional coefficients are only equal at infinite dilution and the value of f_S increases more rapidly than f_D as the concentration increases.

The autocorrelation function of the intensity fluctuation of light scattered from dextran solutions, together with the angular dependence of the integrated intensity of the scattered light, has been measured by means of single-photon counting.[165] The data were analysed by considering the molecular weight distributions of the samples obtained by gel permeation chromatography. The molecular weight dependence of the diffusion coefficient, and the relation between the radii of gyration and the hydrodynamic radii thus obtained, proved that dextran molecules behave as random coils in aqueous solution. The molecular-weight distribution curve measured by gel permeation chromatography has been demonstrated to be useful in the analysis of light-scattering experiments for polydisperse macromolecular solutions.

A low-molecular-weight fraction of a partially acid hydrolysed D-fructan (levan) from *Streptomyces salivarius* has been examined by small angle X-ray scattering.[166] Although this fraction behaves hydrodynamically as a linear random coil, the

---

[161] E. R. Morris, A. N. Cutler, S. B. Ross-Murphy, D. A. Rees, and J. Price, *Carbohydr. Polymers*, 1981, **1**, 5.
[162] R. M. Alsop and G. J. Vlachogiannis, *J. Chromatogr.*, 1982, **246**, 227.
[163] J. Desbrieres, J. Mazet, and M. Rinaudo, *Eur. Polym. J.*, 1982, **18**, 269.
[164] W. Brown, P. Stilbs, and R. M. Johnsen, *J. Polym. Sci., Polym. Phys. Ed.*, 1982, **20**, 1771.
[165] N. Suzuki, A. Wada, and K. Suzuki, *Carbohydr. Res.*, 1982, **109**, 249.
[166] B. A. Khorramian and S. S. Stivala, *Carbohydr. Res.*, 1982, **108**, 1.

observed higher mass per unit length as compared with the calculated value for linear levan confirms its branched structure.

A comparison of the use of the quaternary ammonium salts cetyltrimethylammonium bromide (CTAB) and the commercial mixture Cetavlon for the isolation of xantham gum from fermentations of *Xanthomonas campestris* indicate that the former is the more efficient complexating agent.[167] An assessment of the use of Cetavlon for the isolation of xantham gum in a recycle procedure showed that an 11.5% loss of precipitant per cycle occurred. In the procedure, the xanthan gum was precipitated as the purified K^+ salt from a dispersion of its quaternary ammonium complex in propan-2-ol. Concentration of the propan-2-ol permitted recovery of the quaternary ammonium salt.

Gel permeation chromatography of xanthan gum has been followed using a refractometric and a light scattering detector.[168] Molecular weight distribution analysis indicates a polydispersity of 1.2 for the native polymer and of 1.7 for partially degraded polymer. Xanthan samples (M_r 0.8×10^5—1.95×10^6) have been studied by light scattering to determine the stiffness of the polysaccharide chain in dilute solution.[169] The results support earlier proposals that xanthan exists in aqueous solution as a double-stranded helix.

Studies on heat-denatured xanthan in 4 M urea suggest that these polymers are rod-like in character, no evidence for 'hindered rotation' being obtained.[170] The equivalent rod lengths lie in the range 0.2—1.6 m. These dimensions are compatible with reported electron-microscopic data. The molecular rigidity is attributed to extension of the polyanion due to charge–charge repulsions or to steric hindrance due to side chains.

Increasing concentrations of aqueous solutions of xanthan gum show a distinctive change in the temporal statistics of light scattering in the form of a photon autocorrelation function from close-to-single exponential to bimodal decay.[171] It is suggested that the rapid change in the transport behaviour of solutions of xanthan is associated with the interpenetration or entanglement of polysaccharide chains.

Gellan gum, an extracellular polysaccharide from *Pseudomonas elodea*, is a linear anionic heteropolysaccharide containing D-glucose and L-rhamnose and uronic acid.[172] X-Ray fibre diffraction techniques show that the polymer adopts a threefold helical structure with an axial repeat of 0.94 nm. The polymer forms weak elastic thermoreversible gels and on deacetylation forms rigid brittle gels.

Lamellar single crystals of the β-D-glucan, nigeran, from *Aspergillus niger* have been subjected to controlled enzymic degradation as an approach to understanding polysaccharide organization in these crystals.[173] Analysis of both reactant and products by various methods argues that a situation exists where the crystals are inaccessible at 20 °C but become increasingly accessible as the temperature

---

[167] J. F. Kennedy, S. A. Barker, I. J. Bradshaw, and P. Jones, *Carbohydr. Polymers*, 1981, **1**, 55.
[168] F. Lambert, M. Milas, and M. Rinaudo, *Polym. Bull. (Berlin)*, 1982, **7**, 185.
[169] G. Paradossi and D. Brant, *Macromolecules*, 1982, **15**, 874.
[170] V. J. Morris, K. l'Anson, and C. Turner, *Int. J. Biol. Macromol.*, 1982, **4**, 362.
[171] G. J. Southick, A. M. Jamieson, and J. Blackwell, *Polym. Prepr., Am. Chem. Soc., Div. Polym. Chem.*, 1981, **22**, 143.
[172] V. Carroll, M. J. Miles, and V. J. Morris, *Int. J. Biol. Macromol.*, 1982, **4**, 432.
[173] R. H. Marchessault, J.-F. Revol, T. F. Bobbitt, and J. H. Nordin, *Biopolymers*, 1980, **19**, 1069.

approaches that of solution for the crystals in water. It was concluded that chain folds are inaccessible to enzymes at 20 °C and become increasingly accessible as the melting temperature is approached. This could be due to increased mobility of the surface chains as temperature is raised, which produces large dynamic loops and makes enzyme degradation possible.

The triple helix of a polysaccharide *Schizophyllum commune* (schizophyllan) of viscosity-average molecular weight 4.8×10^5 g mol^{-1} (in water) was melted at denaturation temperatures between 5 and 60 °C in water–DMSO mixtures. The solutions for different denaturation temperature values were studied by viscometry and ultracentrifugation.[174] As denaturation temperature was increased, the intrinsic viscosity of the mixture containing 12.76% (by weight) of water decreased sharply and, at about 50 °C, it approached the value expected for the single chain of schizophyllan in pure DMSO. Schlieren patterns of the sample 'denatured' at temperatures between 25 and 45 °C in the same mixed solvent showed the presence of two solute species. The fast-sedimenting species dominating at denaturation temperatures of 25 °C almost disappeared at denaturation temperature of 45 °C. From sedimentation coefficient data for the mixture pure water and DMSO, the fast- and slow-sedimenting species could be identified with the triple helix and the single chain of schizophyllan, respectively. In water–DMSO mixtures containing 13% of water, the schizophyllan triple helix melts into single chains in all-or-none fashion with increasing temperature.

Regeneration of protoplasts has been used as a system to study the biosynthesis of cell wall β-D-glucans in *S. commune*.[175] Evidence for a water-soluble $(1 \rightarrow 3)$-β-D-glucan functioning as a precursor for the alkali-insoluble $(1 \rightarrow 3)$-β-D-glucan in a D-glucan–chitin complex has been provided.

The hydrodynamic behaviour of lentinan, an anti-tumour $(1 \rightarrow 3)$-linked β-D-glucan, has been described.[176] Quasielectric light-scattering studies show that the polysaccharide molecules are stable and have little tendency to form aggregates.

The walls of two strains of *Saccharomyces uvarum* (flocculent and non-flocculent) have been isolated from exponential and stationary phase cells after growth in media of different Ca^{2+} and K^+ composition.[177] Flocculation of the walls is identical to that of intact cells. The transition from the non-flocculent to the flocculent state involves alterations of the polysaccharide, protein, and inorganic ion composition of the walls. The walls of the flocculent cells have a higher molar ratio of D-mannose/D-glucose, a lower percentage protein, a lower percentage Ca^{2+}, and a higher percentage K^+.

5 Animal Glycoproteins and Glycosaminoglycans

The preferred conformation in aqueous solutions of oligosaccharides related to the complex-type carbohydrate portion of glycoproteins has been inferred from high-resolution ^{1}H and ^{13}C n.m.r. data, combined with hard sphere *exo*-anomeric

---

[174] T. Sato, T. Norisuye, and H. Fujita, *Carbohydr. Res.*, 1981, **95**, 195.
[175] A. S. M. Sonnenberg, J. H. Sietsma, and J. G. H. Wessels, *J. Gen. Microbiol.*, 1982, **128**, 2667.
[176] N. Suzuki, A. Wada, and K. Suzuki, *Carbohydr. Res.*, 1982, **109**, 295.
[177] M. A. Amri, R. Bonaly, B. Duteurtre, and M. Moll, *J. Gen. Microbiol.*, 1982, **128**, 2001.

calculations.[178] 2-Acetamido-2-deoxy-4-O-β-D-galactopyranosyl-D-glucopyranose (N-acetyllactosamine) units are arranged in such a way that they are widely separated in space. Thus these carbohydrates can cover large surface areas of the attached protein and provide good targets for recognition by other proteins.

The flexibility of bi- and tri-antennary glycans of the N-acetyllactosamine-type has been studied by spin-labelling the neuraminic acid residues in glycopeptides of known structure.[179]

A detailed analysis of the 360 MHz ^{1}H n.m.r. spectral parameters for the anomeric and C–2H resonances of a large number of glycopeptides and oligosaccharides of known structure reveals a general method for the determination of the primary structure of glycopeptides for most currently known classes of structures.[180] A two-dimensional display formed by plotting D-mannosyl C1–H against C2–H chemical shifts demonstrates that these pairs of values are sensitive to long-range perturbation by remote substitution by hexoses as well as to direct substitution effects. A total of 41 of these chemical shift clusters have been defined which characterize unique structural micro-environments. On this basis the sequence and branching pattern for most structures can be derived.

^{13}C n.m.r. data for mono- and di-O-α- and β-D-galactosylated dipeptides composed of L-Thr and Gly are presented.[181] The results conclusively show that peptide-bond formation does not affect the chemical shifts of the attached carbohydrate carbon atoms. In the case of the di-O-glycosylated L-threonyl-L-threonine, no carbohydrate–carbohydrate interactions could be observed. For some of the mono-O-glycosylated dipeptides, the attached glycosyl group appears to have a peculiar effect on the chemical shifts of some of the carbon resonances of the amino-acids.

C.d. spectroscopy in the 170—220 nm range and measurements of the amide proton coupling constants in n.m.r. have been used to investigate the conformation of L-asparagine-linked glycopeptides having complex oligosaccharide chains.[182] The spectra have been explained as the sum of three contributions, the first of which is a pair of large bands of opposite sign resulting from coupling of the adjacent chromophores of the 2-acetamido-1-N-(4-aspartyl)-2-deoxy-β-D-glucopyranosyl-amine linkage. Secondly, the amide chromophore of the adjacent core 2-acetamido-2-deoxy-D-glucose residue substituted at C-4 by a β-D-mannosyl residue contributes a negative band at 210 nm together with a small negative signal in the 180—190 nm region. The amides of the antenna 2-acetamido-2-deoxy-D-glucosyl residues, which are usually substituted at C-4 by β-D-galactosyl residues, contribute a negative band at 210 nm and a strong positive band at 185—190 nm. The c.d. and n.m.r. data and other evidence are taken to imply that the 2-amino-2-deoxy-D-glucosyl-L-asparagine linkage is rigidly fixed in a conformation having the amide protons *trans* to the sugar ring protons. These results are consistent with an extended conformation in the shape of a 'Y' or 'T' for complex type asialo–oligosaccharide chains.

---

[178] K. Bock, J. Arnarp, and J. Lonngren, *Eur. J. Biochem.*, 1982, **129**, 171.
[179] J. Davoust, V. Michel, G. Spik, J. Montreuil, and P. F. Devaux, *FEBS Lett.*, 1981, **125**, 271.
[180] J. P. Carver and A. A. Grey, *Biochemistry*, 1981, **20**, 6607.
[181] K. Dill, R. E. Hardy, M. E. Daman, J. M. Lacombe, and A. A. Pavia, *Carbohydr. Res.*, 1982, **108**, 31.
[182] C. A. Bush, V. K. Dua, S. Ralapati, C. D. Warren, G. Spik, G. Strecker, and J. Montreuil, *J. Biol. Chem.*, 1982, **257**, 8199.

The hydrodynamic behaviour of reduced glycopolypeptides has been studied by gel filtration in guanidine hydrochloride in conjunction with h.p.l.c.[183] Although carbohydrate-rich glycopolypeptides may occasionally yield underestimated values of molecular weights, the method is suitable for the rapid estimation of molecular weights of simple polypeptides and glycopolypeptides.

Changes in topography and lateral translational mobility of concanavalin A receptors on the surface of cultured chick muscle cells during myoblast fusion have been studied, using a resonance energy transfer technique employing pyrene- and fluorescein isothiocyanate–concanavalin A conjugates.[184] During the period of myoblast fusion, concanavalin A receptors undergo a dramatic redistribution on the cell surface. Changes in membrane fluidity observed during muscle differentiation serve to modulate the lateral mobility of these receptors.

Phosphatidyl inositol 4,5-bisphosphate (triphospho-inositide) increases glycoprotein lateral mobility in erythrocyte membranes, and probably acts by disrupting linkages in the erythrocyte membrane skeleton.[185] This effect may have important consequences for such vital red cell characteristics as cell shape and deformability.

The relative mobility of neuraminic acid-linked fluorescent probes has been compared *in situ* and after removal of the labelled glycopeptides from human erythrocytes.[186] A significant motional constraint is lost upon separating the glycopeptides from their membrane milieu. The nature of the constraint is unknown, but it may relate to the high density of oligosaccharide chains on the membrane surface.

Fibronectin is a large extracellular glycoprotein that binds to a variety of macromolecules, including fibrin, collagen, cell surfaces, and glycosaminoglycans. Proteases cleave fibronectin into fragments which have one or more of these binding activities contained in molecular domains. The proteolytic fragments and monoclonal antibodies reacting with different domains have revealed the sequential ordering of the active sites of the molecule. The structural and functional diversity of fibronectin has been reviewed.[187]

Heat denaturation of human plasma fibronectin proceeds through at least three stages, with endothermal denaturing transitions observed at 68, 82, and 119 °C.[188] A three-domain structure for the glycoprotein is proposed in which the domain which unfolds at 68 °C is associated with gelatin binding and cell binding, while at 82 °C a domain appears to be associated with much of the immunological activity. The 119 °C domain has heparin binding activity as well as some immunological activity. The gelatin-binding domain of fibronectin is immunogenic, and antisera against this domain recognize cellular fibronectin–gelatin-binding sites.[189] Inhibition of gelatin binding but not cell spreading by anti-gelatin-binding domain Fab' fragments confirms the hypothesis that fibronectin has separate sites mediating these activities.

---

[183] N. Ui, *J. Chromatogr.*, 1981, **215**, 289.
[184] B. A. Herman and S. M. Fernandez, *Biochemistry*, 1982, **21**, 3275.
[185] M. P. Sheetz, P. Febroriello, and D. E. Koppel, *Nature (London)*, 1982, **296**, 91.
[186] P. S. Low, W. A. Cramer, G. Abraham, R. Bone, and M. Ferguson-Segall, *Arch. Biochem. Biophys.*, 1982, **214**, 675.
[187] E. Ruoslahti, M. Pierschbacher, E. G. Hayman, and E. Engvall, *Trends Biochem. Sci.*, 1982, **7**, 188.
[188] D. G. Wallace, J. W. Donovan, P. M. Schneider, A. M. Meunier, and J. L. Lundblad, *Arch. Biochem. Biophys.*, 1981, **212**, 515.
[189] J. A. McDonald, T. J. Broedelmann, D. G. Kelley, and B. Villiger, *J. Biol. Chem.*, 1981, **256**, 5583.

Human plasma fibronectin has been treated with chymotrypsin.[190] Characterization of the isolated fragments has shown that the order of the functional domains from the N-terminus is staphylococcal binding, fibrin cross-linking and binding sites, gelatin-binding site, cell attachment area, and heparin-binding site.

Human plasma fibronectin and a series of its large proteolytic fragments have been analysed by electron microscopy on carbon and polystyrene films.[191] On carbon, intact fibronectin appears as a randomly coiled strand, while on polystyrene it appears as an elongated structure. Two fragments of fibronectin (M_r 2.05×10^5 and 1.9×10^5) which lack the N-terminal domain but retain collagen-binding, cell attachment, and heparin-binding functions, and a fragment (M_r 1.7×10^5) which retains the collagen-binding and cell attachment functions, are seen as rods with varying degrees of nodularity. A fragment (M_r 1.0×10^5) which only binds to collagen, has two clearcut domains. The results support the existing biochemical evidence that the segregation of the functional activities in the fibronectin molecule is based on distinct structural domains and provides evidence for the existence of an additional structural domain not revealed by biochemical and functional studies.

Tryptic fragments of human plasma fibronectin have been analysed by using monoclonal antibodies to delineate the two immunologically distinct regions of the glycoprotein which contain a free sulphydryl group.[192] The data, taken together with other evidence, indicate that the two free SH-groups are in the C-terminal half of the molecule.

Human plasma fibronectin has been enzymically labelled with dansylcadaverine using Factor XIII (transglutaminase).[193] Fluorescence polarization, dynamic light scattering, and intrinsic viscosity studies indicate that fibronectin has an elongated conformation and a significant degree of chain flexibility under physiological conditions. Partial unfolding of this conformation may take place upon binding to collagen in the tissue matrix.

Further studies on the adsorption of fibronectin to hydrophobic and hydrophilic surfaces have been reported.[194] The binding of anti-plasma fibronectin antibodies to adsorbed fibronectin has been measured as another criterion of conformational changes.

Monoclonal antibodies have been used in mapping biologically active proteolytic fragments onto the fibronectin sub-unit (M_r 2.2×10^5).[195] By this approach an N-terminal heparin-binding and the adjacent gelatin-binding fragments have been characterized and a second heparin-binding site located 3.1×10^4 daltons from the C-terminus of the sub-unit is defined.

A monoclonal antibody directed against the area of human plasma fibronectin containing the cell attachment site has been used to isolate quantities of a peptic fragment that comprises the attachment site.[196] The complete amino-acid sequence

---

[190] E. Ruoslahti, E. G. Hayman, E. Engvall, W. C. Cothran, and W. T. Butler, *J. Biol. Chem.*, 1981, **256**, 7277.
[191] T. M. Price, M. L. Rudee, M. Pierschbacher, and E. Ruoslahti, *Eur. J. Biochem.*, 1982, **129**, 359.
[192] D. E. Smith, D. F. Mosher, R. B. Johnson, and L. T. Furcht, *J. Biol. Chem.*, 1982, **257**, 5831.
[193] E. C. Williams, P. A. Janmey, J. D. Ferry, and D. F. Mosher, *J. Biol. Chem.*, 1982, **257**, 14973.
[194] F. Grinnell and M. K. Feld, *J. Biol. Chem.*, 1982, **257**, 4888.
[195] D. E. Smith and L. T. Furcht, *J. Biol. Chem.*, 1982, **257**, 6518.
[196] M. D. Pierschbacher, E. Ruoslahti, J. Sundelin, P. Lind, and P. A. Peterson, *J. Biol. Chem.*, 1982, **257**, 9593.

of this fragment is reported along with a discussion of its possible functional characteristics.

Plasma fibronectin has been subjected to affinity chromatography on various glycosaminoglycan-substituted gels and on collagen–agarose in the presence of glycosaminoglycans in order to examine the effects of different glycosaminoglycans on the interaction between collagen and fibronectin.[197] Heparan sulphate favours the formation of collagen–fibronectin complexes at low molarity, while hyaluronate is ineffective at low concentrations and prevents the formation of complexes when present at concentrations of 1 mg ml^{-1}. It is suggested that heparan sulphate promotes the formation of complexes which bind with fibronectin, thus producing steric changes that increase the affinity for collagen, while hyaluronate prevents the binding of fibronectin to collagen by a steric exclusion mechanism.

Cell surface heparan sulphate binds to fibronectin and collagen in a pH-dependent manner, and the binding does not require Ca^{2+} and Mg^{2+}.[198] The binding site is probably located in the polysaccharide chains.

Sub-stratum adhesion sites of glioma and neuroblastoma cells are enriched with glycosaminoglycan sequences which have the potential to bind to plasma fibronectin.[199] The fibronectin-dependent adhesion of these neural tumour cells on serum-coated tissue culture sub-strata may be mediated by heparan proteoglycans on their surface.

Proteoglycans from pig aorta have been extracted sequentially with inorganic salt solutions under associative and dissociative conditions, followed by fractionation by gel permeation chromatography in order to characterize and compare their chemical properties and hydrodynamic sizes.[200]

Zonal rate centrifugation in sucrose gradients has been used to study interactions in proteoglycan aggregation.[201] The link protein was shown to interact with the isolated hyaluronic acid-binding region of the proteoglycan monomer. Since it also interacts with isolated hyaluronic acid, the binding of proteoglycans to hyaluronate in the presence of link proteins is trifunctional. The procedure has also been used to separate proteoglycan aggregates from monomers and to determine size distributions of the compounds.[202] The binding properties of cartilage link protein and the hyaluronate-binding region of cartilage proteoglycan to hyaluronic acid immobilized on agarose have been compared and shown to be similar.[203] A gross physical characterization of the hyaluronic acid-binding region of proteoglycan from pig laryngeal cartilage has been achieved using densitometric and small angle neutron studies.[204]

Zonal centrifugation on preformed caesium sulphate gradients has been used in the purification of the aggregate fraction from cartilage proteoglycan samples.[205] The characteristics of aggregating and non-aggregating proteoglycans purified

[197] M. Del Rosso, G. Fibbi, F. Pasquali, R. Cappellett, S. Vannucchi, and V. Chiarugi, *Int. J. Biol. Macromol.*, 1982, **4**, 67.
[198] S. C. Stamatoglou and J. M. Keller, *Biochim. Biophys. Acta*, 1982, **719**, 90.
[199] L. A. Culp and C. Domen, *Arch. Biochem. Biophys.*, 1982, **213**, 726.
[200] M. Breton, J. Picard, and E. Berrou, *Biochimie*, 1981, **63**, 515.
[201] A. Franzen, S. Bjornsson, and D. Heinegard, *Biochem. J.*, 1981, **197**, 669.
[202] A. Franzen, S. Bjornsson, and D. Heinegard, *Anal. Biochem.*, 1982, **120**, 38.
[203] A. Tengblad, *Biochem. J.*, 1981, **199**, 297.
[204] S. J. Perkins, A. Miller, T. E. Hardingham, and H. Muir, *J. Mol. Biol.*, 1981, **150**, 69.
[205] K. Kimata, J. H. Kimura, E. J. M. A. Thonar, H. J. Barrach, S. I. Rennard, and V. C. Hascall, *J. Biol. Chem.*, 1982, **257**, 3819.

from Swarm rat chondrosarcoma have been described using this system. Preformed caesium sulphate zonal gradients in different solvents have been used to study purified proteoglycan aggregates.[206] Increasing concentrations of guanidine hydrochloride cause separation of all three components of the aggregate, proteoglycan monomers, link proteins, and hyaluronic acid. Low-solvent pH dissociates a stable complex of proteoglycan monomer–link protein as a unit from hyaluronic acid.

Fine fragmentation of bovine nasal septum cartilage makes it possible to extract significant levels of intact proteoglycan aggregates at low ionic strength.[207] The proteoglycans may then be made to associate within extracted cartilage fragments and agarose beads. A hypothesis has been developed, indicating that proteoglycans are immobilized within cartilage through the formation of aggregates and suggesting that retention is dependent upon the integrity of the collagen mesh.

A proteoglycan, isolated from a rat yolk sac tumour and characterized as a chondroitin sulphate with a smaller amount of dermatan sulphate, forms complexes with collagen and fibronectin.[208] Treatment of the proteoglycan with alkali to separate the glycosaminoglycan chains from the protein, and digestion of the protein part with papain, greatly reduced the capacity of the proteoglycan to precipitate collagen and fibronectin.

An affinity chromatographic procedure which measures binding of proteoglycans to agarose gels substituted with different forms of dermatan sulphate and chondroitin sulphate has been used to investigate the possible involvement of the dermatan sulphate side chains in the aggregation of proteodermatan sulphate.[209] The effects of different modifications on the carbohydrate and protein portions of the proteoglycans have been studied.

Metastatic tumour cells invade, degrade, and eventually solubilize the bovine aortic sub-endothelial matrix by hydrolysing components such as fibronectin and sulphated proteoglycans.[210] The sulphated proteoglycans of the matrix are composed of large amounts of heparan sulphate with minor amounts of chondroitin 4- and 6-sulphate. After incubation of the matrix with tumour cells, a new low-molecular-weight heparan sulphate-containing fragment is released. It is likely that the tumour cells elaborate a glycosidase capable of cleaving specifically glycosaminoglycans and releasing heparan sulphate-rich fragments.

Electron microscopic studies of bovine nasal cartilage proteoglycans have given direct evidence for the variable length of the chondroitin sulphate-rich region of the proteoglycan sub-unit core protein.[211]

Data have been presented for ^{13}C spin–lattice relaxation times and nuclear Overhauser enhancement values in a range of heparinoid glycosaminoglycans.[212] Significant and consistent differences in relaxation times are observed for different ring carbons, for which simple models do not provide an adequate explanation.

The self-association between heparan sulphate chains has been investigated by using oligosaccharides derived from heparan sulphate for the competitive elution of

---

[206] K. Kimata, V. C. Hascall, and J. H. Kimura, *J. Biol. Chem.*, 1982, **257**, 3827.
[207] L. A. Pottenger, N. B. Lyon, J. D. Hecht, P. M. Neustadt, and R. A. Robinson, *J. Biol. Chem.*, 1982, **257**, 11479.
[208] A. Oldberg and E. Ruoslahti, *J. Biol. Chem.*, 1982, **257**, 4859.
[209] L. A. Fransson, L. Coster, A. Malmstrom, and J. K. Sheehan, *J. Biol. Chem.*, 1982, **257**, 6333.
[210] R. H. Kramer, K. G. Vogel, and G. L. Nicolson, *J. Biol. Chem.*, 1982, **257**, 2678.
[211] J. A. Buckwalter and L. C. Rosenberg, *J. Biol. Chem.*, 1982, **257**, 9830.
[212] T. N. Huckerby and I. A. Nieduszynski, *Int. J. Biol. Macromol.*, 1982, **4**, 269.

Natural Polymers: Polysaccharides and Glycoproteins

^{3}H-heparan sulphate from heparan sulphate–agarose.[213] Highly copolymeric regions of the polysaccharide serve as contact zones for the chain–chain association. The self-interaction between the chains is strongly dependent on the overall molecular conformation.[214] The N-sulphate and carboxylate groups, as well as the integrity of the D-glucopyranosyluronic acid residue, are all essential for maintaining the proper secondary structure.

A heparan sulphate fraction possessing the largest proportion of high-affinity variants for human low-density lipoprotein contains almost equal proportions of the repeating units L-idopyranosyluronic acid (O-sulphate)-2-deoxy-2-sulphonamido-D-glucose and D-glucopyranosyluronic acid-2-acetamido-2-deoxy-D-glucose.[215]

An iduronate-rich dermatan sulphate fraction containing variants with high affinity for human low-density lipoprotein has been examined and shown to have features similar to those observed for heparin-related glycans having high affinity for lipoproteins.[216] Digestion of a dermatan sulphate–lipoprotein complex with chondroitin lyase ABC yielded fragments in which half of the uronate residues were L-iduronate sulphate.

A micro-rolling-ball viscometer has been used to measure the viscosity of complexes between human serum low-density lipoprotein (LDL) with heparin.[217] To obtain additional information on particle radius, surface charge, and binding ratio, the rheological results were confirmed by light-scattering studies, electrophoresis, and ultracentrifugation experiments. Quantitative binding studies revealed that one to two heparin molecules are bound to a single LDL particle. Very-low-density lipoproteins (VLDL) and high-density lipoproteins (HDL) gave no rheologically effective aggregates with heparin. Differential scanning calorimetry showed that LDL treated with trypsin or neuraminidase exhibited the same thermotropic transitions as the native LDL.[218] Thus it is unlikely that the effects of glycosaminoglycans on the structure of the lipid core are not related to the protein or carbohydrate arrangement of the shell of the LDL particle.

Porcine intestinal heparin has been partially degraded to produce an octasaccharide with high affinity for antithrombin III.[219] The octasaccharide exhibited more potent inactivation for thrombin, Factor X_a, and plasma coagulation than the initial heparin. The high-affinity binding of heparin to antithrombin III requires the presence of two consecutive N-sulphated 2-amino-2-deoxy-D-glucosyl residues in specific positions of the antithrombin-binding sequence.[220] Loss of either one of these N-sulphate groups, with or without N-acetylation, results in a distinct and appreciable decrease in binding affinity, and in anticoagulant activity.

Porcine heparin has been cleaved randomly by chemical techniques to produce hexasaccharides, octasaccharides, decasaccharides, and fragments containing 14

---

[213] L. A. Fransson, *Carbohydr. Res.*, 1982, **110**, 127.
[214] L. A. Fransson, *Carbohydr. Res.*, 1982, **110**, 135.
[215] L. A. Fransson and B. Havsmark, *Int. J. Biol. Macromol.*, 1981, **3**, 361.
[216] L. A. Fransson and B. Havsmark, *Int. J. Biol. Macromol.*, 1982, **4**, 50.
[217] A. Mitterer, W. D. Eigner, J. Schurz, G. Jurgens, and A. Holasek, *Int. J. Biol. Macromol.*, 1982, **4**, 227.
[218] M. Bihari-Varga, S. Goldstein, D. Lagrange, and E. Gruber, *Int. J. Biol. Macromol.*, 1982, **4**, 438.
[219] N. Ototani and Z. Yosizawa, *J. Biochem. (Tokyo)*, 1981, **90**, 1553.
[220] J. Riesenfeld, L. Thunberg, M. Hook, and U. Lindahl, *J. Biol. Chem.*, 1981, **256**, 2389.

and 16 residues that are able to complex with protease inhibitor.[221] As a result of studies on the avidity of these fractions for protease inhibitors, the binding to antithrombin and the catalysis of the Factor Xa–antithrombin interaction, it is proposed that heparin possesses multiple discrete structural domains that modulate different functions of antithrombin III. The chemical composition and ^{13}C n.m.r. spectra of heparin octasaccharides having high affinity for antithrombin III and high anti-(factor Xa) activity, prepared by three independent approaches, have been studied and compared with those of the corresponding inactive species.[222] Combined with chemical data, the spectra of the active oligosaccharides and of their fragmentation products afforded information on composition and sequence.

High-affinity heparin has been covalently coupled to antithrombin III and the turnover properties of the complex studied.[223] The main mechanism of disappearance of the anticoagulant activity following intravenous injection of heparin is by removal of free heparin and dissociation of the normal heparin–antithrombin III complex and not by clearing of the intact complex.

The rate-determining step for the heparin-catalysed antithrombin–thrombin reaction is independent of thrombin.[224] Based solely on kinetic analysis, the reaction binding sequence was determined to be heparin binding to antithrombin followed by binding to thrombin.

The heparin-enhanced antithrombin III–thrombin reaction has been studied under a variety of conditions designed to determine the validity of kinetic models for the mechanism of action of heparin.[225] A template model has been derived in mathematical terms and theoretical data calculated to test the possible validity of the model.

A modified two-chain form of bovine antithrombin, cleaved at the active site, is produced by stoicheiometric amounts of thrombin in a reaction that competes with the formation of the inactive antithrombin–thrombin complex.[226] Several properties such as the frictional ratio and the c.d. of the modified antithrombin differ from those of the intact inhibitor in a manner suggesting a conformational change following scission of the active site bond. In addition, the affinity of heparin for the modified inhibitor is much lower than for the intact inhibitor. Conformations of the heparin binding site and the active site of antithrombin are probably intimately linked.

The thermal denaturation of antithrombin III is stabilized by heparin and also by anions such as phosphate, sulphate and EDTA ions.[227] Dithiothreitol modulates antithrombin III activity by sulphated polysaccharides.[228] Kinetic analysis of the heparin-enhanced plasmin–antithrombin III reaction supports the theory that heparin must bind to the enzyme to accelerate the plasmin–antithrombin III

---

[221] G. M. Oosta, W. T. Gardner, D. L. Beeler, and R. D. Rosenberg, *Proc. Natl. Acad. Sci., USA*, 1981, **78**, 829.
[222] B. Casu, P. Oreste, G. Torri, G. Zoppetti, J. Choay, J. C. Lormeau, M. Petitou, and P. Sinay, *Biochem. J.*, 1981, **197**, 599.
[223] R. Ceustermans, M. Hoylearts, M. De Mol, and D. Collen, *J. Biol. Chem.*, 1982, **257**, 3401.
[224] C. H. Pletcher and G. L. Nelsestuen, *J. Biol. Chem.*, 1982, **257**, 5342.
[225] M. J. Griffith, *J. Biol. Chem.*, 1982, **257**, 7360.
[226] I. Bjork and W. W. Fish, *J. Biol. Chem.*, 1982, **257**, 9487.
[227] T. F. Busby, D. H. Atha, and K. C. Ingham, *J. Biol. Chem.*, 1981, **256**, 12140.
[228] W. F. Long and F. B. Williamson, *Biochem. Soc. Trans.*, 1981, **9**, 68.

reaction.[229] The binding of high-affinity heparin to antithrombin III has been investigated following characterization of the protein fluorescence enhancement of L-tryptophan residues in the antithrombin in the presence and absence of heparin.[230] Stopped flow kinetic studies of the binding interactions have also been reported.[231]

α_2-Macroglobulin is a high-molecular-weight glycoprotein and a major component in the circulation of vertebrates. It is an inhibitor of trypsin and plasmin but the active site of these endoproteinases is not blocked after complexation with α_2-macroglobulin. The structural features of human α_2-macroglobulin have been discussed and related to their function.[232] Three important sites in the inhibitor have been recognized, viz, the bait region, the internal thioester, and the receptor-recognition site. The cellular receptor on human fibroblasts is specific for α_2-macroglobulin complexes.[233] The α_2-macroglobulin complexes, but not native α_2-macroglobulin, are internalized by receptor-mediated endocytosis by fibroblasts.[234]

Studies on the interaction of immunoglobulins with carbohydrate-containing antigens continues to receive attention. Antibodies have been raised to numerous glycoproteins, glycolipids, and polysaccharides. Those antibodies known to interact with carbohydrate determinants on the antigen molecule are listed in Table 2.

Table 2 *Antibodies interacting with glycoprotein, glycolipid, or polysaccharide antigens*

Antigen	Determinant(s) on antigen	Antibody type	Ref.
Glycoproteins			
Ectoglycoprotein	D-Glcp, D-Galp, β-D-Galp-(1 → 4)-D-Glc	Polyclonal	235
Histocompatibility antigen (H-2K)	α-D-Manp; Neup5Ac; α,β,-D-Galp; D-Manp	Monoclonal	236
Ma (group 1)	β-D-Galp-(1 → 4)-β-D-GlcpNAc-(1 → 6)-β-D-Galp	Monoclonal	237
Glucoamylase	α-D-Manp	Polyclonal	238
Bovine leukaemia envelope viral glycoprotein	intact oligosaccharide chain	Polyclonal	239
Retrovirus envelope glycoprotein	D-GlcNAc; D-Gal; methyl α-D-Manp	Polyclonal	240
Keratan sulphate proteoglycan	keratan sulphate	Polyclonal	241

[229] R. Machovich, P. I. Bauer, P. Aranyi, E. Kecskes, K. G. Buchi, and I. Horvath, *Biochem. J.*, 1981, **199**, 521.
[230] S. T. Olson and J. D. Shire, *J. Biol. Chem.*, 1981, **256**, 11065.
[231] S. T. Olson, K. R. Srinivasan, I. Bjork, and J. D. Shore, *J. Biol. Chem.*, 1981, **256**, 11073.
[232] F. van Leuven, *Trends Biochem. Sci.*, 1982, **7**, 185.
[233] P. Marynen, F. Van Leuven, J. J. Cassiman, and H. van den Berghe, *J. Immunol.*, 1981, **127**, 1782.
[234] F. Van Leuven, J. J. Cassiman, and H. van den Berge, *Biochem. J.*, 1982, **201**, 119.
[235] J. H. Pazur, *Carbohydr. Res.*, 1982, **107**, 243.
[236] H. C. O'Neill, C. R. Parish, and T. J. Higgins, *Mol. Immunol.*, 1981, **18**, 663.
[237] E. A. Kabat, J. Liao, M. H. Burzynska, T. C. Wong, H. Thøgersen, and R. U. Lemieux, *Mol. Immunol.*, 1981, **18**, 873.
[238] J. H. Pazur, K. R. Forry, Y. Tominaga, and E. M. Ball, *Biochem. Biophys. Res. Commun.*, 1981, **100**, 420.
[239] D. Portetelle, C. Bruck, M. Mammerickx, and A. Burny, *Virology*, 1980, **105**, 223.
[240] M. J. F. Schmerr, J. M. Miller, and M. J. van der Maaten, *Virology*, 1981, **109**, 431.
[241] G. W. Conrad, P. Ager-Johnson, and M. L. Woo, *J. Biol. Chem.*, 1982, **257**, 464.

Table 2 cont.

Antigen	Determinant(s) on antigen	Antibody type	Ref.
Polysaccharides			
Dextran B512		Monoclonal	242
Dextran B1355	α-D-Glcp-(1 → 3)-α-D-Glcp-(1 → 6)-D-Glc; α-D-Glcp-(1 → 6)-α-D-Glcp-(1 → 3)-D-Glc	Polyclonal	243
D-Galactan	β-(1 → 6)-D-galactopyranan	Monoclonal	244
Streptococcal glycan	See a below	Polyclonal	245
Glycolipids			
Lewis[b] blood group	See b below	Monoclonal	246
Lewis[x] blood group	α-L-Fucp-(1 → 4)-D-GlcpNAc	Polyclonal	247
Glycosphingolipids	β-D-Galp-(1 → 4)-D-GlcpNAc; α-L-Fucp-(1 → 2)-β-D-Galp-(1 → 4)-D-GlcpNAc	Monoclonal	248
Lacto-N-fucopentaose III	See c below	Monoclonal	249
Lacto-N-biose I	β-D-Galp(1 → 3)-D-GlcpNAc	Monoclonal	250
Ganglioside	α-Neup5Ac-(2 → 8)-α-Neup5Ac-(2 → 3)-β-D-Galp-(1 → 4)-β-D-Glcp-(1 → 1)ceramide	Monoclonal	251
SSEA-1	β-D-Galp-(1 → 4)-β-D-GlcpNAc	Monoclonal	252

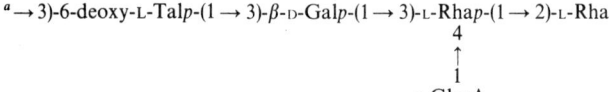

```
                    3
                    ↑
                    1
               α-L-Fucp
```

a → 3)-6-deoxy-L-Talp-(1 → 3)-β-D-Galp-(1 → 3)-L-Rhap-(1 → 2)-L-Rha
```
                        4
                        ↑
                        1
                     D-GlcpA
```

b α-L-Fucp-(1 → 2)-β-D-Galp-(1 → 3)-β-D-GlcpNAc(1 → 3)-β-D-Galp-(1 → 4)-D-Glc
```
                                     4
                                     ↑
                                     1
                                  α-L-Fucp
```

c β-D-Gal-(1 → 4)-β-D-GlcpNAc-(1 → 3)-β-D-Galp-(1 → 4)-D-Glc
```
             3
             ↑
             1
          α-L-Fucp
```

[242] J. Sharon, E. A. Kabat, and S. L. Morrison, *Mol. Immunol.*, 1981, **18**, 831.
[243] M. Torii, B. P. Alberto, and S. Tanaka, *J. Biochem. (Tokyo)*, 1980, **88**, 1855.
[244] A. Roy, B. N. Manjula, and C. P. J. Glaudemans, *Mol. Immunol.*, 1981, **18**, 79.
[245] J. H. Pazur, K. L. Dreher, R. L. Kubrick, and M. S. Erikson, *Anal. Biochem.*, 1982, **126**, 285.
[246] M. Blockhaus, J. L. Magnani, M. Blaszczyk, Z. Steplewski, H. Koprowski, K. A. Karlsson, G. Larson, and V. Ginsburg, *J. Biol. Chem.*, 1981, **256**, 13223.
[247] H. Schenkel-Brunner and P. Hanfland, *Vox Sang.*, 1981, **40**, 358.
[248] W. W. Young, J. Portoukalian, and S. I. Hakomori, *J. Biol. Chem.*, 1981, **256**, 10967.
[249] M. Brockhams, J. L. Magnani, M. Merlyn, M. Blaszczyk, Z. Steplewski, H. Koprowski, and V. Ginsberg, *Arch. Biochem. Biophys.*, 1982, **217**, 647.
[250] M. Terashima, K. Kato, T. Osawa, T. Chiba, and S. Tejima, *Carbohydr. Res.*, 1982, **110**, 345.
[251] M. Y. Yeh, I. Hellstrom, K. Abe, S. I. Hakomori, and K. E. Hellstrom, *Int. J. Cancer*, 1982, **29**, 269.
[252] E. F. Hounsell, H. C. Gooi, and T. Feizi, *FEBS Lett.*, 1981, **131**, 279.

Isoelectric focusing of antibodies in agarose gels has been reported using inexpensive laboratory-synthesized ampholytes.[253] The resolving power of these and commercially available ampholytes are compared and the advantages of substituting agarose gels bonded to plastic films for polyacrylamide gels are discussed. Soluble immune complexes have been studied using a procedure involving binding to immobilized protein A followed by isoelectric focusing to desorb and separate the immune complex.[254] By this approach IgG-type immunoglobulins and putative antigens can be detected in biological fluid from human specimens.

Further studies on the inhibition of the monoclonal anti-Ma (group 1) antibody in quantitative precipitin assays have been reported.[237] It is confirmed that the antibody binds to a portion of an *O-β-D-galactopyranosyl-(1→4)-O-β-D-2-acetamido-2-deoxy-glucopyranosyl-(1→6)-O-β-D-galactopyranosyl* unit of an oligosaccharide structure in a specific conformation. The conformational requirements indicate that the determinant is accepted into the antibody combining site along a hydrophobic portion of the structure which includes the regions about C6–C1′–C3′–O5″–C6″ as represented in Scheme 2. It is assumed that O-3′ is

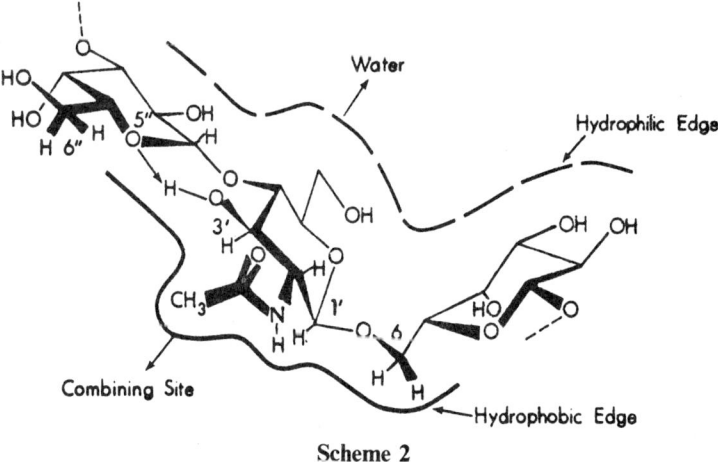

Scheme 2

(Display showing that the information of the IMa antigenic determinant possesses a region along C6–C1′–C3′–O5″–C6″ which can be expected to be compatible with a hydrophobic region for the combining site)

intramolecularly hydrogen bonded to O-5″ and that the other hydroxyl groups of the determinant remain engaged in bonding with solvent water.

Monoclonal antibodies specific for sub-unit 1 of the lectin from *Dolichos biflorus* combine with the C-terminal portion of the sub-unit and may be interacting with the active site or with a determinant conformationally independent of the active

---

[253] S. Binion and L. S. Rodkey, *Anal. Biochem.*, 1981, **112**, 362.
[254] B. W. Maidment, L. D. Papsidero, M. Gamarra, T. Nemoto, and T. M. Chu, *Anal. Biochem.*, 1981, **111**, 336.

site.[255] This antibody, which does not react with another lectin-like protein from the stems and leaves of the plant,[256] can distinguish sub-unit 1 from sub-unit 2 of the seed lectin.[257]

By measuring the affinities at various temperatures between anti-D-galactan immunoglobulin A J539 and anti-dextran immunoglobulin A W3129, and their respective ligands, the enthalpies and entropies of binding have been determined.[258] The interpretation of the results is in agreement with previous work showing that J539 has a groove-type combining region and W3129 has a cavity-type region.

The interaction of saccharide haptens with three homogeneous immunoglobulin A mouse myeloma proteins which have specificity for saccharides has been investigated in a 270 MHz study.[259] Difference spectra show that comparatively few protein resonances are perturbed on binding hapten, suggesting that accompanying conformational changes are limited to the combining site.

---

[255] C. A. K. Borrebaeck and M. E. Etzler, *J. Biol. Chem.*, 1981, **256**, 4723.
[256] M. E. Etzler and C. A. K. Borrebaeck, *Biochem. Biophys. Res. Commun.*, 1980, **96**, 92.
[257] M. E. Etzler, S. Gupta, and C. A. K. Borrebaeck, *J. Biol. Chem.*, 1981, **256**, 2367.
[258] A. K. Bhattacharjee, M. K. Das, A. Roy, and C. P. J. Glaudemans, *Mol. Immunol.*, 1981, **18**, 277.
[259] P. Gettins, J. Boyd, C. P. J. Glaudemans, H. Potter, and R. A. Dwek, *Biochemistry*, 1981, **20**, 7463.

5
Natural Polymers: Nucleic Acids

BY J. T. KNOWLER AND J. P. GODDARD

1 Advances in Methodology

Chemical Synthesis of Oligonucleotides and its Applications.—Advances have been made in both the chemistry of synthesis and the application of synthetic oligonucleotides to molecular biology since our report in the first volume of this series. The major advance in synthesis has been the development of solid-phase methods, used previously for oligopeptide synthesis, to both the phosphite-triester method using a silica gel support[1,2] and the phosphotriester method using a polystyrene or polyacrylamide support.[3,4] Both methods are rapid (a 30-minute cycle for each added nucleotide), reliable, and now sufficiently simplified that even molecular biologists are making the oligonucleotides they require!

The length of oligonucleotides made by this method (usually 15 to 20 bases) is adequate to ensure their specific hybridization under appropriate conditions. However, improvements in solid-support methods have allowed synthesis of longer oligonucleotides (approximately 40 nucleotides long). These may be joined by annealing residues at one end to complementary residues at the opposite end of a second oligonucleotide. The generated gaps may be filled *in vitro* using Klenow fragment DNA polymerase. This method thus reduces the number of building blocks required for a given long block of duplex DNA since complete complementary sequences need not be chemically synthesized.

Khorana's original method (reviewed in Volume 1 of this series), in which relatively short complementary overlapping oligonucleotides are joined by DNA ligase, is, however, still the most extensively used. It has been exploited to synthesize genes for biologically important proteins, including insulin[5] and interferon.[6]

In addition to their use as building blocks for the synthesis of important genes, synthetic oligonucleotides have found increasing application in two fields – those of site-directed mutagenesis (described below) and recovery of specific genes. Both applications rely upon having an oligonucleotide sufficiently long ($\geqslant 14$ nucleotides) that hybrids only occur between the synthetic oligonucleotides and the sequence being studied.

Early cloning of eukaryotic genes for proteins was limited to those which are

---

[1] F. Chow, T. Kempe, and G. Palm, *Nucleic Acids Res.*, 1981, **9**, 2807.
[2] M. D. Matteucci and M. H. Caruthers, *J. Am. Chem. Soc.*, 1981, **103**, 3185.
[3] M. L. Duckworth, M. J. Gait, P. Goelet, G. F. Hong, M. Singh, and R. C. Titmas, *Nucleic Acids Res.*, 1981, **9**, 1691.
[4] H. Ito, Y. Ike, S. I. Kuta, and K. Itakura, *Nucleic Acids Res.*, 1982, **10**, 1755.
[5] M. D. Edge, A. R. Green, G. R. Heathcliffe, P. A. Meacock, W. Schuch, D. B. Scanlong, T. C. Atkinson, C. R. Newton, and A. F. Markham, *Nature (London)*, 1981, **292**, 756.
[6] D. V. Goeddel, D. G. Kleid, F. Bolivar, H. L. Heyneker, D. G. Yansura, R. Crea, T. Hirose, A. Kraszewski, K. Itakura, and A. D. Riggs, *Proc. Natl. Acad. Sci. USA*, 1979, **76**, 106.

abundantly expressed in certain specialist cells, *e.g.*, globin in erythrocytes, and where in consequence the specific messenger RNAs are abundant. Thus copy DNAs (cDNAs) made from mRNA were likely to contain coding sequence for that protein (see Volume 1 of this series). Genes for less common proteins are not readily detected by this method, since the desired mRNA represents only a minute proportion of the total population.

Synthetic oligonucleotides can be used as 'bait' to 'fish' for these rarer genes if all or part of the amino-acid sequence of their encoded proteins is known. In spite of the degeneracy of the code, there are a limited number of nucleotide sequences which can code for a tetra- or penta-peptide sequence, especially if that peptide contains methionine or tryptophan, which have unique codons and/or amino acids which have only two codons. For example, a protein containing the sequence (HN)-Asp-Met-Met-Trp-Gly-(CO) is coded by the mRNA with the sequences 5'GAYAUGAUGUGGGGN where N is any nucleotide and Y is a pyrimidine (C or U). This in turn is encoded by a DNA of sequence 5'CCCCACATCATRTC where R is a purine (A or G). Both of these may be made as a mixture by addition of both A and G blocked precursors on the eleventh cycle of synthesis, but only one will hybridize to the mRNA. This oligonucleotide mixture may be specifically hybridized to the required mRNA without isolating it from the mixed cellular population and the hybrid will then act as a primer for reverse transcription. The cDNA products may then be cloned and/or used as probes to identify the sequences in genomic DNA encoding the rare protein, *e.g.*, the histocompatibility antigen HLA 8B.[3]

Another major contribution of synthetic oligonucleotides is their use to improve methods for rapid determination of sequences in nucleic acids (see Volume 2 of this series). New primers have been synthesized[3] for the 'M13/dideoxy' method of Sanger[7] (see also Volume 2 of this series) or reverse hybridization[8,9] and the versatility increased further by construction of new vectors derived from M13mp2 by insertion of a piece of synthetic DNA at its EcoR1 cloning site. These synthetic oligonucleotides contain the recognition sites of several restriction endonucleases which are not present in, or have been removed from, the remainder of the vector. They thus act as 'multiple cloning sites' (m.c.s.) for insertion or removal of many different restriction fragments. Three of the oligonucleotides most commonly used[10,11] are shown below:

```
  1   2   3   4   5                                                      6   7   8
 THR MET ILE THR ASN            m.c.s. of M13mp7 (ref. 10)              SER LEU ALA
 ATG ACC ATG ATT ACG AAT TCC CCG GAT CCG TCG ACC TGC AGG TCG ACG GATCCG GGG AAT TCA CTG GCC
                  EcoRI         BamHI        PstI              BamHI    EcoRI       HaeIII
                              Sall, AccI,  Sall, AccI,
                              HincII       HincII

  1   2   3   4   5   6                                                  7   8
 THR MET ILE THR ASN SER                 m.c.s. of M13mp8 (ref. 11)      7   8
 ATG ACC ATG ATT ACG AAT TCC CGG GGA TCC GTC GAC CTG CAG CCA AGC TTG GCA CTG GCC
                  EcoRI         BamHI         PstI         HindIII
                              SmaI, XmaI   Sall, AccI
                                           HincII

  1   2   3   4                                          5   6   7   8
 THR MET ILE THR              m.c.s. of M13mp9 (ref. 11) ASN SER LEU ALA
 ATG ACC ATG ATT ACG CCA AGC TTG GCT GCA GGT CGA CGG ATC CCC GGG AAT TCA CTG GCC
                     HindIII     PstI                   BamHI    EcoRI    HaeIII
                                            Sall, AccI        SmaI, XmaI
```

[7] F. Sanger, A. R. Coulson, B. G. Barrell, A. J. H. Smith, and B. H. Roe, *J. Mol. Biol.*, 1980, **143**, 161.
[8] G. F. Hong, *Bioscience Rep.*, 1981, **1**, 243.
[9] D. M. Brown, J. Frampton, P. Goelet, and J. Karn, *Gene*, 1982, **20**, 139.
[10] J. Messing, R. Crea, and P. H. Seeburg, *Nucleic Acids Res.*, 1981, **9**, 309.
[11] J. Messing and J. Vieira, *Gene*, 1982, **19**, 269.

Natural Polymers: Nuclei Acids

The restriction sites present in the m.c.s. of M13mp8 and M13mp9 are identical but their order is reversed. This permits a DNA fragment cloned into one of these vectors to be turned around with respect to the vector simply by recloning into the other vector. Thus both strands (and therefore ends) of the fragment can be sequenced by the 'M13/dideoxy' method.

Site-directed Mutagenesis.—'Classical' genetics depends, in the main, on being derived from random generation of mutants, followed by phenotypic selection and genotypic characterization. These methods have proved very powerful when applied to prokaryotes (*e.g.*, *Escherichia coli* and its phages) but are of much more limited application to eukaryotes. Yeast and *Drosophila* are the most outstanding exceptions to this rule.

With the advent of recombinant DNA techniques it is now possible to construct mutants *in vitro* in which the site or sites of mutation are not randomly distributed but directed to specific segments of DNA (or genetic *loci*). Site-directed mutagenesis may, for convenience of presentation, be divided into three forms on the basis of both the method and product of mutagenesis. In all these methods, recombinant DNA is used so individual mutants can be selected by cloning.

Deletion mutants may be constructed by removal of a segment of DNA from a circular virus or recombinant and religation of the shortened DNA remaining. This deleted segment may simply be a restriction fragment (*e.g.*, a Hae III fragment from the plasma vector pBR322 to generate a new plasmid vector pAT153).[12] Alternatively, a segment of DNA may be sequentially deleted by use of a 3′- and 5′-exonuclease, Bal 31[13] and religation. The deleted fragment can be replaced by insertion of a similar-size fragment of DNA of unrelated sequence, *e.g.*, synthetic linker DNA. Such methods have proved useful in the study of, for example, adenovirus and in establishing the location of the intragenic promoters of 5S RNA and tRNA genes (see Volume 2 of this series).

The production of point mutations by chemical mutagenesis can also be site directed since several chemical mutagens are highly specific for single-stranded DNA. A single-stranded region can be produced in an otherwise duplex DNA by two methods:[14] (*i*) a single strand break can be made by cutting the DNA at one site with a restriction enzyme in the presence of ethidium bromide followed by extension of the gap with exonuclease; (*ii*) a single-stranded DNA can be hybridized to supercoiled duplex DNA such that the segment to be mutated is displaced to form a susceptible, non-base paired loop ('D loop'). The chemical mutagenesis is limited to transition mutations (C → T; A → G). The most commonly used mutagen is bisulphite which catalyses deamination of cytidine residues to give GC → AT transitions. Transversion (purine → pyrimidine) mutations (as well as transitions) can be obtained by formation of mismatches in the region being mutagenized. This may be achieved by one of two methods. A generated gap can be filled in using either mutagenic nucleoside analogues (mutagenic because mistakes cannot be repaired)[15] or an error-prone polymerase.[16] Alternatively, very specific mutagenesis (usually of

---

[12] A. J. Twigg and D. Sherratt, *Nature* (*London*), 1980, **283**, 216.
[13] R. J. Legerski, J. L. Hodnett, and H. B. Gray, *Nucleic Acids Res.*, 1978, **5**, 144.
[14] D. Shortle, D. Di Maio, and D. Nathans, *Annu. Rev. Genet.*, 1981, **15**, 265.
[15] D. Shortle, P. Grisafi, S. J. Benkovic, and D. Botstein, *Proc. Natl. Acad. Sci. USA*, 1982, **79**, 1588.
[16] R. A. Zakour and L. A. Loeb, *Nature* (*London*), 1982, **295**, 708.

one nucleotide) can be achieved by synthesis of an oligonucleotide which is complementary in sequence to the DNA to be changed in all but one nucleotide. The synthetic oligonucleotide is then annealed (with one mismatch) to the recombinant DNA (present in, *e.g.*, the infective 'plus' strand of the viral vector M13) and then extended and ligated to provide a mutant replicative form (RF) of the M13 recombinant. One of the more spectacular applications of this method was the mutation of an anticodon of a human tRNA^{Lys} gene to produce a DNA the transcript of which was able to suppress a UAG (nonsense) codon present in the globin gene of a β^0-thalassaemia.[17]

2 Oncogenes

When the RNA tumour viruses, known as retroviruses, infect a cell, their RNA genome is copied by reverse transcriptase into a double-stranded DNA which becomes integrated into the genomic DNA of their host. The viral genes are thus replicated together with the cellular genes and integration is an obligatory step in the production of virus. The integration of some of the retroviruses into tissue culture cells transforms them, causing them to grow continuously and chaotically, so enhancing virus production. The cause of this transformation was first identified in avian sarcoma virus as the product of a specific viral gene. This, the *src* gene, encodes a protein kinase which phosphorylates tyrosine residues. Its action has been strongly associated with its phosphorylation of a tyrosine residue in vinculin, a protein constituent of cellular adhesion plaques. It appears that such phosphorylation disturbs cell–cell interaction and gives rise to transformed growth. The *src* gene has become known as an oncogene and equivalents have been found in nearly 20 retroviruses and, in at least eight of these, the gene product is a protein kinase which phosphorylates tyrosine residues (for a review, see ref. 18).

By using the techniques of genetic manipulation, equivalents of the retroviral oncogenes have now been found in mammalian cells. Thus, a radioactive *src* DNA probe allowed the identification of a chicken cellular gene, called *c-src* to distinguish it from the viral *v-src* gene. The *c-src* genes, which have since been detected in many vertebrates, are closely related to the *v-src* genes and differ mainly in being split by introns into several exons. Clearly the viral and vertebrate genes must have had the same origins and it is generally assumed that the viruses captured them from their hosts. The lack of introns in the viral genes has led to the suggestion that they were captured as mature mRNA or as a reverse transcriptase-catalysed DNA copy of a mRNA.

The cellular counterpart of oncogenes, often called proto-oncogenes, appear to be active in normal cells. The *c-src* gene gives rise to a protein which is antigenically identical to that encoded by *v-src* but the product does not normally make the cell cancerous. However, proto-oncogenes can be the cause of neoplastic growth if forced to produce their product in excess. One way in which this occurs is by the action of the so-called slow transforming viruses which differ from the avian sarcoma virus (an example of an acute transforming virus) in not containing

[17] G. F. Temple, A. M. Dozy, K. L. Roy, and Y. W. Kan, *Nature (London)*, 1982, **296**, 537.
[18] J. R. Bishop, *Sci. Am.*, 1982, **246** (No. 3), 81.

oncogenes. In the absence of oncogenes, these viruses do not transform tissue culture cells at a detectable rate. They can, however, with a long latent period, induce cancer, and it was first shown by Hayward et al.[19] that they do this by integrating into the host cell genome adjacent to a proto-oncogene so activating it and inducing neoplastic growth. This demonstration was made with another species of oncogene known as *myc*. The viral oncogene of this system (*v-myc*) is the transforming gene of the avian virus MC 29 but again chicken cells have been shown to carry a proto-oncogene equivalent called *c-myc*. Hayward and his collaborators showed that a different virus, the slow transforming avian leukosis virus (ALV), integrated adjacent to the *c-myc* proto-oncogene. Transcription of the gene was enhanced by the viral promotor and this led to lymphoid leukosis.

The oncogenes so far discovered, which currently number about twenty, appear to be widely spread as proto-oncogenes in the DNA of vertebrates. They are strongly conserved through evolution, they exhibit homologies which indicate that they can be grouped into a small number of gene families of common ancestry,[20] and many of them appear to be activated in neoplastic transformation. Thus, the *c-myc* genes, first found in chickens, have also been detected in a variety of human cells. Their transcription is increased in neoplastic tissues[21] and, in the leukaemia cell line, HL-60, increased expression is associated with *c-myc* gene amplification.[22]

The retroviruses do not appear to be major causes of cancer in man. Nevertheless, they may constitute one branch of a common pathway of cancer causation. Thus, cancer may result from insertion of a viral oncogene, activation of a proto-oncogene under the influence of a strong viral promotor, or activation of a proto-oncogene in other ways; this may for instance be one way in which carcinogens exert their effects. It appears that proto-oncogene activation can result from the formation of abnormal chromosomes and again the *c-myc* gene has been implicated. A number of cancers are characterized by chromosome translocations and two of these, Burkitt's lymphoma in man and mouse plasmacytoma, have been linked with *c-myc*. Three simultaneous but independent publications (for a short review, see ref. 23) have shown that the *c-myc* gene is located within a segment of chromosome 8 which is translocated to chromosome 14 in Burkitt's lymphoma. This rearrangement results in the insertion of the *c-myc* gene adjacent to the immunoglobulin heavy chain gene (heavy chain proteins comprise part of an antibody molecule). The translocation is probably a perversion of the normal cellular processes as, during the development of antibody-producing cells, the various segments of the heavy chain gene are brought together by a rearrangement of elements in chromosomes 14, 2, and 22. In Burkitt's lymphoma, the exchange processes also involve chromosome 8 and result in the capture and activation of the *c-myc* gene (refs. 23—25 are a series of mini reviews on these findings).

---

[19] W. S. Hayward, B. G. Neel, and S. M. Astrin, *Nature (London)*, 1981, **290**, 475.
[20] R. A. Weinberg, *Cell*, 1982, **30**, 3.
[21] A. Eva, K. C. Robbins, P. A. Andersen, A. Srinivasan, S. R. Tronick, E. P. Reddy, N. W. Ellmore, A. T. Galen, J. A. Lautenberger, T. S. Papas, E. H. Westin, F. Wong-Staal, R. C. Gallo, and S. A. Aaronson, *Nature (London)*, 1982, **295**, 116.
[22] R. D. Favera, F. Wong-Staal, and R. S. Gallo, *Nature (London)*, 1982, **299**, 61.
[23] P. Newmark, *Nature (London)*, 1983, **301**, 111.
[24] J. D. Rowley, *Nature (London)*, 1983, **301**, 290.
[25] P. Newmark, *Nature (London)*, 1983, **301**, 462.

The link between *c-myc* and mouse plasmacytoma also involves a translocation which places the proto-oncogene in association with the immunoglobulin heavy chain gene. In mouse, the *c-myc* gene is in chromosome 15 and the heavy chain gene is in chromosome 12; a 15:12 translocation brings the two together. Whether the shift of *c-myc* to the immunoglobulin locus is causative or a consequence of the translocation is uncertain. It is tempting to believe that transcription of *c-myc* in its new locus gives rise to the tumour but a minority of both cancer types are not associated with *c-myc* rearrangement. Nevertheless, the above findings have been followed by more evidence for a correlation between cancer associated chromosomal translocations or deletions and the location of proto-oncogenes at the breakpoints. Thus, the *c-mos* proto-oncogene is located at the break point in the 8:21 translocation seen in acute myeloblastic leukaemia.[26] Similarly, *c-abl* is found at the break point of the 9:22 translocation in chromic myelogenous leukaemia[27] and *c-myb* is near the break point of the chromosome 6 long arm deletion seen in acute lymphoblastic leukaemia.[28]

Another way in which a proto-oncogene might be activated is by mutation. Support for this has come with the finding that, in two bladder carcinoma cell lines, transformation appears to be caused not by a change in the expression of the proto-oncogene but by a mutation in its DNA sequence. The gene concerned is the cellular equivalent of Harvey murine sarcoma virus oncogene and is variously known as *c-ras^H*, *c-Ha-ras* or *c-has/bas-1*. Two groups, those of Weinberg[29] and Barbacid,[30] have shown that in bladder carcinoma cells, a single base change in the *c-ras^H* gene leads to the substitution of a valine for a glycine in its 21 000 MW protein product.

With the exception of those encoding a protein kinase, little is known of the normal function of oncogene products. One other has been shown to change in level during embyogenesis and may therefore be involved in differentiation. Very recently, however, the amino terminus of the above 21 000 MW product of *c-ras^H*, known as P21, has been shown to exhibit strong sequence homology with the ATP binding region of mitochondrial ATPase.[31] The mutation in bladder carcinoma cell P21 is in the amino terminus.

The rate of research on oncogenes is such that by the time this review is in print its content will be thoroughly out of date. However, the current position can be summarized as follows. Oncogenes and proto-oncogenes appear to comprise a small, related set of protein encoding genes. Since they have been strongly conserved through evolution, their protein products must have important cellular roles and whatever these are, they are such that, if the proteins are overproduced or modified in their activity, they have a catastrophic effect on the cell and can result in cancer.

---

[26] B. G. Neel, S. C. Jhanwar, R. Chaganti, and W. S. Hayward, *Proc. Natl. Acad. Sci. USA*, 1982, **79**, 7842.

[27] A. de Klein, A. van Kessel, G. Grosveld, C. R. Bartram, A. Hagemeijer, D. Bootsma, N. K. Spurr, N. Heisterkamp, J. Groffen, and J. R. Stephenson, *Nature (London)*, 1982, **300**, 765.

[28] M. E. Harper, M. I. Simon, G. Franchini, R. C. Gallo, and W. Staal, *Proc. Natl. Acad. Sci. USA* (in the press).

[29] C. J. Tabin, S. M. Bradley, C. I. Bargmann, R. A. Weinberg, A. G. Papageorge, E. M. Scolnick, O. Dhar, D. R. Lowy, and E. H. Chang, *Nature (London)*, 1982, **300**, 143.

[30] E. P. Reddy, R. K. Reynolds, E. Santos, and M. Barbacid, *Nature (London)*, 1982, **300**, 149.

[31] N. J. Gay and J. E. Walker, *Nature (London)*, 1983, **301**, 262.

3 The Control of Gene Expression in Eukaryotes

We are here confining our discussion to genes which are transcribed by RNA polymerase II (*i.e.*, those genes which encode proteins).

It has long been established that the regulation of transcription in prokaryotes involves complex nucleic acid–protein interactions. A common mechanism involves regulatory genes which give rise, through mRNA transcripts, to repressor proteins. These repressors control the expression of the genes encoding specific metabolic enzymes, but they themselves are controlled in their activity by the substrates and products of the enzymes concerned.

The mechanisms by which transcription might be controlled in eukaryote cells are far less well understood but similar protein-nucleic acid interactions are likely to occur at several levels:

(*i*) As in prokaryotes, specific protein–DNA interactions are likely to exert an on/off control on the transcription of the protein encoding genes.

(*ii*) Specific factors are likely to regulate the activity of the different eukaryote RNA polymerases.

(*iii*) Chromatin, the association of eukaryote DNA with histones and non-histone proteins, is likely to be important in controlling the availability of genes for transcription. In particular, it is likely that chromatin conformation and structure will play a role in the permanent shut down of genes that ensures that, for instance, proteins specific to a brain cell are not produced in a liver cell.

(*iv*) Many, probably most, gene transcripts are produced as large precursors which are subjected to post-transcriptional processing. Furthermore, the mature sequences must be transported through the nuclear membrane. Both of these events may provide further levels at which gene expression can be controlled.

A detailed understanding of the control of eukaryote gene expression at any of the above levels is still a distant goal. Nevertheless, progress has been made towards elucidating possible mechanisms and the following summarizes the current position.

The Influence of Chromatin Structure on Gene Expressions.—The best evidence that the conformation, packaging, or composition of chromatin exerts an influence on gene transcription comes from studies with the enzyme, deoxyribonuclease I (DNase I). Actively transcribed genes are preferentially digested by this enzyme and, by implication, are part of a chromatin arrangement more available to enzymic interaction.

This increased sensitivity was first described in globin genes which were found to be preferentially digested by DNase I acting on the chromatin of the erythroid cells in which they were expressed but not in other cell types in which they were not expressed.[32] Similar findings have since been reported in many gene systems including genes for enzymes, structural proteins, hormone-induced proteins, and viral proteins (for additional examples, see ref. 33). It appears a general rule that the

---

[32] H. Weintraub and M. Groudine, *Science*, 1976, **193**, 848.
[33] G. M. Lawson, B. J. Knoll, C. J. March, S. Woo, M.-J. Tsai, and B. W. O'Malley, *J. Biol. Chem.*, 1982, **257**, 1501.

sensitivity of a gene to digestion by DNase I is correlated with the ability of the gene to be expressed, *i.e.*, it is related to the differentiated state of a cell and whether it is potentially able to express the gene but not necessarily to actual transcriptional activity. Thus, the γ-globin genes, expressed in foetal erythroid cells, are preferentially sensitive to DNase I but they retain their sensitivity in adult erythroid cells where they are not expressed.[34] Similarly, the globin genes of erythroleukaemia cells, in which globin synthesis is inducible by dimethyl sulphoxide, are still sensitive to DNase digestion in the uninduced state.[35] Also, the ovalbumin genes, expressed in avian oviduct in response to oestrogen, retain sensitivity to DNase after hormone withdrawal.[36]

Studies with an adenovirus transcription unit[37] and an oestrogen-induced genomic domain, which includes the ovalbumin gene and two pseudogenes,[33] indicate that the region of preferential sensitivity to DNase corresponds closely with the transcriptional entities. In the latter case, repetitive DNA elements in opposite orientations to each other are found at or very near the transition regions of DNase sensitivity at either end of the gene domain. The authors suggest that repetitive elements may play a role in defining the ends of DNase I sensitive regions.[38]

Within the genomic regions of DNase sensitivity there are so-called hypersensitive sites, *i.e.*, unique segments that are sensitive to very mild digestion by DNase I. Kene *et al.*[39] have located these at the 5' ends of the genes for each of four heat shock-induced proteins in *Drosophila* and they have also been found in the 5' regions of the DNA encoding avian globin.[40] In contrast to the overall DNase sensitivity of active genes the degree of nuclease sensitivity of these hypersensitive sites may be related to the level of transcription. Chicken α-globin genes become less DNase I sensitive when, during development, they become inactive.[40]

The features of the chromatin of active genes which make them more DNase sensitive are not known in detail but are dependent on the specific occurrence in their constituent nucleosomes of two of the high-mobility group proteins, HMG 14 and HMG 17.[41] Differential DNase sensitivity is lost when non-histone chromatin proteins are removed but is restored by the addition of HMG 14 and HMG 17.[42] Other proteins which have been associated with active chromatin include ubiquitin-conjugated histone H2A (ref. 43) and acetylated histones.[44] Baer and Rhodes[45] have recently shown that RNA polymerase II will selectively bind to 15% of nucleosomes and that these are heavily enriched in active genes and are deficient in histones H2A and H2B. Lawson *et al.*[33] have prepared a model whereby, when a cell differentiates, those genes which will be expressed in the differentiated state are incorporated into

---

[34] N. S. Young, E. J. Benz, J. A. Kantor, P. Kretschmer, and A. W. Neinhuis, *Proc. Natl. Acad. Sci. USA*, 1978, **75**, 5884.
[35] D. M. Miller, P. Turner, A. W. Nienhuis, D. E. Axelrod, and T. V. Gopalakrishnan, *Cell*, 1978, **14**, 511.
[36] R. D. Palmiter, E. R. Mulvihill, G. S. McKnight, and A. W. Senear, *Cold Spring Harbor Symp. Quant. Biol.*, 1977, **42**, 639.
[37] S. J. Flint and H. M. Weintraub, *Cell*, 1977, **12**, 783.
[38] W. E. Stumph, M. Baez, W. G. Beattie, M. J. Tsui, and B. W. O'Malley, *Biochemistry*, 1983, **22**, 306.
[39] M. A. Keene, V. Corces, K. Loewnhaupt, and S. C. R. Elgin, *Proc. Natl. Acad. Sci. USA*, 1981, **78**, 143.
[40] H. Weintraub, A. Larsen, and M. Groudine, *Cell*, 1981, **24**, 333.
[41] S. Weisbrod and H. Weintraub, *Cell*, 1981, **23**, 391.
[42] S. Weisbrod and H. Weintraub, *Proc. Natl. Acad. Sci. USA*, 1979, **76**, 630.
[43] L. Levinger and A. Varshavsky, *Cell*, 1982, **28**, 375.
[44] D. Nelson, J. Covault, and R. Chalkley, *Nucleic Acids Res.*, 1980, **8**, 1745.
[45] B. W. Baer and D. Rhodes, *Nature (London)*, 1983, **301**, 482.

Natural Polymers: Nucleic Acids 141

loops or DNase I sensitive domains. The genes within these domains are not necessarily expressed but can be induced by various modulators. Hypersensitive sites are postulated to occur at key points within the domain such as at the 5' end of genes. The idea of active genes incorporated into exposed loops of chromatin may fit in with a current hypothesis that DNA is arranged in supercoiled loops anchored to the nuclear matrix. The avian ovalbumin gene is preferentially associated with the DNA of the matrix in tissues in which it is expressed but not in transcriptionally inactive tissues.[46]

Base Modification of DNA and the Control of Gene Expression.—A feature of eukaryote DNA which has long held attractions as a possible signal for the control of gene expression is the variable level to which its constituent bases, particularly the cytosine residues, are methylated. The methylation of eukaryote DNA has been the subject of a recent extensive review.[47] The extent of methylation is very variable. It is, for instance, very low in insects and very high in plants while that of vertebrates lies in the middle of the range with approximately one 5-methylcytosine for every 30 cytosines. Clearly, such a variation makes it difficult to assign a generally applicable role to methylation.

There are available pairs of restriction endonucleases which recognize the same nucleotide sequence but are differently affected by its methylation. The most commonly used enzyme pair is Msp I, which cuts the sequence CCGG whether or not the second C is methylated, and Hpa II, which cuts the same sequence provided it is unmethylated. Thus, if the methylation of this sequence within a given gene varies under differing growth conditions, one of these enzymes but not the other will differ in the extent to which it is able to fragment the gene.

Studies with these enzyme pairs have shown that the degree of methylation of specific sites does vary with gene expression. For example, some but not all of the CCGG sequences in globin genes are specifically undermethylated in those tissues, or stages of development, in which they are expressed.[40,48,49] Similarly, the genes for the egg white proteins are relatively undermethylated in oviduct when compared with other tissues in which they are not expressed.[50] Other examples include the loss of methyl groups from immunoglobulin heavy and light chain genes during lymphocyte differentiation,[51,52] the relative hypomethylation of metallothionein genes when induced by cadmium or glucocorticoids,[53] and the correlation between expression and hypomethylation of α foetoprotein genes.[54] The DNA of adenovirus, herpes virus, and retrovirus, which are not methylated in the virion, are also not methylated in productive infection. However, they are methylated in non-productive infections.[55-57] Thus, Jähner et al.[57] showed that when the retrovirus,

[46] S. I. Robinson, B. D. Nelkin, and B. Vogelstein, *Cell*, 1982, **28**, 99.
[47] R. L. P. Adams and R. H. Burdon, *CRC Crit. Rev. Biochem.*, 1982, **13**, 349.
[48] C. J. Shen and T. Maniatis, *Proc. Natl. Acad. Sci. USA*, 1980, **77**, 6634.
[49] L. van der Pleog and R. A. Flavell, *Cell*, 1980, **19**, 947.
[50] J. L. Mandel and P. Chambon, *Nucleic Acids Res.*, 1979, **7**, 2081.
[51] J. Rogers and R. Wall, *Proc. Natl. Acad. Sci. USA*, 1981, **78**, 7497.
[52] M. Yagi and M. E. Koshland, *Proc. Natl. Acad. Sci. USA*, 1981, **78**, 4907.
[53] S. J. Compere and R. D. Palmiter, *Cell*, 1981, **25**, 233.
[54] E. A. Eiferman, P. R. Young, R. W. Scott, and S. M. Tilghman, *Nature (London)*, 1981, **294**, 713.
[55] D. Sutter and W. Doerfler, *Proc. Natl. Acad. Sci. USA*, 1980, **77**, 253.
[56] R. C. Desrosiers, C. Malder, and B. Fleckenstein, *Proc. Natl. Acad. Sci. USA*, 1979, **76**, 3839.
[57] D. Jähner, H. Stuhlman, C. L. Stewart, K. Harbers, J. Löhler, I. Simon, and R. Jaenisch, *Nature (London)*, 1982, **298**, 623.

Moloney murine leukaemia virus, was introduced into pre-implantation mouse embryos, the viral DNA became methylated and was not expressed. When, however, it was injected into post-implantation embryos, it was not methylated and gave rise to viral infection.

As with DNase sensitivity, site specific undermethylation appears to be a prerequisite for gene expression but is not enough to ensure it. A gene can be undermethylated but not expressed. This is perhaps best shown with oestrogen-induced genes in chicken. In response to this hormone, birds produce egg white proteins in the oviduct and the egg yolk precursor protein, vitellogenin, in liver. Wilks et al.[58] have shown that oestrogen causes the specific demethylation of a CCGG sequence near the 5' end of the vitellogenin gene. However, it does this in both target tissues; i.e., in the liver in which it is expressed and in the oviduct where it is not expressed.

The Role of Specific Nucleotide Sequences in the Control of Gene Expression.—A very attractive concept for the control of gene expression is that a regulatory molecule should be able to modulate transcription by interaction with a specific nucleotide sequence associated with the gene or genes under its influence. Evidence is accumulating that such sequences do exist and this has recently been reviewed by Davidson et al.[59] These authors have collected data on five sets or families of genes, each family being subject to a specific regulatory influence. Thus, one set is induced by amino-acid starvation, another by heat shock, a third at a specific stage of insect metamorphosis, and the remaining two by oestrogen and glucocorticoid hormones, respectively. Each gene is characterized by having, in its 5' flanking sequence, a sequence of 9—24 oligonucleotides which is strongly conserved in all the other genes of the same family.

In three of the five families, the evidence for a role of their consensus sequences in their regulation is circumstantial and is limited to the very low probability that several genes could have the same (or almost the same) sequence at similar positions with respect to the coding sequence. The other two families are rather more convincing. The sequence ATGTGACTC occurs twice in the 5' flanking sequence of four genes for amino-acid synthesizing enzymes which are activated when yeasts are starved of amino acids. Mutations that disrupt the sequence no longer show enzyme induction and revertants that re-establish it recover inducibility.[60] The importance of a 15-nucleotide consensus sequence in controlling the expression of eight Drosophila heat shock-induced genes is also supported by mutation data and by the fact that, if the consensus sequence is placed in the 5' sequence of another gene, that too becomes heat-shock-induced.[61]

It would be satisfying if the consensus sequences of the oestrogen and corticosteroid-induced gene sets could be identified with the accepted mode of action of steroid hormones. These compounds enter their target cells and bind to specific receptor proteins. The hormone-receptor complex then binds to chromatin and activates the transcription of hormone-induced genes. It is obviously tempting

---

[58] A. F. Wilks, P. J. Cozens, I. W. Maltaj, and J.-P. Jost, *Proc. Natl. Acad. Sci. USA*, 1982, **79**, 4252.
[59] E. H. Davidson, H. T. Jacobs, and R. J. Britten, *Nature (London)*, 1983, **301**, 468.
[60] T. F. Donahue, R. S. Davis, G. Lucchini, and F. R. Fink, *Cell*, 1983, **32**, 89.
[61] H. R. B. Pelham and M. Bienz, *EMBO J.*, 1982, **1**, 1473.

Natural Polymers: Nucleic Acids

to speculate that the hormone-receptor complex binds to the consensus sequence and this may yet prove to be the case. They can be shown to bind at or close to the genes they induce. Thus, fluorescent derivatives of the insect steroid hormone, ecdysterone, can be visualized bound specifically at the polytene chromosome puffs which correspond to the genes they induce.[62] The family of glucocorticoid-induced genes, shown by Davidson et al.[59] to share a 24-nucleotide consensus sequence, includes as one of its members the genome of the mouse mammary tumour virus, the production of which is enhanced by glucocorticoids. Within the viruses, the consensus sequence occurs upstream of the viral initiation site, 392 nucleotides into the long terminal repeat (LTR) sequence of the retroviral DNA. The glucocorticoid receptor complex has been shown to bind specifically to this LTR and Geisse et al.[63] have shown that binding occurs to the right-hand 400—500 nucleotides of the repeat, a region that would include the consensus sequence. Furthermore, the LTR can be attached to the coding sequences of genes not normally induced by glucocorticoids and, when such recombinant genes are introduced into cells, they become hormone-inducible.[64,65]

The above consensus sequences are well to the 5' side of the genes they appear to control (-486 to -62 nucleotides with respect to the initiation of transcription). Much closer to the initiation site are the so-called promotor sites believed to play a role in the specificity of the initiation of RNA polymerase II (the nature of the promotors for prokaryote and eukaryote polymerases were reviewed in Volume 2 of this series). It is possible that promotors also play a role in the control of transcription. Thus, the mRNAs for salivary and liver amylase are transcribed from the same gene but contain different 5' sequences copied from different genomic sequences. Each 5' sequence is preceded by a different promotor and it has been suggested that the nature of the transcript produced depends on promotor selection.[66]

The Processing of Gene Transcripts in Eukaryotes.—For this review, we are again confining our discussion to those genes transcribed by RNA polymerase II.

The splicing of mRNA to remove the transcripts of introns is the aspect of mRNA processing which is currently attracting most attention and what follows is an update of our review on this subject in Volume 2 of this series.

The role of the nucleotide sequence of the exon–intron boundaries is now clearly established and its nature has been re-examined by Mount.[67] After cataloguing the sequences of 139 exon–intron boundaries (5' end of intron) and 130 intron–exon boundaries (3' end of intron), he derived a consensus sequence which would give rise to the following transcript:

$$5'\text{—exon transcript—}{}^C_A\text{AGGU}{}^A_G\text{AG—intron transcript—}\binom{U}{C}_n N{}^C_U\text{AGG—exon transcript—}3'$$

with the splice point of the exon–intron boundary at AGGU, and the splice point of the intron–exon boundary at AGG.

---

[62] H. Gronemeyer and O. Pongs, *Proc. Natl. Acad. Sci. USA*, 1980, **77**, 2108.
[63] S. Geisse, C. Scheidereit, H. M. Westphal, N. E. Hynes, B. Groner, and M. Beato, *EMBO J.*, 1982, **1**, 1613.
[64] F. Lee, R. Mulligan, P. Gerg, and G. Ringold, *Nature (London)*, 1981, **294**, 228.
[65] A. L. Huang, M. C. Ostrowski, D. Berard, and G. L. Hayer, *Cell*, 1981, **27**, 245.
[66] R. A. Young, O. Hagenbüchle, and V. Schibler, *Cell*, 1981, **23**, 451.
[67] S. M. Mount, *Nucleic Acids Res.*, 1982, **10**, 459.

The sequence immediately 5' to the intron–exon boundary is always pyrimidine rich and free of the dinucleotide AG. The most invariant aspect of the consensus sequence is the GU at the beginning of the intron transcript and the AG at its end (the so-called GT/AG rule when applied to the non-coding strand of the genomic DNA). Exceptions to this rule have been reported but are not rigorously proven (for a review, see ref. 67).

The role of this sequence in splicing is supported by studies of β-thalassaemias, a group of hereditary anaemias in which the production of globin is either diminished (β^+-thalassaemia) or absent (β^0-thalassaemia). In one form of β^0-thalassaemia, lack of β-globin production is due to a mutation in the DNA encoding the GGU splice point at the exon–intron boundary of the large intron of the gene.[68,69] Furthermore, in a β^+-thalassaemia, defective chain synthesis is associated with a single G $\rightarrow$ A mutation which creates a sequence CTATTAG within the large intron. The sequence closely resembles the intron–exon sequence CCGTTAG and competes with it as a splice point to the extent that 90% of transcripts are incorrectly spliced.[70] Similar findings to those with β-thalassaemia have also been reported in one form of α-thalassaemia[71] and with engineered mutants of other genes.

Notwithstanding these findings, an examination of the complete nucleotide sequence of the chicken ovalbumin gene has revealed that sequences similar to the consensus sequence occur throughout the gene but only those at known intron–exon boundaries (the ovalbumin gene has seven introns) are apparently recognized as splice points.[72] There must, therefore, be components in addition to the consensus sequence which are involved in the specificity of splicing.

The components most strongly implicated in the precision of splicing are the small nuclear RNAs (snRNA), particularly the species known as U1. This and other 'U' series snRNAs occur in ribonucleoprotein particles (snRNP) and the proteins of these, together with the proteins associated with pre-mRNA (hnRNP), may also play a role in splicing.

The evidence for a role of snRNP in splicing can be listed as follows:

(1) The 5' end of U1 RNA exhibits extensive sequence complementarity with the consensus sequence of both the exon–intron and the intron–exon boundaries. Thus U1 could hybridize to both ends of an intron and draw the two neighbouring exons into an alignment which would permit their ligation.[73,74]
(2) The sequence of U1 and other 'U' series snRNAs have been strongly conserved through the evolution of eukaryotes from dinoflagelates to man (for a review, see Busch et al.[75]). To judge from their antigenic cross-reaction, some snRNP proteins have also been strongly conserved.[73,76]

---

[68] M. Baird, C. Driscoll, H. Schreiner, G. V. Sciarratta, G. Sansone, G. Niazi, F. Ramirez, and A. Bank, *Proc. Natl. Acad. Sci. USA*, 1981, **78**, 4218.
[69] R. Treisman, N. J. Proudfoot, M. Shander, and T. Maniatis, *Cell*, 1982, **29**, 903.
[70] M. Busslinger, N. Moschonas, and R. A. Flavell, *Cell*, 1981, **27**, 289.
[71] B. K. Felber, S. H. Orkin, and D. H. Hamer, *Cell*, 1982, **29**, 895.
[72] S. L. Woo, W. G. Beattie, J. F. Catterall, A. Dagaiczyk, R. Staden, G. G. Brownlee, and B. W. O'Malley, *Biochemistry*, 1981, **20**, 6437.
[73] M. R. Lerner, J. A. Boyle, S. M. Mount, S. L. Wolin, and J. A. Steitz, *Nature (London)*, 1980, **283**, 220.
[74] J. Rogers and R. Wall, *Proc. Natl. Acad. Sci. USA*, 1980, **77**, 1877.
[75] H. Busch, R. Reddy, L. Rothblum, and Y. C. Choi, *Annu. Rev. Biochem.*, 1982, **51**, 617.
[76] J. C. Wooley, R. D. Conc, D. Tartof, and S.-Y. Chung, *Proc. Natl. Acad. Sci. USA*, 1982, **79**, 6762.

(3) Several groups have shown that a proportion of snRNP is associated with hnRNP. However, when the 5' end of U1 is removed, the particles of which they are a component no longer associate with hnRNP.[73]

(4) The *in vitro* splicing of adenovirus RNA sequences is inhibited when infected cell nuclei are pre-incubated with antiserum raised against snRNP.[77] Furthermore, if antisera are used for the fluorescent labelling of squashed *Drosophila* salivary gland cells, the antigen shows a marked localization among chromosomal puffs.[78] Such puffs are regions of active transcription.

(5) The finding that U1 containing snRNP can be purified on anti-m$^{2.2.7}$G antibody–sepharose columns, indicates that the 5' end of the molecule is available for hybridization.[79] The vulnerability of the 5' end to digestion by RNase further supports these indications.[73,80]

(6) Gross and Cetron[81] have introduced radioactive snRNA into *Drosophila* tissue culture cells. Their data indicate that the RNA molecules remain stable and that they accumulate in the nucleus. Furthermore, at least some of them complex with protein to form snRNP and they base pair with hnRNA [hnRNA includes mRNA precursors (pre-mRNA)].

(7) As a probe of snRNP/pre-mRNA interaction, Mount[82] has prepared ^{32}P labelled transcripts of a cloned section of the β-globin gene which includes the first intron and the flanking sequences. After incubation of these transcripts with purified U1-snRNP, 80% of the labelled RNA could be precipitated with antibodies to snRNP. In further characterization, it was found that the interaction of the particle with the RNA resulted in the protection of a 17-nucleotide sequence from digestion with T1 ribonuclease. This nucleotide sequence occurred at the 5' intron–exon junction. No interaction was observed with the 3' splice junction and no ribonuclease or ligating activities were observed. None of the other 'U' series snRNPs bound to the RNA probe.[82]

The above evidence, although fairly impressive, is largely circumstantial. Much of it strongly suggests a role for snRNP, and, more specifically, for the 5' end of U1 snRNA, in splicing. However, one cannot help feeling that if splicing is as simple as portrayed in the models of Lerner *et al.*[73] and Rogers and Wall[74] it would have been confirmed by now. The fact that intensive research has not led to confirmation is giving rise to wider speculation. Several groups have suggested that hnRNP[83] or hnRNP plus snRNP[84] associate into more complex aggregates. Busch and Reddy[85] have proposed that snRNP are polymerized on moving elements of the nuclear matrix on or near the perinucleolar chromatin.

Potentially of much greater value than the multiplication of splicing models is the recent development of *in vitro* splicing systems. So far, these follow the appearance

---

[77] V. W. Yang, M. R. Lerner, J. A. Steitz, and S. J. Flint, *Proc. Natl. Acad. Sci. USA*, 1981, **78**, 1371.
[78] P. E. Thompson, 1983, personal communication.
[79] J. H. Smith and G. L. Eliceiri, 1982, personal communication.
[80] J. P. Liautard, J. Sri-Widada, J. Brunel, and P. Jeanteur, *J. Mol. Biol.*, 1982, **162**, 623.
[81] R. H. Gross and M. S. Cetron (in the press).
[82] S. M. Mount, 1982, personal communication.
[83] W. M. LeStourgeon, L. Lothstein, B. W. Walker, and A. L. Beyer, in 'The Cell Nucleus', ed. H. Busch, Academic Press, New York, 1981, Vol. IX, pp. 49—87.
[84] C. E. Sekeris and A. Guialis, in 'The Cell Nucleus', ed. H. Busch, Academic Press, New York, 1981, Vol. VIII, pp. 247—259.
[85] H. Busch and R. Reddy, *Biochem. Soc. Bull.*, 1982, **14**(3), 22.

of spliced mRNA from intact nuclei[86] or require whole cell extracts[87–89] and their efficiency is low. Nevertheless, they are capable of splicing viral[87,88] or globin pre-mRNA,[89] and it is to be hoped that their availability will greatly aid the elucidation of mRNA splicing mechanisms.

---

[86] V. W. Yang, M. R. Lerner, J. A. Steitz, and S. J. Flint, *Proc. Natl. Acad. Sci. USA*, 1981, **78**, 1371.
[87] B. Weingartner and W. Keller, *Proc. Natl. Acad. Sci. USA*, 1981, **78**, 4092.
[88] C. J. Goldenberg and H. J. Raskus, *Proc. Natl. Acad. Sci. USA*, 1981, **78**, 5430.
[89] R. Kole and S. M. Weissman, *Nucleic Acids Res.*, 1982, **10**, 5429.

6
Inorganic Polymers

BY K. M. ROCH

1 Introduction

This Report surveys the literature for 1981 and 1982. As in previous Reports appearing in Volumes 1 and 2, the author has summarized those recent developments of significance, within the context of this volume, on those inorganic polymers which are essentially linear in nature, and consist of predominantly inorganic elements. The review is restricted by size to polymers derived from the two most important and active areas of the poly(organosiloxane)s and the poly(organophosphazene)s.

2 Poly(organosiloxane)s

A high level of publication continues on this polymer class. Commercial interest in the 'Silicones' continues to be as active as ever, and is evidenced by the fact that the patent literature on these polymers still outnumbers those publications appearing in the non-patent literature.[1] This review follows the broad classification of publications used in the earlier reviews.

Polymerization.—The ring-opening polymerization of cyclosiloxane monomers with organometallic catalysts has been reviewed.[2] The interesting emulsion polymerization, in aqueous phase, of organosiloxane monomers has been reported. Octamethylcyclotetrasiloxane has been polymerized by such a technique in the presence of a cationic emulsifier, benzyldimethyldodecylammonium halide, in the presence of an alkali. The mechanism of polymerization is postulated to proceed in two stages, first, the ring-opening of the monomer with formation of hydroxyl terminated dimethylsiloxanes, and then further condensation of these to high-molecular-weight polymers.[3] Other cyclic organosiloxanes have also been studied in this system.[4] The polymerization of hexamethylcyclotrisiloxane (D_3) by $\alpha\omega$-bis(tetramethylammonium) oligodimethylsilanolate in the presence of trimethylsilanol has been reported.[5] The polymerization was studied over the temperature range 22—50 °C, and a postulated mechanism was put forward in which solvation of the silanolate groups by silanol groups gives rise to transition state complexes

---

[1] 'Organosilicon Chemistry', CA Selects, 1981/82, American Chemical Society, Washington.
[2] P. Sigwalt, *Angew. Macromol. Chem.*, 1981, **94**, 161.
[3] S. Liu and P. Li, *Gaofenzi Tongxun*, 1981 (4), 257.
[4] X. Zhang, S. Liu, and D. Dai, *Gaofenzi Tongxun*, 1982 (2), 154.
[5] V. M. Kopylov, P. L. Prikhodko, and V. A. Kovyazin, *Vysokomol. Soedin., Ser. A*, 1982, **24**, 1751.

providing mechanistic pathways to polymer formation. The polymerization of D_3 has also been studied with lithium poly(vinyl trimethylsilane). Nearly quantitative conversion to poly(dimethylsiloxane) (PDMS) was observed in the presence of THF at 22—35 °C.[6] The catalysed condensation polymerization of tetramethyldisiloxane-1,3-diol in various solvents has been studied and a mechanism proposed which is based on hydrogen bonding interactions between silanols and solvents.[7]

Gas-phase plasma polymerization of silicon compounds is a rapidly developing area of interest as a technique for producing deposited films of silicon, silica, or organosiloxane, depending on the monomer utilized. A study of the mechanism of silicon deposition in a SiH discharge plasma has been reported, from which it has been possible to predict the deposition of amorphous silicon films with a desired ratio of silicon to hydrogen.[8] The detection of SiH_x radicals by mass spectroscopy has been reported in rf glow-discharge polymerizations, where siloxane films adherent to Si, Al, and glass are formed. Oxygen (2—6%) has been demonstrated by electron microprobe analysis to be present in such films.[9] The separation of silica and siloxane phases in deposited films, produced from $SiMe_4$ and various gas mixtures, has been reported[10] and the characteristics of the deposited silica studied.[11,12] The plasma polymerizations of hexamethyldisiloxane and tetramethyldisiloxane have been reported and the characteristics of the films studied.[13–16] Similarly, the plasma polymerization of methyltrimethoxysilane has been reported and the by-products identified.[17] The plasma polymerization of aromatic silanes has been demonstrated to give polymeric films in which the phenyl group does not undergo fragmentation, which is usual for other aromatic materials.[11]

The technique of gas-phase plasma polymerization has been demonstrated as a useful method of depositing dielectric films on compound semiconductors,[19] and as a quick, versatile method to give siloxane coatings to human implant components such as heart valves.[20]

Molecular Structure.—A review has been published on the structure–property relationships of model polymer networks, using as examples poly(dimethylsiloxane)s (PDMS), having controlled structural characteristics *via* terminal chain-end crosslinking.[21] Important contributions to polymer network theories are arising from studies of model network PDMS, which are readily 'tailored' to specific

---

[6] S. G. Durgaryan, N. K. Gladkova, N. S. Nametkin, and T. Yu. Nikolaeva, *Vysokomol. Soedin., Ser. B*, 1982, **24**, 116.
[7] M. D. Dragojevic and L. J. Bogunovic, *Glas. Hem. Drus. Beograd.*, 1981, **46**, 183.
[8] P. Kocian, *Thin Solid Films*, 1981, **80**, L81.
[9] S. Bourguard, D. Erni, and J. M. Mayor, *Symp. Proc.-Int. Symp. Plasma Chem., 5th*, 1981, **1**, 664.
[10] R. Szeto and D. W. Hess, *J. Appl. Phys.*, 1981, **52**, 903.
[11] R. Szeto and D. W. Hess, *J. Polym. Sci., Polym. Lett. Ed.*, 1981, **19**, 119.
[12] A. C. Adams, F. B. Alexander, C. D. Capio, and T. E. Smith, *J. Electrochem. Soc.*, 1981, **128**, 1545.
[13] G. Akovali and M. Y. Boluk, *Polym. Eng. Sci.*, 1981, **21**, 658.
[14] G. Akovali and M. Y. Boluk, *Polym. Eng. Sci.*, 1981, **21**, 662.
[15] T. Sasuoka and H. Yasuda, *Kobunshi Ronbunshu*, 1981, **38**, 687.
[16] B. U. Tkachuk, V. I. Kalyuzhnyi, A. L. Shustov, and V. A. Tsendrovskii, *Ukn Khim. Zh.*, 1982, **48**, 544.
[17] A. K. Hays, *Thin Solid Films*, 1981, **84**, 401.
[18] I. Haller, *J. Electrochem. Soc.*, 1982, **129**, 180.
[19] Y. Secui, *Symp. Proc. Int. Symp. Plasma Chem., 5th*, 1981, **1**, 265.
[20] W. R. Watson, *Dept. Eng. Sci. Rep. (Univ. Oxford)*, 1981, 1361.
[21] J. E. Mark, *Pure Appl. Chem.*, 1981, **53**, 1495.

network characteristics.[22-26] A study of some network structural effects upon the tear strength of PDMS has been reported and discussed in terms of network theory.[27] Small-strain equilibrium modulus data from 12 different studies on PDMS rubber networks have been compared, and calculations made of important network parameters. The agreement between the various workers is shown to be remarkably good, and the results support the theoretical concept that interactions between siloxane chains contribute directly to the modulus.[28] Good agreement has also been found for small-strain modulus data in PDMS networks, randomly crosslinked by chemical and radiation techniques, and the results add further support to Flory's recent network model.[29,30] Results reported on the tensile and swelling of model PDMS networks, with low extents of crosslinking, are interpreted by assuming significant contributions from trapped chain entanglements.[31] The swelling and elastic properties of PDMS end-linked networks as a function of sol fraction has provided experimental evidence on the concentration of trapped entanglements,[32] and average orientational order of monomeric units.[33] Results of a study of the conformation of PDMS networks have been reported in which small-angle neutron scattering measurements were carried out on deuteriated polymers, and evidence is presented to compare the radius of gyration, in both network and liquid states.[34] Data obtained on the modulus of elasticity in θ solvents, for PDMS networks at increasing dilution, have been discussed.[35] Studies on the bulk and solution crosslinking of PDMS have demonstrated good agreement with the Flory–Huggins interaction parameters.[36,37]

The determination of thermodynamic and physico-chemical properties of liquid PDMS, by using ultrasound measurements, has been reported.[38] Ultrasonic relaxation studies on PDMS and its solutions in toluene, covering a range of molecular weights, ultrasound frequencies and temperatures, gave results in agreement with theoretical predictions.[39] Excellent agreement has also been reported between the theoretical and measured Rayleigh–Brillouin spectra of phenyl containing organosiloxanes.[40-42] Light-scattering studies on cyclic and linear PDMS in good and poor solvents have been published, and the concentration

[22] J. E. Mark and A. L. Andrady, *Rubber Chem. Technol.*, 1981, **54**, 366.
[23] M. A. Llorente, A. L. Andrady, and J. E. Mark, *J. Polym. Sci., Polym. Phys. Ed.*, 1981, **19**, 621.
[24] M. A. Llorente, A. L. Andrady, and J. E. Mark, *Colloid Polym. Sci.*, 1981, **259**, 1056.
[25] Z. M. Zhang and J. E. Mark, *J. Polym. Sci., Polym. Phys. Ed.*, 1982, **20**, 473.
[26] R. Kosfeld and M. Hess, *Angew. Makromol. Chem.*, 1981, **95**, 139.
[27] A. N. Gent and R. H. Tobias, *ACS Symp. Ser.*, 1982, **193**, 367.
[28] M. Gottlieb, C. W. Macosko, G. S. Benjamin, K. O. Meyers, and E. W. Merrill, *Macromolecules*, 1981, **14**, 1039.
[29] M. Gottlieb, C. W. Macosko, and T. C. Lapsch, *J. Polym. Sci., Polym. Phys. Ed.*, 1981, **19**, 1603.
[30] C. W. Macosko and G. S. Benjamin, *Pure Appl. Chem.*, 1981, **53**, 1505.
[31] K. A. Kirk, S. A. Bidstrup, E. W. Merrill, and K. O. Meyers, *Macromolecules*, 1982, **15**, 1123.
[32] S. Candau, A. Peters, and J. Herz, *Polymer*, 1981, **22**, 1504.
[33] J. P. Cohen-Addad, M. Domard, and J. Herz, *J. Chem. Phys.*, 1982, **76**, 2744.
[34] M. Beltzung, C. Picot, P. Rempp, and J. Herz, *Macromolecules*, 1982, **15**, 1594.
[35] L. Z. Rogovina, V. G. Vasilev, and G. L. Slomimski, *Vysokomol. Soedin., Ser. A*, 1982, **24**, 254.
[36] R. W. Brotzman and B. E. Eichinger, *Macromolecules*, 1981, **14**, 1445.
[37] R. W. Brotzman and B. E. Eichinger, *Macromolecules*, 1982, **15**, 531.
[38] K. P. Grinevich, Yu. G. Artemev, D. V. Nagarov, and A. N. Kolosov, *Plast. Massy*, 1981, **3**, 21.
[39] A. R. Eastwood, A. M. North, and R. A. Pethrick, *J. Chem. Soc., Faraday Trans. 2*, 1982, **78**, 95.
[40] G. Fytes, Y. H. Lin, and B. Chu, *J. Chem. Phys.*, 1981, **74**, 3131.
[41] G. Fytas, T. Dorfmueller, Y. H. Lin, and B. Chu, *Macromolecules*, 1981, **14**, 1088.
[42] Y. H. Lin, G. Fytas, and B. Chu, *J. Chem. Phys.*, 1981, **75**, 2091.

dependence of the diffusion coefficients reported to be in good agreement with those measured by the classical boundary spreading technique.[43] The use of light-scattering techniques has also proved valuable in indicating the extent of specific solvent interactions.[44] The proton magnetic relaxation of rotating Me groups has been used to determine the diffusion of long chains of PDMS, and the data obtained are in agreement with previously proposed theoretical models.[45] A study involving use of PDMS has been reported on the influence of the expansion of polymer molecules in solution, at differing shear rates, and in poor and good solvents.[46] Two publications have appeared which describe modified rheometers used to measure the rheological characteristics of PDMS during crosslinking.[47,48]

The effect of thermal prehistory upon the glass transition temperature of crosslinked PDMS has been reported, and indicates that secondary crystallinity changes significantly as chain mobility decreases in the amorphous phase of the polymer.[49,50] Structural investigations of linear and crosslinked PDMS by ^{29}Si n.m.r. have provided valuable data on the microstructure of such polymers,[51] and on the structure of solid organosiloxane polymers.[52]

A comparison of the experimental and theoretical bond length and bond angle variations in silicates and organosiloxanes has been published,[53] as has an electronic structural study, in which the effects of $p\pi$–$d\pi$ bonding on the $\equiv$Si–O–Si$\equiv$ are discussed.[54] The configurational-dependent properties of PDMS have been evaluated, and the allocation of fractional electronic charges to the silicon, oxygen, and carbon atoms and the consequent coulometric charges, yields an essential contribution to the interpretation of the experimental data.[55]

Thermal/Oxidative Stability.—Results have been reported on the study of organo-metal phosphates containing Cu, Cr, Mn, Co, and Ni which have been developed as heat stabilizers for silicone rubbers. The effectiveness of these compounds was found to be Cu > Cr > Mn > Ni = Co.[56] Similar compounds containing Fe and Ti were studied and were shown to increase markedly the thermal stability of poly(dimethylsiloxane)s. The effectiveness at 300 °C was in the order Cu > Fe > Ti, and at 360 °C the order was Ti > Fe > Cu.[57] Stabilization mechanisms are postulated in these two publications. The thermal and oxidative stability of poly(ethylphenylsiloxane)s

---

[43] C. J. C. Edwards, S. Bantle, W. Burchard, R. F. T. Steptoe, and J. A. Semlyen, *Polymer*, 1982, **26**, 865, 869, and 873.
[44] J. P. Herz, S. Boileau, and S. Candau, *Macromolecules*, 1981, **14**, 1370.
[45] J. P. Cohen-Addad, M. Domard, and S. Boileau, *J. Chem. Phys.*, 1981, **75**, 4107.
[46] Y. Ito, S. Shiina, and I. Tokue, *Makromol. Chem.*, 1982, **183**, 505.
[47] J. Unsworth and J. Sendt, *Eur. Polym. J.*, 1982, **18**, 617.
[48] T. Murayama, *J. Appl. Polym. Sci.*, 1982, **27**, 89.
[49] V. A. Martirosov, V. Yu. Levin, A. A. Zhdanov, and G. L. Slonimskii, *Vysokomol. Soedin., Ser. A*, 1981, **23**, 896.
[50] M. F. Bukina and M. D. Parizenberg, *Vysokomol. Soedin., Ser. B*, 1981, **23**, 456.
[51] G. Engelhardt and H. Janke, *Polym. Bull. (Berlin)*, 1981, **5**, 577.
[52] G. Engelhardt, H. Janke, E. Lippmaa, and A. Samoson, *J. Organomet. Chem.*, 1981, **210**, 295.
[53] G. V. Gibbs, E. P. Meagher, M. D. Newton, and D. K. Swanson, *Struct. Bonding Cryst.*, 1981, **1**, 195.
[54] C. A. Ernst, A. L. Allred, M. A. Ratner, M. D. Newton, G. V. Gibbs, J. W. Moskowitz, and S. Topiol, *Chem. Phys. Lett.*, 1981, **81**, 424.
[55] S. Brueckner and L. Malpezzi, *Macromol. Chem.*, 1982, **183**, 2033.
[56] V. D. Lobkov, I. A. Metkin, I. G. Kolokoltseva, T. A. Batkina, E. P. Lebedev, and V. I. Krikunenka, *Prom-st. Sint. Kauch.*, 1981, **1**, 16.
[57] I. A. Metkin, T. V. Dykina, V. D. Lobkov, N. A. Silina, I. G. Kolokoltseva, and P. A. Betkina, *Prom-st. Sint. Kauch.*, 1981, **10**, 11.

Inorganic Polymers 151

have been investigated, together with the electrical properties, at 293 K and 473 K in the presence of a range of metals (Mn, Co, Cu, Cd, Zn, Ni).[58]

Results have been reported of a study on the flammability of low-molecular-weight poly(dimethylsiloxane)s in which the volatile products from the short-term heating at 300 °C in controlled atmospheres were determined. A cyclic oligomer (D_3) was shown to predominate in the oligomeric products and a relationship was established between b.pt. and flash point. D_3 was shown to play a decisive role in the ignition process, and the flash point was found to be improved by the addition of Ce acetylacetonate.[59] Heat aging effects upon the electrical properties of silicone protective coatings, up to temperatures of 300 °C, have been published.[60] Silicon nitride has been studied as a stabilizer for Me_2 and Ph_2 silicone rubber and weight losses at 350 °C (for 24 h under nitrogen) were reduced from 28% to 2% by the use of this stabilizer at a level of 2 phr.[61] The electrical conductivity changes associated with the thermal decomposition of poly(alkylphenylsiloxane)s over the temperature range 25—700 °C have been studied, and the significance of the peaks in the conductivity/temperature curves discussed.[62]

The u.v. degradation of ocatmethylcyclotetrasiloxane, hexamethyldisiloxane, octane, and tetramethylsilane in the presence of ozone have been studied. The disappearance rates, relative to octane, were found to be 3.3, 1.4, 1.0, and 0.55, respectively, indicating the non-accumulation of these compounds in the atmosphere.[63] Mechanisms have been postulated for the degradation, by gamme-ray irradiation, of a series of simple dimethylcyclicsiloxanes, where the by-products were determined by mass spectrometry.[64] An apparatus has been described for the determination of oxygen absorption by organosiloxanes over the temperature range of 150—400 °C, and results have been reported on studies carried out on various dimethyl, methylphenyl, and methylfluoropropyl siloxanes.[65]

Applications.—The applicational areas of the 'silicones' in the bulk form and as additives continues to grow at a high rate, as evidenced both by the patent literature,[1] and by the technical data issued by the manufacturers. As in the earlier volumes, the literature cited in this Section will only refer to significant papers or reviews which have been published dealing with the use of the 'silicones' in particular applicational areas. Such publications have appeared in the areas of dental materials,[66—70] textile hydrophobing,[71—73] medical applications,[74]

[58] A. D. Damaeva, G. G. Mashutina, and E. A. Kirichenko, *Vysokomol. Soedin.*, Ser. A, 1982, **24**, 1684.
[59] S. Yasufuku and T. Umemura, *Proc. Electr./Electron. Insul. Conf.*, *15th*, 1981, 212.
[60] S. O. Izidinov, E. I. Nesyaeva, G. I. Dashintseva, and V. A. Talykov, *Zh. Prikl. Khim. (Leningrad)*, 1982, **55**, 1378.
[61] C. H. Hsien, C. S. Li, and C. T. Wang, *K'O Hsueh T'ung Pao*, 1981, **26**, 737.
[62] R. T. Johnson and R. M. Biefeld, *Polym. Eng. Sic.*, 1982, **22**, 135.
[63] Y. Abe, G. B. Butler, and T. E. Hogan-Esch, *J. Macromol. Sci., Chem.*, 1982, **A16**, 461.
[64] K. S. Maeng, G. J. Trudel, and L. E. St.-Pierre, *Pollimo*, 1982, **98**, 107.
[65] I. A. Metkin, N. A. Silina, and T. V. Dykina, *Prom-St. Sint. Kauch.*, 1981, **8**, 8.
[66] J. N. Ciesco, W. F. P. Malone, J. L. Sandrick, and B. Mazur, *J. Prosthet. Dent.*, 1981, **45**, 89.
[67] A. M. Lacey, T. Bellman, and H. Fukui, *J. Prosthet. Dent.*, 1981, **45**, 209.
[68] A. M. Lacey, H. Fukui, T. Bellman, and M. D. Jendresen, *J. Prosthet. Dent.*, 1981, **45**, 329.
[69] J. A. Stackhouse, *J. Prosthet. Dent.*, 1981, **45**, 146.
[70] W. D. Cooke, *J. Biomed. Mater. Res.*, 1982, **16**, 315.
[71] H. J. Hackmann, *Dtsch. Faerber-Kal.*, 1981, **85**, 242.
[72] W. Gardner, *Shirley Inst. Publ.*, 1981, **S41**, 61.
[73] C. Brooke, *Text. Manuf.*, 1982, March, 47.
[74] R. Van Noort and M. M. Black, *Biocompat. Clin. Implant Mater.*, 1981, **2**, 79.

ophthalmology,[75,76] cosmetics,[77] pharmaceuticals,[78,79] surfactants,[80,81] high vacuum applications,[82] adhesives,[83,84] electroconducting rubbers,[85-87] electrical properties,[88] liquid rubbers,[89-92] polish chemistry,[93] food additives,[94,95] fluorosilicone rubbers,[96] gas permeable membranes,[97] crosslinking of polyolefins,[98] silicone rubbers generally,[99] silane coupling agents,[100] and a general review of silicones (to November 1981).[101].

3 Organosilicon Backbone Copolymers

Continuing activity is reported in the synthesis and structure/property relationships of copolymer systems, particularly block copolymer systems, of linear organosiloxane backbone units and a range of organic backbone units. To date, only those copolymers with polyether or carbaborane units have assumed any significant commercial importance, but the polymer systems which have received study have demonstrated much of interest to polymer scientists. Recently published developments on the more important of these copolymer systems are reported in the following sections. A general area of interest in these systems is the thermal stability/flammability of such copolymers and a study has been reported on the limiting oxygen indices (LOI) of a large number of block copolymers with PDMS, and a synergistic effect upon LOI was evident in several families of the block copolymers. The copolymers studied included the following units: bisphenol A carbonate, styrene, phenylene oxide, and methyl methacrylate.[102]

---

[75] M. Keller, *Pharm. Ztg.*, 1981, **126**, 1327.
[76] M. F. Refojo, *Surv. Ophthalmol.*, 1982, **26**, 257.
[77] A. J. Disapio and M. S. Starch, *Cosmet. Toiletries*, 1982, **96**, 55.
[78] C. E. Creamer, *Pharm. Technol.*, 1982, **6**, 79.
[79] Anon., Manuf. Chem. Aerosol News, 1981, **52**, 25.
[80] J. Ritte, E. G. Dubyaga, O. G. Tarakanov, and H. Hamann, *Plaste Kautsch*, 1981, **28**, 185.
[81] G. Koerner, *Goldschmidt Informiert*, 1982, **56**, 2.
[82] C. B. Whitman, *Ind. Res. Dev.*, 1982, **24**, 157.
[83] O. Cada and N. Smela, *Adhaesion*, 1981, **25**, 162.
[84] F. J. Boerio and J. W. Williams, *Appl. Surf. Sci.*, 1982, **7**, 19.
[85] I. Ya. Poddubnyi, S. V. Averyanov, L. A. Averyanova, and M. P. Grinblat, *Prom-St. Sint. Kauch.*, 1981, **2**, 14.
[86] D. Wolfer, *Kautsch. Gummi Kunstst.*, 1981, **34**, 640.
[87] G. E. Pike, *Energy Res. Abstr.*, 1981, **6**, Abstr. No. 20522.
[88] J. P. Crine, H. Saint-Onge, R. Grob, J. Casanovas, P. Lecollier, and J. Mathieu, *J. Electrost.*, 1982, **12**, 609.
[89] C. Weise, *Kautsch. Gummi Kunstst.*, 1982, **35**, 111.
[90] A. A. Laghi, *Rev. Plast. Mod.*, 1982, **43**, 440.
[91] W. J. Morrow, *Kautsch. Gummi Kunstst.*, 1982, **35**, 585.
[92] J. C. Caprino, *Rubber World*, 1982, **185**, 33.
[93] Anon., *Chem. Times Trends*, 1981, **4**, 40.
[94] W. A. Finzel, *Aust. OCCA Proc. News*, 1981, **18**, 13.
[95] *Fed. Regist.*, 1982, April, **47**, 17986.
[96] C. M. Monroe, *Plast. Rubber Inst.*, 1982, **7**, 105.
[97] N. Ramchandran, A. KDidwania, and K. K. Sirkar, *J. Colloid Interface Sci.*, 1981, **83**, 94.
[98] V. E. Gul, S. V. Genel, and V. Ya. Bulgakov, *Plast. Massy*, 1981, **9**, 20.
[99] R. J. Cush and H. W. Winnan, *Dev. Rubber Technol.*, 1981, **2**, 203.
[100] 'Silane Coupling Agents', ed. E. P. Plueddemann, Plenum Press, New York, 1982.
[101] 'Kirk-Othmer Encyclopedia of Chemical Technology', 3rd Edn., ed. M. Grayson and D. Eckroth, Wiley, New York, 1982.
[102] R. P. Kambour, H. J. Klopfer, and S. A. Smith, *J. Appl. Polym. Sci.*, 1981, **26**, 847.

Carbonate–Siloxane Copolymers.—The synthesis and characterization of polycarbonate–PDMS block copolymers have been reported.[103] The disadvantage in many of the routes to the carbonate–PDMS copolymers stems from the decomposition of the carbonate group, and this has been reported to be overcome by a proposed new synthetic route to these polymers.[104] A study of the strength characteristics of polycarbonate–PDMS demonstrates an almost doubling of tensile strength, from 65—85% PDMS.[105] N.m.r. studies have been reported in the structural investigation of polycarbonate–PDMS block copolymers, particularly as regards end-capping ratio effects.[106] The structure–property relationships for perfectly alternating polycarbonate–PDMS block copolymers have been investigated and reported.[107] The heat aging properties of polycarbonate–PDMS are particularly interesting and these studies have been described, using stress-relaxation and DSC techniques.[108] TGA analysis to 1000 K indicated that there is no simple relationship between composition and thermal stability,[109] and that the flammability resistance of the copolymers indicates a synergistic effect at 50 mol % PDMS.[110]

Styrene–Siloxane Copolymers.—The synthesis of a series of polystyrene and styrene derivatives giving multiblock copolymers with PDMS has been described. The route consisted of a hydrosilylation coupling of $\alpha\omega$-dihydrogen PDMS with $\alpha\omega$-divinylsilylated polystyrene derivatives, including polystyrene, poly(α-methyl styrene), poly(vinyl mesitylene), poly(p-trimethylsilylstyrene), and poly(p-trimethylsilyl-α-methylstyrene). The block copolymers obtained were characterized by gel permeation chromatography, viscometry, and light scattering.[111] The thermal properties of low-molecular-weight polystyrene–PDMS diblock copolymers have been studied by DSC[112] and the glass transition temperatures determined from refractive index–temperature measurements.[113,114] PDMS copolymers with styrene–butadiene diblock and styrene–butadiene–styrene triblock segments, have been studied and the phase changes and glass transition temperatures characterized.[115]

Silphenylene–Siloxane Copolymers.—The synthesis and characterization of alternating block copolymers of tetramethyl-p-silphenylene siloxane with PDMS has been described. The polymers were characterized by n.m.r., molecular weight determinations, and heat of fusion measurements.[116] Exactly alternating

---

[103] J. S. Riffle, *Diss. Abstr. Int. B*, 1981, **42**, 1042.
[104] H. Rosenberg, T. T. Tsai, and N. K. Ngo, *J. Polym. Sci., Polym. Chem. Ed.*, 1982, **20**, 1.
[105] V. Yu. Levin, G. L. Slonimskii, A. A. Zhdanov, and L. I. Makarova, *Vysokomol. Soedin., Ser. B*, 1981, **23**, 163.
[106] E. A. Williams, P. E. Donahue, and J. D. Cargioli, *Macromolecules*, 1981, **14**, 1016.
[107] S. H. K. Tang, *Diss. Abstr. Int. B*, 1981, **42**, 1043.
[108] M. R. Tant and G. L. Wilkes, *Polym. Eng. Sci.*, 1981, **21**, 325.
[109] G. R. Grubbs, M. E. Kleppick, and J. H. Magill, *J. Appl. Polym. Sci.*, 1982, **27**, 601.
[110] R. P. Kambour, *J. Appl. Polym. Sci.*, 1981, **26**, 861.
[111] P. Chaumont, G. Beinert, J. Herz, and P. Rempp, *Polymer*, 1981, **22**, 663.
[112] S. Krause, M. Iskandar, and M. Iqbal, *Macromolecules*, 1982, **15**, 105.
[113] Z. H. Lu and S. Krause, *Macromolecules*, 1982, **15**, 112.
[114] Z. H. Lu, S. Krause, and M. Iskandar, *Macromolecules*, 1982, **15**, 367.
[115] S. Krause, Z. Lu, and M. Iskandar, *Macromolecules*, 1982, **15**, 1076.
[116] Y. Nagase, T. Masubuchi, K. Ikeda, and Y. Sekine, *Polymer*, 1981, **22**, 1607.

silarylene–siloxane copolymers containing *p*-phenylene and *pp'*-diphenyl ether silarylene units have been characterized by n.m.r., i.r., gel permeation chromatography, viscometry, and DSC techniques, and the structures discussed.[117,118] The mechanical properties of rubbers produced from alternating silarylene–siloxane copolymers have been described, and such rubbers have been evaluated at prolonged use in air at 200 °C.[119] Thermal stability investigations have been reported on the *p*-silphenylene copolymers, using a range of analytical techniques, and mechanisms have been proposed for the degradation processes.[120,121]

Other Organic–Siloxane Copolymers.—Results have been published on the surface area measurement of polyacrylate block domains in the copolymers with PDMS.[122] Copolymers of PDMS with phenolphthalein–terephthaloyl chloride polyester have been studied in dilute solutions by light-scattering techniques. The dimensions of the macromolecular associations were found to agree with data obtained by studies using electron microscopy.[123] Copolymers of PDMS with polyethers[124] and with polypeptides[125] have been synthesized, and the results of their characterization reported.

4 Poly(organophosphazene)s

This class of inorganic polymers is of high scientific interest with a major potential for commercialization across a broad area of applications. The volume of publication in the scientific literature is increasing markedly, and, as with the poly(organisiloxane)s, there is high activity demonstrated in the patent literature. The refinement of existing routes to these polymers is evident, with great emphasis being placed upon the development of new synthetic routes to the many novel polymer structures which are possible. The characterization, properties, and applications of the new polymers is proceeding rapidly following their synthesis. Reviews have appeared in which the synthesis, properties and applications of a range of polyphosphazene polymers are surveyed.[126–128]

Synthesis.—The general route to poly(organophosphazene)s has been via the poly(halophosphazene)s synthesis, because while $(NPCl_2)_3$ and $(NPF_2)_3$ react to give open chain high polymers, fully organosubstituted cyclic derivatives generally do not. There are intermediate examples such as mono- and di-organosubstituted

---

[117] P. R. Dvornic and R. W. Lenz, *J. Polym. Sci., Polym. Chem. Ed.*, 1982, **20**, 593.
[118] Y. C. Lai, P. R. Dvornic, and R. W. Lenz, *J. Polym. Sci., Polym. Chem. Ed.*, 1982, **20**, 2277.
[119] M. E. Livingstone, P. R. Dvornic, and R. W. Lenz, *J. Appl. Polym. Sci.*, 1982, **27**, 3239.
[120] B. Zelei, M. Blazso, and S. Dobas, *Eur. Polym. J.*, 1981, **17**, 503.
[121] M. Ikeda, T. Nakamura, Y. Nagase, and K. Ikeda, *J. Polym. Sci., Polym. Chem. Ed.*, 1981, **19**, 2595.
[122] A. N. Ghenkin, N. A. Petrova, and T. V. Evstigneeva, *Vysokomol. Soedin., Ser. A*, 1981, **23**, 329.
[123] S. A. Pavlova, L. V. Dubrovina, E. M. Belavtseva, M. A. Ponomeva, and S. I. Senkevich, *Vysokomol. Soedin., Ser. A*, 1981, **23**, 359.
[124] L. A. Kosenko, V. Ya. Kotomkin, Yu. Yu. Kercha, and E. P. Lebedev, *Vysokomol. Soedin., Ser. A*, 1981, **23**, 2287.
[125] C. M. Kania, H. Nabizadeh, D. G. McPhillimy, and R. A. Patsiga, *J. Appl. Polym. Sci.*, 1982, **27**, 139.
[126] R. E. Singler, G. Hagnauer, and R. W. Sicka, *ACS Symp. Ser.*, 1982, **193**, 229.
[127] H. R. Allcock, *Macromol. Chem. Phys. Suppl.*, 1981, **4**, 3.
[128] H. G. Horn, *Makromol. Chem.*, 1982, **183**, 1833.

cyclotriphosphazenes which yield open chain polymers[129] under certain conditions but, until recently, no general method has been available which allows the incorporation of the desired substituents before polymerization. The synthesis of polyphosphazenes with direct phosphorus–carbon bonds has been possible only in a few cases,[130–131] and a new method with good potential involves the synthesis of suitably constructed N-silphosphinimines, which upon heating eliminate substituted silanes to give poly(organophosphazene)s. This procedure has been used to prepare poly(dimethylphosphazene) and should be capable of extension to other organophosphazenes.[132]

Poly(dichlorophosphazene) presents special problems in polymerization studies because of branching, crosslinking, and cyclization reactions, and in characterization, owing to the polymer's hydrolytic instability. A review of the problem areas and recent advances in the polymerization has been published. The melt, solution, and irradiation polymerization reactions and mechanisms are discussed, together with current research efforts and areas for future investigation.[133] The use of boron halide catalysts has been shown to be particularly effective in polymerizing hexachlorocyclotriphosphazene to poly(dichlorophosphazene) in almost 100% yields. The polymer is linear and crosslinking side-reactions were found to be absent.[134] A study of the polymerization of $(NPCl_2)_3$ with sulphur and selenium as catalysts has also been reported.[135] Chlorine substitution reactions of $(NPCl_2)_3$ with the sodium salts of dihydroxy compounds to produce cyclolinear oligomers has been described, and the i.r. spectra and thermal properties of these oligomers together with the molecular weights were determined.[136] In ring-opening polymerizations of $(NPCl_2)_3$, the catalytic species form chain-end functional groupings which may effect the stability of the polymer. Studies have been reported in which active end-groups have selectively been allowed to react with nucleophilic reagents, giving considerable improvements to the thermal and hydrolytic stability of the $(NPCl_2)_n$.[137,138]

The preparation of poly(difluorophosphazene) as an intermediate in the synthesis of poly(organophosphazene)s has been reviewed. This intermediate is particularly important in facilitating the preparation of a class of poly(organophosphazene)s whose side-groups are linked directly through P–C bonds.[139] The crosslinking reaction occurring when poly(difluorophosphazene) reacts with alkyl-lithium reagents, has been studied by using model cyclic reactants and a mechanism postulated.[140] A route to propenylphosphazenes, by reaction of 2-propenyl-lithium with $(NPF_2)_3$, has been reported, and the copolymerization of these intermediates

---

[129] H. R. Allcock, *Polymer*, 1980, **21**, 673.
[130] H. R. Allcock, T. L. Evans, and D. B. Patterson, *Macromolecules*, 1980, **13**, 201.
[131] R. H. Neilson and O. Wisian-Neilson, *J. Macromol. Sci., Chem.*, 1981, **A16**, 425.
[132] R. H. Neilson and P. Wisian-Neilson, *J. Am. Chem. Soc.*, 1980, **102**, 2848.
[133] G. L. Hagnauer, *J. Macromol. Sci., Chem.*, 1981, **A16**, 385.
[134] J. W. Fieldhouse and D. F. Graves, *ACS Symp. Ser.*, 1981, **171**, 315.
[135] M. Kajiwara and E. Miwa, *Polymer*, 1982, **23**, 495.
[136] B. Laszkiewicz and R. Kotek, *Angew. Makromol. Chem.*, 1981, **99**, 1.
[137] F. Yamada, T. Yasui, and I. Shinohara, *J. Macromol. Sci., Chem.*, 1981, **A15**, 585.
[138] M. Helioni, *Macromol. Chem.*, 1982, **183**, 1137.
[139] T. L. Evans and H. R. Allcock, *J. Macromol. Sci., Chem.*, 1981, **A16**, 409.
[140] T. L. Evans, D. B. Patterson, P. R. Suszko, and H. R. Allcock, *Macromolecules*, 1981, **14**, 218.

with organic vinyl monomers discussed.[141] Copolymer poly(organophosphazene)s of formula [NP(OC$_6$H$_4$CN)$_x$(OCH$_2$CF$_3$)$_{2-x}$] ($x = 0.04 - 2$) have been prepared, and the linear polymers crosslinked to elastomeric networks *via* the 4-cyanophenoxy substituents. The thermal properties of both the linear and crosslinked copolymers were investigated by TGA and DSC techniques.[142]

The synthesis of the first-carborane-substituted cyclophosphazenes, together with the preparation by two different approaches of the first carboborane-substituted phosphazene linear polymers, has been reported, and the properties and structure of the compounds discussed.[143] Synthetic routes enabling phosphazene cyclic and linear polymers to be linked to transition metals or biologically active organic species have been reported, and the commercial potential of such materials indicated.[144-148]

Molecular Structure.—Data have been reported on the unperturbed dimensions of some poly(aryloxyphosphazene)s. A series of copolymers of formula [P(OR)(OR')N]$_n$ with R = phenyl and R' = *p*-ethylphenyl, 2,4-dichlorophenyl, or 2-naphthyl, were separated into molecular weight fractions by fractional precipitation from THF. The fractions were characterized by osmometry and viscometry to give values for the number-average molecular weight, second virial coefficient, and intrinsic viscosity. The values for the unperturbed dimensions, relative to the number of skeletal bonds and the square of their length, showed a remarkably large dependence on the nature of the side groups and structural configurations are proposed to account for this.[149] Results have been described of the fractionation of a sample of poly[bis(*p*-methylanilino)phosphazene] into fractions having M_w from 1.7×10^5 to 1.3×10^7. Characterization of these fractions demonstrated the polymer to be substantially linear with the relationship between [η] and M_w being given by [η] = $5.2 \times 10^{-3} M_w^{0.7}$ mB g^{-1}.[150]

Raman and i.r. spectral studies have been carried out on hexachlorocyclotriphosphazene,[151] octachlorocyclotetraphosphazene,[152] and poly(dichlorophosphazene) and normal co-ordinates calculations presented. The attempts to obtain a common valence force field from the cyclic monomers which could be transferred to molecular models of the linear polymer are discussed.[153] Data obtained from the measurement of the u.v. and fluorescence spectra of poly[bis(phenoxy)phosphazene] leads to the conclusion that the juxtaposition of two phenoxy groups per phosphorus atom does allow facile excimer formation with little or no singlet energy migration. The significance of this work to the photodegradation of

---

[141] C. W. Allen, R. P. Bright, and K. Ramachandran, *ACS Symp. Ser.*, 1981, **171**, 321.
[142] M. Zeldin, W. H. Jo, and E. M. Pearce, *J. Polym. Sci., Polym. Chem. Ed.*, 1981, **19**, 917.
[143] H. R. Allcock, A. G. Scopelianos, J. P. O'Brien, and M. Y. Bernheim, *J. Am. Chem. Soc.*, 1981, **103**, 350.
[144] H. R. Allcock, *ACS Symp. Ser.*, 1981, **171**, 311.
[145] H. R. Allcock, P. J. Harris, and R. A. Nissan, *J. Am. Chem. Soc.*, 1981, **103**, 2256.
[146] H. R. Allcock, P. P. Greigger, L. J. Wagner, and M. Y. Bernheim, *Inorg. Chem.*, 1981, **20**, 716.
[147] H. R. Allcock and T. J. Fuller, *Macromolecules*, 1980, **13**, 1338.
[148] H. R. Allcock and T. J. Fuller, *J. Am. Chem. Soc.*, 1981, **103**, 2404.
[149] A. L. Andrady and J. E. Mark, *Eur. Polym. J.*, 1981, **17**, 323.
[150] G. Pezzin, S. Lora, and L. Busulini, *Polym. Bull. (Berlin)*, 1981, **5**, 543.
[151] P. C. Painter, J. Zarian, and M. M. Coleman, *Appl. Spectrosc.*, 1982, **36**, 265.
[152] J. Zarian, P. C. Painter, and M. M. Coleman, *Appl. Spectrosc.*, 1982, **36**, 272.
[153] M. M. Coleman, J. Zarian, and P. C. Painter, *Appl. Spectrosc.*, 1982, **36**, 277.

poly(organophosphazene) polymers and hence their potential use as photosensitizers is explored.[154]

The rheological characteristics of poly(fluoroalkoxyphosphazene) and poly(aryloxyphosphazene) polymers have been studied both in the bulk and solution states, using the Weissenberg rheogoniometer, and the extensive data obtained fully discussed.[155] The thermomechanical transitions of poly(organophosphazene)s containing a range of different organic side groups have been studied and data obtained over a wide temperature range of -120 to $180\,°C$. The dynamic mechanical response of the polymers was dominated by two transitional regions, namely, (a) a primary softening dispersion region related to T_g and (b) the mesomorphic transition region $T(1)$ that occurs between T_g and T_m.

Properties.—The poly(organophosphazene)s exist in a very wide range of polymer structures whose properties may be 'tailored' to specific applications. Many of the polymers possess high temperature resistance, and the mechanisms of thermal degradation are of prime importance. A study of the thermal decomposition, by mass spectrometric techniques, of a range of poly(organophosphazene)s has been reported and the wide variation in thermal stabilities demonstrated and discussed.[157] Two classes of phosphazene polymers have been developed to the commercial stage, the fluoroelastomers and the aryloxyelastomers, and a large number of polymers are being investigated as fluids and resins, and some of the applicational properties of these polymers have been described.[126] Studies on the methods of linking fire-retardant poly(organophosphazene)s to textiles and paper substrates have been described, and the reactions of a naturally occurring linear aminopolysaccharide (chitosan), with the cyclic poly(dichlorophosphazene)s, have shown promise in this connection.[158] The reactions of organophosphazene polymers with timber have been reported, and by using the techniques of thermogravimetric analysis, toxicity testing, and oxygen index determination, the thermal breakdown of the treated timber has been studied. The data obtained suggest the polymers have useful properties as fire retardants, although pressure impregnation is required to facilitate the timber treatment.[159] Non-flammable adhesives have been synthesized by the reaction of aminoalkoxyphosphazenes with toluene diisocyanate and the products analysed by n.m.r., i.r., and chemical analysis, and results presented on the evaluation of adhesive bond strengths.[160]

The possibility of synthesizing polymer-bound dyes has been reported, using the coupling of phenol and naphthol derivatives to aminophenoxyphosphazenes, and the results obtained showed that the phosphazene skeleton was unaffected by the reduction, diazotization, and diazo coupling processes.[161] Some poly(organophosphazene)s possess important properties as potential carriers for the controlled release of chemotherapeutic agents, stemming from their biocompatibility and facile

---

[154] J. S. Hargreaves and S. E. Webber, *Polym. Photochem.*, 1982, **2**, 359.
[155] P. K. Ho and M. C. Williams, *Polym. Eng. Sci.*, 1981, **21**, 233.
[156] I. C. Choy and J. H. Magill, *J. Polym. Sci., Polym. Chem. Ed.*, 1981, **19**, 2495.
[157] A. Ballisteri, S. Foti, G. Montaudo, S. Lora, and G. Pezzin, *Makromol. Chem.*, 1981, **182**, 1319.
[158] G. G. Allan, E. J. Gilmartin, and H. Struszczyk, *J. Macromol. Sci., Chem.*, 1981, **A15**, 599.
[159] P. J. Lieu, J. H. Magill, and Y. Calarie, *J. Combust. Tosicol.*, 1982, **9**, 65.
[160] M. Kajiwara, *J. Macromol. Sci., Chem.*, 1981, **A16**, 587.
[161] H. R. Allcock, P. E. Austin, and T. F. Rakowsky, *Macromolecules*, 1981, **14**, 1622.

side-group release mechanisms in aqueous media. Side-group hydrolysis of poly(organophosphazene)s is thus an important area of investigation, and model studies have been reported of the Schiff-base coupling of cyclic and linear phosphazenes to aldehydes and amines, to give side-groups which may be easily hydrolysed off *via* the Schiff-base linkage. The physical and chemical properties of the Schiff-base products are discussed.[162] Data have been published describing the catalysed hydrolysis pathways for a large number of aminophosphazenes derived from both model cyclic monomers and high-molecular-weight linear poly(organophosphazene)s.[163]

---

[162] H. R. Allcock and P. E. Austin, *Macromolecules*, 1981, **14**, 1616.
[163] H. R. Allcock, T. J. Fuller, and K. Matsumura, *Inorg. Chem.*, 1982, **21**, 515.

7
Configurations

BY S. B. ROSS-MURPHY

1 Introduction

This Chapter further extends the survey of key literature published on dilute and 'semi-dilute' polymer solutions to cover the period 1981 and 1982. As before, we expect some of these topics also to be discussed, *e.g.*, under neutron scattering (Chapter 10), polymer characterization (Chapter 12), and computer applications (Chapter 19).

For convenience we will closely follow the format previously adopted;[1] where approaches appear without an obvious reference, Yamakawa[2] covers the period up to 1971, and a recent monograph[3] covers the work published in the last decade in some detail. In this Chapter we will also discuss the progress in viscoelasticity of semi-dilute solutions because of the interest generated by recent molecular models.[4,5]

2 Statistics of Model Polymers

Unperturbed Dimensions.—The mean square radius of a flexible linear polymer depends upon the number of chain segments (or molecular weight) in the following way

$$\langle S^2 \rangle \propto n^{2\nu}$$

Here the exponent $\nu = 0.5$ in an ideal (Θ) solvent, close to 0.6 for a very high molecular weight ($n \to \infty$) flexible polymer in a 'good' solvent (excluded volume effects pre-eminent) and $\nu \simeq 0.9$ for an extremely 'stiff' rod polymer. With a strongly attractive potential a chain is said to undergo a coil-globule collapse, with $\nu = 1/3$. The polymer in good solvent excluded volume regime is discussed in the next subsection, and polymers which may be described by rod or wormlike chains follow. Of course, the dimensions decrease in most of the above cases as the concentration is increased; in the present Section we consider dilute and 'infinitely' dilute solutions only.

---

[1] S. B. Ross-Murphy in 'Macromolecular Chemistry', ed. A. D. Jenkins and J. F. Kennedy (Specialist Periodical Reports), The Royal Society of Chemistry, London, 1982, Vol. 2, p. 174.
[2] H. Yamakawa, 'Modern Theory of Polymer Solutions', Harper and Row, New York, 1971.
[3] M. Bohdanecký and J. Kovář, 'Viscosity of Polymer Solutions', Polym. Sci. Libr. II, ed. A. D. Jenkins, Elsevier, Amsterdam, 1982.
[4] M. Doi and S. F. Edwards, *J. Chem. Soc., Faraday Trans. 2*, 1978, **74**, 1789, 1802, and 1818.
[5] C. F. Curtiss and R. B. Bird, *J. Chem. Phys.*, 1981, **74**, 2016 and 2026.

The graph theoretical method of Eichinger[6] for unperturbed dimensions has now been extended,[7] in particular to evaluate the fourth moment $\langle S^4 \rangle_0$ (subscript zero = unperturbed dimensions), for a number of comb-shaped and H-shaped branched polymers, using the Rouse matrix[7] rather than the Kirchoff matrix used by Eichinger.[6] Eichinger[8] has applied his original method to the problem of collapse in networks.

In a number of studies the experimental aspects of unperturbed branched chains have been investigated, including the so-called comb polymers. An extreme comb polymer is the H-shaped polystyrene of Roovers and Toporowski;[9] according to the Zimm–Stockmayer model, the 'shrinkage factor' $g = \langle S^2 \rangle_{0,\text{br}} / \langle S^2 \rangle_{0,\text{l}} < 1$, where the subscripts l and br refer to linear and branched chains, respectively, should be 0.71^2 for this system – experimentally 0.70 was found, in good agreement with theory, and supporting the results obtained from earlier work on lightly branched chains. McCrackin and Mazur[10] have computed g values for comb-polymers on a cubic lattice, and compared the results from their calculations with data obtained in good and Θ solvents. In particular, they find that for highly branched combs the radius of gyration of the comb backbone is greater than that of the equivalent linear chain, in contradiction to the Gaussian sub-chain approximation. Further, in good solvents the chain dimensions increased by less than predicted in contradiction to perturbation theory results (cf. ref. 1). This effect is presumably related to the high intramolecular segment density adjacent to the backbone. Kolinski and Sikorski[11] have also noted this effect in their studies on star branched polymers on a tetrahedral lattice. A number of experimental studies have suggested that the Θ temperature for branched comb and star polymers lies below that for linear polymers. No such effect was observed by Kajiwara and Burchard,[12] who studied 4-, 6-, and 8-arm stars in a Monte Carlo simulation on a five-choice cubic lattice, and confirmed earlier simulation work. They suggest that the Θ point depression previously observed was due to an artefact, viz., the influence of the third virial coefficient,[2] which is not detected in the usual linear extrapolation of concentration dependence.

There are a number of approaches to estimating the unperturbed dimensions of chains by extrapolating to zero molecular weight, where excluded volume effects should disappear. However, the usual methods tend to fail for stiffened chains and when the 'free draining' effect[2,3] (see Section 3) is no longer negligible. Kamide and Saitoh[13] have introduced a new extrapolation plot and applied it to data for cellulose and amylose derivatives. A further plot has been suggested by Tanaka,[14] from an analysis of intrinsic viscosity data – see Section 3.

An area where there has been considerable progress in the last few years is in the synthesis and characterization of macromolecular rings. One incentive for these studies is the recent interest in plasmids – circular rings of DNA – but there is also

[6] B. E. Eichinger, *Macromolecules*, 1980, **13**, 1.
[7] D. S. Pearson and V. Reddy Rau, *Macromolecules*, 1982, **15**, 294.
[8] B. E. Eichinger, *Macromolecules*, 1981, **14**, 1071.
[9] J. Roovers and P. M. Toporowski, *Macromolecules*, 1981, **14**, 1174.
[10] F. L. McCrackin and J. Mazur, *Macromolecules*, 1981, **14**, 1214.
[11] A. Kolinski and A. Sikorski, *J. Polym. Sci., Polym. Lett. Ed.*, 1982, **20**, 177.
[12] K. Kajiwara and W. Burchard, *Macromolecules*, 1982, **15**, 660.
[13] K. Kamide and M. Saitoh, *Eur. Polym. J.*, 1981, **17**, 1049.
[14] G. Tanaka, *Macromolecules*, 1982, **15**, 1028.

considerable progress in flexible synthetic rings, particularly the studies of Semlyen and co-workers on poly(dimethyl siloxanes) (PDMS). Three recent papers[15—17] have investigated the classical diffusion coefficient, D, of rings and chains in a Θ solvent,[15] and also measured diffusion by quasielastic light scattering.[17] Among the principle results are that the ratio of the friction coefficients f_r/f_1 for the higher molecular-weight rings and chains agree quite well with that found from the Kirkwood–Riseman theory[2] in the non-draining limit, $viz., f_r/f_1 = 0.83$ (theory gives 0.85). Partial free draining was observed for shorter rings and chains.

Winnik[18] has reviewed his elegant work on cyclization reactions, especially using the technique of intramolecular fluorescence, in which specific sites along the chain backbone are labelled. Subjecting a dilute solution to u.v. light produces a blue fluorescence, on cyclization the colour changes to green, and by analysing the rate of fluorescence decay the ring–chain equilibria can be studied.[19] Finally, Chen[20] has introduced a new Monte Carlo method to generate undistorted ring polymers[21,22] – the ratio $\langle S^2 \rangle_r / \langle S^2 \rangle_l$ for excluded volume rings and chains is shown to depend slightly upon chain length, and increases to 0.57 for large rings (the Zimm–Stockmayer ratio[2] is 0.50), in agreement with a recent renormalization group calculation of 0.568.[20] The neutron scattering results of Higgins, Dodgson, and Semlyen[23] for PDMS give a ratio of 0.53, under Θ conditions. Des Cloizeaux[24] has also considered excluded volume effects on ring polymers, and distinguished between short rigid and long flexible rings – in the latter case he predicts that the ring polymer should be more soluble, and there should be a slight difference in the Θ temperatures (cf. refs. 15 and 12).

Excluded Volume Effects.—Because of intramolecular segment–segment interactions, $\langle S^2 \rangle$ for a polymer chain in a thermodynamically good solvent is greater than that for a Gaussian chain, and asymptotically approaches an exponent close to 0.6 for extremely high-molecular-weight polymers. The original treatment of the excluded volume effect due to Fixman and to Yamakawa[2] uses a perturbation expansion about $\langle S^2 \rangle_0$ useful only for very small excluded volume. Chikahisa et al.[25] have generalized these results into d-dimensional space, Tanaka[26] has attributed the ln n dependence found for a chain of n segments in $d = 4$ space to a balance between long- and short-range interactions along the chain contour. Increasingly sophisticated techniques have been applied to elucidate the problem of large excluded volume, e.g., the self-consistent field of Edwards,[27] and the analogies between self-avoiding walks (SAW) and the Ising model for the spins in a magnetic

---

[15] C. J. C. Edwards, R. F. T. Stepto, and J. A. Semlyen, *Polymer*, 1982, **23**, 865.
[16] C. J. C. Edwards, R. F. T. Stepto, and J. A. Semlyen, *Polymer*, 1982, **23**, 869.
[17] C. J. C. Edwards, S. Bantle, W. Burchard, R. F. T. Stepto, and J. A. Semlyen, *Polymer*, 1982, **23**, 873.
[18] M. A. Winnik, *Chem. Rev.*, 1981, **81**, 491.
[19] A. E. C. Redpath and M. A. Winnik, *Ann. NY Acad. Sci.*, 1981, **366**, 75.
[20] Y. Chen, *J. Chem. Phys.*, 1981, **75**, 5160.
[21] Y. Chen, *J. Chem. Phys.*, 1981, **74**, 2034.
[22] Y. Chen, *J. Chem. Phys.*, 1982, **75**, 2447.
[23] J. S. Higgins, K. Dodgson, and J. A. Semlyen, *Polymer*, 1979, **20**, 553.
[24] J. des Cloizeaux, *J. Phys. (Paris) Lett.*, 1981, **42**, L433.
[25] Y. Chikahisa, G. Tanaka, K. Solc, and M. Takahashi, *Rep. Prog. Polym. Phys. Jpn.*, 1981, **24**, 33.
[26] G. Tanaka, *Macromolecules*, 1982, **15**, 1525.
[27] K. F. Freed, *Adv. Chem. Phys.*, 1972, **22**, 1.

system. Whittington has published[28] an excellent overview of recent approaches to the excluded volume both for solutions and adsorption of polymers.

Al-Noami[29] has used the original Fixman perturbation, but in dimensionality 4-E, and then calculated v correct to $0(E^2)$, by extrapolating $E \rightarrow 1$. The result found is $\simeq 0.593$, close to the calculation of Le Guillou and Zinn-Justin,[26] viz, 0.588, and the 'classical' value of 0.6. Kosmas[30] has performed another perturbation analysis also in 4-E dimensions (the excluded volume problem is known to be exactly soluble in four dimensions), and produced the estimate $v_d = 4/(4 + d)$, i.e., $v_3 = .571$. The same field theoretic analogue has been used by Elderfield[31] to describe the chain expansion due to excluded volume, an approach which is extended,[32] albeit in a somewhat inconvenient way, to include a polydispersity of chain length. Freed, Oono, and co-workers[33—36] have performed a conformation space renormalization and produced results for the end-to-end distance distribution for chains,[33] and for the so-called internal vector of a 'ring' polymer.[35] The major difference of this method is that calculations are performed using the chain position in space and its contour length as the fundamental variables, rather than rigid 'blobs' as in earlier theories.[37]

Witten and Schafer[38] have reapplied the de Gennes approach to the excluded volume, but re-interpreted in terms of the Fixman expansion method, in particular to describe the fourth moment $\langle L^4 \rangle$ of the end-to-end distance, and the ratio $\langle L^4 \rangle / (\langle L^2 \rangle)^2$. In particular, the latter ratio was found to be around 8% *lower* than in the unperturbed chain. This interesting result has a number of implications, because it contradicts the uniform expansion approximation[2] (i.e., that all moments of the end-to-end distance increase in the same ratio) which is widely applied in polymer solutions theory. The effect is attributed to a 'correlation hole' in the segment–segment distribution function as originally suggested by des Cloizeaux.[39] Witten and Prentis[40] have examined the interpenetration of two chain polymers of different chain length in a good solvent – and in particular obtained an estimate for the so-called 'interpenetration function', $\Psi = A_2 M^2 / 4\pi^{3/2} N \langle S^2 \rangle^{3/2}$, with N, Avagadro's number, lower than found from earlier approaches,[2] but consistent with the results of Miyaki et al.[41] who found Ψ passing through a maximum and then falling to around 0.22. Schafer[42] has reviewed the experimental evidence from different models for the mutually excluded volume, using models from hard sphere repulsion to full interpenetration, but no conclusion can yet be reached as to which

---

[28] S. G. Whittington, *Adv. Chem. Phys.*, 1982, **51**, 1.
[29] G. F. Al-Noaimi, *J. Chem. Phys.*, 1981, **74**, 4701.
[30] M. K. Kosmas, *J. Phys. A*, 1981, **14**, 931.
[31] D. J. Elderfield, *J. Phys. A*, 1981, **14**, 1797.
[32] D. J. Elderfield, *J. Phys. A*, 1981, **14**, 3367.
[33] Y. Oono and K. F. Freed, *J. Chem. Phys.*, 1981, **75**, 993.
[34] Y. Oono and K. F. Freed, *J. Chem. Phys.*, 1981, **75**, 1009.
[35] M. Lipkin, Y. Oono, and K. F. Freed, *Macromolecules*, 1981, **14**, 1270.
[36] Y. Oono, T. Ohta, and K. F. Freed, *Macromolecules*, 1981, **14**, 880.
[37] G. Weill and J. des Cloizeaux, *J. Phys. (Paris)*, 1979, **40**, 99.
[38] T. A. Witten and L. Schafer, *J. Chem. Phys.*, 1981, **74**, 2582.
[39] J. des Cloizeaux, *Phys. Rev. A*, 1971, **10**, 1665.
[40] T. A. Witten and J. J. Prentis, *J. Chem. Phys.*, 1982, **77**, 4247.
[41] Y. Miyaki, Y. Einaga, T. Hirosuye, and H. Fujita, *Macromolecules*, 1977, **10**, 1356.
[42] L. Schafer, *Macromolecules*, 1982, **15**, 652.

is the most suitable. Kosmas[43] has calculated the ratio $\langle S^2 \rangle/\langle R^2 \rangle$ and found it to be a universal ratio of the excluded volume in 4-E dimensions, and Minato and Hatano[44] have calculated the same ratio in three dimensions, and reassessed the effect of the excluded volume on the three principal components of the chain. They find that these ratios $\lambda_1:\lambda_2:\lambda_3$ are 11.3:2.47:1, for the perturbed chain, *i.e.*, slightly more spherical than found by Sŏlc and Stockmayer[45] for the unperturbed chain, but correcting the anomalous result $\lambda_1 \to \infty$ found earlier.[46] Muthukumar and Edwards[47] have produced a series of general expressions for the size of a chain from unperturbed to full excluded volume, and valid for all concentrations. The most interesting results are obtained at the higher concentrations and will be discussed later, but for the isolated chain a closed form of the expansion factor/excluded volume expression is found in agreement with some experimental data, *i.e.*, $\alpha^5 - \alpha^3 = 1.159z$, with z the usual factor in the two-parameter theory.[2]

Alternate approaches to the statistics of perturbed chains include lattice walks, using both exact enumeration and Monte Carlo approaches, off-lattice walks, and analytic treatments of walks on lattice graphs. Domb[48] has made further contributions in the latter area by developing a generating function method for walks with restrictions, *e.g.*, when the end-to-end distance is fixed, without resorting to the Gaussian chain approximation. In this way correction terms for walks of finite step-length may be calculated. The previously mentioned review by Whittington[27] gives a good summary of lattice graph methods. Klein[49] has argued that standard methods for counting the isomers of a tree lattice graph are inappropriate when treating branched polymers with excluded volume. Together with Seitz,[50] he has introduced the intriguingly named 'pulsating amoeba' algorithm for the Monte Carlo construction of branched polymers with excluded volume, and suggested a v exponent of 0.46 for the tree graph with excluded volume. Minato[51] and co-workers have applied a Monte Carlo method to an off-lattice walk, which actually simulates explicitly the renormalization technique applied in the 'blob' model. For large excluded volume, $v \simeq 0.57$, and replacing the repulsive potential by one which is weakly attractive produces $v = 1/3$, in agreement with that predicted for a collapse transition. Baumgartner[52] has extended his off-lattice approach to investigate the effect of excluded volume on a ring polymer, constructed using hard sphere beads. He calculates not only the v exponent (~ 0.59, *i.e.*, effectively the same as for linear chains), but also the static structure factor $S(\mathbf{k})$ for wave vector $\mathbf{k}$. The 'short range' excluded volume is found to have an exponent $\simeq 0.9$, consistent with that found for linear chains.[53] Barr, Brender, and Lax[54] have calculated $\langle S^2 \rangle$ for a self-avoiding walk with span L' limitations,[1] for large L', $v \to 0.6$ is recovered, but $L' \to 0$

---

[43] M. K. Kosmas, *J. Phys. A*, 1981, **14**, 2799.
[44] T. Minato and A. Hatano, *Macromolecules*, 1981, **14**, 1035.
[45] K. Sŏlc and W. H. Stockmayer, *J. Chem. Phys.*, 1971, **54**, 2.
[46] T. Minato and A. Hatano, *Macromolecules*, 1978, **11**, 195 and 200.
[47] M. Muthukumar and S. F. Edwards, *J. Chem. Phys.*, 1982, **76**, 2720.
[48] C. Domb, *J. Phys. A*, 1981, **14**, 219.
[49] D. J. Klein, *J. Chem. Phys.*, 1981, **75**, 5186.
[50] W. A. Seitz and D. J. Klein, *J. Chem. Phys.*, 1981, **75**, 5190.
[51] T. Minato, K. Ideura, and A. Hatano, *Polymer J.*, 1982, **14**, 579.
[52] A. Baumgartner, *J. Chem. Phys.*, 1982, **76**, 4275.
[53] A. Baumgartner, *Z. Phys.*, 1981, **42**, 265.
[54] R. Barr, C. Brender, and M. Lax, *J. Chem. Phys.*, 1981, **75**, 453.

corresponds to a walk on a two-dimensional lattice, and here $v \sim 0.75$. In this indirect way the behaviour of a polymer chain close to a surface, may be investigated. However, the authors comment at the apparent lack of universality, i.e., lattice dependence, of parameters for the two-dimensional model, and also the problems of convergence in changing lattices. Ceperly et al.[55] have simulated both chain statics and dynamics, and again the results for the static $\langle S^2 \rangle$ gives $v \sim 0.6$. Mattice[56] has calculated particle scattering factors $P(\mathbf{k})$ for a real chain RIS (rotational isomeric state) model, but including the excluded volume effect by assuming uniform expansion of sub-chains. At wide scattering angles $P(\mathbf{k})$ is significantly lower than for the unperturbed chain. Finally, Thorpe and Schroll[57] have also applied a more realistic chain model, from which they obtain a closed form for the characteristic function (the Fourier transform of the end-to-end distance distribution), so that higher moments of R are obtained. $\langle R^2 \rangle$ and $\langle R^4 \rangle$ are the same as found for the freely rotating chain, but $\langle R^6 \rangle$ and higher moments are different.

The number of papers devoted to experimental studies of the effect of excluded volume upon chain dimensions is quite limited, perhaps reflecting the greater interest in dynamic light scattering compared with the static technique – more papers now investigate such effects by examining the Stokes radius, R_h^{-1}, measured in the former experiment (but subject to uncertainties about draining behaviour, etc.), rather than the more unequivocal determination of $\langle S^2 \rangle$. However, Oyama et al.[58] have reported a new scattering photometer which splits the light from an incident laser source into two diametrically opposing beams so that the scattering intensity at complementary angles may be directly measured – in this way $\langle S^2 \rangle^{1/2}$ for polystyrene was obtained down to 5 nm (i.e., $M \sim 4 \times 10^4$ in cyclohexane). Fujita and co-workers have extended their series of papers.[59–62] They find[60] that their previously published data for $\langle S^2 \rangle$ of polystyrene and polyisobutylene is well fitted by the Domb–Barrett equation,[63] in terms of the parameter z, but that data for the intrinsic viscosity expansion factor did not produce a single curve for the two systems in different solvents. This casts further light upon the limitations of the two-parameter theory, especially for very high molecular-weight samples. Another paper in this series[61] comments upon the applicability of the modification of the 'blob' theory by Francois et al.,[1] which allows v to increase from 0.5 to 0.6 as the excluded volume increases, but in a way in which v can exceed 0.6 for intermediate exclusion, contrary to experiment. Fujita and Norisuye provide a correction for this anomaly, and from it derive a closed form expression $\alpha^2 = 1.576 z^{2/5}$ close to the asymptotic limit of the Domb–Barrett approach. The other two recent papers in this series are concerned with the coil–globule transition[59] and chain stiffness effects,[62] which are discussed in succeeding sections. Lastly, Chrastova and co-workers[64]

[55] D. Ceperley, M. H. Kalos, and J. L. Lebowitz, *Macromolecules*, 1981, **14**, 1472.
[56] W. L. Mattice, *Macromolecules*, 1982, **15**, 579.
[57] M. F. Thorpe and W. K. Schroll, *J. Chem. Phys.*, 1981, **75**, 5143.
[58] T. Oyama, K. Shiokawa, and K. Baba, *Polymer J.*, 1981, **13**, 821.
[59] Y. Miyaki and H. Fujita, *Polym. J.*, 1981, **13**, 749.
[60] Y. Miyaki and H. Fujita, *Macromolecules*, 1981, **14**, 742.
[61] H. Fujita and T. Norisuye, *Macromolecules*, 1981, **14**, 1330.
[62] T. Norisuye and H. Fujita, *Polymer J.*, 1982, **14**, 143.
[63] C. Domb and A. J. Barrett, *Polymer*, 1976, **17**, 179.
[64] V. Chrastova, D. Mikulasova, J. Lacok, and P. Citovicky, *Polymer*, 1981, **22**, 1054.

have measured $\langle S^2 \rangle$ for very high-molecular-weight polystyrene in Θ (cyclohexane) and good (toluene, benzene) solvents. In the latter, v exponents close to, and slightly above, 0.6 are found, although surprisingly the α' exponents in the Mark–Houwink–Sakurada equation $[\eta] \propto M^{\alpha'}$ were 0.56 in benzene and 0.61 in toluene; somewhat lower than found by other workers, i.e., 0.7—0.8.

The coil–globule collapse transition has stimulated further work both theoretical and experimental, although again the conclusions are somewhat equivocal. The subject has been reviewed, and both mean field and 'blob' theory approaches discussed.[65] Akcasu and co-workers[66] have offered a further modification to the original hypothesis,[1] and to the model criticized by Fujita and Norisuye.[61] The experimental evidence for a sudden chain collapse to $v = 1/3$ is, however, ambiguous.

Nishio and Tanaka et al.[67,68] have pointed out the importance of performing such experiments at extremely high dilution (say, 10^{-6} g ml^{-1}) and for very high-molecular-weight samples. They observe both chain collapse and the concomitant increase in the amplitude of intramolecular motions[68] just below the Θ temperature (34 °C) for polystyrene/cyclohexane. Stepanek and co-workers[69] have also observed the collapse using both static and dynamic light scattering, but for polystyrene in the viscous solvent dioctyl phthalate ($\theta = 22$ °C). In this way intermolecular aggregation is substantially slowed down, allowing measurements to be made well below the θ temperature. They find good agreement with the modified Sanchez formula reported by Sun et al.[67] However, both Miyaki and Fujita[59] and Pritchard and Caroline[70] see little evidence for a drastic collapse. The first paper compares early results by Nierlich et al.[71] (neutron scattering, $M \simeq 3 \times 10^4$) and Slagowski et al.[72] ($M \simeq 4.4 \times 10^7$) with their own results ($M \simeq 1.4 \times 10^6$—3.2×10^7, five samples) and results from scaling theory. Contrary to the latter predictions, no single curve is obtained, and the Miyaki–Fujita results seem to give a limiting v exponent above 1/3. Pritchard and Caroline[70] have also used a range of M, up to 5×10^6, and find a smooth change in dimensions through the θ point. They conclude that perturbation theory provides a useful model for the chain behaviour close to the θ temperature.

As suggested earlier, there has been interest recently in computations for polymer configurations close to a surface; Whittington[28] discusses these and other techniques including exact enumeration. Hammersley et al.[73] discuss two lattice models, one in which walks are allowed to cross a lattice plane (surface), while in the other they are constrained to lie to one side or the other. Both exact enumeration and analytic results are given. Lax, Barr, and Brender[74] have used a scaling method to examine data for exact enumerations for walks in a thin 'slab', while Marqusee

---

[65] C. Williams, F. Brochard, and H. L. Frisch, *Ann. Rev. Phys. Chem.*, 1981, **32**, 433.
[66] A. Z. Akcasu, M. Benmouna, and S. Alkhafaji, *Macromolecules*, 1981, **14**, 147.
[67] S. T. Sun, I. Nishio, G. Swislow, and T. Tanaka, *J. Chem. Phys.*, 1980, **73**, 5971.
[68] I. Nishio, G. Swislow, S. T. Sun, and T. Tanaka, *Nature (London)*, 1982, **300**, 243.
[69] P. Stepanek, C. Konak, and B. Sedlacek, *Macromolecules*, 1982, **15**, 1214.
[70] M. J. Pritchard and D. Caroline, *Macromolecules*, 1981, **14**, 424.
[71] M. Nierlich, J. P. Cotton, and B. Farnoux, *J. Chem. Phys.*, 1981, **69**, 1376.
[72] E. Slagowski, B. Tsai, and D. McIntyre, *Macromolecules*, 1976, **9**, 687.
[73] J. M. Hammersley, G. M. Torrie, and S. G. Whittington, *J. Phys. A*, 1982, **15**, 539.
[74] M. Lax, R. Barr, and C. Brender, *J. Chem. Phys.*, 1981, **75**, 460.

and Deutch[75] have treated a polymer in two dimensions and investigated the effect of allowing the excluded volume to go to zero. Results obtained with their 'real space renormalization group' method are quite close to those found with the classical mean field. Clark and Lal[76] have used the Monte Carlo approach to investigate a chain confined between two plates. The net osmotic pressure/plate separation profile is markedly different when passing from an unperturbed chain to a chain with excluded volume. Klein and Pincus[77] have published an analysis of the interaction between planar surfaces with irreversibly adsorbed polymers under poor-solvent conditions, rationalizing results[78] from the study of mica surfaces with polystyrene adsorbed below the θ temperature. The applicability of the above type of studies to the stabilization of colloidal systems[79] has been well summarized by Silberberg.[80]

Rods and Helices.—Major interest in the theory and experiment of 'stiffened' coils has been stimulated by the interest both in synthetic rod-like polymers and in biopolymers. The former are of use both because of their novel mechanical properties and their tendency to undergo an isotropic–nematic liquid crystal phase transition, and the latter because of the continuing interest in biopolymer structure/property relations. The Kratky–Porod (KP) worm-like chain is characterized only by a persistence length, L, while Yamakawa, Fujii, and co-workers have over the last few years constructed a new worm-like model, which can also describe the helical character of some biopolymers and includes a realistic hydrodynamic cross-sectional diameter (cf. ref. 1). This work has been further extended[81–83] over the last two years, and Shimada et al.[81] have published an analysis of the moments of the radius of gyration of the helical worm-like (HW) chain. Although only approximate solutions can be given, a number of interesting features emerge – in particular when the chain contour length is increased (i.e., as the chain becomes more coil-like). The coil limiting value is reached at lower values of this reduced contour length for the HW chain compared with its limiting case, i.e., the KP model. The shrinkage factor, g, has also been evaluated[82] for branched stars and combs with each branch a HW chain; g is smaller than for branched coil systems, a result previously established by Mansfield and Stockmayer[84] in the KP limit. An alternative approach for real polymer chains is to use the rotational isomeric state model for star branched polymers.[85] Mattice suggests that in most cases the RIS approach is more useful than that for the KP stars. However, inconsistency largely disappears[82] if the full HW chain model is used, although there is still some quantitative uncertainty

---

[75] J. A. Marqusee and J. M. Deutsch, *J. Chem. Phys.*, 1981, **75**, 5179.
[76] A. T. Clark and M. Lal, *J. Chem. Soc., Faraday Trans. 2*, 1981, **77**, 1981.
[77] J. Klein and P. Pincus, *Macromolecules*, 1982, **15**, 1129.
[78] J. Klein, *Nature (London)*, 1980, **288**, 248.
[79] J. W. Goodwin, ed., 'Colloidal Dispersions', Spec. Publ. No. 43, The Royal Society of Chemistry, London, 1982.
[80] A. Silberberg, *J. Macromol. Sci., Phys.*, 1980, **B18**, 677.
[81] J. Shimada, K. Nagasaka, and H. Yamakawa, *J. Chem. Phys.*, 1981, **75**, 469.
[82] M. Fujii, K. Nagasaka, J. Shimada, and H. Yamakawa, *J. Chem. Phys.*, 1982, **77**, 986.
[83] H. Yamakawa and T. Yoshizaki, *Macromolecules*, 1982, **15**, 1444.
[84] M. Mansfield and W. H. Stockmayer, *Macromolecules*, 1980, **13**, 1713.
[85] W. L. Mattice, *Macromolecules*, 1981, **14**, 143.

attributed to perturbations adjacent to the branch point. Fortelny et al.[86] have extended the Yamakawa–Fujii theory for the intrinsic viscosity and sedimentation coefficients of a stiff chain to take account of polydispersity – for short stiff chains this effect produces too small an α' exponent.

For polyelectrolytes, it has been known for many years that such properties as $[\eta]$ are approximately proportional to $I^{-1/2}$, with I the ionic strength. Further, the dependence of $[\eta]$ upon I is much more pronounced for polymers which are known to be intrinsically more flexible. From this qualitative picture Skolnick and Fixman[87] and independently Odijk[88] have proposed that for stiffened polyelectrolytes, the total persistence length, L, is due to an elastic contribution, identical to that for neutral worm-like chains, L_1, and an electrostatic part, L_e, which reflects the reluctance of the polyelectrolyte to bend because of the mutually repulsive coulombic charges arrayed along the chain contour. As the ionic strength of the medium increases the counter-ions screen the charges of the polyion, and the electrostatic contribution to the persistence length decreases. The Odijk–Skolnick–Fixman theory (OSF) agrees reasonably well with experimental data for DNA solutions at low ionic strength, but at high ionic strength produces too low an estimate for L_e. Schurr and Allison[89] have now corrected an error in the Manning theory, and shown it to give results very similar to the OSF model. Now Le Bret[90] and Fixman[91] independently have published a more detailed analysis of the behaviour of L_e with I. In Le Bret's approach the Poisson–Boltzmann equation is solved numerically for a torus whose curvature changes with ionic strength, while in the Fixman model a curved cylinder is used. The numerical difference between the two evaluations is not too great, and agreement with the results of the best current experiments now seems to be very good at all ionic strengths.

Stigter[92] has considered expansion of polyelectrolyte coils and compared light-scattering data for sulphonated polystyrenes with some of the above theories. For flexible coils there is a difference between theory and experiment at low ionic strength. The problem of separating excluded volume effects (which are, of course, relatively greater for flexible coils[67]) from charge effects is overcome by measuring neutral polystyrenes in a Θ solvent. Brender et al.[93] have also investigated polyelectrolyte effects with a dynamic Monte Carlo method, in which the counterions are free; preliminary results are encouraging. Itou et al.[94] have measured M and $[\eta]$ for a series of α-helical polypeptides over the range 3×10^3 up to 5×10^5. Rodlike behaviour was well described by the Yamakawa–Fujii theory adapted for spheroid cylinders ('sausage' model). Similar agreement with results of the Yamakawa–Fujii theory were obtained for stiffened synthetic polymers.[95,96]

---

[86] I. Fortelny, J. Kovar, A. Zivny, and M. Bohdanecky, *J. Polym. Sci., Polym. Phys. Ed.*, 1981, **19**, 181.
[87] J. Skolnick and M. Fixman, *Macromolecules*, 1977, **10**, 944.
[88] T. Odijk, *J. Polym. Sci., Polym. Phys. Ed.*, 1977, **15**, 477.
[89] J. M. Schurr and S. A. Allison, *Biopolymers*, 1981, **20**, 251.
[90] M. Le Bret, *J. Chem. Phys.*, 1982, **76**, 6243.
[91] M. Fixman, *J. Chem. Phys.*, 1982, **76**, 6346.
[92] D. Stigter, *Macromolecules*, 1982, **15**, 635.
[93] C. Brender, M. Lax, and S. Windwer, *J. Chem. Phys.*, 1981, **74**, 2576.
[94] S. Itou, N. Nishioka, T. Norisuye, and A. Teramoto, *Macromolecules*, 1981, **14**, 904.
[95] E. Bianchi, A. Ciferri, J. Preston, and W. R. Krigbaum, *J. Polym. Sci., Polym. Phys. Ed.*, 1981, **19**, 863.
[96] I. Noda, T. Imai, T. Kitano, and M. Nagasawa, *Macromolecules*, 1981, **14**, 1306.

Norisuye and Fujita[62] have compared data for $\langle S^2 \rangle$ for a number of polymers, both flexible and stiff, using the KP chain model. For less than around 50 statistical segments, the KP model fits all the systems examined. This implies that the critical chainlength for the onset of excluded volume is proportional to the total persistence length, but that the proportionality factor is ~ 10 times greater than expected from the perturbation theory for the KP chain.

A technique which has become increasingly important recently is that of transient birefringence, both electrically[97] and magnetically induced,[98] and in which polarizability and rotational diffusion times may be measured.[99] A polyelectrolyte in solution is partially oriented by a pulsed field, and the build-up and decay of the orientation process is followed by monitoring the induced birefringence. For a rodlike molecule the orientation time depends approximately upon (length)^3, so that deviations due to flexibility are easily seen. Elias and Eden[97] have compared data for DNA restriction fragments with the rigid rod model, and other approaches to the polarizability of a macro-ion. A recent Fixman theory[100] is the most successful; finally, Hagerman and Zimm[101,102] have presented a Monte Carlo analysis of rotational diffusion coefficient as a function of persistence length for short wormlike chains, without requiring detailed knowledge of the hydrodynamic radii.

3 Dynamics of Dilute Solutions

For many years the dynamics of dilute polymer solutions have been described by the Rouse–Zimm bead spring model.[2] In terms of the draining parameter, h, the Rouse solution[2] to the bead spring diffusion equation furnished a solution for $h = 0$ – i.e., completely free draining – whereas the Zimm[103] approach corresponds to a non-draining limit, with $h = \infty$. The latter model is a very good approximation for dilute solutions of very long chains; partial draining behaviour is observed for shorter or stiffer chains, or when the concentration of chains is increased. Hydrodynamic interaction is taken into account in the Zimm approach by pre-averaging the elements of the Oseen tensor, as proposed by Kirkwood and Riseman. Recently, however, several workers have attempted to assess the error involved in preaveraging, and to give a more precise treatment. For example, Garcia de la Torre and co-workers[104,105] have introduced the higher-order terms in the interaction tensor as formulated by Rotne and Prager,[106] and then performed a Monte Carlo calculation. In this way the hydrodynamics may be investigated as a function of chainlength; results are quite close to those obtained by Zimm[103] in another recent paper. The Kirkwood–Riseman approach is equivalent to assuming that in certain

---

[97] J. G. Elias and D. Eden, *Macromolecules*, 1981, **14**, 410.
[98] G. Weill, *NATO Adv. Study Int. Ser.*, Ser. B, 1981, **B64**, 473.
[99] J. V. Champion, *Dev. Polym. Charact.*, 1980, **2**, 207.
[100] M. Fixman, *J. Chem. Phys.*, 1980, **72**, 5177.
[101] P. J. Hagerman and B. H. Zimm, *Biopolymers*, 1981, **20**, 1481.
[102] P. J. Hagerman, *Biopolymers*, 1981, **20**, 1503.
[103] B. H. Zimm, *Macromolecules*, 1980, **13**, 592.
[104] J. Garcia de la Torre, A. Jimenez, and J. J. Freire, *Macromolecules*, 1982, **15**, 148.
[105] J. Garcia de la Torre and J. J. Freire, *Macromolecules*, 1982, **15**, 155.
[106] J. Rotne and S. Prager, *J. Chem. Phys.*, 1969, **50**, 4381.

stationary flows a flexible chain polymer moves as if it were a rigid body (cf. ref. 107). Fixman has attempted to simulate polymer chain dynamics, but with the inclusion of fluctuating hydrodynamic interaction, using the Langevin approach which has been widely applied to simple liquids.[108—110] Only preliminary results have so far been obtained, but pre-averaging does not produce a major error. The approach of Wilemski and Tanaka[111] looks to be the most encouraging so far; they derive a new exact formula for $[\eta]$, and comment in depth about the contribution from different terms in their formulation. In particular the bead velocities must still be calculated specifically, because they suggest that Kirkwood–Riseman rigid body assumption is not always appropriate for flexible polymers. This means that, although their formula is exact for all rigid bodies, it can currently only be applied to coils with some further assumptions.

Nevertheless, some particularly interesting results are obtained, notably that pre-averaging does not always provide an upper bound on, e.g., $[\eta]$, as has previously been suspected. Encouraged by these results, Tanaka[14] has employed a Pade approximant to generate a closed form for α_η^2, the excluded volume expansion factor for the intrinsic viscosity, using the perturbation theory coefficients. Results over a wide range of z are good, and suggest a further way of assessing unperturbed dimensions. Fortelny[112] has investigated the error involved in pre-averaging the hydrodynamic interaction for the intrinsic viscosity and for the intrinsic flow birefringence. The error in pre-averaging is $\sim 8\%$, although the overall error in comparing this with other treatments,[113] but neglecting HI, is greater than this.

Ullman[114] has applied a smoothed 'blob' approach to the calculation of $[\eta]$ which seems to be useful for data fitting, although this can predict a MKS α' exponent greater than 0.8 for high-molecular-weight Gaussian coils in a good solvent (cf. ref. 61). Bruns and Bansai[115] have compared the results of different models of polymer chain dynamics, notably the bead-rod and bead-spring models, while Guttman et al.[116] have evaluated the hydrodynamic radius, R_h^{-1}, using a Monte Carlo approach. Extremely slow convergence to the Gaussian chain limit was observed, as has also been appreciated by proponents of the 'blob' theory and much earlier by Stockmayer and Albrecht in 1958.[1,2] Guttman et al. also calculate the ρ parameter $= \langle S^2 \rangle^{1/2} R_h^{-1}$ introduced by Schmidt and Burchard[117,118] and find a value of ~ 1.40, slightly lower than the Kirkwood analytical result of 1.504,[2] but higher than reported in a number of experiments which they have analysed, viz, ~ 1.30. Edwards and co-workers[119] have concentrated upon the effect of partial free draining in short chain polymers (cf. refs. 13 and 15—17) and compared realistic RIS models with experimental data for polymers in the unperturbed state.

[107] B. H. Zimm, Macromolecules, 1982, **15**, 520.
[108] M. Fixman, Macromolecules, 1981, **14**, 1706.
[109] M. Fixman, Macromolecules, 1981, **14**, 1710.
[110] M. Fixman, J. Chem. Phys., 1982, **76**, 6124.
[111] G. Wilemski and G. Tanaka, Macromolecules, 1981, **14**, 1531.
[112] I. Fortelny, Makromol. Chem., 1982, **183**, 193.
[113] K. Tsuda, Rep. Prog. Polym. Phys. Jpn., 1969, **12**, 55.
[114] R. Ullman, Macromolecules, 1981, **14**, 746.
[115] W. Bruns and R. Bansai, J. Chem. Phys., 1981, **75**, 5149.
[116] C. M. Guttman, F. L. McCrackin, and C. C. Han, Macromolecules, 1982, **15**, 1205.
[117] W. Burchard, M. Schmidt, and W. H. Stockmayer, Macromolecules, 1980, **13**, 1265.
[118] M. Schmidt and W. Burchard, Macromolecules, 1981, **14**, 211.
[119] C. J. C. Edwards, D. Rigby, and R. F. T. Stepto, Macromolecules, 1981, **14**, 1808.

The number of workers involved in the technique of quasielastic scattering (QES) continues to grow, including those involved in light-scattering and neutron-scattering experiments. At the same time there is a great interest in theory, especially the evaluation of the dynamic scattering function $S(\mathbf{k}, t)$,[120] and its logarithmic derivative, the observable decay constant $\Gamma(\mathbf{k})$ and Fourier transform, the structure factor $S(\mathbf{k}, \omega)$. The calculation of $S(\mathbf{k}, t)$ purely by using equilibrium averages (i.e., by neglecting the memory function in the projection operation formalism[120]), was carried out by Akcasu and Gurol (AG) in 1976,[121] and has since been used extensively in the study of, e.g., chain branching, etc.[117,118] Some effects are discussed by Stockmayer and Schmidt,[122] especially in polydispersity, branching, and chain stiffness. For example, if the decay constant $\Gamma(\mathbf{k})$ is measured at zero scattering angle, then the result is just $q^2 D_z$, with D_z the z-averaged diffusion coefficient for a polydisperse sample. However, in general

$$\Gamma(\mathbf{k})/k^2 = D_z(1 + Ck^2\langle S^2\rangle_z + \cdots)$$

and the coefficient C depends upon chain structure, polydispersity and goodness of solvent and may be calculated using the AG formalism. For example $C = 13/75$ for non-pre-averaged hydrodynamic interaction, but falls to $10/75$ on pre-averaging. At very short times the AG method may not be correct because the memory function is non-negligible, so that C measured should also be extrapolated to zero sample time.[123] For the same reason the AG approach becomes inapplicable to very stiff chains, and alternative approaches are required. One such is to use the Harris–Hearst model, as has been shown by Schmidt, Stockmayer, and Mansfield.[123] Han and Akcasu[124] have tabulated $S(\mathbf{k}, \omega)$ for different values of a, with a the statistical segment length, with and without hydrodynamic interaction for Gaussian chains. This helps to improve the accuracy of extraction of $\Gamma(\mathbf{k})$ when measuring in the frequency domain; at shorter wavelengths $\Gamma(\mathbf{k})$ becomes proportional to k^3 for an unperturbed chain with hydrodynamic interaction.[125,126] Kajiwara and Burchard[127] have calculated the dynamic scattering behaviour of randomly crosslinked chains using the AG formalism and the cascade theory. A recent review[128] details the power and elegance of these combined techniques in dealing with systems with quite complex chain architecture. Ou and co-workers[129] calculate the structure factor using a different approach, in particular a generalized mean field (cf. ref. 130). Results compare quite well with the known exact result for a free draining chain. Hess et al.[131] use a scaling approach ('blob' model) to calculate the exponent of $\mathbf{k}$, for different distance scales,[132] and to include hydrodynamic and

---

[120] B. J. Berne and R. Pecora, 'Dynamic Light Scattering', Wiley, New York, 1976.
[121] A. Z. Akcasu and H. Gurol, J. Polym. Sci., Polym. Phys. Ed., 1976, **14**, 1.
[122] W. H. Stockmayer and M. Schmidt, Pure Appl. Chem., 1982, **54**, 407.
[123] M. Schmidt, W. H. Stockmayer, and M. L. Mansfield, Macromolecules, 1982, **15**, 1609.
[124] C. C. Han and H. Z. Akcasu, Polymer, 1981, **22**, 1019.
[125] E. Dubois-Violette and P. G. de Gennes, Physics (NY), 1967, **3**, 181.
[126] C. C. Han and H. Z. Akcasu, Macromolecules, 1981, **14**, 1080.
[127] K. Kajiwara and W. Burchard, Polymer, 1981, **22**, 1621.
[128] W. Burchard, Adv. Polym. Sci., 1983, **48**, 1.
[129] J. J. Ou, J. S. Dahler, and M. S. Jhon, J. Chem. Phys., 1981, **74**, 1495.
[130] T. A. Witten, J. Chem. Phys., 1982, **76**, 3300.
[131] W. Hess, W. Jilge, and R. Klein, J. Polym. Sci., Polym. Phys. Ed., 1981, **19**, 849.
[132] A. Z. Akcasu, Macromolecules, 1982, **15**, 1321.

excluded volume interactions, while Escudero and Freire[133] have evaluated $\Gamma(\mathbf{k})$ in the intermediate range of $\mathbf{k}$ for discrete chain lengths – their approach is similar to that given in ref. 129. A number of approaches have been suggested to determine the polydispersity of a real system by inverting experimental data for $S(\mathbf{k}, t)$. In particular, Chu and co-workers[134] and Provencher[135] have recently introduced two techniques. The former method has been criticized by Caroline.[136] In fact, the use of simulated data[137] shows that the analysis of $\Gamma(\mathbf{k})$ as a single exponential is little different from that found using the full cumulants approach for polydispersities up to $M_w/M_n \simeq 1.2$.

As mentioned earlier, a number of experimental determinations of the ratio $\rho = \langle S^2 \rangle^{1/2} R_h^{-1}$ (refs. 116, 118, 138) in a Θ-system have produced the value 1.27—1.30; well outside the range indicated by the Zimm and Kirkwood theories, viz, 1.479 and 1.504, respectively.[118] Although part of the difference may be due to pre-averaging (cf. ref. 103), or to non-asymptotic chain-length behaviour, Schmidt and Burchard suggest that other deficiencies in the model itself, e.g., the use of the Oseen formula, which reduces chain segments to point centres of hydrodynamic resistance, the assumed additivity of these interactions, or the possibility of singularities in the hydrodynamic model, may be to blame. Experimental uncertainties are unlikely, because both components of ρ can now be determined with precision in a simultaneous static and dynamic light-scattering experiment.[139] Improvements in design of software correlators also now allow as many as 512 channels to be accessed.[140]

Dilute solution linear viscoelastic properties have long been used by Ferry, Shrag, and co-workers to probe the effectiveness of the Rouse–Zimm bead-spring model. They have established that, provided the exact eigenvalues are used (to correct for finite chain length) and the draining parameter h is slightly greater than zero, excellent fits to experimental data are obtained.[141] An analogous experiment is that of oscillatory flow birefringence[142,143] (OFB). Recent results confirm that such experiments may be performed close to 'infinite' dilution perhaps with even greater accuracy than the classical viscoelastic techniques. The data of Lodge and Schrag[142] reinforce the results from viscoelastic measurements,[141] over five decades of effective frequency. Interestingly the results obtained from measurements at higher concentrations[143] show no dramatic change in character, corresponding to a c^* transition compared with that observed with the mechanical technique (see below).

---

[133] J. A. Escudero and J. J. Freire, Polymer J., 1982, **14**, 277.
[134] T. Nose and B. Chu, Macromolecules, 1979, **12**, 590.
[135] S. W. Provencher, Makromol. Chem., 1979, **180**, 201.
[136] D. Caroline, Polymer, 1981, **22**, 576.
[137] D. Caroline, Polymer, 1982, **23**, 492.
[138] C. M. Kok and A. Rudin, Makromol. Chem. Rapid Commun., 1981, **2**, 655.
[139] S. Bantle, M. Schmidt, and W. Burchard, Macromolecules, 1982, **15**, 1604.
[140] N. Nemoto, Y. Tsunashima, and M. Kurata, Polymer J., 1981, **9**, 827.
[141] J. D. Ferry, 'Viscoelastic Properties of Polymers', 3rd Edn., Wiley, New York, 1980.
[142] T. P. Lodge, J. W. Miller, and J. L. Schrag, J. Polym. Sci., Polym. Phys. Ed., 1982, **20**, 1409.
[143] T. P. Lodge and J. L. Schrag, Macromolecules, 1982, **15**, 1376.

4 Effect of Increasing Concentration

Dilute and Semi-dilute Solutions.—Most of the discussion in earlier sections has been restricted to polymers in very dilute solution, *i.e.*, so that intermolecular effects are negligible. The transition from dilute solution behaviour (isolated coils/rods) to semi-dilute (interpenetration, entanglement, uniform polymer segment density) usually occurs over a narrow range of concentration, close to the 'critical concentration', c^*, given approximately by $c^* \approx S/[\eta]$, with S, say, 4—8. A number of techniques are suitable for assessing c^* (and S), although the actual transition may be different depending upon the exact technique employed – usually rheological methods suffice, c^* being determined from the change in slope of log η_0 *vs.* loc c plot, with η_0 the zero shear viscosity. (Note some workers distinguish this as a c^{**} transition.) Line-shape analysis from n.m.r. and neutron spin echo[144] may also be employed, but reports[1] that ultraviolet absorbance is also a suitable method have now been shown[145] to be an artefact. Sedimentation methods also seem to suggest a sharp transition from dilute, $S_c \propto c^0$ to semi-dilute, $S_c \propto c^{-1}$ with S_c an effective sedimentation coefficient.[146] These scaling laws have again been verified by Vidakovic *et al.* for polystyrene in cyclohexane.[147] Recent work by Roots and Nystrom,[148] who first introduced the sedimentation technique, have been extended to polyelectrolyte solutions, using the Odijk[149] theory for finite concentrations. Neither the limiting slope predicted for this model nor the sharp transition in c^* are observed, although the former effect may be because of the comparatively low-molecular-weight samples ($\simeq 1 \times 10^6$). Yu *et al.*[150] have performed quasi-elastic scattering experiments in the region between c^* and c^{**}, the former corresponding to $S \simeq 3$. The sharp distinction between these regimes is usually difficult to determine experimentally. Especially when comparing the results of rheological experiments,[151] measurements must be made over a wide range of concentrations and shear rates. A particularly interesting experiment is the forced Rayleigh scattering technique of Leger and co-workers.[152] This involves labelling a few polymer chains with a photochromic probe, and illuminating with a chopped laser beam. In this way a minute 'absorption grating' may be constructed and the self-diffusion of quite concentrated polymer solutions may be followed with a second laser, over realistic timescales. Results obtained for the self-diffusion coefficient, D_{self}, are in very good agreement with scaling predictions, with $D_{self} \propto c^{-7/4}M^{-2}$. D_{self} has previously been measured by several workers with the conventional QES technique, *e.g.*, Amis and Han[153] find support for the scaling prediction.

The Muthukumar–Edwards method[47] provides extrapolation formulae for the average size of a single (labelled) chain at any concentration and excluded volume; one interesting prediction of this remarkably general treatment is that, although the

---

[144] B. Ewen, D. Richter, J. B. Hayter, and B. Lehnen, *J. Polym. Sci., Polym. Lett. Ed.*, 1982, **20**, 233.
[145] C. H. Lee, W. H. Waddell, and E. F. Casassa, *Macromolecules*, 1981, **14**, 1021.
[146] B. Nystrom and J. Roots, *J. Macromol. Sci., Rev. Macromol. Chem.*, 1980, **C19**, 35.
[147] P. Vidakovic, C. Allain, and F. Rondelez, *Macromolecules*, 1982, **15**, 1571.
[148] J. Roots and B. Nystrom, *Polymer*, 1981, **22**, 573.
[149] T. Odijk, *Macromolecules*, 1979, **12**, 688.
[150] T. L. Yu, H. Reihanian, J. G. Southwick, and A. M. Jamieson, *J. Macromol. Sci., Phys.*, 1980, **B18**, 777.
[151] A. M. Jamieson and D. Telford, *Macromolecules*, 1982, **15**, 1329.
[152] L. Leger, H. Hervet, and F. Rondelez, *Macromolecules*, 1981, **14**, 1732.
[153] E. J. Amis and C. C. Han, *Polymer*, 1982, **23**, 1403.

scaling prediction $\langle S^2 \rangle \propto c^{-1/4}$ is recovered at large concentrations, this regime cannot be reached for any $M < 10^{14}$ with reasonable values of the excluded volume. The paper by Lodge and Schrag [143] extends OFB measurements into the range of finite concentrations (although $<c^*$). The relaxation time spectrum is markedly affected by concentration, although agreement with the Muthukumar–Freed[154] concentration dependence is very good.

Dynamic simulations of semi-dilute (and more concentrated) solutions and melts have been carried out by a number of workers, including Ceperley et al.[55,155] Such work is currently restricted to quite small chains and to small numbers of time steps, because of restrictions on computer time. The advent of 'Supercomputers'[156] may well allow more realistic time-dependent simulations to be carried out.

Molecular Rheology of More Concentrated Solutions.—Although it is outside the scope of the present Chapter to consider the behaviour of more concentrated solutions or melts in detail, the advent of recent molecular models should not be ignored, since currently this is an area of great activity.[4,5] In the 'tube' models as polymer concentration becomes increased, the motion of an individual polymer chain becomes so restricted by the environment of the other chains that it can no longer move isotropically, but instead moves down a hypothetical tube formed by these other chains (reptates). Bird[157] has recently reviewed 'kinetic' models of polymer rheology and shown how from them constitutive equations for the time dependence of stress and strain may be derived. Both the models of Doi–Edwards[4] and Curtiss–Bird[5] are examples, although the former is apparently a particular case of the latter.[157] The differences between the two have been well tabulated by Saab, Bird, and Curtiss.[158]

Evans and Edwards[159] have carried out a computer simulation of the dynamics of highly entangled polymers in an attempt to verify that reptation does occur in such a system. The equilibrium dynamics show that such a mechanism does occur, and that the primitive path used by Doi and Edwards[4] can be modelled on the computer and treated as a random walk. From these two results, the simulation may then be used to test that the primitive chain model is applicable to polymer dynamics. Unlike the original scaling argument, the Doi–Edwards approach may be used to predict proportionality constants, some of which have been independently confirmed. Further, this model predicts that, for example, D_r, the diffusion constant of the chain, should be proportional to $\delta^{-3/2} n^{-2}$ in semi-dilute solution, with n the number of statistical segments and δ the number density of monomer units. In the melt D_r is now shown to be $\delta^{-1} n^{-2}$; the diameter of the 'tube' is ca. 2—3 statistical segments. The extension of the simulation to three-arm star polymer dynamics[160] predicts $D_r \propto n^{-3}$ and the tube relaxation time $T \propto n^4$. This is contrary to earlier predictions that a star could not 'reptate', because the simulation shows the star hub to move quite freely and the exponents are in quite good agreement with experimental data.

[154] M. Muthukumar and K. L. Freed, *Macromolecules*, 1978, **11**, 843.
[155] M. Bishop, D. Ceperley, H. L. Frisch, and M. H. Kalos, *J. Chem. Phys.*, 1982, **76**, 1557.
[156] R. D. Levine, *Sci. Am.*, 1982, **246**, 112.
[157] R. B. Bird, *J. Rheol.*, 1982, **26**, 277.
[158] H. H. Saab, R. B. Bird, and C. F. Curtiss, *J. Chem. Phys.*, 1982, **77**, 4758.
[159] K. E. Evans and S. F. Edwards, *J. Chem. Soc., Faraday Trans. 2*, 1981, **77**, 1891, 1913, and 1929.
[160] K. E. Evans, *J. Chem. Soc., Faraday Trans. 2*, 1981, **77**, 2385.

A scaling argument by Evans[161] indicates a general dependence $T \propto n^{f+1}$ for f armed stars, although for highly branched systems entanglement motions themselves would restrict the validity of this result.

Osaki, Kurata, and co-workers[162,163] have continued their careful experimental assessment of the Doi–Edwards theory for concentrated solutions, including measurements of first and second normal stress differences. Except at very short times, the theory is well supported, and a similar exercise has been performed by Menezes and Graessley.[164] The latter workers have measured non-linear viscoelastic parameters, *e.g.*, shear and normal stress growth, and introduce parameters from the molecular theory into a new form of a general constitutive equation.[141] Agreement between theory and experiment is encouraging for a wide range of deformation histories.

The interest in the properties of branched polymers is also evident.[165–168] For example, Roovers and Graessley[168] have reported work on comb polystyrenes, whereas Masuda *et al.* have used six-arm stars.[165] For the combs it was possible to identify three major relaxation mechanisms, *viz*, those at very long and short times corresponding to the slow and fast relaxation of the whole molecule found also in linear systems, and an intermediate process corresponding to the movement of the branches. Finally the Curtiss–Bird kinetic theory has been described in a number of papers, and although currently couched in terms of melt rheology, the results for elongational[169] and shear flows[170] may also be applied to concentrated solutions above c^*.[158] This model, in which a freely jointed bead rod chain has a non-isotropic Stokes law applied to the beads to reflect the entanglement couplings, is less deceptively simple than the tube model, although claimed to be more rigorous.[157] In the CB kinetic theory a parameter ε, the link tension coefficient, varies between zero (the Doi–Edwards equivalent value) and unity. Most experimental data is fitted best by $\varepsilon \simeq 0.4$,[158] including that given in ref. 164. One major advantage of the Curtiss–Bird theory seems to be that for $\varepsilon > 0$ the shear rate reduction of steady shear viscosity, η/η_0, is predicted to be lower than the limiting critical value 0.25 found for the Doi–Edwards theory and in agreement with experiment.[141]

---

[161] K. E. Evans, *J. Polym. Sci., Polym. Lett. Ed.*, 1982, **20**, 103.
[162] K. Osaki, S. Kimura, K. Nishizawa, and M. Kurata, *Macromolecules*, 1981, **14**, 456.
[163] K. Osaki, S. Kimura, and M. Kurata, *J. Polym. Sci., Polym. Phys. Ed.*, 1981, **19**, 517.
[164] E. V. Menezes and W. W. Graessley, *J. Polym. Sci., Polym. Phys. Ed.*, 1982, **20**, 1817.
[165] T. Masuda, Y. Ohta, M. Kitamura, Y. Saito, K. Kato, and S. Onogi, *Macromolecules*, 1981, **14**, 354.
[166] Y. Ohta, Y. Saito, T. Masuda, and S. Onogi, *Macromolecules*, 1981, **14**, 1128.
[167] K. Yasuda, R. C. Armstrong, and R. E. Cohen, *Rheol. Acta*, 1981, **20**, 163.
[168] J. Roovers and W. W. Graessley, *Macromolecules*, 1981, **14**, 766.
[169] R. B. Bird, H. H. Saab, and C. F. Curtiss, *J. Phys. Chem.*, 1982, **86**, 1102.
[170] R. B. Bird, H. H. Saab, and C. F. Curtiss, *J. Chem. Phys.*, 1982, **77**, 4747.

8
The Chemical Microstructure of Synthetic Polymers Investigated by High Resolution Nuclear Magnetic Resonance

BY J. R. EBDON

1 Introduction

This Chapter comprises a selective review of the literature for 1981/82 on the use of nuclear magnetic resonance techniques for the characterization of synthetic and modified natural polymers. Particular emphasis is placed on the use of n.m.r. for microstructural characterization of polymers, both in solution and in the solid state. Rather more complete reviews of the polymer n.m.r. literature for most of the period under consideration may be found in companion publications.[1]

2 Books and Reviews

The work of Bovey and colleagues at the Bell Laboratories on the use of n.m.r. for the characterization of polymers is justly famous. Particularly welcome therefore are reviews of recent work from this source.[2,3] These reviews cover the use, principally, of ^{13}C and ^{19}F n.m.r. for the determination of tacticities, branching, and head–head and tail–tail linkages in such polymers as poly(vinyl chloride), poly(vinyl bromide), polyethylene, poly(vinyl fluoride), and poly(vinylidene fluoride). The second of these reviews lays emphasis on the value of rotational isomeric state models in rationalizing polymer chemical shift data. Work at Bell Laboratories is covered also in a recent symposium report together with, for example, work by Ivin and colleagues on studies of ring-opening polymerizations involving metathesis catalysts; Doskočilova and Schneider on n.m.r. studies of swollen gels; Resing, Garroway, and colleagues, on cured resins; Schaefer and co-workers on solid polymers; Kricheldorf on applications of ^{15}N n.m.r.[4] Harwood has recently reviewed the use of n.m.r. for the structural characterization of polydienes.[5]

The interest in the use of cross-polarization ^{13}C n.m.r. with magic-angle sample spinning (CP MASS) for the structural characterization of solid amorphous and

---

[1] J. R. Ebdon, Chapter 9 in 'Nuclear Magnetic Resonance', ed. G. A. Webb (Specialist Periodical Reports), The Royal Society of Chemistry, London, Vol. 11, 1982, and Vol. 12, 1983.
[2] F. A. Bovey, R. E. Cais, L. W. Jelinski, F. C. Schilling, W. H. Starnes, jun., and A. E. Tonelli, *Polym. Prepr., Am. Chem. Soc., Div. Polym. Chem.*, 1981, **22**, 268.
[3] A. E. Tonelli and F. C. Schilling, *Acc. Chem. Res.*, 1981, **14**, 233.
[4] Main lectures presented at the 22nd Microsymposium on Macromolecules, 'Characterization of Structure and Dynamics of Macromolecular Systems by NMR Methods', Prague, July, 1981, *Pure Appl. Chem.*, 1982, **54**.
[5] H. J. Harwood, *Rubber Chem. Technol.*, 1982, **55**, 769.

crystalline polymers has increased enormously in the last few years. Particularly timely, therefore, are reviews of techniques and applications from Mandelkern,[6] Schaefer and colleagues,[7] McBrierty and Douglass,[8] and Yannoni and co-workers.[9,10]

Also welcome is the appearance of the first volume of an atlas of polymer ^{1}H and ^{13}C n.m.r. spectra.[11] This atlas should prove particularly useful to those contemplating the use of n.m.r. for routine characterization of polymers, perhaps for quality control purposes. It would appear that a second similar enterprise in this direction is also under way.[12]

3 Homopolymers: Tacticity

N.m.r. in general, and ^{13}C n.m.r. in particular, has proved to be a particularly powerful technique for the elucidation of homopolymer tacticity, and publications concerning this topic still regularly appear. Many are devoted to extensions of existing methods to new systems, but it would appear that there are still useful things to be learnt about the tacticities of familiar polymers such as polypropylene, polystyrene, and poly(vinyl chloride). For example, it has been found from the study of ^{13}C and ^{1}H n.m.r. spectra of epimerized isotactic polypropylenes of predictable stereochemistries that small refinements of previously published assignments are necessary.[13] Also, it would appear that a consensus has emerged regarding the stereochemistry of radically prepared polystyrene as determined by ^{13}C n.m.r.[14-16] Sato et al. have examined the ^{13}C n.m.r. spectra of the various diastereomers of styrene pentamer and on the basis of these have more fully assigned the components of the phenyl C-1 and β-methylene carbon signals in the spectrum of atactic polystyrene.[14,15] For the phenyl signals, assignments to configurational pentads can be made, and for the β-methylene signals, assignments to configurational hexads. The configuration of a free radically prepared polystyrene is found to fit Bernoullian statistics with P_m (probability of meso placement) equal to 0.46. In the light of these assignments, Matsuzaki and co-workers have re-examined the tacticities of anionically and cationically prepared polystyrenes.[16] Some revised assignments have appeared also for the ^{13}C n.m.r. spectrum of head-head polystyrene.[17]

Interpretation of the high-field ^{1}H n.m.r. spectrum of atactic polystyrene is greatly facilitated by acetylating the polymer (or its β-dideuterio-derivative) in the *para*-ring

---

[6] L. Mandelkern, *Polym. Prepr., Am. Chem. Soc., Div. Polym. Chem.*, 1981, **22**, 276.
[7] T. R. Steger, J. Schaefer, E. O. Stejskal, R. A. McKay, and M. D. Sefcik, *Ann. NY Acad. Sci.*, 1981, **371**, 106.
[8] V. J. McBrierty and D. C. Douglass, *Macromol. Rev.*, 1981, **16**, 295.
[9] C. S. Yannoni, *Acc. Chem. Res.*, 1982, **15**, 201.
[10] J. R. Lyerla, C. S. Yannoni, and C. A. Fyfe, *Acc. Chem. Res.*, 1982, **15**, 208.
[11] Q. T. Pham and R. Petiaud, 'Proton and Carbon NMR Spectra of Polymers', Heyden, Philadelphia, 1981, Vol. 1.
[12] Y. Fujiwara, K. Hatada, T. Hirano, T. Kawamura, S. Kondo, K. Matsuzaki, A. Nishioka, Y. Tanaka, and B. Tomita, Proc. Int. CODATA Conf., 1981, p. 283.
[13] U. W. Suter and P. Neuenschwander, *Macromolecules*, 1981, **14**, 528.
[14] H. Sato, Y. Tanaka, and K. Hatada, *Makromol. Chem., Rapid Commun.*, 1982, **3**, 175.
[15] H. Sato, Y. Tanaka, and K. Hatada, *Makromol. Chem., Rapid Commun.*, 1982, **3**, 181.
[16] T. Kawamura, T. Uryu, and K. Matsuzaki, *Makromol. Chem., Rapid Commun.*, 1982, **3**, 661.
[17] F. Bangerter, S. Serafini, P. Pino, and O. Vogl., *Makromol. Chem., Rapid Commun.*, 1981, **2**, 109.

position;[18] the methine ^{1}H signals are then more clearly resolved into contributions from isotactic, heterotactic, and syndiotactic triads. For poly(α-fluorostyrene), ^{1}H and ^{13}C n.m.r. are of little use for tacticity determination. However, the high-field ^{19}F n.m.r. spectrum is better resolved[19] and suggests $p_m \sim 0.5$. The stereoregularity of poly(p-isopropyl-α-methylstyrene) also is amenable to analysis by ^{1}H and ^{13}C n.m.r.[20] The polymer is Bernoullian, however prepared, with $P_m = 0.46$ (anionic), 0.07 (cationic), and 0.23 (radical). In some respects, the high-field ^{1}H n.m.r. spectrum of poly(vinyl thiophene) is very like that of polystyrene;[21] assignments to pentad stereosequences have been made for the components of the resonance from the proton in the 3-position on the ring.

The ^{13}C n.m.r. spectrum of poly(vinyl chloride) is very solvent (and temperature) dependent. This feature has complicated the interpretation of the spectrum until recently. Ando and Asakura have carefully monitored the changes in chemical shift over the dielectric constant range 3.7—48.9 in pentachloroethane/dimethyl sulphoxide mixtures and have confirmed some previous assignments.[22] The changes in chemical shift are interpreted in terms of preferential solvation of certain stereosequences rather than in terms of solvent-induced conformational changes. The methine carbon signals for poly(vinyl chloride) appear to be particularly well resolved in 1,4-dioxan.[23] Carbon chemical shifts of model compounds for sequences in poly(vinyl bromide) have been examined by Cais and Kometani.[24] In the polymer itself, methine peak components are assigned to pentads, and methylene components to tetrads. The polymer as normally prepared obeys Bernoullian statistics with $P_m = 0.46$. The ^{13}C and ^{19}F spectra of poly(vinyl fluoride) and related polymers give information on frequencies of head–head and tail–tail linkages as well as on tacticities.[25,26] New assignments to pentads and tetrads have been made also for the ^{13}C spectra of poly(vinyl acetate)s ($P_m = 0.48$).[27]

The high-field ^{13}C n.m.r. spectrum of polyacrylamide has been assigned by analogy with that of poly(vinyl chloride).[28] For this polymer also, Bernoullian statistics are indicated ($P_m = 0.43$). Other nitrogen-containing homopolymers recently examined by n.m.r. include poly(N-vinylcarbazole)[29-31] and poly(N-vinylpyrrolidone).[32] For poly(N-vinylcarbazole), the tacticities of radical polymers as assessed by ^{1}H n.m.r. are markedly temperature dependent[29] with an indication that the polymer, unusually, becomes more syndiotactic as the temperature of preparation is raised.[30] The analysis of ^{13}C n.m.r. spectra of cationic poly(N-vinylcarbazoles) indicates that the configurational statistics in these cases follow a

---

[18] D. L. Trumbo, T. K. Chen, and H. J. Harwood, *Macromolecules*, 1981, **14**, 1138.
[19] R. N. Majumdar, M. K. Niknam, H. A. Nguyen, and H. J. Harwood, *Makromol. Chem., Rapid Commun.*, 1982, **3**, 421.
[20] J. C. Favier, M. Moreau, J. P. Vairon, and J. Leonard, *Polymer*, 1982, **23**, 1501.
[21] D. L. Trumbo, T. Suzuki, and H. J. Harwood, *Polym. Bull. (Berlin)*, 1981, **4**, 677.
[22] I. Ando and T. Asakura, *Makromol. Chem.*, 1981, **182**, 1243.
[23] K. F. Elgert, R. Kosfeld, and W. E. Hull, *Polym. Bull. (Berlin)*, 1981, **4**, 281.
[24] R. E. Cais and J. M. Kometani, *Macromolecules*, 1981, **14**, 1346.
[25] A. E. Tonelli, F. C. Schilling, and R. E. Cais, *Polym. Prepr., Am. Chem. Soc., Div. Polym. Chem.*, 1981, **22**, 271.
[26] A. E. Tonelli, F. C. Schilling, and R. E. Cais, *Macromolecules*, 1982, **15**, 849.
[27] H. N. Sung and J. H. Noggle, *J. Polym. Sci., Polym. Phys. Ed.*, 1981, **19**, 1593.
[28] J. E. Lancaster and M. N. O'Connor, *J. Polym. Sci., Polym. Lett. Ed.*, 1982, **20**, 547.
[29] D. R. Terrell, *Polymer*, 1982, **23**, 1045.
[30] D. R. Terrell and F. Evers, *Makromol. Chem.*, 1982, **183**, 863.

first-order Markov model.[31] For free-radical poly(N-vinyl pyrrolidone)s, triad and tetrad fractions obtained from ^{13}C n.m.r. spectra fit Bernoullian statistics ($P_m = 0.45$), although cationic polymers have increased isotactic contents.[32]

Reports have appeared also on the n.m.r. spectra (^{1}H, ^{13}C, and ^{19}F) of several α-halo acrylates and methacrylates.[33–36] All the radically prepared polymers appear to be predominantly syndiotactic with configurational sequences governed by Bernoullian statistics. The tacticity of poly(chloromethyl methacrylate) as assessed by n.m.r. after hydrolysis and re-esterification with diazomethane appears unusual, however, with almost equal amounts of heterotactic and syndiotactic sequences.[37]

Finally in this section comes a report of the preparation and ^{13}C n.m.r. characterization of a polypropylene sample of very unusual stereochemistry.[38] The polymer was prepared by the hydrogenation of a highly isotactic poly(2-methyl pentadiene) and has an arrangement of asymmetric centres in which alternate pairs are in *meso* configurations but the intervening centres are non-specific. Methyl peaks corresponding to three of the configurational pentads (rrmr, mrmr, and mrmm) are therefore missing as expected. This type of stereochemical arrangement has been termed 'hemitactic'.

4 Homopolymers: Geometrical Isomerism

In many polymers, monomer units can be introduced into the chain in a variety of isomeric forms and in a variety of relative orientations leading to so-called head–head and tail–tail linkages in addition to the normal head–tail arrangements. Polydienes are interesting in this connection and there are three relevant n.m.r. studies to report.[39–42] For polybutadienes initiated with hydrogen peroxide, it has been shown by ^{1}H n.m.r. that short-chain polymers are richer in 1,2-units than are long-chain polymers, indicating a preference for initiation through a 1,2-unit.[39] For a variety of 1,3-polypentadienes, ^{13}C n.m.r. has been used to determine relative amounts of *cis*-1,4- and *trans*-1,4- placements, to distinguish between isotactic and syndiotactic *cis*-1,4,- structures and to identify 1,4–1,4, 1,4–1,2, and 1,4–4,1 linkages.[40] For 1,4-dimethylenecyclohexane it has been shown by ^{1}H and ^{13}C n.m.r. that cyclopolymerization does not occur, but that unsaturated units are incorporated in the polymer.[41] For poly(1,3-cyclohexadiene)s, steric hindrance seems to preclude the formation of long sequences of 1,2- units.[42]

For β-(2-acetoxy ethyl)-β-propiolactone, it has been shown by ^{1}H and ^{13}C n.m.r. that ring-opening polymerization can lead to both poly-β- and poly-δ-esters in

---

[31] T. Kawamura, M. Sakuma, and K. Matsuzaki, *Makromol. Chem., Rapid Commun.*, 1982, **3**, 475.
[32] H. N. Cheng, T. E. Smith, and D. M. Vitus, *J. Polym. Sci., Polym. Lett. Ed.*, 1981, **19**, 29.
[33] R. N. Majumdar and H. J. Harwood, *Polym. Prepr., Am. Chem. Soc., Div. Polym. Chem.*, 1981, **22**, 242.
[34] K. Hatada, T. Kitayama, K. Saunders, and R. W. Lenz, *Makromol. Chem.*, 1981, **182**, 1449.
[35] R. N. Majumdar and H. J. Harwood, *Polym. Bull. (Berlin)*, 1981, **4**, 391.
[36] K. G. Saunders, W. J. Macknight, and R. W. Lenz, *Macromolecules*, 1982, **15**, 1.
[37] M. Ueda, K. Iri, Y. Imai, and C. U. Pittman, jun., *Macromolecules*, 1981, **14**, 1046.
[38] M. Farina, G. di Silvestro, and P. Sozzani, *Macromolecules*, 1982, **15**, 1451.
[39] J.-C. Brosse, M. Bonnier, and G. Legeay, *Makromol. Chem.*, 1982, **183**, 303.
[40] P. Aubert, J. Sledz, F. Schue, and C. Brevard, *J. Polym. Sci., Polym. Chem. Ed.*, 1981, **19**, 955.
[41] A. Sebenik and H. J. Harwood, *Polym. Prepr., Am. Chem. Soc., Div. Polym. Chem.*, 1981, **22**, 15.
[42] Z. Sharaby, M. Marton, and J. Jagur-Grodzinski, *Macromolecules*, 1982, **15**, 1167.

amounts which depend on the catalyst used.[43] In the ring-opening polymerization of cyclo-octadienes, isomerization occurs to give polymers containing mixtures of cis- and trans-units, the distributions of which can be determined by ^{1}H and ^{13}C n.m.r.[44,45] The ring-opening polymerizations of a variety of cyclic olefins by using metathesis catalysts have been studied by Ivin and colleagues. Recent reports concern the use of ^{13}C n.m.r. to determine the proportions of head–head, head–tail, and tail–tail linkages in polymers prepared from 5-methyl- and 5,5-dimethyl-bicyclo[2.2.1]hept-2-enes.[46,47] The distribution of cis- and trans-units in these polymers have also been determined. Head–head and tail–tail structures have been shown to be present also in polymers prepared from 2-methylcyclobutane.[48] The ring-opening polymerization of 2-phenyl-1,3,6,2-trioxaphosphocane has been studied by ^{1}H, ^{13}C, and ^{31}P n.m.r.; the structural units of the polymer are the expected phosphinate groups plus isomerized structures.[49]

Until recently, little was known about the structure of poly(maleic anhydride). However, it has now been demonstrated by ^{1}H and ^{13}C n.m.r. that the repeat unit of the polymer is the expected anhydride, and that the end-groups in polymers prepared with benzoylperoxide are entirely phenyl groups.[50] The cationic polymerization of 4-bromo-6,8-dioxabicyclo[3.2.1]octane occurs with isomerization to give a mixture of axial and equatorial forms. The isomerization can be monitored by ^{13}C n.m.r.[51] ^{1}H and ^{13}C n.m.r. spectra have been used to show that the cyclopolymerization of methacrylic anhydride occurs with both intramolecular head–head and intramolecular head–tail addition to give five- and six-membered rings. The two structures give rise, for example, to different methine, methylene, and carbonyl carbon chemical shifts.[52] Other divinyl polymerizations recently investigated by n.m.r. include those of divinyl formal,[53] diallyl amides and diallyl sulphonamides,[54] and divinyl ferrocene.[55] For divinyl ferrocenes, the polymer structure seems somewhat dependent on the type of initiation used, with cationic polymers showing evidence of residual unsaturation.

Finally in this section there is the extensive work of Schilling and colleagues to report concerning defect structures in various fluorine-containing polymers.[56,57] In connection with this work, a novel ^{13}C-[^{1}H, ^{19}F] triple resonance technique has been developed.[57]

---

[43] T. Araki, S. Hayase, and A. Nakamura, *J. Polym. Sci., Polym. Chem. Ed.*, 1982, **20**, 3337.
[44] A. I. Syatkowsky, T. T. Denisova, E. L. Abramenko, A. S. Khatchaturov, and B. D. Babitsky, *Polymer*, 1981, **22**, 1554.
[45] A. S. Khachaturov, E. L. Abramenko, and A. I. Syatkovskii, *Vysokomol. Soedin., Ser. A*, 1982, **24**, 701.
[46] K. J. Ivin, L.-M. Lam, and J. J. Rooney, *Makromol. Chem.*, 1981, **182**, 1847.
[47] H. T. Ho, K. J. Ivin, and J. J. Rooney, *Makromol. Chem.*, 1982, **183**, 1629.
[48] J. Kops and H. Spanggaard, *Macromolecules*, 1982, **15**, 1200.
[49] S. Kobayashi, M. Y. Huang, and T. Saegusa, *Polym. Bull. (Berlin)*, 1981, **4**, 185.
[50] W. Regel and C. Schneider, *Makromol. Chem.*, 1981, **182**, 237.
[51] M. Okada, H. Sumitomo, and A. Sumi, *Polym. J.*, 1982, **14**, 59.
[52] G. B. Butler and A. Matsumoto, *J. Polym. Sci., Polym. Lett. Ed.*, 1981, **19**, 167.
[53] M. Tsukino and T. Kumitake, *Am. Chem. Soc. Symp. Ser.*, 1982, **195**, 73.
[54] J. H. Hodgkin, S. R. Johns, and R. I. Willing, *Polym. Bull. (Berlin)*, 1982, **7**, 353.
[55] G. C. Corfield, J. S. Brooks, S. Plimley, and A. V. Cunliffe, *Polym. Prepr., Am. Chem. Soc., Div. Polym. Chem.*, 1981, **23**, 270.
[56] A. E. Tonelli, F. C. Schilling, and R. E. Cais, *Macromolecules*, 1981, **14**, 560.
[57] F. C. Schilling, *Polym. Prepr., Am. Chem. Soc., Div. Polym. Chem.*, 1981, **22**, 407.

5 Mechanisms of Polymerization

N.m.r. spectroscopy is being used increasingly to examine polymerization mixtures and the influence of catalyst, solvent, and temperature upon polymer structure with a view to gaining information about polymerization mechanisms. For example, Fink and Rottler have recorded ^{13}C n.m.r. spectra on reaction mixtures containing ^{13}C-enriched ethylene and $Et_2TiRCl/AlEt_nCl_m$-type catalysts.[58] The ^{13}C spectra indicate that there is no preco-ordination of the monomer with the primary complexes and that ethylene insertion occurs unambiguously in the Ti—C bond.

Analysis of oligomers and polymers of 2- and 4-vinyl pyridine produced with living anionic catalysts shows that highly isotactic polymers are produced with *trans* addition of monomer to the propagating anion.[59-61] In the anionic polymerization of styrene, the stereoregularity of the chain as determined by ^{13}C n.m.r. is influenced by the nature of the counterion.[62] These results have been interpreted in terms of the favouring of *meso*-placements by tight ion-pairs. Similar n.m.r. studies have been reported for poly(methoxy-) and poly(methyl-styrenes).[63,64] Schué and co-workers have used ^{1}H n.m.r. to investigate the nature of the living ends in anionic polymerizations of isoprene both in the presence and absence of added amines.[65] The most useful resonance is that of the proton γ to the living end which splits into two components indicative of the *cis–tran* ratio of the terminal unit. Addition of amine produces a shift in the balance of the conformation in favour of the *cis*-structure. In anionic polymerizations of acrylates and methacrylates also, ^{13}C n.m.r. analysis of reaction mixtures and model systems has been used to examine isomer ratios.[66,67] Mechanisms of termination in the anionic polymerization of perdeuteriomethyl methacrylate have been studied with the aid of ^{1}H n.m.r.[68,69]

For the anionic polymerization of acrolein, the use of ^{1}H and ^{13}C n.m.r. has revealed the presence in the polymer of 1,4-, 3,4-, and 1,2-units.[70] To explain these results, a mechanism involving two different living ends has been proposed. ^{13}C n.m.r. has been used recently also to investigate the mechanism of steric control in the cationic ring-opening polymerization of 4-bromo-6,8-dioxabicyclo[3.2.1]-octane.[71]

---

[58] G. Fink and R. Rottler, *Angew. Makromol. Chem.*, 1981, **94**, 25.
[59] S. S. Huang, C. Mathis, and T. E. Hogen-Esch., *Macromolecules*, 1981, **14**, 1802.
[60] T. E. Hogen-Esch and W. L. Jenkins, *Macromolecules*, 1981, **14**, 510.
[61] A. Soum and M. Fontanille, *Am. Chem. Soc. Symp. Ser.*, 1981, **166**, 239.
[62] S. Suparno, J. Lacosta, S. Raynal, J. Sledz, and F. Schue, *Polym. J.*, 1981, **13**, 313.
[63] K. Matsuzaki, T. Kanai, T. Iwamoto, and M. Henmi, *Polym. Prepr., Am. Chem. Soc., Div. Polym. Chem.*, 1981, **22**, 303.
[64] Y. Firat and S. Bywater, *Eur. Polym. J.*, 1982, **18**, 265.
[65] V. Collet-Marti, S. Dumas, J. Sledz, and F. Schue, *Macromolecules*, 1982, **15**, 251.
[66] L. Vancea and S. Bywater, *Macromolecules*, 1981, **14**, 1321.
[67] L. Vancea and S. Bywater, *Macromolecules*, 1981, **14**, 1776.
[68] K. Hatada, T. Kitayama, K. Fumikawa, K. Ohta, and H. Yuki, *Am. Chem. Soc., Symp. Ser.*, 1981, **166**, 327.
[69] K. Hatada, T. Kitayama, S. Okahata, and H. Yuki, *Polym. J.*, 1981, **13**, 1045.
[70] D. Gulino, J. P. Pascault, and Q. T. Pham, *Makromol. Chem.*, 1981, **182**, 2321.
[71] M. Okada, H. Sumitomo, and A. Sumi, *Macromolecules*, 1982, **15**, 1238.

6 End Groups

For polymers of moderate to high molecular weight, end groups are present at such low concentrations as to be indiscernible by n.m.r. However, enriching the end groups relative to the main-chain with, for example, ^{13}C allows them to be detected. This technique has been used successfully by Bevington et al. to examine the structures of end groups derived from azobisisobutyronitrile in several homo- and co-polymers.[72,73] Studies of such end groups are particularly useful since they give information about the relative reactivities of monomers towards the 2-cyano-2-propyl radical.[73] The end-group resonances are also sensitive to the stereochemistries of the neighbouring monomer units. End groups may also be labelled with ^{2}H and ^{15}N, but the discrimination is much less good than with ^{13}C.[74] A very similar approach has been adopted by Moad et al. to investigate end groups derived from benzoyl peroxide in polystyrene.[75] It has been found by ^{13}C n.m.r. using benzoyl peroxide labelled at the carbonyl group with ^{13}C that there is significant head addition and evidence also of termination through transfer to initiator or by primary radicals. ^{13}C-enriched end groups have been monitored by n.m.r. also in polypropylenes prepared with δ-TiCl$_3$–Al(^{13}CH$_3$)$_3$–Zn(^{13}CH$_3$)$_2$.[76,77] The chemical shifts of the terminal methyl groups are influenced by configurational placements up to six bonds away.

For low-molecular-weight polymers, end groups may be discernible by n.m.r. without the need for isotopic enrichment. Such has been the case for low-molecular-weight poly(vinyl chloride)s extracted from bulk polymer with hexane/acetone mixtures.[78,79] For such polymers, ^{1}H n.m.r. can distinguish between $-$CHCl$-$CH$_2-$CH=CH$-$CH$_2$Cl and $-$CH$_2-$CHCl$-$CH=CH$-$CH$_2$Cl ends, and even between cis- and trans- forms of these. End groups in reduced poly(vinyl chloride)s have been examined by ^{13}C n.m.r.; treatment of poly(vinyl chloride) with tributyltin hydride appears to convert some unsaturated end groups to cyclopentane structures.[80] Poly(vinyl acetate)s prepared with a radical initiator in benzene solution show, in their ^{1}H n.m.r. spectra, evidence of incorporation of benzene as end groups, presumably through transfer reactions.[81]

^{1}H n.m.r. analysis of hydroxytelechelic polybutadienes shows hydroxyl groups attached to 1,4- and 1,2- chain ends.[82,83] However, the end groups are significantly richer in 1,2-units than the rest of the chain, indicating an influence of the hydroxyl group on the relative reactivities of carbons 1 and 3 in the allylic radical produced on

---

[72] J. C. Bevington, J. R. Ebdon, T. N. Huckerby, and N. W. E. Hutton, *Polymer*, 1982, **23**, 163.
[73] J. C. Bevington, T. N. Huckerby, and N. W. E. Hutton, *J. Polym. Sci., Polym. Chem. Ed.*, 1982, **20**, 2655.
[74] J. C. Bevington, T. N. Huckerby, and N. W. E. Hutton, *Eur. Polym. J.*, 1982, **18**, 963.
[75] G. Moad, D. H. Solomon, S. R. Johns, and R. I. Willing, *Macromolecules*, 1982, **15**, 1188.
[76] A. Zambelli, M. C. Sacchi, P. Locatelli, and G. Zannoni, *Macromolecules*, 1982, **15**, 211.
[77] A. Zambelli, P. Locatelli, M. C. Sacchi, and I. Tritto, *Macromolecules*, 1982, **15**, 831.
[78] A. Caraculacu, E. C. Buruiana, G. Robila, T. Hjertberg, and E. Sorvik, *Makromol. Chem., Rapid Commun.*, 1982, **3**, 323.
[79] V. Barboiu, G. Robila, E. Buruiana, and A. Caraculacu, *Makromol. Chem.*, 1982, **183**, 2667.
[80] E. H. Starnes, jun., G. M. Villacorta, and F. C. Schilling, *Polym. Prepr., Am. Chem. Soc., Div. Polym. Chem.*, 1981, **22**, 307.
[81] K. Hatada, Y. Terawaki, T. Kitayama, M. Kamachi, and M. Tamaki, *Polym. Bull. (Berlin)*, 1981, **4**, 451.
[82] Q. T. Pham, *Makromol. Chem.*, 1981, **182**, 1167.
[83] L. S. Bresler, E. N. Barantsevich, V. I. Polyansky, and S. S. Ivantchev, *Makromol. Chem.*, 1982, **183**, 2479.

initiation. Similar results are observed in polybutadienes initiated with benzoyl peroxide.[83]

For polymers prepared by the ring-opening polymerization of 2,2-dimethyloxacyclobutane, minor components in the ^{13}C n.m.r. spectra are attributed to end groups rather than to head–head and tail–tail structures.[84] The principle end groups are isopropenyl and hydroxyl; the hydroxyl end group is best seen in the ^{1}H n.m.r. spectrum.

7 Branching

Papers continue to appear dealing with the use of ^{13}C n.m.r. to investigate chain branching in polyethylene and poly(vinyl chloride).[85–90] A consensus seems now to have been reached regarding the interpretation of the most significant features of the spectrum of polyethylene, although it has been suggested that long chain branching could have been seriously overestimated and that there may be more C_6–C_8 branches then previously supposed.[86] It is also now possible to distinguish hexyl branches from others provided that high fields are used.[87] Branching in poly(vinyl chloride)s may be investigated by ^{13}C n.m.r. after first reducing the polymer with tributyltin hydride.[88–90]

8 Copolymers

It is now well established that correlations exist between the ^{13}C chemical shifts of the α and β carbon atoms of vinyl and acrylic monomers and reactivity parameters such as Alfrey–Price Q and e values. Borchardt and Dalrymple have recently used multiple correlation analysis to generate equations relating Q and e values of 63 different monomers to α and β carbon chemical shifts.[91] It remains to be seen whether these equations will have any predictive value. Recent work has also demonstrated that pentad or triad (monomer sequence distribution) analysis by n.m.r. of a single copolymer should be sufficient for the determination of the appropriate reactivity ratios.[92]

Some revised and extended assignments of ^{13}C n.m.r. peaks for ethylene–propylene copolymers have been reported by Randall and Hsieh.[93] The assignments involve the $\alpha\delta^+$ and $\alpha\gamma$ carbons of PPEE and PPEP sequences, which appear upfield from the corresponding carbon signals in EPEE and EPEP sequences in contrast to the behaviour of the analogous carbons in ethylene–but-1-

---

[84] J. Kops and H. Spanggaard, *Macromolecules*, 1982, **15**, 1225.
[85] P. Freche, M.-F. Grenier-Loustalot, F. Metras, and A. Gascoin, *Makromol. Chem.*, 1981, **182**, 2305.
[86] W. L. Mattice and F. C. Stehling, *Macromolecules*, 1981, **14**, 1479.
[87] F. Cavagna, *Macromolecules*, 1981, **14**, 215.
[88] T. Hjertberg and E. Sorvik, *J. Polym. Sci., Polym. Lett. Ed.*, 1981, **19**, 363.
[89] W. H. Starnes, jun., F. C. Schilling, I. M. Plitz, R. E. Cais, and F. A. Bovey, *Polym. Bull. (Berlin)*, 1981, **4**, 555.
[90] D. Braun, G. Holzer, and T. Hjertberg, *Polym. Bull. (Berlin)*, 1981, **5**, 367.
[91] J. K. Borchardt and E. D. Dalrymple, *J. Polym. Sci., Polym. Chem. Ed.*, 1982, **20**, 1745.
[92] A. Rudin, K. F. O'Driscoll, and M. S. Rumack, *Polymer*, 1981, **22**, 740.
[93] J. C. Randall and E. T. Hsieh, *Macromolecules*, 1982, **15**, 1584.

ene and ethylene–hex-1-ene copolymers. The accuracy of the ^{13}C n.m.r. method for determining the compositions of ethylene–propylene copolymers has been checked by radioassay of copolymers prepared with ^{14}C-labelled ethylene and found to be satisfactory.[94] The ^{13}C n.m.r. spectra of ethylene–propylene copolymers have also been analysed with the aid of a second-order Markov model for the overall copolymerization process.[95] Two reports have appeared on the ^{13}C n.m.r. spectra of ethylene–but-1-ene copolymers.[96,97] Both agree that an earlier assignment of a peak to head–head but-1-ene placements was in error and that, in fact, the peak in question (24.5 p.p.m.) arises from BEB triads. Complete assignments of the ^{13}C n.m.r. spectra of ethylene–hex-1-ene copolymers have been made on the basis of studies of model systems (poly–hex-1-ene and ethylene–hex-1-ene copolymers containing only 1.9 mol% hex-1-ene).[98]

Several papers have appeared dealing with n.m.r. spectra of vinyl chloride copolymers.[99–103] The chemical shifts for methine and methylene carbons in ethylene–vinyl chloride are in good agreement with those predicted on the basis of a rotational isomeric state model for the copolymer chain and show that the effects of monomer sequence and stereosequence distributions are resolvable.[99] ^{13}C n.m.r. spectra of copolymers of vinyl chloride with vinyl acetate and vinyl propionate also can be interpreted in terms of both monomer sequence and stereosequence distributions.[100,101] There seems, however, to be some disagreement over whether or not there is a preference for cosyndiotactic placements. For vinyl chloride–sulphur dioxide copolymers, monomer sequence distributions obtained from ^{1}H n.m.r. spectra are interpretable only in terms of a model involving depropagation; penultimate group effects appear not to be significant.[102,103] For vinyl alcohol–vinyl acetate copolymers, monomer sequence distributions have been obtained from methine proton, and methine, methylene, and carbonyl carbon signals;[104] a blocky distribution of hydrolysed units is indicated.

Two reports deal with the ^{13}C n.m.r. spectra of acrylonitrile–vinylidene chloride copolymers.[105,106] Monomer sequence distributions for radical copolymers fit Bernoullian statistics with $r_A r_V = 0.253$ and long acrylonitrile sequences show a preference for syndiotactic placements.[105] Monomer sequence distributions have also been measured by ^{13}C n.m.r. for acrylonitrile–vinyl acetate copolymers; copolymers prepared with cobalt acetylacetonate–triethylaluminium have structures very similar to conventional radical copolymers.[107] For copolymers of acrylonitrile with styrene, ^{13}C n.m.r. analysis has shown that the use of $ZnCl_2$ and

---

[94] M. Kakugo, Y. Naito, K. Mizunuma, and T. Miyatake, *Macromolecules*, 1982, **15**, 1150.
[95] H. N. Cheng, *Anal. Chem.*, 1982, **54**, 1828.
[96] G. J. Ray, J. Spanswick, J. R. Knox, and C. Serres, *Macromolecules*, 1981, **14**, 1323.
[97] E. T. Hsieh and J. C. Randall, *Macromolecules*, 1982, **15**, 353.
[98] E. T. Hsieh and J. C. Randall, *Macromolecules*, 1982, **15**, 1402.
[99] A. E. Tonelli and F. C. Schilling, *Macromolecules*, 1981, **14**, 74.
[100] T. Okada, K. Hashimoto, and T. Ikushige, *J. Polym. Sci., Polym. Chem. Ed.*, 1981, **19**, 1821.
[101] K. Schlothauer and I. Alig, *Polym. Bull. (Berlin)*, 1981, **5**, 299.
[102] R. E. Cais and J. H. O'Donnell, *J. Macromol. Sci., Chem.*, 1982, **17**, 1407.
[103] R. E. Cais, D. J. T. Hill, and J. H. O'Donnell, *J. Macromol. Sci., Chem.*, 1982, **17**, 1437.
[104] G. Van der Velden and J. Beulen, *Macromolecules*, 1982, **15**, 1071.
[105] J. Konig and D. Wendisch, *Angew. Makromol. Chem.*, 1981, **98**, 255.
[106] P. M. Henrichs, J. M. Hewitt, L. J. Schwartz, and D. B. Bailey, *J. Polym. Sci., Polym. Chem. Ed.*, 1982, **20**, 775.
[107] N. Kalyanam, V. G. Gandhi, S. Sivaram, and I. S. Bhardwaj, *Macromolecules*, 1982, **15**, 1636.

cellulose as regulating agents does not lead to perfect alternation of co-monomer units as previously suggested.[108] Sequence distributions for conventional radical acrylonitrile–styrene copolymers are in better accord with a mechanism involving penultimate group effects than one involving the participation in propagation of acrylonitrile–styrene donor–acceptor complexes.[109]

^{13}C n.m.r. spectra of alternating copolymers of styrene with methyl acrylate and with α-methylene-γ-butyrolactone have been studied by Hirai et al.[110–112] Alternating copolymers of styrene and methyl acrylate show only a single carbonyl peak (assigned to SMS triads), whereas random copolymers show three.[110,111] For alternating styrene–α-methylene-γ-butyrolactone copolymers, the carbonyl peak is split, and the components have been assigned to coisotactic, coheterotactic, and cosyndiotactic triads.[112] ^{13}C n.m.r. spectra of alternating copolymers of α-methyl$^{\ominus}$ styrene with methyl methacrylate and methyl acrylate have been investigated most recently by Harwood and colleagues.[113] For both systems, it has been concluded that the tacticities obey simple Bernoullian statistics and therefore that there can be no significant involvement of complexes in chain propagation. Harwood has also investigated, by ^{1}H n.m.r., the cotacticities and monomer sequence distributions of styrene–methacrylic anhydride copolymers by the simple expedient of first converting them to styrene–methyl methacrylate copolymers.[114,115] For copolymers of methacrylic acid with dimethylaminoethyl methacrylate, it has been found that the ^{13}C carbonyl resonances may be interpreted in terms of compositional sequence effects without interference from configurational effects.[116] An interesting observation for these copolymers is that the degree of resolution of (and hence the information obtainable from) the carbonyl signals depends on the pH of the medium. Various copolymers containing maleic anhydride units have also recently been examined by n.m.r.[117–119]

9 Polycondensates and Uncured Resins

Polyamides have been the most extensively studied group of polycondensates with particular interest being shown recently in ^{15}N n.m.r. spectra.[120–122] Moniz and co-workers[120,121] have demonstrated that the use of J cross-polarization techniques when acquiring the ^{15}N spectra of polyamides can lead to a ^{15}N signal enhancement

---

[108] K. Arita, T. Ohtomo, and Y. Tsurami, J. Polym. Sci., Polym. Lett. Ed., 1981, **19**, 211.
[109] D. J. T. Hill, J. H. O'Donnell, and P. W. O'Sullivan, Macromolecules, 1982, **15**, 961.
[110] H. Koinuma, T. Tanabe, and H. Hirai, Macromolecules, 1981, **14**, 883.
[111] T. Tanabe, H. Koinuma, and H. Hirai, J. Polym. Sci., Polym. Chem. Ed., 1981, **19**, 3293.
[112] H. Koinuma, K. Sato, and H. Hirai, Makromol. Chem., Rapid Commun., 1982, **3**, 311.
[113] M. K. Niknam, R. N. Majumdar, F. A. Blouin, and H. J. Harwood, Makromol. Chem., Rapid Commun., 1982, **3**, 825.
[114] D. L. Neumann and H. J. Harwood, Polym. Prepr., Am. Chem. Soc., Div. Polym. Chem., 1981, **22**, 13.
[115] D. L. Neumann and H. J. Harwood, Am. Chem. Soc., Symp. Ser., 1982, **195**, 43.
[116] L. Merle and Y. Merle, Macromolecules, 1982, **15**, 361.
[117] Y. A. Ragab and G. B. Butler, J. Polym. Sci., Polym. Chem. Ed., 1981, **19**, 1175.
[118] W. J. Freeman and D. S. Breslow, Am. Chem. Soc., Symp. Ser., 1982, **186**, 243.
[119] A. L. Smirnov, T. L. Petrova, A. V. Kalabina, V. B. Golubev, and V. P. Zubov, Vysokomol. Soedin., Ser. B, 1982, **24**, 303.
[120] B. S. Holmes, G. C. Chingas, W. B. Moniz, and R. C. Ferguson, Macromolecules, 1981, **14**, 1785.
[121] B. S. Holmes, W. B. Moniz, and R. C. Ferguson, Macromolecules, 1982, **15**, 129.
[122] H. R. Kricheldorf and S. V. Joshi, J. Polym. Sci., Polym. Chem. Ed., 1982, **20**, 2791.

of ~ 10 and a corresponding time saving of a factor of ~ 100 compared with conventional methods. ^{15}N signals from Nylon-6,6 are found to shift down field as the pH of the medium is lowered owing to protonation and/or H-bonding of neighbouring amide carbonyl O atoms.[120] For alternating and random copolyamides, it is found that ^{15}N n.m.r. gives more useful sequence information than does ^{13}C n.m.r. when the constituent amino-acid units differ merely in respect of their substituents rather than their chain lengths.[122] For copolylactams, ^{13}C n.m.r. carbonyl and methylene signals give useful monomer sequence information.[123] For some aliphatic–aromatic polyamide–imides, sequence information may be derived from high field ^{1}H n.m.r. methylene signals.[124]

Two papers have appeared concerning the n.m.r. characterization of unsaturated polyesters.[125,126] ^{1}H and ^{13}C n.m.r. examination of some resins based on maleic acid and glycols has shown that excess glycol may react with free olefinic double bonds to introduce branch points and that the remaining free olefinic double bonds undergo almost quantitative cis–trans isomerization.[125] The lengths of styrene sequences in styrene-cured unsaturated polyesters have been determined by the noval procedure of first hydrolysing the polymer and then determining the styrene sequence lengths in the resulting styrene–fumaric acid copolymer by ^{13}C n.m.r.[126] It is shown that the distribution of styrene sequence lengths may not be random.

Polyurethanes have been examined by both ^{15}N and ^{13}C n.m.r.;[127,128] the use of model compounds has helped in the assignment of resonances.

The use of ^{13}C n.m.r. for the characterization of phenolic and amino resins is now well established, but there are six further studies of interest in this area to report.[129–134] The reaction of phenol with hexamethylene–tetramine has been examined by ^{13}C n.m.r. in situ in the spectrometer; by this means the growth and decay of intermediate hydroxybenzylamines and of bridging methylenes can be conveniently monitored.[129] For the phenol–formaldehyde system, useful model compound data and additivity parameters have been published.[130] For urea–formaldehyde condensates, the conditions necessary for obtaining quantitative data from ^{13}C n.m.r. spectra have been deduced.[132] It has been shown that condensation of urea with formaldehyde in a 1:1 ratio can lead to the formation of linear poly(methylene urea).[133] Both ^{1}H and ^{13}C n.m.r. have recently been used to characterize oligomers of furfuryl alcohol and also to follow the kinetics of the condensation of furfuryl alcohol with formaldehyde.[135,136]

[123] H. R. Kricheldorf, B. Coutin, and H. Sekiguchi, J. Polym. Sci., Polym. Chem. Ed., 1982, 20, 2353.
[124] J. De Abajo, J. G. de la Campa, and J. L. Nieto, Makromol. Chem., Rapid Commun., 1982, 3, 505.
[125] M. Paci, V. Crescenzi, and N. Supino, Makromol. Chem., 1982, 183, 377.
[126] A. W. Birley, J. V. Dawkins, D. Kyriacos, and A. Bunn, Polymer, 1981, 22, 812.
[127] H. R. Kricheldorf and W. E. Hull, Makromol. Chem., 1981, 182, 1177.
[128] C. Delides, R. A. Pethrick, A. V. Cunliffe, and P. G. Klein, Polymer, 1981, 22, 1205.
[129] S. A. Sojka, R. A. Wolfe, and G. D. Guenther, Macromolecules, 1981, 14, 1539.
[130] M. G. Kim, G. T. Tiedeman, and L. W. Amos, Weyerhauser Sci. Symp., 1979, 2, 263.
[131] J. A. Carothers, E. Gipstein, W. W. Fleming, and T. Tompkins, J. Appl. Polym. Sci., 1982, 27, 3449.
[132] R. Taylor, R. J. Pragnell, J. V. McLaren, and C. E. Snape, Talanta, 1982, 29, 489.
[133] I. Ya. Slonim, V. Z. Yashina, V. N. Gorbunov, S. G. Alekseeva, and Ya. G. Urman, Vysokomol. Soedin., Ser. B, 1982, 24, 528.
[134] B. Meyer and R. Nunlist, Polym. Prepr., Am. Chem. Soc., Div. Polym. Chem., 1981, 22, 130.
[135] A. H. Fawcett and W. Dadamba, Makromol. Chem., 1982, 183, 2799.
[136] Z. Laszlo-Hedvig, M. Szesztay, F. Faix, and F. Tudos, Angew. Makromol. Chem., 1982, 107, 61.

The reaction between epoxy resins and aromatic amines up to the gel point has been followed by high field ^{1}H n.m.r.[137] Concentrations of oxirane, primary amine, secondary alcohol, and secondary amine may be recorded as a function of time.

10 Chemical Modification of Polymers

Each year a large number of papers appear concerned with the use of n.m.r. to monitor some feature of reactions involving polymers. However, in many of these n.m.r. is used largely as a confirmatory technique and its use is not central to the problem. Nevertheless, there are some polymer reactions for which n.m.r. has given information unobtainable by other means.[138–144]

Cais and Spencer have shown by ^{13}C n.m.r. that the dechlorination of poly(vinyl chloride) results not simply in 1,3-chlorine elimination to form cyclopropyl rings but also in the introduction of double bonds and methylene units.[138] The structures of dechlorinated polymers can be more easily determined by ^{13}C n.m.r. if they are first reduced with tributyltin hydride. Partial reductive debromination of poly(vinyl bromide) gives ethylene–vinylbromide copolymers, the compositions and microstructures of which can be determined by ^{13}C n.m.r.[139] Removal of bromine by tributyltin hydride is random, but *meso* stereosequences are slightly more reactive than racemic ones. Complete reduction to polyethylene reveals about 1% of OH groups introduced by hydrolysis but negligible amounts of alkyl branching. The chlorination of poly(vinyl chloride) also has been monitored by ^{13}C n.m.r.[140] At chlorine levels up to about 58%, only CH_2 substitution could be observed, but at levels above this, CCl_2 groups are seen.

The distribution of nitrate groups in cellulose nitrates has been measured by ^{13}C n.m.r.[141,142] A non-random distribution of nitrate groups is indicated with the reactivity of a hydroxyl group towards nitration being dependent on the presence or absence of a nitrate group on a neighbouring substituent. For celluloses substituted with 2-hydroxypropyl groups, it is found from ^{13}C n.m.r. studies that O-3 is more reactive than O-2 which, in turn, is more reactive than O-6.[143]

^{13}C n.m.r. analysis of the hydrolysate of bisphenol-A polycarbonate has shown that not all the carbonate groups are equally reactive; the penultimate carbonate group is only half as reactive as the terminal one, which, in turn, is less reactive than the average internal group.[144]

11 Solid Polymers

The last few years have seen a rapid growth in the use of CP-MASS n.m.r. for the characterization of solid linear and cross-linked polymers. Over the past two years,

---

[137] H.-L. Tighzert, P. Berticat, B. Chabert, and Q.-T. Pham, *J. Polym. Sci., Polym. Lett. Ed.*, 1982, **20**, 417.
[138] R. E. Cais and C. P. Spencer, *Eur. Polym. J.*, 1982, **18**, 189.
[139] R. E. Cais and J. M. Kometani, *Macromolecules*, 1982, **15**, 954.
[140] R. A. Komoroski, R. G. Parker, and M. H. Lehr, *Macromolecules*, 1982, **15**, 844.
[141] D. T. Clark, P. J. Stephenson, and F. Heatley, *Polymer*, 1981, **22**, 1112.
[142] D. T. Clark and P. J. Stephenson, *Polymer*, 1982, **23**, 1295.
[143] D. S. Lee and A. S. Perlin, *Carbohydr. Res.*, 1982, **106**, 1.
[144] F. C. Schilling, W. M. Ringo, jun., N. J. A. Sloane, and F. A. Bovey, *Macromolecules*, 1981, **14**, 532.

most of the papers concerned with this technique have described its use for measuring relaxation times and by this means investigating molecular motions and polymer morphology. However, there have also been some applications to studies of the chemical microstructure of solid polymers and 14 such applications are considered here.[145-158]

In two papers, Vander Hart et al. have discussed the various mechanisms by which line-broadening arises in CP-MASS ^{13}C n.m.r.[145] and methods by which ^{13}C chemical shifts for solids may be accurately determined.[146] Chemical shifts may be determined relative to TMS if a capillary of TMS is incorporated in the sample rotor.

Polyimides are generally intractable materials, but they can be investigated by CP-MASS n.m.r.[147] Chemical shift measurements for solid polyimides have shown that there is some conjugation in the polymer backbone. Two conflicting interpretations of the ^{13}C n.m.r. spectra of AsF$_5$ doped polyacetylenes have appeared;[148,149] in one it is claimed that doping leads to cis–trans isomerization of double bonds, while in the other it is claimed that the spectroscopic evidence is that this does not occur. Dilks and colleagues have used CP-MASS ^{13}C n.m.r. to investigate the structures of plasma polymerized hydrocarbons.[150,151] The amount of unsaturation in the polymer is largely determined by the level of unsaturation in the monomer, as expected, but for polymers prepared from ethane and ethylene, about 20% of the C atoms are in methyl groups arising from methyl radicals generated during the polymerization. Structures of some poly(1,4-phenylenevinylene)s have also been investigated by solid state ^{13}C n.m.r.[152]

An interesting recent study concerns the use of CP-MASS ^{13}C n.m.r. to determine the compositions and microstructures of acrylonitrile–butadiene–styrene (ABS) terpolymer blends;[153] the method offers advantages over the more usual infrared methods for this type of characterization.

The structures of crosslinked unsaturated polyester resins have been investigated by Paci et al.[154] Many of the cross-links were found to consist of styrene dyads, although longer styrene sequences are found, as expected, at higher styrene/maleic acid ratios. CP-MASS ^{13}C n.m.r. spectra of cured furfuryl alcohol resins show four main signals.[155] Three of these peaks are from expected structures, but the fourth at around 35 p.p.m. is assigned to methine carbons, which are thought to arise from methylene links that have undergone further reaction. It is suggested that most cross-linking occurs through further reaction of methylene links rather than

---

[145] D. L. Vander Hart, W. L. Earl, and A. N. Garroway, *J. Magn. Reson.*, 1981, **44**, 361.
[146] W. L. Earl and D. L. Vander Hart, *J. Magn. Reson.*, 1982, **48**, 35.
[147] J. R. Havens, H. Ishida, and J. L. Koenig, *Macromolecules*, 1981, **14**, 1327.
[148] T. C. Clarke, J. C. Scott, and C. S. Yannoni, *Polym. Prepr., Am. Chem. Soc., Div. Polym. Chem.*, 1982, **23**, 77.
[149] K. Menke, M. Peo, R. J. Schweizer, and S. Roth, *Polym. Prepr. Am. Chem. Soc., Div. Polym. Chem.*, 1982, **23**, 79.
[150] S. Kaplan and A. Dilks, *Thin Solid Films*, 1981, **84**, 419.
[151] A. Dilks, S. Kaplan, and A. Van Laeken, *J. Polym. Sci., Polym. Chem. Ed.*, 1981, **19**, 2987.
[152] B. Schroter, H.-H. Horhold, and D. Raabe, *Makromol. Chem.*, 1981, **182**, 3185.
[153] L. W. Jelinski, J. J. Dumais, P. I. Watnick, S. V. Bass, and L. Shepherd, *J. Polym. Sci., Polym. Chem. Ed.*, 1982, **20**, 3285.
[154] M. Paci, V. Crescenzi, and F. Campana, *Polym. Bull. (Berlin)*, 1982, **7**, 59.
[155] G. E. Maciel, I. S. Chuang, and G. E. Myers, *Macromolecules*, 1982, **15**, 1218.

through reactions at the 3 and 4 positions on the furan rings. No ether links are evident in the resins.

Finally, there have been three reports of solid-state ^{29}Si n.m.r. studies of silicone resins.[156—158] For some silicone copolymers, monomer sequences up to the pentad level may be determined.[157] In some cases, end-group structures also can be detected.[158]

---

[156] G. Engelhardt, H. Jancke, E. Lippmaa, and A. Samoson, *J. Organomet. Chem.*, 1981, **210**, 295.
[157] G. Engelhardt and H. Jancke, *Polym. Bull. (Berlin)*, 1981, **5**, 577.
[158] G. E. Maciel, M. J. Sullivan, and D. W. Sindorf, *Macromolecules*, 1981, **14**, 1607.

9
Neutron Scattering Studies

BY J. S. HIGGINS

1 Introduction

During the years 1981 and 1982 several improvements in available neutron scattering facilities have come to fruition. Probably most significantly, several small-angle scattering facilities have been commissioned on reactors in the U.S.A. so that an area of activity which had been almost exclusively European has been extended to a large number of new practitioners. Facilities have been provided on medium flux reactors at the University of Missouri[1] and at the National Bureau of Standards[2] and a national centre for small-angle scattering (including a 10-m X-ray camera) has been funded by the National Science Foundation at the Oakridge National Laboratory.[3] At the same time a new generation of pulsed neutron sources[4] is being built and consideration of how best small-angle scattering might be done at such sources is under way. Test facilities exist at the Argonne National Laboratory in the U.S.A.[5] and at KEK, National Laboratory for High Energy Physics, Tsukuba, Japan.[6] This latter facility is already producing publishable data and has thus provided access to the technique to yet another group of scientists.

While the new SANS facilities being built on existing sources cannot compete directly with those at the Institut Laue-Langevin[7] due to reduced source brightness, particularly at long wavelengths, it is clear that pulsed sources are not ideally suited to small-angle scattering either, since their high flux is concentrated at short wavelengths. Nevertheless, viable small-angle scattering spectrometers are being designed and built for such sources and for many experiments somewhat lower fluxes are no hindrance to producing good results. The relaxation of the hectic atmosphere which prevails at the high flux facilities where several hours lost can ruin a year's preparation can even be a positive advantage.

Turning to the literature for these last two years, some changes in emphasis are evident and some new areas are being explored. There has, for example, been a large increase in the investigation by SANS of multicomponent systems, which in turn has necessitated development of more sophisticated data-handling techniques and model calculations. This development is indicative both of the areas of strong

---

[1] D. F. R. Mildner, R. Berliner, O. A. Pringle, and J. S. King, *J. Appl. Crystallogr.*, 1981, **14**, 47.
[2] C. J. Glinka, in Proceedings of 89th AIR Conference, 'Neutron Scattering 1981', 1982, p. 395.
[3] H. R. Child and S. J. Spooner, *J. Appl. Crystallogr.*, 1980, **13**, 259.
[4] C. Windsor, 'Pulsed Neutron Scattering', Taylor and Francis, 1981.
[5] C. S. Borso, J. M. Carpenter, F. S. Williamson, G. L. Holmblad, M. H. Mueller, J. Faber, jun., J. E. Epperson, and S. S. Danyluk, *J. Appl. Crystallogr.*, 1982, **15**, 443.
[6] Y. Ishikawa, ed., KENS Reports I, 1980.
[7] 'Neutron Beam Facilities at the HFR', ed. B. Maier (for information contact Scientific Secretariat, Institut Laue-Langevin, BP 156X, 38042 Grenoble, France).

current interest in the polymer field generally, and in a certain sense of a 'coming of age' of the SANS technique. Its reliability in applications to the understanding of simple systems has been largely demonstrated. This same factor is evident in the growing number of applications to anisotropic scattering from stretched and deformed systems.

There has been a number of publications concerned with the particular problems of SANS experiments on polymeric systems. Tangari et al.[8] and Summerfield[9] have explored the effect on the observed radius of gyration R_g and molecular weight if there are small differences in M_w between the tagged molecules and the surrounding matrix. They show for polystyrene that the scattering can be effectively analysed using a simple formula if the ratio of the molecular weights is closer than a factor of two. Similar results are obtained by Boué et al.,[10] who also look at effects of difference in molecular weight distribution. The group at Michigan has also addressed the problem of multiple scattering from polymeric samples[11] and shown that the ratio of second- to first-order scattering is of order 2% or less for Debye scattering curves up to a value of $QR_g = 10$. Both these results are important in justifying the use of high concentration levels of tagged molecules. At least in simple homogeneous systems the evidence continues to support the conclusion that such high levels may be used with confidence.[12] The use of high tagging levels is particularly important for studying anisotropic samples where radial averaging over the whole area detector can no longer be used to improve counting statistics. This question has been explored by Boué et al. for stretched polystyrene samples[10] and they have shown that chain dimensions can be obtained independent of the tagging concentrations. The authors point out that other technical problems may occur with high levels of deuteration. For example, the net scattering length density in such a sample is much higher than for a largely hydrogenous sample, thus increasing markedly the scattering from voids and other artefacts.

At the same time, the study of partially compatible polymer blends has reopened the question that so troubled the early investigations of crystalline polyethylene – the possible differences between hydrogen and deuterium. The difference in vibrational amplitude of the two isotopes leads to small differences in molar volume. Effects resulting from this are the different crystallization temperatures of deuteriated or hydrogenous polyethylene and the shift of 5° or so in the θ temperature of polystyrene in cyclohexane when one or other component is deuteriated. Prompted by these observations, Buckingham and Heutschel[13] calculated possible demixing effects in mixtures of deuteriated and hydrogenous polymer molecules. It appears to be the case that such effects are only observable near a phase boundary when small effects on the free energy of mixing become important. Polymer blends systems fall into this category, and the situation in terms of calculation and experimental observation are discussed in detail in Section 6. The

---

[8] C. Tangari, J. S. King, and G. C. Summerfield, *Macromolecules*, 1982, **15**, 132.
[9] G. C. Summerfield, *J. Polymer Sci.*, 1981, **19**, 1011.
[10] F. Boué, M. Nierlich, and L. Leibler, *Polymer*, 1982, **23**, 29 and 1035.
[11] J. S. King, P. S. Goyal, and G. C. Summerfield, *Polymer*, 1983, **24**, 131.
[12] G. D. Wignall, R. W. Hendricks, W. C. Koehler, J. S. Lin, M. P. Wai, E. L. Thomas, and R. S. Stein, *Polymer*, 1981, **22**, 886.
[13] A. D. Buckingham and H. G. E. Hentschel, *J. Polym. Sci., Polym. Phys. Ed.*, 1980, **18**, 853.

evidence so far is that problems are unlikely to be encountered in deuteriation of single-component systems away from phase boundaries.

Questions of chain folding in crystalline polymers have recently focused attention on the higher-angle scattering patterns where local molecular conformation dominates. The complementing of experimental results with theoretical calculations particularly noticeable in this area is also occurring in work on inelastic scattering from vibrational motions and on quasielastic scattering from the chain backbone motion in solutions and melts. Experiments in the latter area have been improved by the high resolution of the spin–echo technique.

In this review of the literature for the years 1981 and 1982 no detailed discussion of the neutron scattering technique is attempted. There have been several recent reviews and articles to which the reader is referred for details of the SANS technique,[14] and applications[15,16] of quasielastic and inelastic scattering to polymeric systems.

Readers familiar with the previous two reviews in this Series will inevitably notice a change of emphasis resulting from the change of authorship, although efforts have been made to preserve continuity. The main difference is the omission of a section on biological materials, which has been replaced by the new section on multicomponent systems. This change is justified by the growth in both these latter areas of application. The devotion of a section to work on copolymers and blends was felt to be important in the context of the specialist report. The interested reader is referred to the excellent specialist reviews on applications of neutron scattering to biological systems.[17,18]

2 Amorphous Polymers

The confirmation of the Flory prediction that excluded volume effects should disappear in melts and amorphous solids with a return to θ temperature (random coil) molecular dimensions was one of the first preoccupations of the SANS application to polymers. Investigation of the relationship $R_g = \beta M_w^\alpha$ now occurs only if for some reason a polymer is expected to show deviations from the value $\alpha = 0.5$.

The local stiffness of the polycarbonate[19] molecule, however, causes no observable deviations from the $M_w^{0.5}$ law, but the value of $\beta = 0.458$ is nearly twice as large as for polystyrene chains ($\beta = 0.276$). At higher Q-values, strong deviations from the Debye curve are seen, but these are satisfactorily accounted for by the unperturbed chain model taking account of the local molecular geometry.[20]

---

[14] J. S. King, in 'Methods of Experimental Physics', Academic Press, New York, 1980, Vol. 16A, p. 480.
[15] J. S. Higgins, in 'Developments in Polymer Characterisation – 4', ed. J. V. Dawkins, Applied Science Publishers, Ann Arbor, 1983.
[16] J. S. Higgins, in 'Static and Dynamic Properties of the Solid State', ed. R. A. Pethrick and R. W. Richards, D. Reidel, Dordrecht, 1982.
[17] H. Stuhrmann (and other following contributions), in 'The Neutron and its Applications, 1982', ed. P. Schofield, Conference Series 64, Institute of Physics, London, 1983.
[18] B. Schoenborn, ed., 'Neutrons in Biology', Basic Life Sciences Series, Plenum Press, New York, 1983.
[19] D. G. H. Ballard, A. N. Burgess, P. Cheshire, E. W. Janke, A. Nevin, and J. Schelten, *Polymer*, 1981, **22**, 1353.
[20] D. Yoon and P. J. Flory, *Polym. Bull. (Berlin)*, 1981, **4**, 693.

A similar investigation of polyisobutylene[21] in θ-solution and melt at low and intermediate Q confirms randomness of the configuration down to distances of about 10 Å. In this case a distinction is made between the scattering function of the carbon atoms, which predominates in X-ray scattering, and the hydrogens which dominate in neutron scattering from partially deuteriated samples. The Kratky plots show distinct differences emphasizing the importance of a correct treatment of all the scattering centres in such calculations. This point is also emphasized in recent Monte Carlo calculations of particle scattering factors for poly(ethylene oxide), polymethylene and poly(dimethyl siloxane) chains.[22] Here, calculations employ the exact particle scattering factor for a set of point scatterers on a rotational isomeric chain.

In stretched and drawn samples the molecular conformation may become very anisotropic due to chain entanglements. In polystyrene samples there is no detectable orientation of labelled molecules with M_w less than the entanglement molecular weight,[23] M_e, but high degrees of orientation for $M_w > M_e$. Even with very high draw ratios (up to 10) achieved by solid-state co-extrusion of polystyrene with high-density polyethylene, the molecular dimensions are deformed affinely with the sample dimensions[24] so long as the molecular weight is well above M_e. In these samples, entanglement slippage is evidently reduced by the low draw temperatures (127 °C) allowed by the co-extrusion technique.

In crystalline polymers the molecular deformation is related to the macroscopic sample extension in a much more complicated manner. Drawn polyolefines generally tend to have a high concentration of voids making interpretation of small-angle scattering difficult. Careful sample preparation avoids this problem for polypropylene samples with draw ratios up to 8:1.[25] The molecular deformation is by no means affine with this extension – only increasing by about a factor of 2 parallel to draw axis and diminishing by 40% in a perpendicular direction.

The Kratky plots of both isotropic and drawn samples are interpreted in terms of folded sub-units two lamellae thick connected by long 'amorphous' molecular fragments which scatter relatively little. The effect of drawing is to greatly increase the number of segments two lamella thicknesses in length belonging to the same molecule.

All samples discussed so far have been deformed in the melt state. In the glassy state, different mechanisms might be expected. Lefebvre et al.[26] investigated molecular deformation of polystyrene molecules in shear bands. In such experiments it is important to section samples so that a uniformly deformed segment is observed. In this case the situations of low-temperature–high-strain rate and high-temperature–low shear rate were compared. In the former, the chains deform parallel to the shear but are essentially undeformed in the perpendicular direction, while in the latter no deformation is observed. The authors explain these data in terms of a localized heterogeneous glide mechanism at low temperatures, while at higher temperatures diffusional plasticity allows chain intercalation at the

[21] H. Hiyashi and P. J. Flory, *Physica B*, 1983, **120**, 408.
[22] C. J. C. Edwards, R. W. Richards, and R. F. T. Stepto, submitted to *Macromolecules*.
[23] G. D. Wignall, R. C. Bopp, and R. P. Kambour, to be submitted to *Polymer*.
[24] G. Hadziioannou, L. H. Wang, R. Stein, and R. S. Porter, *Macromolecules*, 1982, **15**, 880.
[25] D. G. H. Ballard, P. Cheshire, E. Janke, A. Nevin, and J. Schelten, *Polymer*, 1982, **23**, 1875.
[26] J. M. Lefebvre, B. Escaig, and C. Picot, *Polymer*, 1982, **23**, 1751.

level of a few monomer units and preserves the randomness of the molecular orientation.

While in principle an SANS experiment could observe directly the recovery of molecular conformation during stress relaxation, the exigencies of limited neutron-beam time and fluxes has meant that in practice a stretch–relax–quench mechanism of sample preparation has been used. So far data are confined to polystyrene.[27—30] Samples are quickly stretched in a time t_s at temperatures above T_g, allowed to relax for a time t_R, and then rapidly quenched below T_g. It is important that $t_s < t_R$. The resulting samples are then examined over the whole range from local conformation to overall chain dimensions. Boué et al.[30] showed that for short t_R (and stretching temperatures close to T_g) the chains are affinely deformed in the direction perpendicular to the stretch axis. The parallel dimension would be too large for SANS measurement for the chains with $M_w > 5 \times 10^5$ used in these experiments. The perpendicular dimensions as a function of t_R were fitted on to a master curve using a time–temperature superposition [equation (1)] where the shift factors agree

$$t_R = t_R a_T/a_{T0} \tag{1}$$

well with those quoted for macroscopic measurements. The data are then compared with the predictions of coil retraction arising from the reptation arguments, but the agreement is poor, a result the authors suggest might be explained by polydispersity effects or by a molecular weight still too low for the theory to apply. Data at higher scattering angles did not obey the time–temperature superposition, but surprisingly followed a time–space superposition proposed for a suddenly deformed gel. It is clear that the understanding of molecular relaxation after deformation is by no means yet complete.

Finally in this section some very recent experiments should be mentioned in which the molecular deformation under flow conditions was investigated.[31] An apparatus has been devised to allow observation of SANS from concentrated polymer solutions in a shear gradient provided by a cylinder rotating round a fixed drum. The initial experiments have mapped out conditions in which molecular deformations can be observed for random coils, polyelectrolytes, and polymeric vanadium oxide. Further investigation into the dependence of molecular and segmental distortion on the shear gradient are to be expected.

3 Crystalline Polymers

The discussion of chain folding in polyethylene is now focused in the intermediate and wide-angle scattering regions, and seems to a disinterested observer to be close to resolution. At times, indeed, the discussion focuses more on semantics than any real physical differences in the proposed models, although some of these do still remain. A calculation by Stamm[32] shows that as few as three adjacent PE stems

---

[27] A. Maconnachie, G. Allen, and R. W. Richards, *Polymer*, 1981, **22**, 1157.
[28] F. Boué, M. Nierlich, G. Jannink, and R. C. Ball, *J. Phys. Lett.*, 1982, **43**, L585.
[29] F. Boué, M. Nierlich, G. Jannink, and R. C. Ball, *J. Phys. Lett.*, 1982, **43**, L593.
[30] F. Boué, M. Nierlich, G. Jannink, and R. C. Ball, *J. Phys. Lett.*, 1982, **43**, 137.
[31] R. C. Oberthür, in 'The Neutron and its Applications', ed. P. Schofield, Conference Series 64, Institute of Physics, London, 1983, p. 321.
[32] M. Stamm, *J. Polym. Sci., Polym. Phys. Ed.*, 1982, **20**, 235.

wihtin the (100) or (010) planes would produce a diffuse scattering halo. Adjacency in the (110) plane would be more difficult to observe. Comparison with data[32,33] then precludes the existence of more than four adjacent stems in the (100) or (010) planes for both melt and solution crystallized samples. Yoon and Flory's calculation[20] of the intermediate-angle scattering also indicates that adjacent re-entry must be rare for melt crystallized samples. They also conclude that the data cannot support a probability of adjacency greater than three.

The problems with the density anomaly which occur for totally random re-entry or switchboard models is overcome following a suggestion of Sadler at the 1979 Faraday Discussion[34] of a 'leap-frogging' or roughening mechanism. In this, chains folding back at the surface of the crystalline lamellae do so at different 'levels' above the surface, thus giving rise to an interface of partial order between crystal and amorphous regions. At the same time the folded chains may return not directly adjacent to the previous segment but a few stems away. The details of the proposed arrangements of 'diluted' adjacent re-entry stems or nearby folding may vary but the essential arrangement now seems to be agreed.[35,36]

There has been considerable discussion about how the molecules achieve these local arrangements as they are cooled from the melt. Schelten and Stamm[37] disprove the pre-existence of back-folding in the melt by showing that this should contribute extra intensity to a peak at wide angles, an effect which is not observed in practice. Moreover, at these angles the scattering from the polymer and an n-alkane molecule (with $C = 36$) are closely similar. The direct freezing-in of random coil statistics when passing from melt to crystal having been disproved, Sadler and Harris[35] propose a model where considerable local rearrangement in the melt allows 'laying down' of sub-units of five stems or so (approximately in rows but not necessarily rigorously adjacent) and with a distribution of sub-units imposed by the Gaussian chain in the melt.

It is well known that if polyethylene samples containing deuteriated molecules are slow-cooled from the melt, a partial isotopic segregation occurs due to differences in the crystallization temperatures of the two species. An ingenious use has been made of this phenomenon to demonstrate that large-scale molecular reorganization takes place under plastic deformation.[38] Slow-cooled samples were prepared and gave the usual anomalous forward scattering intensity. This was shown to be greatly reduced if samples were melted and rapidly quenched. A closely similar reduction was seen when the same slow-cooled samples were subjected to plastic deformation in the temperature range 50—119 °C.

Although having a chemical macrostructure closely similar to polyethylene, the crystallizable polymer, hydrogenated polybutadiene, shows almost negligible serious isotopic segregation effects even when slow-cooled from the melt. In an SANS investigation linear molecules exhibited strictly Gaussian behaviour with the same θ dimensions in melt and crystal, while scattering from a long-chain three-

---

[33] G. D. Wignall, L. Mandelkern, C. Edwards, and M. Glotin, *J. Polym. Sci., Polym. Phys. Ed.*, 1982, **20**, 245.
[34] 'The Organisation of Macromolecules in the Condensed Phase', *Faraday Discuss., R. Soc. Chem.*, 1979, **68**.
[35] D. M. Sadler and R. Harris, *J. Polym. Sci., Polym. Phys. Ed.*, 1982, **20**, 561.
[36] C. M. Guttman, E. A. Di Marzio, and J. D. Hoffman, *Polymer*, 1981, **22**, 597.
[37] J. Schelten and M. Stamm, *Macromolecules*, 1981, **14**, 818.
[38] G. D. Wignall and W. Wu, submitted to *Polymer*.

armed star molecule followed the single-chain scattering function for a random coil star, again with Gaussian statistics.[39]

Isotactic polystyrene is also apparently free from segregation problems during crystallization, but the chain conformation is strongly dependent on the crystallization process.[40,41] In particular, the size of regularly-folded sheets in the (330) plane increases with increasing mobility in the melt prior to crystallization. By observing the conformation of atactic polystyrene molecules in a crystallized isotactic sample, Guenet and Picot were able to address the important question of rejection of non-crystallizable material from crystallizing polymers.[42] The degree of crystallinity was systematically varied by annealing for different times. The atactic chains were excluded from the crystalline regions at low crystallinity, but trapped and extended by the crystals at high crystallinity, with a short-range structure which was far from Gaussian.

The single-chain conformation in polyacetylene has not been determined and this will remain the situation until a mechanism for mixing the deuteriated and hydrogenous molecules is devised. Meanwhile, neutron diffraction is contributing to studies of the crystal structures.[43] Doping with A_5F_5 renders conductive such polymers as polyacetylene and poly(p-phenylene) and this explains the high level of current interest in these intractable materials. The doping appears to occur *via* an intercalation mechanism similar to the doping of graphite.

4 Solutions and Networks

The study of polymer molecular conformation in solution is another area which has benefited from a fruitful interaction between experiment and theory. Richards *et al.*[44-46] have continued investigation of the temperature–concentration phase diagram using SANS results from polystyrene in cyclohexane. This phase diagram has been discussed both in terms of renormalization group theory[47] and mean field theory,[48] and the predictions for the behaviour of radius of gyration and screening length, ξ, are not always identical. Richards *et al.*[44] first investigated the concentration dependence of R_g at 60 °C and of ξ as a function of temperatures above the normal θ point (upper consolute temperature – UCST) in a fairly concentrated solution. From their results they infer the existence of a region in the phase diagram which is referred to as semi-concentrated and which is only covered by the mean theoretical approach. Further results[45] explore the lower consolute temperature which occurs for this system at 213 °C and show that a second phase diagram can be constructed from the behaviour of $R_g(T)$ and $\xi(T)$ which mirrors that at the UCST. Finally, the radius of gyration has been observed over the whole temperature range from 70 to 220 °C for a semi-concentrated solution.[46] While the

---

[39] B. Crist, W. W. Graessley, and G. D. Wignall, *Polymer*, 1982, **23**, 1561.
[40] J. M. Guenet, *Polymer*, 1981, **22**, 313.
[41] J. M. Guenet and C. Picot, *Macromolecules*, 1983, **16**, 519.
[42] J. M. Guenet and C. Picot, *Macromolecules*, 1981, **14**, 309.
[43] M. Stamm, J. Hocker, and A. Axmann, *Mol. Cryst. Liq. Cryst.*, 1981, **77**, 125.
[44] R. W. Richards, A. Maconnachie, and G. Allen, *Polymer*, 1981, **22**, 147.
[45] R. W. Richards, A. Maconnachie, and G. Allen, *Polymer*, 1981, **22**, 153.
[46] R. W. Richards, A. Maconnachie, and G. Allen, *Polymer*, 1981, **22**, 158.

underlying trend of a maximum in $R_g(T)$ is in agreement with the theoretical predictions, the detailed fit of theory to data is poor.

Using a newly available SANS spectrometer on a pulsed neutron source, Okano et al. measured the screening length ξ as a function of concentration and temperature for polystyrene cyclohexane solutions.[49] Analysis of these data allows separation of the effects of two-body and three-body interactions (binary, B_1, and ternary, B_2, cluster integrals). The Flory θ-temperature is defined when B_1 becomes zero. These results show that, while B_1 indeed changes sign at about 40 °C (the Flory θ-temperature), B_2 is almost independent of temperature near θ. At a slightly lower temperature B_1 and B_2 have opposite signs and compensate each other, giving a point at which excluded volume vanishes and the coils become ideal. This corresponds to the observed θ-temperature.

There is a growing interest in star molecules which can now be prepared with well-controlled morphology. Data exist so far for dilute solutions, but Daoud and Cotton[50] use scaling arguments to predict the swelling properties of the star as a function of monomer concentration and solvent quality. By the time of the next specialist report in this Series, experimental data to check these results will no doubt exist.

The scattering from copolymer solutions may be complex functions of inter- and intra-chain contributions. Even in the simplest case where the components differ only in their scattering power (e.g., block copolymers of deuteriated and hydrogenous monomer analogues), peaks are observed in the scattering curves at some concentrations. Koyama[51] explains the peaks observed in scattering from ABA copolymers of this type in terms of an intermolecular radial distribution function assuming inter- and intra-molecular correlations are independent. Benoit, on the other hand,[52] believes that approaches based on radial distribution functions are limited by their failure to take account of correlations between orientation and distance. He shows simply how a peak in the scattering from an AB diblock copolymer may arise as a difference function between correlations in marked and unmarked segments.

The appearance of the maximum as a function of concentration then depends on the relative values of the three light, X-ray, or neutron scattering contrasts: A–B, solvent–A, and solvent–B. The full development[53] to multi-component blocks and allowing the constituents different thermodynamic interactions with each other and with the solvent shows that, in general, the distortions of the scattering functions at low Q which can give rise to maxima in the curves arise from excluded volume effects. Peaks can be made to appear and disappear by adjusting concentration, relative solvent quality or any of the contrast factors.

When the apparent dimensions of block copolymers are obtained from dilute solution experiments, a parabolic variation of R_g as a function of the mean contrast

---

[47] P. G. de Gennes, 'Scaling Concepts in Polymers', Cornell University Press, Ithaca, New York, 1979.
[48] M. Muthukumar and S. F. Edwards, J. Chem. Phys., 1982, **76**, 2720.
[49] K. Okano, K. Kurita, S. Nakajima, E. Wada, M. Furusaka, and Y. Ishikawa, Physica B, 1983, **120**, 413.
[50] M. Daoud and J. P. Cotton, J. Phys., 1982, **43**, 531.
[51] R. Koyama, Macromolecules, 1981, **14**, 1299.
[52] H. Benoit, Jpn. Polym. Soc., 30th Anniversary Lecture.
[53] M. Benmouna and H. Benoit, J. Polym. Sci., Polym. Phys. Ed., 1983, **21**, 1227.

between the two components and the solvent may be observed.[54,55] These curves are strongly affected by polydispersity, but nevertheless information on the state of segregation of the components is obtained.

Swollen network systems have a number of similarities to semi-dilute and concentrated solutions, and, indeed, are discussed in terms of the same theoretical approaches.[56] Information obtained from SANS experiments includes the screening length ξ and the molecular dimensions under various swelling conditions. For polyacrylamide gels the values of ξ agree well with those obtained from quasi-elastic light scattering, via the elastic modulus.[57] The dimensions of tagged end link chains in a swollen poly(dimethyl siloxane) network[58] are close to those in dilute solution and show no apparent effect of functionality or concentration at which the network was formed. This invariance of R_g after end-linking holds even for networks prepared in the melt.

Calculations of the scattering patterns for deformed and linked chains in networks[59] are based on the phantom network model but including both the reduction in cross-link fluctuations arising from chain entanglements and a concept of chain unfolding without deformation. The results are in agreement with the often small effects on chain dimensions observed by SANS from some swollen networks (in contrast to the predictions based purely on a phantom network). When the labelled chains contain several junction points, the calculations[60] show that they will exhibit the anisotropy of the deformation more strongly and that junction fluctuations have less effect. The degree of affinity with the macroscopic deformation is more easy to check. Again the predicted changes in R_g are less than for a purely phantom network, in agreement with experimental results.

5 Charged Molecules

Charged molecules such as polystyrene sulphonates in concentrated aqueous solution also produce peaked structure factors at low Q.[61] By adjusting the contrast factors it is possible to separate the counterion (S_{11}) and polyion (S_{22}) structures. There are peaks in both S_{11} and S_{22} which vary with polyion concentration as $c^{1/2}$. Each polyion is isolated at the centre of its own counterion atmosphere, so that the apparent polyion–polyion structuring is difficult to explain. Moreover, the observed wave vector dependence of $S_{22}(\max)$ cannot be satisfactorily explained in terms of the random phase approximation assuming long-range Coulombic repulsion. Benmouna et al.[62] suggest that the peak is, in fact, another manifestation of the 'correlation hole' argument used to explain the peaks from copolymer

---

[54] L. Ionescu, C. Picot, M. Duval, R. Duplessix, H. Benoit, and J. P. Cotton, J. Polym. Sci., Polym. Phys. Ed., 1981, **19**, 1019.
[55] L. Ionescu, C. Picot, R. Duplessix, M. Duval, H. Benoit, J. P. Lingelser, and Y. Gallot, J. Polym. Sci., Polym. Phys. Ed., 1981, **19**, 1033.
[56] S. Candau, J. Bastide, and M. Delsanti, Adv. Polym. Sci., 1982, **44**, 27.
[57] E. Geissler, A. M. Hecht, and R. Duplessix, J. Polym. Sci., Polym. Phys. Ed., 1982, **20**, 225.
[58] M. Beltzung, J. Herc, and C. Picot, Macromolecules, 1982, **15**, 1594.
[59] R. Ullman, Macromolecules, 1982, **15**, 582.
[60] R. Ullman, Macromolecules, 1982, **15**, 1395.
[61] F. Nallet, J. P. Cotton, M. Nierlich, and G. Jannink, International Conference on Ionic Liquids, Molten Salts and Polyelectrolytes, June 22—25, 1982, Berlin.
[62] M. Benmouna, G. Weill, H. Benoit, and Z. Akcasu, J. Phys., 1982, **43**, 1679.

solutions. In this case the strong repulsive forces exclude other charges from a region surrounding each polyion, so that it is short-range repulsion rather than long-range structure which is responsible for the observed peak in S_{22}.

Both X-ray and neutron scattering exhibit peaks at low Q from charged polymers in the solid state – so-called ionomers.[63,64] In the case of polypentenamer sulphonate ionomers a peak observed in SAXS from 'dry' samples is not seen in SANS, but becomes visible when contrast is increased by addition of D_2O.[64] The peak position is fixed at low water content and thereafter moves to lower Q with an approximately linear relationship between water content and apparent d-spacing. The behaviour is similar to that of perfluoro sulphonate ionomers and is consistent with a two-phase model where water is adsorbed into a separate ionic phase.

When labelled molecules are incorporated into an ionomer sample the observed coil dimensions are increased over the Gaussian value for the precursor polymer.[65] Moreover, R_g increases with increasing ion content. Forsman[66] has used statistical thermodynamic arguments to show that the chain expansion factor α ($\alpha^2 = \langle s^2 \rangle / \langle s_0^2 \rangle$) will be given by equation (2), where m is the number of repeat units per

$$\alpha^2 - 1 = \phi m^{2/3} f^{1/2} [1 + \psi(v_2 f)^{2/3}] \qquad (2)$$

cluster, f is the fraction of repeat units participating in cluster formation, and v_2 is the volume fraction of polymer; ψ is a geometrical factor from the shape of the clusters ($= 0.77$ for spheres) and ϕ is related to the concentration, c, of ionic groups by equation (3), where m_0 is the mass of a repeat unit and K is given by $\langle s^2 \rangle = KM_w$.

$$\phi = (cN_A)^{-2/3} m_0^{-1/3} (2K)^{-1} \qquad (3)$$

The equilibrium value of m is also related to the surface energy/ion pair θ_2 of the cluster formation and the expansion coefficient α. When these calculations are fitted to the observed SANS chain expansion,[67] a reasonable agreement is obtained by using a value for θ_2 of 8.9 kcal/ion pair mole and of m between 12 and 15 ion pairs. This number of ion pairs cannot be accommodated in a spherical arrangement, so that a lamellar cluster is assumed.

6 Multiphase Systems

Variation of the molecular weight of the segments forming regular block copolymers can change the morphology over a wide range of lamellar, microspherical, and rod-like arrangements. These regular arrangements produce intense scattering at low Q of both X-rays and neutrons.[68—70] SANS experiments have been particularly aimed at extracting information about the conformation of tagged molecules within these structures, and hence have confronted the very difficult problem of separating scattering of the molecules from the interference pattern of

---

[63] S. Clough, D. Cortelek, T. Nagbhusham, and J. Salamone, *Bull. Am. Phys. Soc.*, 1982, **27**, 331.
[64] T. R. Earnest, J. S. Higgins, and W. J. MacKnight, *Macromolecules*, 1982, **15**, 1390.
[65] T. R. Earnest, J. S. Higgins, D. L. Handlin, and W. J. MacKnight, *Macromolecules*, 1981, **14**, 192.
[66] W. C. Forsman, *Macromolecules*, 1982, **15**, 1032.
[67] W. C. Forsman, W. J. MacKnight, and J. S. Higgins, *Macromolecules*, in the press.
[68] F. S. Bates, R. E. Cohen, and C. V. Berney, *Macromolecules*, 1982, **5**, 589.
[69] R. W. Richards and J. L. Thomason, *Macromolecules*, 1983, **16**, 982.
[70] C. V. Berney, R. E. Cohen, and F. S. Bates, *Polymer*, 1982, **23**, 8 and 1222.

the structure when the intensity of this interference function is itself altered by the addition of deuteriated material.

An ingenious solution to this problem was achieved by Bates et al.[71] They prepared samples of styrene–butadiene diblock copolymers in which some of the butadiene blocks were deuteriated. The concentration of the deuteriated species was chosen so that assuming they were randomly distributed among the hydrogenous butadiene blocks there was no net contrast between styrene and butadiene microdomains. It was then possible to observe directly the signal from the labelled blocks. However, while for small domain sizes the dimensions of the copolymer blocks were the same as those of the equivalent molecular-weight polymers in a pure polybutadiene melt, and the molecular weights agreed with those determined by gel permeation chromatography; for large domains there was considerable disagreement with the melt values for both R_g and M_w. It is possible that this is a result of chemical differences between the deuteriated and hydrogenous species which increases with molecular weight. The results obtained for blends of polystyrene with polybutadiene which are discussed below also show isotopic effects, so this may well be the correct explanation.

Other workers have relied on differences in the spatial correlation ranges of domain structure and molecular conformation in order to separate the signals.[72] The observed molecular weight of the tagged blocks is a good check on the accuracy of this procedure. Richards and Thomason determined both the radius of gyration of styrene blocks and the interfacial thicknesses in diblock and triblock styrene–isoprene copolymers forming spherical domains. Both the radius of gyration, which is close to the value in a polystyrene melt, and the observed interfacial thickness, agree with theoretical predictions for these copolymer structures. Styrene–isoprene copolymers also form lamellar domains; thus in oriented samples the chain conformation about the interface normal can be determined.[73] The styrene segments show reduced sideways interpenetration compared with that expected if each has the conformation of a Gaussian chain anchored at one end.

Orientation was imposed on a styrene–isoprene triblock system with a face-centred cubic structure of spherical styrene domains by extending the sample.[74] It appeared that the [111] direction is parallel to the direction of stretch but the spacing between the (111) planes does not increase affinely with the extension ratio. The broadening of the diffraction peak indicates that the strain was not uniformly distributed across the planes.

Experiments aimed at investigating the polymer molecular conformation in polymeric micelle systems[75] or in polymers adsorbed on latices[76–79] face the same

[71] F. S. Bates, C. V. Berney, R. E. Cohen, and G. D. Wignall, *Polymer*, 1983, **24**, 519.
[72] R. W. Richards and J. L. Thomason, *Polymer*, 1981, **22**, 581.
[73] G. Hadziioannou, C. Picot, A. Skoulios, M. L. Ionescu, A. Mathis, R. Duplessix, Y. Gallot, and J. P. Lingelser, *Macromolecules*, 1982, **15**, 263.
[74] R. W. Richards and J. L. Thomason, *Polymer*, 1983, **24**, 249.
[75] P. Marie, R. Duplessix, Y. Gallot, and C. Picot, *Macromolecules*, 1979, **12**, 1180.
[76] T. Cosgrove, T. Crowley, and B. Vincent, 'Adsorption from Solution', Academic Press, New York, 1983.
[77] T. Cosgrove, T. L. Crowley, B. Vincent, K. G. Barnett, and Th. F. Tadros, *Faraday Symp. R. Soc. Chem.*, 1982, **16**, 101.
[78] K. G. Burnett, T. Cosgrove, B. Vincent, A. N. Burgess, T. L. Crowley, T. King, J. D. Turner, and Th. F. Tadros, *Polym. Commun.*, 1981, **22**, 283.
[79] K. Burnett, T. Cosgrove, T. L. Crowley, Th. F. Tadros, and B. Vincent, 'The Effect of Polymers on Dispersion Properties', ed. Th. F. Tadros, Academic Press, New York, 1982, p. 183.

problems of separating the relevant scattering from that of the dominant spherical structures. Cosgrove et al. have used both n.m.r. and SANS to investigate the conformation of poly(ethylene oxide) molecules adsorbed onto polystyrene latices and terminally anchored by a polystyrene block 'tail' on the molecule.[76—78] Using deuteriated latices dispersed in D_2O, the density distribution of adsorbed segments was determined normal to the surface. These results indicate that some of the anchored polymers are very extended, giving a segment density extending well into the solvent. The adsorbed molecules, on the other hand, lie close to the surface, with most of the segments contained in loops or trains attached to the surface and a few tails extending into the solvent. Similar results are reported for adsorbed poly(vinyl alcohol).[77,79]

Partially compatible binary polymer mixtures also form microdomains as they phase separate, but the structures are not regular as in block copolymer systems. Experiments have been concentrated on investigation of molecular conformation and interaction parameters in the compatible phase. Even in this case there are two competing scattering functions, one arising from the conformation of labelled molecules, the other from the concentration fluctuations. It is this latter signal which carries information about the interaction parameter. For this reason, a proper separation of the signals is essential. Koberstein[80] gives the most recent discussion of the method which involves a weighted subtraction of the scattering patterns from two samples with different concentrations of labelled molecules. The technique is in principle applicable to any of the two-phase structures. All extra intensity, from incoherent scattering, discussed so far, and from voids, sample containers, *etc.*, must of course first be carefully removed. Methods of analysis of this type are currently being applied to blend systems, but no report has yet appeared in the literature. Some experimenters have assumed that the fluctuation signal will occur at lower Q values than the molecular scattering and for completely phase-separated structures such as polyethylene–polypropylene mixtures this is probably a reasonable assumption, since the domains are large-scale.[81] The observed molecular weight is again a good check on the separation procedure, and shows it works well for this system. The molecular dimensions in the polyethylene domains are close to those in a polyethylene melt.

This crude separation by Q range is more suspect when working in the compatible range since the correlation lengths for fluctuations are much shorter. Thus for labelled poly(vinyl chloride) molecules in various poly(vinyl chloride) blends, the molecular dimensions obtained were taken only as confirmation that true molecular mixing had been achieved.[82] At the limit of low concentration of one component the mixtures can be treated as solid solutions of the deuteriated polymer in the second component and normal solution Zimm analysis yields values of R_g and the virial coefficients, A_2. Measurements on atactic polystyrene in blends with poly(2,6-dimethyl phenylene oxide)[83] and with poly(vinyl chloride) or poly(*o*-chlorostyrene)[84] yield dimensions somewhat expanded from those in θ solvents and the positive virial coefficients characteristic of good solvents.

[80] J. T. Koberstein, *J. Polym. Sci., Polym. Phys. Ed.*, 1982, **20**, 593.
[81] G. D. Wignall, H. R. Child, and R. J. Samuels, *Polym. Lett.*, 1982, **23**, 957.
[82] D. J. Walsh, J. S. Higgins, C. P. Doubé, and J. G. McKeown, *Polymer*, 1981, **22**, 168.
[83] G. D. Wignall, H. R. Child, and F. Li-Aravena, *Polymer*, 1980, **21**, 131.
[84] T. R. Russell and R. S. Stein, *J. Polym. Sci., Polym. Phys. Ed.*, 1982, **20**, 1593.

As noted in the introductory remarks, the thermodynamic properties of polymer blends seem to be particularly sensitive to isotopic differences. Low-molecular-weight polystyrene and polybutadiene are compatible at high temperatures and incompatible at lower temperatures. The compatibility limit is raised in temperature by 20 °C or so when either component is deuteriated.[85] Higher molecular weight polymers usually show an inverted phase diagram, separating as the temperature is raised. When the polystyrene component in polystyrene/poly(vinyl methyl ether) blends is deuteriated the temperature of the compatibility limit again rises[86] (by up to 40 °C), now indicating increased thermodynamic compatibility. Clearly, scattering from samples obtained by adding variable amounts of deuteriated material to such blends must be interpreted with caution. On the other hand, the author's own group has data (as yet unpublished) which indicates that some systems, for example, methoxylated poly(ethylene glycol)/poly(propylene glycol) and poly(methyl methacrylate)/solution chlorinated polyethylene, show little or no effect of deuteriation. The extensive effort in both theory and experiment currently being devoted to polymeric blends will undoubtedly lead to an improved understanding of their molecular behaviour before the appearance of any further report in this Series.

7 Studies of Polymer Dynamics

The first inelastic experiments on polymeric samples published in the sixties investigated optical and acoustic phonons in semi-crystalline samples. At that time the high-resolution spectrometers necessary for investigating the low-frequency main chain motions in solution and melt samples were not yet available. In the last few years there has been a revival of interest in neutron scattering from vibrational modes, partly because careful sample preparation methods have produced better oriented crystalline samples, and partly because theoretical calculations have improved interpretation of the spectra.

Oriented polyethylene has been investigated at lower frequencies using triple axis spectrometers[87] and measurements extended to higher frequencies using a beryllium filter spectrometer.[88] The high-frequency measurements (50—400 mvV) agree will with calculated one phonon density of states. The low-frequency measurements produced dispersion curves for both inter- and intra-molecular modes leading to elastic moduli which agree well with those obtained from bulk measurements. The data were fitted to Urey–Bradley force field calculations, but a better fit was found using atom–atom potential models. Polyoxymethylene can be grown in a single crystal form, thus allowing many of the phonon dispersion curves for inter- and intra-molecular models to be followed.[89] Four of the five independent stiffness constants were determined, but residual disorder precluded detailed fitting of the data to atom–atom potential models.

[85] E. L. Atkin, L. A. Kleint'jens, R. Koningsveld, and L. J. Fetters, *Polym. Bull. (Berlin)*, 1982, **8**, 347.
[86] H. Yang, G. Hadziioannou, and R. S. Stein, *J. Polym. Sci., Polym. Phys. Ed.*, 1983, **21**, 159.
[87] J. F. Twisleton, J. W. White, and P. A. Reynold, *Polymer*, 1982, **23**, 578.
[88] H. Jobic, *J. Chem. Phys.*, 1982, **76**, 2693.
[89] M. R. Anderson, M. B. M. Harryman, D. K. Steinmann, J. W. White, and R. Currat, *Polymer*, 1982, **23**, 569.

The torsional motion of the methyl groups in isotactic polypropylene does not produce an intense peak in neutron inelastic scattering from unoriented samples, unlike poly(propylene oxide) for example. Experiments on stretch-oriented samples and normal co-ordinate calculations both clarify the identification of the torsional mode at 230 cm^{-1} and indicate that coupling between chains may be enhanced by methyl groups on neighbouring molecules leading to a large frequency dispersion for this mode.[90]

The neutron spin–echo technique[91] uses the precession of the neutron spin in a magnetic field as a 'clock' to resolve time-dependent processes. The beams are well collimated, allowing measurements at low Q ($Q \geqslant 0.02$ Å^{-1}) with effective energy resolution of 10^{-9} eV ($\sim 19^7$ s^{-1}). The fact that the neutron spin is followed through the experiment allows separation of coherent and incoherent scattering events.

In dilute solution the observed dynamical motion depends on the value of Q relative to two distance scales. The local structure of the molecule can be considered in terms of a length σ, which might be the persistence length of the molecule, while the overall dimensions are characterized by R_g. For most polymers the value of σ^{-1} falls within the Q range of the spin–echo technique.

In this case, a transitional behaviour for the characteristic decay time Ω or the width function $\Delta\omega$ is observed, with a Q^3 dependence at low Q (characteristic of Zimm–Rouse modes) dropping to Q^2 at higher Q.[92] Such transitional behaviour has been calculated for real polyethylene chains,[93] but since hydrodynamic effects were ignored the low Q behaviour gave $\Delta\omega \propto Q^4$.

For very flexible molecules such as poly(dimethyl siloxane) σ^{-1} moves to higher Q, and, by using low-molecular-weight samples, the effect of molecular dimensions are observed[94] in the spin–echo experiments. In this case $\Omega \propto Q^3$ at the higher Q values within the range and $\Omega \propto Q^2$ at low Q, the transition corresponding to $Q \simeq R_g^{-1}$. Some of the samples explored were ring molecules, and calculations of the transitional behaviour for such small ring molecules indeed show the observed behaviour,[95] though detailed comparisons have not been made.

A different length scale comes into play in concentrated solutions of PDMS molecules.[96,97] Now the dynamic screening length, ξ, defines a value of Q above which single chain (Q^3) behaviour is observed and below which co-operative diffusion is observed with $\Omega \propto Q^2$. A temperature cross-over between two different Q^3 regions is also defined by ξ.[97] Concentrated polyelectrolyte solutions have a complex dynamical behaviour. By using both coherent and incoherent scattering, the diffusive motion of the counterions and the longitudinal collective modes of the polymers have been separately observed.[61]

In polymer melts yet a further length scale is important. The effect of entanglements on the molecular motion has been described in terms of the concept

---

[90] H. Takeuchi, J. S. Higgins, A. Hill, A. Maconnachie, G. Allen, and G. C. Stirling, *Polymer*, 1982, **23**, 499.
[91] 'Neutron Spin-Echo', ed. F. Mezei, Physics 128, Springer-Verlag, Berlin, 1980.
[92] J. S. Higgins, L. K. Nicholson, and J. B. Hayter, *Macromolecules*, 1981, **14**, 836.
[93] G. Allegra and F. Ganazzoli, *J. Chem. Phys.*, 1981, **74**, 1310.
[94] J. S. Higgins, L. K. Nicholson, and J. B. Hayter, *Polym. Prepr., Am. Chem. Soc., Div. Polym. Chem.*, 1981, **22**, 86.
[95] J. L. Viovy and L. Monnerie, *Macromolecules*, 1982, **15**, 1611.
[96] B. Ewen, D. Richter, J. B. Hayter, and B. Lehnen, *J. Polym. Sci., Polym. Lett. Ed.*, 1982, **20**, 233.
[97] B. Ewen, D. Richter, and J. B. Hayter, in 12th European Conference on Macromolecular Physics, Molecular Mobility in Polymer Systems, sponsored by EPS, ed. H. K. Roth, Leipzig, 1981, p. 75.

of reptation within a tube of diameter D. This tube then described the time-averaged limit of local excursions of the polymer molecule. For $D^{-1} < Q < \sigma^{-1}$ Rouse-like motion (Ω or $\Delta\omega \propto Q^4$) is expected, with the same transition to lower power laws as local motion is explored at high Q. At the lowest Q values observed in spin–echo experiment, no reliable deviation from Q^4 behaviour has yet been observed,[98,99] and there has been some discussion about the implications of this for the validity of the reptation model.[98,100] Since, however, the estimates of D from bulk moduli indicate values greater than 30 Å, the neutron experiments may well not be sensitive to the entanglements because of the relatively high Q values explored.[99]

Incoherent scattering experiments made with somewhat lower resolution and at higher Q values, have been used to make qualitative comparisons between different polymeric melts.[101] Here the effect of differing stiffness or σ value is directly observed. Several polymers with low glass transition temperatures show very similar behaviour, but poly(isobutylene) shows a strong effect of hindrance on this local scale, presumably arising from the double methyl substitution. The motion of the junction points in a trifunctional model poly(propylene oxide) network were also slowed down relative to the free chains away from the junctions.[102] Bearing in mind that the measurements were made in the transitional Q-range ($Q \simeq \sigma^{-1}$) there was reasonable agreement with a calculation for Rouse chains[103] which showed that γ-functional junctions should be slowed down by a factor $2/\gamma$.

---

[98] D. Richter, A. Baumgärtner, K. Binder, B. Ewen, and J. B. Hayter, *Phys. Rev. Lett.*, 1981, **47**, 109.
[99] J. S. Higgins, L. K. Nicholson, and J. B. Hayter, *Polymer*, 1981, **22**, 163.
[100] J. M. Deutsch and N. D. Goldenfeld, *Phys. Rev. Lett.*, 1982, **48**, 1694.
[101] J. S. Higgins, A. Maconnachie, R. E. Ghosh, and G. Allen, *J. Chem. Soc., Faraday Trans. 2*, 1982, **78**, 2117.
[102] J. S. Higgins, K. Ma, and R. H. Hall, *J. Phys. C*, 1981, **14**, 4995.
[103] M. Warner, *J. Phys. C*, 1981, **14**, 4985.

10
Polymer Crystallization

BY J. N. HAY

1 Introduction

There have been sustained developments in the study of polymer crystallization behaviour over the period of interest, 1980—1982, particularly with the application and development of new techniques to determining morphology. Such techniques as low angle X-ray scattering (LAXS), various forms of neutron scattering, low angle light scattering (LALS), Fourier transform i.r. spectroscopy, laser Raman spectroscopy, and 'magic angle' solid state ^{13}C n.m.r. spectroscopy all have been increasingly applied to a wider range of polymer systems and a more general idea of polymer morphology, and the part played by molecular parameters in determining it is developing.

2 The Degree of Crystallinity

Molecular order has an important role in determining ultimate material properties and is an important material characteristic. Its measurement, however, is somewhat of a problem and as more techniques are applied it is increasingly apparent that there is no absolute way of defining order, but only relative and defined by the experimental techniques adopted. Amorphous (liquid-like), paracrystalline, liquid crystalline, crystalline, fibre-like, and oriented polymers have been used to represent the different degrees and types of order present and each have been separately defined. Mitchell et al.[1] have outlined the nature of the problems in their study of WAXS of molten polyethylene. This, they observed, contained a defined structure in which there was an essentially random distribution of *gauche* and *trans* conformation sequences, but with local organization with chain segments in an all-*trans* conformation. There was intermolecular correlation over several nm. Similar conclusions[2] were made for liquid crystalline polyesters with a high degree of localized orientation of chain segments with a periodical of 1.5—3.0 nm[3]. Similar latteral biaxiality occurs in polycarbonate and polystyrene.[4] A micro-paracrystalline structure has been suggested for amorphous polymers[5] and an equilibrium size distribution has been calculated which is in good agreement with scattering

---

[1] G. R. Mitchell, R. Lovell, and A. H. Windle, *Polymer*, 1982, **23**, 1273.
[2] G. R. Mitchell and A. H. Windle, *Polymer*, 1982, **23**, 1269.
[3] C. S. Wang and G. S. Y. Yeh, *Polymer*, 1982, **23**, 205.
[4] H. R. Schuback, E. Nagy, and B. Heise, *Colloid Polym. Sci.*, 1981, **259**, 789.
[5] R. Hosemann, W. Schmidt, A. Lange, and M. Hentschel, *Colloid Polym. Sci.*, 1981, **259**, 1161.

behaviour,[6] crystalline polyethylene,[7] and nascent isotactic polypropylene.[8] Various models for the amorphous polymer structure[9] and one- and three-dimensional structures in crystalline polymers[10] are reviewed.

Thermal analysis, especially by differential scanning calorimetry (DSC), continues to be the most popular method of estimating crystallinity and also of characterizing it. Recent developments include the determination of the quality of the crystallinity and the crystallite size distribution from analysing the melting range.[11] This compared favourably with that determined by electron microscopy, LAXS, laser Raman LAM measurements, and gel permeation chromatograph (GPC) evaluation of molecular fragments of the etched crystalline regions.

Booth et al.[12,13] have synthesized single molecular species block oligomers of ethylene oxide and shown from WAXS, Raman and i.r. spectra, DSC and SAXS studies that they are totally crystalline. Similar oligomers with monodispersed ($\bar{M}_w/\bar{M}_n < 1.05$) polyethylene oxides were partially crystalline and contiguous crystallization of the second block was precluded due to a mismatch of the chain segments at the interface. Homogeneity of the chain length was a pre-requisite for achieving complete crystallinity. Glotin and Mandelkern[14] have stressed the importance of the amorphous/crystalline interface in measuring crystallinity of polyethylene by Raman spectroscopy. Lines arise due to crystallinity, the orthorhombic crystalline structure, amorphous content, and also interfacial regions, and differences in the measured crystallinity, by density and calorimetry, are due to the complexity of this morphology. It was concluded that the heat of fusion alone gave a realistic measure of the degree of crystallinity.

WAXS line-widths of polyethylene increase with decreasing temperature close to β and γ relaxations. This is due to stresses set up on the structure, generated by differences in thermal expansion coefficients between amorphous and crystalline regions.[15] Amorphous surface layers have been measured as 1.8 nm for poly(oxymethylene) crystals.[16]

Separating effects due to orientation from those due to crystallinity remains a problem. Several techniques have been used with varying degrees of success. By monitoring the changes in orientation of dried polyethylene gels and isotactic polypropylene, with polarized light microscopy and WAXS, Cannon has obtained a quantitative measure of both.[17] However, while polarization is successful in thin fibres, it fails due to polarization scambling by reflectance and refraction at the surface in thin fibres.[18] Fluorescence polarization microscopy has been used[19] to study the orientation of amorphous regions in crystalline isotactic poylpropylene.

[6] A. M. Hindeleh and R. Hosemann, *Polymer*, 1982, **23**, 1101.
[7] G. I. Asbach and W. Wilke, *Colloid Polym. Sci.*, 1982, **260**, 113.
[8] R. Hosemann, M. Hentschel, E. Ferracini, A. Ferrero, S. Martelli, F. Riva, and M. V. Antisara, *Polymer*, 1982, **23**, 979.
[9] W. Wendorff, *Polymer*, 1982, **23**, 543.
[10] W. Winke, *Colloid Polym. Sci.*, 1981, **259**, 577.
[11] J. N. Hay, in 'Analyses of Polymer Systems', ed. L. S. Bark and N. S. Allen, Applied Science Publishers, London, 1982, Ch. 6.
[12] A. Marshall, R. C. Domszy, W. H. Teo, R. H. Mobbs, and C. Booth, *Eur. Polym. J.*, 1981, **17**, 885.
[13] R. C. Domszy and C. Booth, *Makromol. Chem.*, 1982, **183**, 1051.
[14] M. Glotin and L. Mandelkern, *Colloid Polym. Sci.*, 1982, **260**, 182.
[15] E. Krenzer and W. Ruland, *Colloid Polym. Sci.*, 1981, **259**, 408.
[16] W. D. Warnell, J. P. Runt, and R. Harrison, *J. Polym. Sci., Polym. Phys. Ed.*, 1981, **19**, 1923.
[17] C. G. Cannon, *Polymer*, 1982, **23**, 1123.
[18] D. I. Blower and I. M. Ward, *Polymer*, 1982, **23**, 645.

This depended on morphology, since chains were attached to crystalline regions and were not free to orient.

Measurement of the orientation[20] of isotactic polypropylene during cold-drawing by small angle neutron scattering (SANS), is somewhat of a new development in that chain dimensions were established with changing draw ratio. This is widely used in measuring conformation in crystallizable polymers although segregation of the deuterio-polymer may produce spurious results. This does not appear to be the case in deuteriated polybutadienes with ethyl branches. In the melt and crystalline this polymer has a random coil chain conformation.[21]

Brillouin scattering has been used to measure hypersonic velocity data[22] in draw and transverse directions, i.r. dichroism in polyurethanes,[23] Fourier transform i.r. of polyethylene[24] and polystyrene,[25] birefringence in poly(2,6-dimethyl-1,4-phenylene oxide)–polystyrene blends,[26] and ^{19}F n.m.r. spectroscopy, and in particular the chemical shifts in solid state of uniaxially deformed polymers[27] have all separately been used to measure orientation.

3 Crystallographic Analysis

The crystal structure of the long-chain cycloalkanes has been determined as a model for regular chain folding.[28] Single crystals of cyclohexatriacontane ($C_{36}H_{72}$) have the space group Aa and unit cell, $a = 1.033$, $b = 0.824$, and $c = 4.22$ nm and $\beta = 107.1°$ with four molecules per cell. The sharp fold lies parallel to b-axis and chain conformation in the fold is T(GGTGG)T, with T *trans* and G *gauche*. This structure is representative of the longer cyclic hydrocarbons, *i.e.*, n = 48, 60, and 72, and so representative of a tight chain fold. The cell dimensions of low-molecular-weight polyethylene have been measured from 10 to 300 K, and cell expansion coefficients calculated.[29] A position sensitive proportional counter X-ray measuring system has been developed and used to measure phase transitions in polyethylene with temperature and pressure.[30] An irreversible transformation from hexagonal to orthorhombic occurs at about 350 MPa. This technique has also been applied to formation of extended chain crystals.[31]

The crystal structure of isotactic poly(methyl methacrylate), has been determined, and stabilization energy, of 12.4 kJ monomer mol^{-1}, calculated.[32] The unit cell is

---

[19] F. Pinaud, J. P. Jerry, Ph. Sergot, and L. Monnerie, *Polymer*, 1982, **23**, 1175.
[20] D. G. H. Ballard, P. Cheshire, E. Janka, A. Nevin, and J. Schelten, *Polymer*, 1982, **23**, 1875.
[21] B. Crist, W. W. Graessley, and C. D. Wignall, *Polymer*, 1982, **23**, 1561.
[22] D. B. Cavenaugh and C. H. Wang, *J. Polym. Sci., Polym. Phys. Ed.*, 1981, **19**, 1911.
[23] S. V. Laptij, Yu. S. Lipatou, Yu. Yu. Karcha, L. A. Kosenko, V. N. Vatuleu, and R. Gaiduk, *Polymer*, 1982, **23**, 1917.
[24] K. Holland-Moritz, I. Holland-Moritz, and K. van Wenden, *Colloid Polym. Sci.*, 1981, **259**, 156.
[25] D. Lefebvre, B. Josse, and L. Monnerie, *Polymer*, 1982, **23**, 706.
[26] B. Josse and J. L. Koenig, *Polymer*, 1981, **22**, 1040.
[27] A. J. Brandolini, T. M. Apple, C. Dybowski, and R. G. Pembleton, *Polymer*, 1982, **23**, 39.
[28] T. Trzebiatowski, M. Dräger, and G. R. Strobl, *Makromol. Chem.*, 1982, **183**, 731.
[29] D. J. Burchell and S. L. Hsu, *Polymer*, 1981, **22**, 907.
[30] Y. Maeda, H. Kanetsuma, K. Nagata, K. Matsushige, and T. Takemura, *J. Polym. Sci., Polym. Phys. Ed.*, 1981, **19**, 1313.
[31] Y. Maeda, H. Kanetsuma, K. Tagashina, and T. Takemura, *J. Polym. Sci., Polym. Phys. Ed.*, 1981, **19**, 1325.
[32] F. Bosscher, G. Brinke, E. Eshuis, and G. Challa, *Macromolecules*, 1982, **15**, 1364.

triclinic, pseudo-orthorhombic, with $a = 2.057$, $b = 1.249$, and $c = 1.056$ nm, and $d = 92.9°, \beta = 88.2°$, and $\gamma = 90.6°$. Similar studies have been made on the bacterial polyester poly(hydroxyvalerate),[33] cis-[34] and trans-poly 1,4-butadiene,[35] poly(oxymethylene),[36] and Nylon 5:7.[37]

A series of polyurethane elastomers – diphenyl methane-4,4'-diisocyanate and a range of diols – have a triclinic unit cell[38,39] but the c-axis dimensions are sensitive to chain conformation. With butane diol, the repeat unit is 1.895 nm, and chain adopts an all-*trans* conformation; with propane- and ethanediols, the repeat units are 1.62 and 1.50 nm, respectively, and adopt some *gauche* conformations.[38,39]

The crystallographic structures of *cis*- and *trans*-polyacetylene,[40,41] and iodine-doped polyacetylene,[41] have been determined. WAXS exhibit diffuse rings and contain little structural information. However, model oligomers (trienes and tetraenes) enable various structural analyses to be carried out. Several packing models for the chains are possible which differ little in packing energy and so it is concluded that polymorphs are present.[42]

Poly(oxacyclobutane)[43] has three polymorphs, the first of which has a planar zig-zag conformation stabilized by hydrogen bonding to water since it is a hydrate. The other two have less extended conformations, TTTGTTG and TTGGTTGG, respectively. A fourth form has been produced on uniaxially stretching the second. Fibre diagrams suggest that this contains some extended planar zig-zag regions consistent with repeat period of 0.479 nm. Two crystallographic forms of poly(γ-benzyl-L-glutamate)[44] have been reported. The first is monoclinic with two chain units per cell, $a = 2.90_6$, $b = 1.32_0$, and $c = 2.72_7$ nm, and $\alpha = \gamma = 90°$ and $\beta = 96°$. Adjacent chains alternate up and down. The second form produced on orienting has a nematic-like paracrystalline order, $a = 1.48$—1.52, $b = 1.43$—1.48, and $c = 2.7$ nm with $\gamma = 118°$—$120°$. This structure was attributable to a random placement of the two anti-parallel chains. Poly(hexamethylene terephthalate) has three allotrophic forms,[45] two of which are formed during melt crystallization, and the third (α-form) is produced under stress crystallization. The unit cell parameters are:

α-form (monoclinic): $a = 0.91$ $b = 1.72$ $c = 1.55$ nm
$\alpha = 127.3°$ Density $= 1.284$ g cm^{-3}

β-form (triclinic): $a = 0.48$ $b = 0.57$ $c = 1.57$ nm
$\alpha = 104.4°$ $\beta = 116.0°$ $\gamma = 107.8°$
Density $= 1.262$ g cm^{-3}

---

[33] A. L. Pundsack and T. L. Bluhm, *J. Mater. Sci.*, 1981, **16**, 545.
[34] P. Corradini, R. Napolitano, V. Petraccone, B. Pirozzi, and A. Tuzi, *Eur. Polym. J.*, 1981, **17**, 1217.
[35] G. Bautz, V. Leutz, W. Dollhopf, and P. C. Hazele, *Colloid Polym. Sci.*, 1981, **259**, 714.
[36] A. H. Fawcett, *Polymer*, 1982, **23**, 1865.
[37] J. C. Lin, M. H. Litt, and G. Froyer, *J. Polym. Sci., Polym. Chem. Ed.*, 1981, **19**, 164.
[38] J. Blackwell, M. R. Nagarajan, and T. B. Hoitink, *Polymer*, 1981, **22**, 1584.
[39] J. Blackwell, M. R. Nagarajan, and T. B. Hoitink, *Polymer*, 1982, **23**, 950.
[40] J. C. W. Chein and F. E. Karasz, *Macromolecules*, 1982, **15**, 1012.
[41] Y. Cao, R. Qian, F. Wang, and X. Zhao, *Makromol. Chem., Rapid Commun.*, 1982, **3**, 687.
[42] V. Enkelmann, M. Monkenbusch, and G. Wegner, *Polymer*, 1982, **23**, 1581.
[43] T. Takahashi, Y. Osaki, and H. Tadokoro, *J. Polym. Sci., Polym. Phys. Ed.*, 1981, **19**, 1153.
[44] J. Watanabe, K. Imae, R. Gehani, and I. Uematsu, *J. Polym. Sci., Polym. Phys. Ed.*, 1981, **19**, 653.
[45] I. H. Hall and B. A. Ibrahim, *Polymer*, 1982, **23**, 805.

γ-form (triclinic): $a = 0.53$ $b = 1.39$ $c = 1.55$ nm
$\alpha = 123.6°$ $\beta = 129.6°$ $\gamma = 88.0$
Density = 1.308 g cm^{-3}

Poly(1,4-*trans*-cyclohexanediyl dimethylene terephthalate)[46] also has a triclinic cell for which $a = 0.460$, $b = 0.665$, $c = 1.42$ nm, and $\alpha = 89.4°$, $\beta = 47.0°$, and $\gamma = 114.9°$. The density is 1.266 g cm^{-3}. The flexible *trans*-dimethylene, 1,4-cyclohexane, i.e., $-O-CH_2-C_6H_{10}-CH_2-O-$ has a TGTTTGTTG conformation.

Cellophane[47] has WAXS typical of cellulose-II but electron diffraction pattern of cellulose-I. This has been resolved since both are present but cellulose-I only as very small crystallites, too small to scatter X-rays coherently, and cellulose-II is degraded to amorphous carbon faster than is cellulose-I in the electron beam.

Polyethylene oxide forms complexes with alkali metal salts which conduct ionically. The effect of various ions on the 7/2 helix is discussed and relevance to the conductivity of the complexes.[48–50]

Polytetrafluoroethylene undergoes several transformations. One of these has been studied at 42 K and is sttributed to a rapid increase in translational disorder along the chains. Disorder below this temperature[51] is attributed to excitations of *trans* conformations of 15:7 helix, and above it to *gauche*. Electron diffraction has been used to study the low temperature phase. There are left- and right-handed chain stems placed randomly. Although the structure is ordered, it cannot be ascribed to any classical space group and small structural changes lead to large differences in cell dimensions.[52]

From model structures, the relative stability of the three crystallographic forms of poly(trimethylene oxide) have been calculated and most likely structures determined. These were planar zig-zag conformations $(TTTT)_2$, orthorhombic $(TGGT)_2$, and trigonal (TTTGTTTG').[53] Similar studies have been made on poly(*N*-vinyl pyroline)[54] using semi-empirical functions appropriate to peptide in calculating conformational energies as a function of stereosequences.

Various crystallographic forms of poly(vinylidene fluoride)[55,56] and its copolymers[57–60] have again been extensively studied to correlate structure and piezo- and pyro-electric properties. A ε phase has been postulated;[61] the four previous phases have been well characterized – α-phase, obtained from melt crystallization, has a non-polar structure with chain packed anti-parallel in TGTG

---

[46] B. Remillard and F. Brisse, *Polymer*, 1982, **23**, 1960.
[47] S. Aravindanath, K. M. Paralikar, S. M. Betrabet, and N. K. Chaudhuri, *Polymer*, 1982, **23**, 823.
[48] J. M. Parker, P. V. Wright, and C. C. Lee, *Polymer*, 1981, **22**, 1305.
[49] C. C. Lee and P. V. Wright, *Polymer*, 1982, **23**, 681.
[50] D. R. Payne and P. V. Wright, *Polymer*, 1982, **23**, 690.
[51] Y. Yamamoto and T. Hara, *Polymer*, 1982, **23**, 521.
[52] J. J. Weeks, E. S. Clark, and R. K. Eby, *Polymer*, 1981, **22**, 1480.
[53] M. Ohsaku, T. Izuka, H. Murata, and A. Imamura, *Polymer*, 1981, **22**, 624.
[54] A. E. Tonelli, *Polymer*, 1982, **23**, 676.
[55] A. J. Lovinger, *Macromolecules*, 1982, **15**, 40.
[56] D. T. Grubb and K. W. Choi, *J. Appl. Phys.*, 1981, **52**, 5908.
[57] A. J. Lovinger, G. T. Davis, T. Furakawa, and M. G. Broadhurst, *Macromolecules*, 1982, **15**, 323.
[58] A. J. Lovinger, G. T. Davis, T. Furukawa, and M. G. Broadhurst, *Macromolecules*, 1981, **15**, 329.
[59] T. Yamada and T. Kilayama, *J. Appl. Phys.*, 1981, **52**, 6859.
[60] I. Okawaki, A. Yamazaki, and T. Kitayama, *J. Appl. Phys.*, 1981, **52**, 6856.
[61] A. J. Lovinger, *Macromolecules*, 1981, **14**, 225.

conformation. Poling converts the α to δ phase with the chains parallel and a large dipole. The mechanism by which this occurs has been the centre of much activity.[61,62] The β phase is ferro-electric. The chains are in an all-*trans* conformation. The γ phase has the same unit cell as α phase but with double the repeat unit along the C axis. Some of the difficulty in resolving the crystallographic structures of the phases has been ascribed to statistical disorder in parallel/anti-parallel packing of the chains.[63] Defects built into the crystals assist in phase transformations,[64] from α to γ, either by a flip-flop or crankshaft motion of groups along the chain. A propagation wave along the chain of such a motion is not likely since the calculated energy was greater than observed.[65] The transition occurs randomly from defects. An orthorhombic unit cell has been assigned to α phase, with $a = 0.496$, $b = 0.964$, and $c = 0.462$ nm. The chains pack statistically up and down.[66] The γ phase is monoclinic[67] with $a = 0.496, b = 0.967$, and $c = 0.920$ nm and $\beta = 93°$. The β phase is extensively piezo- and pyro-electric, and is orthorhombic with $a = 0.858$, $b = 0.491$, and $c = 0.256$ nm and the chain extended in an all-*trans* conformation.[68] Stretching the γ phase induces the β phase at 432 K.[69]

Piezo-electric properties have also been reported for γ-form of Nylon-11, but from WAXS studies on the poled and unpoled material it has been deduced that there must be a polar form or γ phase, as well as an unpolar form.[70]

4 Morphology

Different microstructure of crystalline polymers and the part played in mechanical behaviour has been reviewed by Bassett.[71] As a teaching aid, the text is thoroughly recommended, since it gives a comprehensive treatment of a difficult subject in a simple manner.

Solution Crystallization.—Polymer single crystals prepared by crystallizing from dilute solution have a regular folded chain conformation, and occupy a unique portion as a model for the lamellar crystallites produced from melt crystallization. I.r. studies[72,73] have been made on mixtures of deuterio and hydro-polyethylene in ratio 1:40 (ref. 73), respectively, as a function of molecular weight. The CD_2 bending modes were resolved into crystalline and amorphous components, and comparison with n-paraffin mixed crystals show that the crystalline component is incompatible with a switchboard model. Adjacent re-entry folding predominates. WANS of

---

[62] Y. Takahashi, Y. Matsubara, and H. Tadokoro, *Macromolecules*, 1982, **15**, 334.
[63] S. Winhold, M. A. Backmann, M. H. Lilt, and J. B. Lando, *Macromolecules*, 1982, **15**, 631.
[64] A. S. Lovinger, *J. Appl. Phys.*, 1981, **52**, 5934.
[65] J. D. Clark, P. L. Taylor, and A. J. Hopfinger, *J. Appl. Phys.*, 1981, **52**, 5903.
[66] M. A. Backmann and J. B. Lando, *Macromolecules*, 1981, **14**, 40.
[67] A. J. Lovinger, *Macromolecules*, 1981, **14**, 322.
[68] A. J. Lovinger, *Polymer*, 1981, **22**, 412.
[69] A. Servet, D. Broussoux, and F. Micheron, *J. Appl. Phys.*, 1981, **52**, 5926.
[70] J. I. Scheinbeim, *J. Appl. Phys.*, 1981, **52**, 5939.
[71] D. C. Bassett, 'Principles of Polymer Morphology', Cambridge Solid State Science Series, Cambridge University Press, 1981.
[72] T. C. Cheam and S. Krimm, *J. Polym. Sci., Polym. Phys. Ed.*, 1981, **19**, 423.
[73] X. Jing and S. Krimm, *J. Polym. Sci., Polym. Phys. Ed.*, 1982, **20**, 1157.

mixed H-polyethylene and D-polyethylene crystals have been studied[74] in the range $0.68 < K = 4\pi\lambda \sin\theta < 2.02$ Å. For both melt and solution crystallized material, no extra signals were observed in the scattering pattern in their range which placed an upper limit of 4 on the number of stems regularly folded together. An e.s.r. spectroscopic study of free radicals trapped in mats of polyethylene single crystals[75] gives some evidence for the twisted conformations of the methylene groups on the fold surface.

The shape of the LAM-1 band in the Raman spectra[76] has been used to analyse the distribution of straight-chain segments on annealing solution grown polyethylene crystals. Unannealed samples have peak stems of 10.0 nm and half-widths of 2 nm in their distributions. The overall distributions broaden on annealing. Half-width, $\Delta L_{1/2}$, and peak values, L_{max}, are related by equation (1). Fold periods

$$\Delta L_{1/2} = 270[1 - 10.5/L_{max}] \tag{1}$$

measured in dried crystal and suspended in silicone oil are the same,[77,78] even though in dekalin there is an increase due to swelling. In silicone oil, the heat of fusion and melting points increase[78] compared with the dried mat due to the collapse of the original pyrimidal crystals on drying with the production of lattice defects. Similarly, crystals thickened by annealing have a higher heat of fusion.[79] SAXS has been used to determine the amorphous content of dried polyethylene crystal mats[80] by associating it with an amorphous layer on the surface. The amorphous layer was estimated to be 1.2–2.0 nm.

Single crystals of cellulose-IV of different molecular weights have been prepared.[81] Sharp fractions were obtained by gel permeation chromatography of cellulose acetate with subsequent de-acetylation. Well-developed lamellar crystals were obtained only with low degrees of polymerization; higher ones gave only polycrystalline aggregates. Only extended chain single crystals could be prepared and accordingly the presence of chain folding in cellulose could not be demonstrated.

Large single crystals of polydiacetylene have been prepared, and their properties correlated with defects and dislocations built into the crystal during polymerisation.[82] Their Grüneisen constant has been calculated from the coefficient of thermal expansion.[83]

Single crystals of the α and γ forms of poly(vinylidene fluoride) have been prepared and their annealing characteristics measured.[56] Poly(oxymethylene),[84,85]

---

[74] G. D. Wignall, L. Mandelkern, C. Edwards, and M. Glotin, *J. Polym. Sci., Polym. Phys. Ed.*, 1982, **20**, 248.
[75] S. Shimada and H. Kashiwabara, *Polymer*, 1981, **22**, 1385.
[76] R. G. Synder, J. R. Scherer, D. H. Reneker, and J. P. Colson, *Polymer*, 1982, **23**, 1286.
[77] J. Runt, B. O. Hanrahan, and I. R. Harrison, *J. Polym. Sci., Polym. Phys. Ed.*, 1982, **20**, 1683.
[78] K.-H. Illers and G. Kanig, *Colloid Polym. Sci.*, 1982, **260**, 564.
[79] J. H. Magill, *J. Polym. Sci., Polym. Lett. Ed.*, 1982, **20**, 1.
[80] W. D. Varnell, I. R. Harrison, and S. J. Kommiski, *J. Polym. Sci., Polym. Phys. Ed.*, 1981, **19**, 1237.
[81] A. Buleon, H. Chanzy, and P. Froment, *J. Polym. Sci., Polym. Phys. Ed.*, 1982, **20**, 1083.
[82] R. J. Young and J. Petermann, *J. Polym. Sci., Polym. Phys. Ed.*, 1982, **20**, 961.
[83] I. Engeln and M. Meissner, *Colloid Polym. Sci.*, 1981, **259**, 827.
[84] M. R. Anderson, M. B. M. Harryman, D. K. Steinman, J. W. White, and R. Currat, *Polymer*, 1982, **23**, 569.
[85] M. Shimomura and M. Igucha, *Polymer*, 1982, **23**, 509.

Nylon-6,6,[86,87] Nylon-12,[88] and an alternating copolymer single crystals[89] have been prepared and variously studied.

The individual chains of poly(p-xylylene) in the β-form have been directly resolved by electron microscopy.[90] The sole limitation of the procedure is the amount of radiation damage sustained by the polymer and this procedure should be generally applicable to electron resistant polymers. Optical filtering is applied to the high-resolution images to reduce noise. The image, along the *ab* projection, showed the arrangement of the molecules in the crystal. It confirmed the β-modification of these crystals. Dark field electron microscopy with multiple imaging has been used to study beam damage[91] in sensitive materials. This gave information on crystal defects. Electron irradiation, however, produces extensive damage and loss of crystallinity within seconds of exposure. A technique has been developed to circumvent this, monitoring the dark field images for all the diffracted beams simultaneously. Optical transforms have also been applied to the electron diffraction pattern of polyethylene single crystals.[92]

Melt Crystallization.—The nature of the conformation of the chains at the crystalline–amorphous interface in melt crystallized polymer is still in contention. LANS has been applied to melt crystallized polyethylene in order to distinguish between regular adjacent re-entry and random chain folding. The latter is not compatible with the scattering data.[93] Anything approaching a random switchboard model was also rejected because of space-filling requirements from the scattering data; it was deduced that a high proportion of adjacent re-entry – up to 40% – was required. Calculations[94] for several models incorporating various amounts and types of chain foldings with varying degrees of light and regular folds have been reported. The model, which fits the intensity data in the region $\mu = 0.01$–$0.14 = (4\pi/K)\sin\theta/2$, folds along lattice planes in which the stem separation is larger than 0.5 nm. Tight regular folding has a probability of 0.7 which is close to a theoretical limit of 2/3. This model fits the SANS data, the liquid and density requirement, the measured degree of crystallinity, and the measured radius of gyration of the polymer chain straddling crystalline and amorphous regions.

The same authors[95] have modelled the fold surface of semi-crystalline polymers using the classical Gambler's Ruin statistical problem to calculate the ratio of loops and bridges between the same and adjacent lamellae. It was found that the ratio of loops to bridges varied with the thickness of the amorphous layer, and the minimum number of adjacent crystal stems involved in tight folding in semi-crystalline polymers was calculated. For a cubic lattice this is 2/3, but the effect of crystal structure and chain stiffness on this value was discussed.

---

[86] H. W. Starkweather, jun., and G. B. Jones, *J. Polym. Sci., Polym. Phys. Ed.*, 1981, **19**, 467.
[87] J. H. Magill, M. Girolamo, and A. Keller, *Polymer*, 1981, **22**, 43.
[88] M. Kyotani, *J. Polym. Sci., Polym. Phys. Ed.*, 1982, **20**, 345.
[89] J. F. Rabolt, *Polymer*, 1981, **22**, 890.
[90] M. Tsuji, S. Isado, M. Ohara, A. Kawaguchi, and K.-I. Katayama, *Polymer*, 1982, **23**, 1568.
[91] A. J. Lovinger and H. D. Keith, *J. Polym. Sci., Polym. Phys. Ed.*, 1981, **19**, 1163.
[92] A. Kawaguchi, *Polymer*, 1981, **22**, 753.
[93] D. M. Sadler and R. Harris, *J. Polym. Sci., Polym. Phys. Ed.*, 1982, **20**, 561.
[94] C. M. Guttman, E. A. DiMarzio, and J. D. Hoffmann, *Polymer*, 1981, **22**, 597.
[95] C. M. Guttman, E. A. DiMarzio, and J. D. Hoffmann, *Polymer*, 1981, **22**, 1466.

The chain conformation of melt crystallized isotactic polypropylene has been measured by WANS and SANS[96] in the μ range 0.03—0.8 Å^{-1}. Scattering curves, 0.1—0.8 Å^{-1}, were sensitive to fold conformation. Annealing, which increased the crystallinity and stem lengths, had no effect on scattering behaviour. The best model which fitted these results had an equal mixture of single and double strands, and an adjacent re-entry model was not considered appropriate for melt crystallized isotactic polypropylene.

Stamm[97] has re-considered the WANS results from melt crystallized polyethylene and in particular whether the experiments can detect adjacent re-entry folds. Diffuse neutron scattering interactions are sensitive to folding in 100 and 010 planes but not in the 110 plane. Adjacent re-entry folding in 100 and 010 planes with more than five adjacent stems could be excluded. Folding along 110 cannot be excluded nor could adjacent re-entry folds with appreciable amounts of second nearest-neighbour folding. The classical tight adjacent re-entry model of folds was accordingly improbable for melt crystallized polyethylene. This is also consistent with the previous work of Sadler and Keller[98] who concluded that statistical folding predominated. Similar studies[74] place an upper limit of 4 for the average number of adjacent regular folds in 100 and 010 planes.

SANS investigation of crystallized isotactic polystyrene[99] have been interpreted in terms of regularly folded sequences.

Electron microscopy has been widely used to study morphology. Replication and thin sections reveal lamellar fine structure and amorphous thickness.[100] Chlorosulphonation of polyethylene[101] highlights these features and allows the crystalline stem distributions to be measured. There is, however, evidence of distortion of the crystallites and conformational changes.[102] Polyolefin lamellae are also revealed by permanganate etching[102] but artefacts of 10 μm can develop with time which can be eliminated by incorporating orthophosphoric acid into the etching solution.[103] A new method of contrasting and observing directly the microstructure of polyamides in TEM involves staining in a mixture of formaldehyde and osmium tetroxide.[104] This reacts preferentially with the amide group without structural changes. The electron density of the amorphous regions increase. Fuming nitric acid has also been used to etch isotactic polypropylene[105] and polyethylene.[106] Fluorination,[107] and chlorination,[108] occurs in amorphous regions, even after extensive attack.

Stem length distributions have been measured by SAXS. After correcting for background scattering, instrumental broadening, desmearing and the Lorentz factor, many orders of reflections can be measured in polyethylene.[109] This can be

---

[96] M. Stamm, J. Schelten, and D. G. H. Ballard, *Colloid Polym. Sci.*, 1981, **259**, 286.
[97] M. Stamm, *J. Polym. Sci., Polym. Phys. Ed.*, 1982, **20**, 235.
[98] D. M. Sadler and A. Keller, *Science*, 1979, **203**, 263.
[99] J.-M. Guenet, *Polymer*, 1981, **22**, 313.
[100] I. G. Voigt-Martin and L. Mandelkern, *J. Polym. Sci., Polym. Phys. Ed.*, 1981, **19**, 1769.
[101] G. Kanig, *Colloid Polym. Sci.*, 1982, **260**, 356.
[102] B. Bikson, J. Jagur-Grodzinski, and D. Vofsi, *J. Polym. Sci., Polym. Phys. Ed.*, 1981, **19**, 23.
[103] R. H. Olley and D. C. Bassett, *Polymer*, 1982, **23**, 1707.
[104] G. Webber, D. Kuntze, and W. Stix, *Colloid Polym. Sci.*, 1982, **260**, 956.
[105] H. Tanaka, *Colloid Polym. Sci.*, 1982, **260**, 1601.
[106] D. R. Rueda, E. Cagiao, and F. J. Balta-Calleja, *Makromol. Chem.*, 1981, **182**, 2705.
[107] A. Anand, R. E. Cohen, and R. F. Baddour, *Polymer*, 1981, **22**, 361.
[108] B. Bikson, J. Jagur-Grodginski, and D. Vofsi, *J. Polym. Sci., Polym. Phys. Ed.*, 1981, **19**, 361.
[109] W. D. Varnell, I. R. Harrison, and J.-I. Wang, *J. Polymer Sci., Polym. Phys. Ed.*, 1981, **19**, 1577.

Polymer Crystallization

used to determine the order of stacking of the lamellae.[110] Stacking faults[111] have been observed in layer lattice of poly(γ-methyl-L-glutamate). Absolute intensity measurements enable kinetic parameters to be determined for the crystallization.[112] It has been used to measure crystallinity of poly(4-methylpent-1-ene),[113] and domain size in poly(vinyl chloride),[114] although they are not associated with crystallinity but density differences.

The Raman active LAM has also been used to determine lamellae stem distributions.[89] 'Magic angle' ^{13}C n.m.r. spectroscopy of solid poly(phenylacetylene),[115] isotactic polypropylene,[116] and linear polyethylene[117] and broad-line n.m.r.[118,119] spectroscopy have been used to determine details of the crystallographic structure and morphology.

Oriented Crystallization.—The structure of ultra-oriented polyethylene has been reviewed.[120] Fibrils and lamellae are still present even at the highest draw ratios, but a key element in determining the properties of these materials is the development of bridges linking the crystals. The weight average crystal size never exceeds 60 nm even with the highest modulus material, and long crystals alone do not account for the observed properties. Raman LAM vibrations, up to draw ratios of 30 ×, have been measures,[121] and the calculated stem lengths (weight and number average) compare favourably with gel permeation chromatography and LAXS results. At the highest draw ratios, the LAM intensities diminish due to some decrease in crystal size. However, Peterlin and Snyder[122] have interpreted this decrease in terms of defects, *e.g.*, *gauche* conformations which interrupt the all-*trans* conformation of the stem and decouples the longitudinal oscillations on either side of this defect. One defect in five chain stems is required to explain the observed spectrum of drawn polyethylene. SAXS experiments show differences in electron densities between crystalline and amorphous regions and yield little detail on the location and frequency of such defects.

The microfibrillar and lamellar morphologies of cold-drawn, and subsequently annealed, polyethylene has been observed by scanning electron microscopy.[123] Differences in contrast on the fracture surfaces have been interpreted in terms of preferential orientation of inter-fibrillar tie molecules in the plane. In annealed cold-drawn specimens stacks of lamellae were seen with a twinned orientation of inclined lamellae. This 'roof-top' structure[123] was interpreted as due to the stress within the

[110] I. R. Harrison, S. J. Kaziniski, W. D. Varnell, and J.-I. Wang, *J. Polym. Sci., Polym. Phys. Ed.*, 1981, **19**, 487.
[111] J. Watanabe, R. Gehani, and I. Uematsu, *J. Polym. Sci., Polym. Phys. Ed.*, 1981, **19**, 1817.
[112] J. S. Lin, R. W. Hendricks, J. M. Schultz, and M. J. McCready, *J. Polym. Sci., Polym. Phys. Ed.*, 1982, **20**, 1365.
[113] T. Tanigami and K. Miyasaka, *J. Polym. Sci., Polym. Phys. Ed.*, 1981, **19**, 1865.
[114] H. R. Brown, G. M. Musindi, and Z. H. Stachurski, *Polymer*, 1982, **23**, 1508.
[115] T. J. Sanford and R. D. Allendoerfor, *J. Polym. Sci., Polym. Phys. Ed.*, 1981, **19**, 1152.
[116] A. Bunn, M. E. A. Cudby, R. H. Harris, K. J. Parker, and B. J. Soy, *Polymer*, 1982, **23**, 694.
[117] A. Dechter, R. A. Komoroski, D. E. Axelson, and L. Mandelkern, *J. Polym. Sci., Polym. Phys. Ed.*, 1981, **19**, 631.
[118] H. Horii and R. Hitamaru, *J. Polym. Sci., Polym. Phys. Ed.*, 1981, **19**, 109.
[119] D. E. Axelson, *J. Polym. Sci., Polym. Phys. Ed.*, 1982, **20**, 1427.
[120] G. Capaccio, *Colloid Polym. Sci.*, 1981, **259**, 29.
[121] G. Capaccio, M. A. Wilding, and I. M. Ward, *J. Polym. Sci., Polym. Phys. Ed.*, 1981, **19**, 1489.
[122] A. Peterlin and R. G. Synder, *J. Polym. Sci., Polym. Phys. Ed.*, 1981, **19**, 1727.
[123] A. J. Owen, *Colloid Polym. Sci.*, 1981, **259**, 252.

individual microfibrils during micronecking. Light etching with nitric acid was necessary to reveal the individual lamellae.

A statistical mechanics approach[124] has been applied to the morphology and mechanical properties of drawn polymers. A partition function for stacked lamellae of alternating crystalline and amorphous layers was formulated. A random walk problem was used to enumerate the statistical weights of various conformations of the amorphous chains confined by two parallel walls for a body centre cubic lattice.

The chain conformation of drawn isotactic polypropylene has been revealed by neutron scattering[20] up to draw ratios of X8. The chain is extended by factor of 2.

Cold drawing changes the activity of polymer-supported catalyst, in that polyethylene with grafted sulphonated polystyrene units showed an increased activity to dehydrate isopropanol on drawing up to 2.5X.[125] Activity then decreased on further drawing. These effects were attributed to structural changes. In the undrawn materials, the sulphonic acids group are excluded from the crystalline regions and form a network of hydrogen-bonded units. Increased activity on elongation is attributed to the break up of this network with production of free acid groups. This initial deformation occurs in amorphous regions since the hydrogen-bonded network hinders lamellar slip and tilt. At higher elongation, greater structural alterations occur.

Extending previous studies, Keller *et al.*[126] have developed the principles of 'hair-dressing' of shish-kebabs of polyethylene to produce three morphologies – smooth fibres, fibres with closely spaced lamellar overgrowths, and fibres with widely spaced lamellar overgrowths. The spacings of the overgrowths are sensitive to thermal treatment; solution and melt treatments have been carried out and similar morphologies are obtained.

An extended chain morphology of polyethylene has been prepared under high pressure and the pressure and temperature dependence of phase transitions explored,[31] particularly orthorhombic → hexagonal → melt. Two types of extended chain crystallites were observed, ordinary and highly extended.

5 Rate Measurements

Spherulitic and Single Crystals Growth Rate.—The impingement of 2-[127] and 3-dimensional spherulites[128] for athermal, thermal, and combined primary nucleation has been simulated to determine spherulite size distribution and final shape. A pseudo-random number generator was used to determine nuclei co-ordinates. The spherulite size distributions varied with nucleation characterized and the Avrami exponents agreed with theoretical predictions for thermal and athermal nucleation, but mixed nucleation gave a non-integer value. Comparisons with experimentally determined spherulite size distributions were made.

---

[124] K. Itayama, *J. Polym. Sci., Polym. Phys. Ed.*, 1981, **19**, 1873.
[125] C. A. Cooper, B. C. Gates, R. L. McCullough, and J. S. Seferis, *J. Polym. Sci., Polym. Phys. Ed.*, 1982, **20**, 173.
[126] M. J. Hill and A. Keller, *Colloid Polym. Sci.*, 1981, **259**, 335.
[127] A. Galeski, *J. Polym. Sci., Polym. Phys. Ed.*, 1981, **19**, 721.
[128] A. Galeski and E. Piorkowska, *J. Polym. Sci., Polym. Phys. Ed.*, 1981, **19**, 731.

Electron microscopy and WAXS have been used to study the growth of Nylon-12 crystals from solution. SAXS was used to measure the stem lengths and to follow this change on annealing.[88] Similarly, Magill et al.[129] have followed the growth of Nylon-6,6 from solution.

Ethylene oxide–propylene oxide triblock copolymers[130] have been crystallized from ethylbenzene. The size of the crystals produced isothermally increased with the square root of the crystallization time over most of the crystallization range studied, 289—297 K. The solutions self-seeded at the higher temperatures and a decrease of 6.5 K increased the rate 30-fold. The lamellar thicknesses increased with crystallization temperature.

Similarly, trans-1,4-polybutadiene has been crystallized from solution.[131] The annealing of single crystals has been studied in the presence of solvent[132] and the kinetics of the thickening of the crystals determined.[133] The crystal thickness, L_c, increases logarithmically with time, t, in accordance with equation (2). The

$$L(t) = 46 \log (t/t_0) + L_0(t_0) \quad (2)$$

hexagonal phase of polyethylene was studied at 6.1 Kbar at 515 K. A mechanism for annealing has been proposed which involved the coalescence of two crystals and the gradual disappearance of the interface between them. This will proceed at a rate proportional to $(\exp - \alpha L)$ and is dependent on the defects concentrations at the boundaries.

The growth rates of spherulites of poly(ethylene oxide) homogeneously dispersed in poly(methyl methacrylate) have been measured[134] in the range 317—331 K. The rate is reduced, proportional to the concentration of poly(methyl methacrylate) present, consistent with it acting as a diluent. Nucleation control of the growth was always observed.

Nucleation.—The nature of the nuclei[135] in the crystallization of isotactic polypropylene has been examined by isothermal crystallization and also by electron microscopy. Two types of nuclei were effective: (a) metastable nuclei, corresponding to unmelted polymer stabilized by the close proximity to solid heterogeneities, which are only effective after melting at low temperatures and for short melting periods, and (b) stable nuclei, associated with solid heterogeneities, such as catalyst residues, on the surface of which the polymer tends to crystallize.

A kinetic analysis has been presented governing the time dependence of primary nucleation.[136] Solutions of 1-dimensional diffusion equation for the formation of transient embryos (chain-folded or bundle-like nuclei) near the critical point lead to an expression for measuring time dependence. It has shown that these transient processes are important for development of crystallization unless heterogeneous nuclei are present.

[129] J. H. Magill, M. Girolamo, and A. Keller, Polymer, 1981, 22, 43.
[130] M. Droscher and T. L. Smith, Macromolecules, 1982, 15, 442.
[131] S. Tseng, W. Herman, A. E. Woodward, and B. A. Newman, Macromolecules, 1982, 15, 388.
[132] T. H. Magill, J. Polym. Sci., Polym. Lett. Ed., 1982, 20, 1.
[133] T. Asahi, Y. Migamoto, H. Miyaji, and K. Asi, Polymer, 1982, 23, 775.
[134] E. Calahorra, M. Costazar, and G. M. Guyman, Polymer, 1982, 23, 1322.
[135] F. Rybnikar, J. Appl. Polym. Sci., 1982, 27, 1479.
[136] A. J. McHugh, J. Appl. Polym. Sci., 1982, 27, 3663.

The crystal structure of various nucleating agents for the crystallization of poly(n-butane) have been compared to determine the effect of lattice continuity between polymer and nucleating agent in determining nucleation efficiency.[137] While the coarse texture of the crystallized polymer varied with nucleating agent, there was no observable difference in overall degree of crystallinity or in the size of the crystallites. Nucleating agents such as adipic acid, 4-aminobenzoic acid, salicyclic acid, sorbic acid, and sodium benzoate were studied.

Keller et al.[138] have used self-seeding to crystallize polyethylene at very low supercoolings, since this procedure speeded the rate of crystallization by an order of magnitude. Under these conditions lamellar thickening is eliminated. The spherulites produced were uniform in size, and lamellae a constant thickness which increased with temperature. An induction period was observed. Polyethylene was also crystallized below 383 K; normally this is not possible since the rates are too fast. However, by dispersing the polymer as very fine droplets, it can be crystallized at 348 K. Homogeneous nucleation was not present but the concentration of heterogeneities was determined by the polymer–substrate interface. The low-temperature crystallized material has a lamellar morphology and growth rates exceed 1 m s^{-1}. Lamellar thicknesses have been measured as a function of supercooling, thus extending the range available.[139]

Epitaxial crystallization has been used to study the mode of action of the selected nucleating agents. A series[140] of condensed aromatic hydrocarbons and linear polyphenyls as well as graphite[141] have been used to crystallize polyethylene. Structural similarities between polymers and the substrate were required for the most efficient epitaxial agent. The planar zig-zag of the polyethylene chain can deposit in register with the crystallographic units of the substrate; the chain oriented in the 112 direction of the graphite surface layer.[141]

Linear polyesters crystallize epitaxially onto trioxane crystals, as well as on condensed aromatic hydrocarbons,[142] and poly(oxymethylene) on a number of bivalent ionic salts, and other inorganic solids.[143] Substrates with higher surface energies absorb solvent strongly and so inhibit nucleation.[143] Similar studies have been made on the different crystallographic faces of alkali halide crystals.[144]

6 Bulk Crystallization Kinetics

Isothermal.—A theory of chain folding polymer crystallization extended to include reeling-in rate of the chain onto the crystal face from reptation theory has been developed[145] and the temperature dependence of the rate constants derived. The pre-exponential factor incorporates a term $(1/n)$ in which n is the number of

---

[137] M. Cortazar, C. Sarasola, and G. M. Guyman, *Eur. Polym. J.*, 1982, **18**, 439.
[138] R. A. Chivers, P. J. Barham, J. Martinez-Salazar, and A. Keller, *J. Polym. Sci., Polym. Phys. Ed.*, 1982, **20**, 1717.
[139] P. J. Barham, A. Jarvis, and A. Keller, *J. Polym. Sci., Polym. Phys. Ed.*, 1982, **20**, 1733.
[140] J. C. Wittmann and B. Lotz, *J. Polym. Sci., Polym. Phys. Ed.*, 1981, **19**, 1837.
[141] P. R. Baukema and A. J. Hopfinger, *J. Polym. Sci., Polym. Phys. Ed.*, 1982, **20**, 399.
[142] J. C. Wittmann and B. Lotz, *J. Polym. Sci., Polym. Phys. Ed.*, 1981, **19**, 1853.
[143] C. M. Balik, S. K. Tripathy, and A. J. Hopfinger, *J. Polym. Sci., Polym. Phys. Ed.*, 1982, **20**, 2003.
[144] C. M. Balik, S. K. Tripathy, and A. J. Hopfinger, *J. Polym. Sci., Polym. Phys. Ed.*, 1982, **20**, 2017.
[145] J. D. Hoffmann, *Polymer*, 1982, **23**, 656.

monomer units in the chain being reeled by reptation onto the growing face. For the bulk polymer, n is proportional to the molecular weight. The theory predicts the experimentally observed drop in crystallization rate with increasing molecular weight and concludes that the reptation rate is sufficiently fast to allow a significant degree of adjacent re-entry of the chains at the latteral surface. Growth is governed by the activation energy for reptation, Q_p, and the work of chain folding, q. The temperature dependence of the reeling-in rate is then proportional to equation (3) in

$$1/n[A_0(\Delta\delta)\gamma_0 \exp - (Q_p + q)/RT] \quad (3)$$

which γ_0 is the frequency factor, and $A_0(\Delta\delta)$ is the force of crystallization on the pendant chain.

DSC is widely used to measure bulk isothermal crystallization kinetics. The limitations and problems associated with the technique have been reviewed recently,[146] particularly for micro-processor controlled calorimetry. The assumption adopted concerning the constancy of the athermal baseline of the calorimeter has been substantiated from analysis of the heat loss cooling curves which make up a substantial proportion of the initial part of the crystallization isotherms. Correcting for this heat loss, due to cooling to the crystallization, enables the measurement of experimental time from the onset of crystallization to be determined more accurately than previously. Experimental accuracy is improved by taking more data points of improved accuracy.

Despite these improvements in experimental procedure, fractional values of the exponent, n, from the Avrami equation relating the degree of crystallinity, X_t, to time, i.e., $-\ln(1 - X_t) = Zt^n$, with Z an overall rate constant, were observed with linear polyethylene fractions.[146] The n values observed were close to 2 and a spherulitic crystallization involving growth of branching lamellae was considered to be present.

An alternative suggestion[147] is that the Avrami equation can only predict crystallization behaviour in the early stages of crystallization and deviations such as spherulite abutment takes place. Additional corrections also have to be adopted to correct for post-crystallization. A semi-quantitative model has been developed for isothermal crystallization incorporating these corrections which is claimed to fit the experimental data well.[147] The kinetics of bulk crystallization of poly(1,3-dioxalane), $n = 3.0$ by dilatometry[148] and light microscopy,[149] of poly(tetramethylene oxide),[150] of poly(1,3-dioxepane),[151] $n = 3.0$ by dilatometry, of poly(α-methyl-α-N-propyl-β-propiolactone),[152] of polybut-1-ene),[153] $n = 2.0$ by DSC and microscopy, and of rayon in bromine water[154] have been reported. The effect of pressure[155] and of molecular weight[156] on the crystallization of poly(hexamethylene terephthalate) has

[146] P. J. Mills and J. N. Hay, *Polymer*, 1982, **23**, 1380.
[147] W. Dietz, *Colloid Polym. Sci.*, 1981, **259**, 413.
[148] R. Alamo, J. G. Fatou, and J. Guzman, *Polymer*, 1982, **23**, 374.
[149] R. Alamo, J. G. Fatou, and J. Guzman, *Polymer*, 1982, **23**, 384.
[150] F. D. Warner, D. S. Brown, and R. E. Wetton, *Polymer*, 1981, **22**, 1349.
[151] J. Garza, C. Marco, J. G. Fatou, and A. Bello, *Polymer*, 1981, **22**, 477.
[152] D. Grenier, A. Leborgno, N. Spassky, and R. E. Prud'homme, *J. Polymer Sci., Polym. Phys. Ed.*, 1981, **19**, 33.
[153] M. Cortazan and G. M. Guzman, *Makromol. Chem.*, 1982, **183**, 721.
[154] M. Lewin, H. Guttmann, A. Kroll, and D. Derfler, *J. Polym. Sci., Polym. Phys. Ed.*, 1982, **20**, 929.
[155] G. Hinrichsen, H. Krebs, and H. Springer, Colloid Polym. Sci., 1982, **260**, 502.
[156] D. W. Ihm and J. A. Cuculo, *J. Polym. Sci., Polym. Chem. Ed.*, 1982, **20**, 1847.

been measured. In the former case, two separate crystallization stages were observed each with its own n value. These values were low and non-constant, varying with pressure over the range 1.0—1.5 and 0.3—0.5, and were interpreted as being consistent with a change in mechanism or nucleation mode. The effects of pressure on the crystallization kinetics of copolyamides have also been studied.[157]

Poly(ethylene oxide) has been crystallized from various alcohols,[158] and the measured Avrami exponent changed from 1.5 to 2.0 with crystallization temperature. Poly(vinyl alcohol) also crystallized with very low n values which were constant but increased from about 0.7 at 413 K to 1.5 at 463 K; a one-dimensional growth mechanism was assumed.[159] A poly(diacetylene) film[160] which was amorphous to WAXS possessed one-dimensional order as determined from the A exponent determined for an exothermic process. The Avrami exponents were constant but varied between 1.0 and 1.7.

Accordingly, it would appear that the exact meaning of measured Avrami exponents are open to some conjecture. Many systems give meaningful values, others less so.

The kinetics of crystallization of isotactic polypropylene has been followed by i.r. spectroscopic changes using the relative change in the crystalline band at 998 cm^{-1}. The Avrami exponent was found to be 3.0.[161] I.r. bands could also be used to follow the build up of the helices from the glassy state and also from the melt. The processes were second and zero order, respectively.[162]

Calvert and Ryan[163] have used a u.v. fluorescent light microscope to study the development of crystallinity within a spherulite from the melt. As the spherulite-formed impurities were rejected from the spherulite initially, to diffuse back subsequently, the crystallinity within the interior of the spherulites simultaneously increased. There was a non-uniformity of additive concentration across the spherulite diameter.

Several kinetic studies[164—166] have been made on the effect of branches on the crystallization of low-density polyethylene. There is general agreement that the morphology is determined by the branches being excluded from crystalline regions[164] at the lamellae interface.[165] The kinetics of annealing of polymethylene[167] have been studied in order to discover the mechanism of the transport of matter. Previously n-alkanes had been observed to migrate from one crystal to another over macroscopic distances and to involve the plastic 'α' phase. Rotational motion with small translation motions are involved but chains can migrate over long distances. A reptation process is rejected since it requires a large fraction of amorphous regions.

---

[157] S. Gogelewski, *Polymer*, 1981, **22**, 792.
[158] O. Güven, *Colloid Polym. Sci.*, 1982, **260**, 647.
[159] N. A. Peppas and P. J. Hansen, *J. Appl. Polym. Sci.*, 1982, **27**, 4787.
[160] G. N. Patel and Y. P. Khanno, *J. Polym. Sci., Polym. Phys. Ed.*, 1982, **20**, 1029.
[161] A. Wtochowicz and M. Eder, *Polymer*, 1981, **22**, 1285.
[162] M. Glotin, R. R. Rahalker, P. J. Hendra, M. E. A. Cudby, and H. A. Willis, *Polymer*, 1981, **22**, 731.
[163] T. G. Ryan and P. D. Calvert, *Polymer*, 1982, **23**, 877.
[164] R. Kuhn and H. Kromer, *Colloid Polym. Sci.*, 1982, **260**, 1082.
[165] G. R. Strobl, T. Engolke, H. Meier, and G. Urban, *Colloid Polym. Sci.*, 1982, **260**, 394.
[166] T. Pakula, *Polymer*, 1982, **23**, 1300.
[167] G. Zerhi and R. Piazza, *Polymer*, 1982, **23**, 1921.

A similar study has been made with single-crystal mats.[133] The crystal thicknesses increased logarithmically with time, and a mechanism was proposed involving the coalescence of two crystals followed by boundaries diffusing away.

Thermal analysis has been used to follow the annealing of several segmented elastomers as the process was observed to be endothermic.[168]

Non-isothermal.—Non-isothermal crystallization studies are readily measured by DSC because of the ease of following a process at a constant rate of cooling. It gives a ready method of obtaining a rapid overall view of the temperature dependence of crystallization rate, especially when considering the effect of such variables as frozen-in shear and strain.[169] Light scattering has been used to determine morphological changes which occur under non-isothermal conditions.[170] The structure is observed to change systematically and continuously, and not in a discontinuous fashion. These conclusions must have an effect in developing a generalized kinetic equation for non-isothermal studies. A semi-quantitative model for spherulitic crystallization has been derived[147] for non-isothermal crystallization which incorporates the effect of abutment and post-crystallization. The model is claimed to fit the rate data well.

Non-isotropic.—A new thermodynamic model of crystallization in polymer networks subjected to constant deformation has been outlined.[171] Many of the assumptions inherent in previous theories have not been used. The effect of junction displacements is considered and changes in crystallite type, morphology, and size with orientation were allowed. The free energy of the network was calculated from rubber elasticity theory and the equilibrium state of the network found by minimizing with respect to the degree of crystallinity. The average degree of crystallinity and the change in stress due to the equilibrium crystallization were determined.

Stress-induced crystallization of natural rubber[172] has been re-examined using microcalorimetry and photoelasticity. The two techniques were apparently in disagreement unless it was assumed that crystallization occurs at strains less than that detected by calorimetry. The crystallites then formed are highly defective but act as nuclei for the subsequent crystallization at higher strains.

SAXS have been made on the structures which developed from the extended melt of poly(but-1-ene),[173] and in isotactic polypropylene from the melt and on annealing.[174,175] A similar study has been made on poly(hexamethylene terephthalate) in the presence of solvent,[176] and generalized to polymers in the oriented state.[177]

[168] J. W. C. Van Bogart, D. A. Bluemke, and S. L. Cooper, *Polymer*, 1981, **22**, 1428.
[169] K. H. Illers and W. Heckmann, *Colloid Polym. Sci.*, 1981, **259**, 955.
[170] L. Mandelkern, M. Glotin, and R. A. Benson, *Macromolecules*, 1981, **14**, 22.
[171] M. Kosc and A. Ziabicki, *Macromolecules*, 1983, **15**, 1507.
[172] F. DeCandia, G. Romano, R. Russo, and V. Vitterio, *J. Polym. Sci., Polym. Phys. Ed.*, 1982, **20**, 1525.
[173] J. Petermann, R. M. Gohil, J. M. Schultz, R. W. Hendricks, and J. S. Lin, *J. Mater. Sci.*, 1981, **16**, 265.
[174] S. Babajko and J. M. Schultz, *J. Polym. Sci., Polym. Phys. Ed.*, 1982, **20**, 497.
[175] J. Petermann, R. M. Gohil, J. M. Schultz, R. W. Hendricks, and J. S. Lin, *J. Polym. Sci., Polym. Phys. Ed.*, 1981, **19**, 609.
[176] J. Petermann, R. M. Gohil, J. M. Schultz, R. W. Hendricks, and J. S. Lin, *J. Polym. Sci., Polym. Phys. Ed.*, 1982, **20**, 253.
[177] H. Jameal, J. Waldman, and L. Rebenfeld, *J. Appl. Polym. Sci.*, 1981, **26**, 1795.

7 Melting

Thermal analytical techniques have been widely used to determine the melting range of polymers, but precautions have to be taken to eliminate the effect of thermal lags and annealing during melting.[11] Superheating effects[178] occur especially with oriented materials and extended chain crystals. On correcting for these effects, it is generally appreciated that the melting range contains information of the distribution of size and degree of perfection of the crystallites present. There is also a contribution from the melt entropy.[101]

The measured crystallite size distribution in polyethylene compared favourably with that measured by gel permeation chromatography of the etched crystallites, and the melting curves were analysed using the truncated form of the melting-point dependence on degree of polymerization, n, of the model n-alkanes, i.e.,

$$T_m = T_m°[1 - 2RT_m \ln(n)/(n\Delta h)]$$

relating the observed melting point, T_m, to equilibrium melting point $T_m°$, and heat of fusion per monomer unit, Δh. Using this relationship, differences were observed between extended and regular chain folded crystals. Melt crystallized material followed the extended rather than the chain folded crystal – melting point vs. molecular stem length dependence from which it was concluded that the regular chain folded crystal was a poor model for melt crystallized lamellae.[179] Low-molecular-weight polyethylene, with molecular weights about 3000, folded once or twice according to the degree of supercooling,[180] with a change in crystal stability. Consistent with this approach, the thermodynamic parameters of the n-alkanes and linear polyethylene depended on the degree of polymerization, n, and there was a reciprocal dependence of melting point on n. Extrapolation to high n values gave thermodynamic parameters similar to those of polyethylene.[181] These differences, between the n-alkanes and polyethylene, are attributed to chain-end effects, and the entropy and enthalpy of the chain ends have been quantified. However, it should be remembered[13] that the inter-lamellar regions for model alkanes and polymers play an important part in stabilizing the crystals and chain stem distribution determines this interface.

Equilibrium melting points and heats of fusion have been determined by linear extrapolation of the observed melting points against reciprocal stem degree of polymerization, n, i.e., $T_m = T_m°[1 - 2\delta_e/n\Delta h]$, for which δ_e is the lateral surface energy of the lamellae. The melting points of poly(vinylidene difluoride),[56] 1,4-trans-polypentadrene,[182] Nylon-6,6,[129] and Nylon-12 (ref. 88) have been similarly treated. Polyethylene[109] and cross-linked polyethylene[183] similarly have been analysed. The effect of cross-linking and entanglements have been separately treated since:

$$\frac{1}{T_m} - \frac{1}{T_m°} = \frac{R}{\Delta H}[KV_c + 2\delta\varepsilon]$$

---

[178] J. Clements and I. M. Ward, *Polymers*, 1982, **23**, 935.
[179] J. N. Hay, *Chem. Bull.*, 1982, 170; *Polymer*, 1981, **22**, 718.
[180] J. N. Hay and P. R. Fitzgerald, *Polymer*, 1981, **22**, 1003.
[181] W. Dollhopf, H. P. Grossman, and U. Leute, *Colloid Polym. Sci.*, 1981, **239**, 267.
[182] M. Farina and G. D. Silvestro, *Makromol. Chem.*, 1982, **183**, 241.
[183] J. de Boer and A. J. Pennings, *Polymer*, 1982, **23**, 1944.

in which T_m° refers to uncross-linked polyethylene, V_c the concentration of cross-links and ε entanglements in the polymer. K and δ are the corresponding weight factors.

Random and non-random copolymers still represent a problem in correlating their melting points to structure. The composition dependence of the melting points of poly(hexamethylene terephthalate) copolymers and ethylene–propylene copolymers[184] have been studied with a modified form of the Flory equation [equation (4)], where x is the mole fraction of crystallizable monomer.

$$\frac{1}{T_m} - \frac{1}{T_m^\circ} = \frac{R}{\Delta h} \ln(x) \qquad (4)$$

The melting point of poly(ethylene oxide) is depressed by poly(methyl methacrylate), since in the melt the two are compatible, and a similar effect has been observed with copolymer–polycarbonate blends.[186]

8 Mechanical Properties

The drawing of polymers and the production of ultra high modulus and high-strength materials continues to be the centre of massive activity. The morphology of ultra-oriented materials has been reviewed by Capaccio,[120] who collates SEM, WAXS, LAXS, Raman spectroscopic,[121] and DSC results. Microfibrillar and lamellar crystals are present in these materials even at the highest draw ratios but the key to their structures is the crystalline bridges linking adjacent lamellae. WAXS and gel permeation chromatography measurement[187] indicate that the highest modulus is not associated with extremely long crystalline segments, since even the highest draw ratios the weight average crystal size did not exceed 60 nm. The analysis of the deformation process and by analogy to the drawing behaviour of cross-linked polyethylene strongly suggest a network structure[188] in which junctions are provided by crystalline and amorphous entanglements.[120,189]

Highly oriented filaments have been produced by solution spinning of ultra-high-molecular-weight polyethylene and subsequent drawing,[190] and similarly gelation, crystallization, and stretching.[191] A modulus of 100 GPa and tensile strengths of 4 GPa are quoted for such products, about 30% of the theoretical limit. Solid-state extrusion with draw ratios up to 36 × gave modulus of 40 GPa[192] and the effect of the entrance angle, from 10° to 180°, was studied.[193] Annealing leads to a substantial reduction in modulus but there are differences according to whether the material is aged at constant length or unconstrained.[194] These differences become small at high draw ratios. SAXS show that the crystallite thickness decreases as annealing.

[184] N. Tanaka, *Polymer*, 1981, **23**, 647.
[185] M. M. Cortazar, M. E. Calahorr, and G. M. Guzman, *Eur. Polym. J.*, 1982, **18**, 165.
[186] A. S. Barnum, J. W. Barlow, and D. R. Paul, *J. Appl. Polym. Sci.*, 1982, **27**, 4065.
[187] G. Capaccio and I. M. Ward, *J. Polym. Sci., Polym. Phys. Ed.*, 1981, **19**, 667.
[188] L. Fischer, R. Marschberger, R. Ziegeldorf, and W. Ruland, *Colloid Polym. Sci.*, 1982, **260**, 174.
[189] F. De Candia, R. Russo, V. Vittorio, and A. Peterlin, *J. Polym. Sci., Polym. Phys. Ed.*, 1982, **20**, 269.
[190] P. Smith and P. J. Lemstra, *J. Polym. Sci., Polym. Phys. Ed.*, 1981, **19**, 1007.
[191] M. Matsuo and R. St. John Manley, *Macromolecules*, 1982, **15**, 985.
[192] E. S. Sherman, R. S. Porter, and E. L. Thomas, *Polymer*, 1982, **23**, 1069.
[193] W. G. Perkins and R. S. Porter, *J. Mater. Sci.*, 1982, **17**, 1700.
[194] G. Capaccio, J. Clements, P. J. Hine and I. M. Ward, *J. Polym. Sci., Polym. Phys. Ed.*, 1981, **19**, 1435.

Zone annealing has been applied to Nylon-6 to produce a high modulus and high-strength fibre – modulus 10 GPa;[195–198] this is attributed to production of a large number of tie molecules interconnecting the crystallites. Solid-state extrusion has been used to achieve a modulus of about 7 GPa,[199] but a higher value can be reached by using anhydrous ammonia as a plasticizer.[200] Zone annealing has been used to prepare high modulus isotactic polypropylene fibres,[201] zone drawing on pre-oriented polymer,[202] and force quenched polypropylene.[203] Ultra-high modulus liquid crystalline polyesters, based on poly(hexamethylene terephthalate), have been prepared from melt spinning,[204] but the network produced is sensitive to initial molecular weight, as is the draw ratio achieved in cold drawing.[205] Zone annealing has also been used to produce fibres of oriented crystallites.[206] The relationship between the fine structure of poly(hexamethylene terephthalate) and the mechanical properties has been discussed and the technique of zone annealing compared with the effectiveness of others in producing ultra-modulus material.[207]

Tensile creep tests and stress reduction studies during creep have been carried out on polyethylene and polypropylene at 293 K. The creep data could be superimposed with the general equation (5), in which ε_0 is the initial strain on loading, ε_p

$$\varepsilon = \varepsilon_0 + \varepsilon_p[1 - \exp(-K\dot{\varepsilon}_s t)] + \dot{\varepsilon}_s t \tag{5}$$

the primary creep strain, and $\dot{\varepsilon}_s$ the secondary creep ratio.[208] The dynamic shear behaviour[209] of linear oriented polyethylene was measured with reference to the dynamic tensile modulus. The increase in plateau shear modulus at 223 K with draw ratio was due to increase in crystal continuity as measured by the longitudinal thickness, on the basis of the random crystalline bridge model. The creep and the recovery of ultra-high modulus polyethylenes[210] has been studied in range 293–343 K; there was no detectable permanent creep. This behaviour could only be described in terms of two activated processes coupled in parallel.

Amorphous poly(hexamethylene terephthalate) crystallizes on uniaxial stretching. The extent of orientation increases with decreasing temperature, but DSC, SALS, and birefringence studies confirm the development of orientation crystallization.[211] The degree of crystallinity which develops is 40%, and the fibre axis is perpendicular to the draw direction.[212] Instabilities in neck propagation

[195] T. Kunagi, I. Alikayama, and M. Hashimoto, *Polymer*, 1982, **23**, 1193.
[196] T. Kunagi, I. Akiyama, and M. Hashimoto, *Polymer*, 1982, **23**, 1199.
[197] T. Kunagi, *Polymer*, 1982, **23**, 176.
[198] T. Kunagi, T. Ikuta, M. Hashimoto, and K. Matsuzaki, *Polymer*, 1982, **23**, 1983.
[199] T. Shimoda and R. S. Porter, *Polymer*, 1981, **23**, 1124.
[200] T. Karamoto, A. E. Zachariades, and R. S. Porter, *J. Polym. Sci., Polym. Phys. Ed.*, 1982, **20**, 1485.
[201] T. Kunagi, *J. Polym. Sci., Polym. Phys. Ed.*, 1982, **20**, 329.
[202] K. Yamada and M. Takayanagi, *J. Appl. Polym. Sci.*, 1982, **27**, 2091.
[203] K. Yamada, M. Kamezawa, and M. Takayanagi, *J. Appl. Polym. Sci.*, 1981, **26**, 49.
[204] D. Acierno, F. P. La Mantia, G. Polizzolti, A. Ciferri, and B. Valenti, *Macromolecules*, 1982, **15**, 1455.
[205] J. C. Engelacre, J. P. Cavrot, and F. Rietsch, *Polymer*, 1982, **23**, 766.
[206] T. Kunagi, A. Suzaki, and M. Hashimoto, *J. Appl. Polym. Sci.*, 1981, **26**, 395.
[207] A. Suzuki and M. Hashimoto, *J. Appl. Polym. Sci.*, 1981, **26**, 1951.
[208] D. J. Dixon-Stubbs, *J. Mater. Sci.*, 1981, **16**, 389.
[209] A. G. Gibson, S. A. Jawad, G. R. Davies, and I. M. Ward, *Polymer*, 1982, **23**, 349.
[210] M. A. Wilding and I. M. Ward, *Polymer*, 1981, **22**, 870.
[211] T. Terado, C. Sawatari, T. Chigono, and M. Matsuo, *Macromolecules*, 1982, **15**, 998.
[212] M. Matsuo, T. Tamada, T. Terado, C. Sawatari, and M. Niwa, *Macromolecules*, 1982, **15**, 988.

cause self-oscillation at high strain rates. These are not due to heating effects in mechanically deforming the specimens or to the evolved heat of crystallization, but are related to the existence of a critical stress at which the rate of neck propagation changes by orders of magnitude.[213]

Fatigue crack profiles and fracture surfaces have been examined in crystalline poly(vinylidene difluoride), Nylon-6,6, and polyacetal.[214] Cracks were initiated at *trans*-spherulites and at inter-spherulitic tensile crazes. Comprehensive yielding within the plastic zone at the tip of the crack crushes and elongates the spherulites in the direction of crack growth. The morphology in advance of the crack tip is different for the original. Below the glass transition temperature, plastic deformation is limited and fatigue fractures occur with little disruption of the spherulites, the fracture surface reflecting the original micro-structure.

The failure of isotactic polypropylene filled with silicon oxide particles has been similarly studied.[215] The polymer absorbs considerable energy by plastic deformation. Craze initiation and crack propagation required little energy. The fracture toughness was determined by the volume of fraction of filler.

A relationship has been derived between the impact transition temperature and the stress concentration factor for low-density polyethylene.[216] The effect of low-density polyethylene concentration on the ratio of craze initiation and growth in polystyrene–polyethylene blends has been correlated,[217] and explained by assuming that effective crazes are only produced within clusters of polyethylene particles where there is overlapping stress concentration fields. This may be as a result of the changes in morphology near the interface, and different stress states at the interface.

9 Conclusions

The most outstanding problem in polymer crystallization studies is the nature of the chain conformation at the interface between crystalline and amorphous regions in melt crystallized material. The wide application of LANS and WANS to various crystalline systems for which deuterio- and hydro-polymer aggregation is not a problem is beginning to resolve this, and the model which appears to be emerging is one intermediate between adjacent tightfold re-entry and the switchboard model. The continued development of our understanding of the reptation behaviour of chains has led to a development of crystallization rate theories in terms of the 'reeling-in' of chains onto the growing face. Avrami kinetics also appears to be a centre of controversy with reported examples of the exact determination of mechanism by application of the equation to bulk and solution crystallization and an almost equal amount of misleading information being obtained from its application. Perhaps we will soon see this difficulty resolved.

---

[213] T. Pakula and E. W. Fischer, *J. Polym. Sci., Polym. Phys. Ed.*, 1981, **19**, 1705.
[214] P. E. Bretz, R. W. Hertberg, and J. A. Manson, *Polymer*, 1981, **22**, 1272.
[215] W. Friedrich and V. A. Karsch, *J. Mater. Sci.*, 1981, **16**, 2167.
[216] W. Brostow and R. D. Corneliussen, *J. Mater. Sci.*, 1981, **16**, 1665.
[217] S. D. Sjoerdsma, M. E. J. Dekkers, and D. Heikens, *J. Mater. Sci.*, 1982, **17**, 2605.

11
Characterization of Synthetic Polymers

BY J. M. G. COWIE

1 Introduction

The characterization of polymer samples is now achieved by a wide and varied range of techniques, some of which are sufficiently important to be dealt with separately in this Volume. As before, work reviewed here will be confined to synthetic polymers and the more classical approaches to the determination of molar mass, molar mass distribution, and hydrodynamic parameters, and will include a coverage of gel permeation chromatography (GPC). Of general interest, we have the report of a workshop on quasi-elastic light-scattering (QELS) studies of fluids and macromolecular solutions,[1] and Volume 48 of the advances in polymer science series,[2] which contains two extensive articles, by Burchard on static and dynamic light scattering, and Patterson on photon correlation spectroscopy of bulk polymers. Also published is a series on liquid chromatography of polymers edited by Caze,[3] in which three volumes have so far appeared containing a useful selection of symposium papers. The plenary lectures, presented at the symposium on polymer characterization in Durham, have been published, and of immediate interest are those by Dawkins,[4] on high-performance GPC, Hamielec[5] and Inagaki[6] on copolymer analysis by size exclusion chromatography (SEC) and thin-layer chromatography (TLC), Kratochvil,[7] who has reviewed the last ten years of classical light scattering, and Stockmayer[8] on polydispersity and branching analysis using QELS. While these will be of specific interest to readers of this chapter, the complete set of lectures covers other aspects of characterization and are well worth reading.

2 Molar Mass Measurements

A detailed study of the number average molecular weight (M_n) measurement by vapour pressure osmometry (VPO) has been carried out by Marx-Figini and

---

[1] 'Light Scattering in Liquids and Macromolecular Solutions', ed. V. Degiorgio, M. Corti, and M. Giglio, Plenum, New York, 1980.
[2] 'Advances in Polymer Science', Vol. 48, Springer-Verlag, 1983.
[3] 'Liquid Chromatography of Polymers and Related Materials', ed. J. Cazes and X. Delamare, Marcel Dekker, New York.
[4] J. V. Dawkins, *Pure Appl. Chem.*, 1982, **54**, 281.
[5] A. Hamielec, *Pure Appl. Chem.*, 1982, **54**, 293.
[6] H. Inagaki and T. Tanaka, *Pure Appl. Chem.*, 1982, **54**, 309.
[7] P. Kratochvil, *Pure Appl. Chem.*, 1982, **54**, 379.
[8] W. H. Stockmayer and M. Schmidt, *Pure Appl. Chem.*, 1982, **54**, 407.

Figini.[9—11] Using careful calibration procedures, they found that the calibration constant is molecular weight dependent and that the relationship between the reduced bridge imbalance and M_n is $(\Delta v/c)_{c \to 0} = KM^{-a}$ where $a \neq 1$ for the solvents studied. To account for this variable calibration constant they have suggested that there is solute diffusion in the solution drop on the thermistor, and the experimental data presented do appear to support this conclusion. A comparison[12] between an unmodified VPO and a new rotating ebulliometer has been made in which data from benzene solutions of benzil and biphenyl have been recorded for both systems. This could have some bearing on calibration procedures.

A computation of the errors involved in molar mass measurements involving osmometry and viscometry[13] and an analysis of the various factors which influence osmotic pressure measurements has been compiled by Aelenei.[14] In particular, he noted that a new membrane should be allowed to relax for 50—60 h before reliable measurements could be made. Few real developments in molar mass determination have appeared of late, but two novel methods have been developed which are worth further attention. Baumbach[15a] has observed that there is a decrease in the fluorescence intensity of rhodamine B in chloroform solution when a polymer is added to the system and that this varies with the amount of polymer present. If the ratio of the fluorescence intensity without polymer to that with polymer present is plotted as a function of the polymer concentration, then the intercept is proportional to $(M_n/M_0)^{1/2}$ where M_0 is the mass of a main chain segment. The method was tested using polystyrene and polyisobutylene and a single numerical constant could be used for both polymers. A rapid estimation of the molar mass of water-soluble polymers has been achieved by Siano and Bock,[15b] who have noted the fact that the cloud point temperature of a microemulsion containing a small amount of polymer changes monotonically with chain length. The method requires calibration and a suitable choice of polymer concentration, but, once these parameters are established, M is quickly measured with reasonable accuracy over a wide range of chain lengths. The method can be used for polymers such as poly(ethylene oxide), poly(vinyl pyrollidone), dextran, and poly(styrene sulphonate).

The use of the ultracentrifuge for determination of M is not widespread, but an interesting comparison of the precision of measurements based on sedimentation and diffusion to give M_{SD} has been made using standard polystyrene samples.[16] For narrow distribution samples M_W and M_{SD} should be close, but some differences between the values have been noted.

In the previous review it was mentioned that Khan and Bhargava[17] had proposed a general method for determining the M for copolymers using the slope of viscosity–concentration curves. This has been severely criticized by Sen et al.[18] who

---

[9] M. Marx-Figini and R. V. Figini, *Makromol. Chem.*, 1980, **181**, 2401.
[10] M. Marx-Figini and R. V. Figini, *Makromol. Chem.*, 1980, **181**, 2409.
[11] R. V. Figini and M. Marx-Figini, *Makromol. Chem.*, 1981, **182**, 437.
[12] J.-T. Chen, H. Sotobayashi, and F. Asmussen, *Colloid. Polym. Sci.*, 1981, **259**, 1202.
[13] D. Jadraque and J. Pereña, *Angew. Makromol. Chem.*, 1982, **104**, 163.
[14] N. Aelenei, *Eur. Polym. J.*, 1981, **17**, 533.
[15] (a) D. O. Baumbach, *J. Polym. Sci., Polym. Lett. Ed.*, 1982, **20**, 117; (b) D. B. Siano and J. Bock, *ibid.*, 1982, **20**, 151.
[16] V. Petrus, B. Porsch, B. Nyström, and L.-O. Sundelof, *Makromol. Chem.*, 1982, **183**, 1279.
[17] H. U. Khan and G. S. Bhargava, *J. Polym. Sci., Polym. Lett. Ed.*, 1980, **18**, 465.
[18] U. K. Sen, B. M. Mandal, and S. N. Bhattacharyya, *J. Polym. Sci., Polym. Lett. Ed.*, 1981, **19**, 523.

found neither theoretical nor experimental evidence to support this approach and suggest that the original assumptions are invalid.

3 Dilute Solutions

General Characterization.—Modern characterization procedures tend to centre increasingly on the application of gel permeation chromatography, as we shall see in Section 10. In spite of this, or perhaps because of the need for data from alternative sources, a substantial body of work continues to be carried out in which the hydrodynamic properties of polymer solutions are reported. A large number of Mark–Houwink relationships have been measured and the constants in the equation $[\eta] = KM^v$ are summarized in Table 1 for the most important systems.[19–54] While these relationships are extremely useful, they must be treated with due regard to the range of M covered, the particular average used, and the sample polydispersity. Bearing this in mind, Raczek[55,56] has outlined correction

---

[19] W. M. Kulicke and N. Böse, *Polym. Bull.*, 1982, **7**, 205.
[20] Ch. Wandrey, W. Jaeger, and G. Reinisch, *Acta Polym.*, 1982, **33**, 156.
[21] M. G. Vitovskaya, P. N. Lavrenko, O. Y. Okatova, E. P. Astapenko, V. B. Novakovsky, S. V. Bushin, and V. N. Tsvetkov, *Eur. Polym. J.*, 1982, **18**, 583.
[22] N. Hadjichristidis, X. Zhongde, L. J. Fetters, and J. L. Roovers, *J. Polym. Sci., Polym. Phys. Ed.*, 1982, **20**, 743.
[23] M. Sauviat and J.-P. Cohen Addad, *Polymer*, 1981, **22**, 461.
[24] H. Müller, D. Neuray, and A. Horbach, *Makromol. Chem.*, 1981, **182**, 177.
[25] A. Horbach, H. Müller, and L. Bottenbruch, *Makromol. Chem.*, 1981, **182**, 2873.
[26] K. Kishino, T. Kawai, T. Nose, M. Saito, and K. Kamide, *Eur. Polym. J.*, 1981, **17**, 623.
[27] T. C. Amu, *Polymer*, 1982, **23**, 1775.
[28] I. Noda, T. Imai, T. Kitano, and M. Nagasawa, *Macromolecules*, 1981, **14**, 1303.
[29] I. Noda, Y. Yamamoto, T. Kitano, and M. Nagasawa, *Macromolecules*, 1981, **14**, 1306.
[30] K. Kamide, M. Saito, and T. Abe, *Polym. J.*, 1981, **13**, 421.
[31] G. Wenz and G. Wegner, *Makromol. Chem. Rapid Commun.*, 1982, **3**, 231.
[32] F. J. Ansorena, L. M. Revuelta, G. M. Guzman, and J. J. Iruin, *Eur. Polym. J.*, 1982, **18**, 19.
[33] O. V. Kallistov, Yu. E. Svetlov, I. G. Silinskaya, V. P. Skilzkova, V. V. Kudriavtsev, and M. M. Koton, *Eur. Polym. J.*, 1982, **18**, 1103.
[34] L. Mrkvickova and J. Kalal, *Makromol. Chem.*, 1982, **182**, 203.
[35] I. Kössler, M. Netopilik, G. Schulz, and R. Gnauck, *Polym. Bull.*, 1982, **7**, 597.
[36] O. Quadrat, M. Bohdanecky, and L. Mrkvickova, *Makromol. Chem.*, 1981, **182**, 445.
[37] R. Jenkins and R. S. Porter, *Polymer*, 1982, **23**, 105.
[38] N. Hadjichristidis and C. Touloupis, *Makromol. Chem.*, 1982, **183**, 611.
[39] D. Kokkiaris, C. Touloupis, and N. Hadjichristidis, *Polymer*, 1981, **22**, 63.
[40] N. Hadjichristidis, C. Touloupis, and L. J. Fetters, *Macromolecules*, 1981, **14**, 128.
[41] J. W. A. van den Berg, G. van de Ridder, and C. A. Smolders, *Eur. Polym. J.*, 1981, **17**, 935.
[42] S. M. Padaki and J. K. Stille, *Macromolecules*, 1981, **14**, 888.
[43] J. L. Roovers and P. M. Toporowski, *Macromolecules*, 1981, **14**, 1174.
[44] V. Chrastova, D. Mikulášová, J. Lacok, and P. Citovicky, *Polymer*, 1981, **22**, 1054.
[45] L. Sŏltés, D. Mikulášová, and I. Hudec, *Chem. Zvesti*, 1981, **35**, 543.
[46] S. Dragan, S. Ioan, and M. Dima, *Polym. Bull.*, 1982, **7**, 473.
[47] S. V. Bushin, V. N. Tsvetkov, Ye. B. Lysenko, and V. N. Yemel'yanov, *Vysokomol. Soedin., Ser. A*, 1981, **23**, 2494.
[48] L. K. Koopal, *Colloid Polym. Sci.*, 1981, **259**, 490.
[49] D. Engel and R. C. Schulz, *Makromol. Chem.*, 1981, **182**, 3279.
[50] L. A. Tatarova, T. G. Ermakova, N. F. Kedreena, B. A. Kasekeen, D. D. Novekov, and B. A. Lopirev, *Vysokomol. Soedin., Ser. B*, 1982, **24**, 697.
[51] L. Simek, J. Hrnčířik, M. Mládek, and Z. Tuzar, *Angew. Makromol. Chem.*, 1982, **107**, 185.
[52] K. Kamide, Y. Miyazaki, and H. Kobayashi, *Polym. J.*, 1982, **14**, 591.
[53] P. V. Mangalam and V. P. Kalpagam, *Polymer*, 1982, **23**, 991.
[54] S. Miyamoto, Y. Ishii, and H. Ohnuma, *Makromol. Chem.*, 1981, **182**, 483.
[55] J. Raczek, *Eur. Polym. J.*, 1982, **18**, 351.
[56] J. Raczek, *Eur. Polym. J.*, 1982, **18**, 393.

procedures for the molar mass dependence of any function in the general relation $X(M) = K_x M^\alpha$. The suggested methods can be applied either numerically or graphically and the technique has been outlined for application to viscosity, radius of gyration, diffusion, and frictional coefficients.

Aqueous Systems.—Amu[27] has re-investigated some aspects of the dilute solution behaviour of aqueous solutions of poly(ethylene oxide). The polymer coils were observed to contract with increasing temperature when the solvent was pure water and extrapolation indicated that θ conditions would be attained at ~ 381 K. This is in good agreement with the LCST measured independently for this system. The results in aqueous salt solutions were in conflict with results from other workers, however, as the θ temperatures for magnesium sulphate and potassium sulphate solutions reported previously could not be reproduced. In all cases, v values were greater than 0.5, suggesting that the θ temperatures must be lower. The temperature dependence of the unperturbed dimensions was found to be positive which might suggest that these would be larger at the LCST compared with the UCST. The presence of a helical structure has been proposed as a reason for the observed behaviour, but no real evidence for the existence of helices could be found by Brown and Stilbs,[57] who measured the frictional coefficients for short-chain samples in the temperature range 298 to 448 K. If there is orientation of water molecules around the chain, as seems likely, then they do not necessarily force the molecule to assume a helical conformation.

Many dilute solution studies have been published for aqueous polyacrylamide solutions but further work on an aqueous sodium sulphate/polyacrylamide system is justified by the fact that the corrosive qualities of this solution are lower than those containing halogen salts and consequently the pumping system in a GPC should last longer if this is used instead. The unperturbed dimensions of the unhydrolysed[58] and various partially hydrolysed[59] polyacrylamides have been measured in salt/water/methanol mixtures.

Conformational Studies.—Hydrodynamic data have been used to study structural aspects of certain polymeric systems. Roovers et al.[22] have taken stereo-irregular polybutadiene and polyisoprene and found the characteristic ratio C_∞ to be 5.1 in both cases. This value is in good agreement with the Flory–Abe calculations, based on rotational isomer analysis, for polybutadiene, but is higher than the theoretical value for polyisoprene. Values of C_∞ were also measured[34] for the pure *cis* and *trans* polyisoprenes in dioxane at $\theta = 304$ K and as a function of temperature; the rate of change was -1.31×10^{-3} K^{-1}. Measurements of C_∞ for other polymers are collected in Table 1.

Poly(amic acid) chains are known to be relatively rigid structures and an examination of the effect of including an oxygen atom in the backbone has been carried out by Kallistov et al.[33] Synthesis of chains with the 'phenyl–O–phenyl' sequence, rather than an all carbon bonded structure, led to a much more flexible chain when this so-called 'pin joint oxygen' was included. Changes in the rigidity of

---

[57] W. Brown and P. Stilbs, *Polymer*, 1982, **23**, 1780.
[58] T. Schwartz, J. Sabbadin, and J. François, *Polymer*, 1981, **22**, 609.
[59] T. Schwartz and J. François, *Makromol. Chem.*, 1981, **182**, 2757.

Table 1 Dilute solution parameters for synthetic polymers

Polymer	Solvent	Temp./K	$K/\text{cm}^3\,\text{g}^{-1}$	v	$\left[\dfrac{\langle r^2\rangle_0}{M}\right]^{1/2}$ Å	C_∞	σ	Ref.
Polyacrylamide	0.1 M(aq.) Na_2SO_4	298	1.94×10^{-2}	0.70	—	—	—	19
Poly(dimethyl–diallyl ammonium chloride)	1 M(aq.) NaCl		1.12×10^{-2}	0.82	—	—	—	20
Poly(amidobenzimidazole) $M > 10^4$	Dimethyl acetamide	298	1.40×10^{-1}	0.77	—	—	—	21
Poly(amidobenzimidazole) $M > 10^4$	98% H_2SO_4	299	1.30×10^{-1}	0.76	—	—	—	21
Poly(amidobenzimidazole) $M < 10^4$	Dimethylacetamide	298	5.0×10^{-3}	1.11	—	—	—	21
Poly(amidobenzimidazole) $M < 10^4$	98% H_2SO_4	299	3.6×10^{-3}	1.13	—	—	—	21
Poly(butadiene)	Dioxane	299.5	1.78×10^{-1}	0.5	0.892	5.1	—	22
Polyisobutylene	Cyclohexane	298	1.11×10^{-2}	0.76	—	—	—	23
Poly(ε-caprolactam)	96% H_2SO_4	298	3.31×10^{-2}	0.79	1.08	—	—	24
Poly(ε-caprolactam)	Phenol/1,2-dichlorobenzene (1:1)	298	2.29×10^{-2}	0.83	1.04	—	—	24
Bisphenol A polycarbonate	Dichloromethane	298	1.11×10^{-2}	0.82	—	—	2.21	25
3,3′,5,5′-Tetramethyl bisphenol A polycarbonate	Dichloromethane	298	2.58×10^{-2}	0.72	—	—	2.66	25
3,3′-Dimethyl bisphenol A polycarbonate	Dichloromethane	298	2.72×10^{-2}	0.71	—	—	2.52	25
3,5-Dimethyl bisphenol A polycarbonate	Dichloromethane	298	2.58×10^{-2}	0.72	—	—	2.70	25
3,3′,5,5′-Tetrachloro bisphenol A polycarbonate	Dichloromethane	298	2.01×10^{-2}	0.71	—	—	2.68	25
Cellulose disulphate (sodium salt, DS = 1.90)	0.5 M(aq.) NaCl	298	7.17×10^{-4}	0.93	—	17.4	2.90	26
Poly(ethylene oxide)	Water	297.9	6.04×10^{-3}	0.90	—	—	—	27
Poly(ethylene oxide)	Water	307.9	8.62×10^{-3}	0.85	—	—	—	27
Poly(ethylene oxide)	Water	318.5	1.87×10^{-2}	0.78	—	—	—	27
Poly(ethylene oxide)	0.5 M(aq.) $MgSO_4$	298	4.82×10^{-2}	0.65	—	—	—	27
Poly(ethylene oxide)	0.45 M(aq.) K_2SO_4	308	7.03×10^{-2}	0.59	—	—	—	27
Poly(ethylene oxide)	0.39 M(aq.) $MgSO_4$	318	6.4×10^{-2}	0.61	—	—	—	27

Characterization of Synthetic Polymers

Polymer	Solvent	T	col4	col5	col6	col7	Ref
Poly(t-butyl crotonate)	Toluene	298	7.7×10^{-3}	0.82	—	—	29
Cellulose acetate (DS = 0.49)	Dimethyl acetamide	298	1.91×10^{-1}	0.60	1.28	10.0	30
Cellulose acetate (DS = 0.49)	Dimethyl sulphoxide	298	1.71×10^{-1}	0.61	—	—	30
Cellulose acetate (DS = 0.49)	Water	298	2.09×10^{-1}	0.60	2.42	34.7	30
Cellulose acetate (DS = 0.49)	Formamide	298	2.09×10^{-1}	0.60	2.97	52.5	30
Poly(diacetylene)	Chloroform	298	6.4×10^{-3}	0.83	—	—	31
Poly(isoprene) (cis-1,4)	Dioxane	304.2	—	—	—	5.29	32
Poly(isoprene) (trans-1,4)	Dioxane	304.2	—	—	—	7.34	32
Poly(isoprene)						5.1	32
Poly(4,4'-phenylene) pyromellitamic acid	0.1 M LiBr/dimethyl formamide	298	3.78×10^{-2}	1.20	—	—	33
Poly(2,3-epoxypropyl methacrylate)	Dioxane	298	8.32×10^{-3}	0.67	—	—	34
Poly(2,3-epoxypropyl methacrylate)	Tetrahydrofuran	298	1.50×10^{-2}	0.60	—	—	34
Poly(methyl methacrylate)	NN'-Dimethyl formamide	298	2.5×10^{-2}	0.625			35
	Acetonitrile	303	—	0.52			36
		311	—	0.57			36
		318	—	0.59			36
		328	—	0.61			36
	m-Xylene	308	—	0.51			36
	Butyl chloride	318	—	0.53			36
Poly(methyl methacrylate)							
100% Isotactic	Tetrahydrofuran	298	1.66×10^{-2}	0.66	—	10.2	37
92% Isotactic		298	1.69×10^{-2}	0.659	—	10.3	37
83% Isotactic		298	1.72×10^{-2}	0.658	—	9.3	37
35% Isotactic		298	9.84×10^{-3}	0.692	—	7.5	37
23% Isotactic		298	9.58×10^{-3}	0.695	—	7.5	37
15% Isotactic		298	8.89×10^{-3}	0.695	—	7.3	37
Poly(2,4,5-trichlorophenyl methacrylate)	Benzene	295	2.52×10^{-2}	0.50	—	—	38
		303	1.52×10^{-2}	0.54	—	—	38
		308	1.09×10^{-2}	0.57	—	—	38
		313	8.60×10^{-3}	0.59	—	—	38
		318	7.3×10^{-3}	0.61	—	—	38
		323	6.4×10^{-3}	0.62	—	—	38
		328	5.9×10^{-3}	0.63	—	—	38
		333	5.8×10^{-3}	0.63	—	—	38

Table 1 (continued)

Polymer	Solvent	Temp./K	K/cm^3 g^{-1}	v	$\left[\frac{\langle r^2\rangle_0}{M}\right]^{1/2}$ Å	C_∞	σ	Ref.
Poly(phenyl thiol methacrylate)	Tetrahydrofuran	298	7.82×10^{-3}	0.68	—	—	2.26	39
	Benzene	298	1.62×10^{-2}	0.61	—	—	2.26	39
	Toluene	298	1.77×10^{-2}	0.59	—	—	2.26	39
	Butan-2-one	298	3.86×10^{-2}	0.50	—	—	2.26	39
Poly(o-methyl phenyl thiol methacrylate)	Tetrahydrofuran	298	4.07×10^{-3}	0.73	—	—	2.27	39
	Benzene	298	4.75×10^{-3}	0.71	—	—	2.27	39
	Toluene	298	7.12×10^{-3}	0.67	—	—	2.27	39
Poly(cyclohexyl thiol methacrylate)	Tetrahydrofuran	308	4.07×10^{-3}	0.74	—	8.5 (visc)	2.13 (visc)	40
	Cyclohexane	298	8.65×10^{-3}	0.63	—	12.4 (LS)	2.48 (LS)	40
Poly(2,6-dimethyl 1,4-phenylene oxide)	Toluene	298	3.1×10^{-3}	0.70	—	3.5	—	41
Poly(quinoline)	Trichloroethane	298	3.3×10^{-3}	0.70	—	3.5	—	41
Polystyrene – H-shaped	sym.-Tetrachloroethane	298	1.63×10^{-1}	0.58	—	—	—	42
	Toluene	308	7.47×10^{-2}	0.73	—	—	—	43
	Cyclohexane	308	6.65×10^{-2}	0.50	—	—	—	43
Polystyrene ($M_W 1.08 \times 10^7$—2.2×10^7)	Toluene	298	8.52×10^{-2}	0.61	—	—	—	44
	Benzene	298	1.47×10^{-1}	0.56	0.612	—	—	44
Poly(chloromethyl styrene)	Benzene	298	3.17×10^{-2}	0.55	—	—	2.45	46
Poly(phenyl siloxane) (ladder)	Benzene	298	1.2×10^{-4}	1.10	—	—	—	46
Poly(vinyl alcohol)	Water	303	—	—	—	6.6	1.82	48
Poly(2-isopropenyl naphthalene)	Tetrahydrofuran	298	1.43×10^{-2}	0.66	—	—	—	49
	H$_2$O	298	5.44×10^{-2}	0.64	—	—	—	50
Poly(1-vinyl-1,2,4-triazole)	Dimethyl formamide	298	2.27×10^{-1}	0.50	—	—	—	50
Copolymers:								
Segmented polyurethanes								
Poly(butylene adipate)/1,4-butanediol	Dimethyl formamide	298	4.3×10^{-2}	0.70	—	—	—	51
Poly(butylene adipate)/diphenyl methane 4,4'-di-isocyanate	Dimethyl formamide	298	1.10×10^{-1}	0.60	—	—	—	51

Polymer	Solvent	T						Ref.
Poly(acrylonitrile-co-methyl acrylate) [95% AN]	Dimethyl formamide	298	2.13×10^{-2}	0.74	0.88	10.1	2.25	52
	Ethylene carbonate/H_2O [82.5/17.5 wt%]	298	1.52×10^{-1}	0.50	0.86	8.7	2.08	52
	51 wt% HNO_3	298	2.15×10^{-1}	0.50	1.11	14.5	2.68	52
	55 wt% HNO_3	298	1.95×10^{-1}	0.51	1.12	14.6	2.69	52
	67 wt% HNO_3	298	5.88×10^{-2}	0.66	1.30	19.7	3.13	52
	80 wt% HNO_3	298	6.22×10^{-2}	0.68	1.49	25.9	3.59	52
Poly(acrylonitrile-co-methyl methacrylate) Composition [15:85] (0.236 mole fraction AN)	Ethylacetate	303	4.4×10^{-3}	0.73	—	—	1.73	53
		313	1.7×10^{-3}	0.80	—	—	1.69	53
		323	1.0×10^{-3}	0.80	—	—	1.62	53
	Benzene	293	9.12×10^{-3}	0.50	—	—	2.09	53
		303	9.70×10^{-3}	0.67	—	—	1.71	53
		313	4.55×10^{-3}	0.72	—	—	1.58	53
Composition 0.5 mole fraction AN	Acetonitrile	303	4.37×10^{-3}	0.75	—	—	1.73	53
		313	3.40×10^{-3}	0.77	—	—	—	53
		323	2.87×10^{-3}	0.77	—	—	—	53
	Dimethyl sulphoxide	303	9.4×10^{-4}	0.87	—	—	1.48	53
		318	9.4×10^{-4}	0.87	—	—	—	53
		333	5.0×10^{-4}	0.92	—	—	—	53
	(6.5/1) Benzene/dimethyl formamide	303	7.9×10^{-2}	0.50	—	—	1.90	53
Composition 0.74 mole fraction AN	Acetonitrile	303	6.90×10^{-3}	0.70	—	—	1.88	53
		313	3.05×10^{-3}	0.75	—	—	—	53
		323	2.40×10^{-3}	0.75	—	—	—	53
	Dimethyl sulphoxide	303	4.23×10^{-3}	0.60	—	—	1.98	53
Composition 0.74 mole fraction AN	Dimethyl sulphoxide	318	2.07×10^{-2}	0.65	—	—	—	53
		333	1.02×10^{-2}	0.70	—	—	1.95	53
	(1.667/1) Benzene/dimethyl formamide	303	1.12×10^{-1}	0.50	—	—	—	53
Poly(isobutyl vinyl ether-co-maleic anhydride)	Acetone	303	2.10×10^{-1}	0.50	—	—	2.33	54

bisphenol A polycarbonate can be achieved in a different fashion. Insertion of a methyl or chloro substituent in the phenyl rings leads to an increase in chain rigidity as one might expect, because of the greater steric hindrance to rotation about the main chain.[25] A similar effect can be obtained when polystyrene is chloromethylated as shown by an increase in the stiffness parameter σ.[46]

An interesting variation of side chain flexibility is reported by Hadjichristidis et al.[39,40] They have examined several phenyl thiol methacrylates in which a sulphur atom replaces the conventional oxygen atoms in the pendant ester unit. Results suggest that the

$$\begin{array}{c} O \\ \parallel \\ -(C-S-R) \end{array}$$

unit gives a more flexible structure than the polymer with the more normal

$$\begin{array}{c} O \\ \parallel \\ -(C-O-R) \end{array}$$

counterpart. Inclusion of an orthomethyl unit in the phenyl ring left the flexibility essentially unaltered. When the cyclohexyl ring was substituted for the phenyl ring in the thiol structure[40] the flexibility was enhanced even further, but the results were more ambiguous in this case as σ varied between 2.13 and 2.48 depending on the method of measurement. For firm conclusions to be drawn from studies of this kind the comparison of internally consistent data is of prime importance. Thus measurable differences are reported in C_∞ which arise from tacticity differences in a series of poly(methyl methacrylate)s with values ranging from $C_\infty = 10.2$ for 100% isotactic samples to $C_\infty = 7.3$ for polymers with 15% isotacticity. Highly isotactic chains also had a more expanded conformation in tetrahydrofuran as shown by the larger values of $[\eta]$, but as M_W increased beyond 10^6 this effect became much less noticeable. In addition, the polymer solvent interaction was found to be lower for the isotactic chains compared with the more random structures in spite of the more extended form of the former.

Regular H-shaped polystyrenes, with segments having equal molar mass, have been prepared by Roovers and Toporowski.[43] The unperturbed radius of gyration of the branched structure was smaller than its linear analogue by a factor of 0.7 and the properties appeared to be intermediate between those of the three-arm and the four-arm polystyrenes.

Evidence of conformational changes in solution has been observed from $[\eta]$-temperature studies for two systems.[41,60] For atactic poly(methyl methacrylate) samples this occurs between 313 and 323 K, while for the syndiotactic material the temperature range is 283—293 K. Both of these transition ranges are solvent dependent. Similar observations for poly(2,6-dimethyl-1,4-phenylene oxide) in toluene and trichloroethane solutions revealed an inflection point between 308 and 318 K, but in this case it may be associated with crystallization in solution.[41] These observations were substantiated by sedimentation velocity studies.[61]

---

[60] I. A. Katime, P. Guitierrez Cabañas, and C. Ramiro Vera, *Polym. Bull.*, 1981, **5**, 25.
[61] J. W. A. van den Berg, G. van de Ridder, and C. A. Smolders, *Eur. Polym. J.*, 1981, **17**, 935.

Other evidence for conformational changes in solution has come from an evaluation of the Ptitsyn–Eizner flexibility parameter λ at various temperatures. This has been estimated for a wide range of systems by Quintana and co-workers.[62–64] Their general observations were that λ was a function of the polymer chain flexibility, the polymer–solvent interactions, and that it varied linearly with temperature under normal circumstances. In an extensive study of poly(methyl methacrylate) solutions, a discontinuity in the λ-temperature curves was detected which lent support to the existence of a conformational change, which had been located for this polymer by other methods.[63] Values of $\lambda \sim 2.66$ were obtained for poly(methyl methacrylate) which can be compared with $\lambda = 6.31$ measured for the more rigid poly(vinyl carbazole),[64] both sets of values being derived under θ-conditions.

Unperturbed Dimensions.—The variation of the unperturbed dimensions with temperature has been studied for poly(ethylene oxide),[27] poly(methyl methacrylate),[65] poly(diethyleneglycol terephthalate),[66] and poly(2,4,5-trichlorophenyl methacrylate),[65] and the data are summarized in Table 2. The last system is rather peculiar as there is a change of dependence above 313 K, with an apparent

Table 2 *Temperature dependence of unperturbed dimensions*

Polymer	$\left[\dfrac{d \ln \langle r^2 \rangle_0^{1/2}}{dt}\right] K^{-1}$	Temp./K	Ref.
Poly(ethylene oxide)	2.4×10^{-2}	303	27
Poly(2,4,5-trichlorophenyl methacrylate)	-3.0×10^{-3}	295—313	38
	-0.87×10^{-3}	313—333	38
Poly(methyl methacrylate)	$+2.6 \times 10^{-3}$	—	65
Poly(diethylene glycol terephthalate)	$+1.18 \times 10^{-3}$	333	66

contraction in the unperturbed coil size, suggesting that the extended conformation in benzene may have lower rotational barriers.

Measurements of chain dimensions for polystyrene in cyclohexane, over a range of temperatures from below to above the θ temperature, have been made by Oyama et al.[67] They observed a smooth change in dimensions consistent with a coil–globule transition but this contrasts with the abrupt change observed by previous workers on passing through θ-conditions. Huglin[68] has also examined polystyrene chain dimensions, this time in a series of mixed solvents. In cyclohexane/tetralin, Abdel Azim and he established the value of the lattice co-ordination number Z to be 3 or 4, while measurements in 3-methyl cyclohexanol/tetralin indicated that the unperturbed dimensions are independent of the solvent composition.[69]

---

[62] L. M. Leon, J. R. Quintana, A. Martinez, and J. Landabidea, *Eur. Polym. J.*, 1982, **18**, 89.
[63] L. M. Leon, M. C. Gonzalez, J. R. Quintana, F. Zamora, A. Martinez, and G. M. Guzman, *Eur. Polym. J.*, 1982, **18**, 229.
[64] I. Katima and J. R. Quintana, *Polym. Bull.*, 1982, **6**, 455.
[65] O. Quadrat and L. Mrkvickova, *Eur. Polym. J.*, 1981, **17**, 1155.
[66] E. Riande, J. Guzman, and M. A. Llorente, *Macromolecules*, 1982, **15**, 298.
[67] T. Oyama, K. Shiokawa, and K. Baba, *Polym. J.*, 1981, **13**, 167.
[68] A. A. A. Abdel-Azim and M. B. Huglin, *Polymer*, 1982, **23**, 1859.
[69] A. A. A. Abdel-Azim and M. B. Huglin, *Eur. Polym. J.*, 1982, **18**, 735.

The variation of limiting viscosity number with temperature has been measured for aqueous solutions of cellulose ethers[70] and for several methacrylate derivatives.[71] Radic and Gargallo, from their work on methacrylate polymers,[71] concluded that $[\eta]$ will often pass through a maximum at certain temperatures and that this should be considered to be a general phenomenon for all solutions. This is hardly surprising if the existence of both UCST and LCST in polymer systems is also accepted as a general feature common to polymer solutions.

The estimation of unperturbed dimensions for poly(methyl methacrylate) dissolved in acetonitrile has been in some doubt because of the variation in the reported θ temperature for this system. Quadrat et al.[36] have attempted to rectify this situation by making extensive measurements and conclude that $\theta = 301$ K is the best value. Other θ temperatures reported for poly(methyl methacrylate) are $\theta = 271.2$ K in 2-methoxyethanol, $\theta = 300.3$ K in 2-ethoxyethanol and $\theta = 348.2$ K in 2-butoxyethanol.[65]

4 Ultracentrifugation

A minor revival in interest in studies involving the ultracentrifuge is evidenced by a steady flow of publications from a number of laboratories. Elmgren[72-74] has re-examined the concentration dependence of the sedimentation constant, S, and reports that, for compact particles, it is related only to solvent backflow $(1 - \phi)$ and mobility factors (here ϕ is the volume fraction). He has developed a relationship between S and the average mobility of the solvent, located inside and outside the coils, and finds that the buoyancy factor is a product of the true buoyancy, calculated using the solvent density, and $(1 - \phi)$. Hence one should only use the solution density for this correction if no allowance is made for the solvent backflow.[75]

Pressure effects are also important in sedimentation studies and should never be neglected, and this is particularly true if the data are to be used to calculate molar mass distributions. Mulderije[76] has developed an accurate means of reading Schlieren patterns for narrow sedimentation boundaries which enables the operator to determine local values of S more precisely. As the pressure gradient in an ultracentrifuge cell can vary from 1 to 250 atmospheres from meniscus to cell bottom, a reliable analysis of molar mass distribution is impaired unless one can correct for pressure effects with considerable accuracy throughout the cell. By using a non-linear pressure dependence, Mulderije can achieve reliable measurements and also obviate the need to establish a zero time for the sedimentation experiment. Pressure effects must also be considered when flotation systems are encountered and a method to deal with this has been proposed by Van den Berg and Le Grand[77]

---

[70] S. Nagura, S. Nakamura, and Y. Onda, *Kobunshi Ronbunshu*, 1981, **38**, 133.
[71] D. Radic and L. Gargallo, *Polymer*, 1981, **22**, 410.
[72] H. Elmgren, *J. Polym. Sci., Polym. Lett. Ed.*, 1981, **19**, 561.
[73] H. Elmgren, *J. Polym. Sci., Polym. Lett. Ed.*, 1981, **19**, 567.
[74] H. Elmgren, *J. Polym. Sci., Polym. Lett. Ed.*, 1982, **20**, 389.
[75] H. Elmgren, *J. Polym. Sci., Polym. Lett. Ed.*, 1982, **20**, 57.
[76] J. J. H. Mulderije, *Macromolecules*, 1982, **15**, 506.
[77] J. W. A. Van den Berg and P. Le Grand, *Eur. Polym. J.*, 1981, **18**, 51.

Characterization of Synthetic Polymers

using a power series expansion. Also worthy of attention is the first report of an experimental estimation of the complete density profile inside an ultracentrifuge cell which comes from Munk.[78] In contrast to previous reports, Destor and Rondolez[79] have found that S is independent of rotor speed if pressure corrections are carefully applied. However, the corrections required, when temperatures close to θ are used, may only amount to $\sim 2\%$. Pavlov and Frenkel[80] have also reported briefly on the dependence of S on concentration.

An interesting and novel suggestion for a motionless ultracentrifuge has been outlined by Wales and Mir.[81] A semipermeable membrane is placed at the bottom of a liquid column and the polymer solution is layered on top of the membrane. By adjusting the flow rate of liquid down the column, a steady state can be achieved where the diffusion of the polymer is counterbalanced by the pressure of the flowing liquid. This balance of convective and diffusive mass transfer can be used to provide information on M and the sample heterogeneity.

It is often difficult to obtain reliable values of M for high charge density polymers. which often produce strongly coloured solutions. This has been overcome by Bortel et al.,[82] who have used the Archibald procedure to make accurate measurements in less than 30 min.

There are several advantages to be gained by using the ultracentrifuge to study hydrodynamic behaviour and the crossover from dilute to semi-dilute concentration regions. The measurements are insensitive to dust and are always carried out at zero shear. Vidakovic et al.[83] have used this method to examine polystyrene in cyclohexane under θ conditions and found the crossover concentration to be much larger than for thermodynamically better solvents. The effect of penetration of coils under θ conditions on the value of S, and the application of scaling concepts through the semi-dilute regime, have also been examined.[84,85] Poly(ethylene oxide) has also been investigated in semi-dilute aqueous solutions and the data reflect the flexibility of this polymer.[86] On the theoretical front, Zimm[87] has applied a rigid dumbell model to the sedimentation problem.

5 Diffusion

Conventional methods for determining diffusion coefficients in polymer solutions have now been almost wholly replaced by quasi-elastic light-scattering methods. These yield values of D from which one can also estimate the hydrodynamic radius R_H. The type of average value obtained for D must be known and it has been noted that if the method of cumulants is used the D which is obtained is lower than D_Z when measurements are made near a θ temperature.[88]

[78] P. Munk, *Macromolecules*, 1982, **15**, 500.
[79] C. Destor and F. Rondolez, *Polymer*, 1981, **22**, 67.
[80] G. M. Pavlov and S. Ya. Frenkel, *Vysokomol. Soedin., Ser. B*, 1982, **24**, 178.
[81] M. Wales and L. Mir, *J. Polym. Sci., Polym. Lett. Ed.*, 1981, **19**, 321.
[82] E. Bortel, A. Kochanowski, W. Gozdecki, and H. Kozlowska, *Makromol. Chem.*, 1981, **182**, 3099.
[83] P. Vidakovic, C. Allain, and F. Rondolez, *Macromolecules*, 1982, **15**, 1571.
[84] F. A. H. Peeters and H. J. E. Smits, *Bull. Soc. Chim. Belges*, 1981, **90**, 111.
[85] J. Roots and B. Nyström, *J. Polym. Sci., Polym. Phys. Ed.*, 1981, **19**, 479.
[86] B. Nyström, S. Boileau, P. Hemery, and J. Roots, *Eur. Polym. J.*, 1981, **17**, 249.
[87] B. H. Zimm, *Macromolecules*, 1982, **15**, 520.
[88] D. Caroline, *Polymer*, 1982, **23**, 492.

Polystyrene continues to be the most popular polymer for study, particularly when testing current solution theory. As seen in the previous section, a smooth increase in the polystyrene coil size was observed on passing from below to above the θ temperature and this has also been confirmed from measurements of R_H.[89] Forced Rayleigh light scattering has been used to obtain self-diffusion coefficients for polystyrene in benzene solutions, working in the semi-dilute region, and the data are in good agreement with the reptation model and scaling laws.[90] A semi-quantitative agreement between experiment and the Doi-Edwards theory was found for poly(γ-benzyl-L-glutamate) in 1,2-dichloroethane, but the rotational diffusion coefficient was higher than predicted while the translational diffusion coefficient was only 2/3 of the predicted value.[91] The variation of the diffusion coefficient with concentration has also come under scrutiny from both a theoretical[92] and experimental[93] point of view. The effect of pressure on D and the concentration dependence parameter k_D has been measured over a wide range of 1—5000 atmospheres; the observed increase in D was correlated with the pressure dependence of the solvent viscosity, whereas k_D was independent of pressure as long as measurements were in the dilute solution regime.[94] Fluctuation spectroscopy has proved particularly useful for the simultaneous determination of M and D for high-molecular-weight polymers.[95]

The molecular weight dependence of D has been recorded for a number of systems including poly(isopropyl α-methyl styrene),[96] poly(t-butyl crotonate),[29] and poly-(amidobenzimidazole);[21] and other systems which have received attention are poly(α-methyl styrene),[97] cellulose tricarbanilate,[98] and alkanes up to C_{28} in CCl_4 and benzene.[99,100]

6 Light Scattering

While quasi-elastic light scattering is now widely used as a method of studying polymer solutions and has been effectively used to examine the crossover from single chain to multi-chain dynamics in poly(dimethyl siloxane) solutions,[101] classical light scattering still has many adherents. Kratochvil et al.[102] have reviewed the techniques which have proved most reliable for the study of polyethylene solutions by light scattering. Poor solvents are most useful and here diphenyl methane is the one that is recommended. They have also emphasized the care that must be taken to account for the possible presence of supermolecular particles in

---

[89] M. J. Pritchard and D. Caroline, *Macromolecules*, 1981, **14**, 424.
[90] L. Léger, H. Hervet, and F. Rondolez, *Macromolecules*, 1981, **14**, 1752.
[91] K. M. Zero and R. Pecora, *Macromolecules*, 1982, **15**, 87.
[92] A. Ziya Akcasu, *Polymer*, 1981, **22**, 1169.
[93] C. C. Han and A. Ziya Akcasu, *Polymer*, 1981, **22**, 1165.
[94] J. Roots and B. Nyström, *Macromolecules*, 1982, **15**, 553.
[95] K. Ohbayshi, M. Minoda, and H. Utiyama, *Macromolecules*, 1981, **14**, 1031.
[96] D. Landheer and S. L. Malhotra, *J. Macromol. Sci., Chem.*, 1981, **16**, 1349.
[97] J. C. Selser, *Macromolecules*, 1981, **14**, 346.
[98] F. Fried, G. M. Searby, M. J. Seurin-Vellutini, S. Dayan, and P. Sixon, *Polymer*, 1982, **23**, 1755.
[99] J. G. de la Torre, A. Jiménez, and J. J. Freire, *Macromolecules*, 1982, **15**, 148.
[100] J. G. de la Torre and J. J. Freire, *Macromolecules*, 1982, **15**, 155.
[101] B. Ewen, D. Richter, J. B. Hayter, and B. Lehner, *J. Polym. Sci., Polym. Lett. Ed.*, 1982, **20**, 233.
[102] J. Stejskal, J. Horska, and P. Kratochvil, *J. Appl. Polym. Sci.*, 1982, **27**, 3929.

solutions, and ways of interpreting data obtained when these are present have been discussed in general by Dautzenberg.[103] An investigation of branching in polyethylene samples of both low and high density in which light scattering and viscosity have been compared, has been reported by Arndt et al.[104] Also of interest are papers on the characterization of compositional heterogeneity in copolymers using isorefractive solvents,[105] a study of cellulose diacetate solutions,[106] and the use of light scattering to follow the thermal polymerization of styrene.[107] An extensive examination of the second virial coefficients measured in t-butyl acetate solutions of polystyrene has been reported by Wolf and Adam.[108] This is an interesting system to study as it exhibits both UCST and LCST features at reasonable temperatures, 238.5 and 382.5 K respectively, and is also athermal at 318 K.

In the non-classical area, Burchard and his co-workers[109] have developed a modified photon correlation spectrometer, capable of recording both static and dynamic data, and this should eventually find wide application. Molar mass distributions can also be derived from dynamic light-scattering measurements, but one must be able to eliminate or correct for intramolecular interference effects. Raczek[110,111] has developed such a method in which no assumptions are necessary nor is it restricted to low molar mass material. The procedure gave good agreement with GPC results when tested for polystyrene in toluene.

7 Osmotic Pressure

Data from the recent literature, on the osmotic pressure of dilute and semi-dilute polymer solutions, have been treated by Schafer[11] using renormalization group calculations. For polymers in good solvents he found that universal scaling behaviour applied, with good agreement between theory and experiment up to polymer weight fractions of ~ 15. Again, good agreement with scaling laws has been obtained form poly(α-methyl styrene) solutions[113] where the reduced osmotic pressure was observed to be related to the degree of coil overlap and also proportional to $(C/C^*)^{1.32}$

8 Viscosity

The two-parameter theory of polymer solutions has been tested for poly(isobutylene) and polystyrene, in several solvents, by Miyaki and Fujita.[114] Discrepancies were found where viscosity data did not always agree with the

[103] H. Dautzenberg, *Acta Polym.*, 1982, **33**, 158.
[104] K. F. Arndt, E. Schröder, and A. Korner, *Acta Polym.*, 1981, **32**, 620.
[105] T. Tanaka, M. Omoto, and H. Inagaki, *Makromol. Chem.*, 1981, **182**, 2889.
[106] H. Sukuzi, Y. Muraoka, M. Saito, and K. Kamide, *Eur. Polym. J.*, 1982, **18**, 831.
[107] B. Chu and G. Fytas, *Macromolecules*, 1982, **15**, 561.
[108] B. A. Wolf and H. J. Adam, *J. Chem. Phys.*, 1981, **75**, 4121.
[109] S. Bantle, M. Schmidt, and W. Burchard, *Macromolecules*, 1982, **15**, 1604.
[110] J. Raczek, *Eur. Polym. J.*, 1982, **18**, 847.
[111] J. Raczek, *Eur. Polym. J.*, 1982, **18**, 863.
[112] L. Schäfer, *Macromolecules*, 1982, **15**, 652.
[113] I. Noda, N. Kato, T. Kitano, and M. Nagasawa, *Macromolecules*, 1981, **14**, 668.
[114] Y. Miyaki and H. Fujita, *Macromolecules*, 1981, **14**, 742.

theoretical treatment and it was suggested that even for flexible coils such as these, there may be a contribution from partial free draining effects. A great many equations have been proposed in the past in which viscosity data may be used to elicit information on the unperturbed dimensions of polymer coils in dilute solution. Yet another has been proposed by Tanaka[115] [equation (1)], which is similar to the

$$\left(\frac{[\eta]}{M^{1/2}}\right)^{5/3} = K_\theta^{5/3} + 0.627\Phi_0^{5/3}(\langle r^2\rangle_0/M)BM^{1/2} \tag{1}$$

Stockmayer–Fixman equation but has different exponents. The excluded volume calculations from $[\eta]$ have been reassessed[116] on the basis that the equality $\alpha_\eta = \alpha_s$ was a reasonable assumption to make.

An investigation of the concentration at which the onset of the semi-dilute regime occurred has been made for poly(methyl methacrylate) dissolved in a range of solvents.[117] It was found that crossover depended on polymer chain length and solvent quality. A theory, covering this extended concentration range and based on the conception of a 'local viscosity effect', has been developed by Budtov,[118] whereas a more empirical approach has been outlined by Zhoung[119] using equation (2),

$$(\eta_{sp/C}) = [\eta] + (k_1[\eta]\eta_{sp}/\eta_{rel}^\gamma) \tag{2}$$

which is linear up to 10% concentration when $\gamma = 0.28$. Other theoretical studies have considered the effects of pre-averaging of the hydrodynamic interactions[120] and the application of the Bixon–Zwanzig treatment of excluded volume.[121]

The pressure dependence of both $[\eta]$ and the Huggins constant has been measured for polystyrene solutions up to 4000 atmospheres,[122] and assumptions concerning the Huggins constant have been made in order to estimate unperturbed dimensions from viscosity data.[123] While this has given reasonable results for the two systems tested, the unreliability of this constant makes the method one to be treated with some caution. Viscosity–molar mass relations for polystyrene with $M > 10^7$ have been established in good solvents and the v values are smaller than for shorter chains.[44,45] Both polystyrene and poly(methyl methactylate) have been dissolved in a range of viscous solvents and their viscosity behaviour examined.[124]

9 Prediction of Hydrodynamic Parameters

Calculation of hydrodynamic parameters making use of the UNIFAC method has been attempted by several groups. The approach is based on the concept of a

---

[115] G. Tanaka, *Macromolecules*, 1982, **15**, 1028.
[116] T. M. Aminbhavi, R. C. Patel, K. Bridger, and E. S. Jayadevappa, *J. Macromol. Sci., Chem.*, 1982, **17**, 1283.
[117] I. Hernandez-Feuntes, M. G. Prolongo, R. M. Masegosa, and A. Horta, *Eur. Polym. J.*, 1982, **18**, 29.
[118] V. P. Budtov, *Eur. Polym. J.*, 1981, **17**, 191.
[119] C.-S. Zhoung, *Polym. J.*, 1982, **14**, 501.
[120] I. Fortelny, *Makromol. Chem.*, 1982, **183**, 193.
[121] H. R. Berger and E. Straube, *Acta Polym.*, 1982, **33**, 291.
[122] J. R. Schmidt and B. A. Wolf, *Macromolecules*, 1982, **15**, 1192.
[123] K. K. Chee, *J. Appl. Polym. Sci.*, 1982, **27**, 1675.
[124] R. A. M. van Brandwijk, M. G. L. Hesse, H. L. Jalink, and F. A. H. Peeters, *Bull. Soc. Chim. Belges*, 1981, **90**, 105.

solution as a collection of groups so that solution properties are then obtainable from a set of functional group contributions. It is necessary to include volume and surface area parameters which are readily obtained, but, in addition, it is necessary to have an estimate of an interaction parameter a_{mn} describing the interplay of group m with group n. This can be obtained from experimental data and is independent of the particular system used. The χ interaction parameter has been calculated, using this technique, for poly(dimethyl siloxane) solutions and comes within 10% of the experimental values over a range of concentrations.[125] It could prove an attractive procedure for deriving information when experimental results are scarce and has also been applied to concentrated solutions with some success.[126]

Some years ago Rudin derived a relationship between polymer coil dimensions and the concentration of the solution. Briefly, the hydrodynamic volume of a polymer molecule dissolved in a solvent is assumed to be the product of the volume of the unsolvated molecule, v, and a swelling factor, ε. This is related to the viscosity by equation (3), where N_A is the Avogadro constant and the critical concentration,

$$v\varepsilon = \frac{4\pi[\eta]M}{9.3 \times 10^{24} + 4\pi N_A C([\eta] - [\eta]_\theta)} \tag{3}$$

at which polymer coils attain their unperturbed dimensions, is C_x for $\varepsilon = 1$ or defined more fully as $C_x = 9.3 \times 10^{24}/(4\pi N_A[\eta]_\theta)$ with $[\eta]$ expressed in units of g cm^{-3}. Using this model, Rudin has demonstrated that he can predict osmotic pressures from M and $[\eta]$ data;[127] second virial coefficients, in some cases better than the Kurata–Yamakawa theory;[128] sedimentation coefficients, which were reasonable except for high M;[129] Flory–Huggins interaction parameters, where agreement was within 2% of experimental values at infinite dilution;[130] and also account for concentration effects in GPC.[131] Kok and Rudin[132] have also reviewed the relationship between the hydrodynamic radius of a coil R_H, measured from QELS, and the radius of gyration R_G, thereby concluding that a best value relation of $R_H = 0.77 R_G$ can be used for data in θ solvents although this is not in complete agreement with a recent theoretical prediction of the proportionality constant of 0.664.

Second virial coefficients have been described by Tanaka and Solc[133] using the Pade approximation.

10 Polymers with Rigid Chains

Semi-flexible polymers, such as poly(t-butyl crotonate) have relatively stiff backbone chains with persistence lengths around 50–60 Å and very low expansion

---

[125] M. Gottlieb and M. Herskowitz, *Macromolecules*, 1981, **14**, 1468.
[126] M. T. Ratzsch and D. Glindemann, *Acta Polym.*, 1979, **30**, 57.
[127] C. M. Kok and A. Rudin, *J. Appl. Polym. Sci.*, 1981, **26**, 3575.
[128] C. M. Kok and A. Rudin, *J. Appl. Polym. Sci.*, 1981, **26**, 3583.
[129] C. M. Kok and A. Rudin, *J. Appl. Polym. Sci.*, 1982, **27**, 3357.
[130] C. M. Kok and A. Rudin, *J. Appl. Polym. Sci.*, 1982, **27**, 353.
[131] C. M. Kok and A. Rudin, *Makromol. Chem.*, 1981, **182**, 2801.
[132] C. M. Kok and A. Rudin, *Makromol. Chem., Rapid Commun.*, 1981, **2**, 655.
[133] G. Tanaka and K. Solc, *Macromolecules*, 1982, **15**, 791.

factors ($\alpha \sim 1.1$).[28] These general features are shared with sodium cellulose disulphate[26] and lightly substituted cellulose acetate.[29]

Stille[42] has reported on the properties of 'cardo' polyquinolines, which are semi-flexible aromatic polymers with a pendant ring where one carbon atom in the ring is also a member of the polymer backbone chain. These are amorphous materials with good thermal stability and are soluble in common organic solvents, all these properties being enhanced by the 'cardo' structure. The particular polymer examined was derived from 9,9'-bis(4-acetylphenyl)fluorene and 4,4'-diamino 3,3'-dibenzoyl diphenylether.

Viscosity, sedimentation, diffusion, and persistence lengths have been measured for the ladder structure poly(diphenyl siloxane)[47] and the worm-like chain model was found to account adequately for the observed behaviour of these rigid chains. This model was also used to assess data for poly(isophthaloyl-*trans*-2,5 dimethyl-piperazine) measured in *m*-cresol and trifluoroethane.[134] For $M < 10^5$ g mol^{-1} large persistence lengths and typical stiff chain behaviour were observed, but as M increased the Mark–Houwink plot curved slowly towards a lower slope and the polymer began to demonstrate all the features expected of a flexible coil. A similar dependence of dilute solution behaviour on M has been reported for poly(amido-benzimidazole)[21] which is rod-like when M is less than 10^4. This change in behaviour with chain length is thought to be due to excluded volume effects which come into play at high M and this has been examined by Norisuye and Fujita.[135] They found that excluded volume effects cannot be neglected in the worm-like chain model when the number of Kuhn statistical units exceeds 50. Poly(acetylene), which might be expected to have a very rigid structure, appears to behave as a semi-flexible coil in solution.[31]

As it is necessary to account for sample polydispersity in these studies, an analysis of the corrections needed for stiff chain polymers has been given by Fortelny *et al.*[136] It is also evident that partial free draining effects require attention when dealing with semi-flexible or rigid chains and this is important in many cellulose derivatives. Kamide and Saito[137,138] have published a general method for measuring unperturbed dimensions using frictional coefficients derived for non-gaussian coils which require such corrections. In the semi-dilute solution range, screening effects for rod-like molecules can be accounted for using an extended version of the Doi–Edwards theory and introduction of a screened hydrodynamic interaction tensor gave good agreement with experimental observations.[139]

11 Cyclic Structures

Cyclized poly(butadienes) have been examined in mixed θ-solvents of cyclohexane and dioxane (30/70) at 303 K and the hydrodynamic measurements indicate that these samples behave like linear polymers.[140] For cyclic poly(dimethyl siloxane), free

---

[134] T. Sadanobu, T. Norisuye, and H. Fujita, *Polym. J.*, 1981, **13**, 75.
[135] I. Norisuye and H. Fujita, *Polym. J.*, 1982, **14**, 143.
[136] I. Fortelny, J. Kovav, A. Zivny, and M. Bohdanecky, *J. Polym. Sci., Polym. Phys. Ed.*, 1981, **19**, 181.
[137] K. Kamide and M. Saito, *Eur. Polym. J.*, 1981, **17**, 1049.
[138] K. Kamide and M. Saito, *Eur. Polym. J.*, 1982, **18**, 661.
[139] S. Fesciyan and J. S. Dahler, *Macromolecules*, 1982, **15**, 517.
[140] K. Yamamoto, N. Bessho, T. Shiibashi, and E. Maekawa, *Polym. J.*, 1981, **13**, 555.

draining effects have been observed at low M and the second virial coefficients for both linear and cyclic structures are similar when $M_w < 1000$ g mol^{-1}.[141,142]

12 Chromatographic Methods

Gel permeation chromatography is now one of the prime methods for polymer characterization because of the amount of relevant information which can be extracted from each measurement. For reliable results to be obtained, it is essential that certain precautions and corrections be observed and many publications are still devoted to an assessment of calibration procedures and data analysis. In addition to the books on the subject mentioned previously, two substantial reviews have appeared. Ouano[143] deals with recent advances up to 1980 and covers aspects of data acquisition and processing, high-pressure GPC, and detectors. Janca[144] has produced a most comprehensive coverage of applications and quotes 492 references, which makes it an ideal source of material from the late 1960s on.

Calibration Methods.—The application of polydisperse samples in calibration procedures is now becoming quite common. Kubin[145,146] has developed a general method, using standards with a known M_w and M_n but with an unknown distribution shape, which allows the simultaneous determination of the variation of M and the spreading factor with elution volume. It is believed that this method could be extended to aqueous systems where the calibration standards are normally polydisperse. Szewczyk[147] has also employed the technique of using a standard with known M_w and M_n, but has assumed the formalism that the average elution volume corresponds to the average molar mass of the polydisperse sample. This leads to single valued equations from which the calibration parameters can be derived by iterative methods, and the procedures he uses are simpler and faster than those previously reported.

The development of aqueous GPC has tended to be hindered by the paucity of suitable, monodisperse, water-soluble standards. It has been suggested by Bose *et al.*[148] that polyelectrolytes might be suitable as standards if there was an understanding of the effect of ionic strength on the conformation of the polymer. Poly(styrene sulphonate) and dextrans were selected for study and it was found that while the dextran elution volume decreased with increasing ionic strength, that of the poly(styrene sulphonate) actually increased, so a universal calibration could not be used. The Coll–Prusinowski procedure did seem to work but only for high ionic strengths. This area still requires further fundamental work. The need to be able to change rapidly from one solvent system to another has been emphasized by Gilding *et al.*,[149] who highlight this as an essential development in order to achieve rapid

---

[141] C. J. C. Edwards, R. F. T. Stepto, and J. A. Semlyen, *Polymer*, 1982, **23**, 865.
[142] C. J. C. Edwards, R. F. T. Stepto, and J. A. Semlyen, *Polymer*, 1982, **23**, 869.
[143] A. C. Ouano, *Rubber Chem. Tech.*, 1981, **54**, 535.
[144] J. Janča, *J. Liq. Chromatogr.*, 1981, **4** (Suppl. 1), 1.
[145] M. Kubin, *J. Appl. Polym. Sci.*, 1982, **27**, 2933.
[146] M. Kubin, *J. Appl. Polym. Sci.*, 1982, **27**, 2943.
[147] P. Szewczyk, *J. Appl. Polym. Sci.*, 1981, **26**, 2727.
[148] A. Bose, J. E. Rollings, J. M. Caruthers, M. R. Okos, and G. T. Tsao, *J. Appl. Polym. Sci.*, 1982, **27**, 795.
[149] D. K. Gilding, A. M. Reed, and I. N. Askill, *Polymer*, 1981, **22**, 505.

characterization of industrial or medically important polymers and copolymers. They have discussed the use of several calibration methods and have found it possible to change effectively from one solvent to another in approximately 4 h.

Corrections for axial dispersion have been solved mathematically, but the situation is complicated if a molar mass detector is used in series, on line, with a concentration detector. Miniaturization of the detector helps when used in conjunction with high-speed columns but corrections must still be applied and Netopilik[150] has presented a solution to this problem. A simultaneous method for the determination of dispersion and calibration curve has been proposed by Andreeta and Figini.[151] They have used the Gauss method of iteration which does not require monodisperse standards, only the experimental operations and data necessary for an overall calibration.

The universal calibration method is still applied, of course, to great effect, making use of viscosity data,[152] but the nature of the averaging process has been examined in some detail by French and Naufiett.[153] They have suggested that $[\eta]M$ can be replaced by KM^{v+1} as a representation of the hydrodynamic volume, when the polydispersity of the whole sample is large, where K and v are the appropriate Mark–Houwink constants. Methods involving the ratio of $[\eta]$ for two samples as a way of calibrating GPC and estimating the Mark–Houwink constants have been described by Dobbin et al.,[154] and Zhongde et al.;[155] in the latter case three equations are involved but no iteration is required.

The effect on elution volumes of the presence of a second polymer, has been investigated by Burns et al.[156] using mixtures of polystyrene and polybutadiene. As the second polymer was added in increasing amounts, the elution volume of the first increased. It was suggested that the repulsive forces caused contraction of the coils to take place thereby increasing the elution volume. Concentration effects for single polymer systems were also examined[156] and it was noted that a threshold concentration exists below which the peak elution volume was essentially independent of the polymer concentration. This decreased as M increased but was negligible for low M or monodisperse samples. It has also been found that the effect decreases as the heterogeneity becomes large but extrapolation to infinite dilution compensates for these effects.[157]

Flow rate effects,[158] peak broadening,[159] calibration procedures for poly(methyl acrylate),[160] and the relation between elution volume and weight average[161] have also been reported.

Columns and Techniques.—The search for improved column packings and techniques leading to rapid measurements continues. Among the reported new

[150] M. Netopilik, *Polym. Bull.*, 1982, **7**, 575.
[151] H. A. Andreeta and R. V. Figini, *Angew. Makromol. Chem.*, 1981, **93**, 143.
[152] S. F. Sun and E. Wong, *J. Chromatogr.*, 1981, **208**, 253.
[153] D. M. French and G. W. Naufiett, *J. Liq. Chromatogr.*, 1981, **4**, 197.
[154] C. J. B. Dobbin, A. Rudin, and M. F. Tchir, *J. Appl. Polym. Sci.*, 1982, **27**, 1081.
[155] X. Zhongde, S. Mingshi, N. Hadjichristidis, and L. J. Fetters, *Macromolecules*, 1981, **14**, 591.
[156] V. Narasimhah, R. Y. M. Huang, and C. M. Burns, *J. Appl. Polym. Sci.*, 1981, **26**, 1295.
[157] W. L. Elsdon, J. M. Goldwasser, and A. Rudin, *J. Polym. Sci., Polym. Lett. Ed.*, 1981, **19**, 483.
[158] G. Yu Shu, Y. Mei-Ling, L. Hui-Jian, D. Yu Kang, and Y. Qi-Cong, *J. Liq. Chromatogr.*, 1982, **5**, 1241.
[159] V. V. Nesterov, Ye. V. Chubarova, and L. Z. Vilenchik, *Vysokomol. Soedin, Ser. A*, 1981, **23**, 463.
[160] M. Szesztay and F. Tüdös, *Polym. Bull.*, 1981, **5**, 429.
[161] U. Hoechst, *Eur. Polym. J.*, 1982, **18**, 273.

packings is a new series of polystyrene/divinyl benzene microparticulates which have good mechanical stability and low solvent–gel interactions.[162] Silica gels, modified by reaction with octadecyltrichlorosilane, which was meant to suppress adsorption effects, have been prepared by Qi-Jian et al.,[163] but it is difficult to penetrate and modify the internal pores in the silica, with the result that while the larger molecules were not adsorbed the smaller molecules tended to suffer adsorption and had longer retention times than expected. Silica modified by cyanoethyl groups has proved successful in the analysis of random copolymers of styrene and acrylonitrile.[164] Very short columns could be used which reduced spreading and tedious cross fractionation techniques could be avoided.

High-performance GPC using short columns and microparticulate packings is receiving close attention. Microsphere silica packings,[165] with a mean particle size of only 10 μm, give accurate results in 75 cm columns if conditions are correctly selected. The short analysis time simplifies the calculations and broad distribution samples are easily handled.

Dawkins and Yeadon[166] have used similar microparticulate silica columns and have derived a simple expression relating chromatogram broadening to solute mass transfer. Their procedure allows one to estimate sample polydispersity from experimental plate-height data.

Attempts to correlate the average network size of a gel with the maximum polymer size which can permeate through, leads to a linear relation between network size and the size of the molecule according to Hirayama et al.[167] The effect of varying the internal pore volume in gels based on glycidyl methacrylate–ethylene dimethacrylate copolymers has been examined and the decrease in pore volume can be directly related to the decrease in retention volumes in this system.[168] A comparison of elution volumes using two different column packings was made for polystyrene and poly(ethylene glycol),[169] where the solvents were benzene and tetrahydrofuran. For crosslinked polystyrene gels the results were identical for both polymers, but when porous glass was used the elution volume was smaller for polystyrene. This indicates that the poly(ethylene glycol) was adsorbed more strongly on the glass packing than the polystyrene but an increase in temperature reduced this effect for both polymers.

As mentioned in the previous section, Burns [156] has analysed polymer mixtures and commented on the behavioural change in these systems. Kok and Rudin[170] have demonstrated that the reported effects can be accounted for using Rudin's simple model (see Section 9) in which the reduction in the hydrodynamic volumes of the polymer coils can explain the observed concentration effects. No polymer–polymer interactions need be invoked and a quantitative prediction of peak shifts has been achieved.

[162] R. Wernicke and F. Eisenbeiss, *Chromatography*, 1982, **15**, 347.
[163] H. Qi-Jian, L. Xiu-Zhen, and D.Y. Meng, *J. Liq. Chromatogr.*, 1982, **5**, 1321.
[164] M. Danielewicz, M. Kubin, and S. Vozka, *J. Appl. Polym. Sci.*, 1982, **27**, 3629.
[165] O. Chiantore, S. Pokorny, M. Bleha, and J. Janča, *Angew. Makromol. Chem.*, 1982, **105**, 61.
[166] J. V. Dawkins and G. Yeadon, *J. Chromatogr.*, 1981, **206**, 215.
[167] C. Hirayama, M. Iida, and Y. Motozato, *Polymer*, 1981, **22**, 1561.
[168] J. Lukas, M. Bleha, E. Votavova, F. Svec, and J. Kalal, *J. Chromatogr.*, 1981, **210**, 255.
[169] K. Nakamura and R. Endo, *J. Appl. Polym. Sci.*, 1981, **26**, 2657.
[170] C. M. Kok and A. Rudin, *Makromol. Chem.*, 1981, **182**, 2801.

The use of very small columns has been successfully applied by Kever et al.[171] to polystyrene in the range 10^3 to 2×10^6 g mol^{-1}.

Data Analysis and Detectors.—The importance of accurate data analysis is stressed by several authors. Busnel[172] has discussed the points requiring particular attention for the attainment of accurate results, such as precise elution volume control, careful selection of column sets, and good calibration procedures. Tchir et al.[173] have used computer simulation of GPC data to gauge the effects of data and baseline uncertainties, and concluded that imprecise baselines present the most serious problem. Imperfect resolution, when non-linear calibration curves are used, can be corrected for, using general analytical solutions described by Hamielec et al.[174] who have also considered the corrections necessary when on-line detectors are used.

The extension of GPC to systems for which universal calibration techniques may not be valid, again presents difficulties for accurate data acquisition. Averaging methods using the π-theorem have been outlined by Kalfus[175–177] and, while an analytical solution to the equations derived is not possible, an iterative procedure can be used with some effect.

The actual processing of data has been dealt with by Navas,[178] who describes the use of a pocket calculator for this purpose; Mukherji and Ishler,[179] who have interfaced a 3 K microcomputer with a Waters 200 GPC, and Burns et al.,[180] who have carried out a similar operation with an Apple II and describe the necessary hardware and software.

Detector systems are constantly under review, particularly with the move towards copolymer analysis as an extension of the GPC method. Here, compositional differences can present problems for detection by refractive index and ultraviolet systems and Elsdon et al.[181] have assessed a densimeter as an alternative detector. The main drawback appears to be lack of sensitivity which can be partially offset by injecting samples with a higher than normal concentration. It should be recognized, however, that this too can lead to spurious results, especially when the molar mass of the sample is high. Low angle laser light scattering (LALLS) has proved a most useful addition to the arsenal of detection systems and has been successful in detecting microgel in ethylene–propylene copolymers which can cause problems in analysis.[182] The method has also been employed to monitor the influence of flow rate on the elution of polystyrenes with $M > 10^6$, and shear degradation at flow rates of 0.5 to 1.5 cm^3 min^{-1} has been detected for $M \geqslant 3 \times 10^6$ when microparticulate silica was used as column packing.[183] An article on the

---

[171] J. J. Kever, B. G. Belenkii, E. S. Gankina, L. Z. Vilenchik, O. I. Kurenbin, and T. P. Zhmakina, *J. Chromatogr.*, 1981, **207**, 145.
[172] J. P. Busnel, *Polymer*, 1982, **23**, 137.
[173] W. J. Tchir, A. Rudin, and C. A. Fyfe, *J. Polym. Sci., Polym. Phys. Ed.*, 1982, **20**, 1443.
[174] A. E. Hamielec, H. J. Ederer, and K. H. Ebert, *J. Liq. Chromatogr.*, 1981, **4**, 1697.
[175] M. Kalfus, *Acta Polym.*, 1981, **32**, 479.
[176] M. Kalfus, *Acta Polym.*, 1981, **32**, 648.
[177] M. Kalfus, *Acta Polym.*, 1981, **32**, 695.
[178] A. A. Navas, *J. Liq. Chromatogr.*, 1982, **5**, 413.
[179] A. K. Mukherji and J. M. Isher, *J. Liq. Chromatogr.*, 1981, **4**, 71.
[180] V. Narasimhan, A. R. Telfer, R. Y. M. Huang, and C. M. Burns, *J. Appl. Polym. Sci.*, 1982, **27**, 3461.
[181] W. L. Elsdon, J. M. Goldwasser, and A. Rudin, *J. Polym. Sci., Polym. Chem. Ed.*, 1982, **20**, 3271.
[182] B. J. R. Scholtens and T. L. Welzen, *Makromol. Chem.*, 1981, **182**, 269.
[183] W. G. Rand and A. K. Mukherji, *J. Polym. Sci., Polym. Lett. Ed.*, 1982, **20**, 501.

Characterization of Synthetic Polymers

benefit of using LALLS for measuring molar mass distributions directly comes from Martin,[184] which is a most helpful introduction to the principles involved. Extension to aqueous systems is both inevitable and desirable and a combined DRI/LALLS detector system, as applied to aqueous dextran solutions,[185] has been described. Examination of M_n from a GPC/LALLS combination showed that the value measured was higher than the actual one and a correction had to be applied,[186] but on balance it is an advantageous system to use.

The extension of the technique to provide automated on-line polymer analysis on an industrial scale has been shown to be a feasible proposition and is likely to become of greater use with the development of remote sampling techniques to deal with viscous samples, which require dilution prior to pumping some distance to the analyser.[187]

Non-aqueous Systems.—A large number of specific polymer systems have been subjected to GPC and HPLC analysis; these include oligomers of epichlorohydrin,[188] phenolformaldehyde,[189] urea–formaldehyde,[190] poly(ethylene glycols),[191] and styrene.[192] Extremely high molar mass polystyrenes have been characterized by Soltes et al.,[193] while use has been made of pyridine and N-methyl pyrrolidone as solvents for polystyrene and poly(2-vinylpyridine).[194] The application of GPC to study the effect of solvent dielectric constant on methacrylate polymerization has been reported by Boudevska et al.[195] A mixed solvent system of tetrahydrofuran/butyl chloride proved effective for poly(methyl methacrylate) as reported by Hattori and co-workers,[196] and cyclohexanone/1,2 dichloroethane mixtures have been used with this polymer and polystyrene in an attempt to reduce adsorption effects, but with only limited success.[197] Polyolefins,[198] poly(ethylene terephthalate)[199] in which the cyclic trimer content could also be studied, various rubbers,[200] and phenol–formaldehyde resins[201] have all been subjected to GPC analysis.

HPLC and GPC have also been effective in separating hydroxy-terminated methacrylate polymers based on the number of functional groups per molecule[202] and hydroxy terminated polyisoprenes.[203] The separation was unaffected by absorption in the latter system, but it was found that mono- and di-functional OH-terminated polyisoprenes could be separated from each other using silica gel

[184] M. Martin, *Chromatography*, 1982, **15**, 426.
[185] C. J. Kim, A. E. Hamielec, and A. Benedek, *J. Liq. Chromatogr.*, 1982, **5**, 425.
[186] Zhi-Duan He, Kian-Chi Zhang, and Rong-Shi Cheng, *J. Liq. Chromatogr.*, 1982, **5**, 1209.
[187] E. N. Fuller, G. T. Porter, and L. B. Roof, *J. Chromatogr. Sci.*, 1982, **20**, 120.
[188] A. I. Kuzaev and O. M. Ol'Khova, *Vyskomol. Seodin., Ser. A*, 1982, **24**, 2197.
[189] M. Cornia, G. Sartori, G. Casnati, and G. Casiraghi, *J. Liq. Chromatogr.*, 1981, **4**, 13.
[190] S. Katuocak, M. Tomas, and O. Schressl, *J. Appl. Polym. Sci.*, 1981, **26**, 381.
[191] R. Murphy, A. C. Selden, M. Fisher, E. A. Fagan, and V. S. Chadwick, *J. Chromatogr.*, 1981, **211**, 160.
[192] H. Sato, K. Saito, K. Miyashita, and T. Tanaka, *Makromol. Chem.*, 1982, **182**, 2259.
[193] L. Soltes, D. Berek, and D. Mikulasova, *Colloid Polym. Sci.*, 1980, **258**, 702.
[194] W. G. Rand and A. K. Mukherji, *J. Chromatogr. Sci.*, 1982, **20**, 182.
[195] H. Boudevska, C. Brutchkov, S. Platchkova, and J.-P. Pascault, *Makromol. Chem.*, 1981, **182**, 3257.
[196] S. Hattori, H. Nakahara, and T. Kamata, *Kobunshi Ronbunshu*, 1981, **38**, 555.
[197] T. Spychaj and D. Berek, *Acta Polym.*, 1982, **33**, 477.
[198] J. L. Vidal, P. Crouzet, and A. Martens, J. Chromatogr. Sci., 1982, **20**, 252.
[199] S. A. Jabarin and D. C. Balduff, *J. Liq. Chromatogr.*, 1982, **5**, 1825.
[200] J. Chih-An Hu, *J. Chromatogr. Sci.*, 1981, **19**, 634.
[201] H. Much and H. Pasch, *Acta Polym.*, 1982, **33**, 366.
[202] G. D. Andrews and A. Vatvars, *Macromolecules*, 1981, **14**, 1603.
[203] S. Pokorny, J. Janča, L. Mrkvickova, O. Turickova, and J. Trekoval, *J. Liq. Chromatogr.*, 1981, **4**, 1.

modified with $-NH_2$ groups and mixtures of hexane/dichloroethane as the mobile phase.

Ethylene–propylene copolymers have been studied by Ivan et al.,[204] who showed that, contrary to previous reports, the composition of the fractions was not molar mass dependent. The Mark–Houwink parameters were established[205] for ethylene–propylene copolymers of variable composition in o-dichlorobenzene using the universal calibration curve, but this required a trial and error method eventually.

It has been demonstrated that sequence length distribution in SBR can be established by analysis of the ozonolysis products in which styrene and styrene 1,2-sequences could be identified.[206]

Aqueous Systems.—Solute–gel interactions are a particularly bad problem in aqueous GPC, and if polyelectrolytes are to be studied, electrostatic adsorption effects are also present and require correction. Use of specially treated columns can minimize the errors and a number of these are currently available. Polycations such as poly(ethylene imine), poly(vinyl acetamide), and poly(vinyl amine) can be studied with PW(Toyo Soda) hydrophilic columns and a 0.2 M NaCl mobile phase.[207] Aqueous acetic acid containing various salts as the mobile phase has been used for dextrans and various polyanions with some success.[208] A three-detector system (i.r., viscometry, conductimetric) has been found necessary when dealing with aqueous polyelectrolytes using salt solutions to screen the electrostatic forces.[209] Also studied have been dextran-g-polyacrylamides,[210] starch hydrolysates,[211] and poly(4-vinyl-benzyl trimethyl ammonium chloride).[212]

A set of operating conditions to reduce peak broadening and improve resolution in aqueous systems has been established by Omoordian et al.,[213] using samples of poly(sodium styrene sulphonates). Porous glass packing was found to be a suitable stationary phase at pH 6, and an ionic strength of 0.05 was sufficient for electrostatic screening. It was also found that addition of a small amount of a neutral surfactant reduced absorption and improved resolution.

Branching.—GPC is an effective means of studying branching in polymers; natural rubber,[214] polycarbonate,[215] and poly(vinyl acetate)[216] have all been characterized in this way. Short chain branching in polyethylene was analysed accurately using pyrolysis-hydrogenation glass capillary chromatography and the description of how to achieve high resolution thermograms is also given.[217]

---

[204] B. Ivan, Z. Laszlo-Hedvig, T. Kelen, and F. Tüdös, *Polym. Bull.*, 1982, **8**, 311.
[205] Ke Qiang Wang, Shi-Yu Zhang, Jia Xu, and Yang Li, *J. Liq. Chromatogr.*, 1982, **5**, 1899.
[206] Y. Tanaka, H. Sato, and Y. Nakafutami, *Polymer*, 1981, **22**, 1721.
[207] P. L. Dubin and I. J. Levy, *J. Chromatogr.*, 1982, **235**, 377.
[208] Y. Kato and T. Hashimoto, *J. Chromatogr.*, 1982, **235**, 539.
[209] J. Desbrieres, J. Mazet, and M. Rinaudo, *Eur. Polym. J.*, 1982, **18**, 269.
[210] C. L. McCormick and L. S. Park, *J. Appl. Polym. Sci.*, 1981, **26**, 1705.
[211] J. E. Rollings, M. R. Okos, G. T. Tsao, and A. Bose, *J. Appl. Polym. Sci.*, 1982, **27**, 2281.
[212] Y. Higo, Y. Kato, M. Itoh, N. Kozuka, I. Noda, and M. Nagasawa, *Polym. J.*, 1982, **14**, 809.
[213] S. N. E. Omoordian, A. E. Hamielec, and J. L. Brash, *J. Liq. Chromatogr.*, 1981, **4**, 1903.
[214] J. L. Angulo-Sanchez and P. Caballero-Mata, *Rubber Chem. Tech.*, 1981, **54**, 34.
[215] Z. Dobkowski and J. Brzezinski, *Eur. Polym. J.*, 1981, **17**, 537.
[216] S. H. Agarwal, R. F. Jenkins, and R. S. Porter, *J. Appl. Polym. Sci.*, 1982, **27**, 113.
[217] Y. Sugimura, T. Usami, T. Nagaya, and S. Tsuge, *Macromolecules*, 1981, **14**, 1787.

Miscellaneous.—A number of specific techniques have been developed which extends the characterization capabilities. Greschner[218] has continued his study of phase distribution chromatography in which the separation effect is opposite to that in GPC, and a brief review of the application of thin-layer chromatography to polymer polydispersity studies has appeared.[219] A method which looks useful for semicrystalline polymers when combined with GPC is called the 'analytical temperature rising elution technique'. Nakano and Goto[220] have reported on the instrumental design and data handling for the method which analyses the relation between M and crystallizability using an automatic cross-fractionation procedure. Polyethylene was used as the test polymer. A report on the modification of HPLC for use with a supercritical mobile phase and the application to the separation of styrene oligomers has been made by Schmitz and Klesper,[221] and a means of chromatographic separation of macromolecules by centrifugation through a gel medium has been described by Andersen and Vaughan.[222]

The problems associated with trying to establish the compositional heterogeneity of random copolymers by GPC has been discussed by Mori and Suzuki[223] for u.v.–i.r. dual detector systems and styrene–methylmethacrylate copolymers. This problem is better dealt with by gradient column adsorption chromatography according to a report by Danielewicz and Kubin[224] who have used silica columns for styrene–acrylic copolymer analysis.

The application of GPC to the determination of intrinsic viscosity,[225] polymer coil size,[226] and reactivity ratios[227] in a copolymerization have all been discussed.

---

[218] G. S. Greschner, *Makromol. Chem.*, 1981, **182**, 2845.
[219] E. S. Gankina and B. G. Belenkii, *J. Liq. Chromatogr.*, 1982, **5**, 1509.
[220] S. Nakano and Y. Goto, *J. Appl. Polym. Sci.*, 1981, **26**, 4217.
[221] F. P. Schnitz and E. Klesper, *Polym. Bull.*, 1981, **5**, 603.
[222] K. B. Andersen and M. H. Vaughan, *J. Chromatogr.*, 1982, **240**, 1.
[223] S. Mori and T. Suzuki, *J. Liq. Chromatogr.*, 1981, **4**, 1685.
[224] M. Danielewicz and M. Kubin, *J. Appl. Polym. Sci.*, 1981, **26**, 951.
[225] A. K. Mukherji, *J. Liq. Chromatogr.*, 1981, **4**, 741.
[226] P. C. Squire, *J. Chromatogr.*, 1981, **210**, 433.
[227] A. Revillon and T. Hamaide, *Polym. Bull.*, 1982, **6**, 235.

12
Thermodynamics of Solutions and Mixtures

BY J. W. KENNEDY

1 Introduction

Scope of Report.—This Report covers the four-year period 1979/1982 since the previous Report.[1] Because of the rapidly increasing number of publications in this topic, as elsewhere in science, it has only been possible to remain within reasonable space constraints by adopting a policy of highlighting *new* activities concerned with thermodynamic phenomena that result from mixing polymeric systems and the methods of measurement and analysis of parameters associated with the phenomena. Material covered previously is not repeated here and, instead, citations to the earlier Report[1] are employed to give access to relevant background material. Thermodynamic aspects of processes other than mixing are once again only mentioned in as much as they relate to information about mixing polymeric systems. These processes are covered in more detail elsewhere in this volume and cross-references are given, where appropriate, by citing chamber number. This especially relates to crystalline and related morphology of polymers (Chapter 10), to glass transitions and to polymer chain dimensions (Chapter 7). The previous volume[2] did not contain a chapter on thermodynamics of mixing but did cover some of the ground under polymer engineering in so far as the processing and performance of polymers was concerned. This is certainly worth consulting. A worthwhile supplement to the material in this Report can be found in ref. 3, where Barker and Henderson offer a very useful overview of the present state of molecular dynamics as it applies to fluid phases of matter.

Importance and General Overview.—The usefulness of thermodynamic information on polymer solutions and mixtures was discussed previously.[1] This has been further emphasized by activities covered here which have pushed our examination of systems and their phase behaviour to new areas. Stiff-chain polymers exhibiting liquid crystal phase behaviour and phase separation in gels and in systems during polymerization represent important new areas of progress. The combined demands of technology and economics have encouraged the examination of ever more polymer mixtures to find systems that can be blended in useful ways as an alternative to devising new and expensive polymers.

---

[1] J. W. Kennedy in 'Macromolecular Chemistry', ed. A. D. Jenkins and J. F. Kennedy (Specialist Periodical Reports), The Royal Society of Chemistry, London, 1980, Vol. 1, p. 296.
[2] 'Macromolecular Chemistry', ed. A. D. Jenkins and J. F. Kennedy (Specialist Periodical Reports), The Royal Society of Chemistry, London, 1981, Vol. 2.
[3] J. A. Barker and D. Henderson, *Sci. Am.*, 1981, **245**(5), 94.

Thermodynamics of Solutions and Mixtures 249

This Report has been organized like the first[1] with three main sections covering progress in theoretical aspects of the subject, methods for obtaining thermodynamic data, and information concerning particular types of polymer systems. The final section again concludes with a source table for new data on individual polymer systems which it is hoped is as complete as possible. However, this Reporter would be pleased to receive notices of any omissions and of future work that should be included in future Reports.

2 Theoretical Aspects

Introduction.—Theoretical aspects of the subject were outlined previously[1] with a more extensive discussion of statistical mechanical models for polymer solutions and mixtures which have been useful in the practical interpretation of thermodynamic data. The review by Casassa[4] offers further substance to the remarks in ref. 1. He gives special attention to phase equilibria in polymer systems and details theoretical models and thermodynamic relationships that are useful in the practical study of polymer fractionation.[5] Casassa, in general, concludes that, despite their deficiencies, models derived from that of Flory–Huggins (lattice-graph models) (see below) are perhaps the most useful in a practical sense although one must generally abandon attempts to offer detailed physical interpretations to the parameters in these models. From a mainly theoretical point of view, Kurata[6] has surveyed polymer solutions and Solc[7] has edited a survey of polymer mixtures.

Phase Behaviour.—Binder *et al.*[8] have discussed the kinetics of phase separation with some emphasis on the general theoretical aspects of spinodal decomposition. Polymer blends in thin films[9] show that phase separation occurs by spinodal decomposition, apparently with decreasing dimensionality as film thickness is reduced. One theoretical consequence of this is an increase in phase separation temperature.

Phase separation in flowing polymer solutions[10] is of interest while microphase structures that have been observed in block copolymers (see Section 4) are presently receiving considerable attention.

Liquid–Liquid Phase Separation. Charlet *et al.*[11,12] have extensively studied lower critical solution behaviour of olefin polymers in hydrocarbon solvents at high temperatures. Phase diagrams for four new ternary (polymer + solvent + solvent) systems have been presented by Staikos *et al.*[13] Blends of poly(methyl methacrylate)

---

[4] E. F. Casassa, in ref. 5, p. 2.
[5] L. H. Tung, ed., 'Fractionation of Synthetic Polymers (Principles and Practices)', Dekker, New York, 1977.
[6] M. Kurata, 'Thermodynamics of Polymer Solutions', MMI Press, Midland, Michigan, 1982.
[7] K. Solc, ed., 'Polymer Compatibility and Incompatibility', MMI Press, Midland, Michigan, 1982.
[8] K. Binder, C. Billotet, and P. Mirold, *Zt. Phys. B*, 1978, **30**, 183.
[9] S. Reich and Y. Cohen, *J. Polym. Sci., Polym. Phys. Ed.*, 1981, **19**, 1255.
[10] B. A. Wolf and H. Kramer, *J. Polym. Sci., Polym. Lett. Ed.*, 1980, **18**, 789.
[11] G. Charlet and G. Delmas, *Polymer*, 1981, **22**, 1181.
[12] G. Charlet, R. Ducasse, and G. Delmas, *Polymer*, 1981, **22**, 1190.
[13] G. Staikos, P. Skondras, and A. Dondos, *Makromol. Chem.*, 1982, **183**, 603.

with chlorinated polyethylene exhibit both upper and lower critical solution behaviour[14] and offer an interesting system for theoretical analysis.

An interesting new development lies with the phase separation in polymer systems during polymerization, which has been studied experimentally (Section 4). Such systems are unlikely to be in thermodynamic equilibrium, rendering their theoretical analysis a more difficult task but one which has important industrial consequences. In some cases the polymerization process leads to network formation by one of the components and this poses additional interesting problems (cf. 'Phase Equilibria in Gel Polymer Systems', below).

Multiphase Separation. The critique of Tompa's mechanism for three phase separation in a ternary (solvent + polymer$_1$ + polymer$_2$) system made by Chermin (see ref. 39 of ref. 1) has been called to question by Solc.[15] In particular, Solc proves that the critical point in a simple, *i.e.*, concentration independent interaction function, Flory–Huggins system is thermodynamically unstable rather than metastable. He argues that three-phase separation in quasi-ternary systems cannot be inferred from methods advanced for the treatment of strictly ternary systems and offers instead a more general approach.

Phase Interfaces. Leiber[16,17] proposes a formalism for the interfacial properties of 'nearly compatible' mixtures of polymers. A general theoretical formalism for inhomogeneous multicomponent polymer systems[18–20] in which interfacial energy plays a dominant role has interesting consequences when applied to immiscible polymer systems to which are added block copolymers. The latter can behave similarly to soap molecules at an oil–water interface. Careful selection of the type and molecular weight of the block copolymer facilitates phase separation of incompatible polymers into uniformly dispersed microdomains.

Phase Equilibria in Gel Polymer Systems (see also 'Gel Swelling', p. 264). Dusek[21] has surveyed phenomena of phase separation (or syneresis) in gels produced during copolymerization in the presence of diluents. Physically, phase separation of two types is observed: macrosyneresis, in which the gel being crosslinked deswells to continuous liquid and gel phases; and microsyneresis, in which the gel and separated liquid phases form a dispersion. Dusek makes the point that crosslinking in the presence of an inert diluent is sometimes desirable and has practical consequences in the formation of networks that are porous in the dry or swollen state. He offers a good theoretical review of the thermodynamics of macrosyneresis accompanied by illustrative data for real systems. An important conclusion is that the transparency of network systems is not a sufficient guarantee for their

---

[14] D. J. Walsh, S. Lainghe, and C. Zhikuan, *Polymer*, 1981, **22**, 1005.
[15] K. Solc, *J. Polym. Sci., Polym. Phys. Ed.*, 1982, **20**, 1947.
[16] L. Leiber, *Macromolecules*, 1980, **13**, 1602.
[17] L. Leiber, *Macromolecules*, 1982, **15**, 1283.
[18] J. Noolandi and K. M. Hong, *Ferroelectrics*, 1980, **30**, 117.
[19] K. M. Hong and J. Noolandi, *Macromolecules*, 1981, **14**, 727.
[20] J. Noolandi and K. M. Hong, *Macromolecules*, 1982, **15**, 482.
[21] K. Dusek in 'Polymer Networks: Structural and Mechanical Properties', ed. A. J. Chompf, Plenum, New York, 1971, p. 245.

Thermodynamics of Solutions and Mixtures

homogeneity. Tanaka[22] has considered syneresis in some theoretical detail in terms of Flory mean-field theory and he illustrates this with numerous phase diagrams. Rietsch *et al.*[23] have extended these studies to networks contaminated with polymers and describe the effects due to molecular weight.

Syneresis during the formation of epoxy networks modified by elastomers[24] and in poly(ethylene oxide) hydrogels[25] have raised interesting questions. Interpenetrating polymer networks add further complexity to these phase behavioural phenomena that are under study.[26,27]

In preformed gels swollen with solvents, two polymer phases can coexist, differing in their chain conformation and concentration of polymer segments. This was predicted theoretically[28] and has now been demonstrated experimentally in polyacrylamide[29,30] and in poly(vinyl acetate)[31] gel networks.

Flory–Huggins (Lattice-Graph) Models.—Casassa[4] includes an extensive discussion of models of this type with special reference to phase equilibria measurements and polymer fractionation, which adds useful detail to the discussion in ref. 1.

In the lattice-gas derivatives of lattice-graph models,[1] vacant lattice sites or holes of varying concentrations are introduced as a means of quantitatively dealing with effects of pressure. Koningsveld and Kleintjens have explored these models in more detail, especially for small molecule systems[32] where they are able to deal with supercritical phase behaviour[33] and to deduce equations of state from phase behaviour.[34] They have also used lattice-gas models to good effect in dealing with mixtures of *n*-alkanes with polyethylene.[35]

As was noted earlier,[1] the free energy of mixing for Flory–Huggins models can easily be constructed for ternary (or quasi-ternary) systems. The function then contains a ternary interaction term (g_{123}) that describes the deviations of the system from being a strict superposition of binary systems. For many cases the ternary term may be omitted. Pouchly and Zivny,[36] however, have had to include such a term when analysing data for preferential sorption in 22 polymer + solvent + solvent systems. They conclude that all interaction functions have to depend on temperature and concentration but that the theoretical problem of preferential sorption needs to be more rigorously treated. Galin and Rupprecht[37] likewise conclude that their results on interactions in di- and tri-block copolymer solutions in general fall within the scope of classical Flory–Huggins (random lattice-graph)

---

[22] T. Tanaka, *Polymer*, 1979, **20**, 1404.
[23] F. Rietsch, F. Dambrine, and J. Morcellet, *Makromol. Chem.*, 1981, **182**, 2087.
[24] T. T. Wang and H. M. Zupko, *J. Appl. Polym. Sci.*, 1981, **26**, 2391.
[25] N. B. Graham, N. E. Nwachuka, and D. J. Walsh, *Polymer*, 1982, **23**, 1345.
[26] J. Michel, S. C. Hargest, and L. H. Sperling, *J. Appl. Polym. Sci.*, 1981, **26**, 743.
[27] J. M. Widmaier and L. H. Sperling, *Macromolecules*, 1982, **15**, 625.
[28] A. R. Kohohlov, *Polymer*, 1980, **21**, 376.
[29] M. Havsky, *Polymer*, 1981, **22**, 1687.
[30] M. Ilavsky, *Macromolecules*, 1982, **15**, 782.
[31] M. Zrinyi, T. Molnar, and E. Horvath, *Polymer*, 1981, **22**, 429.
[32] L. A. Kleintjens and R. Koningsveld, *Sep. Sci. Technol.*, 1982, **17**, 215.
[33] L. A. Kleintjens and R. Koningsveld, *J. Electrochem. Soc.*, 1980, **127**, 2352.
[34] L. A. Kleintjens and R. Koningsveld, *E.F.C.E. Publ. Ser.*, 1980, **11**, 24.
[35] L. A. Kleintjens and R. Koningsveld, *Colloid Polym. Sci.*, 1980, **258**, 711.
[36] J. Pouchly and A. Zivny, *Makromol. Chem.*, 1982, **183**, 3019.
[37] M. Galin and M. C. Rupprecht, *Macromolecules*, 1979, **12**, 506.

models with three interaction terms (g_{12}, g_{13}, g_{123}). Altena and Smolders[38] detail calculation of binodals for liquid–liquid phase separation in a polymer + solvent + non-solvent mixture with good agreement between Flory–Huggins theory and experiment for cellulose acetate. Aminabhavi and Munk,[39] on the other hand, neglect the ternary term in polymer + solvent$_1$ + solvent$_2$ systems and apply their results to obtain the binary interaction function g_{12} which they interpret in terms of preferential solvent adsorption.

Fujita and Teramoto[40] have derived exact expressions for spinodal, critical conditions and separation factor for a ternary system comprising solvent and two polymer homologues. The interaction function depends separately on the two polymer concentrations rather than on the total polymer concentration as is more often used in quasi-ternary systems. Further work on the separation factor for this case is given in a later report,[41] where it is shown that theoretical prediction of separation factors requires very accurate knowledge about interaction parameters as a function of individual polymer concentrations.

Wolf and Schuch[42] have explored the effect of end groups on phase separation in oligomer mixtures and have contrasted the predictions of Flory–Huggins models with the conclusions that can be drawn from solubility parameter theory (see p. 258).

Koningsveld et al.[43] have discussed the compatibility of polymer mixtures based on new phase volume ratio experiments they have been able to perform. They find deviations from simple lattice-graph theory and discuss possible molecular origins of these.

Equation-of-state Models.—An equation-of-state model (*cf.* ref. 1) has been developed[11,12] and applied to interpret lower critical solution behaviour of polyolefins in hydrocarbons with some success.

Perturbed Hard Chain Theory.—Beret and Prausnitz[44] introduced a model that differs only in detail from those of Flory and Prigogine (see refs. 67 and 76 in ref. 1). This model is summarized by Liu and Prausnitz,[45] who have applied the ideas to a theoretical study of polymer compatibility. They find important effects result from what they term component contact agility, c/q (where c is one-third of the number of degrees of freedom per molecule and q is proportional to the surface area of a molecule). Lower critical solution behaviour is exhibited when the contact agility for polymers are different even though all energy parameters are the same. For quasi-ternary (polymer–polymer–solvent) systems the effects of solvent contact agility are predicted to differ in the upper critical and lower critical solution regions.

---

[38] F. W. Altena and C. A. Smolders, *Macromolecules*, 1982, **15**, 1491.
[39] T. M. Aminabhavi and P. Munk, *Macromolecules*, 1979, **12**, 607.
[40] H. Fujita and A. Teramoto, *J. Polym. Sci., Polym. Phys. Ed.*, 1982, **20**, 893.
[41] J. Hashizume, A. Teramoto, and H. Fujita, *J. Polym. Sci., Polym. Phys. Ed.*, 1981, **19**, 1405.
[42] B. A. Wolf and W. Schuch, *Makromol. Chem.*, 1981, **182**, 1801.
[43] R. Koningsveld, L. A. Kleintjens, and M. H. Onclin, *J. Macromol. Sci., Phys.*, 1980, **18**, 363.
[44] S. Beret and J. M. Prausnitz, *AIChE J.*, 1975, **21**, 1123.
[45] D. D. Liu and J. M. Prausnitz, *Macromolecules*, 1979, **12**, 454.

Huggins Molecular Models.—Until right up to his death in December 1981, Huggins[46—50] continued to develop his ideas for re-expressing phenomenological parameters of polymer solution theory in terms of molecular quantities that can be independently measured (see ref. 1). It is to be hoped that the challenge of this ambitious project will be taken up by others.

A scheme for modelling polymer solutions which is closely related to the ideas applied by Huggins to small molecules was developed by Prausnitz and others[51,52] under the name UNIFAC (universal quais-chemical functional groups activity coefficients). This has been further explored by Ratzsch et al.,[53] who have compared its predictions with a variety of experimental data.[54,55]

Renormalization Group and Scaling Models.—The study of renormalization group and scaling models has been widespread and continues to be controversial, provoking views about its utility that cover extremes. The excellent and highly readable introduction to renormalization group scaling theory by Wilson[56] points out that the renormalization group was introduced for dealing with problems that have multiple scales of length. It is not a descriptive theory of nature, but a general method of constructing theories that can be applied to various types of critical phenomena where fluctuations become extreme. A more detailed introduction to the topic is offered by Pfeuty and Tolouse;[57] for a comprehensive treatment, see the book by de Gennes.[58]

De Gennes[58] calims that scaling theory has given polymer scientists a new perspective on interpreting measurements made in semi-dilute or moderately concentrated polymer solutions predicting both the static and dynamic properties of molecules. Less convinced are Gordon et al.[59—63] who levy heavy criticism at the use of scaling theories since 'true' critical exponents cannot in principle be checked against experiment. They prefer mean-field theories that fit phase diagrams more generally and not merely at the critical point. Gordon and Irvine[59] comment that 'perhaps the only safe hypothesis of universality is that, without exception, each case must be examined on its merits'.

---

[46] M. L. Huggins, *Contemp. Top. Polym. Sci.*, 1978, **1**, 99.
[47] M. L. Huggins and J. W. Kennedy, *Polym. J.*, 1979, **11**, 315.
[48] M. L. Huggins, *Colloid Polym. Sci.*, 1980, **258**, 477.
[49] M. L. Huggins, *J. Macromol. Sci., Phys.*, 1980, **18**, 403.
[50] M. L. Huggins, *ACS Abstract*, New York, 1981, Sept.
[51] A. Fredenslund, R. S. Jones, and J. M. Prausnitz, *AIChE J.*, 1975, **21**, 1086.
[52] T. Oishi and J. M. Prausnitz, *Ind. Eng. Chem., Process Des. Dev.*, 1978, **17**, 333.
[53] M. Ratzsch and D. Glindemann, *Acta Polym.*, 1979, **30**, 57.
[54] M. T. Ratzsch, M. Opel, and Ch. Wohlfarth, *Acta Polym.*, 1980, **31**, 217.
[55] M. T. Ratzsch, D. Glindemann, and E. Hamann, *Acta Polym.*, 1980, **31**, 377.
[56] K. G. Wilson, *Sci. Am.*, 1979, **241**(2), 140.
[57] P. Pfeuty and G. Toulouse, 'Introduction to the Renormalisation Group and to Critical Phenomena', Wiley, New York, 1977.
[58] P. G. de Gennes, 'Scaling Concepts in Polymer Physics', Cornel University Press, Ithaca, New York, 1979.
[59] M. Gordon and P. Irvine, *Polymer*, 1979, **20**, 1450.
[60] M. Gordon and J. A. Torkington, *Ferroelectrics*, 1980, **30**, 237.
[61] P. Irvine and M. Gordon, *Proc. R. Soc. London, Ser. A*, 1981, **375**, 397.
[62] M. Gordon and J. A. Torkington, *Pure Appl. Chem.*, 1981, **53**, 1461.
[63] R. G. Cowell, M. Gordon, and P. Kapadia, *Polym. Prepr. (Jpn.)*, 1982, **31**, 29.

Chu and Nose[64,65] take an intermediate view. A general scaling theory for the thermodynamics of polymer solutions at finite concentrations,[66] obtained without recourse to renormalization group methods or analogies with magnets or critical phenomena, represents a fresh approach and major advance to the problem of concentrated polymer solution behaviour. However, the relations are derived from asymptotic analysis of long chains in the immediate neighbourhood of a critical point not easily accessible to experiment. Elsewhere, deviations from simple scaling relations are to be expected. They give experimental evidence to indicate the approximate nature of scaling theories under realistic experimental conditions. Akcasu and Han[67] tend to agree. They have experimentally studied the dependence of radius of gyration (R_G) and hydrodynamic radius (R_H) of polystyrene in various solvents. R_G is a static property and R_H a dynamic property of polymer coils but, for both, scaling arguments predict a power law dependence on the statistical chain length N and in the asymptotic limit of large N. They conclude that R_H cannot be represented by a simple power law $R_H \sim N^{v'}$ in the range of N accessible to experiment. Most of their data do satisfy $R_G \sim N^v$; however, the data requires that $v' < v$, whereas scaling theory predicts that $v' = v$ asymptotically. Vidakovic et al.[68] arrive at similar conclusions.

A prediction of de Gennes[69] that in good solvents and in the semi-dilute concentration range the sedimentation coefficient (s) should be independent of molecular weight and vary as the $-\frac{1}{2}$ power of concentration (c), has been experimentally examined by Pouyet and Dayantis.[70] They find that while a power relationship does exist between s and c, the exponent does not fit the predicted value. Dobashi et al.[71,72] encounter similar difficulties in reconciling theoretical predictions with their experimental observations on coexistence curves for polystyrene in methyl cyclohexane. Schafer[73] is likewise unable to reconcile scaling theories with observed polydispersity dependences of osmotic pressure and second virial coefficient for shorter chains.

Des Cloizeaux and Noda[74] show that theoretical predictions from renormalization theory agree with experiment for osmotic pressure of long-chain polymers with one exception, the ratio of third to the square of the second virial coefficient. This discrepancy they ascribe to polydispersity and to the fact that the polymer chain is of finite length. They conclude that 'more work is needed to obtain a really precise description of the universal function that describes the asymptotic behaviour of osmotic pressure'. Amirzadeh and McDonnell[75] examine the excess chemical potential of 'monodisperse' polystyrene in toluene and show that it asymptotically follows the predictions of scaling theory, describing their data more satisfactorily than Flory–Huggins type models in which the interaction function is concentration

---

[64] B. Chu and T. Nose, *Macromolecules*, 1979, **12**, 347.
[65] T. Nose and B. Chu, *Macromolecules*, 1979, **12**, 590.
[66] M. K. Kosmas and K. F. Freed, *J. Chem. Phys.*, 1978, **69**, 3647.
[67] A. Z. Akcasu and C. C. Han, *Macromolecules*, 1979, **12**, 276.
[68] P. Vidakovic, C. Allain and F. Rondelez, *Macromolecules*, 1982, **15**, 1571.
[69] P. G. de Gennes, *Macromolecules*, 1976, **9**, 587 and 594.
[70] G. Pouyet and J. Dayantis, *Macromolecules*, 1979, **12**, 293.
[71] T. Dobashi, M. Nakata, and M. Kaneko, *J. Chem. Phys.*, 1980, **72**, 6685.
[72] T. Dobashi, M. Nakata, and M. Kaneko, *J. Chem. Phys.*, 1980, **2**, 6692.
[73] L. Scafer, *Macromolecules*, 1982, **15**, 652.
[74] J. des Cloizeaux and I. Noda, *Macromolecules*, 1982, **15**, 1505.
[75] J. Amirzadeh and M. E. McDonnell, *Macromolecules*, 1982, **15**, 927.

dependent. Daoud–Jannink scaling theory[76] suggests that binodals in polymer solutions should obey a corresponding states principle. Dayantis[77] shows that an equivalent result follows for *both* binodals and spinodals from a Flory–Huggins model in which the interaction function is independent of concentration. Adler and Freed[78] have derived dynamic scaling relations for the diffusion coefficient, viscosity, and dynamic structure factor for polymer chains at infinite dilution.

Models for Stiff-chain Polymers.—Matheson and Flory[79] have added further detail to the statistical mechanics of stiff-chain polymers (see refs. 96 and 97 in ref. 1). The subject has been explored in depth in the large number of papers by Aharoni and various co-authors (see ref. 80). Aharoni notes[81] that the statistical mechanical model of Flory for stiff-chain polymers, when applied to a ternary system, solvent + polymer$_1$ + polymer$_2$, predicts that the phase behaviour depends on the rigidity of the polymers. When (1) is rigid and (2) is flexible, then the solution is isotropic at concentrations below θ_2^*, the critical concentration for appearance of an anisotropic phase remains. The formation of liquid crystal phases in solutions and exist. When both polymers are rigid two phases co-exist above θ_2^*, but at even higher polymer concentrations the isotropic phase disappears and a single anitotropic phase remains. The formation of liquid crystal phases in solutions and mixtures of stiff-chain polymers offers a very interesting development in phase behavioural studies,[82–92] including ternary stiff-polymer + flexible-polymer + solvent systems,[93] where there are useful consequences for high-strength fibre production.[94–97]

Interesting calculations of phase behaviour for model molecular weight distribution of stiff-chain polymers in solution have been made[98,99] and could usefully be applied to the variation in flexibility of amylose tricarbanilate with solvent and temperature observed by dielectric measurements.[100] Khokhlov and Semenov[101]

---

[76] M. Daoud and G. Jannink, *J. Phys. (Paris)*, 1976, **37**, 973.
[77] J. Dayantis, *Macromolecules*, 1982, **15**, 1107.
[78] R. S. Adler and K. F. Freed, *J. Chem. Phys.*, 1979, **70**, 3119.
[79] R. R. Matheson and P. J. Flory, *Macromolecules*, 1981, **14**, 945.
[80] S. M. Aharoni, *J. Macromol. Sci., Phys.*, 1982, **21**, 287.
[81] S. M. Aharoni, *Macromolecules*, 1979, **12**, 537.
[82] S. M. Aharoni and E. K. Walsh, *Macromolecules*, 1979, **12**, 271.
[83] J. S. Aspler and D. G. Gray, *Macromolecules*, 1979, **12**, 562.
[84] S. M. Aharoni, *Macromolecules*, 1979, **12**, 94.
[85] A. K. Gupta, H. Benoit, and E. Marchal, *Eur. Polym. J.*, 1979, **15**, 285.
[86] S. M. Aharoni, *Polymer*, 1980, **21**, 21.
[87] S. M. Aharoni, *Ferroelectrics*, 1980, **30**, 227.
[88] S. M. Aharoni, *J. Polym. Sci., Polym. Phys. Ed.*, 1980, **18**, 1303.
[89] S. M. Aharoni, *J. Polym. Sci., Polym. Phys. Ed.*, 1980, **18**, 1439.
[90] J. K. Moscicki, G. Williams, and S. M. Aharoni, *Polymer*, 1981, **22**, 571.
[91] J. K. Moscicki, G. Williams, and S. M. Aharoni, *Polymer*, 1981, **22**, 1361.
[92] J. K. Moscicki, G. Williams, and S. M. Aharoni, *Macromolecules*, 1982, **15**, 642.
[93] E. Bianchi, A. Ciferri, and A. Tealdi, *Macromolecules*, 1982, **15**, 1268.
[94] C. Balbi, E. Bianchi, A. Ciferri, and W. R. Krigbaum, *J. Polym. Sci., Polym. Phys. Ed.*, 1980, **18**, 2037.
[95] H. Chanzy and A. Pegny, *J. Polym. Sci., Polym. Phys. Ed.*, 1980, **18**, 1137.
[96] M. Takayanagi, T. Ogata, M. Morikawa, and T. Kai, *J. Macromol. Sci., Phys.*, 1980, **17**, 591.
[97] P. Navard, J. Haudin, S. Dayan, and P. Sizou, *J. Polym. Sci., Polym. Lett. Ed.*, 1981, **19**, 379.
[98] J. K. Moscicki and G. Williams, *Polym. Commun.*, 1981, **22**, 1451.
[99] J. K. Moscicki and G. Williams, *Polymer*, 1982, **23**, 558.
[100] A. K. Gupta, E. Marchal, W. Burchard, and B. Pfannemuller, *Macromolecules*, 1979, **12**, 281.
[101] A. R. Khokhlov and A. N. Semenov, *Macromolecules*, 1982, **15**, 1272.

have studied theoretical aspects of the influence of applied external fields on liquid crystal ordering in stiff-polymer solutions.

Dilute Solution Models (and Intermediate Concentrations).—Despite considerable theoretical effort on the second virial coefficient of polymer solutions (see ref. 1) there has been no satisfactory theory of A_2 even for a mixture of two monodisperse polymers differing only in molecular weight. Recent work[102,103] applying Pade approximants based on the first few coefficients of the perturbation expansion for monodisperse linear polymers has given good results on A_2 over a wide range of experimental data. Tanaka and Solc[104] have extended this approach to heterogeneous polymers with some success.

The theory of Manning[105,106] for polyelectrolyte solutions based on the Mayer[107] clustering theory of ionic solutions has been refined. The original work and most subsequent developments (see, *e.g.*, refs. 108 and 109) treat the polyion as being infinitely long and so neglect end effects. Thus recent efforts[100–113] have been directed at developing the detail so that the cluster integral approach applies to polyelectrolytes of finite molecular weight. The main experimental studies covering dilute solutions are based on intrinsic viscosity[114,115] or diffusion.[116]

There has been activity in the context of scaling models and bridging models that is relevant to the dilute solution regime and the topic relates closely to that of polymer chain dimensions.

Bridging Models (and Intermediate Concentrations).—Sanchez and Lohse[117] have considered the various attempts to 'bridge the thermodynamic gap' between theories for dilute and for concentrated polymer solutions (see ref. 1), including the later refinements in mean-field approaches,[118] scaling theories (see, *e.g.*, ref. 58), and self-consistent field theories,[119,120] for the intermediate region. None of these attempts has been completely satisfying, thus Sanchez and Lohse formulate a new cell model for polymer solutions which takes account of concentration inhomogeneities on size scales equal to or larger than the average size of a polymer chain. This generalized cell model is used to develop a new polymer solution theory that works well in both the dilute region and in concentrated solutions. However, it

---

[102] G. Tanaka, *J. Polym. Sci., Polym. Phys. Ed.*, 1979, **17**, 305.
[103] G. Tanaka, *Macromolecules*, 1980, **13**, 1513.
[104] G. Tanaka and K. Solc, *Macromolecules*, 1982, **15**, 791.
[105] G. S. Manning, *J. Chem. Phys.*, 1969, **51**, 924.
[106] G. S. Manning, *J. Chem. Phys.*, 1969, **51**, 3249.
[107] J. E. Meyer, *J. Chem. Phys.*, 1950, **18**, 1426.
[108] G. S. Manning, *Acc. Chem. Res.*, 1979, **12**, 443.
[109] M. Fixman, *J. Chem. Phys.*, 1979, **70**, 4995.
[110] M. T. Record and T. M. Lohman, *Biopolymers*, 1970, **17**, 159.
[111] E. L. Elson, I. E. Scheffler, and R. L. Baldwin, *J. Mol. Biol.*, 1970, **54**, 401.
[112] J. Skolnick and E. K. Grimmelmann, *Macromolecules*, 1980, **13**, 335.
[113] C. P. Woodbury and G. V. Ramanathan, *Macromolecules*, 1982, **15**, 82.
[114] C. M. Kok and A. Rudin, *J. Appl. Polym. Sci.*, 1981, **26**, 3583.
[115] C. M. Kok and A. Rudin, *J. Appl. Polym. Sci.*, 1982, **27**, 353.
[116] B. Chu and E. Gulari, *Macromolecules*, 1979, **12**, 445.
[117] I. C. Sanchez and D. J. Lohse, *Macromolecules*, 1981, **14**, 131.
[118] P. Irvine and M. Gordon, *Macromolecules*, 1980, **13**, 761.
[119] S. F. Edwards and P. Singh, *J. Chem. Soc., Faraday Trans. 2*, 1979, **75**, 1001.
[120] S. F. Edwards and E. F. Jeffers, *J. Chem. Soc., Faraday Trans. 2*, 1979, **75**, 1020.

Thermodynamics of Solutions and Mixtures

fails in the intermediate concentration range for reasons that are at least partly understood although correcting for them may be difficult.

Elsewhere, semi-dilute polymer solutions offer much of interest from both theoretical and experimental viewpoints, especially in relation to scaling models.[70,121] The semi-empirical analysis of Chee[122] suggests useful correlations between critical concentration and unperturbed dimensions of polymer coils which will be helpful in the construction of bridging models.

Pressure Dependence.—Kleintjens[123] has made a comprehensive theoretical and experimental study of the effects of pressure on the thermodynamic properties of polymer solutions. Several authors have examined pressure dependence of thermodynamic parameters[124–127] by viscometric methods (cf. 'Viscometric Methods', p. 262) in work that has an important role in relation to polymer additives to motor oils and mineral oil recovery.

Molecular Weight/Composition Dependence and Fractionation.—Truncation theorems for spinodal and critical point when the free energy of mixing for a polymer solution is expanded in terms of the moments of the polymer molecular weight distribution (see ref. 1) have now been proved.[61] The results have interesting practical consequences and have been used[128,129] to create efficient computational procedures for phase loci in polydisperse polymer solutions using modern matrix algebra techniques. Kleintjens[123,130] has studied the effects of chain branching on thermodynamic properties, including phase loci, of polymer solutions.

In ref. 1 the topic of polymer fractionation which results when a heterodisperse polymer solution separates into distinct phases was briefly discussed. Individual polymer species partition themselves between phases to an extent that depends on (a) their molecular weight and (b) their chemical composition (in the case of branched or copolymers). Theoretical treatment of the subject is straightforward in principle. Thermodynamic equilibrium is established when, for each species, its chemical potential is the same in every phase. Unfortunately, except for a few very special polymer (molecular weight/composition) distributions, no closed mathematical expressions are available to describe the situation and one is forced into approximate computational simulations or other devices (cf. remarks on cloud point curves in ref. 1). On the other hand, preparative polymer fractionation does offer more closely defined polymer samples which can be subjected to further experimental study. The collection of seven papers edited by Tung[5] brings the subject of polymer fractionation up to date, covering both theoretical and practical aspects, including methods to achieve it (cf. 'Fractionation Methods', p. 263). Molecular weight distribution data is now routinely obtained using fractionation

---

[121] B. Chu and T. Nose, *Macromolecules*, 1979, **12**, 599.
[122] K. K. Chee, *J. Appl. Polym. Sci.*, 1981, **26**, 4299.
[123] L. A. L. Kleintjens, Ph.D. Thesis, University of Essex, 1979.
[124] K. Kubota and K. Ogino, *Macromolecules*, 1979, **12**, 74.
[125] K. Kubota and K. Ogino, *Polymer*, 1979, **20**, 175.
[126] J. R. Schmidt and B. A. Wolf, *Macromolecules*, 1982, **15**, 1192.
[127] H. Geerissen, J. R. Schmidt, and B. A. Wolf, *J. Appl. Polym. Sci.*, 1982, **27**, 1277.
[128] P. Irvine and J. W. Kennedy, *Macromolecules*, 1982, **15**, 473.
[129] H. Galina, *Macromolecules*, 1982, **15**, 680.
[130] L. A. Kleintjens, R. Koningsveld, and M. Gordon, *Macromolecules*, 1980, **13**, 303.

methods and 'notable progress has been made on interpreting combined molecular weight/composition data for various copolymers', a topic which has not previously received much attention in the literature.[131]

Solubility Parameter Theory.—An approximate method for assessing solubility of one component in another involves comparing their so-called solubility parameters. The cohesive energy density (CED) of a compound is defined in terms of the molar energy of vaporization, E_v, and the molar volume, V (both at the same reference temperature); thus CED $= E_v/V$. Values for CED can be deduced from a variety of physical measurements, such as refractive index, viscosity, *etc.*, to which it is related. Alternatively, it is to a good approximation an additive function over contributions from parts of the molecule and can thus be estimated from molecular formulae and tables of additivity contributions.[132] The solubility parameter is defined as the square root of the CED. If two materials have similar solubility parameters and no specific interactions intervene then the two materials are likely to form homogeneous mixtures. For further details see references in ref. 1 or consult the detailed early discussion by Burrell.[133,134] The approach continues to be useful industrially where approximate but rapid guidance is required. It has recently been of help in assessing the compatibility of polymers[135,136] with a range of results of significance to mechanical and processing properties of blends.

Wolf and Schuch[42] have detailed the use of solubility parameter theory to predict phase separation in oligomer mixtures. Solubility parameters have been related to, and hence derived from, viscometric[137] and inverse gas chromatographic[138] measurements (see below).

3 Methods for Obtaining Thermodynamic Data

Introduction.—A few new techniques and procedures for obtaining thermodynamic data have been added to the discussion in ref. 1. For the most part these have concerned extending existing methods to a wider range of variables or to a wider range of systems, with special emphasis on polymer mixtures. In the latter case, it is again emphasized that, from a practical point of view, phase morphology has to be studied over a wide range of size scales from microscopic to macroscopic, which necessitates the use of several different methods. Thermodynamically speaking, of course, this is not the case: true phase behaviour can only be adduced from thermodynamic functions. Kinetically, however, homogeneous systems may well survive long enough to be useful even though they are thermodynamically unstable; thus glass transition measurements again dominate among techniques used to study polymer mixtures. Excimer fluorescence offers an interesting ajunct to methods previously available.

[131] G. Riess and P. Sallot, in ref. 5, p. 416.
[132] R. F. Banks, *Polym. Plast. Technol. Eng.*, 1977, **8**, 13.
[133] H. Burrell, *Off. Dig., Fed. Soc. Paint Technol.*, 1955, 726.
[134] H. Burrell, *Off. Dig., Fed. Soc. Paint Technol.*, 1957, 1069.
[135] J. Piglowski and W. Laskawski, *Angew. Makromol. Chem.*, 1979, **82**, 157.
[136] J. Piglowski and T. Skowronski, *Angew. Makromol. Chem.*, 1980, **88**, 165.
[137] E. Perez, M. A. Gomez, A. Bello, and J. G. Fatou, *J. Appl. Polym. Sci.*, 1982, **27**, 3721.
[138] G. DiPaola-Baranyi, *Macromolecules*, 1982, **15**, 622.

Thermodynamics of Solutions and Mixtures

Calorimetric Methods.—Calorimetric methods, when applied to polymer mixtures, are, in fact, more related to glass transition measurements from which phase behaviour is inferred (see 'Glass Transition and Related Relaxation Measurements', p. 263). Experimental details of differential scanning calorimetry (DSC) techniques to study phase separation in polymer blends have been described.[139,140] DSC is one of a number of methods that have been used to study the formation of microphases in block copolymers[141,142] (cf. 'Block Copolymers and their Solutions', p. 266).

Thermal analysis coupled with modulus measurements has been used by Hu et al.[143] to demonstrate the complexity of phase separation processes in polyurethanes. Kwei[144] has studied segmented polyurethanes. Brunette et al. have applied the technique to study hydrogen bonding in polyurethanes.[145]

Centrifugal Methods.—Aminabhavi and Munk[39] have used a new synthetic boundary diffusion technique[146] to study preferential adsorption of solvents in polymer + solvent$_1$ + solvent$_2$ systems.

Further details of the centrifugal homogenizer (see ref. 156 in ref. 1) have been made available,[43,147] together with its use to perform phase volume measurements in polymer blends where high viscosity has previously made experiments of this type difficult.

Gas–Liquid Chromatography (GLC).—Further developments in the use of GLC with a polymer stationary phase to obtain thermodynamic data are described by Ratzsch et al.[55] Lau et al.[148] detail the extraction of thermodynamic data from GLC data in a method that gives accurate results on vapour–liquid equilibria of relevance to process design such as polymer de-volatilization. Some other GLC systems are discussed in refs. 149 and 150.

Inverse gas chromatographic methods are described by Aspler and Gray[83] to measure solvent activity over polymeric liquid-crystalline phases, by DiPaola-Baranyi[138] to obtain solubility parameters for polymers and blends of polymers and by Walsh and McKeown[151] to measure polymer–solvent and polymer–polymer interactions. In general, inverse GLC offers a fast and convenient way to obtain thermodynamic data for concentrated polymer systems.[152]

Scattering Methods.—*Turbidimetry*. Visual observation of cloud points has been used to obtain data for various mixtures of oligomers.[42,153]

---

[139] J. J. Beres, N. S. Schneider, C. R. Desper, and R. E. Singler, *Macromolecules*, 1979, **12**, 566.
[140] S.-F. Lau, J. Pathak, and B. Wunderlich, *Macromolecules*, 1982, **15**, 1278.
[141] S. Krause, M. Iskandar, and M. Iqbal, *Macromolecules*, 1982, **15**, 105.
[142] S. Krause, Z.-H. Lu, and M. Islandar, *Macromolecules*, 1982, **15**, 1076.
[143] C. B. Hu, R. S. Ward, and N. S. Schneider, *J. Appl. Polym. Sci.*, 1982, **27**, 2167.
[144] T. K. Kwei, *J. Appl. Polym. Sci.*, 1982, **27**, 2891.
[145] C. M. Brunette, S. L. Hsu, and W. J. MacKnight, *Macromolecules*, 1982, **15**, 71.
[146] P. Munk, R. G. Alleng, and M. E. Halbrook, *J. Polym. Sci., Polym. Symp.*, 1973, **42**, 1013.
[147] M. Gordon and B. W. Ready, US P. 4 131 369.
[148] W. R. Lau, C. J. Glover, and J. C. Holste, *J. Appl. Polym. Sci.*, 1982, **27**, 3067.
[149] Y. S. Lipatov, A. E. Nestrov, and T. D. Ignatova, *Eur. Polym. J.*, 1979, **15**, 775.
[150] G. DiPaola-Baranyi and P. Degre, *Macromolecules*, 1981, **14**, 1456.
[151] D. J. Walsh and J. G. McKeown, *Polymer*, 1980, **21**, 1335.
[152] G. DiPaola-Baranyi, S. J. Fletcher, and P. Degre, *Macromolecules*, 1982, **15**, 885.
[153] B. A. Wolf and G. Blaum, *Makromol. Chem.*, 1979, **180**, 2591.

Elias[154] reviews turbidimetry under two main headings: (i) cloud-point titrations determine the first detectable turbidity in a solution, which he calls the cloud-point, either by addition of a precipitant to a polymer solution (cloud composition titration) or by altering the temperature (cloud temperature titrations), and (ii) turbidity titrations follow the development of turbidity beyond the cloud-point either isothermally, by addition of a precipitant (turbidity composition titration), or by variation of temperature (turbidity temperature titration). The discussion is a useful contribution to the subject but the caution given earlier[1] regarding the errors that can be introduced when attempting to detect phase boundaries on entering a two-phase region are recalled here. It is much safer to follow the disappearance of turbidity as the cloud point is approached from just within the two-phase region. Tung[5] includes turbidity methods among fractionation techniques, but, unless the main purpose is to obtain polymer fractions for subsequent study rather than phase loci, we prefer to include them among scattering methods.[1]

Turbibimetric techniques have been described for copolymer structure studies,[155] to obtain interesting phase diagrams for some new ternary systems,[13] and to follow the onset of phase separation during polyurethane polymerization processes.[24,156] Other interesting developments extend the method to the study of phase separation in thin films of polymer blends,[9] in flowing polymer solutions,[10] and to high temperatures to explore the lower critical solution behaviour of polyolefin solutions.[11,12]

A 'melt-titration' procedure and apparatus for studying phase separation in polymer mixtures has been detailed[157,158] in which a polymer is cycled through a screw extruder fitted with a special slit die and optical system to detect turbidity in the extruded film. Polymer is returned to the extruder hopper to which the second polymer component is slowly added.

Light Scattering. Suzuki[159,160] has developed a new method for dealing with angular dependence in conventional light scattering primarily for better characterization of polymers of very high molecular weight. Indirectly these methods should prove valuable for obtaining thermodynamic information on high-molecular-weight systems. A low-angle laser light-scattering technique has been used to measure excess chemical potentials of monodisperse polystyrene in toluene.[75] Interesting results on phase separation in polymer blends have been obtained[161] by observing the changing small-angle light-scattering profile with time. As a blend approaches a temperature at which it undergoes phase separation, a light-scattering maximum is observed which grows in intensity and moves to smaller angles with time.

The refractive index and refractive index increment for a polymer in a solvent is a characteristic essential for analysis of light-scattering data. The data available for

---

[154] H.-G. Elias, in ref. 5, p. 346.
[155] T. A. Strivens, *Polymer*, 1981, **22**, 1391.
[156] J. M. Castro, F. Lopez-Serrano, R. E. Camargo, C. W. Macosko, and M. Tirrell, *J. Appl. Polym. Sci.*, 1981, **26**, 2067.
[157] R. H. Somani and M. T. Shaw, *Macromolecules*, 1981, **14**, 886.
[158] R. H. Somani and M. T. Shaw, *Macromolecules*, 1981, **14**, 1549.
[159] H. Suzuki, *Br. Polym. J.*, 1979, **11**, 35.
[160] H. Suzuki, *Br. Polym. J.*, 1979, **11**, 41.
[161] J. Gilmer, N. Goldstein, and R. S. Stein, *J. Polym. Sci., Polym. Phys. Ed.*, 1982, **20**, 2219.

Thermodynamics of Solutions and Mixtures

polyethylene[162,163] and polyacrylamide in various solvents have been significantly improved. An improved method for obtaining refractive index increments for polyelectrolytes has also been noted.[164] Hammel et al.[165] describe a new interferometer to obtain accurate refractive index data at elevated pressures. They use this to obtain data[166] for polystyrene–*trans*-decalin up to 500 bar between 20 and 60 °C. They find that the experimental data follow none of the refractive index mixture rules.

Pulse-induced Critical Scattering (PICS). The PICS technique, which is useful for locating phase loci (see ref. 1) has been explored as a tool for polymer characterization[167] especially in relation to the 'high tail' of a polymer molecular weight distribution. Spinodal curves and, even more so, cloud-point curves, depend on high moments of the molecular weight distribution which are, of course, important in polymer processing and performance.

Wong and Knobler[168,169] have given details of an apparatus which employs pressure (rather than temperature) pulses to induce liquid–liquid phase separation then examined by light scattered using a He–Ne laser. They have not yet applied the method to polymer systems though it would clearly be interesting in this connection.

X-Ray Scattering. X-Ray scattering has been adapted[170] to the study of porosity in gel copolymers (*cf.* 'Phase Equilibria in Gel Polymer Systems', p. 250) and used to study polymer mixtures.[171]

Excimer Fluorescence Spectroscopy.—Frank and Gashgari[172] offer evidence that excimer fluorescence provides a new and powerful probe for molecular interactions in polymer blends and complements the range of methods already available. An excimer is an excited molecular complex which is formed between two identical aromatic rings, one of which is in a singlet excited state. A Schoeffel spectrofluorometer with xenon arc lamp illumination is used to measure the excimer to monomer intensity ratio and clear correlations are noted between this ratio and the difference in solubility parameters between the guest and host polymers. Further work will be required but the technique has already been used to obtain meaningful thermodynamic information on polymers[173—179] and small molecule models for

---

[162] J. Horska, J. Stejskal, and P. Kratochvil, *J. Appl. Polymer Sci.*, 1979, **24**, 1845.
[163] J. Stejskal, J. Horska, and P. Kratochvil, *J. Appl. Polym. Sci.*, 1982, **27**, 3929.
[164] M. Mandel, F. A. Varkevisser, and C. J. Bloys van Treslong, *Macromolecules*, 1982, **15**, 675.
[165] G. L. Hamel, G. V. Schulz, and M. D. Lechner, *Makromol. Chem.*, 1981, **182**, 1829.
[166] G. L. Hamel, G. V. Schulz, and M. D. Lechner, *Makromol. Chem.*, 1981, **182**, 1835.
[167] H. Galina, M. Gordon, P. Irvine, and L. A. Kleintjens, *Pure Appl. Chem.*, 1982, **54**, 365.
[168] N.-C. Wong and C. M. Knobler, *J. Chem. Phys.*, 1978, **69**, 725.
[169] N.-C. Wong and C. M. Knobler, *Phys. Rev. Lett.*, 1979, 1733.
[170] J. Baldrian, B. N. Kolarz, and H. Galina, *Collect. Czech. Chem. Commun.*, 1981, **46**, 1675.
[171] J. H. Wendorff, *J. Polym. Sci., Polym. Lett. Ed.*, 1980, **18**, 439.
[172] C. W. Frank and M. A. Gashgari, *Macromolecules*, 1979, **12**, 163.
[173] P. Ander and M. K. Mahmoudhagh, *Macromolecules*, 1982, **15**, 213.
[174] Y. Iwaya and S. Tazuke, *Macromolecules*, 1982, **15**, 396.
[175] S. Tazuke, H. Ooki, and K. Sato, *Macromolecules*, 1982, **15**, 400.
[176] F. C. De Schryver, J. Vandenreissche, S. Toppet, K. Denmeyer, and N. Boens, *Macromolecules*, 1982, **15**, 406.
[177] G. E. Johnson and T. A. Good, *Macromolecules*, 1982, **15**, 409.
[178] R. Gelles, and C. W. Frank, *Macromolecules*, 1982, **15**, 741.
[179] R. Gelles and C. W. Frank, *Macromolecules*, 1982, **15**, 747.

polymers,[180] including the study of spinodal decomposition in polymer blends.[181]

Viscometric Methods.—*Dilute Solution Viscosity.* Munk and Gutierrez[182] have attempted to remove some of the vagaries of interpreting intrinsic viscosity data in terms of thermodynamic quantities[1] by means of Monte Carlo modelling. A polymer solution is simulated by an ensemble of self-avoiding random walks. Previous work along these lines has usually neglected the effects of polymer–solvent interactions and hence been limited to athermal solutions. The extensive Monte Carlo calculations of McCracken *et al.*,[183] which allow for non-bonded interactions, are used[182] to interpret quantitatively the Mark–Houwink–Sakurada relation, the temperature dependence of its parameters, and hence (indirectly) other thermodynamic quantities. A model has been developed for predicting dilute solution parameters for polymer solutions from intrinsic viscosity data[114,184] and, elsewhere, the feasibility of computing binary and ternary interaction functions from intrinsic viscosity data alone has been discussed.[185] In fact, solvent interactions provide the motivation for several viscometric studies[186,187] including polymer–solvent interactions in mixed solvents.[188,189]

Viscometric methods have been adapted to study phase behaviour in stiff-chain polymers[80,89] and to obtain thermodynamic data for dilute solutions at high pressures.[124–126] Meyerhoff and Appelt[190] have designed a new low-shear automated recording viscometer for polymer solutions that obviates the need for cathetometer measurements over extended periods of time. Perez *et al.*[137] have obtained tables of solubility parameters from intrinsic viscosity measurements.

Rotary Viscometers. Schmidt and Wolf[191] continue to study the thermodynamic basis for Newtonian viscosity coefficients (see ref. 1) and have obtained further information from its pressure dependence for polystyrene in *t*-butyl acetate over the entire temperature range between lower and upper critical solution demixing. Elsewhere, they describe a rolling ball viscometer which they have used to study thermodynamic properties of dilute[126] and moderately concentrated[127] polymer solutions.

A cone and plate viscometric technique is employed by Castro *et al.*[156] for studying phase separation during polyurethane polymerization. Viscometry has been used to study phase behaviour during polymerization of other systems.[24]

---

[180] F. C. De Schryver, L. Moens, M. van der Auweraer, N. Boens, L. Monnerie, and L. Bokobza, *Macromolecules*, 1982, **15**, 64.
[181] R. Gelles and C. W. Frank, *Macromolecules*, 1982, **15**, 1486.
[182] P. Munk and B. O. Gutierrez, *Macromolecules*, 1979, **12**, 467.
[183] F. L. McCrackin, J. Mazur, and C. M. Guttman, *Macromolecules*, 1973, **6**, 859.
[184] C. M. Kok and A. Rudin, *J. Appl. Polym. Sci.*, 1981, **26**, 3575.
[185] J. W. A. Van den Berg and F. W. Altena, *Macromolecules*, 1982, **15**, 1447.
[186] L. M. Leon, J. Galaz, L. M. Garcia, and M. S. Anasagasti, *Eur. Polym. J.*, 1980, **16**, 921.
[187] J. Y. Olauemi, *Makromol. Chem.*, 1982, **183**, 2547.
[188] V. Bottiglione, M. Morcellet, and C. Loucheux, *Makromol. Chem.*, 1980, **181**, 485.
[189] M. Apostolopoulos, M. Morcellet, and C. Loucheux, *Makromol. Chem.*, 1982, **183**, 1293.
[190] G. Meyerhoff and B. Appelt, *Macromolecules*, 1979, **12**, 968.
[191] J. R. Schmidt and B. A. Wolf, *Markomol. Chem.*, 1979, **180**, 517.

Glass Transition and Related Relaxation Measurements.—In their review of polymer mixtures Paul and Barlow[192] have taken a single glass transition as indicative of compatibility, arguing that this is most meaningful for engineering purposes. Results of Matzner et al.[193] strongly support this in a newly found system. Other papers[194–198] likewise consider a single T_g intermediate between those of polymer components as sufficiently indicative of their compatibility.

Fractionation Methods.—Methods for polymer fractionation are founded on the thermodynamics of polymer systems, with special reference to phase separation. However, in the previous Report[1] these methods were covered only incidentally rather than as a class of techniques in their own right. The treatise edited by Cantow[199a] has been elegantly updated in a book edited by Tung,[5] which clearly shows how these methods employ solution thermodynamics to obtain vital information about polymer molecular weight and composition distributions, property–processability–structure relationships, and polymerization mechanisms. Various fractionation techniques are described and are clearly important for polymer characterization. Because of the difficulty with directly interpreting data associated with phase separation boundaries in thermodynamic terms (cf. 'Molecular Weight', p. 257), the methods are less useful for obtaining thermodynamic information directly. However, they do yield closely fractionated samples which can subsequently be subjected to other types of experiment.

Batch Fractionation. Kamide[200] notes that, despite the rapid advance of other fractionation methods, batch fractionation remains popular because the method is straightforward to perform and uses simple apparatus. He reviews the various types of batch fractionation and their theoretical background, discussing the theoretical difficulties and hence the need for computer simulation. The effects of molecular weight have been illustrated in some detail with a large selection of phase diagrams.[106]

Column Fractionation. In this technique, reviewed by Barrall et al.,[201] a solid polymer is first precipitated from solution by evaporation or dilution of a good solvent onto an inert support; a process usually performed outside the chromatograph. Ideally, the support is uniformly coated with a thin layer of polymer which varies in molecular weight from the surface of the support granule outward, with the highest molecular weight closest to the support surface. A portion of this coated support is placed at the solvent inlet of a column on the same inert support material. The coated material is then contacted by progressively more powerful solvent, establishing a solvent gradient which washes off first the more soluble and last the

---

[192] D. R. Paul and J. W. Barlow, *J. Macromol. Sci., Rev. Macromol. Chem.*, 1980, **18**, 109.
[193] M. Matzner, L. M. Robeson, E. E. Wise, and J. E. McGrath, *Makromol. Chem.*, 1982, **183**, 2871.
[194] S. Akiyama and K. Miasa, *Polymer J.*, 1979, **11**, 157.
[195] E. Roerdink and G. Challa, *Polymer*, 1980, **21**, 1161.
[196] M. Aubin and R. E. Prud'homme, *J. Polym. Sci., Polym. Phys. Ed.*, 1981, **19**, 1245.
[197] G. Belorgey and R. E. Prud'homme, *J. Polym. Sci., Polym. Phys. Ed.*, 1982, **20**, 191.
[198] J. S. Chiou, D. R. Paul, and J. W. Barlow, *Polymer*, 1982, **23**, 1543.
[199] (a) M. J. R. Cantow, ed., 'Polymer Fractionation', Academic, New York, 1967; (b) *ibid.*, pp. 466–487.
[200] K. Kamide, in ref. 5, p. 104.
[201] E. M. Barrall, J. F. Johnson, and A. R. Cooper, in ref. 5, p. 268.

least soluble polymer. Eluent from the column is collected for further evaluation. In some cases a thermal gradient is imposed on the column as well as a solvent elution gradient. In either case, the solvent gradient is usually established by blending a non-solvent with a good solvent and for this purpose solvent pairs such as are employed in cloud-point determinations are directly applicable. Cantow[199b] has tabulated many such pairs. Although column chromatographic fractionation of polymers is a tedious method, for many years it was the only way to obtain narrow polymer fractions. Recently there has been some success in dealing with the inherent problems of the technique.[201]

Gel Permeation Chromatography (GPC). This is not really a method for obtaining thermodynamic data directly, although Hashizume et al.[41] have used GPC to obtain isothermal binodal and separation factor data for a ternary system comprising solvent and two 'monodisperse' polymer homologues. The technique is certainly an important technique for the fractionation of polymers. All aspects of the subject, including instrumentation, theory, data analysis anomalies, and special applications of the method, are reviewed in detail by Tung and Moore.[202]

Thin Layer Chromatography (TLC). This is again a useful technique for polymer fractionation and hence characterization rather than a method for obtaining thermodynamic information directly. We mention it here only for completion and note that the method is reviewed by Inagaki[203] covering applications to both homopolymers and copolymers.

Miscellaneous Other Methods.—*Vapour Pressure.* Ratzsch et al.[54] have made vapour pressure measurements of concentrated solutions of polystyrene in various solvents.

Gel Swelling. Gel swelling and porosity data is of increasing interest as a technique. There has been progress both in theoretical[21] and in experimental[170,204] understanding. The principles of gel swelling are, of course, of special relevance to the phenomena of phase separation in swollen gel networks that have been demonstrated (see 'Phase Equilibria in Gel Polymer Systems', p. 250), and new techniques have been discussed[30,205] in this context.

Potentiometric Titrations. This well-known technique has recently been adapted to explore polymer–solvent interactions in mixed solvents,[206] polyelectrolyte properties,[30,206] and the formation of compact structures in some acrylic polymer solutions.[207]

4 Polymer Solutions and Mixtures

Introduction.—Since the previous Report[1] the number of polymer systems that have been the subject of thermodynamic study has increased impressively (see Table 1). This is especially true for phase behavioural studies in polymer mixtures which has

[202] L. H. Tung and J. C. Moore, in ref. 5, p. 545.
[203] H. Inagaki, in ref. 5, p. 649.
[204] H. Galina and B. N. Kolarz, *Polym. Bull.* (Berlin), 1980, **2**, 235.
[205] J. Hrouz, M. Ilavsky, I. Havlicek, and K. Dusek, *Collet. Czech. Chem. Commun.*, 1979, **44**, 1942.
[206] G. Vorreux, M. Morcellet, and C. Loucheux, *Makromol. Chem.*, 1982, **183**, 711.
[207] J. Morcellet-Sauvage, M. Morcellet, and C. Loucheux, *Makromol. Chem.*, 1981, **182**, 949.

been spurred on by the commercial interest in finding compatible systems. Among new types of phase behaviour that have captured interest, phase separation in gels and in polymerizing systems must be highlighted. However, pride of place during the period of this Report must go to the tendency for stiff-chain polymers and copolymers to form and exhibit liquid crystal phases, which has increasingly been the subject of theoretical and experimental study.[82]

Polymer Solutions.—Chu and Nose[121] have studied low-molecular-weight polystyrene in a theta solvent so as to obtain contracted polymer coils which exhibit an extended semi-dilute region. They observe the presence of gel-like behaviour at concentrations less than that at which significant coil overlap occurs, a fact which holds interest for theoretical bridging between dilute and concentrated polymer solution regimes. Some polymers, such as poly(methyl methacrylate), poly(methacrylic acid), and poly(vinyl chloride), undergo conformational changes of unknown origin but thought to be associated with alterations in the motions of polymer segments. Olayemi[187] obtains evidence from dilute solution viscometry that the conformational changes of poly(ethyl methacrylate) in aliphatic organic esters are due to changes in the polymer–solvent interactions. Formation of compact structures in solutions of methacrylic acid copolymers has also been studied.[188]

Appelt and Meyerhoff[190,208] compare a variety of techniques, some thermodynamic, to provide an interesting study on polystyrene of extremely high molecular weight in several solvents. Polymers have been increasingly studied in solvent mixtures,[13,36,188,189,206] and in this connection the data obtained by Aminabhavi and Munk[39] for polystyrene in mixed solvents is interesting. It shows that inversion of the preferentially adsorbed solvent, as the solvent composition is varied, is a common phenomenon which persists even in mixtures of good solvents with marginal ones.

Solutions of stiff-chain polymers have been widely studied,[80,81] partly with a view to testing the model of Flory (see 'Models for Stiff-chain Polymers', p. 255) for these systems but perhaps more so because of the predominance of polymeric liquid crystal behaviour in stiff-chain polymer systems.[209] Dielectric relaxation studies on stiff-chain systems[92] have been especially informative of their character. Much of the interest in liquid crystal forming mixtures of stiff-chain and flexible-chain polymers in solution[93] has resulted from new technologies associated with liquid crystals[209] and with ultra-high modulus polymers.[210] All-*para* aromatic polyamides[94] and many cellulose derivatives[83,95,97,211] form liquid crystalline phases in solvents and these can yield very high-strength fibres when spun from solutions in the appropriate phase state. Some significant aspects of the topic are surveyed by Takayanagi *et al.*[96] A polyamic acid system which does not exhibit liquid crystallinity but does undergo reversible phase separation is discussed by Orwoll *et al.*[212]

---

[208] B. Appelt and G. Meyerhoff, *Macromolecules*, 1980, **13**, 657.
[209] A. Ciferri, W. R. Krigbaum, and R. B. Meyer, ed., 'Polymeric Liquid Crystals', Academic, New York, 1982.
[210] A. Ciferri and I. M. Ward, ed., 'Ultra-High Modulus Polymers', Applied Science Publishers, London, 1979.
[211] S. Dayan, P. Maissa, M. J. Vellutini, and P. Sixou, *J. Polym. Sci., Polym. Lett. Ed.*, 1982, **20**, 33.
[212] R. A. Orwoll, *J. Polym. Sci., Polym. Phys. Ed.*, 1981, **19**, 1385.

Ionic Polymer Solutions.—Onda et al.[213] have studied ionic solutions of polyacrylamide principally to characterize the polymer, but most other studies[92,189,206,214—218] have their focus in understanding interactions in solutions of relevance to biological molecules.

Copolymers and Their Solutions.—Riess and Callot[131] have provided us with one of the few reviews on the fractionation of copolymers. They cover both theoretical and practical aspects of fractionating regular (block) and random copolymers. Interactions of alternating copolymers with mixed aqueous solvents have helped[189] to explain some interesting conformational changes that occur with changing concentration. Dissolution of copolymers in mixed solvents creates possibilities for strongly preferential solvation (adsorption) of polymer segments. Though data of this type exists, its interpretation in theoretical terms still leaves much undone. The topic is of especial interest in polar systems and underlies of the behaviour of biological molecules. Feyereise et al.[219] have provided results and discussion on polypeptides linking preferential interactions to solvent-induced helix–coil transitions.

Phase separation in gel copolymer systems has been surveyed by Dusek[21] (see also 'Phase Equilibria in Gel Polymer Solutions', p. 250). New data have been obtained and interpreted for these systems,[22,31,170,204] including poly(ethylene oxide) hydrogels[25] which have long been of interest in the pharmaceutical and textiles industries but have not yet been fully understood theoretically.

Random Copolymers and Their Solutions. In the previous Report[1] random copolymer solutions were considered together with solutions of polymer mixtures. It has seemed more systematic to alter this format here.

Block Copolymers and Their Solutions. Tuzar and Kratochvil[220] gave a very useful review of block and graft copolymer miscelle solutions that was overlooked in ref. 1. Ikemi et al.[221] have obtained data for such systems. Galin and Rupprecht[37] examine the effect of copolymer structure on interactions in di- and tri-block copolymers and contrast these with mixtures of homopolymers. Mehta and Dole[222] have made a useful study of gas solubility in homopolymers and block copolymers.

Microphase structure has been noted in a variety of block copolymers and has been the subject of several experimental studies.[141,142,223—228] These systems are included separately in Table 1.

---

[213] N. Onda, K. Furusawa, N. Yamaguchi, and S. Komuro, *J. Appl. Polym. Sci.*, 1979, **23**, 3631.
[214] Y. Inoue, N. Teraoka, Y. Suzuki, and R. Chajo, *Makromol. Chem.*, 1981, **182**, 1819.
[215] M. C. Le Bret, *CR Hebd. Seances Acad. Sci., Ser. B*, 1981, **292**, 291.
[216] C. Kienzle-Sterzer, D. Rodriguez-Sanchez, and C. Rha, *J. Appl. Polym. Sci.*, 1982, **27**, 4467.
[217] G. E. Boyd and D. P. Wilson, *Macromolecules*, 1982, **15**, 78.
[218] R. L. Cleland, J. L. Wang, and D. M. Detweiler, *Macromolecules*, 1982, **15**, 382.
[219] C. Feyereisen, M. Morcellet, and C. Loucheux, *Macromolecules*, 1979, **12**, 613.
[220] Z. Tuzar and P. Kratochvil, *Adv. Colloid Interface Sci.*, 1976, **6**, 201.
[221] M. Ikemi, N. Odagiri, S. Tanaka, I. Shinohara, and A. Chiba, *Macromolecules*, 1982, **15**, 281.
[222] M. Mehia and M. Dole, *Macromolecules*, 1982, **15**, 376.
[223] J. M. Cowie, D. Lath, and I. J. McEwen, *Macromolecules*, 1976, **12**, 52.
[224] S. Krause and M. Iskandar, *Adv. Chem. Ser.*, 1979, **176**, 205.
[225] B. Morese-Segela, M. St. Jacques, J. M. Renaud, and J. Prud'homme, *Macromolecules*, 1980, **13**, 100.
[226] Z.-H. Lu and S. Krause, *Macromolecules*, 1982, **15**, 112.
[227] M. Shibayama, H. Hasegawa, T. Hasimoto, and H. Kawai, *Macromolecules*, 1982, **15**, 274.
[228] R. E. Cohen and D. E. Wilfong, *Macromolecules*, 1982, **15**, 370.

Polymer Mixtures and Their Solutions.—Activity on polymer mixtures is increasing rapidly,[229] motivated mainly by the prospects of improving the processing and performance characteristics of polymer blends by better understanding of their phase behaviour and covering ever more diverse systems. Several new reviews[192,230—232] have appeared. DiPaola-Baranyi et al.[152] point out that much of this explosive growth of research interest in miscible polymer blends is due to the fact that they 'may represent a valuable and economical alternative to the use of copolymers'[230,231] even though truly miscible blends are rather infrequent since, as Liu and Prausnitz[45] remind us, the entropy of mixing for a polymer–polymer mixture is much smaller than for a mixture of ordinary liquids, so that a small positive enthalpy of mixing is sufficient for phase separation. Koningsveld et al.[43] also take the view that phase stability in polymer blends can only be rigorously discussed in thermodynamic terms.

On the other hand, Paul and Barlow[192] have reviewed polymer blends (or alloys) from an engineering point of view and with this stance it may be that true thermodynamic compatibility represents too severe a constraint on the selection of systems for practical purposes. Perhaps this has encouraged Matzner et al.[193] to remark that the number of well-documented cases of miscibility is increasing rapidly. Poly(vinyl chloride) is one of the most cited materials in this regard and compatible systems with polyethylene can be obtained by copolymerizing the latter with NN-dimethylacrylamide. In this way, systems with highly attractive mechanical properties can be produced. Similarly, Kozlowski and Lawkawski[233—235] have studied rheological and mechanical properties of blends of poly(vinyl chloride) with copolymers of vinyl chloride with vinyl acetate and with propylene, assessing compatibility by glass transition measurements and microscopic observations.[235] Their results suggest that blends are homogeneous in confirmation of other observations.[233,234] From various indirect methods that examine mechanical and thermodynamic characteristics, Piglowski and Laskawski argue[133,236,237] that blends of poly(vinyl chloride) and polyurethane are incompatible. Beres et al.[139] likewise conclude from DSC data on aryloxy phosphazene homo- and copolymers that crystalline phases, and in some cases amorphous phases also, are incompatible.

Castro et al.[136] have experimentally studied the onset of phase separation during urethane polymerization, a topic of considerable importance to urethane molding operations and for which several models have recently been proposed (see, e.g., ref. 87). Using both light transmission and viscometric measurements, they conclude that phase separation occurs at very early stages in the reaction. Optical and viscometric data appear to be sensitive to different aspects of the phase separation process with the viscometric transition occurring earlier. The results are interpreted in terms of a multiplicity of size scales of the phase separated domains. Quite clearly there are interesting thermodynamic problems to be solved in

---

[229] C. M. Kok and A. Rudin, *Makromol. Chem.*, 1981, **182**, 2801.
[230] D. R. Paul and S. Newman, 'Polymer Blends', Academic, New York, 1978, Vols. 1 and 2.
[231] O. Olabisi, L. M. Robeson, and M. T. Shaw, 'Polymer–Polymer Miscibility', Academic, New York, 1979.
[232] D. Klempner and K. C. Frisch, ed., 'Polymer Alloys', Plenum, New York, 1980, Vol. II.
[233] M. Kozlowski and W. Laskawski, *Angew. Makromol. Chem.*, 1980, **91**, 1.
[234] M. Kozlowski and W. Laskawski, *Angew. Makromol. Chem.*, 1980, **91**, 17.
[235] M. Kozlowski and W. Laskawski, *Angew. Makromol. Chem.*, 1980, **91**, 29.
[236] J. Piglowski and W. Laskawski, *Angew. Makromol. Chem.*, 1980, **84**, 163.
[237] J. Piglowski, T. Skowronski, and B. Masiulanis, *Angew. Makromol. Chem.*, 1980, **85**, 129.

understanding this system and these are likely to require the use of non-equilibrium thermodynamics. Kwei[144] has contributed useful DSC data on this topic. Brunette et al.[145] have studied hydrogen bonding in polyurethanes.

Mixtures of microphase separated block copolymers with one of the homopolymers is contributing to the understanding of these systems.[238—241] For some molecular weights, block copolymers may be homogeneous, whereas their homopolymer mixtures may not be even though the materials are chemically very similar. Cohen and Wilfong[228,238,239,242] have made a detailed study of the phase diagram for polybutadiene systems in which the phase behaviour depends on subtle effects. Interfacial energies in miscible polymer blends may be greatly altered by addition of block copolymers to the blend (cf. 'Phase Interfaces', p. 250) with interesting practical consequences from the uniformly dispersed microdomain structures that can result.[227]

Phase behaviour in interpenetrating polymer networks[243,244] raises interesting new thermodynamic problems that are under study[27,245,246] and promises to offer an exciting new area for polymeric materials. Phase separation studies on polymer blends in thin films[9] (cf. Section 2) has practical consequences, and in oligomer mixtures[42,153] it offers useful insight into the behaviour of polymeric counterparts.

Recent Data Sources for Specific Systems.—Elias[154] tabulates many theta systems for polymer solutions. Kamide[200] gives a large selection of polymer fractionation diagrams showing the effects of polymer molecular weight in some detail. Paul and Barlow[192] offer many tables of polymer mixtures and key references to source data on these mixtures. The new review volume on polymer blends edited by Klempner and Frisch[232] is also useful as a source of data. Polymeric liquid crystal systems are covered in a review edited by Cifferri et al.[209]

Table 1 Recent Data Sources for Specific Systems

Acrylic Polymer Solutions		Ref.
Poly(t-butyl acrylate)	butanone	114
Poly(methacrylic acid)	water	206
	dioxane	206
Poly(methyl acrylate)	hexane	151
	acetone	151
	butanone	151
	propanol	151
	ethyl acetate	151
	chloroform	151
	acetonitrile	151

[238] R. E. Cohen and A. R. Ramos, Adv. Chem. Ser., 1979, **176**, 237.
[239] R. E. Cohen and A. R. Ramos, J. Macromol. Sci., Phys., 1980, **17**, 625.
[240] H. Hashimoto, M. Fujimura, T. Hasimoto, and H. Kawai, Macromolecules, 1981, **14**, 844.
[241] Z.-H. Lu, S. Krause, and M. Isandar, Macromolecules, 1982, **15**, 367.
[242] A. R. Ramos and R. E. Cohen, Polym. Eng. Sci., 1977, **17**, 699.
[243] Yu. S. Lipatov and L. M. Sergeeva, 'Interpenetrating Polymeric Networks', Naukova Dumka, Kiev, 1979.
[244] L. H. Sperling, 'Interpenetrating Polymer Networks and Related Materials', Plenum, New York, 1981.
[245] J. K. Yeo, L. H. Sperling, and D. A. Thomas, Polym. Prepr., Am. Chem. Soc., Div. Polym. Chem., 1980, **21**, 53.
[246] J. K. Yeo, L. H. Sperling, and D. A. Thomas, Polym. Eng. Sci., 1982, **22**, 190.

Table 1 (continued)

		Ref.
Poly(ethyl acrylate)	hexane	151
	acetone	151
	butanone	151
	propanol	151
	ethyl acetate	151
	chloroform	151
	acetonitrile	151
Poly(propyl acrylate)	hexane	151
	acetone	151
	butanone	151
	propanol	151
	ethyl acetate	151
	chloroform	151
	acetonitrile	151
Poly(butyl acrylate)	hexane	151
	acetone	151
	butanone	151
	propanol	151
	ethyl acetate	151
	chloroform	151
	acetonitrile	151
Poly(hexyl acrylate)	hexane	151
	acetone	151
	butanone	151
	propanol	151
	ethyl acetate	151
	acetonitrile	151
Poly(sodium methacrylate)	water	206
	dioxane	206
Poly(methyl methacrylate)	benzene	36, 70, 115
	toluene	73, 114, 115
	methanol	36
	acetone	36, 73, 104, 114, 115, 184
	butanone	114
	ethyl methyl ketone	104
	4-heptanone	115
	tetrahydrofuran	115, 229
	ethyl acetate	114
	chloroform	36, 115
	tetrachloromethane	36
	butyl chloride	115
	methyl methacrylate	122
	nitroethane	114
Poly(ethyl methacrylate)	hexane	151
	acetone	151
	butanone	151
	propanol	151
	methyl acetate	187
	ethyl acetate	151, 187
	isobutyl acetate	187
	chloroform	151
	acetonitrile	151

Table 1 (*continued*)

		Ref.
Poly(propyl methacrylate)	hexane	151
	acetone	151
	butanone	151
	propanol	151
	ethyl acetate	151
	chloroform	151
	acetonitrile	151
Poly(butyl methacrylate)	hexane	151
	acetone	114. 151
	butanone	114, 151
	propanol	151
	ethyl acetate	151
	chloroform	151
	acetonitrile	151
Poly(pentyl methacrylate)	hexane	151
	acetone	151
	butanone	151
	propanol	151
	ethyl acetate	151
	chloroform	151
	acetonitrile	151
Poly(hexyl methacrylate)	hexane	151
	acetone	151
	butanone	151
	propanol	151
	ethyl acetate	151
	chloroform	151
	acetonitrile	151
Polyacrylamide	water	22, 247
	aq. salt solutions	247
	ethylene glycol	247
	acetone	22
	formic acid	247
	acetic acid	247
	ethanolamine	247
	formamide	213, 247
Poly(N-methacryloyl-L-alanine)	water	207
	methanol	207
	dioxane	151
Cellulose Polymer Solutions		
Cellulose	water	95
	N-methyl-morpholine-N-oxide	95
	trifluoroacetic acid	97
Cellulose acetate	water	38
	ethanol	115
	acetone	38, 211
	dioxane	38, 80, 211
	acetic acid	38, 80
	methylene chloride	115
	trifluoroacetic acid	97, 211
	nitromethane	80
	dimethyl formamide	38
	dimethyl acetamide	80

[247] J. Stejskal and J. Horska, *Makromol. Chem.*, 1982, **183**, 2527.

Table 1 (continued)

		Ref.
Cellulose acetate *continued*	aniline	80
	pyridine	80, 211
	tetrachloroethane	80
	picoline	80
	cresol	80
	dimethyl sulphoxide	38
Hydroxy propyl cellulose	water	83
Nitrocellulose	acetone	80
	methyl ethyl ketone	80
	methyl propyl ketone	80
	pentanone	80
	methyl pentanone	80
	hexanone	80
	octanone	80
	methyl acetate	80
	ethyl acetate	80
	propyl acetate	80
	butyl acetate	80
	pentyl acetate	80

Polyamide Solutions
Poly(p-benzamide)	NN-dimethyl acetamide	94

Polyether Solutions
Poly(ethylene oxide)	benzene	36
	methanol	36
	tetrachloromethane	36

Polyisocyanate Solutions
Poly(propyl isocyanate)	toluene	89
Poly(hexyl isocyanate)	toluene	82, 89, 90, 92, 98
	chloroform	82
	1,1,2,2-tetrachloroethene	82
	chlorobenzene	82
	bromoform	82
Poly(octyl isocyanate)	trichloroethane	81

Polyolefin Solutions
Poly(ethene)	butane	11
	pentane	11, 35
	hexane	11, 35
	heptane	11, 35, 53
	octane	11, 35
	nonane	11
	decane	11, 162
	dodecane	11
	various branched alkanes	11
	cyclopentane	11
	methyl cyclopentane	11
	methyl cyclohexane	11
	tetralin	162
	1,2-dichlorobenzene	162, 163
	1,2,4-trichlorobenzene	162, 163

Table 1 (*continued*)

Polymer	Solvent	Ref.
Poly(ethene) *continued*	biphenyl	162
	1-chloronaphthalene	162, 163
	bromobenzene	162
	dibromobenzene	162
	1-methyl naphthalene	162
	1-bromonaphthalene	162
	diphenyl methane	162, 163
	diphenyl	162
	diphenyl ether	167
Polypropene	various alkanes	11
	cyclopentane	11
	cyclohexane	115
	methyl cyclopentane	11
	methyl cyclohexane	11
	benzene	115
Poly(but-1-ene)	various alkanes	11
	cyclopentane	11
	methyl cyclopentane	11
	methyl cyclohexane	11
Poly(isobutene)	pentane	53
	cyclohexane	53, 104, 184
	benzene	53
Poly(pent-1-ene)	various alkanes	11
	cyclopentane	11
	methyl cyclopentane	11
	methyl cyclohexane	11
Poly(4-methyl pent-1-ene)	various alkanes	11
	cyclopentane	11
	methyl cyclopentane	11
	methyl cyclohexane	11
Polybutadiene	hydrogen	222
	deuterium	222
	cyclohexane	13, 148
	benzene	148
	tetrahydrofuran	229
	dimethylformamide	13
Poly(ethylene sulphonate)	aq. ionic solutions	217
Siloxane Polymer Solutions		
Poly(dimethyl siloxane)	hexane	73
	heptane	37, 73
	octane	73
	nonane	73
	decane	37
	cyclohexane	73
	benzene	36, 37, 115
	toluene	37, 73, 115
	methanol	36
	dioxane	37
	ethyl methyl ketone	115
	chloroform	37
	cyclohexyl bromide	73, 124

Table 1 (*continued*)

Styrene Polymer Solutions		Ref.
Polystyrene	hydrogen	222
	deuterium	222
	heptane	36, 37
	isooctane	180
	decane	37
	cyclopentane	127
	cyclohexane	13, 36, 39, 41, 53, 54, 67, 68, 115, 127, 160, 167, 185, 208
	methyl cyclohexane	71, 72
	benzene	36, 37, 39, 53, 54, 67, 68, 104, 114, 115, 122, 159, 160
	ethyl benzene	53, 115
	toluene	37, 67, 73, 75, 104, 114, 115, 122, 184, 190, 208
	xylene	54
	phenyl decane	127
	decalin	64, 65, 67, 78, 104, 121, 122, 160, 166, 125
	methanol	10, 13, 36
	propanol	36
	butanol	36
	ethyl acetate	39, 185
	t-butyl acetate	78, 126, 191
	acetone	166
	methyl ethyl ketone	104, 115, 184, 208
	butanone	114, 160, 248
	dioxane	37, 184
	tetrahydrofuran	67, 114, 190, 229
	diethyl malonate	127
	di-(2-ethyl hexyl) phthalate	10
	chloroform	37
	dichloroethane	115
	trichloroethane	81
	1-chlorodecane	127
	1-chlorododecane	127
	chlorobenzene	115, 184
	bromobenzene	70
	dimethylformamide	13
Poly(α-methyl styrene)	cyclohexane	104
	toluene	73, 104, 114, 122, 248
Poly(*p*-methyl styrene)	toluene	114
Poly(*p*-chlorostyrene)	toluene	114, 115, 248
	chlorobenzene	115
Poly(*p*-bromostyrene)	toluene	114, 248
Poly(styrene sulphonic acid)	aq. ionic solutions	173

[248] D. Stigter, *Macromolecules*, 1982, **15**, 635.

Table 1 (*continued*)

		Ref.
Poly(styrene sulphonate)	aq. ionic solutions	164, 217, 248
Vinyl Polymer Solutions		
Poly(vinyl acetate)	benzene	53, 115
	acetone	115
	ethyl methyl ketone	115
	butanone	122, 184
	4-methyl-2-pentanone	122
	dioxane	115
	*iso*propyl alcohol	31
	chloropropane	122
Poly(vinyl chloride)	hexane	151
	cyclohexane	122
	acetone	135, 151
	butanone	151
	cyclohexanone	115
	propanol	151
	ethyl acetate	151
	chloroform	151
	chlorobenzene	115
	acetonitrile	151
	dimethylformamide	135
Poly(vinyl alcohol)	water	122
Poly(2-vinyl pyridine)	benzene	36
	ethanol	36
	chloroform	36
Poly(*N*-vinyl carbazole)	chlorobenzene	186
	nitrobenzene	186
Poly(*N*-vinyl-2-pyrrolidone)	ethanol	36
	chloroform	36
	1,2-dichloroethane	36
Poly(vinyl benzene trimethyl ammonium)	aq. ionic solutions	217
Miscellaneous Polymer Solutions		
Amylose tricarbonate	ethyl acetate	100
	dioxane	100
Poly(*bis*[*m*-chlorophenoxy]-phosphazene)	chloroform	116
Polyurethane	acetone	135
	dimethylformamide	135
	monomers	156
Poly(ethylene glycol)	octane	55
	decane	55
	dodecane	55
	cyclohexane	13
	benzene	55
	toluene	55
	xylene	55
	methanol	13
	ethanol	55
	propanol	55
	butanol	55
	heptanol	55

Table 1 (continued)

		Ref.
Poly(chloroprene)	carbon tetrachloride	114
Poly(L-lysine)	aq. ionic solutions	214
Natural rubber	benzene	115
	toluene	115
Chitosan (1–4,2-amino-2-deoxy-b-D-glucan	aq. ionic solutions	216
Polyoxetanes	hexane	137
	heptane	137
	octane	137
	isooctane	137
	decane	137
	dodecane	137
	cyclohexane	137
	methyl cyclohexane	137
	benzene	137
	toluene	137
	tetralin	137
	ethanol	137
	cyclohexanol	137
	acetone	137
	ethyl methyl ketone	137
	dioxane	137
	ethyl acetate	137
	butyl acetate	137
	methylene chloride	137
	chloroform	137
	carbon tetrachloride	137
	1,2-dichloroethylene	137
	trichloroethylene	137
	tetrachloroethane	137
	chlorobenzene	137
	o-dichlorobenzene	137
	acetonitrile	137
	nitromethane	137
	nitrobenzene	137
	triethyl phosphate	137
Sodium hyaluronate	aq. salt solutions	218
Copolymer Solutions		
Copoly(ethene:propene)	various hydrocarbons	11, 12
Copoly(styrene:divinyl benzene) gels	iso-octane	21
	cyclohexane	170, 204
	acetone	170, 204
	methanol	170, 204
	toluene	21, 170, 204
Copoly(styrene:butyl methacrylate)	butanone	114
Copoly(methacrylic acid: benzyl methacrylate)	water	188
	2-chloroethanol	188
Copoly(acrylamide:sodium methacrylate)	aq. ionic solutions	30
Copoly(2-hydroxy ethyl methacrylate:ethylene di-methacrylate) gels	water	21
	butanol	21

Table 1 (*continued*)

		Ref.
Copoly(N-methacryloyl-L-alanine:	water	207
N-phenyl methacrylamide)	methanol	207
	dioxane	207
Copoly(propyl isocyanate hexyl isocyanate)	toluene	89
Copoly(n-butyl isocyanate: n-nonyl isocyanate)	toluene	91, 98
Copoly(butyl isocyanate: p-anisole-2-ethyl isocyanate	trichloroethane	81
Copoly(butyl isocyanate:	toluene	82
p-anisole-3-propyl isocyanate	chloroform	82
	tetrachloroethane	89
	1,1,2,2-tetrachloroethene	82
	chlorobenzene	82
	bromoform	82
Copoly(N^5-(2-hydroxyethyl)-	water	219
L-glutamine:N^5-benzyl-L-glutamine)	2-chloroethanol	219
Diblock(styrene:butadiene)	hydrogen	222
	deuterium	222
Diblock(dimethyl siloxane:	heptane	37
styrene	decane	37
	benzene	37
	toluene	37
	dioxane	37
	chloroform	37
Triblock(2-hydroxyethyl methacrylate: ethylene oxide: 2-hydroxyethyl methacrylate)	aq. ionic solutions	221
Triblock(dimethyl siloxane:	heptane	37
styrene:dimethyl siloxane)	decane	37
	benzene	37
	toluene	37
	dioxane	37
	chloroform	37
Alternating copoly(ethyl vinyl	water	189
ether:maleic acid)	2-chloroethanol	189
Alternating copoly(butyl vinyl	water	189
ether:maleic acid)	2-chloroethanol	189
(Co)polymer + (Co)polymer + Solvent Systems		
Poly(styrene) +		
poly(hexyl isocyanate)	toluene	85
poly(octyl isocyanate)	trichloroethane	81
copoly(butyl isocyanate: p-anisole-2-ethyl isocyanate)	trichloroethane	81
poly(isobutene)	tetrahydrofuran	229
poly(methyl methacrylate)	tetrahydrofuran	229
poly(benzyl-L-glutamate)	dioxane	85
	dimethylformamide	85
Poly(octyl isocyanate) +		
copoly(butyl isocyanate: p-anisole-2-ethyl isocyanate)	trichloroethane	81

Table 1 (*continued*)

		Ref.
Poly(urethane) +		
poly(vinyl chloride)	acetone	135
	dimethylformamide	135
Poly(p-benzamide) +		
poly(terephthalamide of p-amino benzhydrazide)	dimethyl acetamide	93
Poly(hexyl isocyanate) +		
poly(octyl isocyanate)	tetrachloroethane	86
poly(styrene)	tetrachloroethane	86
Copoly(butyl isocyanate:		
p-anisole-2-ethyl isocyanate) + poly(styrene)	tetrachloroethane	86

Polymer Blends

	Ref.
Poly(methyl methacrylate) +	
poly(vinylidene fluoride)	152, 171
chlorinated polyethylene	14
Polystyrene +	
poly(α-methyl styrene)	140
poly(vinyl methyl ether)	9, 178, 179, 181
poly(methyl acrylate)	158
poly(ethyl acrylate)	158
poly(butyl acrylate) networks	27, 158
poly(methyl methacrylate)	158
poly(butyl methacrylate)	138, 150, 158
poly(o-chlorostyrene)	161
Poly(1,4-butadiene) +	
poly(cis-1,4-butadiene)	238, 239, 242
poly(1,2-butadiene)	228
Poly(2-vinyl naphthalene) +	
poly(alkyl methacrylate)	172
Polyurethane blends	144, 145, 156
Polyurethane + poly(vinyl chloride)	135, 236, 237
Poly(urethane urea) blends	143
Poly(vinyl chloride) +	
poly(neopentyl glycol adipate)	249
copoly(butyl acrylate:acrylonitrile)	136
copoly(vinyl chloride:vinyl acetate)	233, 234, 235
copoly(vinyl chloride:propylene)	233, 234, 235
copoly(ethylene:NN-dimethyl acrylamide)	193
Poly(vinyl nitrate) +	
poly(vinyl acetate)	194
copoly(ethylene:vinyl acetate)	194
Poly(dimethyl siloxane) +	
poly(propylene glycol) (oligomer mixtures)	153
poly(propylene glycol diacetate)	42
poly(propylene glycol methyl ether)	42
Poly(ethylene glycol) blends	149
Poly(epichlorohydrin) +	
poly(neopentyl glycol adipate)	249
Epoxy polymers + copoly(butadiene:acrylonitrile)	24
Polycaprolactone +	
chlorinated polyethylene	197

[249] S. H. Goh, D. R. Paul, and J. W. Barlow, *J. Appl. Polym. Sci.*, 1982, **27**, 1091.

Table 1 (*continued*)

	Ref.
Polycaprolactone + *continued*	
poly(chlorostyrene)	250
chlorinated polypropene	250
Polylactones +	
poly(vinyl chloride)	196
poly(vinyl fluoride)	196
poly(vinylidene fluoride)	196
Poly(hydroxy ether of bisphenol-A) +	
polycaprolactone	251
poly(ethylene terephthalate)	251
poly(butylene terephthalate)	251
poly(hexamethylene terephthalate)	251
poly(1,4-cyclohexanedimethanol *iso*-terephthalate	251
poly(1,2-propylene terephthalate)	251
poly(1,3-propylene terephthalate)	251
Poly(aryloxy phosphazene) blends	139
Poly(vinylidene fluoride) + poly(methyl methacrylate)	171
Copoly(styrene:divinyl benzene) gels + polyisoprene	23
Copoly(styrene:acrylonitrile)	
poly(methyl acrylate)	198
poly(ethyl acrylate)	198
poly(ethyl methacrylate)	198
poly(propyl methacrylate)	198
poly(cyclohexyl methacrylate)	198
poly(vinyl acetate)	198
Copoly(aryloxy phosphazene) blends	139
Diblock(styrene:dimethyl siloxane) + polystyrene	241
Diblock(styrene:isoprene) +	
polystyrene	240
polyisoprene	240
Microphase Structures in Copolymers	
Diblock(styrene:dimethyl siloxane)	224, 226, 228
Diblock(styrene:isoprene)	225
Diblock(styrene:butadiene)	142
Diblock(styrene:ethylene oxide)	224
Diblock(1,4-butadiene:*cis*-1,4-butadiene)	228
Diblock(1,2-butadiene:1,4-butadiene)	228
Triblock(styrene:isoprene:styrene)	224
Triblock(styrene:ethylene oxide:styrene)	224
Triblock(styrene:ethene-co-butene:styrene)	223
Triblock(styrene:butadiene:styrene)	143
Triblock(styrene:4-vinylbenzene dimethylamine:isoprene)	227

Acknowledgements.—I would like to thank the Associates of the K-M Research Group for their encouragement and support. Especial thanks are due to Dr. H. Galina of Wroclaw Polytechnic, Poland, for supplying details of some of the material included here, and to the staff of Essex University Library for their help in collecting material for this Report.

---

[250] D. Allard and R. E. Prud'homme, *J. Appl. Polym. Sci.*, 1982, **27**, 559.
[251] J. E. Harris, S. H. Goh, D. R. Paul, and J. W. Barlow, *J. Appl. Polym. Sci.*, 1982, **27**, 839.

13
Engineering and Technology

BY S. M. RICHARDSON, G. AKAY, C. B. BUCKNALL, AND D. G. OLDER

PART I Rheology
by S. M. Richardson

1 Introduction

The years 1981 and 1982 have seen much progress in all aspects of rheology. Three important new texts have appeared. Han[1] has written a text on multi-phase flows in polymer processing operations, with a strong emphasis on the rheology of the flows. This is an especially welcome addition to the literature because it covers an area which is otherwise devoid of texts. Particular attention is paid to dispersed flows and stratified flows, which are relevant to, for example, the rheology of filled polymers and the rheology of coating processes, respectively. Cogswell[2] has written a text on the rheology of polymer melts, aimed particularly at industrial practitioners and based on his many years of experience in the field. The text is somewhat uneven in its treatment, but has much to say that is valuable and thought provoking. Arriving just too late for inclusion in the review of advances in rheology in the years 1979 and 1980 was a translation of the text by Vinogradov and Malkin[3] on the rheology of polymers, which was first published in Russia in 1977. This is an excellent text and is commended to anyone with an interest in any aspect of rheology. It is particularly strong on theoretical aspects, but much experimental data is also included.

The years 1981 and 1982 have also seen much research progress in the areas of constitutive equations, elongational and shear flow rheometry, modelling of polymer processing operations, and numerical solution of the equations of motion of viscoelastic fluids. Each of these areas, and others, will be reviewed in some detail in what follows.

2 Constitutive Equations

No revolutionary new types of constitutive equation have been developed in the last two years. Instead, the existing types have been the subjects of very intensive investigations and have been applied to an ever-wider range of flow problems. Bird[4] has extensively reviewed existing kinetic theories of polymer solutions and melts.

---

[1] C. D. Han, 'Multiphase Flow in Polymer Processing', Academic Press, New York, 1981.
[2] F. N. Cogswell, 'Polymer Melt Rheology', George Godwin, London, 1981.
[3] G. V. Vinogradov and A. Ya. Malkin, 'Rheology of Polymers', Mir, Moscow/Springer, Berlin, 1980.
[4] R. B. Bird, *J. Rheol.*, 1982, **26**, 277.

The relationship between various models (dumb-bells and freely jointed chains for solutions, Gaussian networks and reptating chains for melts) are investigated very thoroughly, as is the relationship through phase–space formulations between models for solutions and models for melts. Related to this latter point, Currie[5] has undertaken a theoretical and experimental investigation of the D–E (Doi–Edwards reptating chain) and C–B (Curtiss–Bird freely jointed chain) models. He has shown that there exists a single potential function, U, which is independent of any material constants and dependent only on the eigenvalues of the Finger strain tensor, and which completely determines the strain-dependent characteristics of both the D–E and C–B models. He has obtained an exact expression for U as well as a good working approximation to it. The results of some shear and elongation experiments on a branched LDPE are also reported. Agreement with the predictions of the D–E model are good in shear (a result corroborated by Osaki[6]) but bad in elongation. It is speculated that the reason for this is the fact that the LDPE molecules are branched; it is to be expected that better agreement would result with molecules that are linear. This point illustrates a general problem in the current state of constitutive equations. It is not difficult to develop a model which applies to a given polymer solution or melt in a particular kinematical situation, such as a shear flow. When that model is applied to a different kinematical situation, such as an elongational flow (or a complicated flow comprising shear and elongation, as occurs in practice), it is usually found that the model fails qualitatively, or at least quantitatively, to predict the rheological behaviour. This is a most unsatisfactory state of affairs, and much effort is being expended on the development of reasonably general models. A rather good attempt has been made by de Cleyn and Mewis.[7] They have developed a constitutive equation based on structural kinetics concepts and have so far applied it reasonably successfully to three different transient shear flows: stress relaxation, stress relaxation with superimposed oscillation, and stress overshoot. What will be found to happen when they apply it to elongational flows, whether transient or otherwise, remains, however, to be seen.

3 Elongational Flows

Interest in elongational flows (and, indeed, in all non-shear flows) has again increased over the last two years. Until recently, there was always some doubt about elongational flow data. There was very little independent corroboration of any data, for example, from different types of rheometer. Now this has changed. Münstedt and Laun,[8] and Münstedt and Middleman,[9] have obtained data using several different types of rheometer. They have reported good agreement in some cases and poor agreement in others. As a result, it seems that one rheometer, based on a bubble-collapse technique, gives incorrect results when used with melts, although it appears to be satisfactory when used with solutions. The reason for this is probably that the time-scale for the collapse of the bubble is so short that a steady state stress field does

[5] P. K. Currie, *J. Non-Newtonian Fluid Mech.*, 1982, **11**, 53.
[6] K. Osaki, *J. Soc. Rheol. Jpn.*, 1981, **9**, 139.
[7] G. de Cleyn and J. Mewis, *J. Non-Newtonian Fluid Mech.*, 1981, **9**, 91.
[8] H. Münstedt and H. M. Laun, *Rheol. Acta*, 1981, **20**, 211.
[9] H. Münstedt and S. Middleman, *J. Rheol.*, 1981, **25**, 29.

not exist, even though steady state kinematics do. Several relatively new elongational flow rheometers have been developed. MacSporran[10] has developed a uniaxial suspended syphon viscometer, which works for elongation rates $\dot\varepsilon$ in the range 10 to 160 s^{-1}. Rhi-Sausi and Dealy[11] have developed a biaxial sheet extension technique, which is useful for rubbers and high viscosity melts. Macosko and Winter[12] and Chatraei, Macosko, and Winter[13] have developed a biaxial lubricated squeezing flow technique, which works for $\dot\varepsilon$ in the range 0.003 to 1 s^{-1}.

4 Shear Flows

In many polymer processing operations, it is common for both the shear rate $\dot\gamma$ and the pressure p to be very high. In injection moulding, $\dot\gamma$ is typically 10^5 s^{-1} and p is typically 10^3 bar. There is, therefore, a need for rheological data, particularly shear flow data, for high $\dot\gamma$ and p. Galvin, Hutton, and Jones[14] and Crowson, Scott, and Saunders[15] have independently developed capillary viscometers capable of giving reliable data under such extreme conditions, and some interesting new data have been obtained. In contrast, MacSporran and Spiers[16,17] have investigated the dynamic response of a tried and tested machine, the model R18 Weissenberg Rheogoniometer, which is a cone-and-plate viscometer. They studied the responses to both small and large amplitude oscillations, with particular reference to the effects of fluid inertia. They have shown that inertial effects can be ignored if the Reynolds number, Re, defined in an obvious way, is less than about 0.1.

5 Converging Flows

Between the two extremes of shear and elongational flows lie converging flows. Such flows are of obvious relevance to, for example, flows into dies during extrusion operations. (It is also often claimed that such flows are important because, by lubrication and appropriate contouring of the converging channel walls, something approaching elongational flows can be obtained. Whether, however, this is in fact the case for anything other than slightly converging flows and hence only rather small extensions is, to say the least, questionable.) A great deal of work has been undertaken in this area in the last two years. Yoo and Han,[18] James and Saringer,[19,20] and Walters and Webster[21] have all conducted experimental investiga-

---

[10] W. C. MacSporran, *J. Non-Newtonian Fluid Mech.*, 1981, **8**, 119.
[11] J. Rhi-Sausi and J. M. Dealy, *Polym. Eng. Sci.*, 1981, **21**, 227.
[12] C. W. Macosko and H. H. Winter, *J. Non-Newtonian Fluid Mech.*, 1982, **11**, 301.
[13] S. Chatraei, C. W. Macosko, and H. H. Winter, *J. Rheol.*, 1981, **25**, 433.
[14] G. D. Galvin, J. F. Hutton, and B. Jones, *J. Non-Newtonian Fluid Mech.*, 1981, **8**, 11.
[15] R. J. Crowson, A. J. Scott, and D. W. Saunders, *Polym. Eng. Sci.*, 1981, **21**, 748.
[16] W. C. MacSporran and R. P. Spiers, *Rheol. Acta*, 1982, **21**, 184.
[17] W. C. MacSporran and R. P. Spiers, *Rheol. Acta*, 1982, **21**, 193.
[18] H. J. Yoo and C. D. Han, *J. Rheol.*, 1981, **25**, 115.
[19] D. F. James and J. H. Saringer, *J. Rheol.*, 1982, **26**, 321.
[20] D. F. James and J. H. Saringer, *J. Non-Newtonian Fluid Mech.*, 1982, **11**, 317.
[21] K. Walters and M. F. Webster, *Philos. Trans. R. Soc. London, Ser. A*, 1982, **308**, 199.

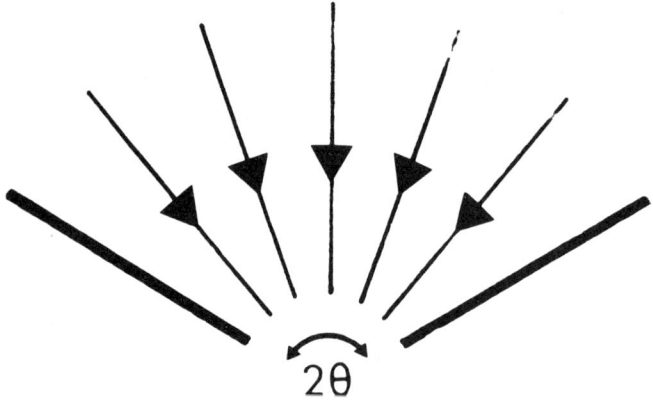

Figure 1 *Converging plane flow*

tions of converging flows of different sorts. Dunn[22] has considered converging plane (or Jeffery–Hamel) flow of a general simple fluid. He has shown that, other than for an 'exceptional subset' of such fluids, a convergent flow with straight (radial) streamlines (see Figure 1) is impossible. Hull[23] has quite independently obtained an exact solution for creeping flow of an upper-convected general linear viscoelastic fluid through a slit, which is a converging flow through a wedge of half-angle $\theta = 90°$ (see Figure 1). The solution is important because it is virtually the only non-trivial exact solution for the flow of any viscoelastic fluid. The solution involves straight streamlines and is thus a member of the 'exceptional subset'. It appears to be a rather isolated solution; no solution exists for $\theta \neq 90°$, for any θ with a co-rotational general linear viscoelastic fluid, or for any θ for a converging flow of an upper-convected general linear viscoelastic fluid through a cone. Thus the 'exceptional subset' is probably only sparsely populated.

6 Numerical Solution of Viscoelastic Flow Problems

Computer simulations of flows of viscoelastic fluids have been reported with increasing frequency over the past two years. The extrudate (or die) swell problem has been tackled using finite-element methods by Caswell and Viriyayuthakorn[24] and Crochet and Keunings[25] for Maxwell fluids, by Crochet and Keunings[26] for an Oldroyd-B fluid, and by Ben-Sabar and Caswell[27] for a Newtonian fluid with a temperature-dependent viscosity. The hole pressure problem – what does a pressure transducer mounted in a hole in a wall, and not flush with the wall, actually

---

[22] J. E. Dunn, Paper G32 presented at the 54th Annual Meeting of the Society of Rheology, held at Evanston, Illinois, on October 24–28, 1982.
[23] A. M. Hull, *J. Non-Newtonian Fluid Mech.*, 1981, **8**, 327.
[24] B. Caswell and M. Viriyayuthakorn, *J. Non-Newtonian Fluid Mech.*, 1983, **12**, 13.
[25] M. J. Crochet and R. Keunings, *J. Non-Newtonian Fluid Mech.*, 1982, **10**, 85.
[26] M. J. Crochet and R. Keunings, *J. Non-Newtonian Fluid Mech.*, 1982, **10**, 339.
[27] E. Ben-Sabar and B. Caswell, *J. Rheol.*, 1981, **25**, 537.

measure? – has also been tackled using finite-element methods by Jackson and Finlayson[28,29] and Richards and Townsend.[30] O'Brien[31] has investigated the related problem of estimation of the second normal stress difference and has used finite-difference methods to solve the equations of motion of a second-order fluid. The problem encountered in all numerical calculations involving viscoelastic fluids when the Deborah number, De, increases to of order unity, has been investigated rather thoroughly by Mendelson, Yeh, Brown, and Armstrong.[32] They examined three possible causes: bifurcation of the solution, non-existence of a steady state solution, and errors in the approximations (in particular, numerical errors in the finite-element approximations) used. They concluded that it is numerical errors alone which cause the problem. These errors arise from a failure to approximate with sufficient accuracy the stress field in those regions of the flow where the stress varies most rapidly, for example, near a die exit or near a corner. Their argument is a compelling one because it is based on calculations for a model problem involving flow of a second-order fluid for which the exact velocity field (and hence the kinematics) is known for all values of De.

7 Polymer Processing

Significant advances have been made in many aspects of polymer processing in the last two years. These advances are in the main a result of the increased use of computer simulation and control. This is especially true in the areas of fibre spinning, which White[33] has comprehensively reviewed, and screw extrusion. The non-isothermal spinning of non-Newtonian fluids has been simulated by Gagon and Denn,[34] Chang, Denn, and Geyling,[35] and Chang, Denn, and Kase.[36] The dynamic modelling and control of screw extruders has been reviewed by Costin, Taylor, and Wright.[37] Hassan and Parnaby[38] have reported what is probably the first true hierarchical automatic (adaptive) optimal control scheme for plastics extruders ever to work completely satisfactorily. The process operator is required merely to set the output melt flow-rate and the melt temperature at the die entry. The rest is handled completely automatically by the computer associated with the extruder. Improved screw designs have been reported by Ingen Housz and Meijer.[39,40] They have presented an analysis of a conventional single-channel screw extruder and followed this with an analysis of a Maillefer two-channel screw extruder. The advantages of the Maillefer extruder, which are primarily avoidance

---

[28] N. R. Jackson and B. A. Finlayson, *J. Non-Newtonian Fluid Mech.*, 1982, **10**, 55.
[29] N. R. Jackson and B. A. Finlayson, *J. Non-Newtonian Fluid Mech.*, 1982, **10**, 71.
[30] G. D. Richards and P. Townsend, *Rheol. Acta*, 1981, **20**, 261.
[31] V. O'Brien, *J. Rheol.*, 1982, **26**, 499.
[32] M. A. Mendelson, P.-W. Yeh, R. A. Brown, and R. C. Armstrong, *J. Non-Newtonian Fluid Mech.*, 1982, **10**, 31.
[33] J. L. White, *Polym. Eng. Rev.*, 1981, **1**, 297.
[34] D. K. Gagon and M. M. Denn, *Polym. Eng. Sci.*, 1981, **21**, 844.
[35] J.-C. Chang, M. M. Denn, and F. T. Geyling, *Ind. Eng. Chem., Fundam.*, 1981, **20**, 147.
[36] J.-C. Chang, M. M. Denn, and S. Kase, *Ind. Eng. Chem., Fundam.*, 1982, **21**, 13.
[37] M. H. Costin, P. A. Taylor, and J. D. Wright, *Polym. Eng. Sci.*, 1982, **22**, 393.
[38] G. A. Hassan and J. Parnaby, *Polym. Eng. Sci.*, 1981, **21**, 276.
[39] J. F. Ingen Housz and H. E. H. Meijer, *Polym. Eng. Sci.*, 1981, **21**, 352.
[40] J. F. Ingen Housz and H. E. H. Meijer, *Polym. Eng. Sci.*, 1981, **21**, 1156.

of break-up of the solids bed and prevention of solids from emerging in the melt output through the die, are then taken further. It is proposed that the depths, and not just the widths, of the channels in the screw be varied, so as to decrease average melt film thicknesses and hence improve heat transfer. Alternatively, or perhaps additionally, it is proposed that a screw with more than two channels should be used. A finite-element simulation of screw extrusion has been reported by Choo, Hami, and Pittman[41] and a simple analytical model, based on somewhat obscure assumptions and hence approximations, has been reported by Mount, Watson, and Chung.[42] Cox, Williams, and Isherwood[43] have investigated experimentally the use of powder, as opposed to granular or chip, feeds to plastics extruders. They found that use of a powder usually results in a lower flowrate of melt through the extruder, lower pressure generation near the die and hence lower power consumption in the motor, compared with the use of granules under otherwise apparently identical conditions. They also found that two different melting mechanisms occurred: the usual one (with the solids bed nearer the downstream side of the screw flight and the melt pool nearer the upstream side) and a reversed one. The reason for this is not clear.

Two relatively new areas in which great advances have been made in the last two years, but in which great advances have yet to be made before anything like satisfactory theories are developed, are mixing of polymers and polymerization during processing. Ng and Erwin[44] have investigated experimentally the mixing of LDPEs of two different colours in laminar shear and elongational flows. Their results seem to corroborate at least qualitatively the theory of Ottino, Ranz, and Macosko,[45] who have used kinematical ideas based on the deformation of contact interfaces between materials or material surfaces. Perhaps the most significant simplification that they made in their theory is complete neglect of interfacial tension effects, which must surely be incorporated in any complete theory of mixing. Recognizing that many polymer processing operations require large energy inputs and machines of very stout construction because of the high viscosities of polymer melts, there has been a great deal of industrial interest in *in situ* polymerization during processing. Yemelyanov, Smetanina, and Vinogradov[46] have measured the shear viscosities of several polymerizing systems, such as methyl methacrylate–poly(methyl methacrylate) and styrene–polystyrene, throughout the course of the polymerization so as to provide data essential for design purposes. Lee[47] has investigated both theoretically and experimentally polyester–styrene copolymerization during compression moulding, *i.e.*, sheet moulding. The copolymerization is an initiation-plus-propagation free-radical chain growth process. Good agreement is claimed at all but high conversions between the theoretical predictions, which are based on finite-difference solution of the energy and species balance equations, and the experimental results, which are obtained using differential scanning calorimetry and a Rheometrics Mechanical Spectrometer. Rojas, Adabbo,

---

[41] K. P. Choo, M. L. Hami, and J. F. T. Pittman, *Polym. Eng. Sci.*, 1981, **21**, 100.
[42] E. M. Mount, J. G. Watson, and C. I. Chung, *Polym. Eng. Sci.*, 1982, **22**, 729.
[43] A. P. D. Cox, J. G. Williams, and D. P. Isherwood, *Polym. Eng. Sci.*, 1981, **21**, 86.
[44] K. Y. Ng and L. Erwin, *Polym. Eng. Sci.*, 1981, **21**, 212.
[45] J. M. Ottino, W. E. Ranz, and C. W. Macosko, *AIChEJ.*, 1981, **27**, 565.
[46] D. N. Yemelyanov, I. Ye. Smetanina, and G. V. Vinogradov, *Rheol. Acta*, 1982, **21**, 280.
[47] L. J. Lee, *Polym. Eng. Sci.*, 1981, **21**, 483.

and Williams[48] have developed a theoretical model of the flow accompanied by curing of thermosets in the nozzle region of an injection moulding machine. Rojas, Borrajo, and Williams[49] have conducted some related experiments. Lindt,[50] Manzione,[51] and Castro, Lipshitz, and Macosko[52] have independently developed theoretical models of reaction injection moulding. The common feature of all of the models of the curing of thermosets and of the reactions during reaction injection moulding is the assumption of nth-order reaction kinetics. Thus, for example, if C denotes the degree or fraction of curing and t denotes time, it is assumed that $\partial C/\partial t$ varies as $(1 - C)^n$. In spite of the fact that the reactions do not have such a simple mechanism in practice – instead, they occur in a step-wise way, so that a Flory kinetics mechanism is more appropriate – this assumption seems to lead to predictions which are in reasonable agreement with experimental results.

---

[48] A. J. Rojas, H. E. Adabbo, and R. J. J. Williams, *Polym. Eng. Sci.*, 1981, **21**, 634.
[49] A. J. Rojas, J. Borrajo, and R. J. J. Williams, *Polym. Eng. Sci.*, 1981, **21**, 1122.
[50] J. T. Lindt, *Polym. Eng. Sci.*, 1981, **21**, 424.
[51] L. T. Manzione, *Polym. Eng. Sci.*, 1981, **21**, 1234.
[52] J. M. Castro, S. D. Lipshitz, and C. W. Macosko, *AIChEJ.*, 1982, **28**, 973.

PART II Engineering and Technology

by G. Akay and C. B. Bucknall

1 Introduction

Developments in engineering and technology of polymers are discussed under three main headings: mechanical properties, melt and solution rheology, and electrical properties. The first two topics are reviewed in some detail, while the newer subject of electrically active polymers is covered briefly.

2 Mechanical Properties

The most active areas of research on mechanical properties are those related to fracture. Considerable progress has been made in studies of crazing, fracture mechanics, and the development of novel polymeric blends, alloys, and composites having improved fracture resistance. These subjects feature prominently in the Proceedings of the Fifth International Conference on 'Deformation, Yielding and Fracture of Polymers', which give a good overview of current work on mechanical properties.[1]

Crazing.—The importance of crazes both as precursors to cracks and as deformation zones contributing to the toughness of plastics is now generally recognized, and there are several active groups studying the morphology, microstructure, growth characteristics, and healing behaviour of crazes. Thin film studies by Donald and Kramer have been particularly fruitful. These authors have shown that polystyrene films over 150 m thick form crazes having the same microstructure as those formed in the bulk polymer,[2,3] and have demonstrated a relationship between the maximum extension ratio of craze fibrils and the molecular chain length between entanglements in a range of glassy polymers.[4,5] This work promises to explain differences in mechanical properties between polymers, that were previously poorly understood. Entanglements also appear to control craze thickening, which occurs by drawing fresh material across the craze–bulk boundary.[6,7] In oriented films, the void content of crazes is a strong function of molecular orientation, falling to a very low value when the applied stress is normal to the direction of molecular orientation.[8] A high volume fraction of voids is also observed in thin polystyrene films when two crazes intersect at right angles, the fibril volume fraction being approximately equal to the product of the volume fractions in the individual crazes.[9] Interactions between

---

[1] 'Deformation, Yielding and Fracture', Proceedings of Conference held at Churchill College, Cambridge, 1982, Plastics and Rubber Institute, London, 1982.
[2] A. M. Donald, T. Chan, and E. J. Kramer, *J. Mater. Sci.*, 1981, **16**, 669.
[3] T. Chan, A. M. Donald, and E. J. Kramer, *J. Mater. Sci.*, 1981, **16**, 676.
[4] A. M. Donald and E. J. Kramer, *Polymer*, 1982, **23**, 461.
[5] A. M. Donald and E. J. Kramer, *J. Polym. Sci., Polym. Phys. Ed.*, 1982, **20**, 899.
[6] A. M. Donald, E. J. Kramer, and R. A. Bubeck, *J. Polym. Sci., Polym. Phys. Ed.*, 1982, **20**, 1129.
[7] P. I. Paredes and E. W. Fischer, *J. Polym. Sci., Polym. Phys. Ed.*, 1982, **20**, 929.
[8] N. R. Farrer and E. J. Kramer, *Polymer*, 1981, **22**, 691.
[9] P. S. King and E. J. Kramer, *J. Mater. Sci.*, 1981, **16**, 1843.

crazes growing parallel to each other have been analysed by Mills,[10] using a dislocation array model. Blunting of crazes through the formation of angled shear bands has been observed in a number of polymers;[11] physical aging raises the yield stress, and allows the formation of simpler craze structures. There are some complex interactions between crazes and existing shear bands, which can act as sites for craze initiation.[12] Active liquid environments are a much more important cause of crazing, and thin film studies show that the void content of the crazes is greatly reduced by the more aggressive liquids.[13] An activated state theory of environmental stress cracking has been proposed by Iannone et al.[14] On a more positive note, crazing is beneficial in rubber-modified plastics; electron microscopy studies of ABS and HIPS have shown a number of interactions between crazes and particles, including cavitation within the particles,[15] with consequent shear yielding in the matrix, debonding at the rubber–matrix interface in poorly-grafted material, and the termination of crazes by large particles.[16,17] The response of HIPS was found to depend strongly on particle size, but to be independent of internal structure; few crazes nucleated at particles below 1 μm in diameter,[18] and in a very dilute sample the equilibrium craze length varied with particle size.

The midrib region has a distinctive structure in many crazes; Doyle has studied crack formation through ductile failure in the midrib in polystyrene.[19] Popli and Roylance have obtained GPC evidence for bond scission during crazing in the same polymer.[20] Wool and O'Connor report healing in crazes over a range of temperatures above and below the glass transition; uniform healing takes place along the length of the craze.[21] The finite element and boundary element approaches to craze micromechanics are compared by Bevan,[22] and a theory of craze extension related to the relaxation of stresses at the tip is advanced by Passaglia.[23]

Deformation.—The second edition of I. M. Ward's well-known book on the 'Mechanical Properties of Solid Polymers'[24] is a welcome addition to the literature on deformation, and provides an excellent review of the subject, which has tended to become overshadowed by work on fracture in recent years. Struik's book is the standard work of reference on the more specialized field of physical aging.[25] The long-term effects of physical aging on the yield strength and fracture behaviour of polypropylene,[26] and of polycarbonate,[27] have been investigated. Another

[10] N. J. Mills, *J. Mater. Sci.*, 1981, **16**, 1332.
[11] A. M. Donald and E. J. Kramer, *J. Mater. Sci.*, 1982, **17**, 1871.
[12] A. M. Donald, E. J. Kramer, and R. P. Kambour, *J. Mater. Sci.*, 1982, **17**, 1739.
[13] M. B. Yaffe and E. J. Kramer, *J. Mater. Sci.*, 1981, **16**, 2130.
[14] M. Iannone, L. Nicolais, L. Nicodemo, and A. T. Dibenedetto, *J. Mater. Sci.*, 1982, **17**, 81.
[15] A. M. Donald and E. J. Kramer, *J. Mater. Sci.*, 1982, **17**, 1765.
[16] M. Dillon and M. Bevis, *J. Mater. Sci.*, 1982, **17**, 1895.
[17] M. Dillon and M. Bevis, *J. Mater. Sci.*, 1982, **17**, 1903.
[18] A. M. Donald and E. J. Kramer, *J. Appl. Polym. Sci.*, 1982, **27**, 3729.
[19] M. J. Doyle, *J. Mater. Sci.*, 1982, **17**, 760.
[20] R. Popli and D. Roylance, *Polym. Eng. Sci.*, 1982, **22**, 1046.
[21] R. P. Wool and K. M. O'Connor, *Polym. Eng. Sci.*, 1981, **21**, 970.
[22] L. Bevan, *J. Appl. Polym. Sci.*, 1982, **27**, 4263.
[23] E. Passaglia, *Polymer*, 1982, **23**, 754.
[24] I. M. Ward, 'Mechanical Properties of Solid Polymers', 2nd Edn., Wiley, New York, 1982.
[25] L. C. E. Struik, 'Physical Aging of Amorphous Polymers', Elsevier, Amsterdam, 1978.
[26] L. C. E. Struik, *Plast. Rubb. Proc. Appl.*, 1982, **2**, 41.
[27] C. Bauwens-Crowet and J. C. Bauwens, *Polymer*, 1982, **23**, 1599.

important theme is the role of shear bands, which Chau and Li have studied in detail in recent years, especially in polystyrene. They report the effects of reverse shearing,[28] of shear bands and layers of HIPS as obstacles to shear band propagation,[29] and fracture in thick shear bands.[30] Some interesting effects of hydrostatic pressure on the ductile–brittle transition in polyethylene are demonstrated by Truss, Duckett, and Ward in torsion tests;[31] the polymer is brittle both at low strain rates and at high strain rates and pressures, because of the pressure dependence of yield stress – the results are correlated by an Eyring model involving two pressure-dependent deformation mechanisms. The Eyring model is also applied by Bucknall and Page to the creep of polypropylene homopolymer and copolymer.[32]

Fracture.—Fracture studies have always been of major importance in the field of plastics. A feature of recent work is the growing understanding of fracture through the application of fracture mechanics and through a detailed analysis of micromechanics. The availability of these techniques, together with an increasing use of plastics in engineering applications, has stimulated work on fatigue. The requirement for high-performance materials having a combination of fracture resistance and other properties has resulted in a rapid growth in research on blends, alloys, and particle- and fibre-filled composites. Useful reviews of the general field of fracture in polymers are to be found in several recent books.[33–35]

Fracture Mechanics.—Linear elastic fracture mechanics (LEFM) is now well established as a technique for characterizing rigid polymers, and has also been applied successfully to fibre composites under restricted conditions. One of the standard methods used for measurements on brittle polymers is the double torsion test, which has been analysed by Leevers,[36] and by Stalder and Kausch.[37] A miniature compact tension specimen suitable for evaluating small quantities of experimental polymers is reported by Lee and Jones.[38] Crack blunting due to adiabatic heating has been observed by Williams and Hodgkinson.[39] The application of fracture mechanics to failure of adhesive joints is covered in a comprehensive review by Kinloch.[40] Practical applications of LEFM have tended to concentrate upon pipes, partly because of the relatively simple geometry, and partly because of the large investment represented by pipe systems. Details of fracture mechanics studies of rigid poly(vinyl chloride) and polyethylene pipes have been published by

---

[28] C. C. Chau and J. C. M. Li, *J. Mater. Sci.*, 1982, **17**, 3445.
[29] C. C. Chau and J. C. M. Li, *J. Mater. Sci.*, 1982, **17**, 652.
[30] C. C. Chau and J. C. M. Li, *J. Mater. Sci.*, 1981, **16**, 1858.
[31] R. W. Truss, R. A. Duckett, and I. M. Ward, *J. Mater. Sci.*, 1981, **16**, 1689.
[32] C. B. Bucknall and C. J. Page, *J. Mater. Sci.*, 1982, **17**, 808.
[33] H. H. Kausch, 'Polymer Fracture', Springer Verlag, Berlin, 1978.
[34] E. H. Andrews, ed., 'Developments in Polymer Fracture', Applied Science, London, 1979.
[35] E. H. Andrews, P. E. Reed, J. G. Williams, and C. B. Bucknall, 'Failure in Polymers', Springer Verlag, Berlin, 1978.
[36] P. S. Leevers, *J. Mater. Sci.*, 1982, **17**, 2469.
[37] B. Stalder and H. H. Kausch, *J. Mater. Sci.*, 1982, **17**, 2481.
[38] C. Y. C. Lee and W. B. Jones, *Polym. Eng. Sci.*, 1982, **22**, 1190.
[39] J. G. Williams and J. M. Hodgkinson, *Proc. R. Soc. Lond., Ser. A*, 1981, **375**, 231.
[40] A. J. Kinloch, *J. Mater. Sci.*, 1982, **17**, 617.

several groups.[41–43] Ductility effects become important in many polymers, especially at higher temperatures, and it is necessary to use yielding fracture mechanics, which is a much less established approach. The applicability of the J integral technique to LDPE,[44] and to ABS has been demonstrated; in the ABS study, results obtained over a range of temperatures have been correlated by means of an Eyring model.[45] The significance of the J integral, and its relationship to the generalized fracture mechanics theory of Andrews is the subject of an interesting discussion.[46,47]

Good reviews of fatigue in polymers are given by Hertzberg and Manson[48] and by Richardson and Sauer.[49] An American Chemical Society conference on fatigue covers most of the topics of current interest.[50] One of the most striking observations concerns the role of crystallization at the crack tip in reducing crack propagation rates in amorphous PET.[51] Short fibres hinder fatigue cracks in nylon, but mineral fillers have the opposite effect.[52] A novel phenomenon has been reported in fatigue tests on polycarbonate and a number of other ductile polymers; the formation of 'epsilon plastic zones', which consist of a craze flanked by a pair of angled shear bands.[53,54]

Composites.—The problem of applying fracture mechanics to fibre composites has already been mentioned briefly. Methods for treating fracture in angled-ply laminates are offered by several groups.[55–57] Modes of failure in carbon[58] and Kevlar[59] reinforced epoxy resins are reported, and a contour mapping method for representing energies and mechanisms of failure in fibre composites has been developed by Wells and Beaumont.[60] Talreja has reviewed mechanisms of fatigue damage in fibre composites, and proposed a pattern for fatigue life diagrams.[61] A general theory for the strength of short fibre composites is proposed by Fukuda and Chou.[62] A general problem for users of composites is how to effect satisfactory repairs; the adverse effects of adsorbed moisture on hot-cured patches are discussed

---

[41] J. F. Mandell, A. Y. Darwish, and F. J. McGarry, *Polym. Eng. Sci.*, 1982, **22**, 826.
[42] G. P. Marshall and M. W. Birch, *Plast. Rubb. Proc. Appl.*, 1982, **2**, 369.
[43] A. Gray, J. W. Mallinson, and J. B. Price, *Plast. Rubb. Proc. Appl.*, 1981, **1**, 51.
[44] J. G. Williams and J. M. Hodgkinson, *J. Mater. Sci.*, 1981, **16**, 50.
[45] N. S. Sridharan and L. J. Broutman, *Polym. Eng. Sci.*, 1982, **22**, 760.
[46] E. H. Andrews, *J. Mater. Sci.*, 1981, **16**, 1705.
[47] J. G. Williams and J. M. Hodgkinson, *J. Mater. Sci.*, 1981, **16**, 1707.
[48] R. W. Hertzberg and J. A. Manson, 'Fatigue in Engineering Plastics', Academic Press, New York, 1981.
[49] J. A. Sauer and G. C. Richardson, *Int. J. Fract.*, 1980, **16**, 499.
[50] C. L. Beatty, ed., 'Polymer Fatigue and Non-Linear Behaviour', *Polym. Eng. Sci.*, 1982, **22**, 921.
[51] A. Ramirez, J. A. Manson, and R. W. Hertzberg, *Polym. Eng. Sci.*, 1982, **22**, 975.
[52] R. W. Lang, J. A. Manson, and R. W. Hertzberg, *Polym. Eng. Sci.*, 1982, **22**, 982.
[53] M. T. Takemori and R. P. Kambour, *J. Mater. Sci.*, 1981, **16**, 1108.
[54] M. T. Takemori and D. S. Matsumoto, *J. Polym. Sci., Polym. Phys. Ed.*, 1982, **20**, 2027.
[55] C. Bathias, R. Esnault, and J. Pallas, *Composites*, 1981, **12**, 195.
[56] T. Nishioka and S. N. Atluri, *Eng. Fract. Mech.*, 1982, **16**, 573.
[57] I. Roman, H. Havel, and G. Marom, *Polym. Compos.*, 1981, **2**, 199.
[58] P. S. Theocaris and C. A. Stassinakis, *J. Compos. Mater.*, 1981, **15**, 133.
[59] R. J. Morgan, E. T. Mones, W. J. Steele, and S. B. Deutscher, *SAMPE Q.*, 1981, **12**(3), 26.
[60] J. K. Wells and P. W. R. Beaumont, *J. Mater. Sci.*, 1982, **17**, 397.
[61] R. Talreja, *Proc. R. Soc. Lond., Ser. A*, 1981, **378**, 461.
[62] H. Fukuda and T. W. Chou, *J. Mater. Sci.*, 1982, **17**, 1003.
[63] S. H. Mayre, J. D. Labor, and S. C. Aker, *Composites*, 1982, **13**, 289.

by Mayre et al.,[63] and the alternative methods of jointing, including bolting, are also covered in a recent conference.[64]

Fracture-resistant Materials.—Commercial interest in both miscible and immiscible blends and alloys remains strong. There is now a range of books covering various aspects of the subject,[65—69] which has been covered by several recent conferences.[70,71] An interesting development is the use of ionic interactions to promote compatibility between phases.[72—74] Some of the most significant advances in the field of composites are related to the use of thermoplastics matrices; short-fibre reinforced materials have been available for many years, and are the subject of a new book,[75] but high-performance long-fibre composites are a more recent development, represented by polyetheretherketone (PEEK) and polyetherimide.[76] These materials are making a serious challenge for the airframe market.

3 Rheology

Flow-induced microstructure is an important theme in current research on polymer rheology, and one that is likely to develop further, since it not only has a bearing on the constitutive equations for polymer melts, but also is fundamental to the study of polymer blends, liquid crystalline polymers, fibre- and particle-filled polymers, emulsions, suspensions, and foams. Microstructural effects may also affect heat and mass transfer.

Constitutive Equations.—Until recently, the only molecular theories of polymer melts and solutions were 'network theories' relating to entanglements of the macromolecules; constitutive equations based on these theories are reviewed by Lodge, Armstrong, Wagner, and Winter.[77] Soong and Shen have developed a kinetic network model which explains time-dependent and non-linear viscoelastic properties in terms of a dynamic equilibrium, involving the formation and loss of entanglements.[78,79] The model has been applied successfully to transient flow.[80]

---

[64] L. N. Phillips, ed., *Composites*, 1982, **13**, 218.
[65] J. A. Manson and L. H. Sperling, 'Polymer Blends and Composites', Plenum Press, New York, 1976.
[66] C. B. Bucknall, 'Toughened Plastics', Applied Science, London, 1977.
[67] D. R. Paul and S. Newman, ed., 'Polymer Blends', Vols. I and II, Academic Press, New York, 1978.
[68] O. Olabisi, L. M. Robeson, and M. T. Shaw, 'Polymer–Polymer Miscibility', Academic Press, New York, 1979.
[69] D. Klempner and K. C. Frisch, ed., 'Polymer Alloys', Plenum Press, New York, 1980.
[70] L. A. Utracki, ed., *Polym. Eng. Sci.*, 1982, **22**, 1107.
[71] 'Polymer Blends', Proceedings of Conference held at Warwick University, 14–16 Sept. 1981, Plastics and Rubber Institute, London, 1981.
[72] A. Eisenberg, P. Smith, and Z. L. Zhou, *Polym. Eng. Sci.*, 1982, **22**, 57.
[73] A. Eisenberg, P. Smith, and Z. L. Zhou, *Polym. Eng. Sci.*, 1982, **22**, 653.
[74] A. Eisenberg, P. Smith, and Z. L. Zhou, *Polym. Eng. Sci.*, 1982, **22**, 1117.
[75] M. J. Folkes, 'Short-Fibre Reinforced Thermoplastics', Wiley, New York, 1982.
[76] Anon., *Mod. Plast. Int.*, 1982, **12**(12), 4.
[77] A. S. Lodge, R. C. Armstrong, M. H. Wagner, and H. H. Winter, *Pure Appl. Chem.*, in the press.
[78] D. S. Soong and M. Shen, *Polym. Eng. Sci.*, 1980, **20**, 1177.
[79] D. S. Soong and M. Shen, *J. Rheol.*, 1981, **25**, 259.
[80] T. Y. Liu, D. S. Soong, and M. C. Williams, *Polym. Eng. Sci.*, 1981, **21**, 675.

Jongchaap[81] has developed a constitutive equation, based on the transient network concept, which reduces to the Marrucci model.[82,83]

The 'reptation' model of Doi and Edwards[84,85] is a totally different approach, which can predict the molecular weight dependence of rheological properties in polymer melts and concentrated solutions. Curtiss and Bird[86] have generalized the model to cover the complete range of polymer concentrations, and the kinetic theory of polymeric fluids has been discussed by Bird.[87] Graessley[88] has analysed the Doi–Edwards model critically, and proposed an improved model, which gives a better prediction of the elasticity of entangled networks.

Predictions of the transient network and reptation models have been compared with experimental results by Dealy and Tsang,[89,90] who found that the characteristic times for re-entanglement after a reduction or reversal of applied stress were much longer than predicted by the Marrucci[82] or Phan Thien-Tanner[83] models. Liu, Soong, and Williams[80] found good agreement between the network model of Soong and Shen[78,79] and experimental results. De Cleyn and Mewis[91,92] have tested the applicability of the Marrucci model to transient flows of concentrated solutions. Soong[93] has reviewed the transient viscoelastic properties of polymeric fluids, and relevant network theories.

Currie[94] has shown that the Doi–Edwards and Curtiss–Bird reptation models involve a potential function that is determined by the first and second invariants of the Finger strain tensor; he concludes that differences between theory and experiments are due to differences in the relative importance of the two invariants between one material and another. Successful applications of the Doi–Edwards theory have been demonstrated by Osai, Kimura, and Kurata,[95] and by Menenzes and Graessley.[96]

Fluids with Orientable Microstructure.—There has been growing interest in fluids having an identifiable microstructure that can be defined by an independent orientation vector, or 'director', lying along a preferred direction. Important examples are liquid crystalline polymers, in which the molecules are oriented, and polymer melts containing short glass fibres. In both cases, there is a distribution of orientations of the individual units.

The most widely used theory of liquid crystal behaviour, advanced by Leslie and Ericksen, is discussed by Wissbrun.[97] The theory has been developed further by

[81] R. J. J. Jongchaap, *J. Non-Newtonian Fluid Mech.*, 1981, **8**, 183.
[82] D. Acierno, F. P. La Mantia, G. Marrucci, and G. Titomanlio, *J. Non-Newtonian Fluid Mech.*, 1976, **1**, 125.
[83] N. Phan-Thien and R. I. Tanner, *J. Non-Newtonian Fluid Mech.*, 1977, **2**, 353.
[84] M. Doi and S. F. Edwards, *J. Chem. Soc., Faraday Trans. 2*, 1978, **74**, 1789, 1802, and 1818.
[85] M. Doi and S. F. Edwards, *J. Chem. Soc., Faraday Trans. 2*, 1979, **75**, 38.
[86] C. F. Curtiss and R. B. Bird, *J. Chem. Phys.*, 1981, **74**, 2016 and 2026.
[87] R. B. Bird, *J. Rheol.*, 1982, **26**, 277.
[88] W. W. Graessley, *Adv. Polym. Sci.*, 1982, **47**, 67.
[89] J. M. Dealy and W. K. W. Tsang, *J. Appl. Polym. Sci.*, 1981, **26**, 1149.
[90] K. W. Tsang and J. M. Dealy, *J. Non-Newtonian Fluid Mech.*, 1981, **9**, 203.
[91] G. DeCleyn and J. Mewis, *J. Non-Newtonian Fluid Mech.*, 1981, **9**, 91.
[92] J. Mewis and G. DeCleyn, *AIChE J.*, 1982, **28**, 900.
[93] D. S. Soong, *Rubber Chem. Technol.*, 1981, **54**, 641.
[94] P. K. Currie, *J. Non-Newtonian Fluid Mech.*, 1982, **11**, 53.
[95] K. Osaki, S. Kimura, and M. Kurata, *J. Rheol.*, 1981, **25**, 549.
[96] E. V. Menenzes and W. W. Graessley, *J. Polym. Sci., Polym. Phys. Ed.*, 1982, **20**, 1817.
[97] K. F. Wissbrun, *J. Rheol.*, 1981, **25**, 619.

Abhiraman and George to provide for a shear rate dependence of the degree of orientation,[98] and by Akay and Leslie[99] to describe unfilled and fibre-filled melts as well as liquid crystals. Doi has advanced a molecular theory of liquid crystalline polymers in concentrated solutions,[100] which can explain concentration and molecular-weight dependence of viscosity.[97] Several recent continuum theories describe the interaction between microstructure and flow field by a microstructure vector or tensor.[101–105] Gupta and Metzner have proposed a thermorheological equation of state in which the rate of change of temperature is incorporated into the stress equation through a constant related to molecular parameters, and obtained good agreement with experiment.[106] The theory is relevant to fibre spinning, film blowing, and injection moulding, in which temperature gradients are often large.

Flow-induced Phenomena.—A large number of flow-induced phenomena, which are a direct result of the microstructure of the fluid, have been recognized in polymer solutions, melts, blends, and fibre-filled or particle-filled melts. Constitutive equations based on structured fluid theories should therefore apply to these systems. This field was pioneered by Eringen,[107,108] and the literature relevant to polymer rheology and processing is reviewed below.

Microstructure Redistribution.—In inhomogeneous velocity gradients, macromolecules migrate away from regions of high shear rate, causing depletion. For example, in capillary flow, migration creates 'slip' at the wall, violating the usually assumed boundary conditions. Kraynik and Scowalter[109] have studied flow through cylindrical tubes, and shown that the amount of slip is affected by the nature of the tube wall; slip reduces extrudate swell and delays the onset of extrudate distortion. Self-oscillations in reservoir pressure are observed with Teflon walls, and attributed to a stick-slip effect. Akay found a similar effect in polypropylene and Nylon-6,6 melts containing short glass fibres, resulting from fibre depletion at the flat capillary entrance and at the capillary wall, over a narrow range of shear rates;[110] when steady flow is restored, the extrudate diameter becomes constant, and equal to that of the capillary. Kanu and Shaw observed slip in blends of EPDM with a lower viscosity fluorinated rubber, which appeared to segregate at the 90° entrance to the capillary, and become concentrated in the recirculating flow zone before feeding along the capillary wall, forming a skin around the EPDM core.[111] Even a minor amount of slip may dramatically affect mass transfer in polymeric fluids; a unified theory has been developed by Mashelkar and Dutta.[112] Akay has investigated both

---

[98] A. S. Abhiraman and W. George, *J. Polym. Sci., Polym. Phys. Ed.*, 1980, **18**, 127.
[99] G. Akay and F. M. Leslie, Cranfield Institute of Technology Report, July, 1982.
[100] M. Doi, *J. Polym. Sci., Polym. Phys. Ed.*, 1981, **19**, 229.
[101] B. R. Duffy, *J. Non-Newtonian Fluid Mech.*, 1978, **4**, 177.
[102] B. R. Duffy, *J. Non-Newtonian Fluid Mech.*, 1980, 7, 107 and 359.
[103] B. R. Duffy, *J. Non-Newtonian Fluid Mech.*, 1981, **9**, 1.
[104] W. E. Vanarsdale, *J. Rheol.*, 1982, **26**, 477.
[105] W. L. Olbricht, J. M. Rallison, and L. G. Leal, *J. Non-Newtonian Fluid Mech.*, 1982, **10**, 291.
[106] R. K. Gupta and A. B. Metzner, *J. Rheol.*, 1982, **26**, 181.
[107] A. C. Eringen, *Int. J. Eng. Sci.*, 1970, **8**, 819.
[108] C. K. Kang and A. C. Eringen, *Bull. Math. Biol.*, 1976, **38**, 135.
[109] A. M. Kraynik and W. R. Schowalter, *J. Rheol.*, 1981, **25**, 95.
[110] G. Akay, *J. Non-Newtonian Fluid Mech.*, in the press.
[111] R. C. Kanu and M. T. Shaw, *Polym. Eng. Sci.*, 1982, **22**, 507.
[112] R. A. Mashelkar and A. Dutta, *Chem. Eng. Sci.*, 1982, **37**, 969.

theoretically and experimentally the effects of flow induced diffusion on the formation of concentration profiles during convective diffusion with or without chemical reactions.[113]

Microstructure Orientation.—Large-scale microstructural orientation occurs in linear polymers, liquid crystalline polymers, and melts containing short glass fibres. Akay concluded from a study of a number of polymers that fibre orientation is mainly parallel to the flow direction if melt elasticity is low, and transverse to the flow direction if it is high.[114,115] Bright and Darlington studied the influence of mould geometry and moulding conditions, and concluded that mould geometry was the most important factor affecting the distribution of fibre orientations.[116] Knutsson, White, and Abbas have shown that the fraction of fibres aligned along the flow direction increases with shear rate.[117]

Liquid crystalline polymers show similarities to fibre-filled melts, as has been mentioned. Wissbrun has reviewed the literature on the unusual rheological properties exhibited by these polymers, both in solution and in the melt, which are due partly to molecular orientation. Jerman and Baird have shown that copolyester liquid crystalline melts have lower melt viscosities than the corresponding homopolymers, and negligible or negative die swell; they attribute these properties to boundary layer orientation of the molecules during flow.[118] The high entrance pressure losses and low die swell values emphasize the similarity with fibre-filled melts. Prasadarao, Pearce, and Han observed first normal stress differences in certain liquid crystalline copolyesters that were either negative or periodic with shear rate; the liquids exhibited a yield stress at low shear rates.[119] Wissbrun reported shear thickening followed by shear thinning with increasing shear rate.[120] These unusual observations can be related to flow induced orientation perpendicular to the flow direction, using the Leslie–Ericksen anisotropic flow theory.[97,114]

Orientation of macromolecules or discrete particles causes anisotropy of heat and mass transfer, effects that are well known in solids. White and Knutsson have developed a theory for anisotropic heat transfer in glass-filled melts.[121] The possibility of anisotropic mass transfer in structure-forming liquids is discussed by Akay.[122]

Rheology of Two-phase Materials.—Han has reviewed the use of capillary rheometers to study filled or multi-phase polymers, and drawn attention to problems of interpreting end-pressure losses, die swell, and normal stress differences.[123] Boger has reviewed circular entry flows in polymer solutions, but his

[113] G. Akay, *Polym. Eng. Sci.*, 1982, **22**, 798.
[114] G. Akay, *Polym. Eng. Sci.*, 1982, **22**, 1027.
[115] G. Akay in 'Interrelations between Processing, Structure, and Properties of Polymeric Materials', ed. J. C. Seferis and P. S. Theocaris, Elsevier, Amsterdam, 1983.
[116] P. F. Bright and M. W. Darlington, *Plast. Rubb. Proc. Appl.*, 1981, **1**, 139.
[117] B. A. Knutsson, J. L. White, and K. B. Abbas, *J. Appl. Polym. Sci.*, 1981, **26**, 2347.
[118] R. E. Jerman and D. G. Baird, *J. Rheol.*, 1981, **25**, 275.
[119] M. Prasadarao, E. M. Pearce, and C. D. Han, *J. Appl. Polym. Sci.*, 1982, **27**, 1343.
[120] K. F. Wissbrun, *Br. Polym. J.*, 1980, Dec., 163.
[121] J. L. White and B. A. Knutsson, *Polym. Eng. Rev.*,1982, **2**, 71.
[122] G. Akay in European Federation of Chemical Engineering Symposium Series, No. 27, 1983.
[123] C. D. Han, *Polym. Eng. Rev.*, 1981, **1**, 363.

conclusions are equally valid for melts.[124] High entrance pressure losses are caused by shear thinning and melt elasticity.[124–126] Utracki and Kamal have compared the rheology of polymer blends with that of model systems, and classified behaviour in terms of deviations from the so-called log-additivity rule.[127] Carley and Crossan have found wide variations in the composition dependence of viscosity in binary blends of thermoplastics, corresponding to all three types of deviation described by Utracki and Kamal.[128] Munstedt reports an increase in yield stress and decrease in die swell with increasing rubber content and decreasing particle size in rubber-modified styrene–acrylonitrile and poly(vinyl chloride).[129] White, Plochocki, and Tanaka have reviewed the shear and elongational flow behaviour of two-phase polymer melts, including rubber-modified plastics, emphasizing the consistency of data from different instruments, and dependence upon composition.[130]

Utracki and Fisa have reviewed the rheology and processing behaviour of fibre- and flake-filled thermoplastics and thermosets, including discussions of surface coatings, interactions between filler and matrix, and filler–filler effects,[131] which are also reviewed by Morrell.[132] Han et al. show that the viscosity of polypropylene melts filled with $CaCO_3$ or glass beads may be reduced significantly with the aid of suitable coupling agents, which also affect the strength and elongation of injection-moulded tensile specimens.[133] Bigg has studied the effects of coupling agents on the viscosity of LDPE containing 15 µm steel spheres.[134] Transient and time-dependent flows and flow instabilities in $CaCO_3$ and fibre-filled polymers have been studied by Akay and related to injection moulding behaviour.[114,115]

Pressure affects the viscosity of melts in several processing operations. Crowson, Scott, and Saunders report a two-stage capillary rheometer for measuring the pressure-dependence of viscosity at high shear rates relevant to injection moulding,[135] and Denn has analysed the capillary flow of a power law fluid having a temperature and pressure dependent viscosity.[136]

Structure Development during Processing.—Molecular orientation and morphology developed during processing have an important effect on mechanical properties, and are the subject of active investigation. Katti and Schultz review microstructure formation in injection-moulded semi-crystalline polymers; a highly oriented skin encloses a largely isotropic core.[137] Orientation development in blow

---

[124] D. V. Boger in 'Advances in Transport Processes', ed. A. S. Mujumdar and R. A. Mashelkar, Wiley, New Delhi, 1982, Vol. 2.
[125] L. Choplin and P. J. Carreau, *J. Non-Newtonian Fluid Mech.*, 1981, **9**, 119.
[126] J. S. Vrentas, J. L. Duda, and S. A. Hong, *J. Rheol.*, 1982, **26**, 347.
[127] L. A. Utracki and M. R. Kamal, *Polym. Eng. Sci.*, 1982, **22**, 96.
[128] J. F. Carley and S. C. Crossman, *Polym. Eng. Sci.*, 1981, **21**, 249.
[129] H. Munstedt, *Polym. Eng. Sci.*, 1981, **21**, 259.
[130] J. L. White, A. P. Plochocki, and H. Tanaka, *Polym. Eng. Rev.*, 1981, **1**, 218.
[131] L. A. Utracki and B. Fisa, *Polym. Compos.*, 1982, **3**, 193.
[132] S. H. Morrell, *Plast. Rubb. Proc. Appl.*, 1981, **1**, 179.
[133] C. D. Han, T. V. D. Weghe, P. Shete, and J. R. Haw, *Polym. Eng. Sci.*, 1981, **21**, 196.
[134] D. M. Bigg, *Polym. Eng. Sci.*, 1982, **22**, 512.
[135] R. J. Crowson, A. J. Scott, and D. W. Saunders, *Polym. Eng. Sci.*, 1981, **21**, 748.
[136] M. M. Denn, *Polym. Eng. Sci.*, 1981, **21**, 65.
[137] S. S. Katti and J. M. Schultz, *Polym. Eng. Sci.*, 1982, **22**, 1000.

moulding[138] and tubular film extrusion[139,140] is described by White and co-workers, and fitted to a biaxial deformation model.[141,142] The influence of injection-moulding conditions on the structure and mechanical properties of polymers has been investigated by Schmidt, Opfermann, and Menges.[143]

Novel processing techniques are reported by a number of groups. Ibar describes a moulding process in which flow is assisted by vibration, the glass transition temperature decreasing with frequency.[144] Zacharides and Economy discuss rotation of the mould cavity during compression or injection moulding, to increase modulus and impact strength through biaxial molecular orientation.[145] Baird and Wilkes describe co-injection of a liquid crystalline copolyester with a filled polyester, adhesion between the copolyester skin and the core being excellent.[146] Akay has studied co-injection of reinforced thermoplastics and shown that mechanical interlocking between phases can be improved by fillers and by an appropriate choice of processing conditions.[147]

Injection moulding and extrusion of melts containing foaming agents have been studied by Han and Yoo,[148] who found that the formation and expansion of gas bubbles during the initial stages of mould filling are strongly influenced by stresses in the melt.[149] Olabisi reports a process in which a gas, usually nitrogen, is injected into the melt in the mould cavity, forming a moulding in which a solid skin encloses an air space reinforced with ribs.[150]

Mathematical Modelling.—Modelling of polymer processing operations requires the simultaneous solution of equations of continuity, momentum, energy, and sometimes diffusion; numerical techniques are therefore necessary. Difficulties in using numerical techniques in non-Newtonian fluid mechanics are discussed by Crochet and Walters.[151] Studies of mathematical modelling and computer simulation have been published relating to injection moulding,[152—154] reaction injection moulding,[155] melt spinning,[156,157] blow moulding,[158,159] single-screw extrusion,[160,161] and wire coating co-extrusion.[162]

[138] J. L. White and J. E. Spruiell, *Polym. Eng. Sci.*, 1981, **21**, 859.
[139] Y. Shimomura, J. F. Spruiell, and J. L. White, *J. Appl. Polym. Sci.*, 1982, **27**, 2663.
[140] K. Choi, J. E. Spruiell, and J. L. White, *J. Polym. Sci., Polym. Phys. Ed.*, 1982, **20**, 27.
[141] J. L. White and A. Agrawal, *Polym. Eng. Rev.*, 1981, **1**, 267.
[142] K. Matsumoto, J. F. Fellers, and J. L. White, *J. Appl. Polym. Sci.*, 1981, **26**, 85.
[143] L. Schmidt, J. Opferman, and G. Menges, *Polym. Eng. Rev.*, 1981, **1**, 1.
[144] J. P. Ibar, *Polym. Plast. Technol. Eng.*, 1981, **17**, 11.
[145] A. E. Zachariades and J. Economy, *Polym. Eng. Sci.*, 1983, **23**, 266.
[146] D. G. Baird and G. L. Wilkes, *Polym. Eng. Sci.*, 1983, **23**, 632.
[147] G. Akay, *Polym. Compos.*, 1983, **4**, 256.
[148] H. J. Yoo and C. D. Han, *Polym. Eng. Sci.*, 1981, **21**, 69.
[149] C. D. Han and H. J. Yoo, *Polym. Eng. Sci.*, 1981, **21**, 518.
[150] O. Olabisi, *Polym. Eng. Rev.*, 1982, **2**, 29.
[151] M. J. Crochet and K. Walters, *Ann. Rev. Fluid Mech.*, 1983, **15**, 241.
[152] T. S. Chung and M. E. Ryan, *Polym. Eng. Sci.*, 1981, **21**, 271.
[153] M. R. Kamal and P. G. Lafleur, *Polym. Eng. Sci.*, 1982, **22**, 1066.
[154] H. van Wijngaarden, J. F. Dijksman, and P. Wesseling, *J. Non-Newtonian Fluid Mech.*, 1982, **11**, 175.
[155] L. T. Manzione, *Polym. Eng. Sci.*, 1981, **21**, 1234.
[156] J. L. White, *Polym. Eng. Rev.*, 1981, **1**, 297.
[157] D. K. Gagon and M. M. Denn, *Polym. Eng. Sci.*, 1981, **21**, 844.
[158] R. K. Gupta, A. B. Metzner, and K. F. Wissbrun, *Polym. Eng. Sci.*, 1982, **22**, 172.
[159] M. E. Ryan and K. Dutta, *Polym. Eng. Sci.*, 1982, **22**, 1075.
[160] H. Fukase, T. Kunio, S. Shinya, and A. Nomura, *Polym. Eng. Sci.*, 1982, **22**, 578.
[161] E. W. Agur and J. Vlachopoulos, *Polym. Eng. Sci.*, 1982, **22**, 1084.
[162] S. Basu, *Polym. Eng. Sci.*, 1981, **21**, 1128.

4 Electronic Properties

Interest in electrically active polymers is at a very high level; the science and technology are making rapid strides and show every sign of developing even more rapidly. There are two main areas of interest: ferroelectric behaviour, especially in poly(vinylidene fluoride), and electrical conductivity in such polymers as polyacetylene. Both subjects are reviewed in a recent book.[163]

Ferroelectrics.—Poly(vinylidene fluoride) exhibits an outstanding combination of electrical, mechanical, and acoustical properties, which make it the only polymeric candidate in piezoelectric and pyroelectric applications, which include microphones, loudspeakers, headphones,[164,165] keyboards, and other electromechanical devices.[166] The mechanisms responsible for these unique properties are discussed in two recent reviews.[167,168] However, alternative ferroelectric polymers may become available in the future; high piezoelectric activity has recently been reported in poled γ-nylon 11, apparently resulting from breaking and reforming of hydrogen bonds.[169—171]

Conductive Polymers.—Developments in conductive polymers are discussed in a recent book.[172] Mechanisms of electrical conduction are the main focus of interest for physicists working in the field, while for chemists the emphasis is on production of novel materials either by synthesis of new polymers or by doping of known polymeric conductors. Doped polyacetylene is the prototype material. Transition-metal salts have been shown to have some advantages over established dopants because of their lower diffusion coefficients.[173]

---

[163] J. Mort and G. Pfister, ed., 'Electronic Properties of Polymers', Wiley, New York, 1982.
[164] G. M. Sessler, *J. Acoust. Soc. Am.*, 1981, **70**, 1596.
[165] A. J. Lovinger in 'Developments in Crystalline Polymers', ed. D. C. Bassett, Applied Science, London, 1982, Vol. 1.
[166] G. T. Pearman and J. L. Hokanson, *Ferroelectrics*, 1980, **28**, 311.
[167] M. G. Broadhurst and G. T. Davis in 'Topics in Modern Physics – Electrets', ed. G. M. Sessler, Springer Verlag, Berlin, 1980.
[168] R. G. Kepler and R. A. Anderson, *CRC Crit. Rev. Solid State Mater. Sci.*, 1980, **9**, 399.
[169] B. A. Newman, P. Chen, K. D. Pae, and J. I. Scheinbeim, *J. Appl. Phys.*, 1980, **51**, 5161.
[170] J. I. Scheinbeim, *J. Appl. Phys.*, 1981, **52**, 5939.
[171] V. Gelfand and D. Katz, *Ferroelectrics*, 1981, **33**, 111.
[172] R. B. Seymour, ed., 'Conductive Polymers', Plenum Press, New York, 1981.
[173] M. Rolland, M. Aldissi, and F. Schue, *Polymer*, 1982, **23**, 834.

PART III Electrical Properties
by D. G. Older

1 Introduction

A large number of papers are published every year on the electrical properties of polymers. The selection of papers for inclusion in this Report has been made on the basis of their potential interest to chemists. This somewhat arbitrary selection has hopefully produced a rather clearer picture of the 'state of the art'. The engineering and technological value of the electrical properties of macromolecules have been historically based on their use as very good insulators. Electrical insulation remains still easily the most commercially important electrical property, and is considered in the first section of this Report. However, not many papers of direct relevance to chemists or chemical technologists have been published on this topic in the period of this review.

The formulation of polymers with various conducting additives to produce conducting polymer blends is now a well-established practice. Research and innovation is still active in this area and is considered in the second section.

The third section deals with the current literature on intrinsically conducting polymers. It seems there is still scope in this newer aspect of electrical properties for the synthesis of novel conducting macromolecules in addition to research into the mechanisms of electrical conduction and the chemical instability of the more conducting polymers. Work on piezoelectric effects is briefly mentioned, as is some work on organic materials for use in light-weight batteries; however, work on photoconduction and magnetic properties has not been included. The patent literature has been excluded from this Report although for a wider view of the background to this topic the reader is recommended to refer to this.

2 Electrical Insulation

A wide variety of different chemical types of macromolecules continue to be used for electrical insulation, in applications ranging from elastomers for power transmission cables to thermoset compounds for microelectronic devices. Much of the development work in these areas is carried out in industrial laboratories, thus published literature on new trends is not readily available. Work is continuing on the reduction of water permeability of insulators and also on the improved fire resistance and suppression of smoke and toxic product emission from cables in fire situations.

An investigation into the dielectric properties of a range of thermoplastic polyurethanes has been reported by Lawandy and Abd-el-Nour.[1] It was concluded that polymers with minimal isocyanate content were the best insulators. Thermoplastic polyurethanes with small amounts of carbon black were found to have better

---

[1] S. N. Lawandy and K. N. Abd-el-Nour, *Eur. Rubber J.*, 1982, **40**, 53.

dielectric properties than the unfilled polymers. The effect of chemical structure on dielectric strength for some epoxy resins made with various ratios of ester and ether bonds is reported by Takahama, Hayashi, and Sato.[2] The mechanism of the electrical breakdown of these polymers at a critical temperature is investigated.

3 Electrically Conducting Fillers in Polymers

Electrical conduction by polymers has for many years past been achieved by the incorporation of electrically conducting additives into insulating polymer matrices. The establishment of optimum concentrations of additives still receives attention. Fowler and Litman[3] have compounded polyacrylonitrile and pitch carbon fibre, metallized glass fibres, aluminium flake, and carbon black into polycarbonate and nylon-6,6 to achieve a critical level of conductive filler while maintaining good mechanical reinforcement.

A study into electrical resistivity, EMI shielding attenuation, and also thermal conductivity for thermoplastic and thermoset matrices, has been made by Briggs and Bradbury.[4] Fillers used included carbon black, metal fibres and flakes, and also metal-coated glass fibres. Sircar and Wells[5] have correlated the temperature dependences of electrical resistance for carbon blacks in an ethylene–vinyl acetate copolymer using a modified differential scanning calorimeter. A further study by Sircar[6] reports on attempts to overcome the processing difficulties introduced with conducting blacks, by modification of the elastomeric phase.

A report (in Italian) by Wallteg[7] describes a polyethylene butyl rubber graft polymer which can accept high loadings of conductive fillers and remain sufficiently pliable for use as a substitute, it is claimed, for metallic conductors in wire and cables. An unusual development is reported from Japan.[8] Metal particles are blended into a silicone elastomer. This material in thin sheet form is found to conduct under the influence of pressure, but to revert to an insulator on removal of pressure. Some applications for this novel switch are mentioned.

A brief report from MIT[9] describes how electrically conductive polyethylene, polystyrene, or polybutadienes containing Ziegler-type catalysts, can be made by impregnation with polyacetylene $(CH)_n$ by direct absorption of acetylene from the gas phase. This blend of polymers is impregnated with iodine to produce the conducting 'doped' polyacetylene phase (see the following section of this Report).

An ISO Standard (2883:1980)[10] for electrical resistance limits for antistatic and conductive elastomers for industrial use has been issued.

[2] T. Takahama, O. Hayashi, and F. Sato, *J. Appl. Polym. Sci.*, 1981, **7**, 2211.
[3] N. E. Fowler and A. M. Litman, *Plast. Eng.*, 1981, **37**, 29.
[4] D. N. Biggs and E. J. Bradbury, *Plast. Chem.*, 1980, **43**, 746.
[5] A. K. Sircar and J. L. Wells, *Polym. Eng. Sci.*, 1981, **21**, 809.
[6] A. K. Sircar, *Rubber Chem. Technol.*, 1981, **54**, 820.
[7] B. Wallteg, *Mater. Plast. Elast.*, 1981, **7/8**, 431.
[8] *Jpn. Chem. Week*, 1982, Sept., 6.
[9] Massachusetts Institute of Technology, *Chem. Week*, 1982, August, 23.
[10] ISO Standard 2883 1980, Geneva, 1980.

4 Electrically Conducting Polymers

The intrinsic electrical conduction by polymers has received great attention over the past few years. Several authoritative reviews,[11–15] an article,[16] and a book,[17] have appeared, setting out the background to these synthetic polymers with marked metallic character. Much of the current literature on the research in this field has been devoted to the study of charge transfer complexes of *trans*-polyacetylene doped with *p*-type electron accepting species. Typically, dopants are AsF_5, Br_2, and I_2, SbF_5, $HClO_4$, or H_2SO_4. The mechanism of electrical conduction in charge transfer complexes of this type is not fully understood. The relationship between unpaired spins (free radicals) and isomerization, doping, and electrical conductivity of polyacetylenes has recently been discussed by Chien.[18] The discussion is based on 'solitons' which occur at the disturbance in *trans*-$(CH)_n$ chains by bond alternation. The structure of polyacetylene doped with iodine has also been investigated by electron microscopy by Chien.[19] The poor oxidative stability of doped polyacetylene is currently seen as a drawback to its commercial exploitation in a wide range of electrical and electronic devices.

Pochan and Gibson[20] have studied the effects of oxygen on the conductivity of $(CH)_n$ and poly(1,6-heptadiyne). In another paper[21] these authors emphasize the need to isolate $(CH)_n$ from oxygen. Wegner[11] is optimistic that these stability problems will be overcome and that useful electrical products such as light-weight batteries, semi-conductors, and light-weight conductors will eventually be developed.

A study of the mechanism of halogen doping and the structural changes it induces in polyacetylene are clearly essential for a better understanding of the causes of this instability. Monkenbusch,[22] Chien *et al.*,[23] and Deits *et al.*,[24] have all used X-ray diffraction to look at structural changes in thin films of I_2 doped $(CH)_n$. Shirakawa *et al.*[25] has reported the use of X-ray photoelectron spectroscopy (XPS), Raman and mass spectrometry to study highly conductive $(CH)_n$ doped with Br_2. Both Monkenbusch and Shirakawa have reported three distinct phases in the relationship between the structure and the electrical conductivity of halogen-doped $(CH)_n$. Conductivity was found by Shirakawa to decrease in the final third stage of the addition of Br_2 to $(CH)_n$.

---

[11] G. Wegner, *Angew. Chem., Int. Ed. Engl.*, 1981, **20**, 361.
[12] R. H. Baughman, *et al.*, *Org. Coatings Plast. Chem.*, 1980, **43**, 762.
[13] J. Ulanski, *et al.*, *Polym. Plast. Technol. Eng.*, 1981, **17**, 139.
[14] K. J. Wynne and G. B. Street, *Ind. Eng. Chem., Prod. Res. Dev.*, 1982, **21**, 23.
[15] K. Seeger, *Angew. Makromol. Chem.*, 1982, **109–110**, 227.
[16] D. Bloor, *New Scientist*, 1982, March 4, 557.
[17] D. A. Seanor, 'Electrical Properties of Polymers', Academic Press, New York, 1982.
[18] J. C. W. Chien, *J. Polym. Sci., Polym. Lett. Ed.*, 1981, **5**, 249.
[19] J. C. W. Chien, *et al.*, *Makromol. Chem., Rapid Commun.*, 1982, **5**, 269.
[20] H. W. Gibson and J. M. Pochan, *Org. Coatings Plast. Chem.*, 1980, **42**, 600.
[21] J. M. Pochan and H. W. Gibson, *Org. Coatings Plast. Chem.*, 1980, **43**, 872.
[22] M. Monkenbusch, *Makromol. Chem., Rapid Commun.*, 1982, **9**, 601.
[23] J. C. W. Chien, *et al.*, *J. Polym. Sci., Polym. Lett. Ed.*, 1982, **2**, 97.
[24] W. D. Deits, *et al.*, *Org. Coatings Plast. Chem.*, 1980, **43**, 867.
[25] H. Shirakawa, *et al.*, *Bull. Chem. Soc. Jpn.*, 1982, **3**, 21.

A study of the structure of Na-doped $(CH)_n$ using optical and e.s.r. spectra has been undertaken by Heeger and MacDiarmid et al.[26] This n-type (electron donor) doping indicates that the 'metallic' stage in Na-doped $(CH)_n$ is similar for both p- and n-type dopant species. No 'free' Na metal was found in the $(CH)_n$ films.

Measurement of the change in dielectric constant and ^1H n.m.r. have been used by DeVreux et al.[27] to study the effect of mechanical rolling on $(CH)_n$. Rolling is found partially to isomerize cis-$(CH)_n$ to the more conductive trans isomer. MacDiarmid and Heeger[28] have discussed the doping and doping procedures for $(CH)_n$ and also the synthesis of substituted polyacetylenes. The photoelectrochemical reactions at $(CH)_n$ interfaces are also discussed. Deits et al.[29] have described the synthesis and characterization of a number of $(CH)_n$ analogues including polyphenylacetylene, polycyanoacetylene, poly(propergyl chloride), and polytrifluormethylacetylene. The electrical conductivities were found to be greatly inferior to that of $(CH)_n$.

Day and Lando[30] have prepared monolayers of the much more stable polydiacetylene

$$\left(\begin{matrix} R' & R'' \\ \mathrm{C-C\equiv C-C} \end{matrix}\right)_n.$$

Doping bilayers of this ordered and highly crystalline polymer increased the current by three orders of magnitude. Photoconductive current increases were also noted. The problems of stability and difficulty with processing for $(CH)_n$ have ensured the continuation of the search for alternative potentially conductive polymeric systems. A variety of chemical types are being investigated; they all have the common features of conjugated bonds in the polymer backbone. A recent brief report[31] outlines some background to the work at IBM on polypyrroles electrolytically doped with p-type perchlorate. This material is stable in air and its conductivity remains substantially intact at temperatures up to 250 °C.

Rabolt et al.[32] have reported the preparation of poly(4-phenylene sulphide hexafluoroarsenate) which is melt or solution processable. Baughman et al.[33,34] have also investigated the electrical and optical properties of this doped poly(4-phenylene sulphide) system. The electrical conductivity of pure poly(4-phenylene sulphide) exponentially changes with doping from a good insulator to a semi-conductor and then to a semi-metal. Clarke et al.[35] have reported some evidence of cross-linking in doped poly(phenylene sulphide hexafluoroarsenate). It is confirmed that this polymer charge-transfer complex is a p-type semi-conductor. Elsenbaumer and Shacklette[36] have reported their work on doping poly(3-phenylene sulphide) derivatives with AsF_5 to produce highly conducting complexes. These authors have found that conduction in this and analogous systems only occurs if C—C bonds are formed intramolecularly between phenyl rings.

---

[26] A. J. Heeger, A. G. MacDiarmid, et al., J. Chem. Phys., 1981, **45**, 5504.
[27] F. DeVreux, et al., J. Polym. Sci., Polym. Phys. Ed., 1981, **19**, 743.
[28] A. G. MacDiarmid and A. J. Heeger, Org. Coatings Plast. Chem., 1980, **43**, 853.
[29] W. Deits, et al., Polym. Prepr., Am. Chem. Soc., Div. Polym. Chem., 1981, **22**, 197.
[30] D. R. Day and J. B. Lando, J. Appl. Polym. Sci., 1981, **26**, 1605.
[31] 'Perspectives', Chem. Br., 1982, Sept., 611.
[32] J. F. Rabolt, et al., Org. Coatings Plast. Chem., 1980, **43**, 772.
[33] R. H. Baughman, et al., Org. Coatings Plast. Chem., 1980, **43**, 768.
[34] R. H. Baughman, et al., J. Chem. Phys., 1981, **75**, 1919.
[35] T. C. Clarke, et al., J. Polym. Sci., Polym. Phys. Ed., 1982, **20**, 117.
[36] R. L. Elsenbaumer and L. W. Shacklette, J. Polym. Sci., Polym. Phys. Ed., 1982, **20**, 1781.

Poly(4-phenylene) itself has been subjected by Tieke et al.[37] to absorption spectroscopy after doping with electron acceptors SbF_5 or anhydrous $AlCl_3$ and electron donors, such as potassium dihydronaphthylide. The spectra indicate that doping can best be explained by a two-phase model, one phase of doped and the other of undoped materials. Gibson et al.[38] have reported the results of doping with I_2 vapour and subsequent oxidation of cyclopolymers of 1,6-heptadiyne. The doping was monitored with e.s.r. spectroscopy. The loss of conductivity found in the absence of I_2 was less than in its presence. It is suggested that a chemical reaction is involved in the loss of conductivity. Oxygen was found to dope the polymer. Also a high rate of oxidation was noted.

The synthesis of poly(2,5-thiophenediyl) and its complexing with AsF_5 is reported by Kossmehl and Chatzitheodorou.[39] Kossmehl and Rohde[40] have reported on the synthesis, characterization, and electrical conductivity of polyesterification products with tetrathiafulvalene units and their complexes with tetracyanoquinodimethane. Watanabe et al.[41] have synthesized highly conductive 7,7,8,8-tetracyanoquinodimethane salts of ionene polymers which have 4,4'-bipyridinium or 1,2 bis(4-pyridinium) ethylene rings. Flexible films of these complexes could be formed from solution. The resistivity of these films was found to increase to about twice its initial value after 300 days.

Huber and Weddigen[42] have attempted to improve the electrical conductivity of bisphenol A-based epoxy resins by substitution of the phthalic anhydride-based hardener with a pyromellitic acid anhydride, then dehydration to form a resin containing C–C bonds. In a further attempt these authors formed a charge-transfer complex by coating the dehydration catalyst grains with a monolayer of tetracyanoethylene. The second method produced an increase in electrical conductivity above 60 °C. Clearly there is still a need for the development of stable organic chemical systems which combine the advantages of high conductivity and low density.

5 Solid State Cells

Watanabe et al.[43] have studied the ionic conductivity of plasticized polyacrylonitrile films filled with $LiClO_4$. These films were found to offer high conductivity and be potentially suitable as electrolytes in Li batteries. Poly(vinylidene fluoride) films with $LiClO_4$ dispersions have been proposed for a similar application by Ohno[44] and co-authors.

---

[37] B. Tieke, et al., Makromol. Chem., Rapid Commun., 1982, **5**, 261.
[38] H. W. Gibson, et al., Polym. Prepr., Am. Chem. Soc., Div. Polym. Chem., 1981, **22**, 35.
[39] G. Kossmehl and G. Chatzitheodorou, Makromol. Chem., Rapid. Commun., 1981, **9/10**, 551.
[40] G. Kossmehl and M. Rohde, Makromol. Chem., 1982, **9**, 2077.
[41] M. Watanabe, et al., Polym. J., 1982, **3**, 189.
[42] R. Huber and G. Weddigen, Colloid Polym. Sci., 1981, **2**, 852.
[43] M. Watanabe, et al., J. Appl. Polym. Sci., 1982, **11**, 4191.
[44] H. Ohno, et al., Polym. Bull., 1982, **5/6**, 271.

6 Piezoelectricity in Polymers

Many papers have been published in this area which investigate the physics or electronic aspects of the well-known piezoelectric effect in poly(vinylidene fluoride). A report by Kato et al.[45] gives details of the piezoelectric properties of cast poly(3-phenylene isophthalamide) films containing 6% $CaCl_2$ which have been stretched in boiling water. A maximum piezoelectric effect was found at stretching draw ratios of 2.

---

[45] K. Kato, et al., 'Gakujutsu Bunken Fujyukai', Tokyo, 1979, p. 343. (Reports on Progress in Polymer Physics in Japan.)

14
Reactions on Polymers

BY D. C. SHERRINGTON

1 Introduction

In Volume 2, Chapter 15, of this series, the author, G. G. Cameron, indicated a shift in emphasis in the area, with a growing proportion of papers being devoted to the preparation and modification of functional polymers for specific applications other than as material substances. This trend has continued and the present review will concentrate on those modifications carried out very often with some subsequent chemical application in mind.

As before, the topics of crosslinking and grafting reactions will not be dealt with comprehensively in identifiable sections, although some reference will be made in passing to specific examples of these.

One area of modification in which considerable interest has been shown and which has developed rapidly in the last two years is the use of phase transfer catalysts for achieving facile chemical change, hitherto achieved only slowly or with poor yield. In view of the potential importance of this development, the results will be dealt with in more detail. The related area of polymers themselves functioning as phase transfer catalysts will also be covered, along with the recent advances in the use of solid polymeric acid and base catalysts and supported metal complex catalysts. The important developments in the use of stoicheiometric quantities of polymeric reagents will be described, with emphasis on those more practical systems capable of recycling or regeneration. The emerging area of the use of modified polymers as selective sorption media will be dealt with, and, although a close relationship exists between this and the use of polymers in chromatographic separations, the latter will not fall within the scope of this review. One final omission which relates closely to many of the modifications and reactions to be described is the area of polymer-modified electrodes. This interfaces intimately with the topic of solid-state devices (both fabrication and operation) and is an aspect of polymer chemistry which should be accounted for perhaps in Volume 3.

Recent review-type literature which readers should be aware of as forming a foundation for this Report include, with respect to chemical modification of polymers, the previous Report by Cameron[1] and an ACS Symposium Publication edited by Carraher and Tsuda,[2] and with respect to reactions involving polymeric

---

[1] G. G. Cameron in 'Macromolecular Chemistry', ed. A. D. Jenkins and J. F. Kennedy (Specialist Periodical Reports), The Royal Society of Chemistry, London, 1981, p. 271.

[2] 'Modification of Polymers', ed. C. E. Carraher and M. Tsuda, American Chemical Society, Washington, 1980.

species, texts by Mathur, Narang, and Williams,[3] and by Hodge and Sherrington.[4] Other central reviews with substantial literature citations have also appeares,[5–11] and a discussion of ecological applications with a number of chemical features relevant to this area has been undertaken.[12]

2 Chemical Modification of Polymers using Phase Transfer Catalysts (PTC)

The background to this area and its substantial development has recently been described by Frechet.[13] The introduction of phase transfer catalysts[14] has allowed the fast convenient synthesis of a wide range or low-molecular-weight organic species previously obtained only with difficulty. The facile transportation of reactive species across aqueous/organic and solid/organic interfaces has given rise to new synthetic possibilities, and similar application in polymer synthesis and modification is now well under way.

Nucleophilic Displacements on Polymers.—Most examples so far have involved nucleophilic displacement of Cl^- from chloromethylated polystyrene [reaction (1)].

$$\text{(P)}-\text{C}_6\text{H}_4-CH_2Cl + Nu^- \xrightarrow{PTC} \text{(P)}-\text{C}_6\text{H}_4-CH_2Nu + Cl^- \quad (1)$$

Nu^- = nucleophile

The polymer is dissolved or swollen in an organic solvent and the nucleophile is present as a solid alkali-metal salt or an aqueous solution of the latter. The nucleophile may also be generated *in situ* by removal of an acidic hydrogen on a substrate using concentrated NaOH. In all cases typical phase transfer catalysts, PTC, such as tetrabutylammonium or phosphonium salts, or crown ethers and cryptands have been employed as the catalyst. In the case of linear polymers very often a crosslinking side-reaction can occur. For example, Nishikubo and co-

---

[3] N. K. Mathur, C. K. Narang, and R. E. Williams, 'Polymers as Aids in Organic Chemistry, Academic Press, New York, 1980.
[4] 'Polymer-supported Reactions in Organic Synthesis', ed. P. Hodge and D. C. Sherrington, Wiley, Chichester, 1980.
[5] G. Cainelli, F. Manescalchi, and M. Contento in 'Organic Synthesis Today and Tomorrow', ed. B. M. Trost and C. R. Hutchinson, Pergamon, Oxford, 1981.
[6] A. Akelah and D. C. Sherrington, *Chem. Rev.*, 1981, **81**, 557.
[7] J. M. J. Frechet, *Tetrahedron*, 1981, **37**, 663.
[8] M. Kaneko and E. Tsuchida, *J. Polym. Sci., Macromol. Rev.*, 1981, **16**, 397.
[9] D. C. Bailey and S. H. Langer, *Chem. Rev.*, 1981, **81**, 109.
[10] 'Selective Polymeric Sorbents', ed. B. Sedlacek, *Pure Appl. Chem.*, 1982, **54**, 2077.
[11] Symposium on 'Polymer-supported Reactions in Organic Chemistry', *Nouv. J. Chim.*, 1982, **6**, 605.
[12] J. Klein, *Makromol. Chem. Suppl.*, 1981, **5**, 155.
[13] J. M. J. Frechet in 'Phase Transfer Catalysis in Polymer Chemistry', ed. C. E. Carraher and L. Mathias, Plenum Press, New York, 1983, p. 1.
[14] *e.g.*, E. V. Dehmlow, *Angew. Chem., Int. Ed., Engl.*, 1974, **13**, 170, and W. P. Weber and G. W. Gokel, 'Phase Transfer Catalysis in Organic Synthesis', Springer-Verlag, Berlin, 1977.

workers[15] allowed KOH to react with linear chloromethylated polystyrene to yield an insoluble gel. This arose from the sequence of processes shown in reaction (2). In

$$\text{—CH}_2\text{Cl} \xrightarrow[\text{PTC}]{\text{K}^+\text{OH}^-} \text{—CH}_2\text{OH} \xrightarrow[\text{PTC}]{\text{K}^+\text{OH}^-} \text{—CH}_2\text{O}^- \xrightarrow{\text{ClCH}_2-} \text{—CH}_2\text{—O—CH}_2-} \quad (2)$$

fact, this complication can be overcome by first using acetate ion and a PTC, and then adding OH⁻ once displacement of Cl⁻ is complete.[16] Use of dimethylformamide//aqueous NaOH also appears to overcome the problem.[17]

A great deal of work has also been carried out on crosslinked polystyrenes where the swollen network provides an evidently separate phase, and which has created many problems over the years. In this case, crosslinking side reactions are not so obvious but are undoubtedly more common than in recognized. Nishikubo and his co-workers[15,18] have examined carefully the effectiveness of different PTC in chemical modification and have also contrasted solid/liquid systems with corresponding liquid/liquid ones. In general, crown ethers in polar organic solvents appear most effective with solid nucleophiles, but with aqueous solutions tetrabutylphosphonium and ammonium salts are superior.

Frechet[13] has summarized those groups which have been introduced readily onto polystyrene aromatic residues via this route. The more recent examples are shown in Table 1. Very often reactions are clean and proceed in high yield. For example, by using dioctylphosphine oxide in toluene and 50% aq. NaOH with tetrabutyl-

---

[15] T. Nishikubo, T. Iizawa, K. Kobayashi, and M. Okawara, *Makromol. Chem., Rapid Commun.*, 1981, **2**, 387.
[16] J. M. J. Frechet, M. D. de Smet, and M. J. Farrall, *Polymer*, 1979, **20**, 675.
[17] J. M. J. Frechet, *J. Macromol. Sci., Chem.*, 1981, **15**, 879.
[18] T. Nishikubo, T. Iizawa, K. Kobayashi, and M. Okaware, *Makromol. Chem., Rapid Commun.*, 1980, **1**, 765.

Table 1 Functional groups introduced onto polystyrene aromatic residues using phase transfer catalysed techniques

Group	Ref.	Group	Ref.
$-CH_2CH(CN)_2$	15	$-CH_2$-crown ether	21
$-CH_2CH(CO_2R)_2$	15	$-CH_2I$	22
$-CH_2C(CO_2R)_2$ | CH_3	15	$-CH_2SCH_2CH_2OH$	23
$-CH_2OH$	15	$-CH_2SCH_2CH(OH)CH_2OH$	23, 24

$-CH_2O-\langle\bigcirc\rangle-X$	18	$-CH_2OCH_2-CH-CH_2$
		         | |
		         O O
		          \/
		           $C(CH_3)_2$
			23

$-CH_2OCOCH_2$	18	$-CH_2OCO-\overset{*}{\langle}\text{pyrrolidine-NH}\rangle$	23
$-CH_2SCOCH_3$	18		
$-CH_2SCSN(C_2H_5)_2$	18		

$-CH_2SO_2-\langle\bigcirc\rangle-CH_3$	18	$-CH_2OCH_2-\overset{*}{C}H-Ph$
		                |
		             $N=R$
			23

$-CH_2-N\langle\text{phthalimide}\rangle$	19	$-CH_2-O-\langle\bigcirc\rangle-\overset{*}{C}H(NH_2)CO_2H$	23
$-CH_2SCN$	20	$-CH_2P(O)(nOct)_2$	25
$-CH_2SH$	20		

ammonium bisulphate or kryptofix as the catalyst chloromethylated polystyrene was converted to its di(n-octyl) phosphine oxide derivative in 70—95% yields.[25] Recently, tetrahydrothiophene has been shown to be a very effective catalyst in azide ion displacements on chloromethylated polystyrene.[26] The authors have

---

[19] A. S. Gozdz, *Polym. Bull.*, 1981, **5**, 591.
[20] A. S. Gozdz, *Makromol. Chem., Rapid Commun.*, 1981, **2**, 595.
[21] A. Cheminat, C. Benezra, M. J. Farrall, and J. M. J. Frechet, *Can. J. Chem.*, 1981, **59**, 1405.
[22] A. S. Gozdz and A. Rapak, *Makromol. Chem., Rapid Commun.*, 1981, **2**, 359.
[23] J. M. J. Frechet and co-workers, personal communication.
[24] P. Hodge and J. Waterhouse, *Polym. Prepr., Am. Chem. Soc., Div. Polym. Chem.*, 1982, **23**, 142.
[25] T. D. N'Guyen, J. C. Gautier, and S. Boileau, *Polym. Prepr., Am. Chem. Soc., Div. Polym. Chem.*, 1982, **23**, 143.
[26] M. Takeishi and N. Umeta, *Makromol. Chem., Rapid Commun.*, 1982, **3**, 875.

argued that *in situ* formation of a sulphonium salt gives rise to an intramolecular phase transfer catalyst [reaction (3)]. This situation is very similar to that reported

$$\text{P}-\text{C}_6\text{H}_4-\text{CH}_2\text{Cl} \xrightarrow{\text{S}\diagup\diagdown} \text{P}-\text{C}_6\text{H}_4-\text{CH}_2^+\text{S}\diagup\diagdown \;\; \text{Cl}^- \longrightarrow \text{P}-\text{C}_6\text{H}_4-\text{CH}_2\text{N}_3 + \text{S}\diagup\diagdown \quad (3)$$

$$\text{Na}^+\text{N}_3^-\text{(aq)} \qquad\qquad \text{Na}^+\text{N}_3^-\text{(aq)} \qquad\qquad \text{Na}^+\text{Cl}^-\text{(aq)}$$

previously[27] where it is well demonstrated that reaction of chloromethyl groups with dimethyl sulphide, with formation of the sulphonium salt, facilitates subsequent reaction with highly polar nucleophiles. It is perhaps worth pointing out that the use of PTC in this context can be somewhat overplayed. Thus CN^- displacements are probably best carried out simply using DMF as the solvent, and similarly displacement with I^- occurs readily in acetone. Furthermore, having generated the iodomethyl group, this is itself readily transformed into its malonitrile derivative under classical conditions.[28]

In principle, other halogenated polymers might be modified by similar PTC reactions, and poly(vinyl chloride) in particular is an attractive substrate. Earlier attempts to catalyse displacement reactions were not optimistic but more recent reports of crown ethers and acetate ion being used as the nucleophile are more encouraging.[29—31] Forcing conditions tend to give rise to side reactions and degradation, and similarly with poly(vinyl bromide) some elimination of HBr results in the formation of ~10—20% of double bonds from the original repeating units.[17,32] Nevertheless, significant yields of methyl sulphone polymer [reaction (4)] were achieved by using methyl sulphinate and tetrabutylammonium chloride as PTC.

$$-(\text{CH}_2-\underset{\underset{\text{Br}}{|}}{\text{CH}})_n- \xrightarrow[\text{Bu}_4\text{NCl}]{\text{MeSO}_2^-} -(\text{CH}_2-\underset{\underset{\underset{\text{Me}}{|}}{\underset{\text{SO}_2}{|}}}{\text{CH}})_n- \quad (4)$$

---

[27] *e.g.*, M. A. Petit and J. Jazefonviez, *J. Appl. Polym. Sci.*, 1977, **21**, 2589.
[28] C. Bied-Charreton, M. Frostin-Rio, D. Pijol, A. Gaudemer, R. Audebert, and J. P. Idoux, *J. Mol. Catal.*, 1982, **16**, 335.
[29] J. Lewis, M. K. Naqui, and G. S. Park, *Makromol. Chem., Rapid Commun.*, 1980, **1**, 411.
[30] J. Lewis, M. K. Naqui, and G. S. Park, *Makromol. Chem., Rapid Commun.*, 1981, **1**, 119.
[31] J. Lewis, M. K. Naqui, and G. S. Park, *Polym. Prepr., Am. Chem. Soc., Div. Polym. Chem.*, 1982, **23**, 140.
[32] J. M. J. Frechet, *Org. Coat. Plast. Chem.*, 1980, **42**, 268.

Another potentially attractive polymeric substrate is poly(epichlorohydrin). This is available as a soluble linear polymer and controlled modification of this has the potential for producing many novel species. Unfortunately, once again reactions are not straightforward. Carbazole groups have been substituted up to ~60% with crown ether and cryptand catalysts, but extensive chain degradation occurs. Similarly, even simple anions like acetate and hydroxide give problems,[17] as indeed does halogen exchange using I^- with PTC in butanone.[33] Recently, superoxide ion has also been shown to cause degradation.[34]

A more successful nucleophilic reaction using PTC has been carried out with a range of copolymers of glycidyl methacrylate. Reaction of these with $Na_2SO_3/NaHSO_3$ in the presence of tetrabutylammonium bisulphate readily generates sulphonate groups [reaction (5)]. Copolymers with styrene, acrylonitrile,

$$\text{(P)}-CH-CH_2 \text{ (epoxide)} + HSO_3^- \xrightarrow{PTC} \text{(P)}-CH(SO_3^-)-CH_2OH \quad (5)$$

cellulose, and acrylate monomers have been examined, and the species involving acrylonitrile appears to sulphonate most easily.[35] Most recently poly(vinyl chloroformate) and its copolymers have been modified under PTC conditions.[36]

Nucleophilic Polymers.—A complementary reaction to nucleophilic displacements on a polymeric substrate is the generation of a nucleophile on the polymer and its reaction with an electrophilic reagent [reaction (6)]. In principle, this should allow a

$$\text{(P)}-Nu^- + E^+ \longrightarrow \text{(P)}-Nu-E \quad (6)$$

wide variety of modifications to be achieved and indeed such reactions might also be catalysed by phase transfer species. In practice, this area has received much less attention. Amine-containing polymers of course readily react under classical conditions.[21,37] Much more interesting, however, are polymers carrying alkoxide or thiolate anions since these can be employed widely in nucleophilic displacements and in Michael additions [reactions (7) and (8)].

$$\text{(P)}-OH \xrightarrow[OH^-]{PTC} \text{(P)}-O^- \xrightarrow{RX} \text{(P)}-OR + X^- \quad (7)$$

$$\text{(P)}-SH + R'CH=CHCOR^2 \xrightarrow[OH^-]{PTC} \text{(P)}-S-CH(R')-CH_2C(=O)R^2 \quad (8)$$

---

[33] E. Schacht, D. Bailey, and O. Vogl, *J. Polym. Sci., Polym. Chem. Ed.*, 1981, **16**, 2343.
[34] A. S. Gozdz, *Polym. Bull.*, 1982, **6**, 375.
[35] S. Paul and B. Ranby, *J. Appl. Polym. Sci.*, 1981, **26**, 3927.
[36] S. Bowin, A. Chettauf, P. Hemery, and S. Boileau, *Polym. Bull.*, 1983, **9**, 114.
[37] J. M. J. Frechet, J. J. Farrall, C. Benezra, and A. Cheminat, *Polym. Prepr., Am. Chem. Soc., Div. Polym. Chem.*, 1980, **21**, 101.

Reactions on Polymers

These reactions are now being examined in some detail,[38] and a particularly interesting polymeric substrate is poly(*p*-hydroxystyrene). The synthesis of this has recently been improved *via* the elegant use of *p*-t-butoxycarbonyl styrene. After polymerization of this monomer, the t-BOC protecting group is removed cleanly by thermolysis and is accompanied by the evolution of gaseous products only, CO_2 and isobutene.[39] Phenoxide polymer is readily generated using KOH and a PTC and a wide variety of displacements is currently being studied. Coincidentally, the modification of cellulose using NaOH and tetrabutylammonium iodide followed by reaction with alkyl or aryl halides uses exactly the same synthetic strategy,[40] as indeed does the reaction of pendant sulphinate groups[39,41] in PTC reactions with various electrophiles. Polymeric alkoxides have also been used to achieve elimination reactions on tosylate and bromide substrates where the presence of a PTC is not required.[42]

Use of Wittig Reactions.—A somewhat different PTC reaction which allows facile modification of polymers has been reported by Hodge and his co-workers.[24,43,44] This utilizes the Wittig reaction. A benzyl triphenylphosphonium salt is generated on the polymer and then the phosphorus ylide produced by reaction with 50% aq. NaOH and cetyltrimethylammonium bromide as PTC. Subsequent reaction with an appropriate aldehyde introduces any required group into the polymer [reaction (9)]. In this way ferrocene and crown ether residues have been successfully attached.

$$\boxed{P}-CH_2\overset{+}{P}Ph_3 \;\; Cl^- \xrightarrow{NaOH/PTC} \boxed{P}-CH=PPh_3 \xrightarrow{RCHO} \boxed{P}-CH=CH-R + Ph_3P=O \quad (9)$$

By using a similar approach in which the aldehyde is formaldehyde, vinyl groups have been introduced.[39,45] Also by reaction of an aldehydic polymer with trimethylsulphonium chloride and NaOH with a PTC, epoxide functionalities can be generated[39] [reaction (10)].

$$\boxed{P}-CHO \xrightarrow{Me_3S^+Cl^-/OH^-/PTC} \boxed{P}-CH-CH_2 \text{ (epoxide)} \quad (10)$$

Other PTC Modifications.—Finally in this area the hydrolysis of methylmethacrylate in the presence of oligoethylene glycols and crown ether PTC has been examined,[46] as has the reaction of hydrolysed glycidyl methacrylate with propane sultone in the presence of tetrabutylammonium hydroxide.[47]

---

[38] J. M. J. Frechet and co-workers, personal communication.
[39] J. M. J. Frechet, W. Amaratunga, and J. Halgas, *Nouv. Chim.*, 1982, **6**, 609.
[40] W. H. Daly, J. D. Caldwell, K. V. Phung, and R. Tang, *Polym. Prepr., Am. Chem. Soc., Div. Polym. Chem.*, 1982, **23**, 145.
[41] A. J. Hagen, M. J. Farrall, and J. M. J. Frechet, *Polym. Bull.*, 1981, **5**, 111.
[42] I. Artaud and P. Viout, 'International Symposium on Polymer-supported Reagents in Organic Chemistry', Communications, Lyon, France, July 1982, p. 134.
[43] P. Hodge and J. Waterhouse, *Polymer*, 1981, **22**, 1153.
[44] P. Hodge, B. J. Hunt, E. Khoshdel, and J. Waterhouse, *Nouv. J. Chim.*, 1982, **6**, 617.
[45] J. M. J. Frechet and E. Eichler, *Polym. Bull.*, 1982, **7**, 345.
[46] A. S. Gozdz, *Makromol. Chem., Rapid Commun.*, 1981, **2**, 443.
[47] J. Hradil and F. Svec, *Polym. Bull.*, 1982, **6**, 565.

3 Phase Transfer Catalysed Polycondensations

In the context of chemical modification using PTC it would be inappropriate not to mention the very closely related field of polymerizations involving PTC. This involves essentially the synthesis of polycondensation species using PTC, although, of course, crown ether and cryptands have also been employed in anionic chain polymerizations where significant effects can arise from complexation with the countercation. In principle, most reactions involve NaOH or an equivalent base and a PTC, to generate a reactive dicarboxylate, diphenolate, or dithiolate which subsequently reacts with a di-halo compounds [reaction (11)]. This approach has a

$$\text{HYRYH} \xrightarrow[\text{PTC}]{\text{NaOH}} {}^-\text{YRY}^- \xrightarrow{\text{XR'X}} -(\text{R}-\text{Y}-\text{R}'-\text{Y})_n- \qquad (11)$$

$$Y = O \text{ or } S$$

number of potential advantages. It can allow the use of cheap inorganic bases and can eliminate the need for expensive dipolar aprotic solvents.

Bisphenolates have been caused to react under these conditions with 1,4-dichloro-2-butene,[48] 1,6-dibromohexane,[49] and *m*-xylylene dibromide[50] to form polyethers. Yields of polymer are generally good and ammonium salts appear to be the best catalysts. Unfortunately, product molecular weights are low because the polymers generally precipitate as they are formed. Similar syntheses of polyesters using di-carboxylates are even less satisfactory.[51] Reaction of a dithiolate with an activated dichloro compound[52] gives a poly(ketone sulphide) in yields up to $\sim 100\%$

Perhaps more interesting are the reactions of dithiolates and diphenolates with hexafluorobenzene.[53,54] The latter undergoes nucleophilic substitution relatively easily and hence allows production of polyaryl species with no aliphatic carbon content [reaction (12)].

$$\text{HX}-\langle\!\!\!\bigcirc\!\!\!\rangle-\text{G}-\langle\!\!\!\bigcirc\!\!\!\rangle-\text{XH} + \langle\!\!\!\bigcirc\!\!\!\rangle\text{F} \xrightarrow[\text{crown ether} + \text{solvent}]{K_2CO_3}$$

$$-(\langle\!\!\!\bigcirc\!\!\!\rangle-\text{G}-\langle\!\!\!\bigcirc\!\!\!\rangle-\text{X}-\langle\!\!\!\bigcirc\!\!\!\rangle^F-\text{X})- \qquad (12)$$

$G = -O-, -S-, -SO_2-, -CMe_2-$
$X = O \text{ or } S$

---

[48] T. D. N'Guyen and S. Boileau, *Polym. Prepr., Am. Chem. Soc., Div. Polym. Chem.*, 1982, **23**, 154.
[49] J. I. Jin and J. H. Chang, *Polym. Prepr., Am. Chem. Soc., Div. Polym. Chem.*, 1982, **23**, 156.
[50] G. G. Cameron and K. S. Law, *Makromol. Chem., Rapid Commun.*, 1982, **3**, 99.
[51] G. G. Cameron and K. S. Law, *Polymer*, 1981, **22**, 272.
[52] M. Ueda and R. Takasawa, *Makromol. Chem., Rapid Commun.*, 1982, **3**, 905.
[53] R. Kellman, J. C. McPheeters, D. J. Gerbi, and R. F. Williams, *Polym. Prepr., Am. Chem. Soc., Div. Polym. Chem.*, 1982, **23**, 383.
[54] R. Kellman, D. J. Gerbi, J. C. Williams, and R. F. Williams, *Polym. Prepr., Am. Chem. Soc., Div. Polym. Chem.*, 1982, **23**, 174.

Reactions of diphenolates and dithiolates with dicyclopentadienyl titanium and zirconium dichlorides under PTC conditions [reaction (13)] produces polymers

$$Cp_2MCl_2 + HXRXH \xrightarrow[R, NHCl^-]{OH} -(M(Cp)_2-X-R-X)_n- \quad (13)$$

Cp = cyclopentadienyl
M = Ti and Zr

with the metal atom in the backbone.[55] Similarly, antimony-containing polymers result from reactions of trialkyl antimony dichloride with salts of diacids and dioximes, and also with diamines, again in the presence of a PTC[55] [reaction (14)].

$$R_3SbCl_2 + \begin{Bmatrix} HO_2CR'CO_2H \\ \\ HONC(R^2)-R^3-C(R^2)NOH \\ \\ H_2NR^4NH_2 \end{Bmatrix} \xrightarrow{\text{NaOH} \atop \text{PTC}} \begin{Bmatrix} -(Sb(R_2)-OCR'(O)-C(O)O)_n- \\ \\ -(Sb(R_2)-ONC(R^2)-R^3-C(R^2)NO)_n- \\ \\ -(Sb(R_2)-N(H)-R^4-N(H))_n- \end{Bmatrix} \quad (14)$$

Flame retardant aromatic polyphosphonates have been made by a similar route using phenylphosphonic dichloride[56] [reaction (15)]. The reaction is carried out in

$$Cl-P(O)(Ph)-Cl + HO-C_6H_4-C(Me)_2-C_6H_4-OH \xrightarrow{\text{aq. NaOH} \atop \text{PTC}} -(P(O)(Ph)-O-C_6H_4-C(Me)_2-C_6H_4-O)_n- \quad (15)$$

methylene chloride and aqueous NaOH at 0 °C to minimize hydrolysis of the dichloride. Yields of polymer are in the range 75—100% and with a crown ether as the catalyst the largest molecular weights are obtained. Absence of a PTC results in

---

[55] C. E. Carraher and M. D. Naus, *Polym. Prepr., Am. Chem. Soc., Div. Polym. Chem.*, 1982, **23**, 158.
[56] Y. Imai, N. Sato, and M. Ueda, *Makromol. Chem., Rapid Commun.*, 1980, **1**, 419.

very-low-molecular-weight products only. Salts of dicarboxylic acids instead of the bisphenol yields analogous polyphosphoanhydrides.[57]

Imai and co-workers[58,59] have employed activated methylene-containing nitriles rather imaginatively in polycondensations with dichloro compounds. The acidic hydrogen atoms are removed under standard PTC conditions, and without a catalyst no condensation occurs [reaction (16)]. Equally attractive is the use of PTC Wittig reactions in producing highly conducting polymers[60] [reaction (17)]. In this case the bisphosphonium salt acts as a reactant and its own PTC.

$$RCH_2CN + ClCH_2R'CH_2Cl \xrightarrow[PTC]{aq.\ NaOH} \pm(\underset{\underset{CN}{|}}{\overset{\overset{R}{|}}{C}} - CH_2R'CH_2)_{\overline{n}} \qquad (16)$$

Where R = Ph, $-\overset{O}{\overset{\|}{C}}OCMe_3$

R' = –C₆H₄–, –C₆H₄–O–C₆H₄–, –(CH₂)₆–

and –CH₂–CH=CH–CH₂–

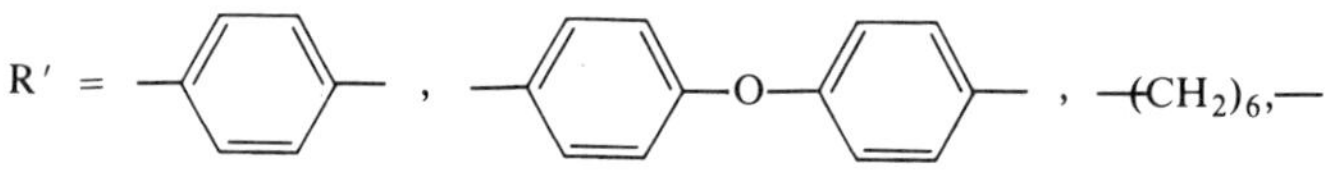

$\overset{+}{C}H_2PPh_3Cl^-$–C₆H₄–$\overset{+}{C}H_2PPh_3Cl^-$ + CHO–C₆H₄–CHO $\xrightarrow{aq.\ NaOH}$ $\left(\!\!\left\langle\!\!\!\begin{array}{c}\end{array}\!\!\!\right\rangle\!\!-CH=CH\right)_n$ (17)

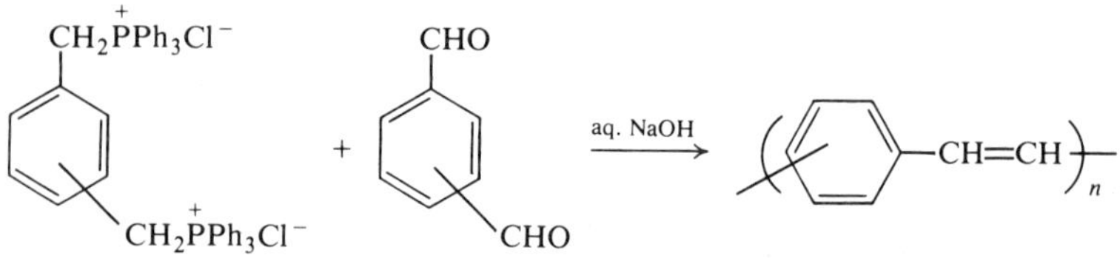

4 Polymer-supported Phase Transfer Catalysts

Polymers can be made and modified using PTC techniques but equally importantly polymers can be used as supports for PTC for application in other reactions. This area has attracted much attention in recent years because of its potential for application in the synthesis of speciality molecules, e.g., pharmaceuticals. Earlier reviews[61,62] have been up-dated more recently.[6,63,64] Typically, a polymer matrix carrying PTC groups is used as a catalyst in liquid/liquid or solid/liquid phase separated reactions. The latter include S_{N^2} displacements, alkylations of active methylene groups, $NaBH_4$ reduction of ketones, Michael additions to α,β

---

[57] C. E. Carraher, J. Raymond, and H. S. Blaxall, *Polym. Prepr., Am. Chem. Soc., Div. Polym. Chem.*, 1982, **23**, 160.
[58] Y. Imai, A. Kameyama, T. Nguyen, and M. Ueda, *J. Polym. Sci., Polym. Chem. Ed.*, 1981, **19**, 2997.
[59] Y. Imai and M. Ueda, *Polym. Prepr., Am. Chem. Soc., Div. Polym. Chem.*, 1982, **1**, 164.
[60] W. A. Feld, A. Ganesan, and D. P. Nymberg, *Polym. Prepr., Am. Chem. Soc., Div. Polym. Chem.*, 1983, **23**, 143.
[61] Ref. 4, Chapter 3.
[62] S. L. Regen, *Angew. Chem., Int. Ed. Engl.*, 1979, **18**, 421.
[63] S. L. Regen, *Nouv. J. Chim.*, 1982, **6**, 629.
[64] A. Akelah and D. C. Sherrington, *Polymer*, 1983, **24**, 1369.

unsaturated ketones and dehydrohalogenations. Polymer bound species which have been employed are quaternary ammonium and phosphonium salts, crown ethers, and cryptands. Closely related linear oligoethers have also been supported and used in the same way.

Considerable debate continues concerning the mechanism of catalysis in liquid/liquid systems[65-67] [reaction (18)]. Reactions certainly occur within the three-

$$\begin{matrix} \text{Aqueous Nucleophile}^- \\ \text{Organic Substrate} \end{matrix} \Big\} - \text{P} - \text{C}_6\text{H}_4 - \text{CH}_2 - \text{PTC} \qquad (18)$$

$\text{PTC} = -\overset{+}{\text{P}}\text{Bu}_3\text{Cl}^-, -\overset{+}{\text{N}}\text{Bu}_3\text{Cl}^-, -\text{crown ether}, -\text{cryptand}$

dimensional networks of polymer resins and ^{13}C n.m.r. spectra have implicated an aqueous solvation shell or organic/aqueous interface.[65] However, analysis of C- and O-alkylation of β-naphthoxide and phenoxide tends to confirm the site of reaction to be an hydrophobic shell around the catalyst.[66] Ford[67] has examined macroporous polymer-supports in some detail and shown them somewhat surprisingly to be no more effective than microporous ones. In the case of CN^- displacements on alkyl bromides he has shown an accumulation of Br^- on phosphonium salt resins, indicating that anion transport is an important problem. Apparently anions experience difficulty in migrating through macropores, which most likely therefore are filled largely with organic solvent rather than water. This would tend to indicate that a more hydrophilic porous structure would be a better support. Clearly, however, some balance is required since cellulose and dextran supports required a surface hydrophobic treatment before they acted as effective PTC in liquid/liquid systems, and in this case penetration by the organic reactant appeared to be the source of the problem.[68] When similar catalysts are used in the absence of water, pre-surface treatment becomes unnecessary.[69]

Catalysts attached by 'spacer arms' to support matrices have in general shown increased activity and recent work with $\text{-(CH}_2\text{)}_4\text{-}$ and $\text{-(CH}_2\text{)}_7\text{-}$ hydrophobic spacers has confirmed this.[70] However, once again the extreme non-polar character of the spacer may prove to be a problem. Furthermore, the synthesis of polymers with crown ethers in the backbone and their effectiveness as catalysts throws considerable doubt on the need for spacers in practical systems.[71,72] In the final analysis where very stable and active systems are required the best PTC will probably be the cryptands, and Manecke[73-75] has continued his elegant work on the development of these. Novel synthetic applications of bound onium catalysts have

[65] N. Oktani, C. A. Wilkie, A. Nigam, and S. L. Regen, *Macromolecules*, 1981, **14**, 516.
[66] F. Montanari, S. Quici, and P. Tundo, *J. Org. Chem.*, 1983, **48**, 199.
[67] W. T. Ford, J. Lee, and M. Tomoi, *Macromolecules*, 1982, **15**, 1246.
[68] H. Kise, K. Arabai, and M. Seno, *Tetramedron Lett.*, 1981, 1017.
[69] A. Akelah and D. C. Sherrington, *Eur. Polym. J.*, 1982, **18**, 301.
[70] M. Tomoi, E. Ogawa, Y. Hosokawa, and H. Kakuichi, *J. Polym. Sci., Polym. Chem. Ed.*, 1982, **20**, 3015 and 3421.
[71] K. Fukunshi, B. Czech, and S. L. Regen, *J. Org. Chem.*, 1981, **46**, 1218.
[72] L. J. Mathias, *Polym. Prepr., Am. Chem. Soc., Div. Polym. Chem.*, 1982, **23**, 187.
[73] G. Manecke and P. Reuter, *J. Mol. Catal.*, 1981, **13**, 355.
[74] G. Manecke and P. Reuter, *Makromol. Chem.*, 1981, **182**, 1973.
[75] G. Manecke, A. Kramer, H. J. Winter, and P. Reuter, *Nouv. J. Chim.*, 1982, **6**, 623.

continued with a report of the selective oxidation of secondary alcohols in the presence of primary ones using $Ca(OCl)_2$ and a commercial anion exchange resin.[76]

Poly(ethylene glycols) may be regarded as 'open chain' crown ethers and these have been used more and more as cheap convenient PTC.[77–80] In spite of the claims of convenience and recoverability,[81] in practice any large-scale application is much more likely to involve these species supported on a rigid polymeric matrix. Regen and his co-workers[82,83] have recently extended the use of such supported systems to dehydrohalogenations and other reactions, where optimum activity arises with the oligoether containing six oxygen atoms, implying a close relationship with 18-crown-6 and its high affinity for K^+. This result is a little surprising in view of other earlier reports showing an increasing activity with chain-lengths beyond this.[84] However, the reactions studied in detail are somewhat different and this may be significant. Evidence has also been presented for the involvement of the terminal hydroxyl group of the catalyst in the reaction.[82] Sherrington and his co-workers[84,85] have examined this effect in great detail in solid/liquid reactions and have recently developed a resin-supported catalyst which is highly loaded with oligoether groups specifically terminally functionalized with 8-quinolyl groups[85] [reaction (19)]. This

$$C_4H_9Br + K^+OPh \xrightarrow[ii]{i} C_4H_9OPh + K^+Br^-$$

Reagents:

(19)

i, (P)—⟨⟩—$CH_2(OCH_2CH_2)_3O$—⟨quinolyl⟩ ; ii, toluene

species is more active than dibenzo-18-crown-6 and represents one of the few examples where a supported species is more reactive than an already highly active low-molecular-weight catalyst. Other polymer-supported 'cosolvents' which have been examined are N-pyrrolidone, N-oxazolidone, and hexamethylphosphoramide.[86]

Chiral polymeric phase transfer catalysts have been reviewed recently.[87] Generally speaking, polymers have been modified with optically active amines to generate a bound chiral ammonium salt,[88] mimicking known low-molecular-weight

---

[76] M. Schneider, J. V. Weber, and P. Fallen, *J. Org. Chem.*, 1982, **47**, 364.
[77] D. G. Lee and H. Karaman, *Can. J. Chem.*, 1982, **60**, 2456.
[78] P. Molina, A. Arques, and M. V. Valcarcel, *Synthesis*, 1982, **11**, 942.
[79] B. G. Zupancic and M. Kokalj, *Synth. Commun.*, 1982, **12**, 881.
[80] Y. Kimura and S. L. Regen, *J. Org. Chem.*, 1982, **47**, 2493.
[81] J. M. Harris, N. H. Hundley, T. G. Shannon, and E. C. Struck, *J. Org. Chem.*, 1982, **47**, 4789.
[82] Y. Kimura and S. L. Regen, *J. Org. Chem.*, 1983, **48**, 195.
[83] Y. Kimura, P. Kirszensztejn, and S. L. Regen, *J. Org. Chem.*, 1983, **48**, 385.
[84] J. G. Heffernan, W. M. MacKenzie, and D. C. Sherrington, *J. Chem. Soc., Perkin Trans. 2*, 1981, 514 and references therein.
[85] J. G. Heffernan and D. C. Sherrington, *Tetrahedron Lett.*, 1983, **24**, 1661.
[86] G. Nee and J. Seyden-Penne, *Tetrahedron*, 1982, **38**, 3485.
[87] D. C. Sherrington in 'Phase Transfer Catalysis in Polymer Chemistry', ed. C. E. Carraher and L. Mathias, Plenum Press, New York, 1983, p. 249.
[88] N. Kobayashi and K. Iwai, *Makromol. Chem., Rapid Commun.*, 1981, **2**, 105.

Reactions on Polymers 315

$$\text{P}-\text{C}_6\text{H}_4-\text{CH}_2\text{Cl} + \text{R}^1\text{R}^2\text{R}^3\text{N} \longrightarrow \text{P}-\text{C}_6\text{H}_4-\text{CH}_2\overset{+}{\text{N}}\text{R}^1\text{R}^2\text{R}^3 \;\; \text{Cl}^-$$

e.g., $\text{R}^1\text{R}^2\text{R}^3\text{N}$ = ephedrine or quinine (20)

catalysts [reaction (20)]. In general, use of these in asymmetric reactions such as displacements, $NaBH_4$ reductions, H_2O_2 epoxidations, and Michael additions has been disappointing[89] and has not lived up to earlier results using supported species.[90] More recently, chiral systems based on other optically active species have been examined. Results using methionine residues[91] were again disappointing. Similarly, Sherrington and co-workers have modified styrene resins with optically active oligoethers, protected saccharides and chiral phosphines,[87,92] and examined their induction properties in a wide range of phase-separated reactions without any success. Almost certainly one fundamental problem involves diffusion through the polymeric supports. Enantiomeric excesses arise only when $\Delta(\Delta G^{\neq})$ is significant for the different enantiomers involved, and for this to be so the inducing chiral species must be involved in the transition state of the rate controlling process. If diffusional barriers intercede in the polymeric reactions no induction will be possible. Thus examination of the polymer structure may help in this context. In the meantime, Manecke and his co-workers[93] have described a polymer modified with a chiral crown ether. Results using this may prove interesting.

5 Polymeric Acids and Bases

Sulphonated and aminated polystyrene resins are well known as ion exchange media, and, indeed, the literature on their use as catalysts is also quite vast.[61] Again, many recent applications have been reviewed.[64] Further examples, however, continue to appear with sulphonic acid species attracting the most interest. These have been used to catalyse the condensation of styrene with formaldehyde,[94] in the dehydrogenation of ethylbenzene,[95] and in synthesis of substituted 1,3-dioxanes using the Prins' reaction[96] [reaction (21)].

$$\text{P}-\text{C}_6\text{H}_4-\text{SO}_3^- \text{H}^+$$

$$\text{Ar}-\text{CH}=\text{CH}-\text{R} + \text{HCHO} \longrightarrow \text{Ar}-\underset{\text{O}}{\overset{\text{R}}{\diagup\!\!\diagdown}}\text{O} \qquad (21)$$

---

[89] E. Chiellini, S. D'Antone, and R. Solaro, *Polym. Prepr., Am. Chem. Soc., Div. Polym. Chem.*, 1982, **23**, 179.
[90] S. Colonna, R. Fornasier, and U. Pfeiffer, *J. Chem. Soc., Perkin Trans. 2*, 1978, 8.
[91] S. Banfi, M. Cinquini, and S. Colonna, *Bull. Chem. Soc. Jpn.*, 1981, **54**, 1841.
[92] D. C. Sherrington and J. Kelly, *Polym. Prepr., Am. Chem. Soc., Div. Polym. Chem.*, 1982, **23**, 177.
[93] G. Manecke and H. Winter, *Makromol. Chem., Rapid Commun.*, 1981, **2**, 569.
[94] M. Delmas and A. Gaset, *J. Mol. Catal.*, 1982, **17**, 35 and 51.
[95] T. Tagawa, K. Iwayama, Y. Ishida, T. Haltori, and Y. Murakami, *J. Catal.*, 1983, **79**, 47.
[96] M. Delmas, P. Kalck, J. P. Gorrichon, and A. Gaset, *J. Chem. Educ.*, 1982, **59**, 700.

A number of mechanistic studies have also been carried out on these systems. The role of water has long been recognized as a very important one,[61] and further information on the reaction of formaldehyde with aryl ketones is now available.[97] The question of diffusion effects in macroporous acid resins has been examined in both the re-esterification of ethylacetate with n-propanol[98] and in the dehydration of methanol.[99] A complex sequence of reactions is involved in the former system (Scheme 1) and, furthermore, acid sites within the dense microparticles of the resins

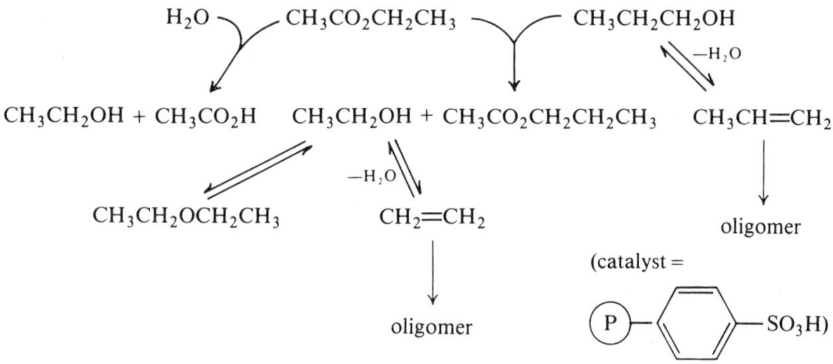

Scheme 1

appear to be more reactive than are surface sites. Higher overall crosslink ratios inhibit diffusion into microparticles, but the sites therein seem more reactive the more crosslinked the environment. In the case of methanol dehydration, macroporous acid resins are less active than conventional gel forms. Attempts have also been made to increase the inherent catalytic activity of sulphonic acid resins by secondary treatment with (i) oleum, (ii) SO_3 gas, and (iii) nitrating and chlorinating reagents.[100] Some success has been achieved, although nitrating and chlorinating reactions tend to induce some simultaneous loss of sulphonic acid sites. A copolymer of vinyl alcohol and styryl sulphonic acid has been shown to be a particularly good catalyst for the hydrolysis of carbohydrates.[101] In this case interaction of the backbone hydroxyl groups with the saccharide seems to play an important role in catalysis.

Chemical modification of macroporous methacrylate-based copolymers has produced a novel range of acid catalysts.[102] In this case the sulphonate group was

---

[97] J. L. Janier-Dubry, M. Delmas, and A. Gaset, *J. Catal.*, 1982, **77**, 16.
[98] K. M. Dooley, J. A. Williams, B. C. Gates, and R. L. Albright, *J. Catal.*, 1982, **74**, 361.
[99] R. B. Diemer, K. M. Dooley, B. C. Gates, and R. L. Albright, *J. Catal.*, 1982, **74**, 373.
[100] Z. Prokop and K. Setinek, *Collect. Czech. Chem. Commun.*, 1982, **47**, 1613.
[101] K. Arai, Y. Ogiwara, and C. Kuwubara, *J. Appl. Polym. Sci.*, 1982, **27**, 1601.
[102] K. Setinek, V. Blazek, J. Hradil, F. Svec, and J. Kalal, *J. Catal.*, 1983, **80**, 123.

introduced by reaction with propane sultone [reaction (22)]. The product in its acid form was an active catalyst in a reesterification reaction but showed no special advantages over normal resins. Novel carboxylic acid resins have been prepared by substitution of polystyrene with aromatic anhydrides[103] [reaction (23)]. In the case

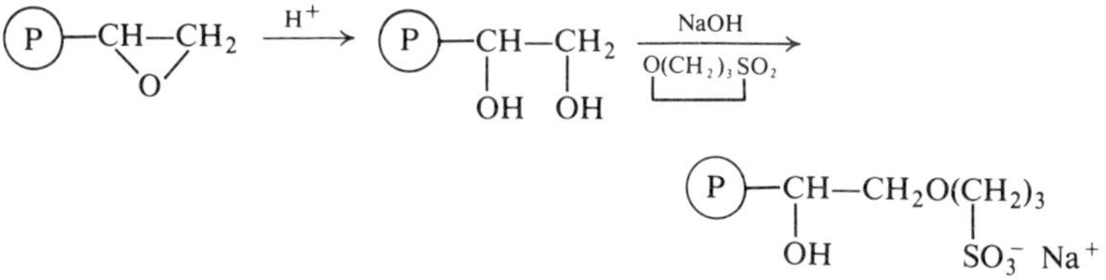

$$\text{(P)}-CH-CH_2 \xrightarrow{H^+} \text{(P)}-CH-CH_2 \xrightarrow[O(CH_2)_3SO_3]{NaOH}$$
$$\underset{OH\quad OH}{}$$

$$\text{(P)}-\underset{OH}{CH}-CH_2O(CH_2)_3 \atop SO_3^- \; Na^+ \quad (22)$$

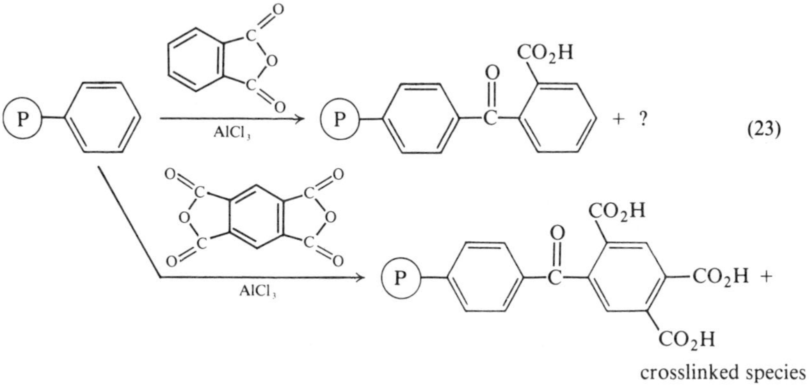

(23)

of pyromellitic dianhydride the authors have recognized the intervention of a crosslinking side-reaction. Parallel work has shown, however, that crosslinking also occurs with phthalic anhydride, although the mechanism by which this occurs is not clear.[104] Polymers carrying pyridinium hydrochloride groups have also been used successfully as acids in catalysing the reaction of ketones and carboxylic acids with alcohols.[105] Finally, under acidic species the trifunctional resin prepared by Gates and his co-workers must be mentioned.[106] In this work a conventional sulphonic acid was partially neutralized to yield bound Hg^{2+} and Fe^{3+} species. The whole system was then employed effectively in the catalysis of the hydration of acetylene to yield acetaldehyde.[106] This synthesis and the catalyst are well documented, but the elegant use of three catalytic species on a single support is a concept well worthy of further development.

New amphoteric resins species have also been described recently.[107] Reaction of salicyclic acid and other hydroxyaromatic acids with epichlorohydrin and ethylene-

---

[103] M. Biswas and S. Chatterjee, *J. Appl. Polym. Sci.*, 1982, **27**, 3851.
[104] D. A. Aitken and D. C. Sherrington, *J. Appl. Polym. Sci.*, 1983, **28**, 2463.
[105] J. Yoshida, J. Hashimoto, and N. Kawabata, *Bull. Chem. Soc. Jpn.*, 1981, **54**, 309.
[106] T. T. Moxley and B. C. Gates, *J. Mol. Catal.*, 1981, **12**, 389.
[107] R. N. Kapadia and A. K. Dalal, *J. Appl. Polym. Sci.*, 1982, **27**, 3793.

diamine yields condensation products with both acidic and basic features. Reductions of poly(β-aminoketones) have also been used to form basic polymers,[108] subsequent alkylation of which yields poly(quaternary salts) [reaction (24)]. Polymers carrying N-benzyl-N-methylamino pyridine residues have also been

$$\text{poly}(\beta\text{-aminoketones}) \xrightarrow{\text{LiAlH}_4} \text{(structure with H, OH, piperazine)} \text{ (and others)}$$

$$\xrightarrow{\text{CH}_3\text{I}} \text{(methylated quaternary salt structure)} \quad (24)$$

synthesized by chemical modification of a non-functional backbone and also *via* an appropriate monomer. These have been used as base catalysts in the acylation of linalool.[109]

6 Polymeric Complexes of Transition Metals

The attachment of transition-metal complexes to polymers and their subsequent use as catalysts is an area which attracts considerable industrial interest for obvious reasons. The field is a very big one and cannot be reviewed here comprehensively. Nevertheless, one or two important developments are worthy of mention.

The vast literature[110—115] generated since 1970 highlighted the potential of these systems in hydrogenations, hydroformylations, and hydrosilylations, and emphasized some shortcomings which must be overcome. Among the latter, not the least is the question of analysis and characterization. This is common to many polymeric systems, and recent developments in this area include the use of mass spectra in analysing polymer–iridium species[116] and the application of 'magic-angle' ^{31}P n.m.r. spectroscopy in the analysis of Ni^{II}, Pd^{II}, and Pt^{III} polymer-attached species.[117] Another vitally important problem concerns the thermal and chemical

---

[108] A. S. Angeloni, P. Ferruti, M. Tramontini, and M. Casolaro, *Polymer*, 1982, **23**, 1693.
[109] M. Tomoi, Y. Akada, and H. Kakiuchi, *Makromol. Chem., Rapid Commun.*, 1982, **3**, 537.
[110] Y. Chauvin, D. Commereuc, and F. Dawans, *Prog. Polym. Sci.*, 1977, **5**, 95.
[111] F. R. Hartley and P. N. Vezey, *Adv. Organomet. Chem.*, 1977, **15**, 189.
[112] C. U. Pittman, Chapter 5 in ref. 4.
[113] F. Ciardelli, G. Braca, C. Carlini, G. Sbrana, and G. Valentini, *J. Mol. Catal.*, 1982, **14**, 1.
[114] D. C. Bailey and S. H. Langer, *Chem. Rev.*, 1981, **81**, 109.
[115] M. Kaneko and E. Tsuchida, *J. Polym. Sci., Macromol. Rev.*, 1981, **16**, 397.
[116] J. Azran, O. Buchman, and J. Blum, *J. Mol. Catal.*, 1983, **18**, 105.

stability of the polymer employed to bind metal complexes. Significant advances have been in the use of polypropylene functionalized via ^{60}Co γ-irradiation techniques.[118,119] This has allowed the attachment of phosphine ligands and hence rhodium complexes, which are useful in hydroformylation reactions. Other more stable polymers which have been examined include carboranylphosphazenes.[120]

With regard to the types of complex and nature of the applications which are attracting current interest, one particularly active area is that of solar conversion. Ru^{II} bipyridyl complexes play a central role in this and numerous polymeric systems have been developed and investigated.[121—127] Somewhat related because of its biological origin is the question of nitrogen fixation and activation, and again polymeric systems have made a contribution in these investigations.[128—130] As far as novel industrial catalysts are concerned, an important possibility with polymeric systems is the ability to generate and stabilize transition-metal species which either have no homogeneous analogues, or which have only poor stability as a result of self-oligomerization processes. In this area, cluster compounds naturally have some potential, and a number of these have now been generated on polymeric systems. Molybdenum-based species have been described,[131,132] as have cobalt-[133] and osmium-[134] based ones. Finally, polymeric systems are being utilized under oxidative conditions, particularly in combination with O_2 and H_2O_2. This is a particularly difficult area, of course, because of the ease of oxidation of most organic-based polymers. Nevertheless, with efficient catalysts allowing mild conditions, thiosalts have been oxidized by poly(vinyl pyridine)-bound copper complexes,[135] thiols have been oxidized by H_2O_2 and bound cobaltphthalocyanin species,[136,137] and phenylalanine has been hydroxylated by similar iron-based complexes.[138] If the problem of oxidative degradation of the polymers can be overcome, no doubt further developments will occur.

---

[117] L. Berni, H. C. Clark, J. A. Davies, C. A. Fyfe, and R. E. Wasylishen, *J. Am. Chem. Soc.*, 1982, **104**, 438.
[118] F. R. Hartley, S. G. Murray, and P. N. Nicholson, *J. Polym. Sci., Polym. Chem. Ed.*, 1982, **20**, 2395.
[119] F. R. Hartley, S. G. Murray, and P. N. Nicholson, *J. Mol. Catal.*, 1982, **16**, 363.
[120] H. R. Allcock, A. G. Scopelianos, R. R. Whittle, and N. M. Tollefson, *J. Am. Chem. Soc.*, 1983, **105**, 1316 and 1321.
[121] M. Kaneko, A. Yamada, S. Nemoto, and M. Yokoyama, *Kobunshi Ronbunshu*, 1980, **37**, 685.
[122] S. F. Chan, M. Chou, C. Creutz, T. Matsubara, and N. Sutin, *J. Am. Chem. Soc.*, 1981, **103**, 369.
[123] J. M. Calvert and T. J. Meyer, *Inorg. Chem.*, 1981, **20**, 27.
[124] J. M. Clear, J. M. Kelly, C. M. O'Connell, and J. G. Vos, *J. Chem. Res.*, 1981, **9**, 260.
[125] H. D. Abrulia and A. J. Bard, *J. Am. Chem. Soc.*, 1982, **104**, 2641.
[126] C. Creutz, A. D. Keller, N. Sutin, and A. P. Zipp, *J. Am. Chem. Soc.*, 1982, **104**, 3618.
[127] J. M. Calvert, J. V. Caspar, R. A. Brinstead, T. D. Westmoreland, and T. J. Meyer, *J. Am. Chem. Soc.*, 1982, **104**, 6620.
[128] M. Koide, T. Masubuchi, Y. Kurimura, and E. Tsuchida, *Polym. J.*, 1980, **12**, 793.
[129] M. Koide, E. Tsuchida, and Y. Kurimura, *Makromol. Chem.*, 1981, **182**, 749.
[130] A. Bossi, F. Barbassi, G. Petrini, and L. Zanderighi, *J. Chem. Soc., Faraday Trans. 1*, 1982, **78**, 1029.
[131] C. Ungurenascu, C. Cotzur, and V. Harabagiu, *Polym. Bull.*, 1980, **19**, 219.
[132] C. G. Francis, H. Huber, and G. A. Ozin, *Inorg. Chem.*, 1980, **19**, 219.
[133] J. G. Gressier, G. Levesque, and H. Patin, *Polym. Bull.*, 1982, **8**, 55.
[134] J. B. N'Guini, J. Lieto, and J. P. Aune, *J. Mol. Catal.*, 1982, **15**, 367.
[135] M. Chanda, K. F. O'Driscoll, and G. L. Rempel, *J. Mol. Catal.*, 1980, **7**, 389.
[136] J. H. Schutten and T. P. M. Beelen, *J. Mol. Catal.*, 1980, **10**, 85.
[137] W. M. Brouwer, P. Piet, and A. L. German, *Polym. Bull.*, 1982, **8**, 245.
[138] T. Shimidza, T. Iyoda, and N. Kanda, *J. Chem. Soc., Chem. Commun.*, 1981, 1206.

7 Stoicheiometric Reactions of Polymers

The modification of polymers to introduce functional groups capable of acting as stoicheiometric reagents has been reviewed previously[139] and some new developments have also been described recently.[39,44,140,141] A much clearer understanding of polymer-bound Wittig reagents has been provided by Ford and his co-workers[142] who have examined in particular the effect of various resin parameters. Polymer-bound carbonate and acetate ions have been used to hydrolyse iodo-aminoalcohols to amino diols,[143] while three different pyridine-based brominating species[144] have been used with success in the stereoselective bromination of *cis/trans* 1-phenyl propenes, and 1-phenyl cyclohexene [reaction (25)]. To some extent all of these reagents are recycleable, as indeed are two polymeric acylating species recently described.[145,146] Two closely related polymer–haem-derived species have been synthesized, one being a reversible O_2 binder,[147] and the other a reversible CN^- exchanger.[148]

$$\text{Ph-cyclohexene} \begin{Bmatrix} \text{A)} & \text{P–C}_5\text{H}_4\text{N–NBr}_2 \\ \text{B)} & \text{P–C}_5\text{H}_4\text{N}^+\text{–O}^-(\text{Br}_2)_x \\ \text{C)} & \text{P–C}_5\text{H}_4\text{N}^+\text{–HBr}_3^- \end{Bmatrix} \longrightarrow \text{Ph(Br)cyclohexane(Br)} + \text{Ph-cyclohexene-Br} \quad (25)$$

Systems Designed for Recycling.—The emphasis on polymeric species capable of facile regeneration or recycling is vitally important if such systems are to find significant technical application. In this context a number of very elegant pieces of work provide a conceptual picture for the future. Thus electrochemical regeneration provides a clean and efficient recycling mechanism. Poly(4-vinylpyridinium hypo-

---

[139] Chapter 2, ref. 4.
[140] A. Patchornik, *Nouv. J. Chim.*, 1982, **12**, 639.
[141] T. W. Hall, S. Greenberg, C. R. MacArthur, B. Khouw, and C. C. Leznoff, *Nouv. J. Chim.*, 1982, **12**, 653.
[142] M. Bernard and W. T. Ford, *J. Org. Chem.*, 1983, **48**, 326.
[143] G. Cardillo, M. Orena, G. Porzi, and S. Sandri, *J. Chem. Soc., Chem. Commun.*, 1982, **22**, 1309.
[144] Y. Johar, M. Zupan, and B. Sket, *J. Chem. Soc., Perkin Trans. 1*, 1982, 2059.
[145] Y. Huang, C. C. Chan, and Q. S. Zhou, *Synth. Commun.*, 1982, **12**, 709.
[146] J. M. J. Frechet and W. Armaratunga, *Polym. Bull.*, 1982, **7**, 361.
[147] E. Tsuchida, H. Nishide, and Y. Sato, *J. Chem. Soc., Chem. Commun.*, 1982, 556.
[148] E. Kakufuta, H. Wanatabe, and I. Nakamura, *Polymer*, 1982, **23**, 1815.

bromite) readily oxidizes secondary alcohols to ketones in the absence of an electric current and without any contaminating reduced product. This oxidant is then reformed on passage of a current[149] [reaction (26)]. This system is quite efficient,

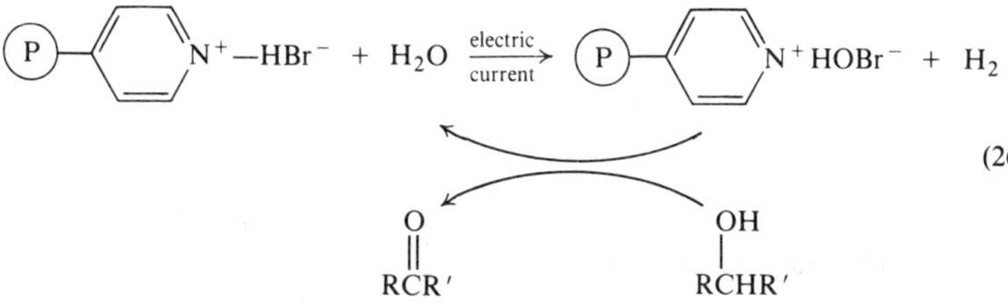

$$\text{(P)}-\text{C}_5\text{H}_4\text{N}^+-\text{HBr}^- + \text{H}_2\text{O} \xrightarrow{\text{electric current}} \text{(P)}-\text{C}_5\text{H}_4\text{N}^+\text{HOBr}^- + \text{H}_2$$

$$\text{RCR'} \;(\text{C=O}) \quad \rightleftharpoons \quad \text{RCHR'} \;(\text{OH})$$

(26)

consuming 2.4 Faradays per mole only slightly above the theoretical required. A more synthetically interesting reaction has been described by the same Japanese group more recently.[150] Olefins electrolysed in dimethylformamide–benzene solutions containing traces of water produce epoxides in yields up to 80% when an alkyl quaternary ammonium bromide resin is present. Yields are virtually zero with Cl^- and NO_3^- resins and significantly reduced with a pyridinium hydrobromide resin. The yield goes through a maximum as the water content is increased because the electrochemical efficiency is much lower; 7.0 Faradays per mole are required (theoretical = 2.0) due to competing hydrolysis of water. In spite of this the reaction is very important because of the significant position of epoxides as chemical intermediates.

Another very interesting development involves supported reduced nicotinamide species. These have been used to reduce ketones to secondary alcohols[151] [reaction (27)]. In the case of the reduction of ethylbenzoyl formate[152] catalysed by Mg^{2+} ions [reaction (28)], the polymeric species was recycled by reaction with an unbound dihydronicotinamide. In addition, yields of ethyl mandelate were significantly enhanced by using a 'spacer arm' to move the reduced nicotinamide away from the

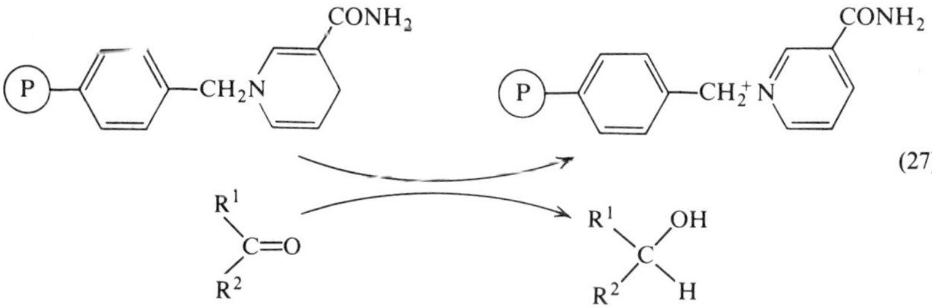

(27)

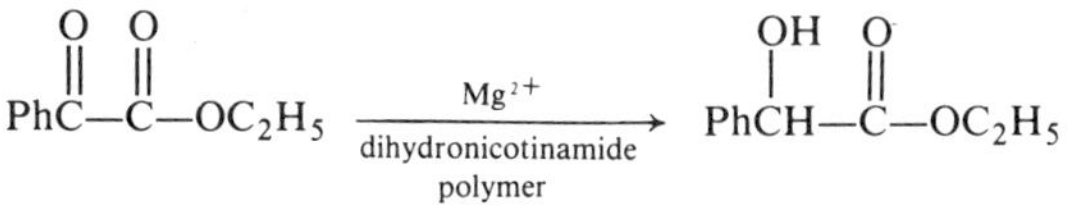

$$\text{PhC(=O)-C(=O)-OC}_2\text{H}_5 \xrightarrow[\text{dihydronicotinamide polymer}]{Mg^{2+}} \text{PhCH(OH)-C(=O)-OC}_2\text{H}_5 \quad (28)$$

---

[149] J. Yoshida, R. Nakai, and N. Kawabata, *J. Org. Chem.*, 1980, **45**, 5269.
[150] J. Yoshida, J. Hashimoto, and N. Kawabata, *J. Org. Chem.*, 1982, **47**, 3575.
[151] G. Dupas, J. Bourguignon, C. Ruffin, and G. Queguiner, *Tetrahedron Lett.*, 1982, **23**, 5141.
[152] T. Endo, T. Takada, and M. Okawara, *J. Polym. Sci., Polym. Chem. Ed.*, 1983, **21**, 603.

polymeric backbone. The synthetic route for achieving the latter is shown in Scheme 2. A similar polymeric species has been used in an ingenious electron transfer from one polymeric species to another using a mobile 'electron carrier'.[153] The three-phase system is depicted in Scheme 3. This clever interchange of electrons

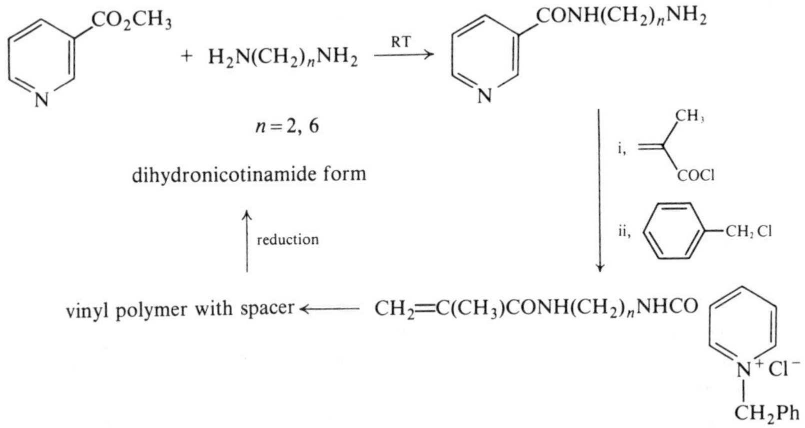

Scheme 2

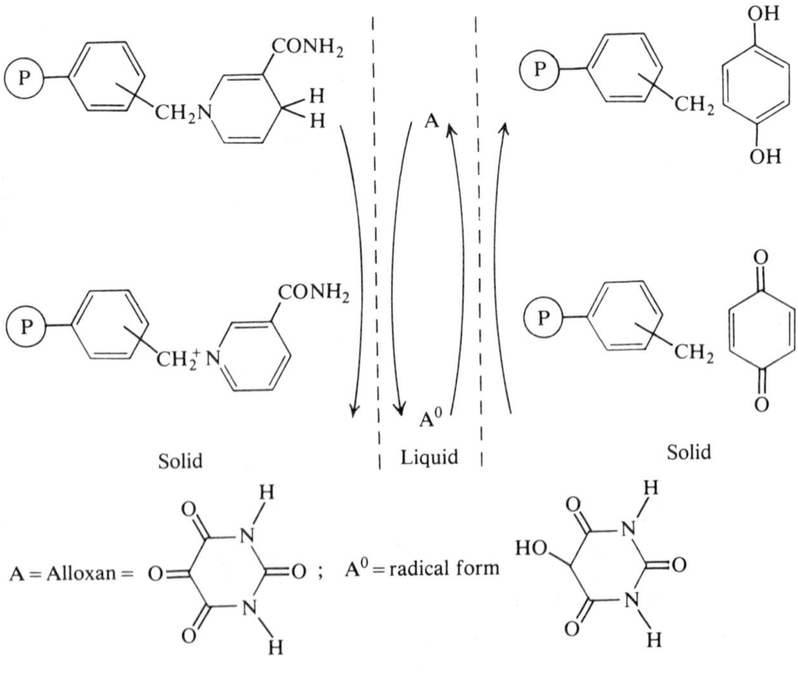

Scheme 3

[153] N. Tsubokawa, E. Endo, and M. Okawara, *J. Polym. Sci., Polym. Chem. Ed.*, 1982, **20**, 2205.

between polymeric species will be important in the application of electronic devices, solid state electrochemistry, and in enzyme and co-enzyme mimics. In this context a polymer-bound model for thiamine-dependent enzymes has also been described.[154] 5-(2-Hydroxyethyl)-4-methylthiazole was attached to a resin and a co-operating basic group introduced subsequently as shown in Scheme 4. The resulting species

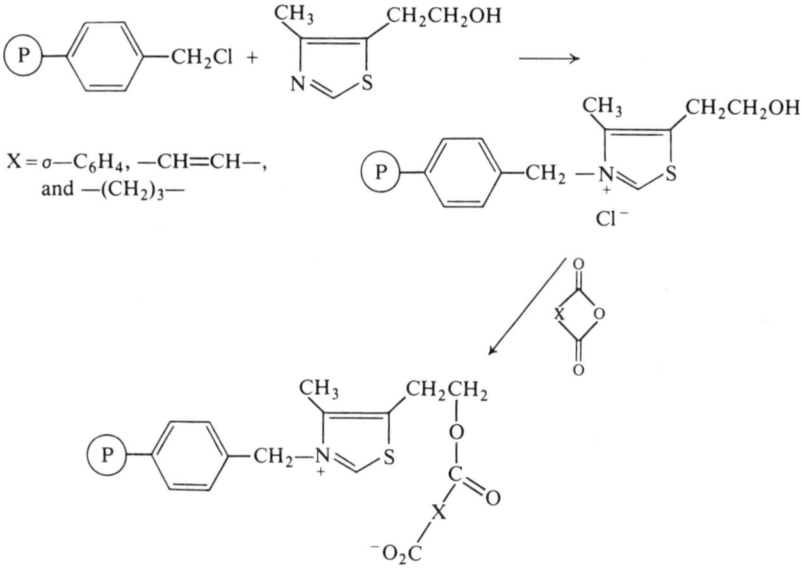

$X = o-C_6H_4$, $-CH=CH-$, and $-(CH_2)_3-$

Scheme 4

readily mediated in the reaction of an aldehyde with an α,β-unsaturated ketone [reaction (29)] and was regenerated in the process. Very significantly if the internal

$$R^1CHO + CH_2=CHCOR^2 \rightarrow R^1COCH_2CH_2COR^2 \qquad (29)$$

carboxylate base was not included but was replaced by an external base source, considerable side reaction occurred.

A final example of a clever autorecycling system has been developed for the oxidation of alcohols to ketones.[155] This involves a polymer-supported pyridodi-

---

[154] C. S. Sell and L. A. Dorman, *J. Chem. Soc., Chem. Commun.*, 1982, 629.
[155] F. Yoneda, H. Yamato, T. Nagamatsu, and H. Egawa, *J. Polym. Sci., Polym. Lett. Ed.*, 1982, **20**, 667.

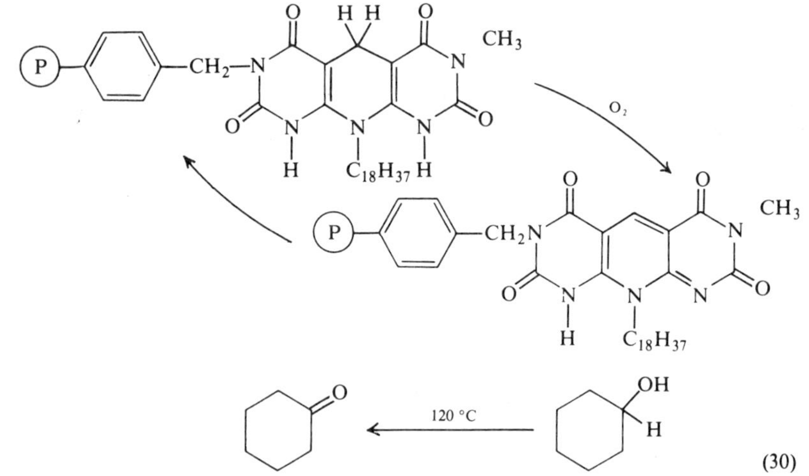

pyrimidine species [reaction (30)]. In this case the dihydro product is spontaneously converted back to the pyrido species by O_2 and the conversion of alcohol to ketone is the reverse of that achieved, for example, in reaction (27).

Chiral Systems.—Another very important use of stoicheiometric polymeric reagents is in the area of protecting groups. Leznoff and his co-workers[139,156] continue to produce some elegant syntheses by this approach and in particular have described routes to unsymmetrical phthalocyanines. Polymer-supported chiral auxiliaries also provide a convenient synthetic tool, again used in stoicheiometric quantities. The bound species is used for temporary attachment of a substrate which is then modified in an enantioselective fashion, before being cleaved from the chiral auxiliary. Again Leznoff and his co-workers have made a valuable contribution.[157] More recently, Frechet and his co-workers[158] have examined the α-methylation of cyclohexanone imines derived from chiral 1-phenylethylamines. By making the latter an integral part of the polymer, isolation of the optically active product is considerably facilitated (Scheme 5). Furthermore, optical yields were somewhat

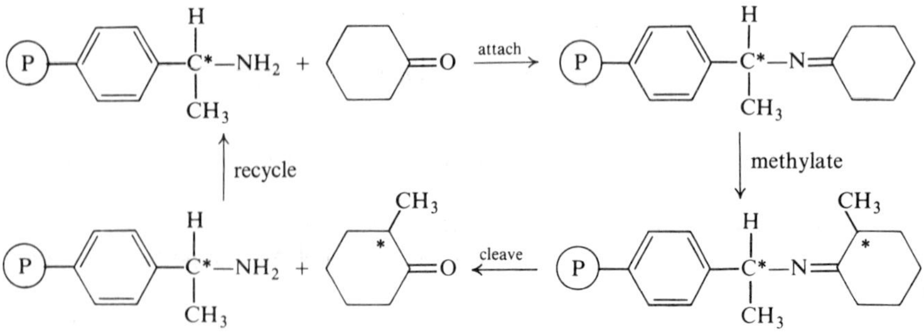

Scheme 5

[156] C. C. Leznoff and T. W. Hall, *Tetrahedron Lett.*, 1982, **23**, 3023.
[157] C. R. McArthur, P. M. Worster, and C. C. Leznoff, *Can. J. Chem.*, 1982, **60**, 1836.
[158] J. M. J. Frechet, J. Halgas, and D. C. Sherrington, *Reactive Polym.*, 1983, **1**, 227.

improved relative to the non-polymeric system. Highly enantioselective epoxidation of olefinic alcohols can be achieved using titanium tetra-isopropoxide, t-butyl hydroperoxide and (+) or (−) tartrate esters. By attaching the latter chiral component to a polymer, however, the entire process is considerably simplified[159] [reaction (31)].

$$\underset{R^3}{\overset{R^1}{\diagdown}}\underset{CH_2OH}{\overset{R^2}{\diagup}} \xrightarrow[\substack{Ti(OCH(CH_3)_2)_4 \\ Bu^tOOH}]{\text{(P)}-CH_2CH_2OCOC^*H(OH)C^*H(OH)CO_2R} \underset{R^3}{\overset{R^1}{\diagdown}}\underset{CH_2OH}{\overset{R^2}{\diagup}}\! O \qquad (31)$$

Another developing area of interest is that of chiral polymeric reagents. Here there is some overlap with chiral catalysts but in this instance reagents are employed in stoicheiometric quantities. Sodium borohydride and aluminium hydride reagents partially deactivated with chiral molecules such as amino-acids and saccharides can be powerful asymmetric reducing agents. Sherrington and his co-workers[160] have synthesized polymeric analogues of a number of chiral borohydrides [reaction (32)]

(Ps)—CH$_2$N⟩ + NaBH$_4$ ⟶ (Ps)—CH$_2$N⟩
      CO$_2$H   H$_3$B CO$_2^-$ Na$^+$

(32)

(C)—OH/CO$_2$H + NaBH$_4$ ⟶ (C)—O—C(=O)—O—BH$_2^-$ Na$^+$

(Ps) = polystyrene; (C*) = polysaccharide

only to find a loss of enantioselectivity in reduction reactions. By using polymer-bound analogous aminols, however, enantioselectivity seems to be retained[161] [reaction (33)]. Why this difference arises is not clear and more work is required. A

(Ps)—CH$_2$N⟩ + BH$_3$ $\xrightarrow{THF}$ (Ps)—CH$_2$N⟩   (33)
      CH$_2$OH   H$_2$B CH$_2$
                                                    \O/

[159] M. J. Farrall, M. Alexis, and M. Trecarter, 'International Symposium on Polymer-supported Reagents in Organic Chemistry', Lyon, France, July, 1982, Communications p. 84.
[160] A. Akelah and D. C. Sherrington, *Polymer*, 1983, **24**, 147.
[161] S. Itsuno, A. Hirao, and S. Nakahama, *Makromol. Chem., Rapid Commun.*, 1982, **3**, 673.

great deal of effort has been directed towards polymeric analogues of chiral aluminium hydride complexes[162] and interesting results will no doubt appear in due course. The generation of 'chiral cavities' in three-dimensional polymeric matrices has recently been reviewed[163] and a move towards their use in asymmetric reactions is clearly in progress.

8 Polymeric Species in Photochemical Processes

Photodegradation and photo-crosslinking of polymeric materials is a vitally important area in both film forming surface treatment and in various lithographic processes. Of special interest at the moment, of course, are the various photo-resist technologies being studied for application in the production of microelectronic devices. This area lies outside the scope of the present review but one development is of interest because of the simple chemistry involved. Most negative resist materials rely on 2 + 2 cycloaddition reactions, *e.g.*, of cinnamate residues, to achieve photo-crosslinking and this is a stoicheiometric photochemical process. Incorporation of a photo-polymerizable vinyl group would allow a chain reaction, *i.e.*, chemical amplification, and indeed this is widely used in forming surface films. Farrell and her co-workers[164] have introduced vinyl groups into linear polystyrene and shown negative resist properties. The routes used are shown in reaction (34). Linear

$$\begin{array}{c} P\text{—}C_6H_4\text{—}CH_2Cl \xrightarrow{\text{i, PPh}_3}_{\text{ii, CH}_2O/\text{NaOH}} \\ \\ P\text{—}C_6H_4\text{—}CHO \xrightarrow{CH_3^+PPh_3Br^-}_{K^+Bu^tO^-} \end{array} \xrightarrow{} P\text{—}C_6H_4\text{—}CH\text{=}CH_2 \quad (34)$$

$$\downarrow \text{benzoyl peroxide, } h\nu$$

$$\text{crosslinking}$$

polymer can, of course, be made directly from divinylbenzene[165] and this material would presumably behave similarly.

Polymers carrying pendant tris(2,2'-bipypridyl) ruthenium(II) dichloride groups have been used in the photoreduction of methyl viologen (1,1'-dimethyl-4,4'-bipyridinium dichloride).[166] This is a reaction of interest in solar conversion mentioned earlier in Section 6. The polymeric system has much better solubility in water and by the use of appropriate copolymers the environment of the complex might be adjusted to control the photochemical reaction.

The use of polymer-supported photo-sensitizers has been reviewed recently by Neckers.[167] Rose Bengal is a useful source of singlet oxygen, O_2^1, and the original

---

[162] J. M. J. Frechet, personal communication.
[163] G. Wulff, R. Kemmerer, J. Vietmeier, and H. G. Poll, *Nouv. J. Chim.*, 1982, **6**, 681.
[164] M. J. Farrall, M. Alexis, and M. Trecarten, *Polymer*, 1983, **24**, 114.
[165] T. Higashimura, S. Aoshima, and H. Hasegawa, *Macromolecule*, 1982, **15**, 1221.
[166] M. Kaneko, A. Yamada, E. Tsuchida, and Y. Kurimura, *J. Polym. Sci., Polym. Lett. Ed.*, 1982, **20**, 593.
[167] D. C. Neckers, *Nouv. J. Chim.*, 1982, **6**, 645.

polystyrene supported species has now been joined by a poly(hydroxyethyl methacrylate)-based species,[168] and also a silica-bound moiety.[169] These have been used successfully in a number of 2 + 2 and 2 + 4 cycloaddition reactions of O_2^1 with alkenes. Also they are useful in furfuryl alcohol photo-oxidations. Recently the photo-oxidation of triphenylphosphine selenide has been described[170] [reaction (35)]. Such supported sensitizers provide advantages in terms of product work-up

$$Ph_3P=Se \xrightarrow[h\nu]{\text{(Ps)}-\text{Rose Bengal}} Ph_3\overset{+}{P}-Se-O-O^-$$

$$\downarrow Ph_3P=Se$$

$$Ph_3P=O \longleftarrow Ph_3P\begin{smallmatrix}Se\\|\\O\end{smallmatrix} \longleftarrow 2Ph_3P=Se=O \quad (35)$$

and isolation, but for prolonged use should not be subject to bleaching. This question seems to be a difficult one, and success in avoiding bleaching may well depend on the ability to produce polymeric systems devoid of contaminants.

Recently, Neckers has described the synthesis and polymerization of p'-vinyl benzoyl-p-t-butylperbenzoate.[171] The polymeric species can be used as a photo-initiator and indeed provide a convenient source of polymeric radicals for any application [reaction (36)]. As a potential extension from polymeric systems the

(P)—C₆H₄—CO—C₆H₄—C(=O)—O—O—Bu^t $\xrightarrow{h\nu}$ polymer bound radicals

photophysics of pyrene and its derivatives encapsulated in polymerized microemulsions has been studied.[172]

9 Polymers as Selective Sorbents

One of the most important areas in this group is that of selective ion exchange polymers. This is a large and well-established area in its own right, and has recently been reviewed.[173] The use of water-soluble metal complexing polymers is a relatively novel idea, and again developments in this field have been reviewed.[174] In general,

---

[168] A. P. Schaap, A. L. Thayer, K. A. Zaklika, and P. C. Valenti, *J. Am. Chem. Soc.*, 1979, **101**, 4106.
[169] S. Tamagaki, C. E. Liesner, and D. C. Neckers, *J. Org. Chem.*, 1980, **45**, 1573.
[170] S. Tamagaki and R. Akatsuka, *Bull. Chem. Soc. Jpn.*, 1982, **55**, 3037.
[171] I. Gupta, S. N. Gupta, and D. C. Neckers, *J. Polym. Sci., Polym. Chem. Ed.*, 1982, **20**, 147.
[172] S. S. Atik and J. K. Thomas, *J. Am. Chem. Soc.*, 1982, **104**, 5868.
[173] A. Warshawsky, *Angew. Makromol. Chem.*, 1982, **109**, 171.
[174] K. Geckeler, G. Large, H. Eberhardt, and E. Bayer, *Pure Appl. Chem.*, 1980, **52**, 1883.

the approach has been to attach to polymers ligands with known complexing ability for particular metal ions and some success has derived from this. Novel systems continue to be described[175,176] and no doubt the elegance of selective extraction will continue to improve. The use of supported crowns, cryptands, and related species in this context has been reviewed recently.[177,178] Of growing interest are species capable of trapping toxic heavy metal ions, and also those capable of application in nuclear fuel reprocessing.[179] Uranium in its UO_2^{2+} ion remains the only metal extracted on any significant scale by resin-supported species and recently U^{VI} has been trapped on a polyheteroatomic chain substituted with bis(di-n-butyl phosphate) groups.[180]

Synthesis of resins with a 'memory' for a particular metal ion (or indeed other moiety) is a novel approach. The idea here is to make a highly crosslinked network of a chelating polymer with the metal ion present as a template, to form a preferred conformational arrangement of ligands. Removal of the template ion then allows subsequent selective re-extraction of the same ion.[181,182] A similar approach has been used to form methacrylate resins capable of selectively absorbing particular dye molecules.[183] The concept has been taken to its most sophisticated, however, by Wulff and his co-workers[184,185] in constructing specific receptor sites.

An area of growing interest, arising from ecological constraints and the evolution of biotechnological processes, is the separation and isolation of organic substances using ion exchange and other resin species. Recently the extraction of cephalosporins on an Amberlite resin has been described[186] and also the isolation of proteins using a polymeric adsorbent containing azobenzene groups.[187] Some theoretical considerations of such systems have also been published.[188] The difficulty of extracting relatively small concentrations of organic materials from large volumes of water is one of extreme importance in the growth of biotechnology, and considerable scope exists for imaginative application of polymer-supported systems.

The work of Frechet and his co-workers[39,189] in separating allergens from essential oils is an excellent example of scientific application. Many allergens are, in fact, α-methylene-γ-butyrolactones and these can be trapped by reaction with a polymeric support carrying either primary amino groups or sulphinate groups. Appropriate treatment with a suitable reagent can later release the allergen for

---

[175] R. A. A. Muzzarelli and F. Tanfani, *Pure Appl. Chem.*, 1982, **54**, 2141.
[176] F. Vernon, *Pure Appl. Chem.*, 1982, **54**, 2151.
[177] E. Balasius and K. P. Janzen, *Pure Appl. Chem.*, 1982, **54**, 2115.
[178] J. Smid, *Pure Appl. Chem.*, 1982, **54**, 2129.
[179] J. Schon and W. Ochsenfeld, *Angew. Makromol. Chem.*, 1982, **109**, 215.
[180] E. Archelas, G. Buono, and B. Waegall, *Polyhedron*, 1982, **1**, 683.
[181] A. A. Efendiev and V. A. Kabanov, *Pure Appl. Chem.*, 1982, **54**, 2077.
[182] S. N. Gupta and D. C. Neckers, *J. Polym. Sci., Polym. Chem. Ed.*, 1982, **20**, 1609.
[183] R. Arshady and K. Mosbach, *Makromol. Chem.*, 1981, **182**, 687.
[184] G. Wulff, *Pure Appl. Chem.*, 1982, **54**, 2093.
[185] A. Sarhan and G. Wulff, *Makromol. Chem.*, 1982, **183**, 85.
[186] C. M. Pirotta, *Angew. Makromol. Chem.*, 1982, **109**, 197.
[187] K. Ishihara, S. Kato, and I. Shinohara, *J. Appl. Polym. Sci.*, 1982, **27**, 4273.
[188] E. M. Savitskaya, L. F. Yakhontova, and P. S. Nys, *Pure Appl. Chem.*, 1982, **54**, 2169.
[189] J. M. J. Frechet, A. J. Hagen, C. Benezra, and A. Cheminat, *Pure Appl. Chem.*, 1982, **54**, 2181.

further study or application (Scheme 6). In a variation on this the allergen is first converted to its diethylamino derivative, then the latter trapped as a copper complex on a suitable polymer (Scheme 7).

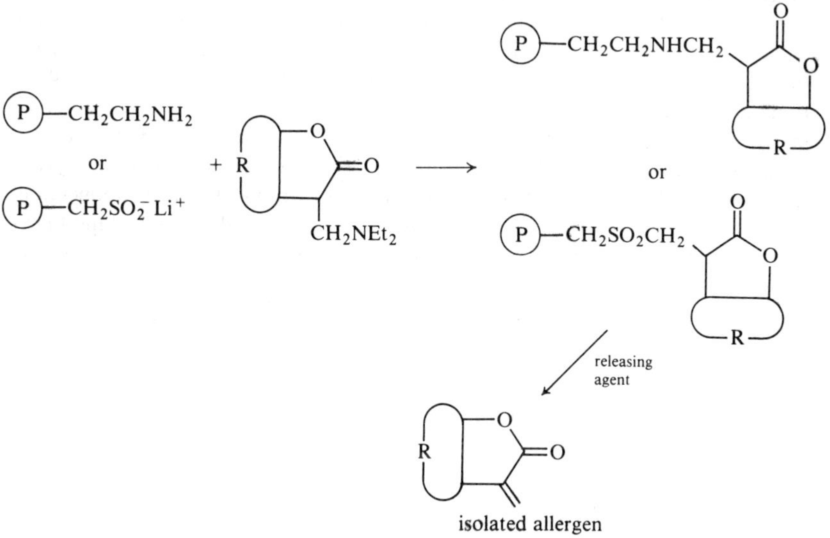

Scheme 6

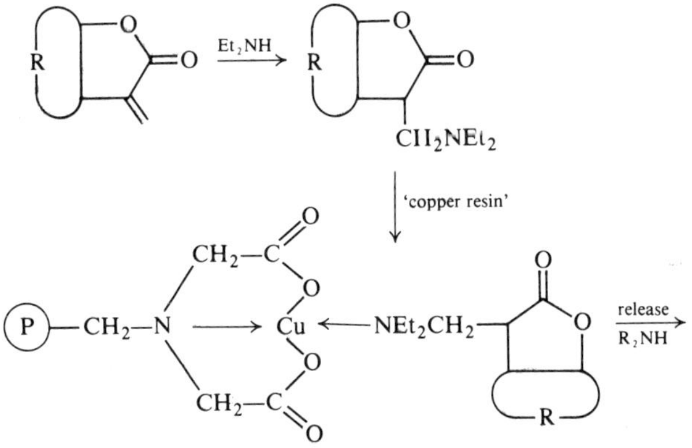

Scheme 7

10 Miscellaneous

The introduction of a wide variety of novel functional groups into polymers continues to occur, and at the same time new and more efficient routes to well-known modifications are also being developed. Thus, benzhydrylamine modified species have been described and used in solid phase synthesis and a new route to these has recently been disclosed.[190] New solid phase supports have also been synthesized by grafting amide-based monomers onto inert hydrocarbon polymers.[191] Macroporous resins have also been similarly modified.[192] Many sophisticated organic transformations of polymers are now being attempted using the most up-to-date organic chemical methodology.[39,193] Some of these modifications are aimed at producing new polymeric substances for application as materials in their own right, or in the fabrication of devices. Others, such as the introduction of nucleotide bases into poly(vinyl amines)[194,195] and saccharides into styrene copolymers,[90,196] are destined for a chemical or biochemical application. Irrespective of the motive or objective, the proliferation of the chemical modification of polymers will continue, as will the variety of applications of the products which ensue.

[190] A. Hirao, S. Itsuno, I. Hattori, K. Yamaguchi, S. Nakahama, and N. Yamazaki, *J. Chem. Soc., Chem. Commun.*, 1983, 25.
[191] J. G. Heffernan, S. B. Kingston, and D. C. Sherrington, *J. Appl. Polym. Sci.*, 1983, **28**, 3137.
[192] T. Brunelet, M. Bartholin, and A. Guyot, *Angew. Makromol. Chem.*, 1982, **106**, 79.
[193] J. M. J. Frechet, J. M. Farrall, and C. G. Wilson, *Polym. Bull.*, 1982, 7, 567.
[194] C. G. Overberger and S. Kikyotani, *J. Polym. Sci., Polym. Chem. Ed.*, 1983, **21**, 525.
[195] K. A. Brandt and C. G. Overberger, *Nouv. J. Chim.*, 1982, **6**, 673.
[196] K. Kobayashi, H. Sumitomo, and Y. Ina, *Kobunshi Ronbunshu*, 1982, **39**, 723.

15
Polymer Degradation

BY J. R. McCALLUM AND W. W. WRIGHT

PART I Photo and Photo-oxidative Degradation
by J. R. McCallum

1 Introduction

The number of papers published on a variety of aspects of the aging of polymers continues to grow. Inevitably in a review of two years' output, a balance between extent of subject coverage and space available dictates that a number of valuable contributions just cannot be mentioned.

2 General

A number of reviews of photodegradation have been published,[1-4] including one devoted to examining changes in mechanical behaviour.[5] The mechanism of operation of stabilizers has received some attention ranging from ultraviolet absorbers,[6,7] phenolic antioxidants,[8] hindered amine light stabilizers[9,10] to a new class of radical and singlet oxygen stabilizers.[11] The application of new techniques in following the course of photo-oxidation has been highlighted by papers describing the use of ESCA[12] and chemiluminescence.[13] The process of oxidation has been put to advantageous use by using rubber as an O_2 absorbent in packaging and storage of O_2 sensitive materials.[14] The most aptly titled review of the period was that produced by Wiles and Carlsson – 'Stop Photodegradation'.[15]

---

[1] M. Tanaka and M. Tsunooka, *Plast. Age*, 1981, **27**, 97.
[2] V. Ya. Shlyapintokh, *Dev. Polym. Photochem.*, 1981, **2**, 215.
[3] D. Gilead and G. Scott, *Dev. Polym. Photochem.*, 1982, **5**, 71.
[4] W. B. Hardy, *Dev. Polym. Photochem.*, 1982, **3**, 287.
[5] J. Pabiot and J. Verdu, *Polym. Eng. Sci.*, 1981, **21**, 32.
[6] A. Gupta, D. Kliger, and W. G. Scott, *ACS Symp. Ser.*, 1981, **151**, 27.
[7] D. A. Tirrell, *Polym. News*, 1981, **7**, 104.
[8] J. Pospisil, *Dev. Polym. Photochem.*, 1981, **2**, 53.
[9] B. Felder, R. Schumacher, and F. Sitek, ACS Symp. Ser., 1981, **151**, 65.
[10] N. S. Allen, *Dev. Polym. Photochem.*, 1981, **2**, 239.
[11] J. Arct, Z. Golulski, J. F. Rabek, and B. Ranby, *Eur. Polym. J.*, 1982, **18**, 81.
[12] D. T. Clark and A. Dilks, *J. Polym. Sci., Polym. Chem. Ed.*, 1981, **19**, 2847.
[13] G. A. George, *Dev. Polym. Degradation*, 1981, **3**, 173.
[14] M. L. Rooney, *Chem. Ind. (London)*, 1982, 197.
[15] D. J. Carlsson and D. M. Wiles, *Chemtech.*, 1981, **11**, 158.

3 Polyethylene

Identification of the initiating species in the photo-oxidation of polyethylene has caused problems. A most interesting observation has been made using ESCA which showed no changes in the surface composition of high- and low-density polymer after photo-oxidation.[16] Luminescence emission from high-density polyethylene has been attributed to the presence of contaminants which could be removed by solvent extraction.[17] Secondary hydroperoxides formed either photochemically or thermally did not appear to contribute to the overall oxidation of polyethylene,[18] a process which has been shown to be controlled by the rate of oxygen diffusion into the sample.[19-21]

Photodegradation has been shown to be enhanced by application of strain[22] and the changes in mechanical properties of polyethylene during photo-oxidation have been measured.[23] The degradation behaviour of blends of polyethylene with poly(vinyl chloride)[24] and styrene–butadiene copolymers[25] has been reported. The effect of additives has been extensively studied. Substituted anthraquinones[26] induce crosslinking in the early stages of photo-oxidation, as also does xanthone.[27] Probing the role of metal-containing complexes in photostabilization has involved investigation of ferrocent derivatives,[28,29] copper salts,[30] and cadmium sulphide.[31] It has been shown that nitroxyl radicals derived from hindered piperidines are very effective in breaking the oxidative chain mechanisms.[32] It was also concluded that the hydroxylamine derivatives were much more effective as stabilizers than the nitroxyl radicals.

4 Polypropylene

One of the ultimate goals of kinetic study of polymer photo-oxidation is the derivation of a function which will predict the service life of the material under given climatic conditions. Papers have been published[33,34] in which predictive equations

---

[16] D. T. Clark and J. Peeling, *Polym. Degradation Stab.*, 1981, **3**, 177.
[17] H. Kuroda and Z. Osawa, *J. Polym. Sci., Polym. Lett. Ed.*, 1982, **20**, 577.
[18] R. Arnaud, J. L. Gardette, J. M. Ginhac, and J. Lemaire, *Makromol. Chem.*, 1981, **182**, 1017.
[19] A. Davis, G. C. Fumlaux, and K. J. Ledbury, *Polym. Degradation Stab.*, 1981, **3**, 431.
[20] A. V. Cunliffe and A. Davis, *Polym. Degrad. Stab.*, 1982, **4**, 17.
[21] K. Perenyi and J. Verdu, *Mater. Tech.* (Paris), 1981, **69**, 69.
[22] D. Banachour and C. E. Rogers, *ACS Symp. Ser.*, 1981, **151**, 263.
[23] O. N. Karpukhin, T. V. Magomedova, V. G. Protasov, and E. M. Slobodetskaya, *Vysokomol. Soedin.*, Ser. A, 1982, **24**, 249.
[24] A. Ghaffar, C. Sadrmohaghegh, and G. Scott, *Polym. Degradation Stab.*, 1981, **3**, 341.
[25] C. David, R. A. Jacobs, and F. Zabeau, *Polym. Eng. Sci.*, 1982, **22**, 912.
[26] L. J. Taylor and J. W. Tobias, *J. Appl. Polym. Sci.*, 1981, **26**, 2917.
[27] A. A. Kachan, N. I. Litsov, and P. V. Zamotaev, *Vysokomol. Soedin.*, Ser. B, 1982, **24**, 577.
[28] M. Z. Borodulina, N. I. Kondrashkina, B. I. Sazhin, and T. N. Zelenkova, *Zh. Prikl. Khim.* (Leningrad), 1982, **55**, 1196.
[29] V. N. Bachischche, E. A. Kalennikov, and V. S. Yuran, *Dokl. Akad. Nauk., USSR*, 1982, **26**, 252.
[30] N. Kurishu, H, Kuroda, and Z. Osawa, *Polym. Degradation Stab.*, 1981, **3**, 265.
[31] R. Arnaud, J. Lemaire, and G. Penot, *Polym. Photochem.*, 1982, **2**, 39.
[32] R. Bagheri, K. B. Chakraborty, and G. Scott, *Polym. Degradation Stab.*, 1982, **4**, 1.
[33] O. N. Karpukhin, E. M. Mostovaya, B. V. Novozhilov, and E. M. Slobodetskaya, *Plast. Massy*, 1981, 24.
[34] E. M. Mostovaya, *Kinet. Katal.*, 1982, **23**, 805.

are evolved. An important observation has been made in that accelerated tests underestimate the effects of peroxide decomposers.[35] The selection of stabilizers has been reviewed.[36,37]

Hindered amine light stabilizers continued to undergo experimental investigation. The mechanism of operation of such substances is not fully understood and frequently the available experimental data is confusing. Several facts which give useful guides have emerged. It has been shown that bifunctional hindered amines become attached to the polymer chain, having reacted *via* the nitroxyl radical intermediate.[38] This conclusion is consistent with the proposal that substituted piperidines and their derived nitroxides associate with oxidation products[39] and induce rapid hyeroperoxide decomposition.[40] The mulfifunctional piperidines were found to be less likely to be lost by migration to the surface of the sample. The conclusion that polymeric hindered amines were less effective than the low-molecular-weight analogues[41] highlights the importance of the association step, an occurrence which would be less likely for polymeric materials. The reactivity of a number of different chemical functions has been examined. The influence of unsaturation,[42,43] carbonyl and hydroperoxide groups,[44,45] processing,[46,47] polychromatic irradiation,[48,49] dienones,[50,51] and pigments have been reported.[52,53]

E.s.r. spectroscopy[54] and chemiluminescence[55] measurements have been made on, respectively, distribution of radicals and hydroperoxide decomposition. Studies of photo-oxidation of blends with polyethylene[56] and olefin rubbers,[57] and thin films of copolymers with ethylene,[58] have been published.

5 Polybutadiene

Oxidation of unsaturated polymers inevitably involves the lowest energy excited singlet state of oxygen, $^1O_2$.[59] Ni^{II} chelates have been found to quench $^1O_2$.[60] The

---

[35] F. R. Mayo and J. Rose, *Macromolecules*, 1982, **15**, 948.
[36] P. Richters, *Plastica*, 1981, **34**, 2.
[37] F. Gugumus and H. Linhart, *Chem. Vlakna*, 1982, **32**, 94.
[38] D. K. C. Hodgeman, *J. Polym. Sci., Polym. Chem. Ed.*, 1981, **19**, 807.
[39] D. J. Carlsson, K. H. Chan, and D. M. Wiles, *ACS Symp. Ser.*, 1981, **151**, 51.
[40] D. J. Carlsson, K. H. Chan, J. Durmis, and D. M. Wiles, *J. Polym. Sci., Polym. Chem. Ed.*, 1982, **20**, 575.
[41] F. Gugumus, *Res. Discl.*, 1981, **209**, 357.
[42] N. S. Allen, K. O. Fatinikum, J. Lemaire, and J. Luc-Gardette, *Polym. Degradation Stab.*, 1982, **4**, 95.
[43] N. S. Allen, K. O. Fatinikun, and T. J. Henman, *Polym. Degradation Stab.*, 1982, **4**, 59.
[44] N. S. Allen, *Polym. Photochem.*, 1981, **1**, 243.
[45] N. S. Allen and K. O. Fatinikun, *Polym. Degradation Stab.*, 1981, **3**, 327.
[46] N. S. Allen, J. Lemaire, and J. Luc-Gardette, *Polym. Photochem.*, 1981, **1**, 111.
[47] N. S. Allen, J. Lemaire, and J. Luc-Gardette, *Polym. Degradation Stab.*, 1981, **3**, 199.
[48] N. S. Allen, J. H. Appleyard, and A. Chirinos-Padron, *Polym. Degradation Stab.*, 1982, **4**, 223.
[49] N. S. Allen, J. Lemaire, and J. Luc-Gardett, *J. Appl. Polym. Sci.*, 1982, **27**, 2761.
[50] J. Kovarova, J. Pospisil, and J. Sedlar, *Polym. Photochem.*, 1981, **1**, 25.
[51] J. Kovarova, J. Pospisil, and J. Sedlar, *Polym. Photochem.*, 1982, **2**, 349.
[52] R. Arnaud and J. Lemaire, *Dev. Polym. Photochem.*, 1981, **2**, 135.
[53] R. Arnaud, J. M. Ginhac, and J. Lemaire, *Makromol. Chem.*, 1981, **182**, 1229.
[54] G. Balint, L. Jokay, T. Kelen, A. Rockenbauer, and F. Tudos, *Polym. Photochem.*, 1981, **1**, 139.
[55] Z. Fodor, M. Iring, L. Matisova-Rychla, and J. Rychly, *Polym. Degradation Stab.*, 1981, **3**, 371.
[56] C. Sadrmoheghegh, G. Scott, and E. Setoudeh, *Polym. Degradation Stab.*, 1981, **3**, 469.
[57] J. Thomas, *Rev. Gen. Caoutch. Plast.*, 1981, **609**, 53.
[58] G. Geuskens and M. S. Kabamla, *Polym. Degradation Stab.*, 1982, **4**, 69.
[59] S. H. Etaiw, W. M. Khalifa, S. E. Morsi, M. A. Salem, and A. B. Zaki, *Polymer*, 1981, **22**, 942.
[60] D. Lala and J. F. Rabek, *Polym. Degradation Stab.*, 1981, **3**, 383.

role of hydroperoxides in the mechanism of photo-oxidation of poly(cis-1,4-butadiene) has been investigated.[61]

Copolymers of butadiene with styrene have been stabilized by addition of Cu^{II} compounds.[62,63] Photo-oxidative breakdown of acrylonitrile–butadiene–styrene rubbers was found to take place in the polybutadiene component resulting in loss of impact strength.[64,65] The mechanism of stabilization achieved by adding polymer bound stabilizers has been investigated.[66,67] The photolytic degradation of poly(but-1-ene) has been studied in some detail.[68,69]

6 Polystyrene

The mechanism of photo-oxidation of polystyrene has been reviewed.[70] Details of the mechanism of photo-initiated decomposition continue to be probed. Study of a model compound in fluid solutions demonstrated quantitative sensitized decomposition of hydroperoxides,[71] although the extension of the interpretation of these experiments to the solid state must be made with caution. The importance of diffusion of additives from the bulk to the surface of a sample has been demonstrated.[72] The effect both of introducing chemical change during processing,[73,74] and by addition of carbonyl functions has been investigated.[75,76]

In solution the significance of molecular complexes, particularly those involving a halogen atom, on the rate of initiation of photo-oxidation has been clearly demonstrated.[77–81] Stabilization of polystyrene has been achieved by adding antioxidants,[82,83] and by blending with poly(2,6-dimethyl-1,4-phenylene oxide).[84,85]

---

[61] D. Lala and J. F. Rabek, *Eur. Polym. J.*, 1981, **17**, 7.
[62] R. P. Singh, *Polym. Photochem.*, 1982, **2**, 331.
[63] R. P. Singh, *Polym. Bull. (Berlin)*, 1981, **6**, 175.
[64] M. Ghaemy and G. Scott, *Polym. Degradation Stab.*, 1981, **3**, 233.
[65] E. I. Kirillova, S. V. Zuznetsova, L. I. Lugova, and G. P. Malakhova, *Vysokomol. Soedin., Ser. B*, 1982, **24**, 296.
[66] E. G. Kolawole and G. Scott, *J. Appl. Polym. Sci.*, 1981, **26**, 2581.
[67] M. Ghaemy and G. Scott, *Polym. Degradation Stab.*, 1981, **3**, 253.
[68] R. P. Singh, *Polym. Bull. (Berlin)*, 1981, **5**, 443.
[69] R. Chandra and R. P. Singh, *Polym. Photochem.*, 1982, **2**, 257.
[70] G. Geuskens, *Dev. Polym. Degradation*, 1981, **3**, 207.
[71] G. Geuskens and Q. Lu-Vinh, *Eur. Polym. J.*, 1982, **18**, 307.
[72] V. B. Ivanov, E. L. Lozovskaya, and V. Y. Shlyapintokh, *Polym. Photochem.*, 1982, **2**, 55.
[73] P. Bostin, G. Geuskens, Q. Lu-Vinh, and M. Rens, *Polym. Degradation Stab.*, 1981, **3**, 295.
[74] P. Bastin and G. Geuskens, *Polym. Degradation Stab.*, 1982, **4**, 111.
[75] C. David, G. Delaunois, G. Geuskens, Q. Lu-Vinh, and W. Piret, *Eur. Polym. J.*, 1982, **18**, 387.
[76] T. Bogdantsaliev, V. Kabaivanov, and G. Nenkov, *Doil. Bolg. Akad. Nauk*, 1981, **34**, 667.
[77] W. Schnabel, Y. Tabata, S. Tagawa, and M. Washio, *Radiat. Phys. Chem.*, 1981, **18**, 1087.
[78] M. J. Easton and J. R. MacCallum, *Polym. Degradation Stab.*, 1981, **3**, 229.
[79] F. Gonzalez and R. Sastre, *Polym. Photochem.*, 1981, **1**, 153.
[80] J. Kowal and M. Nowakowska, *Polymer*, 1982, **23**, 281.
[81] M. Kryszewski, B. Paradowska, and B. Wandelt, *Polym. Bull. (Berlin)*, 1982, **8**, 377.
[82] J. Kowal and M. Nowakowska, *Makromol. Chem.*, 1982, **183**, 1701.
[83] E. G. Kalawole, *J. Appl. Polym. Sci.*, 1982, **27**, 3437.
[84] D. J. S. Birch, R. E. Imhof, M. Kryszewski, A. M. North, R. A. Pethrick, and B. Wandelt, *Polymer*, 1982, **23**, 924.
[85] J. P. Jensen, J. Kops, and J. P. Tovborg, *J. Polym. Sci., Polym. Chem. Ed.*, 1981, **19**, 2765.

7 Poly(vinyl chloride)

Poly(vinyl chloride) is being increasingly used in outdoor applications, and interest in understanding the mechanism of photodecomposition of this polymer grows. A number of reviews have been written.[86-88] Investigation of the initiation process for dehydrochlorination has involved using e.s.r. spectroscopy in simulated sunlight.[89] However, it has been shown that evolution of HCl takes place only when the polymer is subject to irradiation with wavelength less than 300 mm.[90] It has been shown that the rate of dehydrochlorination is proportional to the intensity of radiation absorbed.[91] Study of emission spectroscopy of degraded poly(vinyl chloride) indicated the importance of α-chloroketone structures,[92] and the significance of the product HCl on the mechanism has been discussed.[93]

Photo-oxidation proceeds by a complex mechanism. Plasticized and unplasticized polymers produce HCl and CO_2 as main volatile products.[94] It has also been shown that photo-oxidation of polyenes results in increasing the rate of evolution of HCl, with the primary products of photo-oxidation being the formation of carbonyl and hydroperoxide groups.[95] Chlorination of poly(vinyl chloride) markedly destabilizes the polymer.[96]

Additives can stabilize the polymer, preventing either dehydrochlorination or subsequent photo-oxidation of the resulting polymers. A number of metal acetonates have been shown to act as stabilizers, probably by interaction with α-chloroketones in the polymer.[97] A wide range of stabilizers has been investigated with varying levels of effectiveness: dibutyl tin compounds;[98] organotin and antimony;[99] diaryloxamides;[100] tribasic lead sulphate,[101] phenolic antioxidant combined with dibutyltin maleate;[102] zinc, barium, and calcium carboxylates;[103] blended polybutadiene;[104] blended methyl methacrylate–styrene copolymer.[105]

Two very important aspects of sample preparation have been investigated, the effect of residual solvent in cast films[106] and the effect of thermal processing which

---

[86] W. H. Starnes, *ACS Symp. Ser.*, 1981, **151**, 197.
[87] A. L. Gonzalez and A. J. M. Mendiola, *Metal Electr.*, 1981, **45**, 15.
[88] E. D. Owen, *Dev. Polym. Photochem.*, 1982, **3**, 165.
[89] K. Fueki, A. Torikai, and H. Tsuruta, *Polym. Photochem.*, 1982, **2**, 227.
[90] R. S. Davidson and R. R. Meek, *Polym. Photochem.*, 1982, **2**, 1.
[91] C. Decker and M. Blandier, *J. Photochem.*, 1981, **15**, 213.
[92] M. Aiba and Z. Osawa, *Polym. Photochem.*, 1982, **2**, 397.
[93] E. D. Owen, *ACS Symp. Ser.*, 1981, **151**, 217.
[94] H. Kawaguchi and T. Veda, *Zairyo*, 1981, **30**, 414.
[95] C. Decker and M. Blandier, *Polym. Photochem.*, 1981, **1**, 221.
[96] C. Decker and M. Blandier, *Makromol. Chem.*, 1982, **183**, 1263.
[97] A. Aiba and Z. Osawa, *Polym. Photochem.*, 1982, **2**, 447.
[98] V. Bellenger, L. B. Carette, J. Verdu, Z. Vymazal, and Z. Vymazalova, *Polym. Degradation Stab.*, 1982, **4**, 303.
[99] D. Dieckmann and C. W. Fletcher, *J. Vinyl Technol.*, 1981, **3**, 130.
[100] S. N. Blagova, L. S. Kudryavtseva, M. S. Loyinskii, L. S. Nikitina, I. K. Piguta, and B. I. Shteiselbein, *Plast. Massy*, 1981, 51.
[101] A. L. Gonzalez, *Rev. Plast. Mod.*, 1982, **43**, 167.
[102] B. B. Cooray and G. Scott, *Eur. Polym. J.*, 1981, **17**, 229.
[103] Z. Pokorski and J. Wypych, *J. Appl. Polym. Sci.*, 1981, **26**, 1735.
[104] D. Lala, B. Raanby, and J. F. Rabek, *Polym. Degradation Stab.*, 1981, **3**, 307.
[105] A. Kaminska, *Angew. Makromol. Chem.*, 1982, **107**, 43.
[106] M. Aiba and Z. Osawa, *Polym. Photochem.*, 1982, **2**, 339.

introduces hydroperoxides.[107] Finally, Mössbauer spectroscopy has been used to study chemical changes undergone by stabilizers derived from tin compounds.[108]

8 Polymethacrylates and Polyketones

Interest in the photochemical decomposition of these two groups of polymers has declined in the past two years. In the case of the methacrylates the mechanism seems to have been elucidated, and the decrease in demand for degradable polymers has diminished the amount of research conducted in the field of photolysis of ketone containing polymers.

Vacuum and oxidative photolysis of poly(methyl methacrylate) have been studied by e.s.r. spectroscopy.[109] The same technique has been used in an examination of the photo-induced decomposition of poly(n-butyl acrylate).[110] An interesting investigation of the photodegradation of dilute solutions of butyl methacrylate–naphthyl methacrylate copolymers has demonstrated the stabilizing effect of excimer formation.[111] Interaction of chain segments has also been found in a study of the photodegradation of methyl methacrylate-α-chloroacrylonitrile copolymers.[112]

The photolysis of poly(vinyl benzophenone) proceeds by a random scission reaction.[113] In the photodegradation of poly(acrylophenone), tacticity had no effect on the rate of chain scission.[114]

9 Polyamides

Changes in the dynamic viscoelastic properties of nylon-6 induced by photo-oxidation have been reviewed.[115] Sunlight degradation has been studied in detail with the conclusion that nylon-6 was more sensitive than nylon-66.[116] High levels of delustrant were found to speed up breakdown.[117] Accelerated aging tests of yarns yielded the interesting observation that highly degraded samples showed no change in their i.r. absorption spectra.[118] The u.v. absorption spectra of solutions of degraded fabrics has been used to monitor photodegradation.[119] The 2-hydroxyphenyl benzotriazole class of stabilizers appeared to operate partially as screening

---

[107] B. B. Cooray and G. Scott, *Polym. Degradation Stab.*, 1981, **3**, 127.
[108] D. W. Allen, J. S. Brooks, R. W. Clarkson, M. T. J. Mellor, and A. G. Williamson, *Polym. Degradation Stab.*, 1982, **4**, 359.
[109] K. Fueki and A. Torikai, *Polym. Photochem.*, 1982, **2**, 297.
[110] A. Gupta, R. H. Liang, and F. D. Tsay, *Macromolecules*, 1982, **15**, 974.
[111] T. Kagiya, S. Nishimoto, and K. Yamamoto, *Macromolecules*, 1982, **15**, 1180.
[112] N. Grassie and A. S. Holmes, *Polym. Degradation Stab.*, 1981, **3**, 145.
[113] R. Knoesel and G. Weill, *Polym. Photochem.*, 1982, **2**, 167.
[114] J. E. Guillet, T. Kilp, Y. Merle, and L. Merle-Aubry, *Macromolecules*, 1982, **15**, 60.
[115] S. Yano, *Sen'i Gakkaishi*, 1982, **38**, 89.
[116] G. A. Horsfall, *Textilia*, 1981, **57**, 11.
[117] G. A. Horsfall, *Text. Res. J.*, 1982, **52**, 197.
[118] I. Auerbach, R. H. Ericksen, J. W. Mead, and K. E. Mead, *Ind. Eng. Chem., Prod. Res. Dev.*, 1982, **21**, 158.
[119] I. S. Polikarpov and N. N. Simkovich, *Tekst. Prom-st (Moscow)*, 1982, 6.

Polymer Degradation

agents and partly as quenchers of triplet states.[120] The mechanism of breakdown of polyamides copolymerized such that α-diketone links were in the main chain has been discussed.[121]

10 Polyesters

A study of the photo-oxidation of poly(ethylene terephthalate) showed that stability increased with increasing degree of orientation.[122]

11 Polyurethanes

The photodegradation and stabilization of polyurethanes has been reviewed.[123] The effect of wavelength and condition of irradiation have been examined and mechanisms proposed.[124] Oxygen uptake measurements have been used to study the kinetics of photo-oxidation of crosslinked samples[125] and e.s.r. spectroscopic examination of products has also been reported.[126]

12 Poly(2,6-dimethyl-1,4-phenylene oxide)

This polymer has been reported to undergo random scission during photodegradation. The rate of the process was controlled by the diffusion of oxygen.[127] The effect of added stabilizers has been examined with the conclusion that only u.v. absorbers are effective.[128] A similar experimental investigation has been carried out independently with some variations in mechanism being reported;[129,130] this apparent contradiction can probably be explained by the differences in energy of radiation used by the two groups.

13 Miscellaneous

Two papers have been published on aging of composites.[131,132] This is an area which will grow in importance in spite of the rather difficult experimental problems posed

---

[120] H. E. A. Kramer, T. Werner, and G. Woessner, *ACS Symp. Ser.*, 1981, **151**, 000.
[121] F. Akutsu, M. Miura, K. Nagakubo, S. Sato, and T. Uchiyama, *Polymer*, 1982, **23**, 342.
[122] A. N. Neverov and T. M. Savehuck, *Vysokomol. Soedin., Ser. A*, 1982, **24**, 1009.
[123] Z. Osawa, *Dev. Polym. Photochem.*, 1982, **3**, 209.
[124] J. L. Gardette and J. Lemaire, *Makromol. Chem.*, 1982, **183**, 2415.
[125] E. Y. Davydov, E. V. Davydova, M. I. Karyakina, V. V. Lukyanov, and A. V. Uvarov, *Vysokomol. Soedin., Ser. A*, 1981, **23**, 854.
[126] E. M. Lipskerova and M. Y. Melnikov, *Fizikokhim. Poliuretanov*, 1981, 33.
[127] J. Jackowicz, M. Kryszewski, and B. Wandelt, *Acta Polym.*, 1981, **32**, 637.
[128] Z. Slama, *Chem. Prum.*, 1981, **31**, 185.
[129] R. Chandra and B. P. Singh, *Eur. Polym. J.*, 182, **18**, 199.
[130] R. Chandra and B. P. Singh, *Eur. Polym. J.*, 1982, **18**, 289.
[131] D. L. Fanter, M. A. Grayson, and C. J. Wolf, *Govt. Rep. Announce. Index (U.S.)*, 1982, **32**, 2585.
[132] C. Giori and T. Yamauchi, *Sci. Tech. Aerosp. Rep.*, 1982, 20.

by such materials. The composites examined had relatively low u.v. stability producing a complex range of volatile products. ESCA offers a powerful tool for probing surface oxidation and it has been applied to investigation of the photo-oxidation of poly(phenylene oxide), polysulphone,[133] and polycarbonate.[134] The kinetics and mechanism of photodecomposition of poly(vinyl butyral) have been investigated.[135-137] This polymer is unstable and breaks down by a free-radical process. Vacuum u.v. radiation of polytetrafluoroethylene[138] and weathering of polyoxymethylene[139] have been reported. The oxidative processes involving unsaturated polymers and singlet oxygen still receive attention.[140] An accelerated weathering tester has been described.[141]

[133] D. T. Clark and J. Peeling, *J. Appl. Polym. Sci.*, 1981, **26**, 3761.
[134] D. T. Clark and H. S. Munro, *Polym. Degrad. Stab.*, 1982, **4**, 441.
[135] M. Navratil, V. Reinohl, and J. Sedlar, *Polym. Photochem.*, 1981, **1**, 165.
[136] N. V. Fok, M. Y. Melnikov, O. M. Mikhailik, and E. N. Seropegina, *Eur. Polym. J.*, 1981, **17**, 1011.
[137] N. V. Fok, M. Y. Melnikov, O. M. Mikhailik, and E. N. Seropegina, *Dokl. Akad. Nauk. SSSR*, 1981, **257**, 943.
[138] Y. A. Enhov, B. I. Khrushch, A. N. Lyulichev, A. M. Pshisukha, G. D. Strizhko, and V. F. Udovenko, *Khim. Vys. Energ.*, 1981, **15**, 520.
[139] A. Davis, *Polym. Degradation Stab.*, 1981, **3**, 187.
[140] J. Lucki, B. Rasnby, J. F. Rabek, and S. K. Wu, *Polym. Photochem.*, 1982, **2**, 125.
[141] D. M. Grossman, *Br. Ink Maker*, 1981, **24**, 21.

PART II Thermal and Thermo-oxidative Degradation
by W. W. Wright

1 Introduction

Coverage is once again restricted to papers containing a substantial amount of information on the kinetics or mechanisms of polymer degradation. During the last 2 years there have been no changes in the total number of papers abstracted (approximately 290), in their overall content (almost a third are still devoted to studies of the polyolefins, polystyrene, and poly(vinyl chloride), or in the countries of origin. Because of space restrictions the selection of papers has had to be more rigorous than previously and only 164 have been referenced.

2 General

There appears to be a growing interest in the combustion of polymers and two papers review the role of oxygen in pyrolysis and combustion[1] and the mechanisms of thermal degradation and burning.[2] The latter paper also considers flame retardation and flame proofing. The extinction of diffusion flames has been studied in stagnation-point flow above combusting polymers and the extinction data used to obtain overall rate parameters for gas-phase combustion of these materials.[3] It is claimed that volatile products analysed by pyrolysis/gas chromatography correlate reasonably well with limiting oxygen index values[4] and that the rates of weight loss in air of thin films of various polymers can only be accurately described by mechanisms involving sets of four or five pseudo-first-order reactions.[5]

The use of pyrolysis/mass spectrometry to investigate the structure and selective thermal degradation of polymers has been reviewed.[6] Another paper describes the pyrolysis of a number of polymers directly in the ion source of a chemical ionization mass spectrometer.[7] The evolution of gases has also been quantitatively detected using flowing-afterglow spectroscopy. This is said to be useful for understanding polymer stability and lifetimes in specified environments.[8]

A method has been presented for calculating the temperature of the initial intense thermal degradation of a polymer based on structural parameters of the repeating unit. It is claimed that the influence of various groups and atoms on the thermal stability can be evaluated quantitatively.[9] In particular, molecular mobilities of polymers permit the estimation of the temperatures for beginning of degradation at

---

[1] R. Delbourgo, *Oxid. Commun.*, 1982, **2**, 207.
[2] L. Hostejn, *Chem. Listy*, 1982, **76**, 575.
[3] S. H. Sohrab and F. A. Williams, *J. Polym. Sci., Polym. Chem. Ed.*, 1981, **19**, 2955.
[4] M. R. MacLaury and A. L. Schroll, *J. Fire Flammability*, 1981, **12**, 203.
[5] A. D. Baer, *J. Fire Flammability*, 1981, **12**, 214.
[6] I. Luederwald, *Pure Appl. Chem.*, 1982, **54**, 255.
[7] D. C. Conway and R. Marak, *J. Polym. Sci.*, Polym. Chem. Ed., 1982, **20**, 1765.
[8] M. L. Matuszak and G. W. Taylor, *J. Appl. Polym. Sci.*, 1982, **27**, 461.
[9] Y. I. Matveev, A. A. Askadskii, I. V. Zhuravleva, G. L. Slonimskii, and V. V. Korshak, *Vysokomol. Soedin., Ser. A*, 1981, **23**, 2013.

3 Polyolefins

A number of authors have reported results of thermal and thermo-oxidative degradation of polyethylene and polypropylene using the same technique for both polymers. The complexity of the volatile decomposition products, hydrocarbons, alcohols, aldehydes, ketones, acids, cyclic ethers, cyclic esters, and hydroxy-carboxylic acids has been illustrated by gas chromatographic/mass spectrometric analysis.[11,12] The kinetics of pyrolytic gasification have been studied and it has been found that the kinetic parameters are not appreciably affected by the tacticity of polypropylene,[13] or by molecular weight in the case of polyethylene provided that the value is greater than 10 000.[14] The technique of factor jump thermogravimetry has also been used to determine the overall activation energies for thermal degradation.[15,16] Somewhat lower values [214—257 kJ mol^{-1}] were obtained for polypropylene than for polyethylene [257—275 kJ mol^{-1}]. In both polymers initial breakdown was considered to be chain scission β to allyl groups. The rates of oxidation and formation of unsaturated compounds have been measured for both polymers in chlorobenzene solutions.[17] The number of primary oxidation steps resulting in C—C bond cleavage was higher in polypropylene than in polyethylene but, whereas each C—C bond scission in polyethylene resulted in the formation of a C=C bond, the rate of formation of double bonds in polypropylene was approximately twice the rate of C—C bond cleavage. Another paper comparing degradation of polyethylene in the melt and in solution claims that the main features were the same except for the conversion dependence of the hydroperoxide concentration and the nature of the temperature dependence of the maximum hydroperoxide concentration.[18] The ratio of the different pyrolysis products [C_{2-23} saturated and unsaturated hydrocarbons] obtained from polyethylene did not vary greatly under a wide range of oxidation conditions, including flaming combustion, but this was not true of the oxidation degradation products.[19] Japanese workers have continued their extensive studies of the pyrolysis of polyethylene under continuous flow stirred reactor conditions.[20—22]

---

[10] E. N. Zadorina, G. E. Vishnevskii, and Y. V. Zelenev, *Vysokomol. Soedin., Ser. B*, 1981, **23**, 380.
[11] A. Hoff and S. Jacobsson, *J. Appl. Polym. Sci.*, 1981, **26**, 3409.
[12] A. Hoff and S. Jacobsson, *J. Appl. Polym. Sci.*, 1982, **27**, 2539.
[13] T. Sawaguchi, K. Suzuki, T. Kuroki, and T. Ikemura, *J. Appl. Polym. Sci.*, 1981, **26**, 1267.
[14] S. Niikuni, T. Kuroki, and T. Ikemura, *Polymer*, 1981, **22**, 1403.
[15] B. Dickens, *J. Polym. Sci., Polym. Chem. Ed.*, 1982, **20**, 1065.
[16] B. Dickens, *J. Polym. Sci., Polym. Chem. Ed.*, 1982, **20**, 1169.
[17] N. F. Trofimova, V. V. Kharitonov, and E. T. Denisov, *Dokl. Akad. Nauk SSSR*, 1980, **253**, 651 (*Chem. Abstr.*, 1981, **94**, 16 296).
[18] R. Iring, T. Kelen, Z. Fodor, and F. Tudos, *Magy. Kem. Foly.*, 1982, **88**, 363 (*Chem. Abstr.*, 1982, **97**, 145 408).
[19] J. H. Hodgkin, M. N. Galbraith, and Y. K. Chong, *J. Macromol. Sci., Chem.*, 1982, **17**, 35.
[20] K. Murata, K. Sato, and H. Teshima, *Kagaku Koagku Ronbunshu*, 1981, **7**, 64 (*Chem. Abstr.*, 1981, **94**, 104 200).
[21] K. Murata, *Kagaku Kogaku Ronbunshu*, 1982, **8**, 155 (*Chem. Abstr.*, 1982, **96**, 181 767).
[22] K. Murata, K. Sato, and H. Teshima, *Kagaku Kogaku Ronbunshu*, 1982, **8**, 279 (*Chem. Abstr.*, 1982, **97**, 56 578).

The formation and degradation of hydroperoxides in poly(4-methyl-1-pentene) have been studied at 25—190 °C. The tertiary carbon atoms in the side-groups oxidize several times faster than the carbon atoms in the main chain.[23]

4 Polydienes

Pyrolysis/field ion mass spectrometry has been used to study the thermal degradation of a number of polydienes and their vulcanizates over the temperature range 500—900 °C and fragmentation mechanisms have been postulated.[24] Pyrolysis/gas chromatography of cis-1,4-, trans-1,4-, and 1,2-polybutadiene showed that, although the cis and trans forms gave closely similar product distributions at 450—900 °C, the 1,2-polymer yielded much less vinylcyclohexene at the lower temperatures.[25] The yield of butadiene also depended upon the tacticity of the 1,2-polybutadiene. The effect of novolac-resin and asbestos on the microstructural changes occurring during thermo-oxidation of cis-1,4-polybutadiene was assessed by a number of thermoanalytical techniques.[26] Rates of crosslinking, production of volatiles, and interactions between functional groups were all discussed. Vulcanized polyisoprene has been studied by pyrolysis/gas chromatography/mass spectrometry and the dependence of the composition of volatile degradation products upon crosslink density determined.[27] In the case of block copolymers of isoprene and styrene the thermal behaviour was influenced mainly by the microstructure of the polyisoprene block, although all the results indicated that the individual blocks degraded independently.[28]

The thermal decomposition of polyisobutylene has been followed by differential scanning calorimetry and gel permeation chromatography. Weight loss was independent of molecular weight and oligomers of average degree of polymerization 12 were important products.[29] A comparison of oligoisobutylenes with carboxy and keto end-groups with polyisobutylene showed that the former had poorer thermal stability, degradation occurring from the chain ends.[30]

5 Polystyrene

Controversy still surrounds the so-called weak links in polystyrene. Evidence has been presented[31] that these are not copolymerized peroxide groups. (The same school has earlier shown that the weak points are not head-to-head groups, branch

---

[23] L. L. Yasina and V. S. Pudov, *Vysokomol. Soedin., Ser. A*, 1982, **24**, 491.
[24] G. Czybulka, H. Dunker, H. J. Duessel, H. Logemann, and D. O. Hummel, *Angew. Makromol. Chem.*, 1981, **100**, 1.
[25] T. S. Radhakrishna and M. R. Rao, *J. Polym. Sci., Polym. Chem. Ed.*, 1981, **19**, 3197.
[26] K. C. Gong and Y. C. Zheng, Rubbercon (81), Int. Rubber Conf., 1981, Vol. 1, B.4.1.
[27] X. G. Jin and H. M. Li, *J. Anal. Appl. Pyrolysis*, 1981, **3**, 49.
[28] H. A. Schneider, N. Hurduc, C. N. Cascaval, G. Riess, and S. Marti, *Makromol. Chem.*, 1981, **182**, 921.
[29] S. L. Malhotra, L. Y. Minh, and L. P. Blanchard, *J. Macromol. Sci., Chem.*, 1982, **18**, 455.
[30] S. A. Nasybullin, A. V. Dyuldeva, I. N. Zaripov, F. G. Biknukhametova, and I. N. Faizullin, *Vysokomol. Soedin., Ser. A*, 1981, **23**, 313.
[31] G. G. Cameron and I. T. McWalter, *Eur. Polym. J.*, 1982, **18**, 1029.

points, or unsaturated structures.) Techniques used extensively in the study of polystyrene have been pyrolysis/gas chromatography[32,33] (sometimes coupled with mass spectrometry)[34] and liquid and gel permeation chromatography.[35,36] By using the former method, product distribution has been determined as a function of temperature,[32,34] time,[32] and heating rate[32] and the monomer yields have been compared from head-to-head and head-to-tail polystyrene and poly(α-methylstyrene).[33] The formation of some of the products can only be explained on the basis of intramolecular transfer reactions.[34] This has been corroborated by gel permeation chromatography and reverse-phase liquid chromatographic experiments.[36] Kinetic studies utilizing gel permeation chromatography have indicated that degradation proceeds by random chain scission with only small amounts of volatile products being formed in the early stages.[37] The rate constant for this random scission reaction has been measured by differential scanning calorimetry[38] and was found to undergo a drastic change at a critical molecular weight of approximately 45 000. In the case of anionically polymerized polystyrene, however, the evidence suggests that initiation of degradation takes place at unsaturated chain ends, either present originally in the polymer, or formed during the very early stages of breakdown.[39] On the other hand, the thermal degradation of polydisperse polystyrene was best simulated by a model consisting of both random scission and depolymerization, the ratio of the two remaining constant at all temperatures being studied.[40] Poly(4-bromostyrene) was found to decompose by random scission followed by depolymerization.[41]

The volatile degradation products have also been analysed for copolymers of styrene with acrylonitrile[42] in varying amounts and with 2-bromoethylmethacrylate[43] in a 1:1 molar ratio. In the former case the decomposition was very largely affected by the sequence of monomer units in the copolymer. Blends of polystyrene and poly(2,6-dimethyl-1,4-phenylene oxide) have been studied over a range of compositions.[44] The styrene yield on pyrolysis at 500 °C did not depend upon molecular interactions.

6 Poly(vinyl chloride)

Review articles have compared mechanisms of thermal and photochemical degradation,[45] covered non-oxidative degradation, stabilization, and fire

---

[32] J. J. R. Mertens, E. Jacobs, A. Callaerts, and A. Buekens, *Makromol. Chem., Rapid. Commun.*, 1982, **3**, 349.
[33] M. Tanaka, T. Shimono, Y. Yakubi, and T. Shono, *J. Anal. Appl. Pyrolysis*, 1980, **2**, 207.
[34] M. T. S. P. de Amorim, C. Bouster, P. Vermande, and J. Veron, *J. Anal. Appl. Pyrolysis*, 1981, **3**, 19.
[35] K. Saito and M. Tomita, *Hokkaido Kogyo Kaihatsu Shikenshi Hokoku*, 1980, **20**, 17 (*Chem. Abstr.*, 1982, **96**, 528 182).
[36] D. Daoust, S. Bormann, R. Legras, and J. P. Mercier, *Polym. Eng. Sci.*, 1981, **21**, 721.
[37] T. Kuroki, T. Ikemura, T. Ogawa, and Y. Sekiguchi, *Polymer*, 1982, **23**, 1091.
[38] H. K. Toh and B. L. Funt, *J. Appl. Polym. Sci.*, 1982, **27**, 4171.
[39] L. Costa, G. Camino, A. Guyot, M. Bert, and A. Chiotis, *Polym. Degrad. Stab.*, 1982, **4**, 245.
[40] K. H. Ebert, H. J. Ederer, K. O. Ulrich, and A. W. Hamielec, *Makromol. Chem.*, 1982, **183**, 1207.
[41] S. L. Malhotra, P. Lessard, and L. P. Blanchard, *J. Macromol. Sci., Chem.*, 1981, **15**, 1577.
[42] M. Blazso, G. Varhegyi, and E. Jakab, *J. Anal. Appl. Pyrolysis*, 1980, **2**, 177.
[43] N. Grassie, A. Johnston, and A. Scotney, *Polym. Degrad. Stab.*, 1982, **4**, 123.
[44] B. Wandelt, M. Kryszewski, and P. Kowalski, *Polymer*, 1981, **22**, 1236.
[45] D. Braun, E. Bazdadea, and G. Holzer, *Angew. Makromol. Chem.*, 1982, **106**, 47.

retardance,[46] or dealt more generally with both thermal and thermo-oxidative breakdown[47–49] and the experimental methods used to study these phenomena.[48,49] Ab initio molecular orbital theory has been used to assess the stability of bridged and open forms of the haloethyl and halopropyl cations and it is claimed that the results provide strong support for an intermediate cyclic chloronium ion in thermal degradation of poly(vinyl chloride).[50] Experimental studies have, in the main, followed the conventional lines of determining the rates of hydrogen chloride evolution,[51–53] also in dilute solution, or a plasticized state;[54] the rates of crosslinking and chain scission[55,56] and the formation of volatile aromatic products and residual polyene structures.[53,57–59] Further details of the mechanism of formation of the aromatic volatiles were obtained by isotopic labelling, using either deuterium[57] or ^{13}C.[60] The aromatics arose through intramolecular cyclization reactions, or hydrogen transfer. There was no need to invoke crosslinking steps. Other studies have utilized ultraviolet and infrared spectral and molecular-weight distribution changes,[61] model compounds (chloroalkene–vinyl chloride copolymers) containing different types of structural defects,[62] and polymers containing high levels of short internal polyene sequences and allylic chloride prepared by controlled chemical dehydrochlorination of poly(vinyl chloride) itself.[63] The rate of thermal dehydrochlorination was dependent on the allylic chlorine concentration and upon chain-end defects. One more specialized investigation analysed for phosgene as poly(vinyl chloride) was decomposed under various conditions.[64] Significant quantities were obtained if the polymer was subjected to electric arc initiated flaming combustion.

A series of papers have been published on the thermal degradation of copolymers of vinyl chloride with vinyl acetate[65,66] and methyl acrylate.[67] The presence of acetate groups increased the rate of dehydrochlorination. Other copolymer systems studied

---

[46] W. H. Starnes, *Dev. Polym. Degrad.*, 1981, **3**, 135.
[47] F. Tüdös, T. Kelen, and T. T. Nagy, *Kem. Kozl.*, 1981, **51**, 223 (*Chem. Abstr.*, 1981, **95**, 169 803)
[48] D. Braun, *Pure Appl. Chem.*, 1981, **53**, 549.
[49] D. Braun, *Dev. Polym. Degrad.*, 1981, **3**, 101.
[50] K. Raghavachari, R. C. Haddon, and W. H. Starnes, *J. Am. Chem. Soc.*, 1982, **104**, 5054.
[51] J. D. Danforth, J. Spiegel, and J. Bloom, *J. Macromol. Sci., Chem.*, 1982, **17**, 1107.
[52] J. W. Wimberley, A. B. Carel, and D. K. Cabbiness, *Anal. Lett.*, 1982, **15**, 89.
[53] V. Bellenger, L. B. Carette, E. Fontaine, and J. Verdu, *Eur. Polym. J.*, 1982, **18**, 337.
[54] K. S. Minsker, M. I. Abdullin, and G. E. Zaikov, *J. Vinyl Technol.*, 1981, **3**, 230.
[55] T. T. Nagy, B. Iven, B. Turcsanyi, and F. Tüdös, *Polym. Bull. (Berlin)*, 1980, **3**, 613.
[56] T. Kelen, B. Ivan, T. T. Nagy, B. Turcsanyi, F. Tüdös, and J. P. Kennedy, *Magy. Kem. Foly.*, 1981, **87**, 97 (*Chem. Abstr.*, 1981, **95**, 25 994).
[57] R. P. Lattimer and W. J. Kroenke, *J. Appl. Polym. Sci.*, 1982, **27**, 1355.
[58] A. Guyot, M. Bert, P. Burille, M. F. Llauro, and A. Michel, *Pure Appl. Chem.*, 1981, **53**, 401.
[59] G. Vaneso, T. T. Nagy, B. Turcsanyi, T. Kelen, and F. Tüdös, *Makromol. Chem., Rapid Commun.*, 1982, **3**, 527.
[60] R. P. Lattimer, W. J. Kroenke, and R. H. Backderf, *J. Appl. Polym. Sci.*, 1982, **27**, 3633.
[61] B. Pukanszky, T. T. Nagy, T. Kelen, and F. Tüdös, *J. Appl. Polym. Sci.*, 1982, **27**, 2615.
[62] A. Airinei, E. C. Burniana, G. Robila, C. Vasile, and A. Caraculacu, *Polym. Bull. (Berlin)*, 1982, **7**, 465.
[63] B. Ivan, J. P. Kennedy, T. Kelen, and F. Tüdös, *J. Macromol. Sci., Chem.*, 1982, **17**, 1033.
[64] J. E. Brown and M. M. Birky, *J. Anal. Toxicol.*, 1980, **4**, 166.
[65] K. S. Minsker, V. V. Lisitskii, N. V. Davidenko, A. G. Kronman, and M. A. Chekushina, *Vysokomol. Soedin., Ser. A*, 1981, **23**, 1518.
[66] K. S. Minsker, V. V. Lisitskii, and N. V. Davidenko, *Vysokomol. Soedin., Ser. A*, 1982, **24**, 1157.
[67] K. S. Minsker, A. A. Berlin, V. V. Lisitskii, R. B. Pancheshnikova, N. N. Zavodchikova, D. M. Yanovskii, and Y. B. Monakov, *Vysokomol. Soedin., Ser. A*, 1981, **23**, 1636.

include vinyl chloride–allyl chloride,[68] vinyl chloride–phenylacetylene,[69] and vinyl chloride–2-butynedioic acid dimethylester.[69] The rate of dehydrochlorination decreased with increasing allylic chloride content, but increased significantly in the presence of phenyl groups conjugated with double bonds in the chain.

A general paper has compared the thermal degradation of poly(vinyl chloride) with that of chlorinated polyethylene, chlorosulphonated polyethylene, and polyepichlorhydrin.[70]

7 Fluorine-containing Polymers

The kinetics of the thermal and thermo-oxidative decomposition of poly(vinylidene fluoride) have been studied and the orders of reaction and activation energies determined.[71,72] The technique of ultraviolet photoelectron spectroscopy has been used to follow the early stages of the thermal degradation of polytetrafluoroethylene and the copolymers of tetrafluoroethylene with hexafluoropropene and perfluoropropylvinylether.[73] A concentration–time profile for evolution of tetrafluoroethylene was obtained. Other work has dealt with the analysis of the volatile decomposition products from the copolymer of tetrafluoroethylene and hexafluoropropene,[74,75] polytetrafluoroethylene,[75] poly(vinylidene fluoride),[75] the copolymer of hexafluoropropene and ethylene,[75] and perfluorooxyalkylene triazines.[76] The mechanism adduced for the last-named was supported by the study of model compounds.

The yield of hydrogen fluoride from various fluorine-containing elastomers in gumstock and compounded forms has been monitored by using a fluoride ion specific electrode.[77]

8 Polyacrylates

The thermal degradation of poly(alkyl methacrylates) yielded the monomers as the primary product except for poly(amyl methacrylate) and poly(octyl methacrylate) where pent-1-ene and oct-1-ene are also formed, respectively.[78] The homopolymers of ethyl methacrylate, n-butyl methacrylate and 2-hydroxyethyl methacrylate and their copolymers have also been studied by thermovolatilization analysis[79] and the

---

[68] J. Lewis, F. E. Okieimen, and G. S. Park, *J. Macromol. Sci., Chem.*, 1982, **17**, 915.
[69] D. Braun, A. Michel, and D. Sonderhof, *Eur. Polym. J.*, 1981, **17**, 49.
[70] D. Jaroszynske and T. Kleps, Rubbercon (81), Int. Rubber Conf., 1981, Vol. 1, C.5.1.
[71] M. M. Hirschler, *Eur. Polym. J.*, 1982, **18**, 463.
[72] M. D. Pukshanskii, V. T. Shirinyan, A. P. Bobrovskii, and A. G. Sirota, *Zh. Prikl. Khim. (Leningrad)*, 1981, **54**, 2608 (*Chem. Abstr.*, 1982, **96**, 123 453).
[73] D. Betteridge, N. R. Shoko, M. E. A. Cudby, and D. G. M. Wood, *Polymer*, 1980, **21**, 1309.
[74] C. H. Chen and Y. L. Tuan, *Ko Hsueh Tung Pao*, 1981, **26**, 924 (*Chem. Abstr.*, 1981, **95**, 187 798).
[75] N. E. Shadrina, M. S. Kleshcheva, N. N. Loginova, N. K. Podlesskaya, and S. G. Sannikov, *Zh. Anal. Khim.*, 1981, **36**, 1125 (*Chem. Abstr.*, 1981, **95**, 151 305).
[76] V. N. Shelgaev, S. P. Krukovskii, A. A. Yarosh, and V. A. Ponomarenko, *Vysokomol. Soedin., Ser. B*, 1982, **24**, 211.
[77] G. J. Knight and W. W. Wright, *Polym. Degrad. Stab.*, 1982, **4**, 465.
[78] Y. Hosaka, T. Kojima, and S. Kudo, Proc. 6th Int. Conf. Therm. Anal., 1980, Vol. 6(2), p. 393.
[79] M. S. Choudhary and K. Lederer, *Eur. Polym. J.*, 1982, **18**, 1021.

relationships between monomer yield and their proportion in the copolymers determined. Most other work has been on copolymers except for investigations of polycyanoacrylates[80] and the sodium, calcium, and magnesium salts of polyacrylates.[81] A series of papers has been published on the thermal decomposition of copolymers of 2-bromoethyl methacrylate with methyl methacrylate,[82] methyl acrylate,[83] or acrylonitrile,[84] as well as on blends of some of the homopolymers.[83] In all cases the principal degradation products were determined and integrated reaction sequences proposed. Blends of poly(t-butyl acrylate) and polyethyleneimine have been investigated by measuring the rates of isobutylene evolution.[85] The presence of polyethyleneimine decreased the overall rate constant for degradation, but did not affect the initial rates. Complex polymers containing up to four different monomers (taken from acrylonitrile, butyl acrylate, vinylidene chloride, styrene, and methacrylic acid) have been thermally degraded and the results compared with those for polyacrylonitrile, special attention being paid to the oligomerization of the nitrile groups.[86] Cyclic structural units also have an influence on the thermal decomposition of copolymers of methyl methacrylate and 2,6-dimethoxycarbonyl-1,6-heptadiene.[87]

9 Polyacrylonitrile

The kinetics and mechanism of polyacrylonitrile degradation have been studied at 30—800 °C by using a special pyrolysis/gas chromatographic unit.[88] The yields of eight products were measured as a function of temperature and the respective first-order rate constants determined. The low-temperature oxidation of polyacrylonitrile has again been investigated[89,90] because of its importance in the production of carbon fibres, but little fresh information has been forthcoming. A series of papers[91] has appeared describing the use of Fourier transform infrared spectroscopy to study the thermal degradation of copolymers of acrylonitrile with vinyl acetate, methacrylic acid, or acrylamide, particular attention being again given to the cyclization reaction of the acrylonitrile units.

10 Other Addition Polymers

A kinetic equation describing the isothermal degradation of poly(vinyl acetate) could be obtained assuming that catalysed and uncatalysed deacetylation reactions

---

[80] J. M. Rooney, *Br. Polym. J.*, 1981, **13**, 160.
[81] J. Hetper, W. Balcerowick, and J. Beres, *J. Therm. Anal.*, 1981, **20**, 345.
[82] N. Grassie, A. Johnston, and A. Scotney, *Eur. Polym. J.*, 1981, **17**, 589.
[83] N. Grassie, A. Johnston, and A. Scotney, *Polym. Degrad. Stab.*, 1981, **3**, 349, 365, and 453.
[84] N. Grassie, A. Johnston, and A. Scotney, *Polym. Degrad. Stab.*, 1982, **4**, 173.
[85] A. D. Litmanovich, V. O. Cherkezyan, and T. N. Khromova, *Vysokomol. Soedin., Ser. B*, 1981, **23**, 645.
[86] M. Tomescu, I. Demetrescu, and E. Segal, *J. Appl. Polym. Sci.*, 1981, **26**, 4103.
[87] G. Camino, L. Costa, G. Devalle, and M. Guaita, *Polym. Degrad. Stab.*, 1982, **4**, 459.
[88] R. S. Lehrle, J. C. Robb, and J. R. Suggalt, *Eur. Polym. J.*, 1982, **18**, 443.
[89] M. M. Kanovich and A. P. Rudenko, *Khim. Volokna*, 1982 (3), 19 (*Chem. Abstr.*, 1982, **97**, 40 189).
[90] H. Mellottee, *Eur. Polym. J.*, 1982, **18**, 1041.
[91] M. M. Coleman and G. T. Sivy, *Carbon*, 1981, **19**, 123, 127, 133, and 137.

occurred simultaneously.[92] Studies of other polymers, poly(vinylidene chloride),[93] polyacetylene,[94] and copolymers of vinyl acetate and vinyl alcohol,[95] have concentrated upon analysis of the volatile decomposition products. Several papers have been published[96,97] on the degradation of polyoxymethylene, underlining once again the importance of the end-groups in determining the thermal stability of this polymer. The rates and mechanism of thermal breakdown of nine different poly(olefinsulphones) have been studied at 150 and 200 °C. The rates of degradation showed a moderate correlation with the ceiling temperature for monomer–polymer equilibrium and also with the number of β-hydrogen atoms, but neither of these measures was completely adequate as an explanation.[98] The kinetics of thermal decomposition of poly(but-1-ene sulphone) have also been examined in more detail.[99] A study of polyurea by mass analysed ion kinetic energy spectrometry is claimed to demonstrate the potentiality of the technique for the analysis of mixtures produced by direct pyrolysis of polymers in a mass spectrometer.[100]

11 Polyamides

The aliphatic polyamides ranging from nylon-4 to nylon-12 have been examined.[101–105] In the majority of cases attention has been concentrated upon the volatile decomposition products, with special emphasis placed upon toxic products such as hydrogen cyanide.[104,105] A series of papers has also been published on the thermal degradation of poly(1,3-phenyleneisophthalamide) and poly-(1,4-phenyleneterephthalamide) using e.s.r. spectroscopy,[106] pyrolysis/gas chromatography/mass spectrometry,[107] and a set of model compounds to help elucidate mechanisms of breakdown.[108] Another paper covers similar ground in the use of model compounds and concludes that both homolytic and heterolytic cleavage of amide bonds occurs.[109] Papers outside the main streams of activity have

---

[92] J. M. Mazon-Arechederra, M. S. Chaves, F. Arranz, and J. M. Barrales-Rienda, *An. Quim., Ser. A*, 1982, **78**, 189 (*Chem. Abstr.*, 1982, **97**, 216 840).
[93] A. Ballistreri, S. Foti, P. Maravigna, G. Montaudo, and E. Scamporrino, *Polymer*, 1981, **22**, 131.
[94] J. C. W. Chien, P. C. Uden, and J. L. Fan, *J. Polym. Sci., Polym. Chem. Ed.*, 1982, **20**, 2159.
[95] C. Vasile, C. N. Cascaval, and P. Barbu, *J. Polym. Sci., Polym. Chem. Ed.*, 1981, **19**, 907.
[96] H. Zimmermann and J. Behnisch, *Thermochim. Acta*, 1982, **59**, 1.
[97] G. Opitz, *Plaste Kautsch*, 1981, **28**, 15.
[98] T. N. Bowner and J. H. O'Donnell, *Polym. Degrad. Stab.*, 1981, **3**, 87.
[99] M. J. Bowden, L. F. Thompson, W. Robinson, and M. Brols, *Macromolecules*, 1982, **15**, 1417.
[100] S. Foti, A. Liguori, P. Maravigna, and G. Montaudo, *Anal. Chem.*, 1982, **54**, 674.
[101] T. Konomi, C. Toyoki, W. Mizukami, and C. Shimizu, *Seni Gakkaishi*, 1982, **38**, T200 (*Chem. Abstr.*, 1982, **97**, 24 328).
[102] H. Ohtani, T. Nagaya, Y. Sugimura, and S. Tsuge, *J. Anal. Appl. Pyrolysis*, 1982, **4**, 117.
[103] I. V. Yatsenko, Y. A. Shlyapnikov, and M. S. Akatin, *Izv. Vyssh. Uchebn. Zaved. Khim. Khim. Tekhnol.*, 1981, **24**, 94 (*Chem. Abstr.*, 1981, **94**, 140 541).
[104] J. Michal, J. Mitera, and J. Kubat, *Fire Mater.*, 1981, **5**, 1.
[105] H. H. G. Jellinek, and S. R. Dunkle, *J. Polym. Sci., Polym. Chem. Ed.*, 1982, **20**, 85.
[106] J. R. Brown and D. K. C. Hodgeman, *Polymer*, 1982, **23**, 365.
[107] J. R. Brown and A. J. Power, *Polym. Degrad. Stab.*, 1982, **4**, 379.
[108] J. R. Brown and A. J. Power, *Polym. Degrad. Stab.*, 1982, **4**, 479.
[109] Y. P. Khanna, E. M. Pearce, J. S. Smith, D. T. Burkitt, H. Njuguna, D. M. Hindenlang, and B. D. Forman, *J. Polym. Sci., Polym. Chem. Ed.*, 1981, **19**, 2817.

covered polythioamides,[110] poly(amide esters),[111] and the 4,4'-fluoren-9-ylidenedianiline—terephthalic acid copolymer.[112]

12 Polyesters

All the relevant papers deal with poly(ethylene terephthalate) or poly(butylene terephthalate) and are concerned with the volatile products of decomposition.[113–116] The thermal decomposition of poly(ethylene terephthalate) has been reviewed.[117]

13 Polyurethanes

The thermal degradation of flexible polyurethane foams in fire situations has been reviewed.[118] The kinetics of degradation have been studied by thermogravimetry[119] and the volatile breakdown products analysed.[120] Specific attention has been paid to hydrogen cyanide as a decomposition product[121] and to the primary fragmentation mechanisms by direct pyrolysis of polyurethanes into a mass spectrometer.[122] Different structures give quite different decomposition routes. To aid in the elucidation of mechanisms a number of studies have been made of the pyrolysis of model compounds,[123,124] including deuteriated material.[125]

14 Polyphenylene-type Polymers

A variety of polymers of this type have been investigated. In the case of poly(oxy-2,6-dimethyl-1,4-phenylene) the extent of degradation was followed by monitoring the concentration of hydroperoxy, hydroxyl, carbonyl, and quinone groups, as well as oxygen absorption, weight change, and carbon dioxide formation.[126] Mass spectrometry was used to elucidate the thermal decomposition of a series of

---

[110] J. C. Gressier and G. Levesque, *Eur. Polym. J.*, 1980, **16**, 1175.
[111] V. G. Shelgaeva, A. K. Mikitaev, and V. N. Shelgaev, *Vysokomol. Soedin.*, 1981, **23**, 2099.
[112] S. A. Pavlova, P. N. Gribkova, T. N. Balykova, T. V. Polina, L. G. Komarova, N. I. Bekasova, and V. V. Korshak, *Vysokomol. Soedin.*, 1982, **24**, 1712.
[113] S. A. Motov, L. I. Danilina, and A. N. Pravednikov, *Vysokomol. Soedin.*, 1980, **22**, 2330.
[114] P. A. Matusevich, Y. I. Matusevich, L. V. Soloveva, and A. I. Kumachev, *Vesti. Akad. Navuk CSSR, Ser. Khim. Nauk*, 1981 (2), 108 (*Chem. Abstr.*, 1981, **95**, 44 058).
[115] C. T. Vijayakumar and J. K. Fink, *Thermochim. Acta*, 1982, **59**, 51.
[116] C. T. Vijayakumar, E. Ponnusamy, T. Balakrishnan, and H. Kothandaraman, *J. Polym. Sci., Polym. Chem. Ed.*, 1982, **20**, 2715.
[117] H. Zimmermann, *Plaste Kautsch*, 1981, **28**, 433.
[118] S. J. Grayson, J. Hume, and D. A. Smith, *Plast. Rubber Process Appl.*, 1982, **2**, 111.
[119] F. Gaboriaud and J. P. Vantelon, *J. Polym. Sci., Polym. Chem. Ed.*, 1981, **19**, 139.
[120] F. Gaboriaud and J. P. Vantelon, *J. Polym. Sci., Polym. Chem. Ed.*, 1982, **20**, 2063.
[121] H. H. G. Jellinek and S. R. Dunkle, *J. Polym. Sci., Polym. Chem. Ed.*, 1982, **20**, 2313.
[122] S. Foti, P. Maravigna, and G. Montaudo, *J. Polym. Sci., Polym. Chem. Ed.*, 1981, **19**, 1679.
[123] J. Chambers, J. Jiricny, and C. B. Reese, *Fire Mater.*, 1981, **5**, 133.
[124] P. I. Kordomenos and J. E. Kresta, *Macromolecules*, 1981, **14**, 1434.
[125] K. J. Voorhees and R. P. Lattimer, *J. Polym. Sci., Polym. Chem. Ed.*, 1982, **20**, 1457.
[126] Z. Slana, E. Svejdova, and J. Majer, *Markomol. Chem.*, 1980, **181**, 2449.

structurally related polyureas in which the N–H groups were partially, or totally, replaced by methyl groups[127] and of short chain poly(1,2-acenaphthene-diylidene), poly(1,2-acenaphthylenylene), and poly(9,10-anthracenediylidene).[128] Kinetic studies of phenolphthalein polycarbonate show that very complicated degradation reactions occur involving rearrangement, Friedel–Crafts acylation, decarboxylation, random chain scission, crosslinking, and hydrolysis.[129]

15 Polyimides

The application of thermal methods to the study of the degradation of polyimides has been reviewed.[130] The bulk of the recent experimental work on these polymers has been carried out by Russian workers. Features examined have been the kinetics of breakdown,[131] the formation of free radicals,[132] and the volatile products of decomposition.[133–135] The vexed question of the mechanism of evolution of carbon dioxide has been further investigated.[136] It is claimed to involve scission of N–C bonds with the unpaired electron on the nitrogen atoms of the imide ring leading to formation of CO_2.

16 Other Heterocyclic Polymers

Russian activity on polyimides has been paralleled by studies on other heterocyclic polymers. This has included polybenzoxazole,[137] polyphenylquinoxaline,[138] polytriazine,[139,140] and the semi-ladder poly(naphthoylene-bis-benzimidazole).[141,142] With the polybenzoxazole derived from 3,3'-dihydroxy-4,4'-diaminodiphenylmethane and isophthalic acid dichloride, low temperature oxidation occurred mainly in the aliphatic groups and oxidation of the aromatic nuclei only occurred at temperatures greater than 260 °C.[137] The introduction of bromine into the poly(phenyl quinoxaline) lowered the decomposition temperature and changed the composition of the gaseous decomposition products.[138] With the triazines it was claimed that ionic reactions predominated, catalysed by degradation products such as phenols and

[127] S. Caruso, S. Foti, P. Maravigna, and G. Montaudo, *J. Polym. Sci., Polym. Chem. Ed.*, 1982, **20**, 1685.
[128] D. Price, G. J. Milnes, C. Lukas, and I. Schopov, *Dyn. Mass. Spectrom.*, 1981, **6**, 291.
[129] M. S. Lin, B. J. Bulkin, and E. M. Pearce, *J. Polym. Sci., Polym. Chem. Ed.*, 1981, **19**, 2773.
[130] W. W. Wright, *Dev. Polym. Degrad.*, 1981, **3**, 1.
[131] A. V. Khabenko, V. S. Levshanov, and S. A. Dolmatov, *Vysokomol. Soedin.*, Ser. A, 1981, **23**, 1135.
[132] O. A. Ledneva, G. B. Pariiskii, V. V. Trezvov, and D. Y. Toptygin, *Vysokomol. Soedin.*, Ser. A, 1982, **24**, 361.
[133] G. A. Kalinkevich, V. L. Mikov, T. P. Morozova, I. M. Lukashenko, and R. A. Khmelnitskii, *Izv. Timiryazevsk S-Kh. Akad.*, 1981 (2), 164 (*Chem. Abstr.*, 1981, **94**, 157 621).
[134] G. A. Kalinkevich, V. L. Mikov, T. P. Morozova, I. M. Lukashenko, and R. A. Khmelnitskii, Kinet. Khim. Reakts. Mater. Vses. Simp. Goreniyn. Vzryvu, 1980, 6th, p. 101 (*Chem. Abstr.*, 1981, **94**, 157 626).
[135] A. Toirov, Y. N. Sazanov, L. A. Shibaev, L. M. Shcherbakova, T. M. Muinov, and Z. A. Kabilov, *Dokl. Akad. Nauk Tadch. SSR*, 1981, **24**, 173 (*Chem. Abstr.*, 1981, **95**, 81 726).
[136] J. Kurakowska-Orszagh and T. Chreptowicz, *Eur. Polym. J.*, 1981, **17**, 877.
[137] V. A. Isaeva, V. V. Guryanova, Y. A. Chernikhov, V. V. Korshak, M. P. Noskova, Y. I. Kotov, and B. M. Kovarskaya, *Vysokomol. Soedin.*, Ser. A, 1980, **22**, 1923.
[138] V. V. Korshak, S. A. Pavlova, P. N. Gribkova, I. V. Vlasova, M. N. Belomoina, Y. S. Krongauz, H. Raubach, B. Falk, and H. H. Oelert, *Vysokomol. Soedin.*, Ser. A, 1981, **23**, 789.

bisphenols.[139] The oxidative stability of the bis-benzimidazole was determined by oxygen absorption measurements.[141] Non-Russian work has concentrated upon analysis of the volatile decomposition products from polybenzimidazole,[143] polybenzoxazole,[144] and polyquinoxaline.[145]

17 Other Condensation Polymers

Phenol–formaldehyde resins, epoxy resins, and unsaturated polyesters crosslinked with styrene have all been the subject of study. A series of papers has been published[146–148] on the thermal decomposition of novolacs and resols up to 1200 °C. The kinetics of the non-isothermal degradation were analysed using the Erofeev equation. A variety of chromatographic techniques has been used to analyse qualitatively and quantitatively the phenols, cresols, xylenols, and non-phenolic aromatics produced on pyrolysis of cured phenolic resins.[149] The thermal breakdown of low-molecular-weight compounds modelling the structure of epoxides crosslinked with aromatic, or aliphatic, amines proceeded by a radical chain mechanism involving cleavage of $C_\beta - C_\alpha N$ bonds.[150,151] Two first-order rate constants were observed in the thermo-oxidative degradation of styrene–polyester copolymers. The first involved cross-link, or weak link, scission with liberation of free linear chains and the second random scission of these free linear chains.[152,153]

18 Silicon-containing Polymers

A Russian book has been published on the thermal and thermo-oxidative degradation of poly(organosiloxanes).[154] The majority of experiments have utilized infrared spectroscopy and pyrolysis–gas chromatography to elucidate the mechan-

[139] V. V. Korshak, S. A. Pavlova, P. N. Gribkova, M. W. Tsingiladze, V. A. Pankratov, and S. V. Vinogradova, *Vysokomol. Soedin., Ser. A*, 1980, **22**, 1714.
[140] O. V. Smirnova, D. F. Kutepov, V. P. Bolmosova, and A. A. Atrushkevich, *Tr. Mosk. Khim. Tekhnol. Inst. im D. I. Mendeleeva*, 1980, **110**, 50 (*Chem. Abstr.*, 1982, **97**, 6894).
[141] I. A. Serenkova, E. A. Kazantseva, B. I. Liogonkii, and Y. A. Shlyapnikov, *Vysokomol. Soedin., Ser. B*, 1981, **23**, 600.
[142] V. V. Korshak, S. A. Pavlova, P. N. Gribkova, L. A. Mikadza, A. L. Rusanov, A. M. Berlin, and S. K. Fidler, *Vysokomol. Soedin., Ser. A*, 1981, **23**, 96.
[143] D. A. Chatfield and I. N. Einhorn, *J. Polym. Sci., Polym. Chem. Ed.*, 1981, **19**, 601.
[144] E. G. Jones and I. J. Goldfarb, *Thermochim. Acta*, 1982, **54**, 131.
[145] H. J. Duessel, A. Recca, J. Kolb, D. O. Hummel, and J. K. Stille, *J. Anal. Appl. Pyrolysis*, 1982, **3**, 307.
[146] T. V. Komarova, *Freiberg Forschungsh. A*, 1980, **625**, 33 (*Chem. Abstr.*, 1981, **95**, 8 104).
[147] T. V. Komarova, N. A. Shamkina, S. D. Fedoseer, and G. B. Kortunova, *Khim. Topl. (Moscow)*, 1981 (6), 110 (*Chem. Abstr.*, 1982, **96**, 123 454).
[148] O. M. Petrova, T. V. Komarova, and S. D. Fedoseev, *Zh. Prikl. Khim. (Leningrad)*, 1982, **55**, 1629 (*Chem. Abstr.*, 1982, **97**, 145 615).
[149] D. Braun and R. Steffen, *Fresenius' Z. Anal. Chem.*, 1981, **307**, 7 (*Chem. Abstr.*, 1981, **95**, 44 014).
[150] L. A. Zhorina, L. S. Zarkhin, A. N. Zeleneskii, E. I. Karakozova, L. V. Karmilova, E. N. Kumpaneko, V. P. Melnikov, E. M. Nechvoloclova, and E. V. Prut, *Vysokomol. Soedin., Ser. A*, 1981, **23**, 2799.
[151] T. S. Zarkhina, A. N. Zeleneskii, L. S. Zarkhin, L. V. Karmilova, E. V. Prut, and N. S. Enikolopyan, *Vysokomol. Soedin., Ser. A*, 1982, **24**, 584.
[152] A. N. Das, *Combust. Flame*, 1981, **40**, 1.
[153] A. N. Das and S. K. Baijal, *J. Appl. Polym. Sci.*, 1982, **27**, 211.
[154] N. P. Kharitonov and V. V. Ostrovskii, Nauka Leningradsloe Otdelenie Leningrad, 1982.

isms of decomposition. Plasma-polymerized hexamethylcyclotrisilazane broke down by methyl group abstraction and scission of Si—C bonds.[155,156] Abstraction of methyl and phenyl substituents was confirmed by study of oligocyclosilazanes.[157] The initial cleavage of methyl groups and the later scission of main chain Si—Ph bonds were also observed in poly[4-phenylenebis(dimethylsiloxane)].[158] Other systems investigated have been poly(dimethyl-4-silphenylene) and poly(tetramethyl-4-silphenylene siloxane)[159] by infrared spectrometry and pyrolysis–gas chromatography, block copolymers of poly(dimethylsiloxane) and bisphenol A-polycarbonate[160] by thermogravimetry and copolymers of vinyl siloxanes with methyl acrylate[161] by product analysis.

19 Phosphorus-containing Polymers

Pyrolysis–mass spectrometry has been performed on poly(aryloxyphosphazene) and four polyaminophosphazenes.[162] A wide range of different pyrolysis mechanisms were observed and it is claimed that this is unique to the phosphazene structure. Poly[bis(β-naphthyloxy)phosphazene] gave cyclic oligomers, the poly-[bis(arylamine)phosphazenes] decomposed by a two-stage process involving amine evolution, and in poly(dipiperidinophosphazene) there was complete destruction of the polymer backbone with formation of ammonia and elemental phosphorus.

20 Cellulose

The techniques of differential thermal analysis, thermogravimetry, thermomechanical analysis, evolved gas analysis, and infrared spectrometry have been used to study the pyrolysis of cellulose in air or nitrogen.[163,164] The original polymer structure persisted up to 330 °C, but above this temperature many C=C and C=O bonds were formed.

---

[155] A. M. Wrobel and M. Kryszewski, Symp. Proc. Int. Symp. Plasma Chem., 5th, 1981, Vol. 1, p. 278.
[156] A. M. Wrobel, J. Kowalski, J. Grebowicz, and M. Kryszewski, J. Macromol. Sci., Chem., 1982, 17, 433.
[157] A. I. Gaponova, V. S. Osipchik, V. V. Kazakova, and G. V. Kotrelev, Izv. Vyssh. Uchebn. Zaved. Khim. Khim. Tekhnol., 1981, 24, 495 (Chem. Abstr., 1981, 95, 81 743).
[158] M. Ikeda, T. Nakamura, Y. Nagase, K. Ikeda, and Y. Sekine, J. Polym. Sci., Polym. Chem. Ed., 1981, 19, 2595.
[159] B. Zelei, M. Blazso, and S. Dobos, Eur. Polym. J., 1981, 17, 503.
[160] G. R. Grubbs, M. E. Kleppick, and J. H. Magill, J. Appl. Polym. Sci., 1982, 27, 601.
[161] O. D. Sutina, S. S. Zislina, L. M. Terman, and G. A. Razuvaev, Vysokomol. Soedin., Ser. A, 1981, 23, 322.
[162] A. Ballistreri, S. Foti, G. Montaudo, S. Lora, and G. Pezzia, Makromol. Chem., 1981, 182, 1319.
[163] D. Dollimore and J. M. Hoath, Proc. Eur. Symp. Therm. Anal., 2nd, 1981, Vol. 2, p. 576.
[164] D. Dollimore and J. M. Hoath, Thermochim. Acta, 1981, 45, 87 and 103.

16
Reactions in Macromolecular Systems

BY M. I. PAGE AND D. A. CROMBIE

1 Introduction

There have been several reviews on the simulation of biological phenomena, particularly catalysis and specificity,[1] transport,[2] and drug-receptor interactions[3] by macromolecules. Quantitative aspects of binding and recognition,[4] internal motion,[5] and synthetic utility[6] of macromolecules have also been discussed.

2 Cyclomalto-oligosaccharides (Cyclodextrins)

Interest in cyclomalto-oligosaccharides as hosts and potential catalysts continues to expand. In the last two years there have been 22 reviews and two books[7] on the subject and the more important ones only are listed.[8] The role of cyclomalto-oligosaccharides in the complexation of drugs for improving their bioavailability and stabilization have been discussed[9] together with their role as preservatives and flavour-improving agents.[10]

---

[1] R. Breslow, *Science (Washington, DC)*, 1982, **218**, 532; T. H. Crawshaw, D. A. Laidler, J. C. Metcalfe, R. B. Pettman, J. F. Stoddart, and J. B. Wolstenholme, *Stud. Org. Chem.*, 1982, **10**, 49; F. M. Menger, *ibid.*, p. 138; Y. Murakami, *Kagaku Kogyo*, 1982, **33**, 677; G. P. Royer, *Adv. Catal.*, 1980, **29**, 197; S. Shinkai, *Prog. Polym. Sci.*, 1982, **8**, 1; S. P Spragg, 'The Physical Behaviour of Macromolecules with Biological Functions', 1980; J. F. Stoddart, *Actual. Chim.*, 1982, 17; I. Tabushi, *Front. Chem., Plenary Keynote Sect. IUPAC Congr.*, 1981, 275, and *Kagaku Zokan (Kyoto)*, 1981, 71; F. Voegtle, 'Topics in Current Chemistry', Vol. 98, 'Host Guest Complex Chemistry', Pt. 1, 1981.

[2] T. M. Fyles, V. A. Malik-Diemer, and D. M. Whitfield, *Can. J. Chem.*, 1981, **59**, 1734; R. M. Izatt, B. L. Nielsen, J. J. Christensen, and J. D. Lamb, *J. Membr. Sci.*, 1981, **9**, 263; J. D. Lamb, R. M. Izatt, D. G. Garrick, J. S. Bradshaw, and J. J. Christensen, *Ibid.*, p. 83; J. P. Behr and J. M. Lehn, *Colloq. Ges. Biol. Chem.*, 1981, **32**, 24.

[3] J. Pitha and J. W. Kusiak, 'Controlled Release Pestic.: Pharm.' (Proc. Int. Symp.), 7th, 1982, p. 67; H. Batzer, *C. R. Mag.*, 1982, 22.

[4] I. Tabushi, *Kagaku (Kyoto)*, 1981, **36**, 868; S. Ferguson-Miller and W. H. Koppenol, *Trends Biochem. Sci.* (Pers. Ed.), 1981, **6**, IV-VII.

[5] O. Jardetzky, *Acc. Chem. Res.*, 1981, **14**, 291.

[6] T. Osa and J. Anzai, *Kagaku (Kyoto)*, 1981, **36**, 884.

[7] J. Szejtli, 'Cyclodextrins and their Inclusion Complexes', Akad. Kiado, Budapest, Hungary, 1982, and in 'Proc. 1st Int. Symp. Cyclodextrins', 1982.

[8] I. Tabushi, *Acc. Chem. Res.*, 1982, **15**, 66; W. Saenger, *Colloq. Ges. Biol. Chem.*, 1981, **32**, 33; I. Tabushi, *Stud. Org. Chem.*, 1982, **10**, 275; M. Komiyama, *Kagaku Kogyo*, 1981, **32**, 1107; M. Komiyama and H. Hidefumi, *ibid.*, 1982, **33**, 726; T. Ikeda and F. Toda, *ibid.*, p. 721; I. Tobushi and T. Nabeshima, *Gendai Kagaku*, 1982, **134**, 22; F. Cramer, 'Proc. 1st Int. Symp. Cyclodextrins', 1981, p. 3.

[9] K. H. Froemming, 'Proc. 1st Int. Symp. Cyclodextrins', 1981, p. 367; A. Stadler-Szoke and J. Szejtli, *ibid.*, p. 377; J. Szejtli, *Starch (Weinheim)*, 1981, **33**, 387; K. Uekama, *Yakugaku Zasshi*, 1981, **101**, 857; T. Nagai, 'Proc. 1st Int. Symp. Cyclodextrins', 1981, p. 15.

[10] G. Venczel, 'Proc. 1st Int. Symp. Cyclodextrins', 1981, p. 481; J. Szejtli, *ibid.*, p. 469.

The hydrolysis of 4-nitrophenyl 1-adamantaneacetate, $R(CH_2)_nCO_2C_6H_4NO_2$ (I, $n = 1$) is catalysed by cyclomaltoheptaose and cyclomalto-octaose, whereas that of the analogous substrate (I, $n = 0$) is retarded.[11] This is a good example which demonstrates that substrate structure, namely the presence or absence of a methylene group, governs the rate effects of cyclodextrins.

The cyclomalto-oligosaccharide catalysed hydrolysis of 4-nitrophenyl esters of long-chain alkanoic acids in 1:1 dimethyl sulphoxide:water shows, except for acetate, Michaelis–Menten kinetics. Stabilities and catalytic efficiencies of the inclusion complexes increase with increasing chain-length of the substrate.[12]

The magnitude of the acceleration of the cleavage of phenyl acetates by cyclomaltohexaose in 1:1 dimethyl sulphoxide:water increases with the decreasing distance between the carbonyl carbon of the substrate and the O-2 of the cyclomaltohexaose, as determined by ^1H n.m.r.[13]

The substrate binding constant for 2-naphthyl acetate and cyclomaltoheptaose, but not for cyclomaltohexaose and cyclomalto-oxaose, decreases with increasing pressure. The rate constants for the activation step all increase with pressure, giving volumes of activation for acylation of -13 to -17 cm^3 mol^{-1}. The differences are incautiously interpreted in terms of strongly solvated tight and weakly solvated loose transition states.[14]

The hydrolysis of (+)-amino-acid phenyl esters, $H_2NCHRCO_2Ar$, is more efficiently catalysed by cyclomaltoheptaose and cyclomaltohexase than that of the (−)-enantiomers.[15]

The first example of general base catalysed ester hydrolysis by cyclomalto-oligosaccharide has been reported for 2,2,2-trifluoroethyl-4-nitrobenzoate and cyclomaltohexaose (1). The secondary alkoxide ion of the saccharide also acts as a nucleophilic catalyst but mostly (80%) as a general base.[16]

Freezing the internal rotation of the acrylate side chain of the good substrate 3-*trans*-ferrocenyl propenoate by bridging the ferrocene nucleus gives a substrate (2) which is acylated by cyclomaltoheptaose at saturation, 3×10^6 faster than it is

[11] M. Komiyama and S. Inove, *Bull. Chem. Soc. Jpn*, 1980, **53**, 3266.
[12] Y. Hui, S. Wang, and X. Jiang, *J. Am. Chem. Soc.*, 1982, **104**, 347.
[13] M. Komiyama and H. Hidefumi, *Chem. Lett.*, 1980, 1471.
[14] Y. Taniguchi, S. Makimoto, and K. Suzuki, *J. Phys. Chem.*, 1981, **85**, 3469.
[15] M. Yamamoto, H. Kobayashi, M. Kitayama, H. Nakaya, S. Tanaka, K. Naruchi, and K. Yamada, *Kogakubu Kenhyer Hokoku* (*Chiba Daigaku*), 1981, **33**, 89.
[16] M. Komiyama and S. Inove, *Bull. Chem. Soc. Jpn*, 1980, **53**, 3334.

hydrolysed under the same conditions. Not surprisingly there is a discrimination (20-fold) between the two enantiomers of (2).[17] The rate acceleration approaches the maximum of 10^8 predicted by entropy changes in the absence of strain effects.[18] The hydrolysis of phosphate esters is retarded by 1:2 complexes of cyclomaltohexaose and cyclomaltoheptaose with Cu^{II}.[19] The effect of substituents on the alkaline hydrolysis of 7-substituted coumarins is modified by inclusion of the substrate within the cyclomalto-oligosaccharide cavity.[20] The deprotonation of oxazolones (3) is catalysed by cyclomaltohexaose and cyclomaltoheptaose but with little enantioselectivity. Hydrolysis of (3) is also catalysed but with higher enantioselectivity. Inclusion of the substrate is attributed mainly to the phenyl substituent.[21]

(3)   (4)

The rate of hydrolysis of the vinyl ether of prostaglycin is retarded by cyclomaltohexaose and cyclomaltoheptaose by competitive inhibition.[22] The first example of catalysis of electrochemical cleavage by cyclomalto-oligosaccharides has been reported. Cyclomeltoheptaose with an electrophilic group chemically bonded to the periphery cleaves benzyl esters at a modest potential associated with the electrophore. Electron transfer or reaction probably occurs between the reduced electrophore and the complexed ester.[23]

Complexation of electroactive substrates with cyclomaltoheptaose can profoundly alter the course of their reactions. The cathodic reduction of complexes ethyl cinnamate, benzaldehyde, and benzophenone proceeds by very efficient protonation of the radical-anions whereas a previously unobserved reductive coupling is found for the acetophenone complex.[24]

Nicotinamide bonded to cyclomeltaheptaose (4) is more efficient (40—60-fold) at reducing ninhydrin than NADH.[25] Proton transfer reactions of 5-(3- and 4-nitrophenylazo) salicylic acids are accelerated in the presence of cyclomaltohexaose

[17] G. L. Trainer and R. Breslow, *J. Am. Chem. Soc.*, 1981, **103**, 154.
[18] M. I. Page and W. P. Jencks, *Proc. Natl. Acad. Sci. USA*, 1971, **68**, 1678; M. I. Page, *Angew. Chem., Int. Ed. Engl.*, 1977, **16**, 449.
[19] K. Mochida, Y. Ozoe, H. Miyazaki, and Y. Matsui, *Shimane Daigaku Nogakubu Kenkyu Hokoku*, 1980, 158.
[20] K. Uekama, C.-L. Lin, F. Hirayama, M. Otagiri, A. Takadate, and S. Goya, *Chem. Lett.*, 1981, 563.
[21] V. Daffe and J. Fastrez, *Stud. Org. Chem.*, 1982, **10**, 298.
[22] K. Vekama, F. Hirayama, T. Wakuda, and M. Otagiri, *Chem. Pharm. Bull.*, 1981, **29**, 213.
[23] C. Z. Smith and J. H. Putley, *J. Chem. Soc., Chem. Commun.*, 1981, 792.
[24] C. Z. Smith and J. H. Putley, *J. Chem. Soc., Chem. Commun.*, 1981, 492.
[25] M. Kojima, F. Toda, and K. Hattori, *Tetrahedron Lett.*, 1980, **21**, 2721.

and cyclomaltoheptaose. The first example of an intramolecular Diels–Alder reaction catalysed by cyclomaltoheptaose has been reported for a substituted furan and a dienophile separated by three atoms. The yield of cyclization product depends on the substituents present on the connecting chain.[27]

The formation of dichlorocarbene within the cyclomaltoheptaose cavity affords a selective synthesis of 2,5-cyclohexadienones.[28] In the presence of cyclomaltoheptaose the E2 elimination from 1-(β-bromoethyl)-naphthalene to give 1-vinylnaphthalene is favoured over the substitution reaction to give 1-(β-hydroxyethyl)-naphthalene.[29] 4-Hydroxybenzoic acid is selectively formed (96% yield) from phenol, carbon tetrachloride, and copper powder in the presence of cyclomaltoheptaose as catalyst.[30] Rotaxanes where cyclomalto-oligosaccharides are threaded by α,ω-diaminoalkanes co-ordinated to cobalt(III) complexes (5) may be prepared in high yield.[31]

(5)

The formation of channels is one of the general strategies adopted in biological systems to achieve specific ion transport. One approach to the synthesis of artificial channels is to prepare a molecule with appropriate hydrophilic and hydrophobic parts similar in length to a lipid molecule and with a suitable number of ionophilic sites. A cyclomalto-oligosaccharide having four hydrophobic tails and three metal binding sites has now been synthesized as a 'half-channel' and is soluble in organic solvents but insoluble in water. Artificial liposome modified with this derivative transported Co^{2+} much faster than the rate for the unmodified liposome. The channel is designed to align along lecithin molecules and to bind the metal ion moderately by several approximately equally spaced binding sites (6).[32] See also Section 7.

Cyclomalto-oligosaccharides with two different functional groups may be unsymmetrically introduced by using a new type of N-oxide capped cyclomaltooligosaccharide which shows half-of-the-sites reactivity.[33] The enhancement of catalytic hydrolysis of nitrophenyl esters by hydroxamate containing cyclomaltoheptaose relative to cyclomaltoheptaose is sensitive to substrate structure, par-

[26] M. Kojima, F. Toda, and K. Hattori, *J. Chem. Soc., Perkin Trans. 1*, 1981, 1647; N. Yoshida and M. Fujimoto, *Bull. Chem. Soc. Jpn.*, 1982, **55**, 1039.
[27] D. D. Sternbach and D. M. Rossana, *J. Am. Chem. Soc.*, 1982, **104**, 5853.
[28] M. Komiyama and H. Hirai, *Macromol. Chem., Rapid Commun.*, 1981, **2**, 177; *Bull. Chem. Soc. Jpn.*, 1981, **54**, 2053.
[29] I. Tabushi in 'Advances in Solution Chemistry', 5th Ed., ed. I. Bertini, L. Lunazzi, and A. Dei, Plenum, New York, 1980, p. 221.
[30] M. Komiyama and H. Hirai, *Macromol. Chem., Rapid Commun.*, 1981, **11**, 661.
[31] H. Ogino, *J. Am. Chem. Soc.*, 1981, **103**, 1303.
[32] I. Tabushi, Y. Kuroda, and K. Yokota, *Tetrahedron Lett.*, 1982, **23**, 4601.
[33] I. Tabushi, T. Nabeshima, H. Kitaguchi, and K. Yamamura, *J. Am. Chem. Soc.*, 1982, **104**, 2017.

```
HO                                                    OH
  ⟋⎯⎯⎯⟍  S-----NHCO--    --OCHN-----S  ⟋⎯⎯⎯⟍
 |   S-----NHCO--    --OCHN-----S   |
 |   S-----NHCO--    --OCHN-----S   |
  ⟍⎯⎯⎯⟋  S-----NHCO--    --OCHN-----S  ⟍⎯⎯⎯⟋
```

(±)∿∿∿∿∿∿∿∿∿   ∿∿∿∿∿∿∿∿∿(±)

(6)

ticularly the location of the nitro group(s).[34] Enzyme catalysed reactions can be inhibited by the formation of an inclusion complex of the substrate with cyclomaltoheptaose. Examples are the hydrolysis of 4-nitrophenyl β-D-glucopyranoside by β-glucosidase and that of 4-nitrophenyl phosphate by alkaline phosphatase.[35]

Physical studies on cyclomalto-oligosaccharide complexes have included the location of binding sites using ^{13}C (ref. 36) and the thermodynamics of binding.[37] Significantly large deviations are found between the observed geometries of inclusion complexes and those calculated from van der Waals interaction energies.[38]

The suggestion that the cyclomeltahexaose 3-hydroxyls are inherently unreactive and therefore not involved in catalytic hydrolysis has been questioned by the determination of both the static and dynamic aspects of binding of guests to methylated and unmethylated cyclohexaamyloses.[39] There have been several studies of photolytic reactions in the presence of cyclomalto-oligosaccharides.[40]

3 Crown Ethers and Cryptands

There have, again, been 22 reviews of this topic and only the major ones are listed.[41]

The rate of the hydrogen ion catalysed hydrolysis of crown ether acetals (7) is decreased on complexing with alkali-metal cations, whereas a rate increase is observed for the alkaline hydrolysis of the crown ether ester (8). This is

---

[34] I. Tabushi, Y. Kuroda, and Y. Sakata, *Heterocycles*, 1981, **15**, 815.
[35] M. Kodaka, *Kobunshi Ronbunshu*, 1980, **37**, 803.
[36] R. I. Gelb, L. M. Schwartz, B. Cardelino, H. S. Fuhrman, R. F. Johnson, and D. A. Laufer, *J. Am. Chem. Soc.*, 1981, **103**, 1750; M. Komiyama and H. Hirai, *Bull. Chem. Soc. Jpn.*, 1981, **54**, 828.
[37] M. R. Eftink and J. C. Harrison, *Bioorg. Chem.*, 1981, **10**, 338; J. C. Harrison and M. R. Eftink, *Biopolymers*, 1982, **21**, 1153.
[38] Y. Matsui, *Bull. Chem. Soc. Jpn.*, 1982, **55**, 1246.
[39] R. T. Bergeron and P. S. Burton, *J. Am. Chem. Soc.*, 1982, **104**, 3664.
[40] A. Ueno, R. Saka, K. Takawashi, and T. Osa, *Heterocycles*, 1981, **15**, 671; A. Ueno, K. Takawashi, and T. Osa, *J. Chem. Soc., Chem. Commun.*, 1981, **3**, 94; M. Hoshino, M. Imamura, K. Ikehara, and Y. Hama, *J. Phys. Chem.*, 1981, **85**, 1820; I. Tabushi and L. C. Yuan, *J. Am. Chem. Soc.*, 1981, **103**, 3574.
[41] S. Patai, ed., *J. Chem. Educ.*, 1982; J. M. Lehn, *Recherche*, 1981, **12**, 1213, and in 'Front. Chem., Plenary Keynote Sect. IUPAC Congr., 28th', 1981, p. 265; J. F. Stoddart, *Spectrum*, 1981, **19**, 45; I. O. Sutherland, *Chem. Ind.*, 1981, 421; F. Voegtle, E. Weber, and V. Elben, *Kontakte (Darmstadt)*, 1980, 36; E. Weber and F. Voegtle, *ibid.*, 1981, 24; E. Weber, *ibid.*, 1982, 24; F. Voegtle, H. Sieger, and W. M. Mueller, *Top. Curr. Chem.*, 1981, **98**, 107; F. Voegtle, *Pure Appl. Chem.*, 1980, **52**, 2405.

(7) $n = 2-7$ (8)

understandable on a simple electrostatic model if the charge interaction is similar to that in the bulk solvent.[42] The dissociation of metal ions from cryptate complexes is general acid catalysed and kinetic isotope effects suggest that the rate-limiting step involves proton transfer.[43]

The rate of racemization of chiral bipyridyl crown ethers may be increased 10^6-fold in the presence of $PdCl_2$. This is attributed to the increase in binding between dipyridyl and the metal in the coplanar transition state.[44] The rate constant for the decomposition of arene diazonium ions complexed with crown ethers decreases with increasing stability of the complex.[45]

Crown ethers are powerful complexing agents for alkali-metal cations which, in turn, produce highly reactive anions. For example, the alkylation of anions in the Williamson and Gabriel syntheses occurs readily in the presence of cryptands.[46] However, under conditions of rate-limiting nucleophilic attack, the rate constant for displacement of the methane sulphonate group is similar for bulky and less bulky ligands. This is presumably because the anion separation is already optional in the less bulky cryptate, which would indicate that anion activation obtained with [2.2.2]-cryptates soluble in organic non-polar media is probably the highest actually obtainable in solution.[47] Other aspects of phase-transfer catalysis are reviewed in Section 6 of this Chapter.

The effect of crown ethers on the steric hindrance to base approach in bimolecular elimination reactions indicate that the clump aggregate model of ion-paired alkoxide bases is incorrect.[48] Highly selective nitrating agents for electrophilic aromatic substitution may be obtained with nitronium tetrafluoroborate–crown ether complexes in methylene chloride.[49] Crown ethers form molecular complexes with bromine that may be used as reagents for the bromination of alkenes and alkynes. There is very little difference between the formation constants of the complex or between rates of bromination as the ethers are varied indicating that bromine is not encapsulated by the macrocycle. However, the stereoselectivity of addition is significantly changed as a function of the ether.[50]

[42] D. S. Baker and V. Gold, *J. Chem. Soc., Chem. Commun.*, 1982, 1401.
[43] B. G. Cox, W. Jedral, P. Firman, and H. Schneider, *J. Chem. Soc., Perkin Trans. 2*, 1981, 1486.
[44] J. Rebel, jun., T. Costello, and R. V. Wattley, *Tetrahedron Lett.*, 1980, **21**, 2379.
[45] H. Nakazumi, I. Szele, and H. Zollinger, *Tetrahedron Lett.*, 1981, **22**, 3053.
[46] M. A. Pasquini, R. le Goaller, and J. L. Pierre, *Tetrahedron*, 1980, **36**, 1223; P. Viout, *J. Mol. Catal.*, 1981, **10**, 231.
[47] D. Landini, A. Maia, F. Montawari, and F. Rolla, *J. Chem. Soc., Perkin Trans. 2*, 1981, 821.
[48] M. Pankova and J. Zavada, *Collect. Czech. Chem. Commun.*, 1980, **45**, 3150.
[49] B. Masci, *J. Chem. Soc., Chem. Commun.*, 1982, 1262.
[50] K. H. Pannell and A. J. Mayr, *J. Chem. Soc., Perkin Trans. 1*, 1982, 2153.

Chiral crown ethers complexes to potassium bases catalyse, with high turnover numbers, the Michael additions of a β-keto ester to methyl vinyl ketone to give products of 60—99% optical purity. The configurational bias may be rationalized by steric effects.[51] 1,4-Dihydropyridines contained in chiral macrocycles are useful asymmetric reducing agents in the presence of Hg^{2+} ions. For example, variation of the amino-acid, length, and form of the bridge in (9) shows that hydride is always transferred preferentially from (9) derived from L-amino acids to the *re*-face of the ketone carbonyl group to be reduced. Ether oxygens in the bridge contribute minimally to the enantiomeric excess, indicating that metal ion binding at this site is not important for hydride transfer. It is suggested that a ternary complex is formed in which the reactive carbonyl group is bound to Mg^{2+} complexed with the macrocycle amide group with the fit dictated by steric requirements. Quite remarkably, the length and shape of the bridge in (9) can be varied extensively without decreasing enantiomeric excesses below 83%.[52]

Cytochrome P_{450} monooxygenase enzymes selectively oxygenate organic substrates by simultaneously activating dioxygen and undergoing a hydrophobic guest–host association between the enzyme and the target substrate. Bicyclic ligand (10) has sufficiently commodious persistent voids to engulf many potential organic substrates and its cobalt(II) and iron(II) complexes exhibit exceptional O_2-carrying capacities.[53]

Macromonocyclic polyamines, because of their potential positive charge and hydrogen-bonding abilities, are suitable model receptors for biochemically important polyoxyanions. For example, they form 1:1 complexes with carbonate ion and interestingly there is a liberation of a free proton at pH 7 when bicarbonate binds to polyamines in the H_3L^{3+} form to give complexes presumably of the type

---

[51] D. J. Cram and G. D. Y. Sogah, *J. Chem. Soc., Chem., Commun.*, 1981, 625.
[52] P. Jouin, C. B. Troostwijk, and R. M. Kellog, *J. Am. Chem. Soc.*, 1981, **103**, 2091.
[53] K. J. Takeuchi and D. H. Busch, *J. Am. Chem. Soc.*, 1981, **103**, 2421.

(11). This may relate the production of gastric HCl, secretion of H^+ in kidney tubules, and the 'chloride shift' in red blood cells to the acid–base equilibrium of carbonic acid.[54]

Shinkai and co-workers have continued their elegant studies of photoresponsive crown ethers. The ion binding functions of the ethers may be controlled by an on–off light-switch mechanism. For example, ion transport across a liquid membrane may be facilitated by photoconversion of *trans* to *cis* azo linkages in the crown ether carrier molecule (12).[55]

The competitive alkali-metal transport from an alkaline aqueous source phase through a chloroform phase to an acidic aqueous receiving phase is facilitated by crown ethers with pendant carboxylic acid groups. Transport selectivity if controlled by the size of the polyether cavity of the carrier. Increasing the lipophilicity of the carrier enhances the transport rate but does not affect the selectivity. There is poor agreement between the results of competitive transport and the behaviour anticipated from single cation transport studies.[56] See Section 7 for other aspects of ion transport.

The first example of the formation of a crystalline complex of an uncharged macrocyclic crown host with water as a guest molecule has been reported. The water guest is bound exclusively in the centre of the crown cavity by hydrogen-bonding with the crown ether oxygens.[57]

The macrocyclic cryptand (13) binds one or two protons either outside or inside its intramolecular cavity. Thermodynamically it is a very strong base but kinetically an extremely slow one. The pK_a values for the externally mono- and di-protonated species are 7 and ~1, respectively, while the corresponding values for the internally protonated derivatives are ⩾18 and ~6. The rates of proton transfer into and out of the cavity are very slow. One of the two protons in the internally diprotonated species may only be removed by base under vigorous conditions.[58]

The direction and extent of chiral recognition of amino-acids and their esters by

---

[54] E. Kimura, A. Sakonaka, and M. Kodama, *J. Am. Chem. Soc.*, 1982, **104**, 4984.
[55] S. Shinkai, T. Kouno, Y. Kusano, and O. Manabe, *J. Chem. Soc., Perkin Trans. 1*, 1982, 2741; S. Shinkai, K. Shigematsu, M. Sato, and O. Manabe, *ibid.*, 1982, 2735; S. Shinkai, K. Shigematsu, Y. Kusano, and O. Manabe, *ibid.*, 1981, 3279; S. Shinkai, T. Minami, Y. Kusano, and O. Manabe, *J. Am. Chem. Soc.*, 1982, **104**, 1967; S. Shinkai, T. Minami, and Y. Kusano, *Tetrahedron Lett.*, 1982, **23**, 2581; S. Shinkai, T. Minami, T. Kouno, Y. Kusano, and O. Manabe, *Chem. Lett.*, 1982, 499.
[56] J. Strzelbicki and R. A. Bartsch, *J. Membr. Sci.*, 1982, **10**, 35.
[57] G. R. Newkome, H. C. R. Taylor, F. R. Fronczek, T. J. Delord, D. K. Kohli, and F. Voegtle, *J. Am. Chem. Soc.*, 1981, **103**, 7376.
[58] P. B. Smith, J. L. Dye, J. Cheney, and J. M. Lehn, *J. Am. Chem. Soc.*, 1981, **103**, 6044.

(13)   (14)

chiral crown ether can be rationalized by molecular model structures of the diastereoisomeric complexes.[59] There have been other reports of enantiomeric selectivity by chiral crown ethers.[60] A cryptand (14) with a tetrahedral recognition site has a high binding complementarity to the ammonium ion. The cation is bound by a tetrahedral array of N^+-H ... $N-H$ bonds to the four sites and by 12 electrostatic interactions to the oxygens of the ligands.[61]

The co-ordination radii of Group I and II metal ions in the cavity of a crown ether or cryptand are significantly ($\sim 35\%$) larger than the corresponding Pauling crystal radii because of the differences in the charge densities in the crystal and ligand environments.[62] N.m.r. correlation times of cryptates indicate that complementarity (steric and electronic fit) is reflected in dynamic fit between receptor and substrates.[63] Thermodynamic studies[64] and molecular mechanics calculations[65] of cryptate formation have been reported.

Cyclophanes.—The water-soluble heterocyclophane (15) shows substrate specificity in the catalysed hydrolysis of esters and is more discriminating than cetyltrimethyl-ammonium bromide micelles or cyclodextrins.[66] The copper(II) complex of a [10:10] paracyclophane bearing imidazolyl side chains catalyses the hydrolysis of esters via acylation of the imidazolyl groups by the bound substrate. Deacylation of the monoacylcyclophane intermediate is facilitated by the imidazole-bound copper(II).[67] Azaparacyclophanes with alkyl chains as branches of the skeleton incorporate both neutral and anionic hydrophobic substrates.[68] A [20]-paracyclo-

---

[59] D. S. Lingenfelter, R. C. Helgeson, and D. J. Cram, *J. Org. Chem.*, 1981, **46**, 393.
[60] W. Bussmann, J. M. Lehn, U. Oesgh, P. Plumere, and W. Simon, *Helv. Chim. Acta*, 1981, **64**, 657; F. Dietl, A. Merz, and R. Tomahogh, *Tetrahedron Lett.*, 1982, **23**, 5255; D. J. Chadwick, I. A. Cliffe, I. O. Sutherland, and R. F. Newton, *J. Chem. Soc., Chem. Commun.*, 1981, 992.
[61] E. Graf, J. P. Kintzinger, J. M. Lehn, and J. Lemoigne, *J. Am. Chem. Soc.*, 1982, **104**, 1672.
[62] R. T. Myers, *Inorg. Nucl. Chem. Lett.*, 1980, **16**, 329.
[63] J. P. Kintzinger, F. Kotzyba-Hibert, J. M. Kehn, A. Pagelot, and K. Saigo, *J. Chem. Soc., Chem. Commun.*, 1981, 833.
[64] M. H. Abraham and H. C. Ling, *Tetrahedron Lett.*, 1982, **23**, 469; B. G. Cox, P. Firman, J. Garcia-Rosas, and H. Schneider, *ibid.*, 1982, **23**, 3777.
[65] G. Wipff, P. Weiner, and P. Kollman, *J. Am. Chem. Soc.*, 1982, **104**, 3249.
[66] I. Tabushi, Y. Kimura, and Z. Yamamura, *J. Am. Chem. Soc.*, 1981, **103**, 6486, and *Stud. Org. Chem.*, 1982, **10**, 328.
[67] Y. Murakami, Y. Aoyama, and M. Kida, *J. Chem. Soc., Perkin Trans. 2*, 1980, 1665.
[68] Y. Murakami, A. Nakano, K. Akiyoshi, and K. Fukuya, *J. Chem. Soc., Perkin Trans. 1*, 1981, 2800.

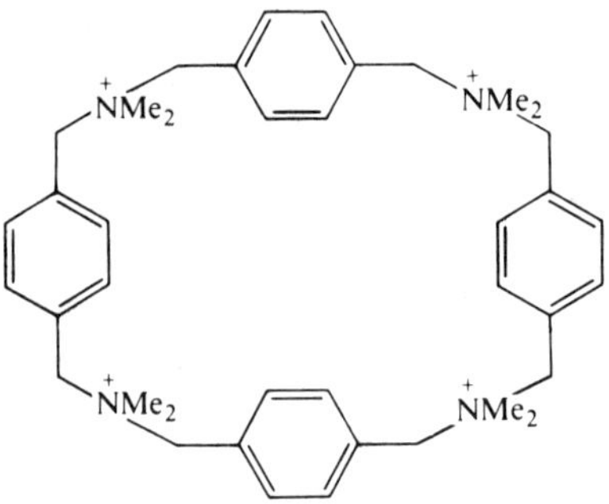

(15)

phane oxime shows a small specificity for hydrophobic substrates in its catalytic hydrolysis of esters and is more effective than cycloheptaamylose as a catalyst.[69]

A [20]-paracyclophane bearing a 1,4-dihydronicotinamide on the benzene ring (16) binds zinc(II).[70] The 2:1 complex in which dihydronicotinamide is not co-ordinated to the metal is 7-fold more reactive in the reduction of hexachloroacetone than metal-free (16).[71] [6,6]-Anthracene-haem-cyclophane binds carbon monoxide and oxygen about 200-fold less effectively than the [7,7] derivative.[72] Selectivity of guests by water-soluble paracyclophanes has been reported.[73]

4 Synthetic Polymers

Polymer-supported reagents for organic synthesis,[74] functional group containing polymers,[75] enzyme-like activities of polymers,[76] soluble polymer-enzyme adducts,[77] polymer controlled release of macromolecules,[78] and functional polymer membranes[79] have been reviewed.

Poly(ethylenimine)s.[80]—Poly(ethylenimine) provides a remarkably versatile macromolecular matrix for the introduction of environments that might accelerate specific

---

[69] H. Okamoto, D. Horiguchi, H. Kondo, and J. Sunamoto, *Nagasaki Daigaku Kogakubu Kenkyu Hokoku*, 1981, 81.
[70] Y. Murakami, J. Aoyama, and J. Kikuchi, *J. Chem. Soc., Perkin Trans. 1*, 1981, 2809.
[71] Y. Murakami, Y. Aoyama, and J. Kikuchi, *Bull. Chem. Soc. Jpn.*, 1982, **55**, 2898.
[72] T. G. Traylor, M. J. Mitchell, S. Tsuchiya, D. H. Campbell, D. V. Stynes, and N. Koga, *J. Am. Chem. Soc.*, 1981, **103**, 5234.
[73] K. Odashima, T. Soga, and K. Koga, *Tetrahedron Lett.*, 1981, **22**, 5311; K. Odashima and K. Koga, *Kagaku (Kyoto)*, 1982, **37**, 396.
[74] P. Hodge and D. C. Sherrington, *J. Am. Chem. Soc.*, 1982, **104**, 3548; J. M. J. Frechet, *Tetrahedron*, 1981, **37**, 663; F. Ciardelli, G. Braca, C. Carlini, G. Sbrana, and G. Valentini, *J. Mol. Catal.*, 1982, **14**, 1; J. Evans, *Chem. Soc. Rev.*, 1981, **10**, 159.
[75] A. Ledwith, *Chem. Ind.*, 1981, 358; A. Peterlin, *Macromol. Rev.*, 1980; M. Okawara, *Nippon Setchaku Kyokaishi*, 1982, **18**, 165.
[76] I. Tabuse, *Iwanami Koya Gendai Kagaku*, 1980, **20**, 87—108 and 205.
[77] A. Abuchowski and F. F. Davis, *Enzymes Drugs*, 1981, 367.
[78] R. Langer, *Chemtech*, 1982, **12**, 98; R. Langer, D. S. T. Hsieh, A. Peil, R. Bawa, and W. Rhine, *AIChE Symp. Ser.*, 1981, **77**, 10; R. Langer, *Methods Enzymol.*, 1981, **73**, 57.
[79] A. Nakajima, *Nippon Zairyo Kyodo Gakkaishi*, 1980, **15**, 119.
[80] E. Kimura, *Kagaku No Ryoiki*, 1981, **35**, 865.

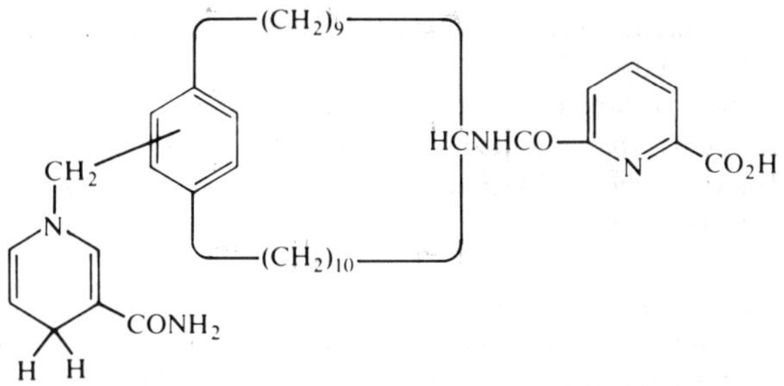

(16)

reactions. Alkylated cationic derivatives of the polymer are more effective nucleophilic catalysts for ester hydrolysis than are corresponding alkyl cationic micelles. Similarly the polymer is more effective than micelles in influencing donor–acceptor electron transfer.[81]

Substituted aminopyridines attached to poly(ethyleneimine)s catalyse the hydrolysis of 4-nitrophenyl caproate 50—2000-fold more effectively than in the absence of polymer.[82] Stereoselectivity in the hydrolysis of amino-acid 4-nitrophenyl esters can be introduced by using graft polymers of poly(ethyleneimine)–histidine.[83] The largest rate enhancements using dodecane-block-poly[ethylenimine-graft-4,(5)-methylimidazole] and derivatives as catalysts for the hydrolysis of anionic esters are obtained when both electrostatic and a polar interactions are evident. Michaelis–Menten kinetics are observed and the catalyst can be inhibited.[84]

Poly(ethylenimine) containing an isolated apolar binding site (17) is a more effective catalyst for the hydrolysis of long-chain 4-nitrophenyl esters than for the shorter chains. This is attributed to the substrate binding to the apolar block.[85]

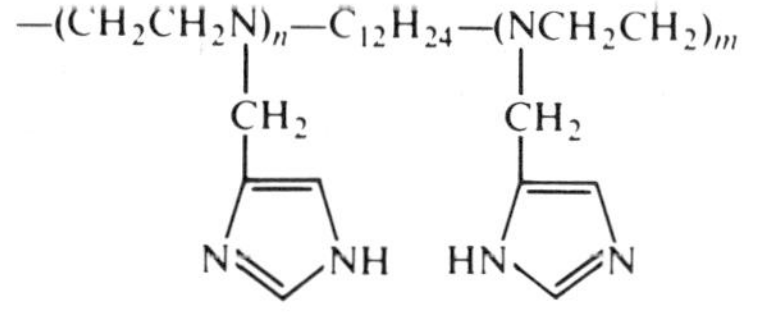

(17)

The Brønsted β-value for nucleophilic attack of pendant primary amino groups of poly(ethylenimine) dodecyl iodide is 0.81 based on the apparent change in pK_a of the polymer amino groups with pH. The attachment of imidazole, pyridine, or 2-aminopyridine residues to the polymer did not change the pH dependency which is

---

[81] I. M. Klotz, E. N. Drake, and M. Sisido, *Bioorg. Chem.*, 1981, **10**, 63.
[82] E. Delaney, L. E. Wood, and I. M. Klotz, *J. Am. Chem. Soc.*, 1982, **104**, 799.
[83] M. Nango, H. Kozuka, Y. Kimura, N. Kuroki, Y. Ihara, and I. M. Klotz, *J. Polym. Sci., Polym. Lett. Ed.*, 1980, **18**, 647.
[84] J. A. Pavlisko and C. G. Overberger, *J. Polym. Sci., Polym. Chem. Ed.*, 1981, **19**, 1757, and *Org. Coat. Plast. Chem.*, 1980, **42**, 537.
[85] J. A. Pavlisko and C. G. Overberger, *J. Polym. Sci., Polym. Chem. Ed.*, 1981, **19**, 1621.

compatible with similar electrostatic effects on nucleophilicity. There is no evidence for bifunctional catalysis.[86]

The catalytic constant for proton transfer between ethyl nicroacetate and a poly(amidoamine) is that expected from a Brønsted plot of non-polymeric bases.[87] Metal complexes of poly(ethylenimine) catalyse the decomposition of hydrogen peroxide,[88] the reduction of water to hydrogen,[89] and cyclohexene.[90]

Poly(4-vinylpyridine) Derivatives.—The decarboxylation of 6-nitrobenzisoxazole-3-carboxylate anion is catalysed by quaternized poly(4-vinylpyridines). The effectiveness increases with increasing hydrophobicity of the polymer domain as indicated by the average side-chain length.[91] *Anti*-stereoselective bromination of alkenes may be obtained by using complexes of bromine and cross-linked poly(styrene-4-vinylpyridine) beads.[92] The addition of alcohols to dihydropyran is catalysed by poly(vinyl pyridine) tosylate salts.[93]

The rate of α,β-elimination of tryptophan, but not that of serine or threonine, catalysed by pyridoxal phosphate and copper(II) ions is enhanced by quaternized poly(4-vinyl pyridine) bearing hydrophobic dodecyl groups. This is attributed to hydrophobic interaction between the indole of tryptophan and the polymer.[94] A poly(vinyl pyridine) co-ordinated cluster of molybdenum and iron may be used repeatedly as a catalyst for nitrogen fixation.[95]

Poly(vinylimidazole)s.[96]—The catalytic effect of poly(3-alkyl-1-vinylimidazolium) salts on the hydrolysis of esters increases with increasing hydrophobicity of the polymer–substrate complex.[97]

The catalytic effect of copolymers of poly(vinylimidazole) on solvolysis reactions increases with increasing ethanol concentration but decreases with increasing rigidity of the copolymer compared with the homopolymer.[98] Enantioselectivity is observed in the hydrolysis of chiral amino-acid esters catalysed by optically active imidazole-containing poly(iminomethylenes).[99] Enhanced activities, compared with low-molecular-weight analogs of histidine, are observed for the imidazole-containing polymer catalysed hydrolysis of hydrophobic esters but not of lower-molecular-weight substrates.[100]

---

[86] C. S. Lege and J. A. Deyrup, *Macromolecules*, 1981, **14**, 1629, 1634.
[87] P. De Maria, A. Fini, P. Ferruti, M. C. Beni, and R. Barbucci, *Macromolecules*, 1982, **15**, 679.
[88] J. H. Park and T. S. Cho, *Yaehan Hwakakhoe Chi*, 1981, **25**, 394.
[89] E. R. Buyanova, L. G. Matvienko, A. I. Kokorin, G. L. Elizarova, V. N. Parmon, and K. I. Zamaraev, *React. Kinet. Catal. Lett.*, 1981, **16**, 309.
[90] I. V. Patsevich, I. A. Ogorodnikov, A. I. Fridman, V. I. Nefedov, and Y. V. Salyn, *Vysokomol. Soedin., Ser. A*, 1982, **24**, 1559.
[91] S. Shinkai, S. Hirakawa, M. Shimomura, and T. Kunitake, *J. Org. Chem.*, 1981, **46**, 868.
[92] Y. J. M. Zupan and B. Sket, *J. Chem. Soc., Perkin Trans. 1*, 1982, 2059.
[93] F. M. Menger and C. H. Chu, *J. Org. Chem.*, 1981, **46**, 5044.
[94] H. Nakano, T. Yagi, O. Sangen, and Y. Yamamoto, *J. Polym. Sci., Polym. Lett. Ed.*, 1982, **20**, 23.
[95] C. Sun, S. Li, Q. Huang, and S. Niu, *Gaodeng Xuepiao Huapue Xuebao*, 1982, **3**, 398.
[96] J. A. Pavlisko and C. G. Overberger, *Polym. Sci. Technol.*, 1981, **14**, 257.
[97] S. C. Israel, K. I. Papathomas, and J. C. Salamone, *Polym. Prepr., Am. Chem. Soc., Div. Polym. Chem.*, 1981, **22**, 221.
[98] S. Hayama, T. Okamura, N. Watanabe, and S. Osawa, *Yamagata Daigaku Kiyo Kogaku*, 1980, **16**, 27.
[99] J. M. Van der Eijk, R. J. M. Nolte, V. E. M. Richters, and W. Drenth, *Recl. Trav. Chim. Pays-Bas*, 1981, **100**, 222.
[100] I. Cho and J. S. Shin, *Bull. Korean Chem. Soc.*, 1982, **3**, 34.

Poly(amino acids).—A review of catalysis by cyclopeptide macrocycles has appeared.[101]

The asymmetric addition of HCN to benzaldehyde catalysed by cyclo(L-phenylalanyl-L-histidine) gives an enantiomeric excess of 90% in the early stages of reaction which decreases with time.[102]

Other Polymers.—Poly-(R)-2-ethylaziridinoacetic acid specifically catalyses the hydrolysis of phenyl esters of L-amino acids with Michaelis–Menten kinetics.[103] Polyaziridines may be used as chiral catalysts for the asymmetric addition of methanol to phenyl methyl ketene.[104] Asymmetric synthesis in the epoxidation of alkenes may be achieved by using quinium salts anchored by polystyrene matrix.[105]

Ternary complexes have been proposed to account for acceleration of metal ions on the poly(acrylic acid) catalysed hydrolysis of esters.[106] Other studies have included the dehydration of methanol catalysed by sulphonated poly(styrene–divinylbenzene),[107] the proteolytic activity of copolymers of vinyl benzene, maleic anhydride with mixtures of histidine and other amino-acids,[108] the Michael condensation catalysed by polystyrene anchored metal acetylacetonates,[109] and the polymer catalysed cation-radical Diels–Alder reaction.[110]

Two-stage reactions in which a soluble reagent reacts first with one insoluble polymeric reagent and the product with the second polymeric reagent have several advantages over analogous reactions in solution. This is exemplified by the actylation of carbon acids where the simultaneous use of a polymeric strong base and a polymeric acylating reagent.[111] Macroporous polymer catalysts consist of aggregates of gelular microparticles interdispersed with macropores. The sites on the surfaces of the gel microparticles are less active than the internal sites in the transesterification of ethyl acetate catalysed by macroporous sulphonated cross-linked polystyrene. The more highly cross-linked polymers offer greater resistance to diffusion in the microparticles but have more active catalytic sites within the microparticle.[112]

Pendant hydroxamic acid groups in water-soluble microgels catalyse the hydrolysis of esters more effectively than monomeric or linear polymeric hydroxamic acids. This is attributed to an increased nucleophilicity of the hydroxamate ions in a negatively charged microenvironment rather than to an increased accessibility of the microgel architecture as the pH is raised.[113] Chiral spaces in the microgel bead allow discrimination between chiral substrates.[114]

---

[101] B. Sarkar, *Prog. Macrocyclic Chem.*, 1981, **2**, 251.
[102] J. Oku and S. Inove, *J. Chem. Soc., Chem. Commun.*, 1981, 229.
[103] N. Yahiro, K. Asakawa, and K. Tsuboyama, *Kobunshi Ronbunshu*, 1982, **39**, 549.
[104] T. Yamashita, H. Mitsui, H. Watanabe, and N. Nakamura, *Polym. J. (Tōkyō)*, 1981, **13**, 179.
[105] N. Kobayashi and K. Iwai, *Macromol. Chem., Rapid Commun.*, 1981, **2**, 105.
[106] M. Takeishi, T. Watanabe, S. Niino, and S. Hayama, *J. Polym. Sci., Polym. Chem. Ed.*, 1980, **18**, 3081.
[107] R. B. Diemer, jun., K. M. Dooley, B. C. Gates, and R. L. Albright, *J. Catal.*, 1982, **74**, 373.
[108] V. M. Belikov, V. K. Latov, and M. I. Fastovskaya, *J. Mol. Catal.*, 1980, **8**, 443.
[109] C. P. Fei and T. H. Chan, *Synthesis*, 1982, 467.
[110] N. L. Bauld and D. J. Bellville, *Tetrahedron Lett.*, 1982, **23**, 825.
[111] B. J. Cohen, M. A. Kraus, and A. Patchornik, *J. Am. Chem. Soc.*, 1981, **103**, 7620.
[112] K. M. Dooley, J. A. Williams, B. C. Gates, and R. L. Albright, *J. Catal.*, 1982, **74**, 361.
[113] K. A. Stacey, R. H. Weatherhead, and A. Williams, *Makromol. Chem.*, 1980, **181**, 2517; R. H. Weatherhead, K. A. Stacey, and A. Williams, *ibid.*, p. 2529.
[114] M. G. Harun and A. Williams, *Polymer*, 1981, **22**, 946.

Template Polymerization.[115]—By using D-glyceric acid as a template, an amino and a boronic acid group have been introduced in a defined steric arrangement into divinylbenzene-based polymers. Removal of the template gives a polymer capable of resolving D,L-glyceric acid.[116] Some of these polymers bind glyceric acid by a covalent amide bond or an electrostatic interaction.[117] Similarly chiral cavities in polymers have been used to resolve mandelic acid,[118] diols, monoalcohols, and amines.[119]

Polymerization of methacrylate in the presence of template dyes gives only marginal compartmental affinity. The size of the template molecules dictates the size of the compartments, but the role of polymer composition is more fundamental. The presence of carboxyl groups in the polymer increases non-specific binding but also greatly increases the specificity of compartments for the dyes.[120] Polymer complexons to be used as selective sorbents for metal ions have been developed using template synthesis.[121]

Template polymers accelerate the polycondensation of esters with diamines.[122] The use of poly(deoxyribosecytosine) containing 3-methylcytosine as a template for DNA polymerase inhibits DNA chain elongation and does not induce any mispairing under high fidelity conditions.[123] Template-induced molecular assembly could have led to the instructive evolution of proteins before tRNAs existed.[124]

Polymers as Supports and Protecting Groups.[125]—Insoluble polymer supported thiazolium salts are active catalysts for benzoin condensations virtually in any solvent, even those in which conventional thiazolium salts are insoluble.[126] Linear and 1% cross-linked polystyrenes have been mercuriated and thallated and then converted to polymer-supported alkylboranes and alkylboranic acids which can be used as polymer-supported protecting groups for diols.[127]

Polystyrene-bound benzyltri-n-butylphosphonium salts catalyse the reactions of bromoalkanes with aqueous sodium cyanide more effectively as the speed of stirring increases, as catalyst particle size decreases, with increasing swelling power of the solvent and as the amount of cross-linking in the polymer decreases. These observations may be rationalized by mass transfer and intraparticle diffusion limitations on the reaction rates.[128]

---

[115] C. H. Bamford, *Chem. Aust.*, 1982, **49**, 341; S. Shinkai and T. Kunitake, *Kagaku (Kyoto)*, 1981, **36**, 76; T. Kunitake, *Kagaku Sosetsu*, 1982, **35**, 56; M. Maciejewski, *J. Macromol. Sci., Chem.*, 1982, **17**, 689.
[116] A. Sarhan and G. Wulff, *Makromol. Chem.*, 1982, **183**, 85.
[117] A. Sarhan and G. Wulff, *Makromol. Chem.*, 1982, **183**, 1603.
[118] A. Sarhan, *Makromol. Chem., Rapid Commun.*, 1982, **3**, 489.
[119] G. Wulff, *Pure Appl. Chem.*, 1982, **54**, 2093.
[120] R. Arshady and K. Mosbach, *Macromol. Chem. Phys.*, 1981, **182**, 687.
[121] A. A. Efendiev and V. A. Kabanov, *Pure Appl. Chem.*, 1982, **54**, 2077; B. Thulin and F. Voegtle, *J. Chem. Res.*, 1981, 256.
[122] H. Nakagawa, M. Muraki, Y. Miura, and M. Kinoshita, *Makromol. Chem.*, 1982, **183**, 2065.
[123] S. Boiteux and J. Laval, *Biochimie*, 1982, 64.
[124] G. D. Wassermann, *J. Theor. Biol.*, 1982, **96**, 77.
[125] J. M. J. Frecher, *Tetrahedron*, 1981, **37**, 663; S. Mazur, P. Jayalekshmy, J. T. Andersson, and T. Matusinovic, *ACS Symp. Ser.*, 1982, **192**, 43; S. J. Teichner, V. C. Hoang, and M. Astler, *Stud. Surf. Sci. Catal.*, 1982, **11**, 121.
[126] C. Segura and M. Pascual, *Ser. Univ. - Fund. Juan March*, 1980, **140**, 57.
[127] N. P. Bullen, P. Hodge, and F. G. Thorpe, *J. Chem. Soc., Perkin Trans. 1*, 1981, 1863.
[128] M. Tomoi and W. T. Ford, *J. Am. Chem. Soc.*, 1981, **103**, 3821 and 3828.

Polystyrene-bound organo-titanium complexes catalyse the isomerization of alkenes with a different specificity from that when an analogous homogeneous system is used.[129] The cyclopropane derivative (18) is formed from the Michael addition of PhCHClCO$_2$Me to the polystyrene ester of acrylic acid. This presumably arises from stereospecific ring closure (19).[130] Other aspects of selectivity have been reported for polymer-supported nucleophiles[131] and the alkylation of enolates by using polymer-bound hexanethylphosphotriamide.[132]

(18)   (19)

The hydroformylation of alkenes has been studied with polymer-bound ruthenium[133] and rhodium complexes.[134] The use of a chiral platinum catalyst leads to asymmetric hydroformylation.[135] The exploration of Wilkinson-type complexes anchored to solid supports continues but the effects of substrate diffusion into support and swelling of the latter often obscure accurate kinetic studies. The indications are that the mechanism of alkene hydrogenation is essentially the same as in the homogeneous case.[136] Linear polymers highly loaded with rhodium complexes have a higher activity in the catalysed hydrogenation of alkenes compared with the corresponding monomers which have a tendency to dimerize.[137]

The polymer support environment of ruthenium complexes induces some selectivity in alkene hydrogenation which is also affected by the extent of metal loading on the polymer and solvent composition.[138] Polymer-supported ruthenium complexes have also been used for investigating ligand substitution[139] and alkene isomerization.[140] The importance of the primary and secondary structures of the polymers in determining the distribution of metal species and the isomerization and hydrogenation catalytic properties of heterogeneous systems has been demonstrated.[141]

Selective hydrogen transfer from formic acid to α,β-unsaturated ketones may be obtained with polystyrene-anchored complexes of iridium.[142] Asymmetric organic synthesis by using transition metals with optically active phosphine ligands

---

[129] D. E. Bergbreiter and G. L. Parsons, *J. Organomet. Chem.*, 1981, **208**, 47.
[130] I. Artaud and P. Viout, *Tetrahedron Lett.*, 1981, **22**, 1009.
[131] G. Cainelli, M. Contento, F. Manescalchi, and L. Plessi, *J. Chem. Soc., Chem. Commun.*, 1982, 725.
[132] G. Nee, Y. Leroux, and J. Seyden-Penne, *Tetrahedron*, 1981, **37**, 1541; G. Nee and J. Seyden-Penne, *ibid.*, 1982, **38**, 3485.
[133] C. V. Pittman, jun., and G. M. Wilemon, *J. Org. Chem.*, 1981, **46**, 1901.
[134] H. Hirai, S. Komatsuzaki, S. Hamasaki, and N. Toshima, *Nippon Kagaku Kaishi*, 1982, 316.
[135] C. V. Pittman, jun., Y. Kawabata, and L. I. Flowers, *J. Chem. Soc., Chem. Commun.*, 1982, 473.
[136] M. H. J. M. De Croon and J. W. E. Coenen, *J. Mol. Catal.*, 1981, **11**, 301; M. Bartholin and C. Graillat, *ibid.*, 1981, **10**, 361.
[137] A. J. Naaktgeboren, R. J. M. Nolte, and W. Drenth, *J. Mol. Catal.*, 1981, **11**, 343.
[138] C. P. Nicolaides and N. J. Coville, *J. Organomet. Chem.*, 1981, **222**, 285.
[139] S. Torroni, G. Innorta, A. Foffani, A. Modelli, and F. Scagnolari, *J. Organomet. Chem.*, 1981, **221**, 309.
[140] A. Zoran, Y. Sasson, and J. Blum, *J. Org. Chem.*, 1981, **46**, 255.
[141] G. Valentini, G. Sbrana, and G. Braca, *J. Mol. Catal.*, 1981, **11**, 383.
[142] J. Azran, O. Buchman, and J. Blum, *Tetrahedron Lett.*, 1981, **22**, 1925.

attached to polymers is often less effective than with homogeneous catalysts.[143] If the polymer contains optically active pendent alcohols the enantiomeric excess of product depends upon the structure of the alcohol.[144] Polymer-supported metal complexes have been used as catalysts for the oxidation of thiosalts by molecular oxygen,[145] for the oxidative coupling of phenols,[146] and the dehydrogenation of alcohols.[147]

The photolytic production of hydrogen from water is catalysed by colloidal platinium–poly(vinyl alcohol) and an aqueous solution containing tris(2,2'-bipyridine) ruthenium(II) and methyl viologen at pH 5.[148] Excited states within metallopolymers undergo oxidative and reductive electron-transfer quenching. Lifetimes and quantum yield data are related to analogous monomeric metal complexes.[149]

5 Micelles[150]

During 1981 and 1982 there were over 150 papers describing reactions taking place in micelles. The following discussion reviews some of the important studies and is organized by the type of reaction.

The hydrolysis of esters within micelles is the most popular reaction studied, with most emphasis on stereoselectivity. The largest selectivity so far observed by an assembly composed on single molecular species is a factor of 4.4 for the hydrolysis of N-(benzyloxycarbonyl)-L and D-phenylalanine 4-nitrophenyl esters by the vesicular assembly of NN-didodecyl-$N\alpha$-(6-trimethylammoniumhexanoyl) histidinamide bromide which acts as a functionalized membrane.[151] However, a factor of 5.7 has been observed for the hydrolysis of 4-nitrophenyl esters of L- and D-amino acids with optically active hydroxamic acids in the presence of cetyltrimethylammonium bromide micelles.[152]

Ihara's laboratory has been very active in this area and demonstrated that deacylation of optically active 4-nitrophenyl esters of amino-acids is rate limiting when the hydrolysis is catalysed by optically active micellar catalysts containing a

---

[143] G. L. Baker, S. J. Fritschel, J. R. Stille, and J. K. Stille, *J. Org. Chem.*, 1981, **46**, 2954; J. K. Stille, *Proc. China–US Bilateral Symp. Polym. Chem. Phys.*, 1979, 90.
[144] G. L. Baker, S. J. Fritschel, and J. K. Stille, *J. Org. Chem.*, 1981, **46**, 2960.
[145] M. Chanda, K. F. O'Driscoll, and G. L. Rempel, *J. Mol. Catal.*, 1980, **7**, 389; *ibid.*, 1980, **8**, 339; *ibid.*, 1981, **11**, 9.
[146] P. J. Verlaan, J. P. C. Bootsma, and G. Challa, *J. Mol. Catal.*, 1981, **14**, 211.
[147] W. K. Rybak and J. J. Ziolkowski, *J. Mol. Catal.*, 1981, **11**, 365; N. Takamiya, M. Takano, N. Miyata, and H. Shoji, *Nippon Kagaku Kaishi*, 1981, 326.
[148] M. Gratzel and J. Kiwi, Ger. Offen., 3 032 303.
[149] J. M. Calvert, J. V. Caspar, R. A. Rinstead, T. D. Westmoreland, and T. J. Meyer, *J. Am. Chem. Soc.*, 1982, **104**, 6620.
[150] T. Kunitake and S. Shinkai, *Adv. Phys. Org. Chem.*, 1980, **17**, 435; J. H. Fendler, 'Membrane Mimetic Chemistry', 1982; Y. Murakami and J. Sunamoto, 'Enzyme Chemistry: Chemistry of Biomembrane Models', 1981; T. Kunitake, *Polym. Prepr., Am. Chem. Soc., Div. Polym. Chem.*, 1979, **20**, 1079; J. Burgess, *Inorg. React. Mech.*, 1981, **7**, 287; H. Chaimovich, R. M. V. Aleixo, I. M. Cuccovia, D. Zanette, and F. H. Quina, *Solution Behav. Surfactants: Theor. Appl. Aspects* [*Proc. Int. Symp.*], 1980, 949; P. Linda, F. Rubessa, and G. Savelli, *Chim. Ind. (Milan)*, 1981, **63**, 333; D. G. Whitten, J. C. Russell, and R. H. Schmehl, *Tetrahedron*, 1982, **38**, 2455; R. A. Moss and Y. S. Lee, *Stud. Org. Chem.*, 1982, **10**, 200.
[151] Y. Murakami, A. Nakano, A. Yoshimatsu, and K. Fukuya, *J. Am. Chem. Soc.*, 1981, **103**, 728.
[152] S. Ono, H. Shosenji, and K. Yamada, *Tetrahedron Lett.*, 1981, **22**, 2391.

$$Me(CH_2)_8CONH-CH-CO_2H$$
$$|$$
$$CH_2$$

[imidazole ring structure: HN—N]

(20)

histidine residue (20). Stereoselectivity is observed in both acylation and deacylation of the imidazole functionality[153] and the reactivity may be enhanced by the incorporation of a carboxyl group which may act by intramolecular hydrogen bonding.[154]

The diastereoselectivity of the hydrolysis of some peptide esters is lower in thiol functionalized vesicles compared with analogous micelles which is attributed to a greater molecular ordering in vesicular systems.[155] Although nucleophilic catalysed hydrolysis of 4-nitrophenyl esters by imidazole groups is normally observed, comicelles of peptide surfactants bearing both histidyl and aspartyl residues exhibit predominantly general base catalysed hydrolysis. The normal nucleophilic mechanism is suppressed in the tight micelles of peptide surfactants which prevents the close proximity of reactants. The reaction mechanism can thus be changed by changing the aggregation properties of the micellar phase.[156] Similarly a bifunctional surfactant containing imidazole and hydroxyl groups has been compared with the α-chymotrypsin catalysed hydrolysis of esters.[157]

The hydrolysis of 4-nitrophenyl cyanoacetate proceeds by an ElcB mechanism and cationic micelles of cetyltrimethylammonium bromide act as catalysts for the reaction by stabilizing the carbanion intermediate. However, at pH values above the pK_a of the ester the micelle inhibits the spontaneous decomposition of the carbanion.[158] The negative volumes of activation for the hydrolysis of 4-nitrophenyl esters of pentanoic acid within micelles of cetyltrimethylammonium bromide cause the rate to increase with increasing pressure. However, binding of the ester to the micelle is inhibited by increased pressure.[159]

The alkaline hydrolysis of 5,5-dithiobis (2-nitrobenzoic acid) is 15- and 1500-fold faster in cetyltrimethylammonium bromide micelles and dioctadecyldimethylammonium chloride surfactant vesicles, respectively. Rate enhancements are a consequence of increased concentrations of hydroxide and ester in the micelles and surfactant vesicles.[160] The catalytic action of imidazole on the hydrolysis of 4-nitrophenyl acetate in water solubilized in non-ionic surfactants in carbon tetrachloride is enhanced with the decrease in polarity of the interior of the aggregates.[161] Inverted micelles absorbed on platinum catalyse the hydrolysis of

[153] Y. Ihara, Y. Kimura, M. Nango, and N. Kuroki, *Makromol. Chem., Rapid Commun.*, 1982, **3**, 521; Y. Ihara, *J. Chem. Soc., Perkin Trans. 2*, 1980, 1483; Y. Ihara, N. Kunikiyo, T. Kunimasa, M. Nango, and N. Kuroki, *Chem. Lett.*, 1981, 667; Y. Ihara and R. Hosako, *Bull. Chem. Soc. Jpn.*, 1982, **55**, 1979.
[154] Y. Ihara, R. Hosako, M. Nango, and N. Kuroki, *J. Chem. Soc., Chem. Commun.*, 1981, 393.
[155] R. A. Moss, T. Taguchi, and G. O. Bizzigotti, *Tetrahedron Lett.*, 1982, **23**, 1985.
[156] Y. Murakami, A. Nakano, A. Yoshimatsu, and K. Matsumoto, *J. Am. Chem. Soc.*, 1981, **103**, 2750.
[157] L. Anoardi, R. Fornasier, and V. Tonellato, *J. Chem. Soc., Perkin Trans. 2*, 1981, 260.
[158] H. Al-Lohedan and C. A. Bunton, *J. Org. Chem.*, 1981, **46**, 3929.
[159] Y. Taniguchi, S. Makimoto, and K. Suzuki, *J. Phys. Chem.*, 1981, **85**, 2218.
[160] J. H. Fendler and W. L. Hinze, *J. Am. Chem. Soc.*, 1981, **103**, 5439.
[161] K. Kon-No, T. Inove, T. Hanada, K. Nakamura, and A. Kitahara, *Yukagaku*, 1980, **29**, 670.

ethyl benzoate to a degree dependent on the surface area of the platinum, the number of layers of surfactant and the potential on the platinum.[162]

The stereoselective deacylation of long-chain amino-acid esters by chiral micelles containing histidine residues is enhanced by metal ions. This is attributed to either the metal-bound ester being incorporated into the micelle or to the co-ordination of the ester to the metal-bound micelle. Stereoselectivity increases with the hydrophobic nature of the ester.[163] Cationic surfactant vesicles accelerate the rate of thiolysis of 4-nitrophenyl octanoate by n-heptyl mercaptan several million-fold in the pH range 4—6, i.e., up to 6 pH units below the pK_a of the thiol. However, this is attributed to predominantly a concentration effect rather than enhanced dissociation and reactivity of the nucleophile at the vesicle surface.[164]

It is suggested that the aminolysis of esters in alkylammonium carboxylate reversed micelles proceeds by general base catalysis by the carboxylate groups. Support for this mechanism is that there is no evidence for nucleophilic catalysis. However, the authors also showed that the rate-limiting step involved expulsion of phenoxide from the tetrahedral intermediate which is inconsistent with general base catalysis.[165]

Enantioselectivity in the aminolysis of optically active esters in chiral reversed micelle forming surfactants is due to differences in the binding constants of the esters to the aggregates.[166] Cationic micelles inhibit the spontaneous hydrolysis of anhydrides and the degree of inhibition increases with hydrophobicity of the substrate.[167] However, the alkaline hydrolysis is enhanced and the data fit a model in which the distribution of hydroxide between the aqueous and micellar pseudophases depends on the concentration of hydroxide.[168] The hydrolyses of carbonate esters and alkyl halides have been used to support a porous cluster model for micelles in which water-filled regions bind guests hydrophobically.[169]

The hydrolysis of activated amides catalysed by micelles and comicelles of functional surfactants containing hydroxyl and/or imidazole groups show different kinetic effects from the hydrolysis of esters. This is attributed to a change in the rate-limiting step.[170,171]

The alkaline hydrolysis of the β-lactam of benzylpenicillin is increased ca. 50-fold in the presence of micelles of cetyltrimethylammonium bromide. The rate of reaction is inhibited by increasing concentrations of hydroxide ion and the penicillin anion. It is assumed that both reactants have to be bound to the micelle for reaction to occur.[172] The binding of alkylpenicillins (21) to the micelle increases with

---

[162] T. C. Franklin and M. Iwunze, *J. Am. Chem. Soc.*, 1981, **103**, 5937.
[163] K. Ohkubo and N. Matsumoto, *J. Mol. Catal.*, 1981, **12**, 393.
[164] I. M. Cuccovia, F. H. Quina, and H. Chaimovich, *Tetrahedron*, 1982, **38**, 917; K. Ohkubo, K. Kawazoe, M. Toyoda, and R. Ueoka, *J. Mol. Catal.*, 1980, **9**, 219.
[165] M. I. El-Seoud, R. C. Vieira, and O. A. El-Seoud, *J. Org. Chem.*, 1982, **47**, 5137; M. I. El-Seoud and O. A. El-Seoud, *J. Org. Chem.*, 1981, **46**, 2686.
[166] K. Konno, M. Tosaka, Y. Saratani, and A. Kitahara, *Nippon Kagaku Kaishi*, 1982, 543; K. Konno, M. Tosaka, and A. Kitahara, *J. Colloid Interface Sci.*, 1981, **79**, 581.
[167] H. Al-Lohedan, C. A. Bunton, and M. M. Mhala, *J. Am. Chem. Soc.*, 1982, **104**, 6654.
[168] H. Al-Lohedan and C. A. Bunton, *J. Org. Chem.*, 1982, **47**, 1160.
[169] F. M. Menger, Y. Yoshinaga, K. S. Vekatasubban, and A. R. Das, *J. Org. Chem.*, 1981, **46**, 415.
[170] R. Fornasier and V. Tonellato, *J. Chem. Soc., Perkin Trans. 2*, 1982, 899.
[171] T. J. Broxton, D. R. Fernando, and J. R. Rowe, *J. Org. Chem.*, 1981, **46**, 3522.
[172] N. P. Gensmantel and M. I. Page, *J. Chem. Soc., Perkin Trans. 2*, 1982, 147.

(21) structure: RCONH group attached to β-lactam fused with thiazolidine ring bearing CO_2^-

increasing chain length and reaches a maximum value with heptylpenicillin. The free energy of transfer of a methylene group from water to the micelle shows a maximum value of 0.7 kcal mol^{-1}.[173] Surfactants have no effect upon the neutral hydrolysis of the antibiotics but anionic micelles enhance the acidic hydrolysis.[174]

Anionic and cationic micelles inhibit the water-catalysed hydrolysis of benzoyl triazoles.[175] It has been estimated that the rate of the acid catalysed hydrolysis of N-(trifluoroacetyl) indole is ca. 30-fold smaller in anionic micelles than in bulk water.[176] There have been several studies of the hydrolysis of phosphate esters in micelles. Comicelles of cetyltrimethylammonium bromide and surfactants containing oximate and hydroxamate ions dephosphorylate 4-nitrophenyl diphenyl phosphate with second-order rate constants which are very similar to those in water.[177] The rate of dephosphorylation by benzimidazole is increased over 1000-fold by cetyltrimethylammonium bromide micelles.[178]

As found for the hydrolysis of anhydrides, the pseudophase ion-exchange model of micellar catalysis is not suitable for the alkaline hydrolysis of phosphate esters. The charge on the micelle decreases, i.e., the concentration of hydroxide ion in the micelle increases with increasing addition of hydroxide.[179] The hydrolysis of dextrin in reversed micelles in benzene is about 300-fold faster than in aqueous solution. Unlike sucrose hydrolysis the first-order rate constant decreases with increasing dextrin concentration.[180] The decarboxylation of 6-nitrobenzisoxazole-3-carboxylate is very sensitive to solvent effects and the reaction on micelles of 1-methyl-4-dodecylpyridinium iodide is increased ca. 160-fold compared with that in water due to the reduced micropolarity in the Stern layer.[181]

Cationic micelles of cetyltrimethylammonium bromide inhibit acid catalysed transimination reactions; the degree of inhibition increases with substrate hydrophobicity.[182] The transamination of pyridoxal 5′-phosphate with N-dodecyl-L-alaninamide is enhanced 230-fold in imidazole functionalized single wall vesicles. The isomerization of the aldimine Schiff-base to the corresponding ketimine is rate-limiting.[183] The rate of hydration of 1,3-dichloroacetone and dehydration of the hydrate is enhanced in reversed micelles of the non-ionic surfactant Triton X-100 in

---

[173] N. P. Gensmantel and M. I. Page, J. Chem. Soc., Perkin Trans. 2, 1982, 155.
[174] A. Tsuji, E. Miyamoto, M. Matsuda, K. Nishimura, and T. Yamana, J. Pharm. Sci., 1982, 71, 1313.
[175] N. Fadnavis and J. B. F. N. Engberts, J. Org. Chem., 1982, 47, 152.
[176] A. Cipiciani, P. Linda, G. Savelli, and C. A. Bunton, J. Org. Chem., 1981, 46, 911.
[177] C. A. Bunton, F. H. Hamed, and L. S. Romsted, J. Phys. Chem., 1982, 86, 2103; C. A. Bunton, S. E. Nelson, and C. Quan, J. Org. Chem., 1982, 47, 1157.
[178] C. A. Bunton, Y. S. Hong, L. S. Romsted, and C. Quan, J. Am. Chem. Soc., 1981, 103, 5784.
[179] C. A. Bunton, L. H. Gan, J. R. Moffatt, L. S. Romsted, and G. Savelli, J. Phys. Chem., 1981, 85, 4118.
[180] K. Arai and Y. Ogiwara, Bull. Chem. Soc. Jpn., 1982, 55, 838.
[181] L. A. M. Rupert and J. B. F. N. Engberts, J. Org. Chem., 1982, 47, 5015.
[182] I. A. K. Reddy and S. S. Katiyar, Tetrahedron, 1981, 37, 585, 655.
[183] Y. Murakamiki, A. Nakano, and K. Akiyoshi, Bull. Chem. Soc. Jpn., 1982, 55, 3004.

carbon tetrachloride. It is suggested that the surfactant acts as a general base catalyst for water attack.[184]

The micellar catalysed proton abstraction from carbon acids has been reported for acetone[185] and tetranitromethane.[186] There have been several reports of the effect of micelles on nucleophilic substitution reactions. The reactions of stabilized carbonium ions with anions are enhanced by cationic micelles of cetyltrimethylammonium bromide.[187] Micellar dediazonization is independent of bromide concentration and, although the rate is similar to the non-micellar reaction, which gives carbonium ions trapped by solvent, the micellar reaction gives products resulting from bromide ion reacting with the cation intermediate.[188] Nucleophilic aromatic substitution in microemulsions is insensitive to the composition of the pseudo-phase and proceeds at similar rates to those in micelles.[189] Although the reactivity of nucleophiles in deacylation and aliphatic substitution reactions is usually similar in micellar and aqueous phases, the second-order rate constant for aromatic nucleophilic substitution with azide ion is much larger in the micellar pseudo-phase than in water.[190] Nitrite ions are also effective nucleophiles in S_NAr reactions catalysed by micelles of cetyltrimethylammonium bromide.[191]

Reductive desulphonation of 2,4,6-trinitrobenzene sulphonate by N-dodecyl-1-benzyldihydronicotinamide bound to sodium dodecyl sulphate micelle in D_2O gives 1,3,5-trinitrobenzene containing 5% deuterium. The micelle surface presumably is capable of dissociating the radical ion-pair.[192] The formation of cationic radicals in anionic micelles is favoured but anionic radicals are also stabilized by the alkali-metal counterion.[193] The rate of reaction of N-methylphenothiazine cationic radicals with anions is enhanced in the presence of cationic micelles.[194] The oxidation of unsaturated fatty acid micelles by the superoxide free-radical anion may occur by a chain oxidation process. Tetranitromethane, which reacts rapidly with O_2^-, protects the micelle from oxidation.[195]

During 1981 and 1982 there have been nearly 50 reports of photochemical reactions in micellar and related systems.[196] Although the same products are formed in the photo-oxidation of protoporphyrin IX derivatives in neutral and charged micelles and vesicles as in homogeneous organic solvents, the product distributions are quite different. Electron transfer from excited porphyrin to generate superoxide is exclusively intramicellar.[197] Singlet oxygen is implicated as the primary oxidizing agent in the photo-oxidation of sulphides formed by the intramicellar recombina-

[184] O. A. El-Seoud and G. J. Vidotti, *J. Org. Chem.*, 1982, **47**, 3984.
[185] M. Maruthamuthu and N. Lakshmikanthan, *Int. J. Chem. Kinet.*, 1981, **13**, 695.
[186] T. Harada, N. Nishikido, Y. Moroi, and R. Matuura, *Bull. Chem. Soc. Jpn.*, 1981, **54**, 2592.
[187] S. K. Srivastava and S. S. Katiyar, *Int. J. Chem. Kinet.*, 1982, **14**, 1007; S. S. Katiyar and K. L. Patel, *Indian J. Chem., Sect. A*, 1981, **20**, 1065.
[188] R. A. Moss, F. M. Dix, and R. Romsted, *J. Am. Chem. Soc.*, 1982, **104**, 5048.
[189] V. Athanassakis, C. A. Bunton, and F. de Buzzaccarini, *J. Phys. Chem.*, 1982, **86**, 5002; C. A. Bunton and F. de Buzzaccarini, *ibid.*, 1981, **85**, 3142, and 1982, **86**, 5010.
[190] C. A. Bunton, J. R. Moffat, and E. Rodenas, *J. Am. Chem. Soc.*, 1982, **104**, 2653.
[191] T. J. Broxton, *Aust. J. Chem.*, 1981, **34**, 969.
[192] S. Sinkai, T. Suno, and O. Manabe, *J. Chem. Soc., Chem. Commun.*, 1982, 592.
[193] L. McIntire and H. N. Blount, *Solution Behav. Surfactants: Theor. Appl. Aspects [Proc. Int. Symp.]*, 1980, **2**, 1101.
[194] D. Lardet, E. Laurent, M. Thomalla, and M. Genies, *Nouv. J. Chim.*, 1982, **6**, 349.
[195] J. M. Gebicki, *Arch. Biochem. Biophys.*, 1982, **214**, 1.
[196] D. G. Whitten, J. C. Russell, and R. H. Schmehl, *Tetrahedron*, 1982, **38**, 2455.
[197] S. G. Cox, M. Kreig, and D. G. Whitten, *J. Am. Chem. Soc.*, 1982, **104**, 6930.

tion of the transient 10-methylphenothiazine cation radical–superoxide anion radical ion pair.[198]

Irreversibly reduced cytochrome-C is observed when the oxidized form is photoreduced in reversed micelles.[199] Photohydrolysis of substances poorly soluble in water may be performed very efficiently in aqueous micellar solution.[200] Micellar aggregates are insufficiently organized to influence the radiation-induced polymerization of 3-n-dodecyl-1-vinylimidazolium iodide.[201]

Micelle catalysed reactions can be controlled by light. For example, photoresponsive surfactants containing the azo group form micelles which have different catalytic activity in the light and dark. This is attributed to the partitioning of the *trans* and *cis* isomers in the micellar phase.[202]

Different potentials, created by the high charge densities on the surface of synthetic surfactant vesicles may be exploited for enhanced energy and electron transfer. The application to photochemical solar energy conversion has been discussed.[203] Single compartment vesicles formed with a double-chain amphiphile greatly enhance the stability of charge-transfer complexes, presumably due to the higher structured environment provided.[204]

Anionic micelles enhance the rate of reaction between aqueous nickel(II) ions and bidentate ligands.[205] The association constants for several metal ions to anionic micelles are not related to metal ion charge.[206]

Water soluble enzymes may be stabilized against the inactivating action of organic solvents by incorporating them in reversed micelles.[207] For example, α-chymotrypsin binds anilide substrates more tightly in the water pools compared with bulk water, but the pH-rate profile is shifted to higher pH. This is probably due to a modification of the pK_a of the enzyme's active groups. It is suggested that the enzyme has a more rigid conformation in the reversed micelle.[208]

The so-called hexapus (22) has six hydrophobic chains covalently linked which can associate to form an adjustable void capable of incorporating small organic molecules. It is not surface active but does aggregate in water with an immeasurably low critical micelle concentration.[209] There have been several theoretical studies of diffusion and reactions in micelles.[210] Studies on the porosity of micelles continues.

[198] M. C. Hovey, *J. Am. Chem. Soc.*, 1982, **104**, 4196.
[199] M. P. Pileni, *Chem. Phys. Lett.*, 1981, **81**, 603.
[200] T. Wolff, *J. Photochem.*, 1982, **18**, 285.
[201] C. M. Paleos, S. Voliotis, G. Margomenov-Leonidopoulou, and P. Dais, *J. Polym. Sci., Polym. Chem. Ed.*, 1980, **18**, 3463.
[202] S. Shinkai, K. Matsuo, M. Sato, T. Sone, and O. Manabe, *Tetrahedron Lett.*, 1981, **22**, 1409; S. Shinkai, K. Matsuo, A. Aarada, and O. Manabe, *J. Chem. Soc., Perkin Trans.*, 1982, 1261.
[203] M. S. Tunuli and J. H. Fendler, *ACS Symp. Ser.*, 1982, **177**, 53; J. H. Fendler, *J. Photochem.*, 1981, **17**, 303; M. S. Tunuli and J. H. Fendler, *J. Am. Chem. Soc.*, 1981, **103**, 2507.
[204] Y. Marakami, Y. Aoyama, J. Kikuchi, K. Nishida, and A. Nakano, *J. Am. Chem. Soc.*, 1982, **104**, 2937.
[205] J. R. Hicks and V. C. Reinsborough, *Aust. J. Chem.*, 1982, **35**, 15; P. D. I. Fletcher and V. C. Reinsborough, *Can. J. Chem.*, 1981, **59**, 1361.
[206] H. Ziemiecki and W. R. Cherry, *J. Am. Chem. Soc.*, 1981, **103**, 4479.
[207] K. Martinek, A. V. Levashov, N. L. Klyachko, V. I. Pantin, and I. V. Berezin, *Biochim. Biophys. Acta*, 1981, **657**, 277; F. J. Bonner, R. L. Wolff, and L. Pier, *J. Solid-Phase Biochem.*, 1980, **5**, 255.
[208] S. Barbaric and P. L. Luisi, *J. Am. Chem. Soc.*, 1981, **103**, 4239.
[209] F. M. Menger, M. Takeshita, and J. F. Chow, *J. Am. Chem. Soc.*, 1981, **103**, 5938.
[210] H. Sano and M. Tachiya, *J. Chem. Phys.*, 1981, **75**, 2870; G. E. Amidon, W. I. Higuchi, and N. F. H. Ho, *J. Pharm. Sci.*, 1982, **71**, 77; K. L. Patel and S. S. Katiyar, *Indian J. Chem., Sect. A*, 1981, **20**, 788; M. D. Hatlee and J. J. Kozak, *J. Chem. Phys.*, 1981, **74**, 1098; M. D. Hatlee, J. J. Kozak, and M. Graetzel, *Ber. Bunsenges. Phys. Chem.*, 1982, **86**, 157.

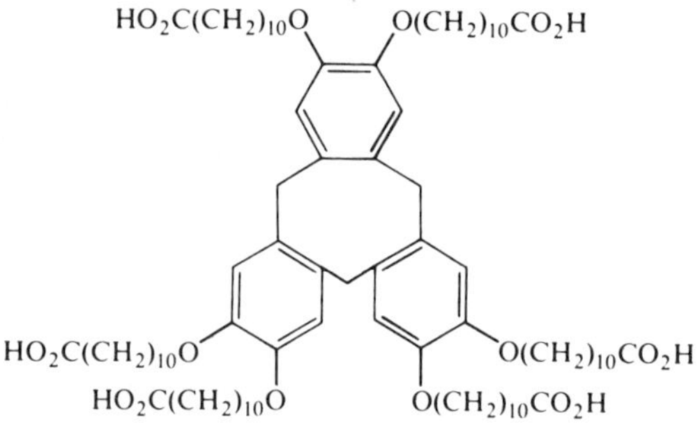

$HO_2C(CH_2)_{10}O$ $O(CH_2)_{10}CO_2H$

$HO_2C(CH_2)_{10}O$ $O(CH_2)_{10}CO_2H$
$HO_2C(CH_2)_{10}O$ $O(CH_2)_{10}CO_2H$

(22)

Despite recent claims to the contrary, the evidence seems to support the suggestion that micelles possess water-filled pores and 'fatty patches'.[211] Water molecules at the micelle surface reorient anisotropically and two to three times slower than in bulk water but remain there for only 6—37 ns.[212]

Polymerized surfactant vesicles are considerably more stable and less permeable and have reduced rates of turbidity compared with their unpolymerized counterparts.[213]

6 Phase-transfer Catalysis[214]

The first examples of the free-radical polymerization of vinyl monomers under phase transfer conditions have been reported using solid $K_2S_2O_8$ in the presence of crown ethers as the initiator system. A correlation is found between the rate of polymerization and the complexing ability of the crown ether for potassium ions.[215]

The first example of a mononuclear Co carbonyl complex in a phase-transfer catalysed process has been described. The carbonylation reactions give different results from single-phase reactions.[216] The negatively charged phase-transfer catalyst $Na^+[3,5-(CF_3)_2C_6H_3]_4B^-$ produces a large rate acceleration in the coupling of diazonium salts.[217] The conversion of halides and sulphonate esters to alkanes by using sodium borohydride in a two-phase system with phosphonium or ammonium salts as phase-transfer catalysts may be executed without decomposition of the catalyst.[218]

The dephosphonylation of 4-nitrophenyl diphenylphosphate by benzimidate

---

[211] F. M. Menger and J. M. Bonicamp, *J. Am. Chem. Soc.*, 1981, **103**, 2140; W. Reed, M. J. Politi, and J. H. Fendler, *J. Am. Chem. Soc.*, 1981, **103**, 4591.
[212] B. Halle and C. Goeran, *J. Phys. Chem.*, 1981, **85**, 2142.
[213] P. Tundo, D. J. Kippenberger, P. K. Klahn, N. E. Prieto, T. C. Jao, and J. H. Fendler, *J. Am. Chem. Soc.*, 1982, **104**, 456; P. Tundo, D. J. Kippenberger, M. J. Politi, P. Klahn, and J. H. Fendler, *ibid.*, p. 5352.
[214] E. V. Dehmlow and S. S. Dehmlow, Monographs in Modern Chemistry, Vol. 11: 'Phase Transfer Catalysis', Verlag Chemie, 1980; F. Montanari, D. Landini, and F. Rolla, *Top. Curr. Chem.*, 1982, **101**, 147; H. Alper, *Adv. Organomet. Chem.*, 1981, **19**, 183; R. Gallo, H. J. M. Dou, and P. Hassanaly, *Bull. Soc. Chim. Belg.*, 1981, **90**, 849; T. Ando and J. Yamawaki, *Yuki Gosei Kagaku Kyokaishi*, 1981, **39**, 14.
[215] J. K. Rasmussen and H. K. Smith, *J. Am. Chem. Soc.*, 1981, **103**, 730.
[216] S. Gambarott and H. Alper, *J. Organomet. Chem.*, 1981, **212**, C23.
[217] H. Kobayashi, T. Sonoda, H. Iwamoto, and M. Yoshimura, *Chem. Lett.*, 1981, 579.
[218] F. Rolla, *J. Org. Chem.*, 1981, **46**, 3909.

$nC_{18}H_{37}-\overset{+}{N}\diagup\diagdown\overset{+}{N}-nC_{18}H_{37}$

(23)

anion in the presence of tri-n-octylethylammonium mesylate proceeds by the incorporation of both reactants within aggregates of the phase-transfer agent.[219] Quaternary ammonium salts control anomer formation in glycosylation reactions.[220] Nucleotide phosphates can be synthesized in a hydrophobic medium by using the rigid diammonium salt (23) which has a high specificity for the pyrophosphate residue.[221] Carboxylate ester and ethers have been synthesized by gas–liquid phase-transfer catalysis with quaternary phosphonium salts.[222]

Interest in the use of chiral catalysts in phase-transfer reactions in order to achieve asymmetric synthesis continues. Quaternary ammonium salts whose structures demand that the anion be associated with a specific face of the tetrahedral nitrogen have been synthesized[223] and the errors which must be avoided in evaluating chiral phase-transfer catalysts described.[224]

Chiral ion-pair intermediates formed between the achiral carbanion and the chiral ammonium cation are thought to be responsible for enantiomeric excesses produced from the reaction of carbon acids and alkyl halides in the presence of ephedrine or alkaloid derivatives.[225] The same laboratory has described asymmetric epoxidation using poly(amino acids) in a triphase system.[226]

There have been many reports of the use of phase-transfer catalysts supported on insoluble polymers or inorganic carriers such as alumina, silica gel, or molecular sieves. Catalysts based on a cross-linked polystyrene carrier of polar and dipolar functional groups of the type of quaternary ammonium and phosphonium salts and thioester, sulphone, and sulphoxide groups are active in the three-phase (liquid–solid liquid) reaction of potassium cyanide with bromobutane. There is a difference in catalytic activity of functional groups which are pendant on the polymer chain and those which are part of the network.[227]

Polymer-bound polyethers have advantages over polymer-bound quaternary anion salts because their activity is higher and they are more stable. However, loss of mechanical properties due to grinding of the polymer matrix remains a problem.[228] Increasing the cross-linking of polystyrene resins with phosphonium pendant groups decreases the efficiency of catalysis as does increasing the amount of ring substitution.[229] Poly(ethylene glycol)s can be precipitated from solvents by the

[219] C. A. Bunton, Y. S. Hong, L. S. Romsted, and C. Quan, *J. Am. Chem. Soc.*, 1981, **103**, 5788.
[220] F. Seela and H. D. Winkeler, *Liebigs' Ann. Chem.*, 1982, 1634.
[221] I. Tabushi and J. Imuta, *Tetrahedron Lett.*, 1982, **23**, 5415.
[222] E. Angeletti, P. Tundo, P. Venturello, *J. Chem. Soc., Perkin Trans. 1*, 1982, **993**, 1137.
[223] J. M. McIntosh, *J. Org. Chem.*, 1982, **47**, 3777.
[224] E. V. Dehmlow, P. Singh, and J. Heider, *J. Chem. Res.*, 1981, 292.
[225] S. Julia, A. Ginebreda, J. Guixer, and A. Tomas, *Tetrahedron Lett.*, 1980, **21**, 3709.
[226] S. Julia, J. Guixer, J. Masana, J. Rocas, S. Colonna, R. Annuziata, and H. Molinari, *J. Chem. Soc., Perkin Trans. 1*, 1982, 1317.
[227] V. Janout and P. Cefelin, *Collect. Czech. Chem. Commun.*, 1982, **47**, 1818.
[228] F. Montanari and P. Tundo, *J. Org. Chem.*, 1981, **46**, 2125.
[229] S. L. Regen, D. Bolikal, and C. Barcelon, *J. Org. Chem.*, 1981, **46**, 2511.

addition of ether which offers an alternative method to immobilizing the phase-transfer catalyst on insoluble polymer backbones.[230]

Phosphonium salts immobilized on silica gel are both micellar and phase-transfer catalysts, being able to exchange anions and catalyse nucleophilic substitution and decarboxylation reactions. Catalysis is not controlled by diffusion but by the regeneration of catalytic centres. The reactions proceed rapidly even in the absence of stirring and are facilitated by a long alkyl chain between the catalytic centre and the matrix.[231] Similar observations have been made by immobilization on alumina.[232] Cross-linked polyamides can act as solid-phase cosolvents under biphasic and triphasic conditions.[233]

7 Ionophores

Ionophores are hydrophobic molecules which enhance the passive transport of ions across cell and organelle membranes. They are divided into two classes: mobile ion carriers and channel formers. They shield the charge of the transported ion so that it can penetrate the hydrophobic interior of the lipid bilayer. As ionophores are not coupled to energy sources they only permit net movement of ions down their electrochemical gradients.[234] Synthetic ionophores are of interest for aiding the study of transport mechanisms.[235]

A cylindrical ionophore in which two diaza-crown ethers are linked by two photoresponsive azobenzene pillars changes its binding ability for poly(methylenediammonium) ions in response to photo-irradiation.[236]

Selective ionophores for lithium ions as components in solvent–polymeric membrane electrodes have increased activity if lipophilic anions are incorporated into the membrane. Bulky spherical polarizable monovalent anions enable the cation to interact more effectively with the ligand binding sites generating a more selective and specific system.[237]

Several types of crown ethers mediate the passive and active transport of amino-acid and peptide anions coupled with potassium ion transport. Their transport efficiencies, selectivities and directions are controlled by the nature of the crown ether used and the cations present.[238]

---

[230] J. M. Harris, N. H. Hundley, T. G. Shannon, and E. C. Struck, *J. Org. Chem.*, 1982, **47**, 4789.
[231] P. Tundo and P. Venturello, *Tetrahedron Lett.*, 1980, **21**, 2581; P. Tundo and P. Venturello, *J. Am. Chem. Soc.*, 1981, **103**, 856.
[232] P. Tundo, P. Venturello, and E. Angeletti, *J. Am. Chem. Soc.*, 1982, **104**, 6551.
[233] S. L. Regen, A. Mehrotra, and A. Singh, *J. Org. Chem.*, 1981, **46**, 2182.
[234] G. R. Painter and B. C. Pressman, *Top. Curr. Chem.*, 1982, **101**, 83; A. E. Shamoo and T. R. Herrmann, *Usp. Sovrem. Biol.*, 1981, **91**, 350; G. Szabo, *Fed. Proc., Fed. Am. Soc. Exp. Biol.*, 1981, **40**, 2196; Y. A. Ovchinnikov, *Int. J. Quantum Chem.*, 1981, **20**, 461.
[235] J. F. Stoddart, *Prog. Macrocyclic Chem.*, 1981, **2**, 173; T. R. Herrman and A. E. Shamoo, *Membr. Transp.*, 1982, **1**, 579; R. Hilgenfeld and W. Saenger, *Top. Curr. Chem.*, 1982, **101**, 1.
[236] S. Shinkai, Y. Honda, Y. Kusano, and O. Manabe, *J. Chem. Soc., Chem. Commun.*, 1982, 848.
[237] U. Olsher, *J. Am. Chem. Soc.*, 1982, **104**, 4006.
[238] H. Tsukube, *J. Chem. Soc., Perkin Trans. 1*, 1982, 2359.

17
Biomedical Applications of Polymers

BY B. J. TIGHE

1 Introduction

This review continues directly from those in 'Macromolecular Chemistry' Volumes 1 and 2. The aim in all three reviews has been to provide a balance between synthetic work and the results of clinical applications. Having established a basis of existing and novel applications in the earlier reviews, more attention has been given in this account to the interesting long-term results that are now appearing. Selectivity has been essential in the face of the large and ever-growing volume of published work in this area. To this end dental materials, which are regularly and expertly reviewed elsewhere, have been omitted except where they impinge on other topics (*e.g.* see ref. 55). An attempt has been made, however, to indicate areas where increased interest is reflected in a growth in literature activity.

2 Biocompatibility Studies

Biocompatibility continues to be a dominant issue. Andrade,[1] Bruck,[2] and Leininger,[3] who are amongst the best-known workers in the field, have expressed their individual views in reviews and discussions. One aspect of the subject that presents many problems is the comparison of biomaterials in contact with blood, and the assessment of their relative compatibility. An *ex-vivo* couvette flow device has been reported which enables the direct comparison of synthetic materials with endothelial tissue in terms of their relative thrombogenicity.[4] Similarly, dynamic techniques have been applied to the direct *in-vivo* study of interactions between flowing blood and artificial surfaces.[5] Blood flow rates represent one variable whose effect on protein and platelet deposition at polymer surfaces has been studied.[6,7] The effects are broadly small but obviously significant. It is more conventional to attempt to design *in-vitro* test techniques that mimic flow rates encountered in

---

[1] R. M. Lindsay, R. G. Mason, S. W. Kim, J. D. Andrade, and R. M. Hawkins, *Trans. Am. Soc. Artif. Intern. Organs*, 1980, **26**, 603.
[2] S. D. Bruck, *Biomaterials*, 1982, **3**, 12.
[3] T. I. Leininger, *Org. Coatings Plast. Chem.*, 1980, **42**, 73.
[4] S. M. Lindenauer, J. S. Schultz, and J. A. Penner, *Trans. Am. Soc. Artif. Intern. Organs*, 1981, **27**, 231.
[5] B. Basse-Cathalinat, C. Baquey, Y. Llabador, and A. Fleury, *Int. J. Appl. Radiat. Isot.*, 1980, **31**, 747.
[6] R. C. Eberhart, M. E. Lynch, F. H. Bilge, and J. A. Arts, *Trans. Am. Soc. Artif. Intern. Organs*, 1980, **26**, 185.
[7] J. T. Christenson, J. Megerman, K. C. Hanel, G. J. L. Ralien, H. W. Strauss, and W. M. Abbott, *Trans. Am. Soc. Artif. Intern. Organs*, 1981, **27**, 188.

normal prosthetic applications. One such facility is used for the evaluation of prosthetic aortic valves. These are sutured into flexible anatomically shaped aortic roots made from silicone rubber in a test rig allowing pulsatile flow.[8]

Specific studies of platelet interaction with polymer surfaces represent an important aspect of this area and a relatively simple *ex-vivo* method has been proposed as a suitable basis for platelet–surface interactions.[9] In this connection a range of polyamides differing in density of hydrogen bonding, hydrophilic character, and functional group content have been synthesized.[10] Those with anionically charged groups adsorbed fewer platelets than those with undissociated carboxylate groups, whereas introduction of piperazine units caused an increase in platelet adhesion. An alternative synthetic approach, employing block copolymers of styrene and 2-hydroxyethyl methacrylate, has been used to examine the effect of hydrophilic and hydrophobic domains on the mode of interaction between blood platelets and polymers.[11] An attempt to couple effects of hydrophilicity and surface rugosity in this respect has also been made with[12] various commercial polymers ranging from Teflon to polyacetal. Under static conditions, surface roughness did not affect platelet adhesion to any of the polymers, whereas under laminar flow an effect could be demonstrated.

The importance of hydrophilic–hydrophobic balance of surfaces in relation to the adsorption of plasma proteins has been recognized for some time, but the diversity of results reported indicates the complexity of this phenomenon. Some attempt to overcome this problem has been made by choosing carefully controlled structures and a range of characterization techniques. Simple copolymers of 2-hydroxyethyl methacrylate and ethyl methacrylate[13] have been used in this way in combined studies of cell adhesion[14] and protein adsorption.[15,16] Among the more difficult parameters to measure is water structuring at the polymer surface, and although attempts have been made to assess this by ESCA[17] much more work remains to be done. Other interesting studies include those with hydrophilic water-soluble polymers either in homogeneous systems[18] or grafted on to polyethylene substances.[19] In the latter case it was reported that adsorption of both bovine serum albumin and fibrinogen decreased as the overall hydrophilicity of the polymer increased. The generally accepted requirement (for enhanced compatibility) of high albumin and low fibrinogen adsorption appears to have been achieved with several clinically significant polymers by alkyl substitution.[20] The technique, which involves

[8] T. R. P. Martin, W. B. Tindale, R. Van Noort, and M. M. Black, *Trans. Am. Soc. Artif. Intern. Organs*, 1981, **27**, 475.
[9] F. Mantorani, W. Marconi, L. Caprino, G. Goglia, and G. Togna, *Int. J. Artif. Organs*, 1980, **3**, 305.
[10] J. L. Brash and S. Uniyal, *Artif. Organs*, 1981, **5** (July), 9.
[11] T. Okano, S. Nishiyama, I. Shinohara, T. Akaiki, Y. Sakurai, K. Kataoka, and T. Tsuruta, *J. Biomed. Mater. Res.*, 1981, **3**, 393.
[12] Q. Zingg, A. W. Neuman, A. B. Strong, O. S. Hum, and D. R. Absdom, *Biomaterials*, 1981, **3**, 156.
[13] B. D. Ratner and A. S. Hoffman, *Org. Coatings Plast. Chem.*, 1979, **40**, 714.
[14] J. J. Rosen and M. B. Schway, *Org. Coatings Plast. Chem.*, 1979, **40**, 636.
[15] T. A. Horbett, *Org. Coatings Plast. Chem.*, 1979, **40**, 642.
[16] T. A. Horbett and P. K. Weathersby, *J. Biomed. Mater. Res.*, 1981, **3**, 403.
[17] D. K. Gilding, R. W. Paynter, and J. E. Castle, *Biomaterials*, 1980, **1**, 163.
[18] H. Ohno, K. Abe, and E. Tsuchida, *Makromol. Chem.*, 1981, **4**, 1253.
[19] Y. Ikada, H. Iwata, F. Horii, T. Matsunaga, M. Taniguchi, M. Suzuki, W. Taki, S. Yamagato, T. Yonekawa, and H. Handa, *J. Biomed. Mater. Res.*, 1981, **5**, 697.
[20] M. S. Munro, A. J. Quattrone, S. R. Ellsworth, P. Kulkarni, and R. C. Eberhart, *Trans. Am. Soc. Artif. Intern. Organs*, 1981, **27**, 499.

successive treatment of the polymer surface with sodium ethoxide and an alkyl halide, is claimed to have been successfully employed with polyamides, polyurethanes, and polyesters.

Despite the fact that protein adsorption is known to have an important part to play, the determinants of thrombus formation at artificial surfaces are still open to some debate and widely discussed.[21,22] In particular, attempts are made to relate surface free energy, ultrastructural morphology, surface charge, and surface molecular motion to thrombogenesis.[23—25] The question of rugosity has been previously raised in connection with platelet adhesion. Similar studies involving *in-vivo* work with dogs and sheep have demonstrated unequivocally the importance of surface roughness in relation to thrombus formation at the implant surface. Although an increase in surface roughness promoted thrombus formation, however, workers found it difficult to distinguish differences in adhesion of the thrombus to the more rugous surfaces from true differences in thrombogenicity.[26,27] The fact that relatively small differences in surface chemistry can affect thrombogenicity is illustrated by work on Biomer polyurethane. It was found that extruded Biomer was more thromboresistant than solution cast Biomer. Surface studies indicated[28] that there was a difference between the two grades and suggested a correlation between thromboresistance and the surface concentration of the soft segment.

Two techniques for improving thromboresistance, referred to in previous reviews, are well represented in the recent literature. The first is the binding of heparin and heparin-like polysaccharides on to polymer surfaces.[29,30] Variations on the conventional ionic binding techniques include initial grafting of chloromethylstyrene with γ-rays, quaternization with pyridine, and finally exchange of heparin anions for chlorine by means of the sodium salt.[31] Although improved thromboresistance is invariably claimed, the mechanism of action of heparin complexes of this sort is still in doubt and under investigation.[32] The second group of techniques is associated particularly with vascular grafts and the need to establish a pseudointima of endothelial cells on the surface of the prosthesis. The existence and consequences of widely different degrees of maturity in this prosthetic pseudointima have been demonstrated.[33] Attempts to improve the rate of endothelial cell growth in this respect have included preliminary seeding of the Dacron poly(ethylene tere-

---

[21] J. S. Schulz, S. M. Lindenauer, J. A. Penner, and S. Barenberg, *Trans. Am. Soc. Artif. Intern. Organs*, 1980, **26**, 279.
[22] C. D. Forbes, *Clin. Haematol.*, 1981, **10**, 653.
[23] M. Thubrikar, T. Reich, and I. Cadoff, *J. Biochem.*, 1980, **13**, 663.
[24] S. A. Barenberg, J. M. Anderson, and K. A. Mauritz, *J. Biomed. Mater. Res.*, 1981, **2**, 231.
[25] W. M. Reichert, F. E. Filisko, and S. A. Barenberg, *J. Biomed. Mater. Res.*, 1982, **3**, 301.
[26] L. Goldberg, P. Bosco, E. Shors, S. Klein, R. Nelson, and R. White, *Trans. Am. Soc. Artif. Intern. Organs*, 1981, **27**, 517.
[27] J. F. Hecker and R. O. Edwards, *J. Biomed. Mater. Res.*, 1981, **2**, 1.
[28] M. D. Lelah, R. J. Stafford, L. K. Lambrecht, B. R. Young, and S. L. Cooper, *Trans. Am. Soc. Artif. Intern. Organs*, 1981, **27**, 504.
[29] R. Larsson, P. Olssen, and U. Lindahl, *Thromb. Res.*, 1980, **19**, 43.
[30] Y. Noishiki, S. Nagaska, T. Kikuchi, and Y. Mori, *Trans. Am. Soc. Artif. Intern. Organs*, 1981, **27**, 213.
[31] J. E. Wilson, R. C. Eberhart, and A. B. Elkowitz, *J. Macromol. Sci., Chem.*, 1981, **4**, 769.
[32] L. K. Lambrecht, B. R. Young, D. F. Mosher, A. P. Hart, W. J. Hammar, and S. L. Cooper, *Trans. Am. Soc. Artif. Intern. Organs*, 1981, **27**, 380.
[33] G. P. Clagett, M. Robinowitz, Y. Maddox, J. M. Langloss, and P. W. Ramwell, *Surgery*, 1982, **91**, 87.
[34] L. M. Graham, W. E. Burkel, J. W. Ford, D. W. Vinter, R. H. Kahn, and J. C. Stanley, *Arch. Surg.*, 1980, **115**, 1289.

phthalate), prior to implantation, with endothelial cells or, in the case of Biomer, precoating with extracellular matrix.[35] Both procedures were reported to be successful.

Studies of cell attachment, morphology, and growth on the surface of synthetic polymers has much wider implications than those referred to above. The ability to control cell spreading has consequences in, for example, wound healing and scar tissue formation. It appeared from studies[36] based on the work of Folkman and Moscona that it is possible to mediate cell growth on poly(2-hydroxyethyl methacrylate) by simply varying film thickness. This has, however, been more properly interpreted in terms of the inhomogenicity of very thin evaporated coatings.[37] Other workers have attempted to relate adhesion to ionic forces[38] or interfacial energy,[39] but as yet no unifying features have emerged.

3 Applications

Soft Tissue Prosthesis.—The results of some interesting longer-term studies of *in-vivo* materials' performance are now becoming available. The conclusions of workers[40] examining prolonged effects of polytetrafluoroethylene in the treatment of velopharyngeal insufficiency were that the injection procedure used is safe and effective and that no significant changes in the polymer occurred over a period of several years. Similarly, the conclusions of a series of papers[41-44] relating to breast reconstruction indicate that although longer-term problems do occur they are not specifically related to silicone rubber which is still the material of choice. Good results over periods of up to 10 years are also reported with composites of Marlex (polypropylene) mesh and poly(methyl methacrylate) in reconstruction of the chest wall.[45,46] In contrast, the use of polypropylene in abdominal wall reconstruction has been found to produce significant long-term problems despite its advantages in providing short-term abdominal integrity, even in the presence of severe infection. One of the major problems encountered was mesh extrusion.[47] In a comparative study in rats involving Gore-Tex, polytetrafluoroethylene, and Marlex mesh, the materials were inserted in a manner that yielded results both intraperitoneally and extraperitoneally.[48] Whereas the polymers produced similar intraperitoneal tissue

---

[35] W. K. Nichols, D. Gospodarowicz, T. R. Kessler, and D. B. Olsen, *Trans. Am. Soc. Artif. Intern. Organs*, 1981, **17**, 208.
[36] A. S. Belmont, F. M. Kendall, and C. A. Nicolini, *Cell Biophys.*, 1980, **2**, 165.
[37] T. W. Minett, B. J. Tighe, M. J. Lydon, and D. A. Rees, *Eur. J. Cell Biol.*, 1983, Suppl. 1, 30.
[38] M. Horisberger, *Physiol. Chem. Phys.*, 1980, **12**, 195.
[39] D. F. Gerson and D. Scheer, *Biochem. Biophys. Acta*, 1980, **602**, 506.
[40] L. T. J. Furlow, W. N. Williams, C. R. Eisenback, and K. R. Bzoch, *Cleft Palate J.*, 1982, **19**, 47.
[41] G. Lemperle, *Acta Chir. Belg.*, 1980, **79**, 159.
[42] M. I. Lejour, H. Eder, A. De May, and W. Mattheiem, *Acta Chir. Belg.*, 1980, **79**, 125.
[43] J. Bostwick, *Acta Chir. Belg.*, 1980, **79**, 125.
[44] I. Prpic, *Acta Chir. Belg.*, 1980, **79**, 103.
[45] P. McCormack, M. S. Bains, E. J. Beattie, jun., and N. Martini, *Ann. Thorac. Surg.*, 1981, **31**, 45.
[46] H. Eschpasse, J. Gaillard, F. Henry, G. Fourimal, F. Berthoumieu, and X. Desrez, *Ann. Thorac. Surg.*, 1981, **32**, 329.
[47] C. R. Voyles, J. D. Richardson, K. I. Bland, G. R. Tobin, L. M. Flint, and H. C. Polk, jun., *Ann. Surg.*, 1981, **194**, 219.
[48] W. Sher, D. Pollack, C. A. Paulides, and T. Matsumoto, *Ann. Surg.*, 1980, **46**, 618.

reactions microscopic examination revealed significant differences. All the Gore-Tex grafts retained their original shape and demonstrated focal adherence to the muscle. In contrast, strands of Marlex showed disorganization in the host in 90% of the specimens and no focal adherence to the muscle. It was concluded that the characteristics of Gore-Tex produced more acceptable abdominal wall prostheses. Other examples of prosthetic and reconstruction work include the use of Proplast, polytetrafluoroethylene, in the reconstruction of the lower sternum[49] and the posterior meatal wall,[50] silicone rubber in carpal bone[51] and penile prostheses,[52,53] and initial results with poly(2-hydroxyethyl methacrylate) sponge as a potential means of repairing cartilage defects.[54] Extensive reviews of the use of polymers in dental and maxillofacial surgery[55] and mentoplasty (chin augmentation)[56] have been compiled.

Joint Prostheses.—Although not without problems, hip-joint prostheses are obviously the most widely performed and arguably the most successful of joint prostheses. Increasing emphasis is now being placed on total knee prostheses both in terms of stress analysis of the various tibial component designs[57] and of material evaluation. In a review[58] of failed knee-joint prostheses one design was found to have a failure rate in three years of over 70%. Failure was apparently largely due either to the generation of wear debris or to the loosening of the polyester femoral component. This extremely unsatisfactory situation reflects the proliferation in the past of designs based on intuitive thinking rather than interdisciplinary research. Experimental techniques to assess deformation and wear of knee-joint components have been developed[59] and carbon-fibre-reinforced ultra-high molecular-weight polyethylene appears to offer promise as a constructional material.[60] A useful overview of the current state of elbow prostheses is given in a recent group of papers.[61—63] Polyethylene is the dominant material.

Ducts and Canals.—Details of an interesting and thorough examination of several silicone-rubber ureteral prostheses after implantation for periods of four to six years in dogs and in humans have been published.[64] This appears to be the longest implantation period so far studied and the deposits appear to be predominantly lipoidal with possible initiation sites for calcium-based incrustations. There are strong indications that progress is being made in understanding the initiation

[49] M. L. Bell, *Plast. Reconstr. Surg.*, 1981, **68**, 795.
[50] A. N. Johns, *J. Laryngol. Otol.*, 1981, **95**, 819.
[51] L. Zolczer, K. Somogyrari, T. Nyari, J. Manninger, and J. Nemes, *Magy. Traumatol. Orthop. Helyreallito Sebesz*, 1981, **24**, 282.
[52] R. J. Krane, P. S. Freedberg, and M. B. Siroky, *J. Urol.*, 1981, **126**, 475.
[53] J. M. Barry, *J. Urol.*, 1980, **123**, 350.
[54] M. Kon and A. C. de Visser, *Plast. Reconstr. Surg.*, 1981, **67**, 288.
[55] D. F. Williams, *Biomaterials*, 1981, **2**, 133.
[56] B. Hagihara, F. Ishibashi, N. Sato, T. Minami, Y. Okada, and T. Sugimoto, *J. Biomed. Eng.*, 1981, **3**, 9.
[57] J. L. Lewis, M. J. Askew, and D. P. Jaycox, *J. Bone Jt. Surg., Am. Vol.*, 1982, **64**, 129.
[58] H. U. Cameron and D. M. Federkow, *Arch. Ortho. Trauma Surg.*, 1980, **97**, 87.
[59] P. S. Walker, M. Ben-Dor, M. J. Askew, and J. Pew, *Eng. Med.*, 1981, **10**, 33.
[60] F. J. Halcomb and D. Bardos, *Trans. Am. Soc. Artif. Intern. Organs*, 1981, **27**, 364.
[61] W. A. Souter, *Eng. Med.*, 1981, **10**, 59.
[62] T. G. Wadsworth, *Eng. Med.*, 1981, **10**, 69.
[63] M. A. Tuke, B. A. Roper, S. A. V. Swanson, and S. O'Riordan, *Eng. Med.*, 1981, **10**, 75.
[64] J. P. Triboulet, *J. Chir. (Paris)*, 1981, **118**, 1.

mechanisms involved in the calcification of elastomers during prolonged use in various body sites.[65–68] The evaluation of microporous polytetrafluoroethylene as a material for tracheal prosthesis is reported,[69] together with a comparative study of latex rubber and silicone rubber for prosthetic oesophageal tubes.[70] A new implantable (hydraulic) sphincter forming the basis of an artificial anus has been developed in silicone rubber.[71] Initial results from a group of seven patients are reported.

Vascular Prosthesis.—Long-term implantation results are again available here. Early Dacron arterial prostheses with up to 15 years' use have been examined and a detailed report of some five hundred presented in which a significant proportion showed eventual fibre deterioration. The conclusions suggest that, in addition to manufacturing problems, factors related to storage and sterilization, direct *in-vivo* degradation is involved. A similar conclusion is reached in the case of a Dacron prosthesis[72] that failed after seven years and in which hydrolytic degradation was found to have initiated the failure process. The most dramatic complication of reconstructive vascular surgery, even after long periods, is considered to be infection. An *in-vitro* infusion system has been developed to assess resistance to bacterial colonization of materials used in arterial prosthesis.[73] Much attention is given to the question of the assessment of the relative merits of fibre-based poly(ethylene terephthalate) and expanded polytetrafluoroethylene prostheses. Although clearly presented, the results do not suggest overwhelming advantages for one or the other. Detailed haemodynamic studies of grafts of diameters less than 8 mm suggested that Dacron produces more disturbance of pulsatile flow[74] whereas direct implantation studies produced higher early failure rates for Gore-Tex prostheses.[75] A canine model is proposed for characterizing arterial prostheses in the above diameter range and an initial comparative evaluation presented of the two polytetrafluoroethylene grafts (Gore-Tex and Impra) currently available.[76] Studies with small-diameter (1—3 mm) grafts indicate that polytetrafluoroethylene is not at present an adequate substitute for autogenous vein in microvascular surgery.[77] Reviews and debates on various aspects of vascular prostheses are presented.[78–80]

[65] E. Pollock, E. J. Andrews, D. Lentz, and K. Sheikh, *Trans. Am. Soc. Artif. Intern. Organs*, 1981, **27**, 405.
[66] D. R. Owen and R. M. Zane, *Trans. Am. Soc. Artif. Intern. Organs*, 1981, **27**, 528.
[67] Y. Nose, H. Harasaki, and J. Murray, *Trans. Am. Soc. Artif. Intern. Organs*, 1981, **27**, 714.
[68] S. D. Bruck, *Biomaterials*, 1981, **2**, 14.
[69] J. R. Bottema, J. Feijen, H. W. fen Hoopen, I. Molenaar, and C. R. Wildevuur, *Trans. Am. Soc. Artif. Intern. Organs*, 1980, **26**, 412.
[70] F. J. Branicki, A. L. Ogilvie, M. R. Willis, and M. Atkinson, *Br. J. Surg.*, 1981, **68**, 861.
[71] G. Szinicz, *Int. J. Artif. Organs*, 1980, **3**, 358.
[72] F. Godard, M. King, R. Guidoin, M. Marois, A. Garton, P. Blais, C. Gosselin, and K. Gunasekera, *J. Mal. Vasc.*, 1981, **6**, 167.
[73] O. Goeau-Brissoniere, J. C. Pechere, R. Goudoin, and H. P. Noel, *J. Chir. (Paris)*, 1980, **117**, 397.
[74] E. J. Santiago, K. Chatamra, and D. E. Taylor, *Ann. R. Coll. Surg., Eng.*, 1981, **63**, 253.
[75] H. Hamann, *Fortschr. Med.*, 1981, **99**, 1457.
[76] K. C. Hanel, C. McCabe, W. M. Abbot, J. Fallan, and J. Megerman, *Ann. Surg.*, 1982, **195**, 456.
[77] D. H. Lidman, B. Faibisoff, and K. Dar, *J. Microsurg.*, 1980, **1**, 447.
[78] J. M. Rocko and K. G. Swan, *Biomaterials*, 1981, **2**, 172.
[79] R. E. Clark, C. B. Anderson, J. L. Kardos, and C. B. Wright, *Trans. Am. Soc. Artif. Intern. Organs*, 1980, **26**, 598.
[80] P. Sabanayagam, A. B. Schwartz, R. R. Soricelli, J. L. Chinitz, and P. Lyons, *Trans. Am. Soc. Artif. Intern. Organs*, 1980, **26**, 573.

Tendons and Ligaments.—An example[81] of the work required to take experimental prostheses to a satisfactory stage for general use is found in the case of a new polyethylene-based ligament (Polyflex). In a group of nine patients the ligament was so unsatisfactory as to require removal in a third of the patients and half of the remaining group complained of pain. Despite the fact that the prosthesis had reached the stage of clinical use, it was concluded that many problems relating to the material and the operation remain to be resolved. A new technique for attachment to soft tissue in tendon and ligament replacement and repair has been described.[82] This entails the use of a filamentous carbon–poly(lactic acid) composite as a tissue scaffold. Following implantation, soft tissue ingrowth into the carbon-fibre network. In a rabbit study the bond developed rapidly and was mechanically secure under physiological loading conditions. Strength tests after 12 weeks demonstrated that breaking strains equivalent to those of the natural system had been reached. An interesting synthetic tendon[83] based on poly(2-hydroxyethyl methacrylate) hydrogel reinforced with poly(ethylene terephthalate) has been reported but as yet no clinical results are available.

Sutures and Adhesives.—During the last decade the use of synthetic absorbable sutures has become widespread. There are two currently available materials, poly(glycollic acid) and a poly(lactic–glycollic) copolymer. Because of their stiffness the polymers are normally produced as sutures in the form of braided filaments, except in very fine diameters. A new flexible monofilament polymer produced by ring opening of 1,4-dioxanone has recently been developed and several reports of its properties and performance have now appeared.[84—87] The mechanical properties of absorbable sutures have been the subject of several papers, especially in comparison with those of non-absorbable sutures. Results are presented of both *in-vitro*[88] and *in-vivo* comparisons, the latter being carried out in various body sites.[89—92] Further detailed studies of the degradation of poly(glycollic acid) and poly(lactic acid) have been presented. These include the effects of enzymes,[93] lipids,[94] pH,[95] and buffer solutions.[96]

The field of surgical adhesives is widely recognized as being in need of a new generation of materials, but as yet there is little sign of innovative work. The strengths of adhesives used in bone surgery have been examined under *in-vitro* and *in-vivo* conditions.[97] The study is claimed to form the basis of development of a new

[81] M. Scharling, *Acta Orthop. Scand.*, 1981, **52**, 575.
[82] J. Aragona, J. R. Parsons, H. Alexander, and A. B. Weiss, *Clin. Orthop.*, 1981, **160**, 268.
[83] J. Kolarik, C. Migharesi, M. Stol, and L. Nicholais, *J. Biomed. Mat. Res.*, 1981, **15**, 147.
[84] J. A. Ray, N. Doddi, D. Regula, J. A. Williams, and A. Melveger, *Surg. Gynaecol. Obst.*, 1981, **153**, 497.
[85] D. Beurton, D. Gonties, S. Terdjman, S. H. Abraham, and A. Dana, *J. Urol. (Paris)*, 1981, **87**, 295.
[86] R. S. Bartholmew, *Ophthalmologica*, 1981, **183**, 81.
[87] A. R. Berry, M. C. Wilson, J. W. Thomson, and T. J. McNair, *J. R. Coll. Surg., Edinb.*, 1981, **26**, 170.
[88] C. C. Chu, *Ann. Surg.*, 1981, **193**, 365.
[89] D. C. Birdsell, G. E. Gavelin, G. M. Kemsley, and K. S. Hein, *Plast. Reconst. Surg.*, 1981, **68**, 742.
[90] G. T. Rodehearer, J. G. Thacker, and R. F. Edlich, *Surg. Gynaecol. Obst.*, 1981, **153**, 835.
[91] B. Valltars, H. A. Hansson, and J. Svenson, *Neurosurgery*, 1981, **9**, 407.
[92] G. Ross, C. Parlides, F. Lang, A. Kusaba, M. Perlman, and T. Matsumoto, *Ann. Surg.*, 1981, **47**, 541.
[93] D. F. Williams, *Eng. Med.*, 1981, **10**, 5.
[94] C. P. Sharma and D. F. Williams, *Eng. Med.*, 1981, **10**, 8.
[95] C. C. Chu, *Ann. Surg.*, 1982, **195**, 55.
[96] C. C. Chu, *J. Biomed. Mater. Res.*, 1981, **15**, 19.
[97] G. Giebel, M. Rumpler, and L. Borchers, *Biomed. Tech. (Berlin)*, 1981, **26**, 170.

adhesive for use in this field. Elsewhere[98] poly(butyl methacrylate) has been considered as an alternative bone cement to poly(methyl methacrylate) and various aspects of the volumetric behaviour of bone cements have been studied as a function of the development of transient and residual strength.[99] Cyanoacrylate adhesives are now becoming the subject of wider clinical evaluation with varying degrees of success. Typical examples of use include middle ear[100] and intracranial[101,102] surgery, re-attachment of tendons, and reconstruction of the frontal sinuses.[104]

Ophthalmic Applications.—The problems associated with intraocular lens implantation have been widely discussed. A group of collected papers provides a good source of comments on the present situation.[105] Poly(methyl methacrylate) is the material generally used for the lens itself with nylon, polypropylene[107] or poly(ethylene terephthalate)[108] fixation loops. Evidence suggests that the suture materials cause more problems than the methacrylate[107] itself. Silicone polymers[109] and poly(2-hydroxyethyl methacrylate)[110] have been evaluated as alternative lens materials and there is clearly room for more specific polymer design work in this area. The use of extended wear contact lenses as an alternative to intraocular lenses in the treatment of aphakia has attracted more attention recently.[111] Long-term (up to eight years) studies of extended wear contact lens use, particularly in the treatment of myopia[112] and aphakia,[113] are now appearing. The results are generally good. Overviews[114,115] of the many presently available types of contact lens materials have been presented together with a discussion[116] of ocular compatibility of polymers in relation to the more widely studied problems encountered in blood contact devices.

Artificial Skin and Wound Dressings.—There is considerable activity here. The small number of reports of work with Hydron[117,118] continue, on balance, to suggest that whereas it has no adverse effect on healing it produces no significant

---

[98] A. Crugnola, E. J. Ellis, R. M. Rose, and E. L. Rodin, 'Plastics – Creating Value through Innovation', 39th ANTEC Proceedings, SPE, 1981, p. 253.
[99] A. M. Ahmed, W. Pak, D. L. Burke, and J. Miller, *J. Biochem. Eng.*, 1982, **104**, 21.
[100] K. H. Siedentop, *Am. J. Otol.*, 1980, **2**, 77.
[101] D. Samson, Q. M. Ditmore, and C. W. Beyer, jun., *Neurosurgery*, 1981, **8**, 43.
[102] D. Samson, Q. M. Ditmore, and C. W. Beyer, jun., *Neurosurgery*, 1982, **8**, 52.
[103] J. Endroodi, A. J. Simanka, K. Barabas, G. Kiss, and G. Dosa, *Magy. Traumatol. Orthop. Helyreallito Sebesz*, 1980, **23**, 296.
[104] J. Wiecko, P. Karasiewicz, and J. Fruba, *Polim. Med.*, 1980, **10**, 165.
[105] J. E. Blaydes, *Contact Intraoc. Lens Med. J.*, 1981, **7**, 67.
[106] R. L. Knonenthal, *Ophthalmology (Rochester)*, 1981, **88**, 965.
[107] A. W. Tuberville, M. A. Galin, H. D. Perez, D. Banda, R. Ong, and I. M. Goldstein, *Invest. Ophthalmol. Vis. Sci.*, 1982, **22**, 727.
[108] N. S. Jaffe, *Ophthalmology (Rochester)*, 1981, **88**, 955.
[109] B. S. Kassar and E. D. Varnell, *J. Am. Intraoc. Implant Soc.*, 1980, **6**, 344.
[110] R. B. Packard, A. Garner, and E. J. Arnott, *Br. J. Ophthmol.*, 1981, **65**, 585.
[111] E. Gruber, *Trans. Ophthalmol. Soc. UK*, 1980, **100**, 231.
[112] W. J. Stark and N. F. Martin, *Arch. Ophthalmol.*, 1981, **99**, 1963.
[113] G. P. Kracher, W. J. Stark, and L. W. Hirst, *Am. J. Optom. Physiol. Opt.*, 1981, **58**, 467.
[114] B. J. Tighe, *Chem. Ind.*, 1981, 796.
[115] B. J. Tighe, *Manuf. Opt. Internat.*, 1983, **36**(3), 10.
[116] D. Baker and B. J. Tighe, *Contact Lens J.*, 1981, **10**, 3.
[117] J. H. Kronman and M. Goldman, *J. Endod.*, 1981, **7**, 441.
[118] D. P. Dressler, W. K. Barbee, and R. Sprenger, *J. Trauma*, 1980, **20**, 1024.

improvement. The same polymer, poly(2-hydroxyethyl methacrylate), has been produced in the form of a composite laminate with polybutadiene.[119] This is primarily intended as a burn cover, permitting oxygen transport while preventing water diffusion. A similar type of active surface is presented to the wound in the poly(vinyl alcohol)–formaldehyde foam dressing which has been described and taken to the stage of animal and clinical work.[120] Several reports of polyurethane-based systems have appeared. A bilaminar foam (SYSpur-derm) has behaved extremely well in clinical trials as a temporary skin substitute.[121] The new and similar product Opsite has also given good results[122–124] in surgery and on burns. *In-vitro* and *in-vivo* work has been reported with a new synthetic skin consisting of a silicone rubber membrane laminated to 6,6-nylon looped velour and the results are encouraging.[125] The most complex of the developments in this area consists of a silicone-rubber top layer and a porous collagen layer which has been crosslinked with chondroitin sulphate.[126,127] The collagen surface is seeded with epithelial cells from the patient before application and successful use for periods up to 46 days has been recorded.

Artificial Organs.—The major single expansion in this area has been in publications relating to the total artificial heart. A useful collection of papers serves as a state of the art review and describes various aspects of its development and use.[128] In particular, various aspects of design and construction of the currently evaluated systems are discussed together with details of the animal work involved.[129–132] Maximum survival periods still seem to be uniformly less than one year with any system.

Whereas interest in haemoperfusion for treatment of drug overdose[133–136] continues to grow, clinical activity in its use as a basis for an artificial liver support

---

[119] C. Migharesi, C. Carfagna, and L. Nicholais, *Biomaterials*, 1980, **1**, 205.
[120] W. Mutschler, C. Burri, and E. Plank, *Helv. Chir. Acta*, 1980, **47**, 163.
[121] E. Rose, J. Riedeberger, and P. F. Mahnke, *Z. Exp. Chir.*, 1980, **13**, 70.
[122] A. C. Eaton, *Br. J. Surg.*, 1980, **67**, 857.
[123] Y. Cavlak, *Akt. Trauma.*, 1980, **10**, 311.
[124] L. F. Nahas and B. L. Swartz, *Plast. Reconstr. Surg.*, 1981, **67**, 791.
[125] Sin-Daw Lin, E. C. Robb, and P. Nathan, *Trans. Am. Soc. Artif. Intern. Organs*, 1981, **27**, 522.
[126] I. V. Yannas, J. F. Burke, M. Warpehoski, P. Stasikelis, E. M. Skrabut, D. Orgill, and D. J. Giard, *Trans. Am. Soc. Artif. Intern. Organs*, 1981, **27**, 19.
[127] J. F. Burke, I. V. Yannas, W. C. Quinty, jun., C. C. Bandoc, and W. K. V. Jung, *Ann. Surg.*, 1981, **194**, 413.
[128] Various authors in *Trans. Am. Soc. Artif. Intern. Organs*, 1981, **27**, 71—191.
[129] T. Mochizuik, J. H. Lawson, D. B. Olsen, H. Fukumasu, N. Daitch, R. Jarvik, T. R. Kessler, A. B. Pons, L. Hastings, K. J. Razzeca, S. D. Nelson, and W. J. Kolff, *Artif. Organs*, 1981, **5**, 125.
[130] J. Vasku, J. Gerng, P. Hanzkelka, E. Urbanek, M. Dostal, P. Urbanek, J. Filkuta, H. Haneckora, O. Sotalova, B. Hartmannova, P. Guba, V. Pavlicek, L. Krcek, T. Sladek, E. Sotakova, S. Dolezel, V. Krcma, and K. Cidl, *Artif. Organs*, 1981, **5**, 388.
[131] R. K. Janik, T. R. Kessler, L. D. McGill, D. B. Olsen, W. C. De Vries, J. Deneris, J. T. Blaylock, and W. J. Kolff, *Trans. Am. Soc. Artif. Intern. Organs*, 1981, **27**, 90.
[132] K. Atsumi, I. Fujumasa, K. Imachi, H. Miyake, N. Takido, M. Nakajima, A. Kouno, T. Ono, S. Yuasa, T. Mori, S. Nagaoka, S. Kawase, and T. Kikuchi, *Trans. Am. Soc. Artif. Intern. Organs*, 1981, **27**, 77.
[133] B. Sangster, A. N. van Heijst, and J. J. Sixma, *Arch. Toxicol.*, 1981, **47**, 269.
[134] D. Baggish, S. Gray, P. Jatlow, and M. J. Bia, *Yale J. Biol. Med.*, 1981, **54**, 147.
[135] T. M. Chang, J. F. Winchester, J. Rosenbaum, A. Saito, S. Stefoni, C. Shu, and H. L. Gurland, *Trans. Am. Soc. Artif. Intern. Organs*, 1980, **26**, 593.
[136] B. Gosselin, D. Mathieu, C. Chopin, F. Wattel, B. Dupius, J. M. Hagnenoer, and M. Desprez, *Clin. Toxicol.*, 1980, **17**, 439.

system seems to be limited to the Williams' group.[137] The position in artificial kidney research appears to be similar. The field has been reviewed[138] with some emphasis on future developments but little active progress (in portable systems for example) has been reported. Questions of molecular weight selectivity and its importance,[138,139] the relative ineffectiveness of heparin in preventing deposition at membrane surfaces,[140] and the dispersion of silica particles from dialysis (silicone-rubber) tubing into various body organs[141] are discussed. A significant interest in the question of an implantable artificial pancreas is shown in pieces of recent work which approach the problem from different standpoints.[142—144]

Polymeric Drugs and Drug Delivery Systems.—The immunoregulatory capacity of maleic anhydride–vinyl ether copolymers has been studied fairly widely. The background literature has been summarized and effects such as molecular weight and the role of the pyran ring discussed.[145—148] A relatively simple application of the slow release concept, and one which has proved to be extremely effective, is the release of gentamycin from bone cement.[149] Similar techniques are now extended to poly(ethylene terephthalate) vascular prostheses.[150] A novel extension of the principle has been used with polytetrafluoroethylene vascular grafts. This entails the use of benzalkonium chloride as a 'cationic anchor' and simultaneous ionic binding of both antibiotics and heparin are reported.[151] Results are said to be good. Many workers have presented findings on various aspects of the slow release of metoprolol in comparison to multiple dosage administration.[152—155]

A much more sophisticated group of delivery techniques, which are of great potential value but still at the stage of exploratory work, are those related to absorbtive pinocytosis. Studies are currently designed to determine the factors relating the size and nature of natural and synthetic macromolecules to their uptake in peritoneal macrophages.[156—159] Microencapsulation techniques (which would be

[137] R. D. Hughes, P. G. Langley, and R. Williams, *Int. J. Artif. Organs*, 1980, **3**, 277.
[138] S. Jorstrad, L. C. Smeby, and T. E. Wideroe, *Artif. Organs*, 1981, **4**, 98.
[139] L. C. Smeby, S. Jorstrad, T. E. Wideroe, and T. M. Strartaas, *Artif. Organs*, 1981, **4**, 104.
[140] P. Stratta, C. Canavese, G. Mangiariotti, A. Pacitti, C. Tetta, R. Coppo, R. Ragni, and A. Vercellone, *Proc. Eur. Dial. Transplant Assoc.*, 1981, **18**, 269.
[141] J. Bommer, R. Waldherr, and E. Ritz, *Proc. Eur. Dial. Transplant Assoc.*, 1981, **18**, 731.
[142] S. P. Bessman, L. J. Thomas, H. Kojima, D. F. Sayler, and E. C. Layne, *Trans. Am. Soc. Artif. Intern. Organs*, 1981, **27**, 7.
[143] T. Kondo, H. Kojima, K. Ohkura, S. Ikeda, and K. Ito, *Trans. Am. Soc. Artif. Intern. Organs*, 1881, **27**, 250.
[144] A. Sun, G. O'Shea, and Y. Leung, *Artif. Organs*, 1981, **5**, 69.
[145] M. A. Chinigos, *Exp. Path.*, 1981, **19**, 19.
[146] W. Regelson, *Pharmacol. Ther.*, 1981, **15**, 1.
[147] A. R. Zander, G. Spitzer, D. S. Vorma, S. Ginzberg, and K. A. Dicke, *Biomedicine*, 1980, **33**, 69.
[148] D. S. Breslow, *Polym. Prepr., Am. Chem. Soc., Div. Polym. Chem.*, 1981, **22**, 24.
[149] G. Josefsson, L. Lindberg, and B. Wiklander, *Clin. Orthop.*, 1981, **159**, 194.
[150] W. S. Moore, M. Chrapil, G. Seiffert, and K. Keown, *Arch. Surg.*, 1981, **116**, 1403.
[151] R. A. Harvey and R. S. Greco, *Ann. Surg.*, 1981, **194**, 642.
[152] T. J. Bloem and P. C. Lindner, *Ann. Clin. Res.*, 1981, **13**, 61.
[153] L. Oro, *Ann. Clin. Res.*, 1981, **13**, 58.
[154] J. Asphind and P. Ohman, *Ann. Clin. Res.*, 1981, **13**, 30.
[155] R. C. Browning, V. A. John, R. J. Till, and W. Theobald, *Br. J. Clin. Pharmacol.*, 1981, **12**, 600.
[156] R. Duncan, M. K. Pratten, H. C. Cable, H. Ringsdorf, and J. B. Lloyd, *Biochem. J.*, 1981, **196**, 49.
[157] T. Kooishra, M. K. Pratten, and J. B. Lloyd, *Biosci. Rep.*, 1981, **1**, 587.
[158] R. N. Rowland and J. F. Woodley, *Biosci. Rep.*, 1981, **1**, 399.
[159] M. K. Pratten, P. C. Millard, and J. B. Lloyd, *Biosci. Rep.*, 1981, **1**, 125.

an obvious route by which pinocytosis might be exploited) continue to find widespread use.[160–163] Biodegradation is the major release mechanism involved in microcapsular delivery and several useful reviews, outlining the current situation in the field of biodegradable polymers for drug release, are contained in a recent monograph.[164] These, together with other timely reviews, present a good overview of biomaterials in drug release systems.[165,166] An interesting contrast in release mechanisms for medium-molecular-weight molecules from polyesters[167] and silicone rubber[168] (which systems are of increasing importance for veterinary applications[169]) can be drawn from recent studies.

Enzyme-related Applications.—Immobilization of enzymes with polymer substrates is now a routine activity either by covalent bonding, adsorption, or entrapment. Covalent bonding is frequently achieved by coupling agents such as cyanogen bromide or glutaraldehyde. Thus urokinase, immobilized directly on to a collagen–polyolefin composite, produces a potentially useful biomaterial for artificial organs.[170] Several interesting methods for the production of more specific coupling sites have been described.[171–173] Nitro groups are frequently used. Work on entrapment in particular polymers includes accounts of the inhibitory effect of collagenous capsules formed around the active particles after implantation,[174] and of the use of fluorescence microscopy[175] to determine the distribution of enzyme within the particles. Related subject areas are those involving immunoassay techniques in which specific antibodies are linked to polymeric substrates[176,177] or specimens are embedded for histological sectioning in polymeric media.[178] The development of novel polymers both as embedding media and as carriers is becoming increasingly significant. Thus, although epoxies,[179–181] poly(ethylene glycol),[182] and poly(2-hydroxyethyl methacrylate)[183,184] are used as embedding media, none is without disadvantage.

[160] L. R. Beck and D. Cowsar, *Acta Eur. Fertil.*, 1980, **11**, 139.
[161] M. Ndong-Nkoume, P. Labrude, J. C. Humbert, B. Teisseire, and C. Vigneron, *Ann. Pharm. Fr.*, 1981, **39**, 247.
[162] K. Bala and P. Vasuderan, *J. Macromol. Sci., Chem.*, 1981, **16**, 819.
[163] Y. Morimoto, M. Okumara, K. Sugibayashi, and Y. Kato, *J. Pharmacobiodyn.*, 1981, **4**, 624.
[164] Various authors in National Institute for Drug Abuse Research Monograph Series, 1981, Vol. 28.
[165] C. E. Carraher, D. J. Giron, D. R. Cerutis, S. Tsuji, T. J. Gehrke, R. S. Venkatachalam, and H. S. Bloxall, *Org. Coatings Plast. Chem.*, 1981, **44**, 1.
[166] R. S. Langer and N. A. Peppas, *Biomaterials*, 1981, **4**, 201.
[167] J. Heller, D. W. H. Penhale, R. F. Helwing, and B. K. Fritzinger, *Polym. Eng. Sci.*, 1981, **21**, 727.
[168] T. J. Roseman, L. J. Larion, and S. S. Butler, *J. Pharm. Sci.*, 1981, **70**, 562.
[169] H. A. Turner, R. L. Philips, M. Vavra, and D. C. Young, *J. Anim. Sci.*, 1981, **52**, 939.
[170] S. Watanabe, Y. Shimizu, T. Teramatsu, T. Murachi, and T. Hino, *J. Biomed. Mater. Res.*, 1981, **15**, 553.
[171] G. Manecke and H. G. Vogt, *Biochemie*, 1981, **62**, 603.
[172] G. Manecke and D. Polakowski, *J. Chromatog.*, 1981, **215**, 13.
[173] F. Abdel-Hay, C. G. Beddows, and J. T. Guthrie, *Makromol. Chem.*, 1981, **182**, 717.
[174] P. Edman and I. Sjoholm, *J. Pharm. Sci.*, 1981, **70**, 684.
[175] M. Yoshida and I. Kaetsu, *J. Appl. Polym. Sci.*, 1981, **26**, 687.
[176] E. Ishikawa, Y. Hamaguchi, M. Imagawa, M. Inada, H. Imura, N. Nakazawa, and H. Ogawa, *J. Immunoassay*, 1980, **1**, 385.
[177] A. R. Neurath and N. Strick, *J. Virol. Meth.*, 1981, **3**, 155.
[178] H. C. Bickley, G. von Hagens, and F. M. Townsend, *Arch. Pathol. Lab. Med.*, 1981, **105**, 674.
[179] W. D. Kuhlmann and R. Krischan, *Histochemistry*, 1981, **72**, 377.
[180] M. Tulliez and J. Diebold, *Ann. Pathol.*, 1981, **1**, 163.
[181] Y. le Charpentier, A. Galien, A. Gaste, and L. Herbe, *Ann. Pathol.*, 1981, **1**, 160.
[182] F. T. Bossman and P. M. Go, *Histochemistry*, 1981, **73**, 195.
[183] P. Felman and P. A. Bryon, *Ann. Pathol.*, 1981, **1**, 156.
[184] D. Wynford-Thomas, B. Stringer, and G. R. Newman, *Med. Lab. Sci.*, 1981, **38**, 121.

4 Synthetic Work and the Development of New Materials

A useful collection of 'current status' reports and papers on synthetic work involving sulphur dioxide copolymers, polyamidoamines, fluorinated acrylic monomers, organometallics, and several other topics has recently been published.[185] Work in the hydrogel field has included the study of 3-alkoxy-2-hydroxy propylacrylates,[186] 2-hydroxyethyl methacrylate[187] copolymers (as scleral buckling agents), and derivation of vinyl pyridine-1-oxide (as drug release matrices).[188] The incorporation of a hydroxyl-terminated allyl amylose oligomer into terpolymers of potential value as bioerodible materials provides an alternative approach to drug release.[189] Many reports have appeared on the synthesis of linear polymers of potential pharmacological activity, usually achieved by coupling drugs to active sites within the parent polymer. The use of mercapto,[190] imidazole,[194] quinone,[195] and hydroxyl[196] groups in this respect is described. The synthesis of potential artificial haemoglobins has been reported. This involves the copolymerization of the vinyl group in hemin with vinyl monomers such as styrene derivatives and 2-aminoethylmethacrylate. The amounts of hemin incorporated and the effect of the polymeric environment on its properties are discussed.[197] Two novel membrane materials and their use in charcoal microencapsulation have been reported. One of these is based on sulphohexyl methacrylate[198] and the other on a functionalized vinyl chloride–ethylene–vinyl acetate terpolymer on to which hydrophilic monomers were grafted.[199] Plasma polymerization is used in production of thin biocompatible coatings,[200] for example in articles for implantation.[201] A useful review of its application in the biomedical field has been compiled.[202] One of the most novel proposals in the recent literature is the concept of a membrane acting as a chemical valve with change in conformation (through reversible complexation) causing dilation and contraction. Some preliminary results based on a poly(methacrylic acid) network in conjunction with poly(ethylene glycol) and albumin–haemoglobin solutions are reported.[203]

[185] Various authors in *Org. Coatings Plast. Chem.*, 1980, **42**.
[186] Chenxur Lu, Naichun Chen, Zhang Wei Gu, and Zind Feng, *J. Polym. Sci., Polym. Chem. Ed.*, 1980, **18**, 3403.
[187] M. F. Refojo and F. L. Leong, *J. Biomed. Mater. Res.*, 1981, **15**, 497.
[188] V. N. Hasirci, *Biomaterials*, 1981, **2**, 3.
[189] K. S. Lee, V. T. Stanett, and R. D. Gilbert, *J. Polym. Sci., Polym. Chem. Ed.*, 1982, **20**, 997.
[190] J. Pitha, S. Zawadzki, and B. A. Hughes, *Makromol. Chem.*, 1982, **183**, 781.
[191] T. Gehrmann and W. Vogt, *Makromol. Chem.*, 1981, **182**, 3069.
[192] C. Lu, J. Yang, L. Leng, Z. Feng, and D. Li, *J. Polym. Sci., Polym. Chem. Ed.*, 1981, **19**, 3333.
[193] T. X. Neenan and H. R. Allock, *Biomaterials*, 1982, **3**, 78.
[194] G. Manecke and W. Beier, *Angew. Makromol. Chem.*, 1981, **97**, 23.
[195] M. Kuhn, K. Pommerening, P. Mohr, M. Benes, and J. Stamberg, *Angew. Makromol. Chem.*, 1981, **97**, 161.
[196] G. Bauduin, D. Bondon, J. Martel, Y. Pietrasanto, B. Pucci, J. J. Serrano, and C. Francois, *Makromol. Chem.*, 1981, **182**, 2589.
[197] L. C. Dickinson and J. C. W. Chien, *Polym. Prepr., Am. Chem. Soc., Div. Polym. Chem.*, 1981, **22**, 134.
[198] W. Y. Chen, B. Z. Xu, and X. D. Feng, *J. Polym. Sci., Polym. Chem. Ed.*, 1982, **20**, 547.
[199] Y. Mori, S. Nagaoka, H. Tanzawa, T. Kikuchi, Y. Yamanda, M. Hagiwara, and Y. Idezuki, *J. Biomed. Mater. Res.*, 1982, **16**, 17.
[200] A. S. Chawla, *Biomaterials*, 1981, **2**, 83.
[201] A. W. Hahn, H. K. Yasuda, W. J. James, M. F. Nichols, R. K. Sadhir, A. K. Sharma, O. A. Pringle, D. H. York, and E. J. Charlson, *Biomed. Sci. Instrum.*, 1981, **17**, 109.
[202] H. Yasuda and M. Gazicki, *Biomaterials*, 1982, **3**, 68.
[203] Y. Osada and Y. Takeuchi, *J. Polym. Sci., Polym. Lett. Ed.*, 1981, **19**, 303.

18
Computer Applications

BY A. H. FAWCETT

1 Polymer Kinetics

The stepgrowth formation of thermosetting resins from phenol and formaldehyde under both acidic and basic conditions was simulated by Ishida *et al.* by a Monte Carlo procedure which provided information on the types of methylene bridges and methylol groups as well as the oligomer distribution as a function of the extent of reaction.[1] Gupta *et al.* have computed by solving the appropriate differential equations of the concentrations of particular types of aromatic ring recognized as forming and subsequently reacting during resole[2] and novolac[3] formation. In the latter case the number of structural units used was smaller, so that it was possible to include the reverse reactions. Elsewhere, treatments of novolac production have shown how three consecutive ideal tank reactors might achieve the same conversion as a batch reactor[4] but that the polydispersity and degree of branching would then differ.[5] More general explorations of reversible polycondensation processes dealt with the molecular-weight distribution[6] and the influence of the geometry of the gas phase through which the condensation byproduct escaped.[7] One consequence of allowing reversibility in AB-type is a diminishment in the difference between the molecular-weight distribution curves for even and odd homologues found when monomers are less reactive than the oligomers.[8]

The formations of nylon-6[9–12] and nylon-66[13,14] by step growth and poly-(ethylene terephthalate)[15] by transesterification in different types of reactor have been modelled as an aid to process design and control. Brown's algorithm was found

---

[1] S.-I. Ishida, Y. Tsutsumi, and K. Kaneko, *J. Polym. Sci., Polym. Chem. Ed.*, 1981, **19**, 1609.
[2] P. K. Pal, A. Kumar, and S. K. Gupta, *Polymer*, 1981, **22**, 1699.
[3] A. Kumar, S. K. Gupta, and B. Kumar, *Polymer*, 1982, **23**, 1929.
[4] P. M. Frontini, T. R. Cuadrado, and R. J. J. Williams, *Polymer*, 1982, **23**, 267.
[5] A. Kumar, U. K. Phukan, and S. K. Gupta, *J. Appl. Polym. Sci.*, 1982, **27**, 3393.
[6] A. Kumar, P. Rajora, N. L. Agarwalla, and S. K. Gupta, *Polymer*, 1982, **23**, 222.
[7] S. K. Gupta, A. Kumar, and K. K. Agrawal, *Polymer*, 1982, **23**, 1367; S. K. Gupta, N. L. Agarwalla, and A. Kumar, *J. Appl. Polym. Sci.*, 1982, **27**, 1217.
[8] S. K. Gupta, N. L. Agarwalla, P. Rajora, and A. Kumar, *J. Polym. Sci., Polym. Phys. Ed.*, 1982, **20**, 933.
[9] S. K. Gupta, A. Kumar, and K. K. Agrawal, *J. Appl. Polym. Sci.*, 1982, **27**, 3089.
[10] K. Tai, Y. Arai, and T. Tagawa, *J. Appl. Polym. Sci.*, 1982, **27**, 731.
[11] S. K. Gupta, A. Kumar, P. Tandon, and C. N. Naik, *Polymer*, 1981, **22**, 481.
[12] S. K. Gupta, C. D. Naik, P. Tandon, and A. Kumar, *J. Appl. Polym. Sci.*, 1981, **26**, 2153.
[13] A. Kumar, R. K. Agarwal, and S. K. Gupta, *J. Appl. Polym. Sci.*, 1982, **27**, 1759.
[14] A. Kumar, S. Kuruville, A. R. Raman, and S. K. Gupta, *Polymer*, 1981, **22**, 387.
[15] K. Ravindranath and R. A. Mashelkar, *J. Appl. Polym. Sci.*, 1981, **26**, 3179; 1982, **27**, 471 and 2625; A. Kumar, S. K. Gupta, B. Gupta, and D. Kunzru, *ibid.*, 1982, **27**, 4421.

to be more efficient than the Gauss–Jordan for the solution of the non-linear algebraic equations of the reversible polymerization in a homogenous continuous flow stirred tank reactor.[13] An attempt was made, by using an equilibrium expression, to include in one study the possibility of cyclization.[12]

Polymer network formation has been simulated with a Monte Carlo procedure[16] in order to find that, for a trifunctional monomer, many of the results are similar to those obtained by the cascade substitution route. The recursive approach has been adopted to calculate the characteristic properties of chains pendant within a formed network.[17] These network characteristics were found, together with the proportion of loops and other features, in a simulation of the formation of random elastic networks from stars with terminal reactive groups.[18]

Monte Carlo algorithms developed to compute the distribution of monomer sequences in copolymers and the stereo sequences of vinyl homopolymers have been reviewed and certain of the Memory programs have been listed.[19] By such methods reactivity ratios have been obtained from n.m.r. data on diads, triads, and pentads.[20] Two programs are available which treat terpolymerizations at low and high concentrations as first- or second-order Markov processes.[21]

The Tromsdorf effect, encountered at high conversion in free radical polymerizations, is associated with a reduction in the value of the termination rate constant, for which reptation expressions have been developed and tested.[22] At present, results which are nearly as well in agreement with experiment may be achieved with semiempirical models such as those devised by Hamielec,[23] O'Driscoll[24] and others[25–27] and applied to the polymerization of styrene,[23,25] methyl methacrylate,[24] and vinyl acetate.[26] Soh and Sundberg, in common with Hamielec, recognize a stage subsequent to that concerned with the gel effect when reptation ceases.[28] A weak gel effect was detected in a simulation of the free radical polymerization of methyl methacrylate, in which the program REMECH was employed to verify the role of certain oligomers.[29] IBM's Continuous System Modeling Program (CSMP) appears to be straightforward to apply to reactions whose kinetics follow a complex system of differential equations,[30,31] for programming the equations themselves is as simple as preparing the data for other programs![30]

For numerical solution of the Smith–Ewart differential equations of emulsion

[16] J. Mikes and K. Dusek, *Macromolecules*, 1982, **15**, 93.
[17] M. I. Bibbo and E. M. Valles, *Macromolecules*, 1982, **15**, 1293.
[18] B. E. Eichinger, *J. Chem. Phys.*, 1981, **75**, 1964.
[19] 'Monte Carlo Applications in Polymer Science', ed. W. Bruns, I. Motoc, and K. F. O'Driscoll, Lecture Notes in Chemistry, Vol. 27, Springer-Verlag, Berlin, 1981.
[20] A. Rudin, K. F. O'Driscoll, and M. S. Rumack, *Polymer*, 1981, **22**, 740.
[21] Y. Kodaira and H. J. Harwood in 'Computer Applications in Applied Polymer Science', ed. T. Provder, ACS Symposium Series, **197**, 1982.
[22] T. J. Tulig and M. Tirrell, *Macromolecules*, 1981, **14**, 1501; 1982, **15**, 459.
[23] F. L. Marten and A. E. Hamielec, *J. Appl. Polym. Sci.*, 1982, **27**, 489.
[24] K. F. O'Driscoll, *Pure Appl. Chem.*, 1981, **53**, 617.
[25] L. A. Cutter and T. D. Dresler, in ref. 21.
[26] W. Baade, H. U. Moritz, and K. H. Reichert, *J. Appl. Polym. Sci.*, 1982, **27**, 2249.
[27] S. Balke, L. Garcia, and R. Patel, *Polym. Eng. Sci.*, 1982, **22**, 777.
[28] S. K. Soh and D. C. Sundberg, *J. Polym. Sci., Polym. Chem. Ed.*, 1982, **20**, 1299, 1315, and 1331; p. 27 of ref. 21.
[29] M. Stickley and G. Meyerhoff, *Polymer*, 1981, **22**, 929.
[30] P. Bataille, B. T. Van, and Q. B. Pham, *J. Polym. Sci., Polym. Chem. Ed.*, 1982, **20**, 795 and 811.

polymerization in the non-steady state, Birtwhistle and Blackey have devised an iterative procedure which allows any number of radicals to occupy a particular particle.[31] Under steady-state conditions a 'closure-recursion' method provides the average number of radicals per particle.[32] According to one model of a batch reactor the particle size distribution is determined by events during the nucleation stage,[33] but another found that during the interval II of the reaction new particles did form, to widen the particle size distribution.[30] A reaction model of emulsion copolymerization has been applied to styrene and acrylonitrile in an azeotropic composition,[34] and to styrene and methyl methacrylate.[35] The latter case deviated from the Smith–Ewart model (because of radical desorption) which has been developed by others for the general case during intervals II and III.[36] Simulations have been described of emulsion polymerizations in a tubular reactor[37] and of a seeded continuous stirred tank reactor.[38] The removal of oscillations in the polymer and particle properties and other commercial advantages can be obtained by altering the configuration of a reactor system.[39]

Other models with industrial applications include one of a high-pressure polyethylene reactor[40] and a controlled batch copolymerization reactor.[41] Expressions have been derived for modelling continuous flow reactors as a continuous time Markov chain.[42]

2 Polymer Characterization and Spectroscopy

The time for the attainment of an equilibrium sedimentation may be substantially reduced by a two-step initial loading, according to a simulation dependent upon the finite element method.[43] Elsewhere it has been shown by iterative computations when measurements of the compressibility of proteins will become sensitive to the influence of high pressures on the composition density distribution.[44] New methods of calibrating gel permeation chromatography curves using non-fractionated polymer samples of known $\bar{M}_n$ and $\bar{M}_w$ have been devised and tested.[45,46] An Apple II has been linked to a gel permeation chromatographic system to provide an analysis of the chromatographs,[47] the precision of which, according to a modelling study, may influence strongly the values of derived parameters (such as $\bar{M}_n$).[48] Some

---

[31] D. T. Birtwistle and D. C. Blackley, *J. Chem. Soc., Faraday Trans. 2*, 1981, **77**, 1351.
[32] M. J. Ballard, R. G. Gilbert, and D. H. Napper, *J. Polym. Sci., Polym. Lett. Ed.*, 1981, **19**, 533.
[33] D. A. Cauley and R. W. Thompson, *J. Appl. Polym. Sci.*, 1982, **27**, 363.
[34] C.-C. Lin, H.-C. Ku, and W. Y. Chiu, *J. Appl. Polym. Sci.*, 1981, **26**, 1327.
[35] M. Nomura, K. Yamamoto, I. Horie, K. Fujita, and M. Harada, *J. Appl. Polym. Sci.*, 1982, **27**, 2483.
[36] M. J. Ballard, D. H. Napper, and R. G. Gilbert, *J. Polym. Sci., Polym. Chem. Ed.*, 1981, **19**, 939.
[37] D. Lynch and C. Kaparissides, *J. Appl. Polym. Sci.*, 1981, **26**, 1283.
[38] C.-C. Lin and W. Y. Chiu, *J. Appl. Polym. Sci.*, 1982, **27**, 1977.
[39] M. Pollock, J. F. MacGregor, and A. E. Hamielec, in ref. 21, p. 209.
[40] S. Goto, K. Yamamoto, S. Furui, and M. Sugimoto, *J. Appl. Polym. Sci.*, 1981, **36**, 21.
[41] A. F. Johnson, B. Khaligh, and J. Ramsay, in ref. 21, p. 117.
[42] R. Nassar, J. R. Too, and L. T. Fan, *J. Appl. Polym. Sci.*, 1981, **26**, 3745.
[43] G. P. Todd and R. H. Haschemeyer, *Biopolymers*, 1982, **21**, 17.
[44] D. S. Sharp and J. B. Ifft, *Biopolymers*, 1982, **21**, 1127.
[45] M. Kubin, *J. Appl. Polym. Sci.*, 1982, **27**, 2933 and 2943.
[46] P. Szewczyk, *J. Appl. Polym. Sci.*, 1981, **26**, 2727.
[47] V. Narasimhan, A. R. Telfer, R.Y.-M. Huang, and C. M. Burns, *J. Appl. Polym. Sci.*, 1982, **27**, 3461.
[48] W. J. Tchir, A. Rudin, and C. A. Fyfe, *J. Polym. Sci., Polym. Phys. Ed.*, 1982, **20**, 1443.

methods of treating experimental[49] and simulated[50,51] photon correlation data have been compared, and the empirical histogram method defended[52] and developed.[49] A new solution has been reported for inversion of the Laplace transform, to provide molecular-weight distributions, based upon its recently discovered eigenvalues and eigenfunctions.[51] Full details have appeared of the methods of approximating molecular-weight distributions by a number of Δ-functions for the purpose of computing spinodals and other characteristics of phase equilibria.[53]

Standard linear and polynomial regressional analysis techniques have been applied to volume–temperature data on polymers, to show how an examination of the residuals helps the development of a correct algebraic description and the discovery of the transition temperatures.[54] Transitions within the liquid state have been identified by this means. Annealing has been incorporated in a model which reproduces differential scanning calorimetric data in the region of the glass transition temperature.[55] Factor analyses of the i.r. spectra of samples of polyethylene found orthorhombic crystallites together with disordered regions resembling the monoclinic phase.[56] A computer model which generates a polymer paracrystal has upheld a theory of the equilibrium size distribution of microparacrystals in polymers.[57]

Two books collecting articles on SAXS[58] and fibre diffraction methods[59] record the impact in these fields of computers through model building, data acquisition, and analysis. A similar insight into the art of refining protein structures has been provided in a volume from Daresbury.[60] One method of refinement from data of low resolution which is given elsewhere has been developed for use with a molecular graphics system,[61] while another, a modification of the real space method, utilizes the potential energy function UNICEPP.[62]

Methods for and accounts of the computer analysis and manipulation of the primary and secondary structure of nucleic acids have been collected in a single volume of *Nucleic Acids Research*.[63] The crystal structures of certain proteins have been examined from two points of view: the shape and surface features of globular proteins in relationship to the various residues present,[64] and the incidence of certain four residue structures within a differential geometric representation, which

---

[49] A. DiNapoli, B. Chu, and C. Cha, *Macromolecules*, 1982, **15**, 1174.
[50] D. Caroline, *Polymer*, 1982, **23**, 492.
[51] N. Ostrowsky, D. Sornette, P. Parker, and E. R. Pike, Opt. Acta, 1981, **28**, 1059.
[52] E. Gulari, E. Gulari, and B. Chu, *Polymer*, 1982, **23**, 649.
[53] P. A. Irvine and J. W. Kennedy, *Macromolecules*, 1982, **15**, 473.
[54] R. F. Boyer, R. L. Miller, and C. N. Park, *J. Appl. Polym. Sci.*, 1982, **27**, 1565; R. F. Boyer, *Macromolecules*, 1982, **15**, 1498.
[55] I. M. Hodge and A. R. Berens, *Macromolecules*, 1982, **15**, 762.
[56] D. J. Burchell and S. L. Hsu, *Polymer*, 1981, **22**, 907.
[57] A. M. Hindeleh and R. Hoseman, *Polymer*, 1982, **23**, 1101.
[58] 'Small Angle X-ray Scattering', ed. O. Glatter and O. Kratky, Academic Press, London, 1982.
[59] 'Fibre Diffraction Methods', ed. A. D. French and K. H. Gardner, *ACS Symp. Ser.*, 1980, **141**.
[60] 'Refinement of Protein Structure', ed. P. A. Machin, J. W. Campbell, and M. Elder, Science and Engineering Research Council, Daresbury Lab., 1981.
[61] D. S. Moss and A. J. Morffew, *Comput. Chem.*, 1982, **6**, 1.
[62] S. Fitzwalter and H. A. Scheraga, *Proc. Natl. Acad. Sci. USA*, 1982, **79**, 2133.
[63] 'The Application of Computers to Research on Nucleic Acids', ed. D. Sol and R. J. Roberts, IRL Press, Oxford, 1982; *Nucleic Acids Res.*, 1982, No. 1.
[64] M. Prabhakaran and P. K. Ponnuswamy, *Macromolecules*, 1982, **15**, 314.

through their stability may be considered as nucleation sites for extended and compact types of secondary structure.[65]

Sturm has devised a Monte Carlo technique to simulate equilibrium binding on a one-dimensional molecule with neighbouring group effects, which proved useful in a kinetics study.[66] Multiple closure approximations and matrix iteration methods have been developed for the kinetics of ligand binding which require less computational time than the Monte Carlo approach if the ligand binds at several sites.[67] In fitting the melting profiles of certain DNAs, an improvement was obtained by associating the stability parameter with near nearest-neighbour doublet base pairs, as a variation from the Fixman–Freiere treatment.[68]

3 Intermolecular Potentials and Force Fields

Three related aspects of the modelling of the interaction of enzymatic polypeptides with their substrates have been reviewed.[69-71] Warshell concentrates on the electrostatic basis of protein function,[69] Pincus and Scheraga deal with substrate recognition,[70] while Momany is concerned with the design of peptides, which, by having a particular configuration and the capacity for a particular interaction, might function as a drug.[74] By using a 'third generation' force field program, AMBER, designed to function in either Cartesian or internal co-ordinates as well as with interaction graphics,[72] the differential binding of several thyroid analogues to albumin have been simulated. Approximately 250 atoms within the protein were included.[73] In a similar manner have been computed the reaction paths for the hydrolysis of a polynucleotide by staphylococcal nuclease[74] and the peptide–protein interactions responsible for the catalytic effect of chromotrypsin.[75] A major part of the different behaviour of D and L substrate isomers in this case comes from differences in hydrogen bonding.

An account has appeared of the use of a high-speed array processor to search more extensively than allowed previously for the stable conformations of a polypeptide.[76] The initial problem, of locating the global minimum of many local minima on the potential energy surface, has been surmounted for particular cases such as small cyclic polypeptides[77] and the α-helix,[78] but remains for globular proteins.[77] A spacefilling model of bovine pancreatic trypsin inhibitor provided a set

[65] S. Rackovsky and H. A. Scheraga, *Macromolecules*, 1982, **15**, 1340.
[66] J. Sturm, *Biopolymers*, 1981, **20**, 753; 1982, **21**, 1189.
[67] C. Dates and I. R. Epstein, *Biopolymers*, 1981, **20**, 1651.
[68] O. Gotoh and Y. Tagashira, *Biopolymers*, 1981, **20**, 1033 and 1043.
[69] A. Warshell, *Acc. Chem. Res.*, 1981, **14**, 284.
[70] M. R. Pincus and H. A. Scheraga, *Acc. Chem. Res.*, 1981, **14**, 299.
[71] F. A. Momany in 'Crystal Cohesion and Conformational Energies', ed. R. M. Mutzger, Springer-Verlag, Berlin, 1981.
[72] P. K. Weiner and P. A. Kollman, *J. Comput. Chem.*, 1981, **2**, 287.
[73] J. M. Blaney, P. K. Weiner, A. Dearing, P. A. Kollman, E. C. Jorgensen, S. J. Oatley, J. M. Burridge, and C. C. F. Blake, *J. Am. Chem. Soc.*, 1982, **104**, 6424.
[74] J. A. Deiters, J. A. Gallucci, and R. R. Holmes, *J. Am. Chem. Soc.*, 1982, **104**, 5457.
[75] D. Los F. DeTar, *J. Am. Chem. Soc.*, 1981, **103**, 107.
[76] D. C. Rappoport and H. A. Scheraga, *Macromolecules*, 1981, **14**, 1238.
[77] H. A. Scheraga, *Biopolymers*, 1981, **20**, 1877.
[78] C. M. Venkatachalam, M. A. Khaled, H. Sugano, and D. N. Ury, *J. Am. Chem. Soc.*, 1981, **103**, 2372.

of initial structures distinguished by the spacial geometric arrangements of the loops (SGAL) created by the cystine links. Procedures which allowed small- and large-scale adjustments of both main-chain and side-chain dihedral angles did not find the native conformation even when the proper SGAL was chosen, but an improvement in energy was obtained when further short-ranged restrictions were introduced.[79] The strategy of building up the chain with judicial combinations of small peptides gave reasonable structures for the membrane portion of melittin[80] and for a hydrophobic hexapeptide portion of murine κ light chain.[81] The success of these procedures owed much to the actual structures occurring in a non-polar medium.[82] For the three-stranded collagen II a complete analysis of all possible conformations was performed using a single variable derived from a set of simultaneous equations.[83] Others have examined the conformational preferences of the side chains of collagen.[84] Cyclic oligopeptides have been studied as models for β-bends[85] and a spiral version of elastin.[86] Calculations have also been performed on the geometry of one type-II β-turn,[87] and of the influence of the sugar moiety on that of a glycosylated turn.[88]

Aspects of the conformations of DNA that have been examined by modelling include the coupling between helix twisting and sugar repuckering,[89] the relative energies of Z- and B-helices,[90] and of A- and B-helices,[91] the force constants for large-scale motions of the A- and B-forms,[92] and the design of intercalating agents.[93] Further illustration of the plasticity of the double helix is provided by Olson's transformation of the left-handed helix to a right-handed form.[94,95] Inadequacies have been identified in estimates of the potentials which govern pseudorotation of both the ribose and deoxyribose rings,[96] prior to the presentation of a force field which predicts barriers to pseudorotation of 6 to 8 kcal mol^{-1}. The single chain of poly(ribouridylic acid) has three principle conformational sequences.[97]

In their book,[98] Painter, Coleman, and Koenig show how the vibrational spectra may be treated as arising from various interatomic forces and other structural features, the success of which depends much upon the ability of computers to

---

[79] H. Meirovitch and H. A. Scheraga, *Macromolecules*, 1981, **14**, 1250; *Proc. Natl. Acad. Sci. USA*, 1981, **78**, 6584.
[80] M. R. Pincus, R. D. Klansner, and H. A. Scheraga, *Proc. Natl. Acad. Sci. USA*, 1982, **79**, 5107.
[81] M. R. Pincus and R. D. Klasner, *Proc. Natl. Acad. Sci. USA*, 1982, **79**, 3413.
[82] J. M. Dungan and T. M. Hooker, *Macromolecules*, 1981, **14**, 1812.
[83] V. G. Tumangan and N. G. Espiora, *Biopolymers*, 1982, **21**, 475.
[84] G. Nemethy and H. A. Scheraga, *Biopolymers*, 1982, **21**, 1535.
[85] G. Nemethy, R. J. McQuie, M. S. Pottle, and H. A. Scheraga, *Macromolecules*, 1981, **14**, 975.
[86] C. M. Venkatachalam and D. W. Ury, *Macromolecules*, 1981, **14**, 1225.
[87] B. V. Prasad, H. Balaram, and P. Balaram, *Biopolymers*, 1982, **21**, 1261.
[88] C. A. Bush, *Biopolymers*, 1983, **21**, 535.
[89] P. Kollman, J. W. Klesper, and P. Weiner, *Biopolymers*, 1982, **21**, 2345.
[90] P. Kollman, P. Weiner, G. Quigley, and A. Wang, *Biopolymers*, 1982, **21**, 1945.
[91] V. I. Poltev, L. A. Milova, B. S. Zhorov, and V. A. Govyrin, *Biopolymers*, 1981, **20**, 1.
[92] B. F. Putnam, E. W. Prohofsky, and L. L. Van Zandt, *Biopolymers*, 1982, **21**, 885.
[93] K. J. Miller and D. D. Newlin, *Biopolymers*, 1982, **21**, 633.
[94] W. K. Olson in 'Biomolecular Stereodynamics, I', ed. R. H. Sarma, Adenine Press, New York, 1981.
[95] C. K. Mitra, M. H. Sarma, and R. H. Sarma, *J. Am. Chem. Soc.*, 1981, **103**, 6727.
[96] W. K. Olson and J. L. Sussman, *J. Am. Chem. Soc.*, 1982, **104**, 270; W. K. Olson, ibid., p. 278.
[97] B. E. Hingerty, S. B. Broyde, and K. W. Olson, *Biopolymers*, 1982, **21**, 1167.
[98] P. C. Painter, M. M. Coleman, and J. L. Koenig, 'The Theory of Vibrational Spectroscopy and its Application to Polymeric Materials', Wiley, New York, 1982.

perform the necessary numerical calculations. Putnam and Van Zandt describe a procedure for obtaining macromolecular force constants which does not suffer from the instability associated with the large number of factors in the force field.[99] An alternative to the usual GF matrix method has been described by Tasumi et al. It treats a polypeptide as a sequence of more tractable subchains, as is shown by the example polypeptide with 29 residues, glucagon.[100] Complete analyses of the i.r. and Raman spectra of normal and partially deuteriated polyglycine in the I(α) and II(β) crystalline forms and for β-poly(L-alanine) have been reported.[101] If adjacent chains of polyglycine II were antiparallel, details in the fine structure were better reproduced. A dipeptide model for the type-II β-bend has also been examined.[102] An illustration of the scope of the method is provided by the current papers on isotactic poly(t-butyl acrylate)[103] and isotactic polystyrene,[104] which are both 3_1 helices, and on poly(vinylidene fluoride) in the phase III.[105] Tadokoro et al. have identified the nature of the disorders in the chain conformation which may be found in samples of this phase.[106]

Mansfield, in computing the potential energy surface of fragments of the poly(alkene sulphone) chains, has explored the dipole–dipole electrostatic interactions, developed a mean force theory of side-chain interactions, and showed how both factors tend to support a helix conformation in solution when each repeat unit carries one side chain.[107] There has been reported a conformational analysis of the crystalline states of syndiotactic polypropylene and 1,2-poly(1,3-butadiene).[108] Other studies on particular crystalline polymers have dealt with the packing of trans-polyacetylene, as this determines both the crystal structure[109] and the incidence of solitons (defects in double bond alternation),[110] with the packing of polytetrafluoroethylene in and near to the 13/6 helix,[111] and with the attraction between poly(ethylene sulphide) chains which cause a rather high melting point.[112] Chain packing in poly(cis-1,4-butadiene)[113] and in β-sheets of poly(L-valine)[114] does influence the main chain conformation. A new method of performing packing calculations for chains in fixed conformation has been utilized for polyethylene and for phase I of poly(vinylidene fluoride), in which dipolar forces are treated in a novel way.[115]

---

[99] B. F. Putnam and L. L. Van Zandt, *J. Comput. Chem.*, 1982, **3**, 297.
[100] M. Tasumi, H. Takeuchi, S. Ataka, A. M. Dwivedi, and S. Krimm, *Biopolymers*, 1982, **21**, 711.
[101] A. M. Dwivedi and S. Krimm, *Macromolecules*, 1982, **15**, 177 and 186; *Biopolymers*, 1982, **21**, 2377.
[102] F. R. Maxfield, J. Bandekar, S. Krimm, D. J. Evans, J. S. Leach, G. Nemethy, and H. A. Scheraga, *Macromolecules*, 1981, **14**, 997.
[103] N. N. Aylwood and S. S. Ti, *J. Polym. Sci., Polym. Phys. Ed.*, 1981, **19**, 1805.
[104] R. W. Snyder and P. C. Painter, *Polymer*, 1981, **22**, 1633.
[105] M. A. Bachmann and J. L. Koenig, *J. Chem. Phys.*, 1981, **74**, 5896.
[106] K. Tashiro, M. Kobayashi, and H. Tadokoro, 1981, **14**, 1757.
[107] M. Mansfield, *Macromolecules*, 1982, **15**, 1587.
[108] P. Corradini, R. Napolitano, V. Petraccone, B. Pizzio, and A. Tuzi, *Macromolecules*, 1982, **15**, 1207.
[109] V. Enkelmann, M. Monkenbusch, and G. Wegner, *Polymer*, 1982, **23**, 1583.
[110] R. H. Baughman and G. Moss, *J. Chem. Phys.*, 1982, **77**, 6321.
[111] B. L. Farmer and R. K. Eby, *Polymer*, 1981, **22**, 1487.
[112] B. Bhaumik and J. E. Mark, *Macromolecules*, 1981, **14**, 162.
[113] P. Corradini, R. Napolitano, V. Petraccone, B. Pirozzi, and A. Tuzi, *Eur. Polym. J.*, 1981, **17**, 1217.
[114] C.-K. Chou and H. A. Scheraga, *Proc. Natl. Acad. Sci. USA*, 1982, **79**, 7047.
[115] S. K. Tripathy, A. J. Hopfinger, and P. L. Taylor, *J. Phys. Chem.*, 1371, **85**, 1981.

4 Rotational Isomeric State Calculations

Tonelli and Schilling have reviewed[116] the application of rotational isomeric state matrix multiplication techniques to the interpretation of the tacticity and conformation related fine structure in the ^{13}C n.m.r. spectra of linear polymers. It is usual to have a single spectral parameter, an upfield shift of 2 to 5 p.p.m. for carbon atoms subject to a γ-*gauche* configuration interaction by an alkane carbon, or by Br, Cl, or F, but recently it has been found that ^{19}F nuclei in these 3-bond *gauche* arrangements experience much larger (10 to 30 p.p.m.) upfield shifts.[117] The conditional probability matrix featured, together with a model for chain packing in a treatment of the wide angle X-ray scattering of liquid polyethylene[118] which confirmed conventional understanding, and in a model of the free radical polymerization of ethylene, where the production of side chains by back-biting processes is governed by chain conformations.[119] The disorder longitudinal modes (D-LAM), which produce low-frequency vibrations in the Raman spectra of liquid n-alkanes and perfluoro-n-alkanes have also been treated.[120]

The theory of stereochemical equilibria in vinyl polymers has been generalized by Suter, to permit the successful calculation of the population of sequences up to the pentad level, within the five-state model, for polypropylene diastomeric structures equilibrated over Pd.[121] Others have discussed the ability of this and other models to analyse vicinal H···H coupling constants.[122] Such coupling constants in models of poly(alkyl propenyl ethers), together with ^{13}C shifts, have been computed, allowance being made for side-chain rotational states, and compared with experimental values obtained in a selection of solvents at different temperatures.[123]

In their treatment of the experimental dipole moment characteristics of poly(vinyl chloride), Riande *et al.* suggest that the CHCl group dipole moment lies 10 to 15° away from the C—Cl bond because of the presence of a C—H moment.[124] For simple models of this polymer and poly(vinylidene chloride), Boyd and Kesner found small induced moments from conformationally sensitive polarization effects, before proceeding to the statistical mechanical study.[125] In one of two R.I.S. models of poly(vinyl bromide) that have been described,[126,127] copolymers with ethylene were also examined.[126]

The experimental molar Kerr constants and dipole moments have been used to determine the tacticity of partially *p*-halogenated polystyrene.[128] Earlier peculiarities in the Kerr effect of poly(oxyethylene glycols) now seem to have been rationalized.[129] The success of their treatment of the configurational properties

---

[116] A. E. Tonelli and F. C. Schilling, *Acc. Chem. Res.*, 1981, **14**, 233.
[117] A. E. Tonelli, F. C. Schilling, and R. E. Cais, *Macromolecules*, 1982, **15**, 849.
[118] G. R. Mitchell, R. Lovell, and A. H. Windle, *Polymer*, 1982, **2**, 1273.
[119] W. L. Mattice and F. C. Stehling, *Macromolecules*, 1981, **14**, 1479.
[120] R. G. Snyder, *J. Chem. Phys.*, 1982, **76**, 3921.
[121] U. W. Suter, *Macromolecules*, 1981, **14**, 523; U. W. Suter and P. Neuenschwander, *ibid.*, p. 529.
[122] D. R. Ferro and M. Ragazzi, *Macromolecules*, 1981, **14**, 1830.
[123] K. Matsuzaki, H. Morii, T. Kanai, and Y. Fujiwara, *Macromolecules*, 1981, **14**, 1004; K. Matsuzaki, H. Morii, N. Inoue, T. Kanai, Y. Fujiwara, and T. Higashimura, *ibid.*, p. 1008.
[124] F. Blasco Cantera, E. Riande, J. P. Almendo, and E. Saiz, *Macromolecules*, 1981, **14**, 138.
[125] R. H. Boyd and L. Kesner, *J. Polym. Sci., Polym. Phys. Ed.*, 1981, **19**, 375 and 393.
[126] A. E. Tonelli, *Macromolecules*, 1982, **15**, 290.
[127] E. Saiz, *Macromolecules*, 1982, **15**, 1152.
[128] G. Kanarian, R. E. Cais, J. M. Kometani, and A. E. Tonelli, *Macromolecules*, 1982, **15**, 866.
[129] G. Khanarian and A. E. Tonelli, *Macromolecules*, 1982, **15**, 145.

which depend upon the optical anisotropy of the groups in the polycarbonate of diphenyl propane was attributed by Flory et al. to the lack of strong mutual inductive effects because of good separation of the polar groups.[130] Kanarian has developed the R.I.S. theory of the perturbations in conformations in a non-Gaussian dipolar chain molecule caused by an external electric field to give the magnitude of the non-linear dielectric effect for several polymers and oligomers.[131] For similar systems the direct current electric field induced second-harmonic generation has been calculated following the derivation of the necessary equations.[132]

Models with two rotational states for the main chains are described for poly(N-vinyl pyrrolidone)[133] and poly(methyl acrylate).[134] The side-chain dipoles were respectively pre-averaged and fixed in orientation within the reference frame of the main chain atoms. The configurations of the main chains of poly(3,3-dimethyl oxetane),[135] poly(1-oxa-3-thiacyclopentane),[136] altcopoly(pentamethylene sulphide ethylene sulphide),[137] poly(3,3-dimethyl thietane),[138] poly(diethylene glycol terephthalate),[139] and poly(diethylene glycol terephthalate)[140] have all been computed and compared with one or more experimental properties (proton coupling constant,[140] unperturbed dimensions including the temperature coefficient,[135,139] and dipole moments including temperature coefficients[135—138]).

The expansion in dimensions and other related effects in an otherwise unperturbed linear and branched polymethylene chain which would be caused by placing like charges at each end have been examined in media of different dielectric constants.[141] If these and other types of perturbation are expressed in the generator method by Mattice, the probability of *trans* conformations is unaltered, but their distribution on the backbone is modified through the second-order statistical weights, which become dependent upon bond location. When comparing this approach to the more time-consuming Monte Carlo method, the conventional excluded volume effect was obtained by causing perturbations at the centre of the chain to allow a study of the asymmetry and overall expansion of finite chains and of sub-chains in an infinitely long chain.[142] The method has been used to show that the location of a sub-chain in a linear molecule influences the angular light-scattering function at low angles,[143] and that chain asymmetry may be sensibly affected by attaching an end to a wall, though the perturbing effect upon *trans* conformational weight persists merely for about 20 bonds along the chain.[144] Numerical analysis of

---

[130] B. Erdman, D. C. Marvin, P. A. Irvine, and P. J. Flory, *Macromolecules*, 1982, **15**, 664; B. E. Wu, P. A. Irvine, D. C. Martin, and P. J. Flory, *ibid.*, p. 670.
[131] G. Khanarian, *Macromolecules*, 1982, **15**, 1429; *J. Chem. Phys.*, 1982, **76**, 3186.
[132] G. Khanarian, *J. Chem. Phys.*, 1982, **77**, 2684.
[133] A. E. Tonelli, *Polymer*, 1982, **23**, 676.
[134] T. M. Birshtein, A. A. Merkireva, and A. N. Goryunov, *Polym. Sci. USSR*, 1981, **22**, 889.
[135] L. Garrido, E. Riande, and J. Guzman, *J. Polym. Sci., Polym. Phys. Ed.*, 1982, **20**, 1805.
[136] E. Riande and J. Guzman, *Macromolecules*, 1981, **14**, 1234 and 1511.
[137] E. Riande, J. Guzman, E. Saiz, and J. de Abajo, *Macromolecules*, 1981, **14**, 608.
[138] E. Riande, J. Guzman, and M. A. Llorente, *Macromolecules*, 1982, **15**, 298.
[139] J. S. Roman, J. Guzman, and E. Riande, *Macromolecules*, 1982, **15**, 609.
[140] J. S. Roman, J. Guzman, E. Riande, J. Santora, and M. Rico, *Macromolecules*, 1982, **15**, 609.
[141] W. L. Mattice and J. Skolnick, *Macromolecules*, 1981, **14**, 863 and 1463.
[142] W. L. Mattice, *Macromolecules*, 1981, **14**, 1485 and 1491.
[143] W. L. Mattice, *Macromolecules*, 1982, **15**, 579.
[144] W. L. Mattice and D. H. Napper, *Macromolecules*, 1981, **14**, 1066.

the partition function in a theory of helix–coil transitions of two-chained coiled coils and an expression of the co-operativity of conformational transitions in polysaccharide derivatives depended upon matrix algebra techniques.[145] For a treatment of the tropomyosin dimer, Mattice and Skolnick have formulated the partition function to express the effect of crosslink on the co-operativity of the helix ordering process.[146] Freed et al. have provided for random copolymers an alternative method of calculating the conformational free energy and characteristic ratio to the use of a Monte Carlo sequence of generator matrices.[147] For the free energy the extrapolations involved seem to be algebraically simpler and to have less uncertainty. It has been shown how the alternative Fourier method of configurational statistics may be applied to chains of finite length and of non-uniform structure.[148]

5 Monte Carlo and Molecular Dynamic Simulations

To examine the theory of reptation for polymer melts, Evans and Edwards have modelled Rouse chains on a lattice within fixed obstacle meshes. The dynamic behaviour of a simple linear chain was as expected (e.g., $D_R \sim N^{-2}$), and the ideas of Doi and Edwards of the static and dynamic properties of the 'primitive chain' in their theory of the viscoelastic properties of polymer melts were verified.[149] In contrast, the results for a three-branched star molecule were closer to experimental behaviour than certain experimental expectations, which would not permit such a molecule to reptate, so that the discovery, $D_R \sim N^{-3}$ has yet to be explained.[150] Evidence for reptation of a linear chain was found in dynamic models by Baumgartner and Binder[151] and by Bishop et al.[152] if the concentration was relatively high and the temperature was lowered, or some other strategy used, to freeze the environmental molecules. Such studies are the culmination of many years' development, and tax the capacity of the computers presently available to achieve the time scales greater than that of the slowest motion when many molecules are present.[153] With a similar model, with excluded volume interactions, the screening length of the 'blob hypothesis' was found to diminish as concentration rose.[154] Chain statistics in the bulk were studied by Mansfield, whose method of producing new configurations – by breaking and reforming chains – allowed every lattice site to be occupied.[155] Though chains extended to 512 beads, the M exponent v, characterizing mean chain dimensions, remained 2 to 4% above the expected value of 0.50.

The mutual orientational and conformational ordering of flexible polymers by

---

[145] J. Skolnick and A. Holtzer, *Macromolecules*, 1982, **15**, 303 and 812.
[146] W. L. Mattice and J. Skolnick, *Macromolecules*, 1982, **15**, 1088.
[147] A. L. Kholodenko and K. F. Freed, *Macromolecules*, 1982, **15**, 899.
[148] S. Bruckner, *Macromolecules*, 1981, **14**, 449.
[149] K. E. Evans and S. F. Edwards, *J. Chem. Soc., Faraday Trans. 2*, 1981, **77**, 1891, 1913, and 1929.
[150] K. E. Evans, *J. Chem. Soc., Faraday Trans. 2*, 1981, **77**, 2385.
[151] A. Baumgartner and K. Binder, *J. Chem. Phys.*, 1981, **75**, 2994.
[152] M. Bishop, D. Ceperley, H. L. Frisch, and M. H. Kalos, *J. Chem. Phys.*, 1982, **76**, 1557.
[153] M. Bishop, D. Ceperley, H. F. Frisch, and M. Kalos, in 'Supercomputers in Chemistry', ed. P. Lykos and I. Shavitt, *ACS Symp. Ser.*, **173**, 1981.
[154] P. G. Khalatur, S. G. Pletneva, and Y. G. Papulov, *J. Phys. Lett.*, 1982, **43**, L683.
[155] M. Mansfield, *J. Chem. Phys.*, 1982, **77**, 1555.

strong intermolecular interactions may be expected in concentrated solutions.[156] The lively discussion[157] on the types of loops and the incidence of tie molecules between lamellae in non-equilibrium bulk crystallized polyethylene has been developed in a Monte Carlo treatment of the 'Gambler's Ruin' method, which used a 'real chain', to find that a large fraction ($\sim \frac{2}{3}$) of the chains reaching the lamella surface must be in a tight fold, in order to maintain normal amorphous densities.[158]

Lebowitz et al., using a bead-spring model including excluded volume,[159] have analysed the simulation by a special Monte Carlo procedure of the static and dynamic properties of a single polymer chain. Others have used a molecular dynamics model of a bead-spring chain in a solvent near its triple point.[160] Equilibrium properties of a single chain were also sampled by a reptation Monte Carlo dynamic technique in the region of the θ-point, and at lower temperatures when the chain collapses to a globule.[161] Under similar circumstances, according to a lattice study, multibranch chains may undergo the transition more readily.[162] This and related topics received the attention of Khalatur, who generated polymethylene chains off lattice.[163] Star-branched polymers on a tetrahedral lattice[164] and comb-branched polymers on a cubic lattice[165] show deviations from the ideal random flight behaviour, even when θ-conditions are obtained. This and earlier work has verified that whatever model is used the mean-square dimensions of a single linear chain scale with $M^{6/5}$ in a good solvent.[166]

When a chain is confined in two dimensions, its collapse to a globule upon decreasing the temperature appears to be qualitatively different from that of a chain in three dimensions, in terms of the behaviour of the chain configurations[167] and the thermodynamic functions.[168] Contraction is also obtained by increasing chain concentration,[169] when a segregation tendency is also seen.[170] Further tests of scaling theory were also performed in studies of the influence of a strongly adsorbing surface on the configurations of one chain,[171] and in an investigation of span limitations.[172] The dynamics of adsorption and desorption of a macromolecule on a surface has been modelled by Birshtein et al.[173] The stabilization and flocculation of colloidal suspensions by dissolved polymers has been simulated by placing an excluded volume chain between two parallel plates, which tend to be pushed apart

---

[156] T. M. Birshtein, A. A. Sariban, and A. M. Skvortsov, *Polymer*, **23**, 1481.
[157] P. J. Flory, *J. Chem. Soc., Faraday Discuss.*, 1979, **68**, 489.
[158] C. M. Guttman and E. A. DiMarzio, *Macromolecules*, 1982, **15**, 525.
[159] D. Ceperley, M. H. Kalos, and J. L. Lebowitz, *Macromolecules*, 1981, **14**, 1472.
[160] W. Bruns and R. Bansal, *J. Chem. Phys.*, 1981, **74**, 2064; **75**, 5149.
[161] I. Webman, J. L. Lebowitz, and M. H. Kalos, *Macromolecules*, 1981, **14**, 1495.
[162] K. Kajiwara and W. Burchard, *Macromolecules*, 1982, **15**, 660.
[163] P. G. Khalatur, *Polym. Sci. USSR*, 1981, **22**, 2247.
[164] A. Kolinski and A. Sikorski, *J. Polym. Sci., Polym. Lett. Ed.*, 1982, **20**, 177; *Polym. Chem.*, 1982, **20**, 3147.
[165] F. L. McCrackin and I. Mazur, *Macromolecules*, 1981, **14**, 1214.
[166] W. Bruns, in ref. 19, p. 105.
[167] J. Tobochnik, I. Webman, J. L. Lebowitz, and M. H. Kalos, *Macromolecules*, 1982, **15**, 549.
[168] A. Baumgartner, *J. Phys.*, 1982, **43**, 1407.
[169] M. Bishop, D. Ceperley, H. L. Frisch, and M. H. Kalos, *J. Chem. Phys.*, 1981, **75**, 5538.
[170] A. Baumgartner, *Polymer*, 1982, **23**, 282.
[171] E. Eisenriegler, K. Kremer, and K. Binder, *J. Chem. Phys.*, 1982, **77**, 6296; T. Ishinabe, *ibid.*, 1982, **76**, 5589.
[172] R. Barr, C. Bender, and M. Lax, *J. Chem. Phys.*, 1981, **75**, 453 and 461.
[173] T. M. Birshtein, V. N. Gridnev, and A. M. Skvortsov, *Polym. Sci. USSR*, 1981, **23**, 330.

by entropy effects provided that the energy of adsorption and the width of the gap do not promote the formation of bridges.[174]

Van Gunsteren and Karplus have progressively constrained bond lengths then bond angles in molecular dynamics simulations of bovine pancreatic trypsin inhibitor, to discover how computer requirements and the motions of the atoms are affected.[175] A similar investigation of the stochastic dynamics of n-alkanes was elsewhere reported by Van Gunsteren et al.,[176] who found the behaviour of n-decane, in contrast to that of n-butane, to lack a sensitivity to solvent packing effects. The n.m.r. relaxation parameters of such molecules and of alkane side chains of a macromolecule have been obtained from a stochastic dynamics simulation,[177] while the molecular dynamics simulation of an α-helix has been compared with analytical results and experimental values,[178] and the transition state theory for rotational isomerism of a tyrosine side chain has been evaluated.[179]

The review by Winnik[180] deals with the cyclization of hydrocarbon chains, experimental studies of which are often tested against Monte Carlo models, as at present when exciplex formation between chain ends is controlled by chain conformation.[181]

---

[174] A. T. Clark and M. Lal, *J. Chem. Soc., Faraday Trans. 2*, 1981, **77**, 981; in 'The Effect of Polymers on Dispersion Properties', ed. Th. F. Tadros, Academic Press, London, 1982.
[175] W. F. Van Gunsteren and M. Karplus, *Macromolecules*, 1982, **15**, 1528.
[176] W. F. Van Gunsteren, H. J. C. Berendsen, and J. A. C. Rullman, *Mol. Phys.*, 1981, **44**, 69.
[177] R. M. Levy, M. Karpous, and J. A. McCammon, *J. Am. Chem. Soc.*, 1981, **103**, 994 and 5998.
[178] R. M. Levy, D. Perahia, and M. Karplus, *Proc. Natl. Acad. Sci. USA*, 1982, **79**, 1346.
[179] S. H. Northrup, M. R. Pear, C.-Y. Lee, and J. A. McCammon, *Proc. Natl. Acad. Sci. USA*, 1982, **79**, 4035.
[180] M. A. Winnik, *Chem. Rev.*, 1981, **81**, 491.
[181] A. Mar, S. J. Frazer, and M. A. Winnik, *J. Am. Chem. Soc.*, 1981, **103**, 4941; S. J. Frazer and M. A. Winnik, *J. Chem. Phys.*, 1981, **75**, 4683; M. Sisido and Y. Imanishi, *Biopolymers*, 1982, **21**, 1613.

19
Selected Topics in the Photochemistry of Polymers

BY K. L. PETRAK AND M. D. PURBRICK

1 Introduction

In this article it is our intention to cover material complementary to that dealt with in the Specialist Periodical Reports on Photochemistry.[1] Other more general reviews are also available.[2-4] Because of space limitations, we have restricted this review to three topics. Two of these – photoconductivity and resist materials – have found extensive technological applications. The third topic, energy transfer, encompasses material of relevance to both these areas, and precedes them in this article.

2 Energy Transfer and Related Topics

Empirical calculations of energy transfer in copolymers of N-vinylcarbazole with other vinyl aromatic monomers have been carried out by Cabaness.[5] It was demonstrated that pendant phenyl groups had sufficient mobility to interact with the aromatic rings of the carbazole. However, the insertion of a carbonyl spacing group between the polymer backbone and the naphthyl substituent precluded similar interaction, giving rise to exciton traps and short migration lengths. Fluorescence depolarization measurements by Guillet et al.[6] on poly-[(9-phenanthryl)methyl methacrylate] and its copolymers with (9-anthryl)methyl methacrylate and methyl methacrylate gave evidence, at 77 K, for substantial singlet energy migration among the phenanthrene chromophores. In fluid solutions, the efficiency of intrachain singlet energy transfer from phenanthrene to anthracene units was measured as a function of the mole fraction of the anthracene component. The same authors also described[7] transient fluorescence decay measurements on these polymers and the terpolymer, poly[methyl methacrylate-co-(9-phenanthryl)-methyl methacrylate-co-(9-anthryl)methyl methacrylate]. Singlet energy migration

---

[1] 'Photochemistry', ed. D. Bryce-Smith (Specialist Periodical Reports), The Royal Society of Chemistry, 1970—1982, Vol. 1—12.
[2] N. S. Allen, *Photochemistry*, 1982, **12**, 475.
[3] W. Schnabel, *Dev. Polym. Photochem.*, 1982, **3**, 237.
[4] J. L. R. Williams and M. F. Molaire, 'Kirk-Othmer Encyclopedia of Chemical Technology', 3rd Ed., ed. M. Grayson and D. Eckroth, 1982, **17**, 680.
[5] W. R. Cabaness, *Polym. Prepr., Am. Chem. Soc., Div. Polym. Chem.*, 1980, **21**, 261.
[6] D. Ng and J. E. Guillet, *Macromolecules*, 1982, **15**, 724.
[7] D. Ng and J. E. Guillet, *Macromolecules*, 1982, **15**, 728.

was significant, and increased with the phenanthrene contents of the polymers. At 77 K in a 2-methyltetrahydrofuran–tetrahydrofuran glass (35:65 V/V), the migration coefficient of the singlet excitation in the poly(9-phenanthryl)methyl methacrylate homopolymer was *ca.* 1.5×10^{-5} cm^2 s^{-1}.

A one-dimensional random walk model,[8] developed to describe singlet exciton migration in isolated chains of aromatic vinyl polymers, successfully explained the molecular-weight dependence of intramolecular excimer fluorescence from poly(2-vinylnaphthalene) in dilute solution at low temperature. In the model, the polymer chain was considered to contain two types of site, namely traps – consisting of *trans, trans meso* dyads, which could lead to excimer fluorescence – and non-traps – single repeat units or monomer sites. A similar model has been applied to the study of energy migration in polymers of *N*-vinylcarbazole incorporating nitrated substituents as excimer-forming sites,[9] fluorescence spectra being obtained for a series of such polymers with different degrees of nitration. The analysis of the ratio of the maximum intensities of fluorescence and phosphorescence at 4.2, 77, and 293 K was interpreted[10] as indicating that energy transfer involved singlet excitons.

The results of a study of singlet excitation energy migration in vinyl and methacrylate polymers containing the 1,3,5-triphenyl-2-pyrazoline chromophore as a pendant unit were discussed in terms of a random migration model.[11] At 0 °C, it was shown that the singlet excitation energy migrated over 48 pendant groups in poly[1,3-diphenyl-5-(*p*-vinylphenyl)-2-pyrazoline] within the lifetime of the excited state (*ca.* 5 ns). The consequences of the steric effects exerted by a tertiary butyl group on intramolecular energy transfer from naphthalene to anthracene units have been reported.[12] Intramolecular singlet energy transfer in poly(2-vinylnaphthalene), poly(2-t-butyl-6-vinylnaphthalene) and in copolymers of each of the monomers with 2-vinylanthracene (0.2—3.6 mole%) indicated that steric interactions not only decreased excimer formation, and therefore donor lifetime, but also suppressed energy migration from the donor chromophores and so significantly reduced the overall transfer efficiency.

The experimental evidence for singlet energy migration in poly(vinylarenes) has been critically examined by MacCallum.[13] In particular, he rejects Klöppfer's[14] explanation, based on singlet energy transfer, for the delayed fluorescence observed with poly(*N*-vinylcarbazole) films containing perylene. Instead, MacCallum proposed a mechanism involving migrating triplet energy leading to triplet–triplet annihilation. His general contention[13] was that there was no conclusive evidence to support the proposal that singlet energy migration was a common phenomenon in polymers having aromatic substituents, and that triplet energy migration was in many cases a more likely process. It was the long lifetime and favourable absorption properties of the triplet state of polymers and copolymers of *p*-methoxyacrylo-

---

[8] P. D. Fitzgibbon and C. W. Frank, *Macromolecules*, 1982, **15**, 733.
[9] V. V. Slobodyanik, A. N. Faidysh, V. N. Yashchuk, and L. N. Fedorova, *Zh. Prikl. Spektrosk.*, 1982, **36**, 309.
[10] V. V. Slobodyanik, A. N. Faidysh, V. N. Yashchuk, and L. N. Fedorova, *Ukr. Fiz. Zh. (Russ. Ed.)*, 1982, **27**, 506.
[11] F. Iinuma, H. Mikawa, and Y. Shirota, *Mol. Cryst. Liq. Cryst.*, 1981, **73**, 309.
[12] T. Nakahira, T. Sasaoka, S. Iwabuchi, and K. Kojima, *Makromol. Chem.*, 1982, **183**, 1239.
[13] J. R. MacCallum, *Eur. Polym. J.*, 1981, **17**, 209.
[14] W. Klöpffer, *J. Chem. Phys.*, 1969, **50**, 2337.

phenone which commended their use in laser flash photolysis studies of energy transfer and migration processes.[15] The homopolymer and copolymers with styrene, methyl methacrylate, or phenyl vinyl ketone were used. Intermolecular quenching by oxygen, 1-methylnaphthalene, and conjugated dienes was examined. It was concluded that efficient energy migration played a significant role in triplet state processes and proceeded at a hopping frequency of 4.3×10^{11} s^{-1}. The influence of polymer tacticity on triplet energy migration in poly(acrylophenone) has been investigated.[16] The quenching of phosphorescence intensity in solid solutions at 77 K was described by the Perrin model. Active spheres of quenching differed for the isotactic and atactic polymers, being 2.25 nm and 2.03 nm, respectively.

Fluorescence emission investigations of energy transfer processes and excimer/exciplex kinetics in macromolecular systems may serve as useful probes for polymer miscibility. Their use in the elucidation of polymer–polymer interactions has been reviewed by Ledwith.[17] Serious doubts have recently been voiced[18—21] regarding the applicability of Birks'[22] kinetic analysis to intramolecular excimer formation in macromolecules. The use of laser excitation[23] upon a series of 1-vinylnaphthalene/methyl methacrylate copolymers of varying composition allowed the individual rate constants governing intramolecular excimer formation to be determined over a temperature range 205—295 K. The proposed kinetic scheme involved the excimer, D^*, and two kinetically distinct 'monomer' entities, M_1^* and M_2^*. Of these, only M_1^* could interact with D^* by exciton diffusion. M_2^* was excluded from this process, and was involved in excimer formation solely through energy transfer to M_1^*. The photoluminescence characteristics of styrene/2-vinylnaphthalene copolymers have been studied under photostationary and transient conditions in compiling a kinetic analysis of excimer emission.[24] Excited state properties of poly(2-vinylnaphthalene) containing pyrene have also been investigated.[25] The mode of incorporation of the pyrene influenced the rate of quenching of its fluorescence; as an end-group, this was essentially unchanged from the value for 3-methylpyrene, and double that observed as a pendant group.

Iwaya and Tazuke[26] have described work on the exciplex emission behaviour of a polymethacrylate with 2-[(1-pyrenyl)methyl]-2-[4-(dimethylamino)benzyl]ethyl substituents. Exciplex emission intensity was concentration dependent, indicative of the presence of both intra- and inter-molecular species. Of these, the latter were solely responsible for the observed molecular weight dependence. An optimum molecular weight for exciplex formation was apparent, resulting from the counterbalancing factors of increased zipping effect (favourable) and increased chain entanglement and shrinking (unfavourable) as molecular weight rose. Excimer

---

[15] J. S. Scaiano and J. C. Selwyn, *Macromolecules*, 1981, **14**, 1723.
[16] T. Kilp and J. E. Guillet, *Macromolecules*, 1981, **14**, 1680.
[17] A. Ledwith, *Pure Appl. Chem.*, 1982, **54**, 549.
[18] D. Phillips, A. J. Roberts, and I. Soutar, *Polymer*, 1981, **22**, 293.
[19] D. Phillips, A. J. Roberts, and I. Soutar, *Eur. Polym. J.*, 1981, **17**, 101.
[20] D. Phillips, A. J. Roberts, and I. Soutar, *Polymer*, 1981, **22**, 427.
[21] F. C. De Schryver, K. Demeyer, M. Van der Anweraer, and E. Quanten, *Ann. N Y Acad. Sci.*, 1981, **366**, 93.
[22] J. B. Birks, 'Photophysics of Aromatic Molecules', Wiley, London, 1970, Chap. 7.
[23] D. Phillips, A. J. Roberts, and I. Soutar, *J. Polym. Sci., Polym. Phys. Ed.*, 1982, **20**, 411.
[24] S. Ito, M. Yamamoto, and Y. Nishijima, *Polym. J. (Tokyo)*, 1981, **13**, 791.
[25] J. S. Hargreaves and S. E. Webber, *Macromolecules*, 1982, **15**, 424.
[26] Y. Iwaya and S. Tazuke, *Macromolecules*, 1982, **15**, 396.

formation by polyesters and by their dimer model compounds was also studied.[27] Excimer formation by the polymers was up to ten times more efficient than that by the dimer models.

Energy transfer is frequently studied in blended or doped polymer systems. For example, excimer fluorescence was used to study the conformational properties of, and energy migration in, polystyrene dispersed at low concentration in blends with poly(methyl vinyl ether).[28] Fluorescence spectra of monodisperse polystyrene samples of varying molecular weights were taken at 286—323 K and analysed with the one-dimensional random walk model described earlier.[8] The molecular weight dependence of the ratio of excimer to monomer emission was consistent with the predictions of the model up to 303 K. Slight deviations encountered above this temperature were attributed to unfavourable thermodynamic interactions as the blend approached the lower critical solution temperature. The same workers[29] developed a three-dimensional random walk energy migration model to explain the dependence of the ratio of excimer to monomer fluorescence on polystyrene concentration in these blends. The lifetime of phosphorescence emission from triphenylene and coronene suspended in a range of styrene/methyl methacrylate copolymers was measured at 77 K by MacCallum et al.[30] The triplet lifetimes of both polycyclic additives decreased as the mole fraction of styrene in the copolymer increased. It was proposed that this was due to the quenching of the triplet state of the additive by the polymer matrix, presumably by an energy transfer mechanism, with the styrene unit being a more efficient participant than the methacrylate. Singlet energy transfer in polystyrene–polyphenylene oxide blends has been investigated.[31] Efficient quenching of excited polystyrene by polyphenylene oxide and its monomer, xylenol, was thought to contribute to the stability against photo-oxidative degradation shown by the blends. This stability, despite the greater ease of localized motion in polyphenylene oxide,[32,33] was greater than that of polystyrene alone.

The energy transfer process to guest molecules in doped poly(N-vinylcarbazole) films was directly measured using a picosecond streak camera system.[34] In a film containing dimethyl terephthalate, exciplex formation with the carbazole was thought to occur through the trapping of the migrating terephthalate singlet excited state. The methylene blue (MB) photosensitized isomerization of cis-p-(phenylazo)-phenyltrimethylammonium iodide (cis-PTA) was used to investigate the effect of polyanions on triplet excitation energy transfer between cationic dyes in aqueous media.[35] The efficiency of energy transfer was particularly enhanced in the presence of poly(styrenesulphonate) and poly(vinyl sulphate) anions. These polyanions were also used in a study on the effect of dye aggregation on MB–cis–PTA energy transfer

---

[27] S. Tazuke, H. Ooki, and K. Sato, Macromolecules, 1982, 15, 400.
[28] R. Gelles and C. W. Frank, Macromolecules, 1982, 15, 741.
[29] R. Gelles and C. W. Frank, Macromolecules, 1982, 15, 747.
[30] A. N. Jassim, J. R. MacCallum, and T. M. Shepherd, Eur. Polym. J., 1981, 17, 125.
[31] M. Kryszewski, B. Wandelt, D. J. S. Birch, R. E. Imhof, A. M. North, and R. A. Pethrick, Polymer, 1982, 23, 924.
[32] B. Wandelt, Polym. Bull., 1981, 4, 199.
[33] B. Wandelt and M. Kryszewski, Eur. Polym. J., 1980, 16, 583.
[34] H. Masuhara, N. Tamai, and N. Mataga, Chem. Phys. Lett., 1982, 91, 209.
[35] M. Shirai, M. Ohyabu, and M. Tanaka, J. Polym. Sci., Polym. Chem. Ed., 1981, 19, 1847.

conducted by the same workers.[36] It was found that the formation of highly aggregated MB reduced the efficiency of energy transfer.

As the final topic under this heading, the influence of polymers as reaction media upon energy transfer and electron transfer will be reviewed. The energy transfer efficiencies of free and polymer-bound acetophenone and benzophenone have been compared by Catalina et al.[37] and found to be higher in the free state than when polymer-bound. The concentration dependence of the phosphorescence of benzil, biacetyl, or benzophenone in polymers or solvents indicated that energy transfer, in the deactivation of the excited triplet, involved charge transfer.[38] The process was slower in the polymer than in solution and was diffusion controlled. Amphiphilic copolymers of 2-acrylamido-2-methylpropanesulphonic acid (AMPS) with styrene,[39] 9-vinylphenanthrene, and 1-vinyl-pyrene[40] were prepared as potential media for photo-induced electron transfer reactions. The fluorescence spectrum of the AMPS–9–vinylphenanthrene copolymer in aqueous solution indicated quenching *via* excimer formation by phenanthryl groups. Also, the fluorescence of the copolymers in aqueous solution was quenched far more effectively by bis(2-hydroxyethyl)terephthalate, an amphiphilic quencher, than by the hydrophilic quencher, fumaric acid.

3 Photoconductivity

Extrinsic charge carrier photogeneration in poly(N-vinylcarbazole) has been studied by Yokoyama et al.[41,42] Electric field-induced quenching of exciplex[41,42] and charge-transfer complex[42] fluorescence has been observed with polymer films doped with weak electron acceptors, such as dimethyl terephthalate, and stronger acceptors, such as tetracyanobenzene, respectively. Based upon analyses of field-induced fluorescence quenching experiments, a general mechanism was advanced for the sensitizing of carrier generation through the addition of electron acceptors. It was proposed that carrier photogeneration in solid-state excited donor–acceptor systems involved field-assisted thermal dissociation of an ion pair. An attempt was made by the same workers[43,44] to estimate the trap-free drift mobility in poly(N-vinylcarbazole) from the transient photocurrent curves by modelling the thermal equilibrium between trapped and free carriers. Charge carrier photogeneration in pure and sensitized poly(N-epoxypropylcarbazole)[45] was studied by measuring exciplex fluorescence and surface conductivity. Sensitization was effected by incorporation of a rhodamine dye. The influence of oxygen was examined, its

---

[36] M. Shirai, M. Ohyabu, Y. Ono, and M. Tanaka, *J. Polym. Sci., Polym. Chem. Ed.*, 1982, **20**, 555.
[37] F. Catalina, R. Martinez-Utrilla, and R. Sastre, *Polym. Bull.*, 1982, **8**, 369.
[38] M. V. Encina, E. A. Lissi, and F. A. Olea, *Bol. Soc. Chil. Quin.*, 1982, **27**, 200.
[39] Y. Morishima, Y. Itoh, and S. Nozakura, *Makromol. Chem.*, 1981, **182**, 3135.
[40] Y. Itoh, Y. Morishima, and S. Nozakura, *J. Polym. Sci., Polym. Chem. Ed.*, 1982, **20**, 467.
[41] M. Yokoyama and H. Mikawa, *Photogr. Sci. Eng.*, 1982, **26**, 143.
[42] M. Yokoyama, S. Shimokohara, A. Matsubara, and H. Mikawa, *J. Chem. Phys.*, 1982, **76**, 724.
[43] M. Fujino, H. Mikawa, and M. Yokoyama, *Polym. J. (Tokyo)*, 1982, **14**, 81.
[44] M. Fujino, H. Mikawa, and M. Yokoyama, *Photogr. Sci. Eng.*, 1982, **26**, 84.
[45] D. I. Kadyrov, B. M. Rumyantsev, I. A. Sokolik, and E. L. Frankevich, *Polym. Photochem.*, 1982, **2**, 243.

participation in charge-transfer complex formation with polymer molecules being identified as an intermediate step in free charge carrier generation.

Intramolecular excimeric states were deemed responsible[46] for intrinsic excitonic photogeneration in a series of solid polymers including poly(N-vinyl-7H-benzo[c]-carbazole), its copolymer with octyl methacrylate, poly(N-vinylcarbazole)-co-(N-vinyl-5H-benzo[b]carbazole), and poly(N-vinylcarbazole) itself. Both energy transfer in films of poly(N-vinylcarbazole) at low temperatures, due to singlet and triplet excitons whose migratory range was limited by excimer-forming sites, and delayed fluorescence, arising from triplet–triplet annihilation between triplet excimer and triplet exciton, were described by an exciton hopping model.[47] This predicted concentrations of excimer-forming sites of $ca.$ 10^{-5} and 10^{-4} mol (mol monomer unit)$^{-1}$ respectively for the cationically- and radically-formed polymers. The model also described satisfactorily the dependence of fluorescence, phosphorescence, and delayed fluorescence intensities on excitation intensity. Enthalpy of charge-transfer complex formation and photocurrent values were measured for sandwich samples containing nickel–poly(N-vinylcarbazole) (sensitized) and gold–electron acceptor complexes in efforts to elicit a relationship between the photoconductivity of the complexes and their charge-transfer behaviour.[48] p-Chloranil, p-fluoroanil, tetracyanoethylene, and 7,7′,8,8′-tetracyanoquinonedimethane were among the electron acceptors used.

Intrinsic photoconduction induced by vacuum ultraviolet light was studied in polyethylene, poly(vinylidene fluoride), and polystyrene.[49] The photoconduction response observed in high-density polyethylene differed from that in the low-density polymer and indicated the possibility of ionic carrier migration in the latter at higher temperatures. Photoconduction and vacuum ultraviolet photoelectron spectroscopy (UPS) were used to probe the electronic structure of poly(p-xylylene) (PPX).[50] Photoconduction analysis revealed two modes of carrier formation–electron-hole pair generation via exciton excitation and the photo-injection of holes from metal electrodes. A detailed study of the latter determined the interfacial barrier height (Au–PPX–Au) and the photoelectron emission threshold as 2.26 and 6.84 eV, respectively. The value for the photoelectron emission threshold obtained by UPS was in good agreement, at 6.9 eV.

The photoconductivity of organic polymers can be greatly enhanced and extended into the visible region by the incorporation of small amounts of additives such as electron acceptors. These effects were discussed in terms of photoredox processes and the influence of donor–acceptor complexes in a review of the mechanisms of doping and spectral sensitization of polymer photoconductors, with particular emphasis on poly(N-vinylcarbazole).[51] The introduction of chemically bound methylene blue sensitizing dye into poly(N-vinylcarbazole) films has been effected[52] via sulphonation of the polymer followed by ion exchange. Photogenera-

---

[46] V. V. Slobodyanik, V. P. Naidenov, V. Ya. Pochinok, and V. N. Yashchuk, *Chem. Phys. Lett.*, 1981, **81**, 582.
[47] W. Klöpffer, *Chem. Phys.*, 1981, **57**, 75.
[48] G. I. Orlov, V. P. Bezzubaev, and M. I. Cherkashin, *Dokl. Akad. Nauk. SSSR*, 1981, **260**, 374.
[49] K. Yoshino, K. Nojima, I. Kitani, S. Iwakawa, and Y. Inuishi, *Technol. Rep. Osaka Univ.*, 1981, **31**, 49.
[50] Y. Takai, A. Kurachi, T. Mizutani, M. Ieda, K. Seki, and H. Inokuchi, *J. Phys. D*, 1981, **15**, 917.
[51] H. Meier, W. Albrecht, E. Zimmerhackl, N. Geheeb, and U. Tschirwitz, *Polym. Bull.*, 1982, **7**, 505.
[52] J. M. Pochan and H. W. Gibson, *J. Polym. Sci., Polym. Phys. Ed.*, 1982, **20**, 2059.

tion in a series of such surface-sulphonated, dye-exchanged films was found to be modified according to the depth of the resultant sensitizing layer, and to become less efficient with increasing sulphonation. It was proposed that the generation of photo-holes in sensitized poly(N-vinylcarbazole)–zinc oxide systems required a strong electronic interaction between the sensitizing dye molecule and the zinc oxide surface.[53] This was based on the classification of a series of dyes as sensitizers either for zinc oxide or the polymer. Photohole generation occurred with dyes in the former category but not the latter. The photosensitivity and photocurrent kinetics of zinc oxide–polymer binders have also been studied[54] as functions of polymer concentration, incident light intensity, and applied field.

The measurement of the photoconductivity of some polymeric cyanine dyes has been described in some detail.[55] Values were quoted for the dark conductivities ($\sigma \leqslant 2.5 \times 10^{-9}$ ohm cm^{-1}) and photoconduction ($I_{\text{phot.}} \leqslant 2.0 \times 10^{-10}$ A). The results of steady state photoconductivity experiments with *cis*- and *trans*-polyacetylene were interpreted[56] in terms of charged solitons, photogenerated either directly or indirectly through coupling of the lattice to electron-hole pair excitations. The transient photocurrent, after laser pulse excitation, decayed as a power law, $I(t) \sim t^{-0.6}$, suggesting dispersive transport of the photogenerated carriers. Finally, two reviews, entitled 'Photoconduction Processes in Polymers'[57] and 'Photoelectronic Properties of Photoconducting Polymers',[58] have recently been published.

4 Polymeric Resist Materials

The expanding microelectronics industry has stimulated interest in polymeric resist materials, which play an important role in the manufacture of semiconductor devices. A good resist must satisfy the many conflicting requirements for microcircuit processing such as resolution, sensitivity, adhesion, etch resistance, and freedom from defects. As the demand for high packing densities in integrated circuits increases, the lithographic techniques are required to produce finer lines with better linewidth control. Recent reviews covering the progress up to 1981 are available.[59—61] Since currently resolution in the sub-micron region is required, the overall performance of the photoresist cannot be ignored.[62—65] Indeed, the performance of the resist polymer plays a deciding role in which of the major

---

[53] N. C. Khe, I. Shimizu, and E. Inoue, *Photogr. Sci. Eng.*, 1981, **25**, 254.
[54] P. C. Meheudru, S. Radhakrishnan, and M. N. Kamalasanan, *Indian J. Pure Appl. Phys.*, 1982, **20**, 90.
[55] G. Kossmehl and P. Bocionek, *Makromol. Chem.*, 1981, **182**, 3445.
[56] S. Etemad, T. Mitani, M. Ozaki, T. C. Chung, A. J. Heeger, and A. G. MacDiarmid, *Solid State Commun.*, 1981, **40**, 75.
[57] Y. Takai and M. Ieda, *Dev. Polym. Photochem.*, 1982, **3**, 93.
[58] J. Mort and G. Pfister, *Electron. Prop. Polym.*, 1982, 215.
[59] G. E. Green, B. P. Stark, and S. A. Bahir, *J. Macromol. Sci., Rev. Macromol. Chem.*, 1981/82, (**21**)2, 187.
[60] S. Mitra, *J. Appl. Photogr. Eng.*, 1981, **7**, 37.
[61] S. Tazuke, *Dev. Polym. Photochem.*, 1982, **3**, 53.
[62] (*a*) A. N. Broers in 'Microcircuit Engineering 80', ed. R. P. Kramer, Delft Univ. Press, 1981, pp. 9—18; (*b*) A. N. Broers, *IEEE Trans. Electron. Devices*, 1981, **28**, 1268.
[63] S. Wittekoek in 'Microcircuit Engineering 80', ed. R. P. Kramer, Delft Univ. Press, 1981, pp. 155—170.
[64] M. J. S. Bowden, *J. Electrochem. Soc.*, 1981, **128**, 195C.
[65] E. A. Chandross, E. Reichmanis, C. W. Wilkins, jun., and R. L. Hartless, *Solid State Technol.*, 1981, **24**, 81.

Photoresists.—Relatively few novel photoresist polymers have been reported during the period under review. Copolymers of indenone and methyl methacrylate (MMA) were evaluated as positive u.v. photoresists.[67] Sensitivities of 0.02 and 0.1 J cm^{-2} at 248 and 280 nm were found for polymers containing 7% and 18% indenone, respectively.

Similarly, copolymers of methyl methacrylate and 3-oximino-2-butanone methacrylate worked as positive deep u.v. resists as the degradation of these copolymers was activated by the 3-oximino-2-butanone entity.[68] Several novel resists based on novolak resins have been described. Grant et al.[69] sensitized cresol–formaldehyde resins with 2,2-dimethyl-4,6-dioxo-5-diazo-1,3-dioxolane to get a deep u.v. (254 nm) resist that showed some unusual dissolution kinetics. Similarly, poly(p-vinylphenol) sensitized with 3,3'-diazido-diphenyl sulphone gave resists capable of recording steep profile images of 1 μm linewidth in 1 μm-thick films.[70-72] Telomers and block cotelomers were prepared by redox catalysis from 2-hydroxyethyl acrylate and telogens, e.g., CCl_4, CCl_3CO_2Me, CF_3CCl_3, and 2,4,4,4-tetrachloro-2-methyl-butyric acid). Polyols with 2—20 primary OH groups were thus obtained and esterified with acrylic acid to give photosensitive compositions.[73] A novel synthesis of polymers containing azido-nitrobenzyl groups was described.[74] In the photochemical u.v. initiated reaction, the reactivity of the azide group in poly[2-(2-azido-5-nitrobenzoyloxy)ethyl methacrylate] was affected by the presence of a spacer in the polymer chain. Poly(N-alkyl-o-nitroamides), prepared from 3,3'-dinitro-4,4'-di-N-methylaminodiphenyl ether isophthaloyl chloride or 3-nitro-4-(N-methylamino)benzoyl chloride,[75] contain photosensitive anilides in the backbone that degrade on u.v. light exposure by a mechanism analogous to the decomposition of o-nitroanilides. The reaction not only reduced the molecular weight of the polymer but also converted a hydrophobic disubstituted amide to a carboxylic acid. Water-soluble photoresists derived from poly(vinyl alcohol) (PVA) have been prepared. Thus, PVA was treated with 2,2-dialkoxyethoxystyrylpyridinium or quinolinium salts.[76] The high photosensitivity of these polymers arose from the aggregation of photofunctional groups in the polymer matrix resulting in cyclodimerization on exposure to u.v. light.[77] Photosensitivity of polymers prepared by esterifying PVA with chloroacetic acid increased in order tri- < di- < monochloroacetylated

---

[66] P. R. Thornton, *Adv. Electron. Electron. Phys.*, 1980, **54**, 69.
[67] R. L. Hartless and E. A. Chandross, *J. Vac. Sci. Technol.*, 1981, **19**, 1333.
[68] T. N. Bowmer, E. Reichmanis, C. W. Wilkins, jun., and M. Y. Hellman, *J. Polym. Sci., Polym. Chem. Ed.*, 1982, **20**, 2661.
[69] B. D. Grant, N. J. Clecak, R. J. Twieg, and C. G. Willson, *IEEE Trans. Electron. Devices*, 1981, **28**, 1300.
[70] T. Iwayanagi, T. Kohashi, S. Nonogaki, T. Matsuzawa, K. Douta, and H. Yanazawa, *IEEE Trans. Electron. Devices*, 1981, **28**, 1306.
[71] T. Matsuzawa and H. Tomioka, *IEEE Trans. Electron. Devices*, 1981, **28**, 1284.
[72] E. Gipstein, A. C. Ouano, and T. Tompkins, *J. Electrochem. Soc.*, 1982, **129**, 201.
[73] B. Boutevin, M. Maliszewicz, and Y. Pietrasanta, *Makromol. Chem.*, 1982, **183**, 2333.
[74] T. Nishikubo, T. Iizawa, I. Imagawa, and K. Kobayashi, *J. Polym. Sci., Polym. Chem. Ed.*, 1981, **19**, 2705.
[75] S. A. MacDonald and C. G. Willson, *ACS Symp. Ser.*, 1982, **184**, 73.
[76] K. Ichimura, *J. Polym. Sci., Polym. Chem. Ed.*, 1982, **20**, 1411.
[77] K. Ichimura and S. Watanabe, *J. Polym. Sci., Polym. Chem. Ed.*, 1982, **20**, 1419.

derivative.[78] The need for optimizing the photochemical and the physical properties of polymers is reflected in the number of copolymers used as photoresists. Photopolymers with low surface energy were obtained by copolymerizing β-p-azidobenzoyloxyethyl methacrylate with 2,2,3,4,4,4-hexafluorobutyl methacrylate, 2,2,3,3,4,4,5,5-octafluoropentyl acrylate, or perfluoro-octylethyl methacrylate.[79] The surface energies of these photopolymers were from 11.3 to 32.8 dynes cm^{-1}. Copolymers of MMA with p-substituted (H, Me, or MeO) phenyl isopropenyl ketone have an aromatic ring conjugated to a carbonyl group, shifting the absorption of the copolymers to a longer wavelength.[80] Poly[(MMA)-co-(3-oximino-2-butanone methacrylate)] underwent a rapid photolytic decomposition involving cleavage of the N—O bond, yet showed sufficient thermal stability.[81] The photosensitivity of PMMA was significantly enhanced, and terpolymerization with methacrylonitrile increased the sensitivity still further. A negative photoresist based on a glycidyl methacrylate–styrene polymer modified by cinnamic acid was reported.[82] Polyacrolein and acrolein–butadiene block copolymers, and graft copolymers based on poly(styrene-co-p-vinylbenzophenone-p'-t-butyl perbenzoate) were also examined as possible photoresist materials.[83]

Interfacial condensation of 2,4-hexadiyne-1,6-diol with diacid chlorides of succinic, glutaric, adipic, azelaic, sebacic, terephthalic, and isophthalic acids gave highly photoreactive, highly crystalline polyesters.[84,85] A more general method of crosslinking a variety of polymers carrying nucleophilic functional groups was described by Wagner and Purbrick.[86] Arene transition-metal carbonyls, either free or polymer bound, were used to effect crosslinking by free-radical and ligand-exchange mechanisms.

Some previously described photoresists have been studied in more detail. Thus, Tsunooka et al. observed that the photocrosslinking of poly(2,3-epithiopropyl methacrylate) was promoted by anthracene in air but not under nitrogen. Crosslinking involved dimerization of sulphenic acid groups produced by the oxidation and rearrangement of episulphide groups, and was inhibited by singlet oxygen quenchers.[87] Poly(vinyl cinnamates) have retained their popularity. The photosensitivity of donor poly(vinyl alcohol-p-isopropyl cinnamate) was increased by the presence of acceptor poly(vinyl alcohol-p-cyanocinnamate) containing 10% of 5-nitro-acenaphthene; the study indicated that the intermolecular crosslinking was promoted by excited energy transfer from donor to acceptor.[88] Similarly, poly(chloromethylstyrenes) containing pendant p-(2-benzoylvinyl)cinnamate were

---

[78] H. Tanaka, C. Azuma, K. Sanui, and N. Ogata, *Kobunshi Ronbunshu*, 1981, **38**, 529.
[79] T. Yamaoka, H. Kato, and T. Tsunoda, *Nippon Insatsu Gakkai Rombunshi*, 1981, **19**, 128.
[80] K. Nate and T. Kobayashi, *J. Electrochem. Soc.*, 1981, **128**, 1394.
[81] E. Reichmanis and C. W. Wilkins, jun., ACS Symp. Ser., 1982, **184**, 29.
[82] A. Ya. Ivainer, K. M. Dyumaev, A. I. Kirilin, N. S. Glybina, and S. V. Skobochkina, *Zh. Prikl. Khim. (Leningrad)*, 1981, **54**, 2357.
[83] I. Gupta, S. N. Gupta, and D. G. Neckers, *J. Polym. Sci., Polym. Chem. Ed.*, 1982, **20**, 147.
[84] A. O. Patil, D. D. Deshpande, S. S. Talwar, and A. B. Biswas, *J. Polym. Sci., Polym. Chem. Ed.*, 1981, **19**, 1155.
[85] A. O. Patil, D. D. Deshpande, and S. S. Talwar, *Polymer*, 1981, **22**, 434.
[86] H. M. Wagner and M. D. Purbrick, *J. Photogr. Sci.*, 1981, **29**, 230.
[87] M. Tsunooka, T. Ueda, S. Tanaka, M. Tanaka, and H. Egawa, *J. Polym. Sci., Polym. Lett. Ed.*, 1982, **20**, 589.
[88] S. Watanabe and K. Ichimura, *J. Polym. Sci., Polym. Chem. Ed.*, 1982, **20**, 3261.

studied by Nishikubo et al.[89] Photocrosslinking of this polymer proceeded by a radical chain mechanism, and was accelerated by energy transfer sensitizers. The same workers related the photosensitivity of cinnamate esters of 2-hydroxyethyl methacrylate copolymers with acrylic compounds containing photosensitive nitroaryl groups to the molecular motion of the polymer chains. A photosensitizer group bonded to the main chain *via* a long side chain showed higher activity than that attached directly to the backbone.[90] The photocrosslinking reactions of polymers containing cyanocinnamylidene-pyridinium and pyridinium dicyanomethylide groups were found to be promoted by the triplet-state sensitizers.[91,92] The absolute quantum yield of crosslink formation in PVA-2,3-diphenyl-2- cyclopropenecarboxylate was found to approach the theoretical maximum for a single-step crosslinking process.[93] In the solid photopolymer matrix, only *ca.* 4% of chromophores were in reactive configurations, and cycloaddition of suitably oriented cyclopropene rings was the primary crosslinking step. The high initial quantum yield and low concentration of reactive sites suggested that extensive energy migration was involved. The response of PMMA and poly(methyl isopropenyl ketone) to deep u.v. light displayed pronounced temperature dependence over the range 20—160 °C. The molecular-weight dispersity of the irradiated resist was found to be influential.[94] Kilp et al. did not find any difference in the quantum yield of main-chain scission by the Norrish type II reaction for isotactic and atactic poly(acrylophenones).[95] The u.v. light initiated photocrosslinking reactions of polymers having pendant vinyloxy groups,[96] and of unsaturated polyesters were also studied.[97] At a more fundamental level, the photo-imaging speed of a photopolymerizable/crosslinkable dry-film system suffering from free-volume dependence was correlated with the glass-transition temperature of the polymeric binder and the monomer, the volume fraction of the monomer, and the temperature of exposure.[98] A general photographic theory of crosslinking resists was presented by Reiser and Pitts.[99] Speed, as measured by the gel-point exposure, was inversely proportional to the quantum yield of the crosslinking reaction, to the extinction coefficient and the molar concentration of the photoreactive chromophore, and to the weight average molecular weight.

The converse of photocrosslinking, *i.e.* photo-etching, also received some attention. Poly(vinyl cinnamate)-5-nitrofluorenone photoresist layer underwent photochemical reactions upon u.v. irradiation that led to the formation of an etched pattern.[100] The laser flash photolysis (at 347 nm) of an alternating copolymer of

---

[89] T. Nishikubo, T. Iizawa, and M. Hasegawa, *J. Polym. Sci., Polym. Lett. Ed.*, 1981, **19**, 113.
[90] T. Nishikubo, T. Iizawa, and K. Tsuchiya, *Makromol. Chem. Rapid Commun.*, 1982, **3**, 377.
[91] C. Roucoux, C. Loucheux, and A. Lablache-Combier, *J. Appl. Polym. Sci.*, 1981, **26**, 1221.
[92] J. J. Cottart, C. Loucheux, and A. Lablache-Combier, *J. Appl. Polym. Sci.*, 1981, **26**, 1233.
[93] M. V. Mijovic, P. J. Beynon, T. J. Shaw, K. L. Petrak, A. Reiser, A. J. Roberts, and D. Phillips, *Macromolecules*, 1982, **15**, 1464.
[94] K. Harada and S. Sugawara, *J. Appl. Polym. Sci.*, 1982, **27**, 1441.
[95] T. Kilp, J. E. Guillet, L. Merle-Aubry, and Y. Merle, *Macromolecules*, 1982, **15**, 60.
[96] T. Nishikubo, T. Iizawa, A. Yoshinaga, and M. Nitta, *Makromol. Chem.*, 1982, **183**, 789.
[97] D. S. Sadafule and S. P. Panda, *Polym. Photochem.*, 1982, **2**, 13.
[98] M. F. Molaire, *J. Polym. Sci., Polym. Chem. Ed.*, 1982, **20**, 847.
[99] A. Reiser and E. Pitts, *J. Photogr. Sci.*, 1981, **29**, 187.
[100] X.-Y. Hong, R.-X. Pei, W.-X. Jin, Q. Zhou, and W.-P. Fu, *Kao Fen Tzu Tung Hsun*, 1981, **1**, 47.

phenyl isopropenyl ketone with styrene,[101] of poly(ethylene terephthalate) films (at 193 nm),[102] of polyethylene, polypropylene, polyisobutylene, polybutadiene, and polyisoprene (at 123.6 and 147 nm),[103] and of PMMA and poly(butene-1-sulphone)[104,105] was reported.

In order to meet the requirements imposed by modern electronics, the performance limits of optical systems must be extended beyond the current limits of about 0.4 to 0.8 μm.[106]

Electron-beam Resists.—In comparison with u.v. radiation, electron-beam (e-beam) offers better focusing and thus higher resolution. The energy of the beam can be readily controlled and the beam can be deflected with great precision. The resolution is not limited by that of the electron optical system, but by scattering of electrons in the resist and back-scattering from the substrate. Various aspects of the e-beam lithography have been discussed.[107—109] The combination of e-beam with optical (photo) lithography was found to be beneficial.[110]

Various polymers including PMMA, poly(acrylic acid) (PAA), poly(dialkylorthophthalate), and chlorinated polystyrenes have been used previously as e-beam resists. Some of the more common negative-working materials were discussed by Hieke and Oldham.[111] More recently, chloromethylated polystyrenes of a wide range of molecular weight (6800—560 000) were made from nearly mono-disperse polystyrenes (PS). The effect of molecular parameters on sensitivity and resolution were investigated.[112] Choong and Kahn[113] used poly(chloromethylstyrene) homopolymer. This resulted in (1) higher homogeneity, (2) easier preparation and control of polymer properties, and (3) higher sensitivity due to the higher content of chloromethyl groups.

Methyl methacrylate–methacrylic acid and glycidyl methacrylate–ethyl methacrylate copolymers showed greater sensitivity to e-beam than PMMA.[114] Etching rates for poly(vinylnaphthalene) were found to be about 2/3 of those for PS, but the polymer gave patterns of higher contrast. The sensitivity of poly(vinylnaphthalene) was insufficient for the current requirements but it could be increased by chloromethylation.[115] Various poly(vinylpyridinium) salts showed sensitivities on exposure to 20 kV e-beam similar to that of PMMA while being able to record

---

[101] I. Naito, A. Kinoshita, and W. Schnabel, *Nippon Insatsu Gakkai*, 1982, **20**, 77.
[102] R. Srinivasan and V. Mayne-Banton, *Appl. Phys. Lett.*, 1982, **41**, 576.
[103] V. E. Skurat and Yu. I. Dorofeev, *Zfl-Mitt.*, 1981, **43b**, 465 (*Chem. Abstr.*, 1981, **96**, 123 468).
[104] P. W. Bohn, J. W. Taylor, and H. Guckel, *Anal. Chem.*, 1981, **53**, 1082.
[105] Y. Kawamura, K. Toyoda, and S. Namba, *Appl. Phys. Lett.*, 1982, **40**, 374.
[106] R. K. Watts and J. H. Bruning, *Solid State Technol.*, 1981, **24**, 99.
[107] P. F. W. Pease, *Contemp. Phys.*, 1981, **22**, 260.
[108] B. Dance, *New Electron.*, 1981, **14**, 61.
[109] V. M. Korsunskii, *Mikroelectronika*, 1982, **11**, 291.
[110] R. L. Maddox, N. Casey, C. Sallee, F. Kinoshita, R. Imerson, and G. Whitcomb, *Solid State Technol.*, 1982, **25**, 240.
[111] E. Hieke and W. G. Oldham in 'Microcircuit Engineering 80', ed. R. P. Kramer, Delft Univ. Press, 1981, p. 395.
[112] S. Imamura, T. Tamamura, K. Harada, and S. Sugawara, *J. Appl. Polym. Sci.*, 1982, **27**, 937.
[113] H. S. Choong and F. J. Kahn, *J. Vac. Sci. Technol.*, 1981, **19**, 1121.
[114] Yu. M. Aleksandrov, K. A. Valiev, L. V. Velikov, B. G. Gribov, S. D. Dushenkov, R. Kh. Makhmutov, D. D. Mozzhukhin, A. S. Pleshivtsev, and G. K. Selivanov, *Mikroelectronica*, 1982, **11**, 483.
[115] Y. Ohnishi, *J. Vac. Sci. Technol.*, 1981, **19**, 1136.

submicron images. A series of such ionic polymers varying in molecular weight, molecular-weight distribution, counterions, alkyl groups, and in the degree of quaternization was prepared.[116] Poly(N-butyl methacrylate) also gave a highly sensitive (10^{-7} C cm^{-2}), high-resolution (0.8 μm) e-beam resist.[117] It appears that no currently available commercial e-beam resist satisfies all the necessary requirements. However, poly(allyl methacrylate)-co-(2-hydroxyethyl methacrylate) prepared by Tan et al. was claimed to be suitable in all respects.[118]

Some new positive e-beam resists have also been reported. Vinyl monomers containing the α-CF_3 group were shown to increase the e-beam sensitivity in copolymers with MMA and methacrylonitrile. For example, the incorporation of about 30% of α-trifluoromethacrylonitrile into PMMA increased sensitivity from 2×10^{-4} to 3×10^{-5} C cm^{-2} (at 2 keV).[119] A similar effect was reported for 2,2,2-trichloroethyl methacrylate, 2,2,2-trifluoroethyl α-chloroacrylate,[120a] and bromoacrylates.[120b] Printed pattern accuracy within 0.2 μm were reported for such materials. Poly(methyl-α-chloroacrylate) was reported to undergo six chain scission events per 100 eV of energy absorbed, significantly higher than 1.6 for PMMA, the current standard e-beam resist.[121] The acrylate resists have been the most popular choice of positive resists for advanced lithography. An overview of the lithographic performance of polyacrylates of the general formula $-CH_2-C(X)(COOY)-$ where X may be an electron donor or acceptor and Y may be alkyl, aryl, epoxy, etc. was given by Moreau.[122]

An understanding of the structure–property relationship is an essential aspect of resist design. For example, the dry etch rates were found to vary significantly for vinyl polymers with different side-chain substituents.[123a] Willson et al.[123b] verified experimentally, using poly(methyl-α-trifluoromethacrylate), a mechanistic hypothesis designed to explain the radiation chemistry of positive-working poly(methyl α-haloacrylate)s. Similarly, the room temperature γ-ray degradation of PMMA, poly(methyl-α-chloroacrylate), poly(methyl-α-fluoroacrylate), and polymethacrylonitrile was studied by e.s.r. and ENDOR to determine degradation mechanisms relevant to e-beam lithography.[124] The compactness of the radiochemical degradation region was related to the potential resolution of the resist. The sensitivity and contrast of PMMA and poly(methyl isopropenyl ketone) on e-beam radiation showed a minor temperature dependence.[94] The methods for predicting the radiation sensitivity of polymers and for rapidly screening polymers have been discussed.[125,126]

[116] K. I. Lee, H. Jopson, and P. Cukor, Proc. SPIE, 1982, 333, 15.
[117] T. Nakayama, K. Kotake, T. Sato, S. Ichikawa, S. Asaumi, A. Yokota, and H. Nakane, Denki Kagaku Oyobi Kogyo Butsuri Kagaku, 1981, 49, 245.
[118] Z. C. H. Tan, C. C. Petropolous, and F. J. Rauner, J. Vac. Sci. Technol., 1981, 19, 1348.
[119] C. U. Pittman, jun., M. Ueda, C. Y. Chen, J. H. Kwiatkowski, C. F. Cook, jun., and J. N. Helbert, J. Electrochem. Soc., 1981, 128, 1758.
[120] (a) T. Fujiyoshi, New Jpn., 1981, 8, 9; (b) B. N. Babu and J. C. W. Chien, Polym. Prepr., Am. Chem. Soc., Div. Polym. Chem., 1982, 23, 85.
[121] J. H. Lai, Proc. SPIE, 1982, 333, 8.
[122] W. M. Moreau, Proc. SPIE, 1982, 333, 2.
[123] (a) J. N. Helbert and M. A. Schmidt, ACS Symp. Ser., 1982, 184, 61; (b) C. G. Willson, H. Ito, D. C. Miller, and T. G. Tessier, 6th International Conference on Photopolymers, Ellenville, NY, 1982, p. 207.
[124] S. Schlick and L. Kevan, J. Appl. Polym. Sci., 1982, 27, 319.
[125] D. O'Sullivan, P. B. Price, K. Kinoshita, and C. G. Willson, J. Electrochem. Soc., 1982, 129, 811.
[126] R. G. Brault and L. J. Miller, J. Electrochem. Soc., 1981, 128, 1158.

Ion-beam Resists.—A collimated beam of ions (e.g., H^+, H_2^+, He^+, O^+, Li^+) can also be used for recording images in polymer layers. As a general rule, resists are more sensitive to ions than to electrons.[127] Further, ions, in contrast to electrons, produce only low-energy secondary electrons, and no high-energy backscattered electrons. This facilitates the formation of images with submicron dimensions. Linewidths of about 400 Å have been recorded in PMMA exposed to 40 keV H^+ ions.[128] A computer simulation of ion-beam exposure profiles in PMMA demonstrated that proximity problems are greatly reduced in comparison with e-beam exposed patterns.[129] The ion-beam exposure technique can be used for mask making as well as for direct exposure of the wafer.[130]

The polymers examined as ion-beam resists include PMMA, poly(dimethylsiloxane), poly(fluoroalkyl methacrylates), poly(methyl isobutyl ketone), poly(glycidyl methacrylate), and poly(glycidyl methacrylate)-co-(ethyl acrylate).[131] Polyimide high-resolution masks for ion-beam exposure were also described.[132] The sensitivity of PMMA to ion-beam exposure in relation to the electronic stopping power was studied. The absorbed energy for H^+ ions was comparable with X-ray lithography data, but heavier ions were more effective. The exposure characteristics of six polymer resists – PS, PMMA, and PMMA blends – to 1.5 MeV H^+, He^+, and O^+ ions and to 20 keV electrons were measured by Hall et al.[133] The deposited energy per unit volume needed to expose a resist was found to be a function of the spatial energy dissipation rate of the ion in the resist. This was rationalized in terms of the nature of the energy distribution around the primary particle path and whether the resist required the activation of a singlet site or two adjacent sites to produce an image.

The theory and the statistical limitations of the ion-beam exposure were discussed.[134—136]

X-ray Resists.—Poly(chloroalkyl acrylates), in particular poly(2,3-dichloro-1-propyl acrylate)/poly(glycidyl methacrylate-co-ethyl acrylate) mixture were found to meet the current sensitivity requirements for X-ray lithography.[137] Polystyrene–tetrathiafulvalene films doped with a haloalkyl acceptor showed good sensitivity with high contrast value and no evidence for the classical swelling phenomena.[138] A study on the X-ray exposure characteristics of glycidyl methacrylate–ethyl acrylate copolymer identified several undesirable effects that limited the resolution.[139] Ultrahigh molecular weight ($>10^7$ g mol^{-1}) and a commercial PMMA (7×10^{-5} g

---

[127] H. Ryssel, K. Haberger, and H. Kranz, J. Vac. Sci. Technol., 1981, 19, 1358.
[128] L. Karapiperis and C. A. Lee, Appl. Phys. Lett., 1979, 35, 395.
[129] L. Karapiperis, I. Adesida, C. A. Lee, and E. D. Wolf, J. Vac. Sci. Technol., 1981, 19, 1259.
[130] G. Stengl, R. Kaitna, H. Loschner, R. Rieder, P. Wolf, and R. Sacher, Solid State Technol., 1982, 25, 104.
[131] H. Ryssel, H. Kranz, K. Haberger, and J. Bosch, in 'Microcircuit Engineering 80', ed. R. P. Kramer, Delft Univ. Press, 1981, p. 293.
[132] N. P. Economou, D. C. Flanders, and J. P. Donnelly, J. Vac. Sci. Technol., 1981, 19, 1172.
[133] T. M. Hall, A. Wagner, and L. F. Thompson, J. Appl. Phys., 1982, 53, 3997.
[134] H. Ryssel, G. Prinke, H. Bernt, K. Haberger, and K. Hoffmann, Appl. Phys. A, 1982, 27, 239.
[135] A. Macrander, D. Barr, and A. Wagner, Proc. SPIE, 1982, 333, 142.
[136] L. Karapiperis, Univ. Microfilm Int., 1982, DA 8210762.
[137] J. M. Moran and G. N. Taylor, J. Vac. Sci. Technol., 1979, 16, 2014.
[138] D. C. Hofer, F. B. Kaufman, S. R. Kramer, and A. Aviram, Appl. Phys. Lett., 1980, 37, 314.
[139] W. D. Buckley and J. A. DalleAve, Proc. Electrochem. Soc., 1979, 78, 458.

mol^{-1}) were exposed to doses ranging from 0.1—10 μC cm^{-2}. When 1:1 methyl ethyl ketone and isopropanol was used to develop the exposed resists, 0.5 μm pattern sizes were formed in both materials. Lower doses (ca. 0.3 ×) could be used with the higher-molecular-weight polymer, since the dissolution rate of the unexposed resist was about 30 times lower than for the commercial PMMA.[140] According to Aleksandrov et al. the rate of photo-etching of poly(hexane sulphone) resist layers decreased with the increase in the irradiation dose, and it increased with the decrease in the wavelength of irradiation (in three spectral ranges – 0.7—2.4; 0.8—2.8; and 2.4—6 nm). The thickness of the etched layer was about 10 times higher as compared with PMMA (at irradiation dose 1 J cm^{-2}).[141] Several reviews of this topic are available.[142–144]

Plasma-developed Resists.—A serious problem in high-resolution lithography using polymeric negative resist materials is the limit to resolution caused by swelling of polymer during solution development. The resolution is generally limited to about 1 μm in 0.5 μm-thick films. In order to avoid the use of solvents during the development process, a combination of post-exposure thermal treatment (often called baking or fixing), and plasma (e.g., O_2, Ar, H_2, CF_4) development can be used to develop negative resist images. The materials used are mixtures of a host polymer and a moderately volatile monomer 'locked' into the host by the impinging radiation.[145] For example, a mixture containing 81 wt.% poly(2,3-dichloro-1-propyl acrylate) and N-vinylcarbazole resolved features as small as 0.3 μm after 1.5 min X-ray exposure.[146] A quinone-sensitized modification of the same mixture has also been reported.[147] Locking of carbazole moiety by cationic polymerization was presumed to proceed via cationic radical intermediates produced on quenching the excited states of N-vinylcarbazole by electron-accepting quinones. Vacuum lithography by using plasma polymerization and plasma etching of MMA and hexafluorobutyl methacrylate was described. The resists were improved further by copolymerizing the methacrylates with styrene or tetramethyltin.[148] A variety of silicon-containing monomers for plasma-developed resists based on chlorine-containing host polymers were prepared. The major problems encountered were monomer incompatibility, high volatility, and low reactivity. Bis-acryloxy- and methacryloxy-alkyltetramethyldisiloxanes, however, met all the requirements. A combination of poly(2,3-dichloro-1-propyl acrylate) with bis-acryloxybutyltetramethyldisiloxane showed optimum properties.[149] Tsuda et al. used poly(methyl isopropenyl ketone) and 4,4'-diazidobiphenylthioether mixture as a plasma-developable negative resist material, fine patterns of 0.5 μm being readily produced

---

[140] D. W. Hess, Report 1981, B2FOSR-TR-81-0729, Order No. AD-A107833, 15 pp.
[141] Yu. M. Aleksandrov, K. A. Valiev, L. V. Velikov, A. M. Prokhorov, and M. N. Yakimenko, Pisma Zh. Tekh. Fiz., 1982, **8**, 577.
[142] J. R. Maldonado, Proc. SPIE, 1982, **333**, 131.
[143] C. R. Fencil and G. P. Hughes, Proc. SPIE, 1982, **333**, 100.
[144] J. Przyluski and M. Zagorska, Elektronika, 1982, **23**, 12.
[145] H. Nakane, M. Tsuda, W. Kanai, S. Oikawa, A. Yokota, I. Hijikata, and A. Uehara, in 'Microcircuit Engineering 80', ed. R. P. Kramer, Delft Univ. Press, 1981, p. 427.
[146] G. N. Taylor and T. M. Wolf, J. Electrochem. Soc., 1980, **127**, 2665.
[147] G. N. Taylor, T. M. Wolf, and M. R. Goldrick, J. Electrochem. Soc., 1981, **128**, 361.
[148] S. Morita, S. Hattori, M. Ieda, J. Tamano, and M. Yamada, Kobunshi Ronbunshu, 1981, **38**, 657.
[149] G. N. Taylor, T. M. Wolf, and J. M. Moran, J. Vac. Sci. Technol., 1981, **19**, 872.

by deep-u.v. lithography.[150] Plasma-developed ion-implanted resists, where an organic polymer was exposed using In$^+$ ions and developed in an oxygen plasma, were also described.[151]

By using plasma etching processing, an 'all-dry' lithography was developed, based, for example, on plasma-polymerized MMA etched with CCl_4 plasma,[152] poly(methacrylonitrile) and acrylonitrile–methacrylic acid copolymer (positive-working), and poly(α-chloroacrylonitrile) (negative-working) etched with oxygen plasma.[153] The quality of the plasma-processed resist depended not only on the plasma-etching conditions but also on the structure of the polymer.[154] The chemical reaction mechanism of the plasma development process of resist materials was theoretically examined.[155,156]

The flow of resist material during the manufacture of electronic devices often caused a reduction in the resolution. Linewidth variations of 0.5—1.5 µm were observed. Treatment of positive and negative resists with a nitrogen plasma prior to postbaking reduced resist flow and enabled the linewidth integrity to be maintained. Minimal resist erosion occurred as a result of plasma treatment. A model was proposed to account for the improved flow characteristics.[157] Similarly, a u.v. hardening process was described that rendered micron-sized images in diazo-oxide photoactive resists resistant to flow at temperatures above 210 °C.[158]

Post-irradiation polymerization is another process that can interfere with the optimum performance of the resists. The resist pattern width and thickness are known to suffer from time dependence after irradiation. Ohnishi et al.[159] presented a theoretical analysis of the 'postpolymerization' mechanism. In the model used for analysis, active radicals were not extinguished and the polymerization proceeded at a rate proportional to the radical concentration (1st-order reaction). A radical lost its activity as a result of collision with another radical (2nd-order reaction). Post-irradiation polymerization was inhibited by adding a radical scavenger to resists.[159] Experiments showed that many more stable radicals were present in irradiated poly(glycidyl methacrylate) as compared with polystyrene. The postpolymerization effects observed in the former polymer were attributed to these stable radicals.[160]

---

[150] M. Tsuda, S. Oikawa, W. Kanai, K. Hashimoto, A. Yokota, K. Nuino, I. Hijikata, A. Uehara, and H. Nakane, *J. Vac. Sci. Technol.*, 1981, **19**, 1351.
[151] T. Venkatesan, G. N. Taylor, A. Wagner, B. Wilkens, and D. Barr, *J. Vac. Sci. Technol.*, 1981, **19**, 1379.
[152] S. Hattori, J. Tamano, M. Yamada, M. Ieda, S. Morita, K. Yoneda, and S. Ishibashi, *Thin Solid Films*, 1981, **83**, 189.
[153] H. Hiraoka, *J. Electrochem. Soc.*, 1981, **128**, 1065.
[154] J. Tamano, S. Hattori, S. Morita, and K. Yoneda, *Plasma Chem. Plasma Process.*, 1981, **1**, 261.
[155] V. M. Dolgopolov, V. V. Gusev, V. I. Ivanov, and E. F. Shelykhmanov, *Zh. Prikl. Spektrosk.*, 1982, **36**, 10.
[156] M. Tsuda and S. Oikawa, *Jpn. J. Appl. Phys.*, 1982, **21**, 135.
[157] J. M. Moran and G. N. Taylor, *J. Vac. Sci. Technol.*, 1981, **19**, 1127.
[158] H. Hiraoka and J. Pacansky, *J. Vac. Sci. Technol.*, 1981, **19**, 1132.
[159] Y. Ohnishi, M. Itoh, K. Mizuno, H. Gokan, and S. Fujiwara, *J. Vac. Sci. Technol.*, 1981, **19**, 1141.
[160] T. Tada, *J. Electrochem. Soc.*, 1982, **129**, 1070.

Author Index

Aarada, A., 371
Aaronson, S. A., 137
Abadie, M. J. M., 52
Abbas, K. B., 84, 293
Abbey, K. J., 70
Abbott, W. M., 375, 380
Abdel-Azim, A. A. A., 233
Abdel-Hay, F. I., 87,387
Abd-el-Nour, K. N., 297
Abdullaev, A. A., 70
Abdullin, M. I., 343
Abe, K., 130, 376
Abe, T., 226
Abe, Y., 151
Abhiraman, A. S., 292
Abraham, G., 123
Abraham, M. H., 359
Abraham, S. H., 381
Abramenko, E. L., 179
Abruhna, H. D., 74
Abrulia, H. D., 319
Absdom, D. R., 376
Abuchowski, A., 360
Acharya, H. K., 34
Acierno, D., 83, 222, 291
Adabbo, H. E., 285
Adam, H. J., 237
Adams, A. C., 148
Adams, R. L. P., 141
Ades, D., 41
Adesida, I., 411
Adler, P., 31
Adler, R. S., 255
Adur, A. M., 32
Aelenei, N., 225
Agarwal, R. K., 387
Agarwal, S. H., 246
Agarwalla, N. L., 387
Ager-Johnson, P., 129
Aggour, Sh. Sh., 87
Agrawal, A., 295
Agrawal, J. P., 84
Agrawal, K. K., 387
Agur, E. W., 295
Aharoni, S. M., 255
Ahmed, A. M., 382
Aiba, M., 335
Aida, T., 83
Airinei, A., 343
Aitken, D. A., 317
Ajo, D., 4
Akada, Y., 318
Akaiki, T., 376
Akatin, M. S., 346

Akatsuka, R., 327
Akay, G., 292, 293 295
Akcasu, A. Z., 165, 170, 236, 254
Akcasu, Z., 197
Akelah, A., 304, 312, 313, 325
Aker, S. C., 289
Akiyama, S., 263
Akiyoshi, K., 359, 369
Akopyan, G. D., 63, 69
Akovali, G., 148
Aksiment'eva, E. I., 72, 75
Akutin, M. S., 84
Akutsu, F., 87, 337
Al-Abidin, K. M. Z., 52
Alamo, R., 31, 217
Alba, J., 77
Al-Bazi, S. J., 97
Alberto, B. P., 130
Albrecht, G., 34
Albrecht, W., 404
Albright, R. L., 316, 363
Aldissi, M., 296
Aleixo, R. M. V., 366
Aleksandrov, Yu. M., 409, 412
Alekseeva, S. G., 185
Alekseeva, T. A., 73, 75
Alexander, F. B., 148
Alexander, H., 381
Alexander, R., 33
Alexis, M., 325, 326
Alger, K. W., 80
Aliev, R. D., 32
Aliev, R. E., 39
Alieva, Z. M., 30
Alig, I., 183
Alijev, R., 37, 85
Alikayama, I., 222
Al-Kass, S., 31
Alkhafaji, S., 165
Allain, C., 172, 235, 254
Allan, G. G., 157
Allard, D., 278
Allcock, H. R., 154, 155, 156, 157, 158, 319, 386
Allegra, G., 4, 202
Allen, C. W., 156
Allen, D. W., 336
Allen, G., 193, 195, 202, 203
Allen, N. S., 331, 333, 399
Allen, P. E. M., 47
Allendoerfor, R. D., 213
Alleng, R. G., 259
Allison, S. A., 167

Al-Lohedan, H., 367,368
Allport, D. C., 96
Allred, A. L., 150
Almendo, J. P., 394
Al-Noaimi, G. F., 162
Alper, H., 372
Al-Shahib, W. A., 66
Alsop, R. M., 119
Altena, F. W., 252, 262
Alvino, W. M., 91
Amaratunga, W., 309, 320
Amerik, Yu. B., 53
Amidon, G. E., 371
Amin, A. F., 86
Aminabhavi, T. M., 238, 252
Amirtaraj, J., 86
Amirzadeh, J., 254
Amis, E. J., 172
Amos, L. W., 185
Amosova, S. V., 35
Amri, M. A., 121
Amu, T. C., 226
Anand, A., 212
Ander, P., 261
Andersen, K. B., 247
Andersen, P. A., 137
Anderson, B. C., 47
Anderson, C. B., 380
Anderson, J. M., 377
Anderson, J. S., 116
Anderson, M. R., 201, 210
Anderson, R. A., 296
Andersson, J. T., 364
Ando, I., 177
Ando, T., 33, 372
Andrade, J. D., 375
Andrady, A. L., 149, 156
Andreeta, H. A., 242
Andresen, A., 27
Andrews, E. H., 288, 289
Andrews, E. J., 380
Andrews, G. D., 47, 245
Andreyeva, M. A., 55
Andreyeva, N. Yu., 55
Andreyeva, Ye. D., 54
Andrianova, Z. S., 85
Andruzzi, F., 36, 75
Angeletti, E., 373, 374
Angeli, G., 73
Angeloni, A. S., 318
Angulo-Sanchez, J. L., 246
Anhang, T., 81
Annuziata, R., 373
Anoardi, L., 367

Author Index

Anosov, V. I., 34
Ansorena, F. J., 226
Antisara, M. V., 205
Anzai, J., 351
Anzur, I., 63
Aoshima, S., 39, 326
Aoyama, J., 360
Aoyama, S., 86
Aoyama, Y., 359, 371
Apostolopoulos, M., 262
Appelt, B., 262, 265
Apple, T. M., 206
Appleyard, J. H., 333
Appokattan, P. S., 111
Apraksina, L. M., 73
Arabai, K., 313
Aragona, J., 381
Arai, K., 66, 316, 369
Arai, M., 66
Arai, Y., 85, 387
Araki, T., 179
Aranyi, P., 129
Aravindanath, S., 208
Archelas, E., 328
Archibald, A. R., 117
Arct, J., 331
Aref'yev, N. M., 88
Arita, K., 184
Armistead, D. M., 86
Armstrong, R. C., 174, 283, 290
Arnarp, J., 122
Arnaud, R., 332, 333
Arndt, K. F., 237
Arnott, E. J., 382
Arques, A., 314
Arranz, F., 346
Arshady, R., 328, 364
Artaud, I., 309, 365
Artemev, Yu. G., 149
Arthur, P., 47
Arts, J. A., 375
Arutyunyan, R. S., 36
Asahi, T., 215
Asai, D., 52
Asai, K., 215
Asakawa, K., 363
Asakura, T., 177
Asami, R., 38
Asaumi, S., 410
Asbach, G. I., 205
Asbeck, A., 65
Askadskii, A. A., 339
Askerov, A. K., 59, 69
Askew, M. J., 379
Askill, I. N., 241
Asmus, R. A., 52
Asmussen, F., 225
Asphind, J., 384
Aspler, J. S., 255
Astapenko, E. P., 226
Astler, M., 364
Astrin, S. M., 137
Ataka, S., 393
Atha, D. H., 128
Athanassakis, V., 370

Atik, S. S., 327
Atkin, E. L., 201
Atkins, E. D. T., 106
Atkinson, M., 380
Atkinson, T. C., 133
Atluri, S. N., 289
Atrushkevich, A. A., 349
Atsumi, K., 383
Aubert, P., 178
Aubin, M., 77, 82, 263
Audebert, R., 307
Auerbach, I., 336
Aune, J. P., 319
Austin, P. E., 157, 158
Averyanov, S. V., 152
Averyanova, L. A., 152
Aviram, A., 411
Axelrod, D. E., 140
Axelson, D. E., 213
Axmann, A., 195
Aylwood, N. N., 393
Azar, A. A., 81
Azran, J., 318, 365
Azuma, C., 407
Azuma, J. I., 104

Baade, W., 56, 388
Baba, K., 164, 233
Baba, Y., 9
Babajko, S., 219
Babitsky, B. D., 179
Babu, B. N., 410
Babu, G. N., 90
Bachhawat, B. K., 115
Bachishche, V. N., 332
Bachmann, M. A., 393
Bachrach, A., 77
Backderf, R. H., 343
Backmann, M. A., 209, 393
Baddiley, J., 117
Baddour, R. F., 212
Baenziger, J. U., 109
Baer, A. D., 339
Baer, B. W., 140
Baer, M., 69
Baez, M., 140
Baggish, D., 383
Bagheri, R., 332
Bahir, S. A., 405
Bahr, H., 114
Baidenok, I. V., 54
Baijal, S. K., 349
Bailey, D., 308
Bailey, D. B., 183
Bailey, D. C., 304, 318
Bailey, F. E., 97
Bailey, W. J., 54
Bain, A. D., 102
Bains, M. S., 378
Bair, H. E., 83
Baird, D. G., 293, 295
Baird, M., 144
Bajaj, B. P., 78
Baker, D., 382
Baker, D. A., 111

Baker, D. S., 356
Baker, G. L., 366
Bakeyev, N. F., 81
Baklagina, Yu. G., 92
Bala, K., 385
Balakrishnan, T., 77, 347
Balaram, H., 392
Balaram, P., 109, 392
Balasius, E., 328
Balasubramaniam, M., 86
Balbi, C., 255
Balcerowick, W., 345
Baldrian, J., 261
Balduff, D. C., 245
Baldwin, R. L., 256
Balik, C. M., 216
Balint, G., 333
Balke, S., 388
Balkin, B. J., 83
Ball, E. M., 129
Ball, L. E., 35
Ball, R. C., 193
Ballard, D. G. H., 191, 192, 206, 212
Ballard, M. J., 61, 68, 389,
Ballistreri, A., 157, 346, 350
Balta-Calleja, F. J., 212
Balyberdina, T. G., 65
Balykova, T. N., 88, 347
Bamford, C. H., 50, 54, 364
Banda, D., 382
Bandekar, J., 393
Bandoc, C. C., 383
Banerjee, M., 62
Banfi, S., 315
Bangerter, F., 176
Bank, A., 144
Banks, R. F., 258
Bansal, R., 169, 397
Banthia, A. K., 82
Bantle, S., 150, 161, 171, 237
Bao, J., 86
Baquey, C., 375
Barabas, E. S., 69
Barabas, K., 382
Baranova, S. A., 54
Barantsevich, E. N., 51, 56, 181
Barbacid, M., 138
Barbaric, S., 371
Barbassi, F., 319
Barbe, P., 4
Barbee, W. K., 382
Barboiu, V., 181
Barbu, G., 346
Barbucci, R., 362
Barcelon, C., 373
Bard, A. J., 319
Bardos, D., 379
Barenberg, S., 377
Barenberg, S. A., 377
Bargmann, C. I., 138
Barham, P. J., 216
Barkalov, I. M., 57
Barker, C., 96
Barker, J. A., 248

Barker, S. A., 120
Barlow, J. W., 82, 84, 85, 221, 263, 277, 278
Barnett, K. G., 199
Barnickel, G., 118
Barnum, A. S., 221
Barnum, R. S., 84
Barondes, S. H., 109, 111
Barr, D., 411, 413
Barr, R., 163, 165, 397
Barrack, H. J., 125
Barrales-Rienda, J. M., 346
Barrall, E. M., 263
Barrell, B. G., 134
Barrett, A. J., 164
Barrie, J. A., 42
Barry, J. M., 379
Barth, H. G., 104
Bartholin, M., 330, 365
Bartholmew, R. S., 381
Bartram, C. R., 138
Bartsch, R. A., 358
Baskent, F. O., 95
Bass, S. V., 187
Basse-Cathlinat, B., 375
Bassett, D. C., 209, 212
Bassi, I. W., 4
Bastelberger, T., 36
Bastide, J., 197
Bastin, P., 334
Basu, D., 111
Basu, P. S., 111
Basu, S., 295
Bataille, P., 61, 63, 388
Bates, F. S., 198, 199
Batey, I. L., 99
Bathias, C., 289
Batkina, P. A., 150
Batkina, T. A., 150
Batley, M., 116
Battistel, E., 113
Batz, H. G., 67
Batzer, H., 351
Bauduin, G., 386
Bauer, C., 114
Bauer, P. I., 129
Baughman, R. H., 299, 300, 393
Baukema, P. R., 216
Bauld, N. L., 363
Baulin, A. A., 7, 19, 20
Baumann, C. M., 111
Baumann, G. F., 95
Baumbach, D. O., 225
Baumgartner, A., 163, 203, 396, 397
Bautz, G., 207
Bauwens, J. C., 84, 287
Bauwens-Crowet, C., 84, 287
Bawa, R., 360
Bayazeed, A. M., 79
Bayer, R., 327
Bazdadea, E., 342
Beato, M., 143
Beattie, E. J., jun., 378
Beattie, W. G., 140, 144

Beaumont, P. W. R., 289
Bechara, I. S., 95
Beck, L. R., 385
Becker, H., 94
Beddows, C. G., 385
Bednas, M. E., 80
Beelen, T. P. M., 319
Beeler, D. L., 128
Beghishev, V. P., 85
Behnisch, J., 346
Behr, J. P., 351
Beier, W., 386
Beileryan, N. M., 60, 63, 69
Beinert, G., 38, 45, 153
Bekasova, N. I., 347
Belavtseva, E. M., 154
Belen'kaya, B. G., 37
Belenkii, B., 44
Belenkii, B. G., 244, 247
Bel'govskii, I. M., 55
Belikov, V. M., 363
Belina, K., 81
Bell, A. T., 57
Bell, J. P., 84
Bell, M. L., 379
Bell, V. L., 91
Bellenger, V., 335, 343
Bellman, T., 151
Bello, A., 217, 258
Bellobono, I. R., 79, 87
Bellsci, P., 84
Bellville, D. J., 363
Belmont, A. S., 378
Belogovodskaya, K. V., 63
Belomoina, M. N., 348
Bélorgey, G., 82, 263
Belton, P. S., 107
Beltzung, M., 149, 197
Bemi, L., 319
BeMiller, J. N., 101
Benachour, D., 332
Bender, C., 397
Ben-Dor, M., 379
Benedek, A., 245
Benedetti, E., 117
Benes, M., 386
Benezra, C., 306, 308, 328
Beni, M. C., 362
Benjamin, G. S., 149
Benkovic, S. J., 135
Benmouna, M., 165, 196, 197
Benoit, H., 196, 197, 255
Ben-Sabar, E., 282
Benson, R. A., 219
Benz, E. J., 140
Berard, D., 143
Berek, D., 245
Berens, A. R., 390
Berendsen, H. J. C., 398
Beres, J., 345
Beres, J. J., 259
Beret, S., 252
Berezin, I. V., 371
Berezin, M. P., 57
Bergbreiter, D. E., 365

Berger, H. R., 238
Bergeron, R. T., 355
Bergozza, M., 4
Berlin, A. A., 343
Berlin, A. M., 349
Berliner, R., 189
Berman, E. L., 36
Bernard, M., 320
Berne, B. J., 170
Berney, C. V., 198, 199
Bernheim, M. Y., 156
Bernt, H., 411
Berrou, E., 125
Berry, A. R., 381
Berry, H. M., 80
Bert, M., 342, 343
Berthoumieu, F., 378
Berticat, P., 186
Bertini, V., 53
Besecke, S., 67
Bessho, N., 240
Bessman, S. P., 384
Bestrabet, S. M., 208
Betteridge, D., 344
Beulen, J., 183
Beurton, D., 381
Bevan, L., 287
Bevington, J. C., 50, 51, 181
Bevis, M., 287
Beyer, A. L., 145
Beyer, C. W., jun., 382
Beynon, P. J., 408
Bezuglyi, V. D., 72, 73, 75
Bezzubaev, V. P., 404
Bhardwaj, I. S., 34, 183
Bhargava, G. S., 225
Bhattacharjee, A. K., 118, 132
Bhattacharyya, L., 111
Bhattacharyya, S. N., 225
Bhaumik, B., 393
Bia, M. J., 383
Biagini, E., 85, 167, 255
Bibbo, M. I., 388
Bickley, H. C., 385
Bidstrup, S. A., 149
Bied-Charreton, C., 307
Biefeld, R. M., 151
Bienz, M., 142
Bigg, D. M., 294
Biggs, D. N., 298
Bihari-Varga, M., 127
Biknukhametova, F. G., 341
Bikson, B., 212
Bílá, J., 55
Bilge, F. H., 375
Biliaderis, C. G., 99, 100
Billotet, C., 249
Binder, K., 203, 249, 396, 397
Bini, D. A., 86
Binion, S., 131
Birch, D. J. S., 334, 402
Birch, M. W., 289
Bird, R. B., 159, 173, 174, 279, 291
Birdsell, D. C., 381

Author Index

Birks, J. B., 401
Birky, M. M., 343
Birley, A. W., 80, 185
Birshtein, T. M., 395, 397
Birtwistle, D. T., 61, 389
Bishop, J. R., 136
Bishop, M., 173, 396, 397
Biswas, A. B., 407
Biswas, M., 30, 34, 90, 317
Bityurin, N. M., 57
Bizzigotti, G. O., 367
Bjork, I., 128, 129
Bjornsson, S., 125
Black, M. M., 151, 376
Blackley, D. C., 61, 71, 389
Blackwell, J., 82, 97, 103, 120, 207
Blagova, S. N., 335
Blais, P., 380
Blake, C. C. F., 391
Blake, D. A., 112
Blanchard, L. P., 341, 342
Bland, K. I., 378
Blandier, M., 335
Blaney, J. M., 391
Blankenship, R. M., 69
Blasco Cantera, F., 394
Blaszczyk, M., 130
Blaum, G., 259
Blaxall, H. S., 312
Blaydes, J. E., 382
Blaylock, J. T., 383
Blazek, V., 316
Blazso, M., 154, 342, 350
Bleha, M., 243
Blin, P., 35
Bloem, T. J., 384
Bloom, J., 343
Bloor, D., 299
Blouin, F. A., 184
Blount, H. N., 370
Blower, D. I., 205
Bloxall, H. S., 385
Bloys van Treslong, C. J., 261
Bluemke, D. A., 219
Bluhm, T. L., 100, 207
Blum, J., 318, 365
Blumstein, A., 78
Blumstein, R. B., 78
Blundell, D. T., 82
Bobbitt, T. F., 120
Bobrovskii, A. P., 344
Bocionek, P., 405
Bock, J., 225
Bock, K., 122
Bocnyk, C. P., 83
Bodor, G., 5
Böhm, L. L., 15, 19
Bömer, B., 53
Boens, N., 261, 262
Boerio, F. J., 152
Boes, J., 32
Böse, N., 226
Bogdantsaliev, T., 334
Boger, D. V., 294

Bogomolova, T. B., 33
Bogunovic, L. J., 148
Bohdanecky, M., 159, 167, 226, 240
Bohmer, V., 86
Bohn, P. W., 409
Boiesan, V., 65
Boiko, G. I., 89
Boileau, S., 53, 150, 235, 306, 308, 310
Boiteux, S., 364
Bokobza, L., 262
Bolgov, S. A., 85
Bolikal, D., 373
Bolivar, F., 133
Bolmosova, V. P., 349
Bolotnikova, L. S., 85
Boluk, M. Y., 148
Bolza, F., 60
Bommer, J., 384
Bonaly, R., 121
Bonamy, A., 64
Bond, A. J., 74
Bondarenko, S. G., 54
Bondon, D., 386
Bone, R., 123
Bone, T., 66
Bonicamp, J. M., 372
Bonner, F. J., 371
Bonner, P., 31
Bonnier, M., 51, 178
Bonora, G. M., 117
Bonsignore, P. V., 96
Bookstein, F. L., 111
Booth, C., 205
Boots, H. M. J., 56
Bootsma, D., 138
Bootsma, J. P. C., 366
Bopp, R. C., 192
Borchardt, J. K., 182
Borchers, L., 381
Bormann, S., 342
Borodulina, M. Z., 332
Boros-Gyevi, E., 49
Borrebaeck, C. A. K., 132
Borso, P. C. S., 189
Borsus, J. M., 95
Bortel, E., 235
Borziau, C., 72
Boscato, J. F., 42
Bosch, J., 411
Bosco, P., 377
Bose, A., 241, 246
Bose, D. T., 81
Bosnyak, C. P., 83
Bossaer, P., 38
Bosscher, F., 206
Bossi, A., 319
Bossman, F. T., 385
Bostin, P., 334
Bostwick, J., 378
Botan, E., 78
Botstein, D., 135
Bottema, J. R., 380
Bottenbruch, L., 226

Bottiglione, V., 262
Bouchal, K., 55
Boucher, D. G., 19
Boudevska, H., 245
Boué, F., 190, 193
Bourguard, S., 148
Bourguignon, J., 321
Bourzem, Y., 72
Bouster, C., 342
Boutevin, B., 51, 406
Bovey, F. A., 79, 84, 175, 182, 186
Bowden, M. J., 346
Bowden, M. J. S., 405
Bowin, S., 308
Bowner, T. N., 346, 406
Boyd, G. E., 266
Boyd, J., 132
Boyd, R. H., 394
Boyer, R. F., 84, 390
Boyle, J. A., 144
Bozek, F., 38, 43
Braca, G., 318, 360, 365
Bradaczek, H., 118
Bradbury, E. J., 298
Bradley, S. M., 138
Bradshaw, I. J., 120
Bradshaw, J. S., 351
Brandolini, A. J., 206
Brandt, K. A., 330
Branicki, F. J., 380
Brant, D., 120
Brasch, D. J., 106
Brash, J. L., 246, 376
Brault, R. G., 410
Braun, C., 88
Braun, D., 51, 182, 342, 343, 344, 349
Brede, O., 32
Brender, C., 163, 165, 167
Bresler, L. S., 51, 181
Breslow, D. S., 27, 184, 384
Breslow, R., 351, 353
Breton, M., 125
Bretz, P. E., 89, 223
Brevard, C., 178
Brewer, C. F., 113
Bridger, K., 238
Briggs, P. J., 96
Bright, P. F., 293
Bright, R. P., 156
Brinke, G., 206
Brinkman, N., 81
Brinstead, R. A., 319
Brisse, F., 81, 208
Brittain, O., 84
Britten, R. J., 142
Broadhurst, M. G., 208, 296
Brochard, F., 165
Brockhaus, M., 130
Broedelmann, T. J., 123
Broers, A. N., 405
Brols, M., 346
Bromnikov, S. V., 90
Brooke, C., 151

Brooks, B. W., 61, 62, 63
Brooks, J. S., 179, 336
Brossas, J., 42
Brosse, J.-C., 41, 51, 178
Brossmer, R., 111
Brostow, W., 223
Brotzman, R. W., 149
Broussoux, D., 209
Broutman, L. J., 289
Brouwer, W. M., 319
Brown, C. W., 60
Brown, D. M., 134
Brown, D. S., 217
Brown, H. R., 213
Brown, J. E., 343
Brown, J. R., 88, 346
Brown, R. A., 283
Brown, R. D., 113
Brown, W., 119, 227
Browning, R. C., 384
Brownlee, G. G., 144
Broxton, T. J., 368, 370
Broyde, S. B., 392
Bruck, C., 129
Bruck, S. D., 375, 380
Brückner, S., 150, 396
Brumá, M., 90
Bruneau, C. M., 79
Brunel, J., 145
Brunelet, T., 330
Brunette, C. M., 259
Bruni, G., 4
Bruning, J. H., 409
Bruns, W., 169, 397
Brutchkov, C., 245
Bruzzone, M., 34
Bryon, P. A., 385
Brzezinski, J., 83, 246
Bu, N., 4
Bubeck, R. A., 286
Buchan, G. M., 77
Buchi, K. G., 129
Buchman, O., 318, 365
Buckingham, A. D., 190
Buckley, W. D., 411
Bucknall, C. B., 288, 290
Buckwalter, J. A., 126
Buday, M., 66
Budesinsky, M., 37, 85
Budtov, V. P., 52, 56, 238
Buekens, A., 342
Buffington, L. A., 104
Bukatov, G. D., 16, 18, 19
Bukatova, Z. K., 12
Bukina, M. F., 150
Bulacovschi, V., 78
Buleon, A., 103, 210
Bulgakov, V. Ya., 152
Bulkin, B. J., 348
Bull, R. A., 74
Bullen, N. P., 364
Bulpin, P. V., 101
Bundle, D. R., 111
Bunel, C., 33, 34, 34, 35
Bunn, A., 80, 185, 213

Bunton, C. A., 367, 368, 369, 370, 373
Buono, G., 328
Burchard, W., 150, 160, 161, 169, 170, 171, 237, 255, 397
Burchell, D. J., 206, 390
Burdon, R. H., 141
Burfield, D. R., 13
Burgess, A. N., 191, 199
Burgess, J., 366
Burille, P., 343
Burillo, G., 57
Burke, D. L., 382
Burke, J. F., 383
Burkel, W. E., 377
Burkhardt, T. J., 10
Burkitt, D. T., 346
Burnett, K. G., 199
Burns, C. M., 242, 244, 389
Burny, A., 129
Burrage, M. E., 44
Burrell, H., 258
Burri, C., 383
Burridge, J. M., 391
Burton, P. S., 355
Buruiana, E. C., 181, 343
Burzynska, M. H., 129
Busby, T. F., 128
Busch, D. H., 357
Busch, H. 144, 145
Bush, C. A., 122, 392
Bush, C. N., 67
Bushin, S. V., 226
Busnel, J. P., 79, 244
Busslinger, M., 144
Bussmann, W., 359
Busulini, L., 156
Butler, G. B., 37, 54, 151, 179, 184
Butler, S. S., 385
Butler, W. T., 124
Buyanova, E. R., 362
Bye, M. L., 95
Byrikhin, V. S., 30, 34
Bywater, S., 44, 180
Bzoch, K. R., 378

Caballero-Mata, P., 246
Cabaness, W. R., 399
Cabbiness, D. K., 343
Cable, H. C., 384
Cada, O., 152
Cadoff, I., 377
Cagiao, E., 212
Cainelli, G., 304, 365
Cais, R. E., 175, 177, 179, 182, 183, 186, 394
Calafate, B. A. L., 75
Calahorra, M. E., 221
Calahorra, E., 215
Calarie, Y., 157
Caldwell, J. D., 309
Calgari, S., 79
Callaerts, A., 342
Calvert, J. M., 74, 319, 366

Calvert, P. D., 218
Camargo, R. E., 260
Camberlin, Y., 97
Cameron, G. G., 77, 303, 310, 341
Cameron, H. U., 379
Camino, G., 342, 345
Campana, F., 187
Campbell, D. H., 360
Canavese, C., 384
Candau, F., 71
Candau, S., 149, 150, 197
Cannon, C. G., 205
Canterberry, J. B., 37
Cao, T., 62
Cao, Y., 207
Capaccio, G., 213, 221
Capio, C. D., 148
Cappallett, R., 125
Caprino, J. C., 152
Caprino, L., 376
Caraculacu, A., 181, 343
Cardelino, B., 355
Cardillo, G., 320
Carel, A. B., 343
Carette, L. B., 335, 343
Carfagna, C., 383
Cargioli, J. D., 83, 153
Carley, J. F., 294
Carlini, C., 318, 360
Carlsson, B. J., 80
Carlsson, D. J., 331, 333
Caroline, D., 165, 171, 235, 236, 390
Carothers, J. A., 185
Carpenter, J. M., 189
Carraher, C. E., 311, 312, 385
Carreau, P. J., 294
Carrier, F., 87
Carriere, F. J., 78
Carroll, F. P., 97
Carroll, V., 120
Cartmill, T. D. I., 117
Caruso, S., 348
Caruthers, J. M., 241
Caruthers, M. H., 133
Carver, J. P., 115, 122
Casanovas, J., 152
Casassa, E. F., 172, 249
Cascaval, C. N., 341, 346
Casey, N., 409
Casiraghi, G., 245
Casnati, G., 245
Casolaro, M., 318
Caspar, J. V., 319, 366
Cassiman, J. J., 129
Castillo, J. L., 74
Castle, J. E., 376
Castro, J. M., 96, 260, 285
Casu, B., 128
Caswell, B., 282
Catala, J. M., 42
Catalina, F., 403
Catterall, J. F., 144
Cauley, D. A., 62, 389

Author Index

Cavagna, F., 182
Cavanaugh, D. B., 81, 206
Cavlak, Y., 383
Cavrot, J. P., 222
Ceccarelli, G., 36
Cefelin, P., 373
Cěgolja, A. S., 78
78
Cella, J. A., 31
Cenini, S., 94
Ceperley, D., 164, 173, 396, 397
Cerrai, P., 33, 73, 75
Cerutis, D. R., 385
Cesca, S., 34
Cetron, M. S., 145
Ceustermans, R., 128
Cha, C., 390
Chabert, B., 186
Chabot, F., 77
Chadwick, D. H., 93
Chadwick, D. J., 359
Chadwick, V. S., 245
Chaganti, R., 138
Chaimovich, H., 366, 368
Chaineaux, J., 41
Chainey, M., 69
Chajo, R., 266
Chakrabarti, P. M., 64
Chakraborty, K. B., 332
Chalkley, R., 140
Challa, G., 206, 263, 366
Chalykh, A. E., 39
Chambat, G., 102
Chamberlain, B. J., 62
Chambers, J., 96, 347
Chambon, P., 141
Champion, J. V., 168
Chan, C. C., 320
Chan, K. H., 333
Chan, M. M., 96
Chan, S. F., 319
Chan, T., 286
Chan, T. H., 363
Chanda, M., 319, 366
Chandra, R., 334, 337
Chandross, E. A., 405, 406
Chang, E. H., 138
Chang, E. P., 88
Chang, J.-C., 283
Chang, J. H., 310
Chang, J. Y., 37
Chang, M., 60
Chang, T. M., 383
Chang, V. S. C., 38, 39
Chanzy, H., 102, 103, 210, 255
Chapelet-Letourneux, G., 51
Chapiro, A., 57
Chapman, J. F., 96
Charlet, G., 249
Charlson, E. J., 386
Chatamra, K., 380
Chatfield, D. A., 349
Chatraei, Sh., 281
Chattergee, S., 317
Chatzitheodorou, G., 301

Chau, C. C., 288
Chaudhuri, N. K., 208
Chaumont, P., 153
Chauvin, Y., 318
Chaves, M. S., 346
Chawla, A. S., 386
Cheah, S. C., 117
Cheam, T. C., 209
Chee, K. K., 238, 257
Chein, J. C. W., 207
Chekushina, M. A., 343
Cheminat, A., 306, 308, 328
Chen, C. H., 344
Chen, C. Y., 410
Chen, J., 48
Chen, J.-T., 225
Chen, N., 386
Chen, P., 296
Chen, T. K., 177
Chen, W. C., 42
Chen, W. Y., 386
Chen, Y., 161
Cheney, J., 358
Cheng, H. N., 178, 183
Cheng, R.-S., 245
Cheng, T. C., 48
Chéradame, H., 53
Chereiskii, Z. Yu., 88
Cherkashin, M. I., 404
Cherkezyan, V. O., 345
Chernikhov, Y. A., 348
Cherry, W. R., 371
Cheshire, P., 191, 192, 206
Chesworth, A. G., 13
Chettauf, A., 308
Chevalier, J., 102
Chiantore, O., 243
Chiarugi, V., 125
Chiba, A., 266
Chiba, T., 130
Chidambareswaran, P. K., 103
Chiellini, E., 53, 78, 315
Chien, J. C. W., 7, 9, 299, 346, 386, 410
Chigono, P., 222
Chikahisa, Y., 161
Child, H. R., 189, 200
Chilvers, G. R., 107
Chin, W.-Y., 68
Chingas, G. C., 184
Chinigos, M. A., 384
Chinitz, J. L., 380
Chiotis, A., 342
Chiou, J. S., 263
Chirinos-Padron, A., 333
Chiu, W.-Y., 70, 389
Chivers, R. A., 216
Cho, I., 36, 54, 362
Cho, T. S., 362
Choay, J., 128
Choi, K., 295
Choi. K. W., 208
Choi, Y. C., 144
Chojnowski, J., 37
Chonde, Y., 68

Chong, Y. K., 340
Choo, K. P., 284
Choong, H. S., 409
Chopin, C., 383
Choplin, L., 294
Chosalow, F. I., 80
Chou, C.-K., 393
Chou, M., 319
Chou, T. W., 289
Choudhary, M. S., 344
Chow, A., 97
Chow, F., 133
Chow, J. F., 371
Chow, L. C., 88
Choy, C. L., 81
Choy, I. C., 157
Chrapil, M., 384
Chrastova, V., 164, 226
Chreptowicz, T., 91, 348
Christel, P., 77
Christensen, J. J., 351
Christenson, J. T., 375
Christianson, D. D., 102
Chu, B., 149, 171, 237, 254, 256, 257, 390
Chu, C. C., 381
Chu, C. H., 362
Chu, T. M., 131
Chuah, C. T., 106
Chuang, I. S., 187
Chubarova, Ye. V., 242
Chumaevskii, N. B., 18
Chumpitazi-Hermoza, B., 102
Chung, C. I., 284
Chung, D. Y. L., 38
Chung, S.-Y., 144
Chung, T. C., 405
Chung, T. S., 295
Chunng, M. J., 100
Chwialkowska, W, 36
Ciardelli, F., 53, 318, 360
Cidl, K., 383
Ciesco, J. N., 151
Ciferri, A., 167, 222, 255, 265
Cihlar, J., 25, 27
Cimerol, J. J., 94
Cinquini, M., 315
Cipiciani, A., 369
Citovicky, P., 164, 226
Clagett, G. P., 377
Clark, A. T., 166, 398
Clark, D. T., 186, 331, 332, 338
Clark, E., 71
Claer, E. S., 89, 208
Clark, H. C., 319
Clark, J. D., 209
Clark, R. E., 380
Clarke, T. C., 187, 300
Clarkson, R. W., 336
Claudy, P., 97
Clear, J. M., 319
Clecak, N. J., 406
Clegg, R. M., 113
Cleland, R. L., 266
Clements, J., 220, 221

Cleveland, T. H., 93
Cliffe, I. A., 359
Clough, S. B., 78, 198
Coenen, J. W. E., 365
Cogswell, F. N., 279
Cohen, B. J., 363
Cohen, J.-P., 226
Cohen, R. E., 174, 198, 199, 212, 266, 268
Cohen, Y., 249
Cohen-Addad, J. P., 149, 150
Cojazzi, G., 87
Coleman, M. M., 82, 156, 345, 392
Collawn, J. F., 111
Collen, D., 128
Collet-Marti, V., 44, 180
Colonna, S., 315, 373
Colson, J. P., 210
Colvin, H. A., 33
Colvin, J. R., 103
Comănită, E., 52
Combs, R. L., 16
Commereuc, D., 318
Compere, S. J., 141
Conc, R. D., 144
Conder, K., 75
Conlon, D. A., 37
Conrad, G. W., 129
Constantin, D., 56
Contento, M., 304, 365
Conway, D. C., 339
Cook, C. F., jun., 410
Cooke, W. D., 151
Cooper, A. R., 263
Cooper, C. A., 214
Cooper, S. L., 97, 219, 377
Cooray, B. B., 335, 336
Coover, H. W., 16
Coppo, R., 384
Corces, V., 140
Cordes, H. G., 27
Corfield, G. C., 54, 179
Corneliussen, R. D., 223
Corner, T., 71
Cornia, M., 245
Corno, C., 34
Corot, J. P., 81
Corradini, P., 207, 393
Cortazar, M. M., 215, 216, 217, 221
Cortelek, D., 198
Cosaveanu, A., 65
Cosgrove, T., 199
Costa, L., 342, 345
Costa Bizzari, P., 86
Costello, T., 356
Coster, L., 126
Costin, M. H., 283
Cothran, W. C., 124
Cottart, J. J., 408
Cotton, J. P., 165, 196, 197
Cotzur, C., 319
Coudane, J., 35
Coulson, A. R., 134

Cousin, P., 82
Coutin, B., 185
Covault, J., 140
Coville, N. J., 365
Cowell, R. G., 253
Cowie, J. M., 266
Cowie, J. M. G., 79
Cowsar, D., 385
Cox, A. P. D., 284
Cox, B. G., 356, 359
Cox, S. G., 370
Cozens, P. J., 142
Cram, D. J., 357, 359
Cramer, F., 351
Cramer, W. A., 123
Crasiorek, M., 88
Crawshaw, T. H., 351
Crea, R., 133, 134
Creamer, C. E., 152
Crescenzi, V., 185, 187
Creutz, C., 319
Crine, J. P., 152
Crist, B., 195, 206
Critchfield, F. E., 97
Crivello, J. V., 31, 32, 37
Crochet, M. J., 282, 295
Crossman, S. C., 294
Crouzet, P., 245
Crowley, J. F., 111
Crowley, T., 199
Crowson, R. J., 281, 294
Crugnola, A., 382
Cser, F., 81
Cuadrado, T. R., 387
Cuccovia, I. M., 366, 368
Cuculo, J. A., 217
Cudby, M. E. A., 213, 218, 344
Cukor, P., 410
Culp, L. A., 125
Cunliffe, A. V., 179, 185, 332
Cunningham, A., 95
Cupples, B. L., 30
Currat, R., 201, 210
Currie, P. K., 280, 291
Curtiss, C. F., 159, 173, 174, 291
Cush, R. J., 152
Cuthbertson, M. J., 50
Cutler, A. N., 119
Cutter, L. A., 388
Czajlik, I., 49
Czech, B., 313
Czerniawska, K., 88
Czlonkowska-Kohutnicka, Z., 84
Czybulka, G., 341

Dadamba, W., 185
Daffe, V., 353
Dagaiczyk, A., 144
Dahler, J. S., 170, 240
Dai, D., 147
Daimon, H., 86
Dais, P., 371
Daitch, N., 383

Dalal, A. K., 317
Dalla Casa, C., 86
Dalleave, J. A., 411
Dalrymple, E. D., 182
Daly, W. H., 309
Damaeva, A. D., 151
Daman, M. E., 122
Dambrine, F., 251
Damusis, A., 96
Damyanov, D., 19
Dana, A., 381
Dance, B., 409
Danforth, J. D., 343
Dani, M., 113
Daniel, J. C., 71
Danielwicz, M., 243, 247
Daniewska, I., 81
Danilina, L. I., 347
D'Antone, S., 315
Danyluk, S. S., 189
Daoud, M., 196, 255
Daoust, D., 342
Dar, K., 380
Darăngă, M., 78
Darlington, M. W., 293
Darwish, A. Y., 289
Das, A. N., 349
Das, A. R., 368
Das, D. P., 51
Das, M. K., 132
Das, N. N., 104
Das, P. K., 111
Das, P. R., 51
Das, S. C., 104
Das, S. K., 90
Dashintseva, G. I., 151
Dates, C., 391
Datta, T. K., 111
Dautzenberg, H., 237
David, C., 332, 334
Davidenko, N. V., 343
Davidjan, A., 44
Davidson, E. H., 142
Davidson, R. S., 31, 335
Davies, G. R., 222
Davies, J. A., 319
Davies, S., 13
Davis, A., 332, 338
Davis, F. F., 360
Davis, G. T., 208, 296
Davis, J. H., 54
Davis, M. A. F., 105
Davis, R. S., 142
Davoust, D., 122
Davtyan, S. P., 85
Davydov, E. Y., 337
Davydova, E. V., 337
Dawans, F., 318
Dawkins, J. V., 80, 185, 224, 243
Dawson, J. R., 95
Day, D. R., 300
Day, M., 80
Dayan, S., 236, 255, 265
Dayantis, J., 254, 255

Author Index 421

Dea, I. C. M., 106
De Abajo, J., 185, 395
Dealy, J. M., 281, 291
de Amorim, M. T. S. P., 342
Dearing, A., 391
Deb, P. C., 56
De Boeck, H., 112, 113
De Boer, J., 220
De Bruyne, C. K., 112, 113
de Buzzaccarini, F., 370
De Candia, F., 219, 221
Decastel, M., 111
Dechter, A., 213
Decker, C., 335
De Cleyn, G., 280, 291
De Croon, M. H. J. M., 365
Deffieux, A., 32
de Gennes, P. G., 56, 170, 196, 253, 254
Degre, P., 259
Dehmlow, E. V., 304, 372, 373
Dehmlow, S. S., 372
Deiters, J. A., 391
Deits, W., 82, 87, 300
Deits, W. D., 299
Dekkers, M. E. J., 223
de Klein, A., 138
de la Campa, J. G., 185
Delaney, E., 361
de la Torre, J. G., 236
Delaunois, G., 334
Delbourgo, R., 339
Delides, C., 185
Dellinger, T. A., 80
Delmas, G., 249
Delmas, M., 315, 316
Delmotte, F. M., 112
Delord, T. J., 358
Delos, S. E., 89
Del Rosso, M., 125
Delsanti, M., 197
Del Vecchio, W. D., 65
De Maria, P., 362
De May, A., 378
Demetrescu, I., 345
Demeyer, K., 401
Demina, M. I., 85
De Mol, M., 128
de Munno, A., 53
Deneris, J., 383
Denisevich, P., 74
Denisov, E. T., 340
Denisova, T. T., 179
Denmeyer, K., 261
Denn, M. M., 283, 294, 295
Derfler, D., 217
Desbrieres, J., 119, 246
De Schryver, F. C., 261, 262, 401
des Cloizeaux, J., 161, 162, 254
Deshpande, D. D., 407
Desjardins, S. B., 83
de Smet, M. D., 305
Desper, C. R., 259
Desprez, M., 383

Desrez, X., 378
Desrosiers, R. C., 141
Destor, C., 235
De Tar, F., 391
Detroy, R. W., 102
Dettenmaier, M., 84
Detweiler, D. M., 266
Deutsch, J. H., 65
Deutsch, J. M., 166, 203
Deutscher, S. B., 289
Devalle, G., 345
Devaux, P. F., 122
Deveux, R., 37
de Visser, A. C., 379
De Vreux, F., 300
De Vries, W. C., 383
Deyey, Yu. S., 85
Deyrup, H. A., 362
Dhal, A. K., 51
Dhar, O., 138
Dhyaneswari, E. S., 72
Diaz, A. F., 74
Diaz, F. R., 86, 90
Di Benedetto, A. T., 84, 287
Di Blasio, B., 117
Dicke, H. R., 63
Dicke, K. A., 384
Dicke, R., 64
Dickens, B., 340
Dickinson, L. C., 386
Didwania, A. K., 152
Diebold, J., 385
Dieckmann, D., 335
Diemer, R. B., jun., 316, 363
Dietl, F., 359
Dietrich, W., 95
Dietz, W., 217
Dijksman, J. F., 295
Dilks, A., 187, 331
Dill, K., 122
Dillon, M., 287
Dima, M., 226
Di Maina, M., 30
Di Maio, D., 135
Di Marzio, E. A., 194, 211, 397
Dimov, K., 77, 86
Dimtirov, D., 86
Di Napoli, A., 390
Ding, Y.-K., 242
Dipaola-Baranyi, G., 258, 259
Dirlikov, S. D., 81
Disapio, A. J., 152
di Silvestro, G., 178
Ditmore, Q. M., 382
Dix, F. M., 370
Dixon-Stubbs, D. J., 222
Dobas, S., 154
Dobashi, T., 254
Dobb, M. C., 89
Dobbin, C. J. B., 242
Dobkowski, Z., 83, 84, 246
Dobó, J., 57
Dobos, S., 350
Doddi, N., 381
Dodgson, K., 161

Doerfler, W., 141
Doi, M., 159, 291, 292
Doi, Y., 12, 22, 23, 24
Doicheva, J., 31
Doiuchi, T., 53
Dole, M., 266
Dolezel, S., 383
Dolgopolov, V. M., 413
Dollhopf, W., 207, 220
Dollimore, D., 350
Dolmatov, S. A., 348
Domard, M., 149, 150
Domareva, N. M., 20
Domb, C., 163, 164
Domen, C., 125
Domszy, R. C., 205
Donahue, P. E., 83, 153
Donahue, T. F., 142
Donald, A. M., 84, 286, 287
Dondos, A., 249
Donescu, D., 71
Donnelly, J. P., 411
Donovan, J. W., 123
Dooley, K. M., 316, 363
Dorfmueller, T., 149
Dorman, L. A., 323
Dorofeev, Yu. I., 409
Dosa, G., 382
Doskočilova, D., 79
Dostal, M., 383
Dou, H. J. M., 372
Doubé, C. P., 200
Douglas, W., 80
Douglass, D. C., 176
Douta, K., 406
Doyle, M. J., 287
Dozy, A. M., 136
Drach, V. A., 56
Dräger, M., 206
Dragan, S., 226
Dragojevic, M. D., 148
Drake, E. N., 361
Dreher, K. L., 130
Drenth, W., 362, 365
Dresler, T. D., 388
Dressler, D. P., 382
Dreyfuss, M. P., 94
Dreyfuss, P., 94
Driscoll, C., 144
Droscher, M., 215
Drusiani, A. M., 87
Dua, V. K., 122
Dubin, P. L., 246
Dubois, J. E., 74
Dubois-Violette, E., 170
Dubrovina, L. V., 154
Dubyaga, E. G., 152
Ducasse, R., 249
Duckett, R. A., 288
Duckworth, M. L., 133
Duda, A., 42, 43
Duda, J. L., 294
Dudchenko, V. K., 18
Duessel, H. J., 341, 349
Duff, C. C., 97

Duffy, B. R., 292
Duk, M., 115
Dumais, J. J., 187
Dumas, S., 44, 180
Dumitriu, S., 52
Duncan, R., 384
Dungan, J. M., 392
Dunker, H., 341
Dunkle, S. R., 346, 347
Dunn, A. S., 61, 62, 66, 71
Dunn, J. E., 282
Dupas, G., 321
Dupius, B., 383
Duplessix, R., 197, 199
Durand, D., 79
Durgaryan, S. G., 148
Durmis, J., 333
Dussek, K., 250, 264, 388
Dushenkov, S. D., 409
Duteurtre, B., 121
Dutt, A. S., 104
Dutta, A., 292
Dutta, K., 295
Duval, H., 197
Dvorak, J., 31
Dvornic, P. R., 154
Dwek, R. A., 132
Dwivedi, A. M., 393
Dyachkovskii, F. S., 27
Dybowski, C., 206
Dye, J. L., 358
Dykina, T. V., 150, 151
Dyuldeva, A. V., 341
Dyumaev, K. M., 407

Earl, W. L., 187
Earnest, T. R., 198
East, G. C., 42, 77
Easton, M. J., 334
Eastwood, A. R., 149
Eaton, A. C., 383
Eaton, D. R., 102
Ebdon, J. R., 50, 51, 175, 181
Eberhardt, H., 327
Eberhart, R. C., 375, 376, 377
Ebert, K. H., 244, 342
Ebisu, S., 112
Eby, R. K., 208, 393
Economou, N. P., 411
Economy, J., 82, 295
Eden, D., 168
Eder, H., 378
Eder, M., 218
Ederer, H. J., 244, 342
Edge, M. D., 133
Edlich, R. F., 381
Edman, P., 385
Edwards, C., 194, 210
Edwards, C. J. C., 150, 161, 169, 192, 241
Edwards, R. O., 377
Edwards, S. F., 159, 163, 173, 196, 256, 291, 396
Efendiev, A. A., 328, 364
Eftink, M. R., 355

Egan, W., 118
Egawa, H., 323, 407
Egglestone, G. T., 88
Egusa, S., 68
Ehlerding, U., 114
Eichinger, B. E., 149, 160, 388
Eichler, E., 309
Eiferman, E. A., 141
Eigner, W. D., 127
Einaga, Y., 162
Einhorn, I. N., 349
Eisenbach, C. D., 78
Eisenback, C. R., 378
Eisenbeiss, F., 243
Eisenberg, A., 290
Eisenriegler, E., 397
Ek, K., 109
Ekstrom, C., 80
El-Aasser, M. S., 69
Elben, V., 355
Elderfield, D. J., 162
Elgert, K. F., 177
Elgin, S. C. R., 140
Elias, H. G., 77, 260
Elias, J. G., 168
Eliceiri, G. L., 145
Eliseeva, V. I., 59, 60
Elizarova, G. L., 362
Elkowitz, A. B., 377
Ellis, E. J., 382
Ellis, H. A., 42
Ellison, M. S., 80
Ellison, S. A., 116
Ellmore, N. W., 137
Ellsworth, S. R., 376
Elmgre, H., 234
El-Rafie, M. H., 87
Elsdon, W. L., 242, 244
Elsenbaumer, R. L., 300
El-Seoud, M. I., 368
El-Seoud, O. A., 368, 370
Elson, E. L., 256
El'yashevich, A. M., 91
Encina, M. V., 403
Endo, J., 322
Endo, R., 243
Endo, T., 54, 321
Endroodi, J., 382
Engbers, J. B. F. N., 369
Engel, D., 226
Engelaere, J. C., 81, 222
Engelhardt, G., 150, 188
Engeln, I., 210
Engolke, T., 218
Engvall, E., 99, 123, 124
Enhov, Y. A., 338
Enikolopyan, N. S., 55, 349
Enkelmann, V., 207, 393
Entelis, S. G., 30
Epperson, J. E., 189
Epstein, I. R., 391
Erdman, B., 395
Ericksen, R. H., 336
Erickson, J. R., 69
Erikson, M. S., 130

Eringen, A. C., 292
Ermakova, T. G., 226
Erman, B., 83
Erni, D., 148
Ernst, C. A., 150
Ershov, A., 64
Erussalimsky, B. L., 40, 44
Erwin, L., 284
Escaig, B., 192
Eschpasse, H., 378
Escudero, J. A., 171
Eshdat, Y., 116
Eshuis, E., 206
Esipora, N. G., 392
Esnault, R., 289
Etaiw, S. H., 333
Etemad, S., 405
Etzler, M. E., 132
Eva, A., 137
Evans, D. J., 393
Evans, J., 360
Evans, K. E., 173, 174, 396
Evans, T. L., 155
Evers, F., 36, 177
Evstigneeva, T. V., 154
Ewen, B., 172, 202, 203, 236
Ezhova, E. A., 30, 34

Faber, J., jun., 189
Fadnavis, N., 369
Fagan, E. A., 245
Faibisoff, B., 380
Faidysh, A. N., 400
Faix, F., 185
Faizullin, I. N., 341
Fajardo, O., 74
Fakirov, S., 81, 89
Falcone, D. R., 83
Fales, F. W., 100
Falk, B., 348
Fallan, J., 380
Fallen, P., 314
Fan, F. R. F., 74
Fan, J. L., 346
Fan, L. T., 389
Fancey, K. S., 107
Fanter, D. L., 337
Farina, M., 178, 220
Farmer, B. L., 393
Farnoux, B., 165
Farrall, J. J., 308
Farrall, J. M., 330
Farrall, M. J., 43, 305, 306, 309, 325, 326
Farrer, N. R., 286
Fastovskaya, M. I., 363
Fastrez, J., 353
Faterpeker, S. A., 79
Fatinikum, K. O., 333
Fatou, J. G., 31, 217, 258
Faulkner, K., 73
Faullimmel, J. G., 33
Favera, R. D., 137
Favier, J. C., 33, 45, 177

Author Index

Fawcett, A. H., 185, 207
Febroriello, P., 123
Federkow, D. M., 379
Fedorova, L. N., 400
Fedoseev, S. D., 349
Fehérvári, A., 49
Fei, C. P., 363
Feijen, J., 380
Feizi, T., 130
Fejgin, J., 37
Felber, B. K., 144
Feld, M. K., 124
Feld, W. A., 312
Felder, B., 331
Fellers, J. F., 295
Fencil, C. R., 412
Fendler, J. H., 366, 367, 371, 372
Feng, X. D., 386
Feng, Z., 386
fen Hoopen, H. W., 380
Ferguson, R., 79
Ferguson, R. C., 184
Ferguson-Miller, S., 351
Ferguson-Segall, M., 123
Fernandez, S. M., 113, 123
Fernando, D. R., 368
Ferracini, E., 205
Ferrari, A., 4
Ferrero, A., 205
Ferro, D. R., 394
Ferruti, P., 318, 362
Ferry, J. D., 124, 171
Fesicyan, S., 240
Fetters, L. J., 40, 44, 48, 201, 226, 242
Feyereusen, C., 266
Fibbi, G., 125
Fichera, A., 4, 87
Fidler, S. K., 349
Fieldhouse, J. W., 155
Fiete, D., 109
Figini, R. V., 79, 225, 242
Filisko, F. E., 377
Filkuta, J., 383
Fini, A., 362
Fink, F. R., 142
Fink, G., 26, 180
Fink, J. K., 49, 80, 347
Finlayson, B. A., 283
Finzel, W. A., 152
Firat, Y., 180
Firman, P., 356, 359
Firon, N., 116
Fisa, B., 294
Fischer, E. W., 84, 223, 286
Fischer, L., 221
Fish, W. W., 111, 128
Fisher, L. D., 80
Fisher, M., 245
Fisher, M. C., 47
Fitch, R. M., 60, 62
Fitzgerald, P. R., 220
Fitzgibbon, P. D., 400
Fitzwalter, S., 390

Fixman, M., 167, 168, 169, 256
Flanders, D. C., 411
Flavell, R. A., 141, 144
Fleckenstein, B., 141
Fleming, W. W., 185
Fletcher, C. W., 335
Fletcher, P. D. I., 371
Fletcher, S. J., 259
Fleury, A., 375
Flint, L. M., 378
Flint, S. J., 140, 145, 146
Florián, S., 49
Florquin, S. M., 31
Florshiem, R., 35
Flory, P. J., 83, 191, 192, 255, 395, 397
Flowers, H. M., 105
Flowers, L. I., 365
Fodor, Z., 333, 340
Földes-Berezsnich, T., 49, 58
Foffani, A., 365
Fok, N. V., 338
Folkes, M. J., 290
Folonari, C., 73
Fontaine, E., 343
Fontaine, F., 80
Fontanville, M., 41, 45, 47, 180
Foord, S. A., 106
Forbes, C. D., 377
Force, C. G., 65
Ford, J. W., 377
Ford, W. T., 313, 320, 364
Forman, B. D., 86, 346
Fornasier, R., 315, 367, 368
Forry, K. R., 129
Forsman, W. C., 198
Fortelny, I., 167, 169, 238, 240
Fortunato, B., 77
Foti, S., 96, 157, 346, 347, 348, 350
Fouassier, J. P., 50, 64, 71
Fourimal, G., 378
Fowler, N. E., 298
Fradet, A., 94
Fraih, M., 54
Frampton, J., 134
Franchini, G., 138
Francik, W. P., 38
Francis, C. G., 319
Francois, C., 386
Francois, J., 227
Frangou, S. A., 106
Frank, C. W., 261, 262, 400, 402
Frank, O., 89
Frank, W. F. X., 79, 81
Frankevich, E. L., 403
Franklin, T. C., 368
Fransson, L. A., 126, 127
Franta, E., 38, 42, 43, 45
Franz, H., 111
Franzen, A., 125
Frazer, S., 398
Freche, P., 182
Frecher, J. M. J., 364

Frechet, J. M. J., 43, 304, 305, 306, 307, 308, 309, 320, 324, 326, 328, 330, 360
Fredenslund, A., 253
Freed, K. F., 161, 162, 254, 255, 396
Freed, K. L., 173
Freedberg, P. S., 379
Freelin, R. G., 82
Freeman, W. J., 184
Freire, J. J., 168, 171, 236
French, D. M., 242
Frenkel, S. Ya., 90, 235
Frénoy, J. P., 111
Freudenberger, V., 80
Fridd, P. F., 51
Fridman, A. I., 362
Fried, F., 236
Friedrich, W., 223
Frigerio, P. F., 90
Frisch, H. L., 165, 173, 396, 397
Frisch, K. C., 93, 95, 96
Fritschel, S. J., 366
Fritzinger, B. K., 385
Froehling, P. E., 80
Froemming, K. H., 351
Frolov, V. G., 85
Froment, P., 103, 210
Fronczek, F. R., 358
Frontini, P. M., 387
Frostin-Rio, M., 307
Froyer, G., 207
Fruba, J., 382
Frunze, T. M., 54, 85
Fu, W.-P., 408
Fueki, K., 335, 336
Fuhrman, H. S., 355
Fujii, M., 49, 166
Fujimoto, M., 354
Fujimura, M., 268
Fujino, M., 403
Fujishige, S., 35
Fujita, H., 121, 162, 164, 237, 240, 252
Fujita, K., 63, 68, 389
Fujiwara, S., 413
Fujiwara, Y., 88, 89, 176, 394
Fujiyoshi, T., 410
Fujumasa, I., 383
Fukase, H., 295
Fukuda, H., 289
Fukui, H., 151
Fukumasu, H., 383
Fukunshi, K., 313
Fukuya, K., 359, 366
Fuller, E. N., 245
Fuller, T. J., 156, 158
Fumikawa, K., 46, 180
Fumlaux, G. C., 332
Funke, W., 62, 71
Funt, B. L., 73, 342
Furakawa, T., 208
Furcht, L. T., 124
Furlow, L. T. J., 378

Furtsch, T. A., 91
Furui, S., 389
Furusaka, M., 196, 266
Furusawa, K., 67
Fusezi, S., 94
Fuwa, H., 99, 102
Fyfe, C. A., 176, 244, 319, 389
Fyles, T. M., 351
Fytas, G., 149, 237

Gaboriaud, F., 347
Gagon, D. K., 283, 295
Gaiduk, R., 206
Gaillard, J., 378
Gait, M. J., 133
Gajardo, I., 35
Gajria, C., 66
Galbraith, M. N., 340
Galen, A. T., 137
Galeski, A., 214
Galien, A., 385
Galin, M., 251
Galin, M. A., 382
Galina, H., 257, 261, 264
Galli, G., 78
Galli, P., 4
Gallo, R., 372
Gallo, R. C., 137, 138
Gallo, R. S., 137
Gallot, Y., 197, 199
Gallucci, J. A., 391
Galvin, G. D., 281
Gamarra, M., 131
Gambarott, S., 372
Gamichon, C., 53
Gan, L. H., 369
Ganazzoli, F., 202
Gandhi, V. G., 183
Ganesan, A., 312
Gankina, E. S., 244, 247
Gankov, N., 86
Gantmakher, A. R., 33
Ganuchak, N. I., 75
Gao, Y.-S., 242
Gaponov, A. I., 350
Garag, S. K., 80
Garbunova, Ye. V., 85
Garcia, L., 388
Garcia de la Torre, J., 168
Garcia-Rosas, J., 359
Gardette, J. L., 332, 337
Gardini, G. P., 74
Gardner, W., 151
Gardner, W. T., 128
Garegg, P. J., 112
Gargallo, L., 234
Garner, A., 382
Garnier, F., 74
Garreau, H., 34
Garrick, D. G., 351
Garrido, L., 31, 36, 395
Garroway, A. N., 89, 187
Garton, A., 380
Garza, J., 217

Gascoin, A., 182
Gaset, A., 315, 316
Gashgari, M. A., 261
Gaste, A., 385
Gatechair, L. R., 31
Gates, B. C., 214, 316, 317, 363
Gaudemer, A., 307
Gaur, N., 111
Gaur, U., 82
Gautier, J. C., 306
Gavelin, G. E., 381
Gavrilov, L. B., 84
Gay, N. J., 138
Gayathri, V., 51
Gaymons, R. T., 86
Gazicki, M., 386
Gebicki, J. M., 370
Geckeler, K., 327
Geerissen, H., 257
Gehani, R., 207, 213
Geheeb, N., 404
Gehrke, T. J., 385
Gehrmann, T., 386
Geisse, S., 143
Geissler, E., 197
Gelb, R. I., 355
Gelfand, V., 296
Gelles, R., 261, 262, 402
Genel, S. V., 152
Genies, M., 370
Genkin, V. N., 57
Gensmantel, N. P., 368, 369
Gent, A. N., 149
George, G. A., 88, 331
George, M. H., 42
George, W., 292
Georgescu, M., 69
Gerbi, D. J., 310
Gerg, P., 143
German, A. L., 319
Gerner, M. M., 75
Gerng, J., 383
Gerok, W., 114
Gerson, D. F., 378
Gettins, P., 132
Geuskens, G., 333, 334
Geyling, F. T., 283
Ghaemy, M., 334
Ghaffar, A., 332
Ghenkin, A. N., 154
Ghosh, P. K., 74
Ghosh, R. E., 203
Gianazza, E., 109
Gianchandoi, J., 89
Giannini, U., 4
Giard, D. J., 383
Gibbs, G. V., 150
Gibson, A. G., 222
Gibson, H. W., 84, 299, 301, 404
Gibson, P. E., 97
Gidley, M. J., 105
Giebel, R., 381
Giesbrecht, P., 118
Gilbert, R. D., 386

Gilbert, R. G., 61, 62, 68, 389
Gilding, D. K., 80, 241, 376
Gilead, D., 331
Gill, W. D., 74
Gillberg, G., 81
Gillette, P. C., 81
Gillot, M. A., 72
Gilmartin, E. J., 157
Gilmer, J., 260
Ginebreda, A., 373
Ginhac, J. M., 332, 333
Ginsburg, B. M., 90
Ginsburg, V., 130
Ginzburg, S., 384
Giori, C., 337
Gipstein, E., 185, 406
Giroloamo, M., 89, 211, 215
Giron, D. J., 385
Girshab, A. M., 77
Gittleman, J. I., 91
Giusti, P., 33
Gladkikh, I. F., 30
Gladkova, N. K., 148
Glassey, K., 117
Glaudemans, C. P. J., 130, 132
Glindemann, D., 239, 253
Glinka, C. J., 189
Glotin, M., 194, 205, 210, 218, 219
Glover, C. J., 259
Glushchenok, I. N., 54
Glybina, N. S., 407
Gnauck, R., 226
Go, K., 14
Go, P. M., 385
Godard, F., 380
Goeau-Brissoniere, O., 380
Goebel, K. H., 62
Goeddel, D. V., 133
Goelet, P., 133, 134
Goeran, C., 372
Goethals, E. J., 31, 37, 38
Goglia, G., 376
Gogolewski, S., 88, 218
Goh, S. H., 277, 278
Goh, S. L., 82
Gohil, R. M., 219
Gokan, H., 413
Gokel, G. W., 304
Gold, V., 356
Goldberg, L., 377
Goldenberg, C. J., 146
Goldenfeld, N. D., 203
Goldfarb, I. J., 349
Goldman, M., 382
Goldman, R. C., 118
Goldrick, M. R., 412
Goldstein, I. J., 111, 112, 114, 382
Goldstein, N., 260
Goldstein, S., 127
Goldwasser, J. M., 69, 242, 244
Golikov, I. V., 57
Golubev, A. V., 184
Golulski, Z., 331

Author Index

Gomez, M. A., 258
Gong, K. C., 341
Gong, M. S., 36, 54
Gonties, D., 381
Gonzale, A. L., 335
Gonzalez, A. L., 335
Gonzalez, F., 334
Gonzalez, M. C., 233
Good, T. A., 261
Goodall, A. R., 60
Goodall, B. L., 5
Goodin, J. W., 31
Goodwin, J. W., 166
Gooi, H. C., 130
Gopalakrishnan, T. V., 140
Gorbatenko, A. N., 85
Gorbunov, V. N., 185
Gordon, M., 253, 256, 257, 259, 261
Gorrichon, J. P., 315
Gorski, D., 38
Goryunov, A. N., 395
Gospodarowicz, D., 378
Gosselin, B., 383
Gosselin, C., 380
Goto, S., 389
Goto, Y., 247
Gotoh, O., 391
Gottlieb, M., 149, 239
Goudoin, R., 380
Govyrin, V. A., 392
Goya, S., 353
Goyal, P. S., 190
Gozdecki, W., 235
Gozdz, A. S., 306, 308, 309
Grabowski, B., 9
Graessley, W. W., 174, 195, 206, 291
Grätzel, M., 366, 371
Graf, E., 359
Graham, L. M., 377
Graham, N. B., 251
Graillat, C., 365
Granozzi, G., 4
Grant, B. D., 406
Grant, D. R., 99, 100
Grant, P. M., 74
Grasdaken, H., 106
Grassie, N., 336, 342, 345
Graves, D. F., 155
Gray, A., 289
Gray, D. G., 81, 255
Gray, H. B., 135
Gray, S., 383
Grayson, M. A., 337
Grayson, S. J., 347
Grebowicz, J., 350
Greco, R. S., 384
Green, A. R., 133
Green, G. E., 405
Greenberg, S., 320
Greenwood, T. D., 86
Greigger, P. P., 156
Grenier, D., 217
Grenier-Loustalot, M.-F., 182

Greschner, G. S., 247
Gressel, J., 105
Gressier, J. C., 86, 319, 347
Grey, A. A., 122
Gribkova, P. N., 88, 91, 347, 348, 349
Gribov, B. G., 409
Gridnev, V. N., 397
Griffin, A. C., 78
Griffith, M. J., 128
Griffiths, P. G., 50
Grigoryan, E. A., 27
Grigoryan, R. G., 36
Grimmelmann, E. K., 256
Grinblat, M. P., 152
Grinevich, K. P., 149
Grinevich, T. V., 30
Grinnell, F., 124
Grisafi, P., 135
Grob, R., 152
Grobler, A., 7
Groeninckx, G., 80
Groffen, J., 138
Gronemeyer, H., 143
Groner, B., 143
Gross, R. H., 145
Grossman, D. M., 338
Grossman, H. P., 220
Grossman, S., 87
Grosveld, G., 138
Groudine, M., 139, 140
Grovu, M., 72
Grubb, D. T., 208
Grubbs, G. R., 153, 350
Gruber, E., 127, 382
Gruentzig, C., 33
Guaita, M., 345
Guba, P., 383
Guckel, H., 409
Guenet, J. M., 195, 212
Guenther, G. D., 185
Guerra, G., 75
Güven, O., 218
Gugumus, F., 333
Guhaniyogi, S. C., 39
Gui, L., 4
Guialis, A., 145
Guidetti, G., 4
Guidoin, R., 380
Guillet, J. E., 16, 54, 336, 399, 401, 408
Guillot, J., 69
Guitierrez Cabañas, P., 232
Guixer, J., 373
Gul, V. E., 152
Gulari, E., 256, 390
Gulino, D., 180
Gunasekera, K., 380
Gupta, A., 331, 336
Gupta, A. K., 255
Gupta, B., 76, 387
Gupta, I., 52, 327, 407
Gupta, R. K., 292, 295
Gupta, S., 132
Gupta, S. K., 76, 85, 387

Gupta, S. N., 52, 53, 327, 328, 407
Gurland, H. L., 383
Gurol, H., 170
Guruvaih, S., 72
Guryanova, V. V., 348
Guseinov, Z. M., 70
Gusev, V. V., 413
Guthrie, J. T., 385
Gutierrez, B. O., 262
Gutierrez, G., 82
Guttman, C. M., 169, 194, 211, 262, 397
Guttmann, H., 217
Guyot, A., 38, 330, 342, 343
Guyot, P., 45
Guzenko, S. I., 34
Guzman, G. M., 215, 216, 217, 221, 226, 233
Guzman, J., 31, 36, 37, 79, 217, 233, 395

Haase, L., 38
Haberger, K., 411
Hackman, H. J., 151
Haddon, R. C., 343
Hadjichristidis, N., 54, 226, 242
Hadziioannou, G., 192, 199, 201
Hageman, H. J., 57
Hagemann, H., 53
Hagemeijer, A., 138
Hagen, A. J., 43, 309, 328
Hagenbüchle, O., 143
Hagerman, P. J., 168
Hagihara, B., 379
Hagino, A., 39
Hagiopol, C., 62
Hagiwara, M., 386
Hagnauer, G. L., 155
Hagnenoer, J. M., 383
Hahan, A. W., 386
Hahn, M. T., 89
Haight, G. P., 73
Hajek, K., 79
Hakomori, S. I., 130
Halasa, A. F., 44
Halbrook, M. E., 259
Halcomb, F. J., 379
Hales, P. D., 108
Halgas, J., 309, 324
Hall, H. K., 36
Hall, H. K., jun., 54
Hall, I., 77
Hall, I. H., 207
Hall, J. E., 44
Hall, R. H., 203
Hall, T. M., 411
Hall, T. W., 320, 324
Halle, B., 372
Haller, I., 148
Hallpap, P., 33
Halman, M., 96
Haltori, T., 315
Ham, G. E., 59

Hama, Y., 355
Hamaguchi, Y., 385
Hamaide, T., 247
Hamann, E., 253
Hamann, H., 152, 380
Hamasaki, S., 365
Hambrecht, J., 65
Hamed, F. H., 369
Hamel, G. L., 261
Hamer, D. H., 144
Hami, M. L., 284
Hamielec, A., 224
Hamielec, A. E., 56, 70, 244, 245, 246, 388, 389
Hamielec, A. W., 342
Hammar, W. J., 377
Hammel, R., 82
Hammersley, J. M., 165
Hamrik, O., 25, 27
Han, C. C., 169, 170, 172, 236, 254
Han, C. D., 279, 281, 293, 294, 295
Han, M. J., 37, 87
Han, X.-Z., 50
Hanada, T., 367
Hancock, I., 117
Handa, H., 376
Handa, T., 100
Handlin, D. L., 198
Haneckora, H., 383
Hanel, K. C., 375, 380
Hanfland, P., 130
Hanna, R., 70
Hanrahan, B. O., 210
Hansen, P. J., 218
Hansson, H. A., 381
Hanusa, L. H., 95
Hanzelka, P., 383
Haq, Z., 66
Hara, T., 208
Harabagiu, V., 319
Harada, K., 408, 409
Harada, M., 61, 63, 389
Harada, T., 86, 370
Harasaki, H., 380
Harbers, K., 141
Hardingham, T. E., 125
Hardy, R. E., 122
Hardy, W. B., 331
Hargest, S. C., 251
Hargreaves, J. S., 157, 401
Hariharan, P. C., 33
Harper, M. E., 138
Harrington, P. C., 114
Harris, J. E., 278
Harris, J. M., 314, 374
Harris, P. J., 156
Harris, R., 194, 211
Harris, R. H., 213
Harris, T. E., 82
Harrison, I. R., 210, 212, 213
Harrison, J. C., 355
Harrison, R., 205
Harryman, M. B. M., 201, 210

Hart, A. P., 377
Hartless, R. L., 405, 406
Hartley, F. R., 318, 319
Hartley, M. D., 95
Hartmannova, B., 383
Harun, M. G., 363
Harvey, R. A., 384
Harwood, H. J., 33, 35, 175, 177, 178, 184, 388
Hasan, S. M., 61, 71
Hascall, V. C., 125, 126
Haschemeyer, R. H., 389
Hasegawa, H., 30, 34, 39, 266, 326
Hasegawa, M., 91, 408
Hashimoto, H., 268
Hashimoto, J., 317, 321
Hashimoto, K., 183, 413
Hashimoto, M., 81, 222
Hashimoto, S., 37
Hashimoto, T., 246
Hashizume, J., 252
Hasimoto, T., 266, 268
Hasirci, V. N., 386
Hassan, G. A., 283
Hassan, S. A., 61
Hassanaly, P., 372
Hastings, L., 383
Haszeldine, R. N., 13
Hatada, K., 46, 176, 178, 180, 181
Hatano, A., 163
Hatlee, M. D., 371
Hattori, I., 42, 330
Hattori, K., 353, 354
Hattori, S., 245, 412, 413
Haudin, J., 255
Havel, H., 289
Havens, J. R., 90, 187
Havlicek, I., 264
Havsmark, B., 127
Haw, J. R., 294
Haward, R. N., 18, 19, 83, 84
Hawkett, B. S., 61, 62
Hawkins, R. M., 375
Hawthorne, J. H., 81
Hay, J. N., 81, 83, 84, 205, 217, 220
Hayama, S., 363
Hayame, S., 362
Hayase, S., 179
Hayashi, J., 103
Hayashi, K., 32
Hayashi, O., 298
Hayashi, S., 67
Hayer, G. L., 143
Hayes, K., 80
Hayman, E. G., 99, 123, 124
Hays, A. K., 148
Hayter, J. B., 172, 202, 203, 236
Hayward, W. S., 137, 138
Hazele, P. C., 207
He, Q.-J., 243
He, Z.-D., 245
Hearn, J., 60, 69

Heathcliffe, G. R., 133
Heatley, F., 186
Hebeish, A., 79, 87
Hecht, A. M., 197
Hecht, J. D., 126
Heckel, M. J., 70
Hecker, J. F., 377
Heckmann, W., 219
Heeger, A. J., 300, 405
Heffernan, J. G., 314, 330
Heider, J., 373
Heikens, D., 223
Hein, K. S., 381
Heinegard, D., 125
Heins, F., 68
Heise, B., 204
Heisterkamp, N., 138
Heitz, W., 51, 63, 64
Helbert, J. N., 410
Helgeson, R. C., 359
Helias, P., 79
Helioni, M., 155
Heller, J., 385
Hellman, M. Y., 83, 406
Hellstrom, I., 130
Hellstrom, K. E., 130
Helmstreit, W., 32
Helwing, R. F., 385
Hémery, P., 53, 235, 308
Henderson, D., 248
Hendra, P. J., 79, 218
Hendricks, R. W., 190, 213, 219
Henman, T. J., 333
Henmi, M., 180
Henrichs, P. M., 183
Henrici-Olivé, G., 13
Henry, F., 378
Hentschel, H. G. E., 190
Hentschel, M., 204, 205
Hepburn, C., 97
Hepner, M., 89
Heranishi, H., 88
Herbe, L., 385
Herc, J., 197
Hergeth., W. D., 71
Herman, B. A., 113, 123
Herman, W., 215
Hernandez-Feuntes, I., 238
Herrmann, T. R., 374
Herskowitz, M., 239
Hert, M., 56
Hertzberg, R. W., 89, 223, 289
Hervet, H., 172, 236
Herwig, J., 27
Herz, J., 149, 150, 153
Herzog-Cance, M. H., 102
Hess, D. W., 148, 412
Hess, M., 149
Hess, W., 170
Hesse, M. G. L., 238
Hetper, J., 345
Hett, E. M., 91
Heublein, G., 31, 33, 34
Heuschen, J., 78
Hewitt, J. M., 183

Author Index

Heyneker, H. L., 133
Hibionado, Y. M., 80
Hicks, J. R., 371
Hidalgo, J., 19, 20
Hidefumi, H., 351, 352
Hieke, E., 409
Higashimura, T., 30, 33, 34, 35, 39, 326, 394
Higgins, J. S., 161, 191, 198, 200, 202, 203
Higgins, T. J., 129
Higo, Y., 246
Higuchi, W. I., 371
Hijikata, I., 412, 413
Hild, G., 54
Hilgenfeld, R., 374
Hill, A., 202
Hill, D. J. T., 56, 183, 184
Hill, M. J., 214
Hindeleh, A. M., 205, 390
Hindenlang, D. M., 346
Hindrickscen, G., 80
Hine, P. J., 221
Hingerty, B. E., 392
Hino, T., 385
Hinrichsen, G., 81, 217
Hinze, W. L., 367
Hirai, A., 102
Hirai, H., 184, 354, 355, 365
Hirakawa, S., 362
Hirami, M., 88
Hirano, T., 176
Hirao, A., 42, 325, 330
Hiraoka, H., 413
Hirayama, C., 243
Hirayama, F., 353
Hirohashi, R., 91
Hirose, T., 133
Hirosuye, T., 162
Hirschler, M. M., 344
Hirst, L. W., 382
Hirte, R., 88
Hisasue, M., 69
Hitamaru, R., 213
Hiyashi, H., 192
Hiza, M., 33
Hjertberg, T., 49, 56, 181, 182
Ho, H. T., 179
Ho, K., 80
Ho, N. F. H., 371
Ho, P. K., 157
Hoang, P. M., 73
Hoang, V. C., 364
Hoath, J. M., 350
Hocker, J., 195
Hodge, I. M., 390
Hodge, J. E., 102
Hodge, P., 306, 309, 360, 364
Hodgeman, D. K. C., 88, 333, 346
Hodgkin, J. H., 179, 340
Hodgkinson, J. M., 288, 289
Hodnett, J. L., 135
Hoechst, U., 242
Hoecker, H., 36, 47

Hoerhold, H. H., 38
Hofer, D. C., 411
Hoff, A., 340
Hoffman, A. S., 376
Hoffmann, J. D., 194, 211, 216
Hoffmann, K., 411
Hogan, J. P., 29
Hogen-Esch, T. E., 151, 180
Hoitink, T. B., 207
Hokanson, J. L., 296
Holasek, A., 127
Holden, D. A., 54
Holland-Moritz, I., 206
Hollensleben, M. L., 79
Holmblad, G. L., 189
Holmes, A. S., 336
Holmes, B. S., 184
Holmes, R. R., 391
Holste, J. C., 259
Holtzer, A., 396
Holzer, G., 182, 342
Homer, L. D., 112
Honda, Y., 374
Hond, G. F., 133, 134
Hong, J., 84
Hong, K. M., 250
Hong, S. A., 294
Hong, Y. S., 369, 373
Hong, X.-Y., 408
Hook, M., 127
Hooker, T. M., 392
Hopfinger, A. J., 209, 216, 393
Horak, J., 69
Horbach, A., 82, 87, 226
Horbett, T. A., 376
Horhold, H.-H., 187
Horie, I., 389
Horiguchi, D., 360
Horii, F., 102, 376
Horii, H., 213
Horisberger, M., 378
Horn, H. G., 154
Horsfall, G. A., 88, 336
Horska, J., 236, 261, 270
Horta, A., 238
Horvath, E., 251
Horvath, I., 129
Hosaka, Y., 344
Hosako, R., 367
Hosemann, R., 204, 205, 390
Hoshino, H., 67
Hoshino, M., 355
Hosokawa, Y., 313
Hostejn, L., 339
Hotzel, H. E., 34
Hounsell, E. F., 130
Hourston, D. J., 69, 82
Hovens, S. J., 78
Hovey, M. C., 371
Hoyleats, M., 128
Hrabák, F., 53, 55
Hradil, J., 309, 316
Hrdlovic, P., 53
Hrncaiŕik, J., 226
Hrouz, J., 264

Hsieh, D. S. T., 360
Hsieh, E. T., 182, 183
Hsieh, K. H., 95
Hsieh, W. C., 32
Hsien, C. H., 151
Hsu, C.-Z., 70
Hsu, L.-C., 73
Hsu, S. L., 81, 206, 259, 390
Hu, A., 69
Hu, A. T., 56
Hu, C., 86, 87
Hu, C. B., 259
Hu, J. C., 245
Hu, M., 36
Huang, A. L., 143
Huang, M. Y., 179
Huang, Q., 362
Huang, R. Y. M., 242, 244, 389
Huang, S. S., 180
Huang, Y., 320
Huang, Y.-W., 70
Huber, H., 319
Huber, R., 301
Huckerby, T. N., 51, 126, 181
Hudec, I., 226
Hudee, P., 25
Hughes, B. A., 386
Hughes, G. P., 412
Hughes, I. D., 82
Hughes, R. D., 384
Huggins, M. L., 253
Huglin, M. B., 87, 233
Huguet, Y., 88
Hui, Y., 352
Hull, A. M., 282
Hull, W. E., 177, 185
Hultberg, H., 112
Hum, O. S., 376
Humbert, J. C., 385
Hume, J., 347
Hummel, D. O., 341, 349
Hundley, N. H., 314, 374
Hung, N. A., 30
Hunt, B. J., 309
Hunt, R. N., 95
Hurduc, N., 341
Hussey, H., 117
Hutton, J. F., 281
Hutton, N. W. E., 51, 181
Hux, R. A., 102
Hynes, N. E., 143

Iannone, M., 84, 287
Ibar, J. P., 295
Ibrahim, A. M., 90
Ibrahim, B. A., 77, 207
Ichikawa, S., 410
Ichimura, K., 406, 407
Ichiro Imanura S., 88
Ideura, K., 163
Idezuki, Y., 386
Idoux, J. P., 307
Ieda, M., 404, 405, 412, 413
Ifft, J. B., 389
Iglesias, J. L., 111

Ignatova, T. D., 259
Igucha, M., 210
Ihara, Y., 361, 367
Ihm, D. W., 217
Iida, M., 243
Iinuma, F., 400
Iizawa, T., 37, 305, 406, 408
Ikada, Y., 376
Ike, Y., 133
Ikeda, K., 153, 154, 350
Ikeda, M., 154, 350
Ikeda, S., 21, 22, 384
Ikeda, T., 351
Ikehara, K., 355
Ikemi, M., 266
Ikemura, T., 340, 342
Ikushige, T., 183
Ikuta, T., 222
Ilavansky, M., 251
Ilavsky, M., 264
Iler, L. R., 91
Illers, K.-H., 210, 219
Imachi, K., 383
Imae, K., 207
Imagawa, I., 406
Imagawa, M., 385
Imai, T., 53, 54, 55, 86, 167, 178, 226, 311, 312
Imajo, H., 90
Imamura, A., 208
Imamura, M., 355
Imamura, S., 409
Imanishi, Y., 398
Imerson, R., 409
Imhof, R. E., 334, 402
Imura, H., 385
Imuta, J., 373
Ina, Y., 330
Inada, M., 385
Inagaki, H., 65, 224, 237, 264
Ingen-Housz, J. F., 283
Ingham, K. C., 128
Inglefield, P. T., 83
Innorta, G., 365
Inokuchi, H., 404
Inoue, E., 405
Inoue, N., 35, 394
Inoue, S., 83, 352, 363
Inoue, T., 36, 38, 367
Inoue, Y., 266
Inuishi, Y., 404
Ioan, S., 226
Ionescu, L., 197
Ionescu, M. L., 199
Iqbal, M., 153, 259
Iri, K., 54, 178
Irig, M., 88
Iring,M., 333
Iring, R., 340
Iruin, J. J., 226
Irvine, P., 253, 256, 257, 261
Irvine, P. A., 83, 390, 395
Isado, S., 211
Isaeva, V. A., 348
Isandar, M., 268

Isgur, I. E., 65
Isher, J. M., 244
Isherwood, D. P., 284
Ishibashi, F., 379
Ishibashi, S., 413
Ishida, H., 33, 90, 187
Ishida, K., 102
Ishida, S.-I., 387
Ishida, Y., 315
Ishido, H., 54
Ishihara, K., 328
Ishii, S. I., 112
Ishii, Y., 226
Ishikawa, E., 385
Ishikawa, S., 86
Ishikawa, T., 69
Ishikawa, Y., 189, 196
Ishinabe, T., 397
Iskandar, M., 153, 259, 266
Ismailov, I., 69
Ismailov, T. I., 70
Isoda, S., 91
Isogai, S.-I., 54
Israel, S. C., 362
Itakura, K., 133
Itayama, K., 214
Ito, H., 133, 410
Ito, K., 51, 85, 384
Ito, O., 54
Ito, S., 401
Ito, T., 89
Ito, Y., 150
Itoh, K., 54
Itoh, M., 246, 413
Itoh, Y., 403
Itou, S., 167
Itsuno, S., 325, 330
Ivan, B., 39, 246, 343
Ivainer, A. Ya., 407
Ivanchev, S. S., 7, 20, 51, 52, 56, 59, 60, 64, 71, 181
Ivanov, M. P., 85,
Ivanov, V. B., 334
Ivanov, V. I., 413
Ivanov, V. M., 86
Ivanova, A. N., 85
Ivanova, S. L., 85
Iven, B., 343
Iverson, T., 112
Ivin, K. J., 179
Iwabuchi, S., 86, 400
Iwai, K., 314, 363
Iwakawa, S., 404
Iwakura, Y., 90, 91
Iwamoto, H., 372
Iwamoto, T., 180
Iwanami, K., 54
Iwata, H., 376
Iwaya, Y., 261, 401
Iwayama, K., 315
Iwayanagi, T., 406
Iwunze, M., 368
Iyoda, T., 319
Izatt, R. M., 351
Izidinov, S. O., 151

Izmailov, B. A., 54
Izuka, T., 208
Izumi, K., 21
Izyminkov, A. L., 37

Jabarin, S. A., 245
Jackowicz, J., 337
Jackson, N. R., 283
Jacobs, E., 342
Jacobs, H. T., 142
Jacobs, R. A., 332
Jacobson, H. W., 47
Jacobsson, S., 340
Jadraque, D., 103, 225
Jaeger, W., 226
Jähner, D., 141
Jaenisch, R., 141
Jaffe, N. S., 382
Jagur-Grodzinski, J., 34, 48, 178, 212
Jain, K., 84
Jakab, E., 342
Jakob, F., 80
Jakobi, R., 51
Jakubowski, J. J., 73
Jalink, H. L., 238
Jameal, H., 219
James, D. F., 281
James, W. J., 386
Jamieson, A. M., 120, 172
Janča, J., 241, 243, 245
Jancke, H., 188
Janier-Dubry, J. L., 316
Janik, R. K., 383
Janke, E., 206
Janke, E. W., 191
Janke, H., 150, 188, 192, 206
Janmey, P. A., 124
Jannink, G., 193, 197, 255
Janout, V., 373
Janović, Z., 71
Janzen, K. P., 328
Jao, T. C., 372
Jardetzky, O., 351
Jarm, V., 69, 71
Jaroszynske, D., 344
Jarvik, R., 383
Jarvis, A., 216
Jassim, A. N., 402
Jatlow, P., 383
Jawad, S. A., 222
Jayadevappa, E. S., 238
Jayalekshmy, P., 364
Jayaraman, T. V., 53
Jaycox, D. P., 379
Jazefonviez, J., 307
Jeanteur, P., 145
Jedral, W., 356
Jefferson, A., 33
Jeffries, M., 108
Jelinski, L. W., 79, 175, 187
Jellinek, H. H. G., 346, 347
Jencks, W. P., 353
Jendresen, M. D., 151
Jenkin, V. B., 96

Author Index

Jenkins, A. D., 58
Jenkins, R., 226
Jenkins, R. F., 246
Jenkins, W. L., 180
Jensen, J. P., 334
Jerguŝová, J., 49
Jerman, R. E., 293
Jërme, R., 78, 95
Jerry, J. P., 206
Jeuck, H., 46
Jhanwar, S. C., 138
Jhon, M. S., 170
Jiang, X., 352
Jilge, W., 170
Jiménez, A., 168, 236
Jin, J.-I., 78, 310
Jin, W.-X., 408
Jin, X. G., 341
Jing, X., 209
Jiricny, J., 96, 347
Jo, B.-W., 78
Jo, W. H., 156
Jobic, H., 201
Johann, C., 46
Johansson, R., 112
Johar, Y., 320
John, V. A., 384
Johns, A. N., 379
Johns, S. R., 50, 179, 181
Johnsen, R. M., 119
Johnson, A. F., 389
Johnson, D. J., 89
Johnson, D. R., 57
Johnson, G. E., 83, 261
Johnson, J. F., 263
Johnson, R. B., 124
Johnson, R. F., 355
Johnson, R. T., 151
Johnson, S. D., 116
Johnston, A., 342, 345
Jokay, L., 333
Jones, A. A., 83
Jones, B., 281
Jones, E. G., 349
Jones, J. B., 211
Jones, P., 120
Jones, R. G., 52
Jones, R. S., 253
Jones, W. B., 288
Jongchaap, R. J. J., 291
Jonte, J. M., 33
Jopson, H., 410
Jordan, F., 114
Jorgensen, E. C., 391
Jorstrad, S., 384
Josefsson, G., 384
Joseleau, J. P., 102
Joseph, E. A., 84
Joshi, M. G., 53
Joshi, S. V., 184
Josse, B., 206
Jost, J.-P., 142
Jouin, P., 357
Jovin, T. M., 113
Joyner, F. B., 16

Julia, S., 373
Jung, W. K. V., 383
Juret, C., 72
Jurgens, G., 127

Kabaivanov, V., 334
Kabamla, M. S., 333
Kabanov, V. A., 328, 364
Kabanov, V. Ya., 32, 39
Kabat, E. A., 129, 130
Kabilov, Z. A., 348
Kachan, A. A., 332
Kadyrov, D. I., 403
Kaetsu, I., 385
Kagawa, K., 32
Kagiya, T., 336
Kahn, F. J., 409
Kahn, R. H., 377
Kai, T., 255
Kaitna, R., 411
Kajiura, T., 100
Kajiwara, K., 160, 170, 397
Kajiwara, M., 155, 157
Kaku, M., 38
Kakufuta, E., 320
Kakugo, M., 183
Kakuichi, H., 313, 318
Kalabina, A. V., 184
Kalal, J., 226, 243, 316
Kalawole, E. G., 334
Kalck, P., 315
Kalennikov, E. A., 332
Kalfus, M., 244
Kalinin, V. N., 54
Kalinkevich, G. A., 348
Kallistov, O. V., 226
Kalnins, K., 41
Kalos, M. H., 164, 173, 396, 397
Kalpagam, V. P., 226
Kalyanam, N., 183
Kalyuzhnyi, V. I., 148
Kamachi, M., 49, 50, 181
Kamal, M. R., 294, 295
Kamalasanan, M. N., 405
Kamata, T., 245
Kambe, H., 91
Kambour, R. P., 83, 152, 153, 192, 287, 289
Kameyama, A., 312
Kameyama, M., 81
Kamezawa, M., 222
Kamide, K., 160, 226, 237, 240, 263
Kamienski, C. W., 4
Kaminska, A., 335
Kaminsky, W., 27, 28
Kamp, H., 86
Kan, Y. W., 136
Kanai, T., 35, 45, 47, 180, 394
Kanai, W., 412, 413
Kanarian, G., 394
Kanazawa, K. K., 74
Kanda, N., 319
Kaneda, T., 86

Kaneko, K., 387
Kaneko, M., 254, 304, 318, 319, 326
Kanetsuma, H., 206
Kang, C. K., 292
Kania, C. M., 154
Kanig, G., 210, 212
Kanovich, M. M., 345
Kantor, J. A., 140
Kanu, R. C., 292
Kapadia, P., 253
Kapadia, R. N., 317
Kaparissides, C., 389
Kapko, J., 86
Kaplan, S., 187
Kaplanová, M., 89
Karakozova, E. I., 349
Karaman, H., 314
Karamoto, T., 222
Karamyan, D. R., 63
Karapiperis, L., 411
Karasiewicz, P., 382
Karasz, F. E., 207
Karayannidis, G., 86
Karayannis, N. M., 11
Karcha, Yu. Yu., 206
Karods, J. L., 380
Karlsson, K. A., 130
Karmilova, L. V., 349
Karn, J., 134
Karpinets, A. P., 72
Karplus, M., 398
Karpukhin, O. N., 332
Karsa, D. R., 64
Karsch, V. A., 223
Kartavykh, V. P., 56
Karyakina, M. I., 337
Kasai, K. I., 112
Kasajima, M., 85
Kase, S., 283
Kasekeen, B. A., 226
Kashiwa, N., 6, 7, 16
Kashiwabara, H., 210
Kaspar, K., 70
Kaspar, M., 43
Kass, R. L., 89
Kassar, B. S., 382
Kast, H., 62
Kataoka, K., 376
Katayama, K.-I., 211
Katime, I., 233
Katime, I. A., 232
Katiyar, S. S., 369, 370, 371
Kato, H., 407
Kato, K., 130, 174, 302
Kato, N., 237
Kato, S., 328
Kato, T., 60
Kato, Y., 246, 385
Katsuki, S., 49
Katsura, T., 86
Katsuta, Y., 69
Katti, S. S., 294
Katuocak, S., 245
Katz, D., 296

Kaufman, F. B., 73, 411
Kaufman, J. J., 33
Kausch, H. H., 84, 288
Kawabata, N., 317, 321
Kawabata, Y., 365
Kawaguchi, A., 211
Kawaguchi, H., 67, 335
Kawahanasaki, N., 86
Kawai, H., 266, 268
Kawai, T., 226
Kawakami, Y., 33
Kawakishi, S., 101
Kawamura, T., 30, 33, 36, 52, 176, 178
Kawamura, Y., 409
Kawase, S., 383
Kawazoe, K., 368
Kazaev, A. I., 36, 55, 245
Kazakova, V. V., 350
Kazantseva, E. A., 349
Kazaryan, G. A., 36
Kazinski, S. J., 213
Kecskes, E., 129
Kedreena, N. F., 226
Keene, M. A., 140
Kehayoglou, A., 86
Kehn, J. M., 359
Keii, T., 12, 14, 22, 23, 24
Keith, H. D., 211
Kelen, T., 33, 246, 333, 340, 343
Kellcher, P. G., 83
Keller, A., 89, 211, 212, 214, 215, 216
Keller, A. D., 319
Keller, J. M., 125
Keller, M., 152
Keller, W., 146
Kellman, R., 310
Kellog, R. M., 357
Kelly, D. G., 123
Kelly, J., 315
Kelly, J. M., 319
Kemmerer, R., 326
Kemp, D., 81
Kempe, T., 133
Kemsley, G. M., 381
Kendall, F. M., 378
Kennedy, J. F., 120
Kennedy, J. P., 30, 31, 33, 35, 38, 39, 343
Kennedy, J. W., 248, 253, 257, 390
Keown, K., 384
Kepler, R. G., 296
Kerber, M. L., 84
Kerber, R., 51
Kercha, Yu. Yu., 154
Kern, W., 86
Kesner, L., 394
Kessler, T. R., 378. 383
Keszler, B., 5, 7
Keunings, R., 282
Kevan, L., 410
Kever, J. J., 244

Khabenko, A. V., 348
Khachaturov, A. S., 179
Khalafov, F. R., 30
Khalatur, P. G., 396, 397
Khaled, M. A., 391
Khalid, N., 52
Khalifa, W. M., 333
Khaligh, B., 389
Khalil, M. I., 87
Khan, A. H., 81
Khan, H. U., 225
Khan, M. I., 109, 115
Khanarian, G., 394, 395
Khanna, D. N., 78
Khanna, Y. P., 86, 218, 346
Kharitonov, N. P., 349
Kharitonov, V. V., 340
Khatchaturov, A. S., 179
Khe, N. C., 405
Khmelnitskii, R. A., 348
Khokhlov, A. R., 251, 255
Kholodenko, A. L., 396
Khomyakov, A. K., 37
Khor, E., 91
Khorramian, B. A., 119
Khoshdel, E., 309
Khouw, B., 320
Khromova, T. N., 345
Krushch, B. I., 338
Khune, G. D., 91
Kian-Chi Zhang, 245
Kida, M., 359
Keinzle-Sterzer, C., 266
Kikuchi, J., 360, 371
Kikuchi, T., 377, 383, 386
Kikyotani, S., 330
Kilyama, T., 208
Kilp, T., 336, 401, 408
Kilzer, L., 91
Kim, C. J., 245
Kim, K. U., 71
Kim, M. G., 185
Kim, S. W., 375
Kimata, K., 125, 126
Kimura, E., 358, 360
Kim'ıra, J. H., 125, 126
Kimura, S., 174, 291
Kimura, Y., 314, 359, 361, 367
Kindt, A., 111
King, J. S., 189, 190, 191
King, M., 380
King, P. S., 286
King, T., 199
Kingston, S. B., 35, 330
Kinloch, A. J., 288
Kinoshita, A., 409
Kinoshita, F., 409
Kinoshita, K., 410
Kinoshita, M., 87, 364
Kinstle, J. F., 30
Kintzinger, J. P., 359
Kiparissides, C., 70
Kippenberger, D. J., 372
Kirchner, D. G., 64
Kiriċhenko, E. A., 151

Kirilin, A. I., 407
Kirillova, E. I., 334
Kirk, K. A., 149
Kirszensztejn, P., 314
Kiryukhin, D. P., 57
Kise, H., 313
Kiseleva, T. M., 90
Kishino, K., 226
Kishore, K., 51
Kiss, G., 382
Kitaguchi, H., 354
Kitahara, A., 367, 368
Kitamaru, R., 102
Kitamura, M., 174
Kitamura, T., 54
Kitani, I., 404
Kitano, K., 38
Kitano, T., 167, 226, 237
Kitayama, M., 352
Kitayama, T., 46, 178, 180, 181, 208
Kiwi, J., 366
Klahn, P. K., 372
Klasner, R. D., 392
Kleid, D. G., 133
Klein, A., 69
Klein, A. J., 84
Klein, D. J., 163
Klein, J., 166, 304
Klein, P. G., 185
Klein, R., 170
Klein, S., 377
Kleiner, G., 96
Kleintjens, L. A., 201, 251, 252, 257, 261
Klemm, E., 38
Klempner, D., 267
Kleppick, M. E., 153, 350
Kleps, T., 344
Kleshcheva, M. S., 344
Klesper, E., 247
Klesper, J. W., 392
Kliger, D., 331
Klimos, D. M., 80
Klimovich, A. F., 72
Klöpffer, W., 400, 404
Klopfer, H. J., 152
Klotz, I. M., 361
Klyachko, N. L., 371
Kminek, J., 43
Knight, G. J., 83, 344
Knobler, C. M., 261
Knoesel, R., 336
Knoll, B. J., 139
Knonenthal, R. L., 382
Knox, J. R., 183
Knox, K., 116
Knutsson, B. A., 293
Kobayashi, H., 226, 352, 372
Kobayashi, K., 305, 330, 406
Kobayashi, M., 393
Kobayashi, N., 314, 363
Kobayashi, T., 407
Kobayashi, S., 37, 38, 179
Kobayashi, Y., 84

Author Index

Koberstein, J. T., 200
Kobmehl, G., 77
Kobori, M., 69
Kochanowski, A., 235
Kochi, M., 91
Kocian, P., 148
Kocourek, J., 109
Kodaira, K., 51
Kodaira, Y., 388
Kodaka, M., 355
Kodama, M., 358
Koehler, W. C., 190
Koenig, J. L., 90, 187, 206, 392, 393
Koenig, S. H., 113
Koenig, T. L., 81
Koerner, G., 152
Kössler, I., 226
Köttgen, E., 114
Kofinov, P., 68
Koga, K., 360
Koga, N., 360
Kohashi, T., 406
Kohli, D. K., 358
Kohn, R., 106
Koide, M., 319
Koinuma, H., 184
Kojima, G., 69
Kojima, H., 384
Kojima, K., 400
Kojima, M., 14, 353, 354
Kojima, T., 344
Kok, C. M., 171, 239, 243, 256, 262, 267
Kokalj, M., 314
Kokkiaris, D., 54
Kokorin, A. I., 362
Kolarik, J., 381
Kolarz, B. N., 261, 264
Kolawole, E. G., 334
Kolb, J., 349
Kole, R., 146
Kolff, W. J., 383
Kolinski, A., 160, 397
Kollman, P., 359, 392
Kollman, R. A., 391
Kollmeier, H., 97
Kollross, P., 89
Kolodziej, P., 96
Kolokoltseva, I. G., 150
Kolosov, A. N., 149
Komarov, V. S., 35
Komarov, L. G., 347
Komarov, T. V., 349
Komatsuzaki, S., 365
Kometani, J. M., 177, 186, 394
Komiya, T., 101
Komiyama, M., 351, 352, 354, 355
Kommiski, S. J., 210
Komoroski, R. A., 186, 213
Komuro, S., 266
Kon, M., 379
Konak, C., 165
Konami, Y., 111

Konar, R. S., 62
Kondeliková, J., 37, 85
Kondo, H., 79, 360
Kondo, S., 52, 54, 176
Kondo, T., 384
Kondrashkina, N. I., 332
Konig, J., 183
Koningsveld, R., 201, 251, 252, 257
Konno, K., 368
Kon-No, K., 367
Konno, M., 66
Konomi, T., 346
Konovalenko, N. A., 34
Konovalenko, V. V., 51, 52
Konstantinov, I. I., 53
Kooishra, T., 384
Koopal, L. K., 226
Koppel, D. E., 118, 123
Koppenol, W. H., 351
Korrowski, H., 130
Kops, J., 36, 179, 182, 334
Kopylov, E. P., 34
Kopylov, V. M., 147
Kordomenos, P. I., 347
Kordowicz, M., 115
Korner, A., 237
Korolev, G. V., 55, 57
Koros, W. J., 91
Korovina, G. V., 30
Koršak, V. V., 88, 90, 91, 339, 347, 348, 349
Korsunskii, V. M., 409
Korzhaven, L. N., 90
Kortunova, G. B., 349
Kosc, M., 219
Kosenko, L. A., 154, 206
Kosfeld, R., 149, 177
Koshijima, T., 104
Koshland, M. E., 141
Kosmas, M. K., 162, 163, 254
Kossmehl, G., 301, 405
Kostanck, A., 79
Kostyushkina, V. Yu., 72
Kotake, K., 410
Kotek, R., 155
Kotel'nikov, V. A., 85
Kothandaraman, H., 77, 347
Koto, T., 84
Kotomkin, V. Ya., 154
Koton, M. M., 90, 226
Kotov, Y. I., 348
Kotrelev, G. V., 350
Kotsareva, N. S., 34
Kotzyba-Hibert, F., 359
Kouno, A., 383
Koutinas, A. A., 88
Kovac-Filipovic, M., 69
Koval'chuk, E. P., 72, 75
Kovař, J., 159, 167, 240
Kovarik, J., 69
Kovarova, J., 333
Kovarskaya, B. M., 348
Koutunenko, L. I., 65
Kouno, T., 358

Kovyazin, V. A., 147
Kouno, T., 358
Kowal, J., 334
Kowalski, A., 69
Kowalski, J., 350
Kowalski, P., 342
Koyama, R., 196
Kozak, J. J., 371
Kozerski, L., 9
Kozlowska, H., 235
Kozlowski, M., 267
Kozuka, H., 361
Kozuka, N., 246
Kracher, G. P., 382
Kralicek, J., 37, 85
Kramer, A., 313
Kramer, E. J., 84, 286, 287
Kramer, H., 249
Kramer, H. E. A., 337
Kramer, R. H., 126
Kramer, S. R., 411
Krane, R. J., 379
Kranz, H., 411
Krasikow, V., 44
Kraszewski, A., 133
Kratochvil, P., 224, 236, 261, 266
Kraus, M. A., 363
Krause, S., 153, 259, 266, 268
Kraynik, A. M., 292
Krcek, L., 383
Krcma, V., 383
Krebs, H., 80, 217
Kreitser, T. V., 70
Kremer, K., 397
Krenzer, E., 205
Kresge, A. N., 30
Kresta, J. E., 95, 347
Kretschmer, P., 140
Kricheldorf, H. R., 78, 184, 185
Krieg, M., 370
Krieger, I. M., 68
Krigbaum, W. R., 167, 255, 265
Krikunenka, V. I., 150
Krimm, S., 209, 393
Krischan, R., 385
Kroenke, W. J., 343
Króliček, J., 85
Kroll, A., 217
Kromer, H., 218
Kronberg, B., 66
Krongauz, Y. S., 348
Kronis, K. A., 115
Kronman, A. G., 343
Kronman, J. H., 382
Krukovskii, S. P., 344
Kryszewski, M., 334, 337, 342, 350, 402
Ku, H.-C., 68, 389
Kubat, J., 346
Kubin, M., 241, 243, 247, 389
Kubisa, P., 36
Kubo, M., 68
Kubota, K., 257
Kubrick, R. L., 130

Kucera, M., 38, 43
Kuchanov, S. I., 59
Kudaikulova, S. K., 89
Kudáček, L., 89
Kudo, S., 344
Kudriavtsev, V. V., 226
Kudryavtsev, V. N., 32, 39
Kudryavtseva, L. S., 335
Kuhlmann, W. D., 385
Kuhn, M., 386
Kuhn, R., 218
Kukushkina, N. P., 63
Kulichikhin, S. G., 85
Kulicke, W. M., 226
Kulkarni, P., 376
Kulp, K., 101
Kul'velis, Y. I., 37
Kumachev, A. I., 347
Kumakura, T., 53
Kumar, A., 76, 85, 387
Kumar, B., 387
Kumbour, R. P., 92
Kumitake, T., 179
Kumpaneko, E. N., 349
Kunii, D., 85
Kunikiyo, N., 367
Kunimasa, T., 367
Kunio, T., 295
Kunitake, T., 31, 33, 362, 364, 366
Kuntze, D., 212
Kunugi, T., 81, 222
Kunzru, D., 76, 387
Kuo, C.-I., 7
Kurachi, A., 404
Kurakowska-Orszagh, J., 348
Kuran, W., 78
Kurashov, V. V., 85
Kurata, M., 171, 174, 249, 291
Kurenbin, O. I., 244
Kurenkov, V. F., 68
Kurganskii, V. S., 53
Kurimura, Y., 319, 326
Kurishu, N., 332
Kurita, K., 196
Kuriyama, A., 56
Kuriyama, I., 89
Kurnosova, L. K., 34
Kuroda, H., 332
Kuroda, Y., 354, 355
Kuroki, N., 361, 367
Kuroki, T., 340, 342
Kuruville, S., 387
Kusaba, A., 381
Kusano, Y., 358, 374
Kusiak, J. W., 351
Kuta, S. I., 133
Kutepov, D. F., 349
Kuwabara, T., 38
Kuwae, Y., 50
Kuwubara, C., 316
Kuznetsava, S. V., 334
Kuznetsova, T. A., 52
Kwiatkowski, J. H., 410
Kyotani, M., 89, 211

Kyriacos, D., 80, 185
Kwak, J. E., 74
Kwei, T. K., 259
Kwita, K., 90

Laaksonen, J., 67
Labes, M. M., 35
Labischinski, H., 118
Lablache-Combier, A., 408
Labor, J. D., 289
Labrude, P., 385
Lacaze, P. C., 74
Lacelle, N., 114
Lacey, A. M., 151
Lacher, K. P., 116
Lachinov, M. B., 57
Lacok, J., 164, 226
Lacombe, J. M., 122
Lacosta, J., 180
Lafleur, P. G., 295
Laghi, A. A., 152
Lagrange, D., 127
Lagunov, V. M., 57
Lai, J. H., 410
Lai, Y. C., 154
Laidler, D. A., 351
Lainghe, S., 250
Laius, L. A., 90
Laker, R., 96
Lakshmikanthan, N., 370
Lal, M., 166, 398
Lala, D., 333, 334, 335
Lam, J. H. W., 32
Lam, L.-M., 179
La Mantia, F. P., 83, 222, 291
Lamb, J. D., 351
Lamb, J. E., 111
Lambert, F., 120
Lambrecht, L. K., 377
La Monica, G., 94
Lancaster, J. E., 177
Landabidea, J., 233
Landheer, D., 236
Landini, D., 356, 372
Lando, J. B., 209, 300
Lando, T. B., 81
Lang, F., 381
Lang, R. W., 89, 289
Lang, W. K., 117
Lange, A., 204
Langer, A. W., 10
Langer, R., 360
Langer, R. S., 385
Langer, S. H., 304, 318
Langley, P. G., 384
Langloss, J. M., 377
l'Anson, K., 120
Laptij, S. V., 206
Lardet, D., 370
Large, G., 327
Larion, L. J., 385
Larrain, R., 86
Larsen, A., 140
Larsen, B., 106
Larson, G., 130

Larsson, R., 377
Lasarova, R., 77
Lashkina, E., 65
Laskawski, W., 258, 267
Lastukhin, Yu. A., 53
Laszkiewicz, B., 155
Laszlo-Hedvig, Z., 185, 246
Lath, D., 266
Latov, V. K., 363
Lattimer, R. P., 343, 347
Lau, S.-F., 259
Lau, W. R., 259
Laufer, D. A., 355
Laun, H. M., 280
Laundon, R. C., 91
Laurent, E., 370
Lautenberger, J. A., 137
Laval, J., 364
Lavrenko, P. N., 88, 226
Lavrova, S. N., 78
Law, K. S., 77, 310
Lawandy, S. N., 297
Lawson, G. M., 139
Lawson, J. H., 383
Lax, M., 163, 165, 167, 397
Layne, E. C., 384
Le, H.-J., 242
Leach, J. S., 393
Leal, L. G., 292
Lebedev, A. V., 59
Lebedev, E. P., 150, 154
Lebedkina, O. K., 72, 73
Leborgno, A., 217
Lebowitz, J. L., 164, 397
Le Bret, M., 167
Le Bret, M. C., 266
Lecayon, G., 72
le Charpentier, Y., 385
Lechert, H., 101
Lechner, M. D., 261
Lecollier, P., 152
Ledbury, K. J., 332
Ledent, T., 80
Lederer, K., 344
Ledneva, O. A., 348
Ledwith, A., 31, 360, 401
Lee, C. A., 411
Lee, C. C., 208
Lee, C. H., 172
Lee, C.-Y., 398
Lee, C. Y. C., 288
Lee, D. G., 314
Lee, D. I., 69
Lee, D. M., 103
Lee, D. S., 186
Lee, F., 143
Lee, J., 313
Lee, J. L., 31, 32, 37
Lee, K. I., 410
Lee, K. S., 386
Lee, L. J., 96, 284
Lee, M. M.-P., 69
Lee, S. S., 11
Lee, W. Y., 74
Lee, Y. S., 366

Author Index

Lee, Y. Y., 37
Leevers, P. S., 288
Lefebvre, D., 206
Lefebvre, J. M., 192
Leffew, K. W., 70
Lege, C. S., 362
Legeay, G., 51, 178
Leger, L., 172, 236
Legerski, R. J., 135
le Goaller, R., 356
Le Grand, P., 234
Legras, R., 342
Le Gressus, C., 72
Lehn, J. M., 351, 355, 358, 359
Lehnen, B., 172, 202, 236
Lehr, M. H., 186
Lehrle, R. S., 345
Leiber, L., 250
Leibler, L., 190
Leidner, C. R., 74
Leininger, T. I., 375
Leive, L., 118
Lejour, M. I., 378
Lelah, M. D., 377
Lelievre, J., 102
Lemaire, J., 88, 332, 333, 337
Lemieux, R. U., 129
Le Moel, A., 72
Lemoigne, J., 359
Lemperle, G., 378
Lemstra, P. J., 221
Lenain, J.-C., 51
Leng, L., 386
Lenka, S., 51, 87
Lentz, D., 380
Lenz, R. W., 33, 77, 78, 154, 178
Leon, L. M., 233, 262
Leonard, J., 33, 41, 177
Leong, F. L., 386
Leong, Y. S., 71
Lepsch, T. C., 149
Leray, J., 77
Lerner, M. R., 144, 145, 146
Leroux, Y., 365
Leslie, F. M., 292
Lessard, P., 342
Lester, P. D., 33
Le Stourgeon, W. M., 145
Letoffe, J. M., 97
Leung, W. P., 81
Leung, Y., 384
Leute, U., 220
Leutz, V., 207
Levashov, A. V., 371
Levelut, A. M., 79
Levesque, G., 86, 319, 347
Levi, G., 111
Levin, V. Yu., 150, 153
Levine, M. J., 116
Levine, R. D., 173
Levinger, L., 140
Levshanov, V. S., 348
Levy, I. J., 246
Levy, R. M., 398

Lewin, M., 217
Lewis, F. D., 32
Lewis, J., 307, 344
Lewis, J. L., 379
Lewis, M. S., 118
Leznoff, C. C., 320, 324
Li, C. S., 151
Li, D., 386
Li, H. M., 341
Li, J. C. M., 288
Li, P., 71, 147
Li, S., 362
Li, X.-Z., 243
Li, Y., 246
Liang, J. N., 105, 106
Liang, R. H., 336
Liao, J., 129
Liao, T., 38, 39
Li-Aravena, F., 200
Liautard, J. P., 145
Licchelli, J. A., 18
Lichti, G., 61
Lidman, D. H., 380
Lidy, W., 97
Liebert, L., 78, 79
Lien, Q. S., 37
Liesner, C. E., 327
Lieto, J., 319
Lieu, P. J., 157
Liguori, A., 346
Lilt, M. H., 209
Lin, C.-C., 68, 70, 389
Lin, C.-L., 353
Lin, J. C., 207
Lin, J. S., 190, 213, 219
Lin, M. S., 83, 348
Lin, S.-D., 383
Lin, Y. H., 149
Lind, P., 124
Linda, P., 366, 369
Lindahl, U., 127, 377
Lindberg, L., 384
Lindenauer, S. M., 375, 377
Lindner, P. C., 384
Lindsay, R. M., 375
Lindsell, M. E., 47
Lindsey, J. J., 94
Lindstroem, T., 103
Lindt, J. T., 285
Ling, H. C., 359
Lingelser, J. P., 199
Lingelser, L., 197
Lingenfelter, D. S., 359
Linhart, H., 333
Liogonkii, B. I., 349
Lipatov, Yu. S., 53, 85, 206, 259
Lipkin, M., 162
Lippmaa, E., 150, 188
Lipshitz, S. D., 285
Lipskerova, E. M., 337
Liratov, Yu. S., 268
Lis, H., 109, 111
Lisitskii, V. V., 343
Lisowska, E., 115

Lissi, E. A., 403
Litman, A. M., 298
Litmanovich, A. D., 345
Litsov, N. I., 332
Litt, M. H., 207
Liu, D. D., 252
Liu, L.-J., 68
Liu, P., 87
Liu, S., 71, 147
Liu, T. Y., 290
Liu, W., 4
Livingstone, M. E., 154
Llabador, Y., 375
Llauro, M. F., 343
Llorente, M. A., 79, 149, 233, 395
Lloyd, J. B., 384
Lobkov, V. D., 150
Locatelli, P., 181
Lodge, A. S., 290
Lodge, T. P., 171
Loeb, L. A., 135
Löhler, J., 141
Lövy, J., 79
Logan, J. A., 74, 82
Logemann, H., 341
Loginov, V. S., 81
Loginova, N. N., 344
Lohman, T. M., 256
Lohr, D. F., 44
Lohse, D. J., 256
Lokaj, J., 53, 55
Lokhande, H. T., 103
Long, W. F., 128
Long, W. P., 27
Lonngren, J., 122
Loontiens, F. G., 112, 113
Lopez-Serrano, F., 260
Lopirev, B. A., 226
Lora, S., 156, 157, 350
Lorenz, K., 101
Lorenz, M. D., 84
Lormeau, J. C., 128
Los, D., 391
Loschner, H., 411
Lothstein, L., 145
Lotz, B., 216
Loucheux, C., 262, 264, 266, 408
Lougnot, D.-J., 50, 64
Loutfy, R. O., 57
Lovell, R., 204, 394
Lovinger, A. J., 208, 211, 296
Lovinger, A. S., 209
Low, P. S., 123
Lowenhaupt, K., 140
Lowy, D. R., 138
Loyinskii, M. S., 335
Lozovskaya, E. L., 334
Lu, C., 87, 386
Lu, M.-K., 70
Lu, Z. H., 153, 259, 266, 268
Lubianez, R. P., 83
Lucchesi, A., 33, 73
Lucchini, G., 142

Luc-Gardette, J., 333
Lucki, J., 80, 338
Luederwald, I., 339
Luft, G., 51
Lugova, L. I., 334
Luisi, P. L., 371
Lukáč, I., 53
Lukas, C., 348
Lukas, J., 243
Lukas, R., 60
Lukashenko, I. M., 348
Lukasheva, N. V., 91
Lukhovitskii, V. I., 70
Lukyanov, V. V., 337
Lund, P. A., 103
Lundblad, J. L., 123
Lundquist, K., 104
Lundstrom, I., 74
Luston, J., 38
Lutz, P., 45
Lu-Vinh, Q., 334
Luxton, A. R., 44
Lydon, M. J., 378
Lyerla, J. R., 176
Lynch, J., 389
Lynch, M. E., 375
Lyon, N. B., 126
Lyons, P., 380
Lysenko, Ye. B., 226
Lyudvig, E. B., 37
Lyulichev, A. N., 338

Ma, K., 203
MacArthur, C. R., 320, 324
McBrierty, V. J., 176
McCabe, C., 380
MacCallum, J. R., 73, 334, 400, 402
McCammon, J. A., 398
McCormack, P., 378
McCormick, C. L., 246
McCrackin, F. L., 160, 169, 262, 397
McCready, M. J., 213
McCullough, R. L., 214
MacDiarmid, A. G., 300, 405
McDonald, J. A., 123
MacDonald, S. A., 406
McDonnel, M. E., 254
McEwen, I. J., 79, 266
McFay, D., 43
McGarry, F. J., 289
McGill, L. D., 383
McGrath, J. E., 82, 263
MacGregor, J. F., 70, 389
Machi, S., 71
Machon, J. P., 56
Machovich, R., 129
McHugh, A. J., 215
Maciejewski, M., 364
Maciel, G. E., 187, 188
McIntire, L., 370
McIntosh, J. M., 373
McIntyre, D., 165
McKay, R. A., 176

McKechnie, M. T., 71
McKenzie, I. D., 13
MacKenzie, W. M., 314
McKeown, J. G., 200, 259
MacKerron, D. H., 73
McKinley, S. V., 56
Mackley, M. R., 82
McKnight, G. S., 140
MacKnight, W. J., 178, 198, 259
McLaren, J. V., 185
McLaury, M. R., 339
McNair, T. J., 381
Macomber, D., 53
Maconnachie, A., 193, 195, 202, 203
Macosko, C. W., 96, 149, 260, 281, 284, 285
McPheeters, J. C., 310
McPhillimy, D. G., 154
McQuie, R. J., 392
Macrander,, A., 411
Macret, M., 54
MacSporran, W. C., 281
McWalter, I. T., 341
Maddox, R. L., 409
Maddox, Y., 377
Madruga, E. L., 50
Maeda, M., 53
Maeda, Y., 206
Maekawa, E., 240
Maeng, K. S., 151
Maeshima, T., 58
Mafezzoni, C., 94
Magdelev, Ye. T., 9
Magill, J. H., 89, 153, 157, 210, 211, 215, 350
Magnani, J. L., 130
Magomedova, T. V., 332
Mah, S., 32
Mahadevan, V., 86, 90
Mahmoudhagh, M. K., 261
Mahnke, P. F., 383
Maia, A., 356
Maidment, B. W., 131
Maidunny, Z. A. B., 41
Mair, C., 47
Maissa, P., 265
Maiti, S., 90
Majer, J., 25, 347
Majerova, F., 43
Majerova, K., 38
Majid, M. A., 42
Majumdar, R. N., 33, 177, 178, 184
Makanjuola, B. O., 61, 62, 63
Makarov, K. A., 72, 73
Makarov, L. I., 153
Makaruk, L., 83
Makawinta, T., 69, 70
Makhmutov, R. Kh., 409
Makhtarulin, S. I., 18
Makimoto, S., 352, 367
Maksimov, V. L., 7
Makuuchi, K., 68

Malakhova, G. P., 334
Malanga, M., 36
Malder, C., 141
Maldonado, J. R., 412
Malhotra, S. L., 236, 341, 342
Malik-Diemer, V. A., 351
Maliszewicz, M., 406
Malkin, A. Ya., 85, 279
Malkov, V. D., 69
Mal'kova, G. Y., 7
Malley, P. J., 50
Mallinson, J. W., 289
Malmstrom, A., 126
Malone, W. F. P., 151
Maloshitskii, A. S., 30
Malovikova, A., 106
Malpezzi, L., 150
Malta, V., 87
Maltaj, I. W., 142
Malumyan, I. V., 71
Malyukova, E. B., 65
Mammerickx, M., 129
Manabe, O., 358, 370, 371, 374
Manaresi, P., 77
Manasek, Z., 38
Manca, F., 113
Mandal, B. M., 225
Mandel, I. D., 116
Mandel, J. L., 141
Mandel, M., 261
Mandelkern, L., 176, 194, 205, 210, 212, 213, 219
Mandell, J. F., 289
Mandric, G., 90
Manecke, G., 313, 315, 385, 386
Manen, J. F., 114
Manescalchi, F., 304, 365
Mangalam, P. V., 226
Mangiariotti, G., 384
Maniatis, T., 141, 144
Manjula, B. N., 130
Mankowski, Z., 57
Manners, D. J., 99
Manning, G. S., 256
Manninger, J., 379
Mano, E. B., 75
Manolova, N., 31
Manson, J. A., 89, 223, 289, 290
Mantorani, F., 376
Mantsch, H. H., 111, 114
Manzione, L. T., 96, 285, 295
Marak, R., 339
Maravigna, P., 96, 346, 347, 348
March, C. J., 139
Marchal, E., 255
Marchenko, A. P., 55
Marchessault, R. H., 120
Marchuk, L. V., 75
Marco, C., 217
Marconi, W., 376

Author Index

Marechal, E., 30, 33, 34, 35, 94
Margolin, A. L., 88
Margomenov-Leondiopoulou, G., 371
Marie, P., 199
Marigo, A., 4
Mariguchi, S., 89
Mark, J. E., 148, 149, 156, 393
Markham, A. F., 133
Marks, T. J., 32
Marois, M., 380
Marom, G., 289
Marqusee, J. A., 166
Mar, A., 398
Marrucci, G., 291
Marshall, A., 205
Marshall, G. L., 80
Marshall, G. P., 289
Marshall, Y. S., 97
Marschberger, R., 221
Martan, M., 34, 48
Martel, J., 386
Martelli, S., 205
Marten, F. L., 56, 388
Martens, A., 245
Marti, S., 341
Martin, A., 19
Martin, D. C., 83, 395
Martin, M., 245
Martin, N. F., 382
Martin, T. R. P., 376
Martin, W. E., 94
Martinek, K., 371
Martinez, A., 74, 233
Martinez, C., 20
Martinez, F., 35
Martinez, F. O., 35
Martinez, H., 35
Martinez Duran, A., 74
Martinez-Salazar, J., 216
Martinez-Utrilla, R., 403
Martini, N., 378
Martirosov, V. A., 150
Marton, M., 178
Martorana, A., 4
Maruthamuthu, M., 370
Maruyama, Y., 100
Marvin, D. C., 395
Marx-Figini, M., 225
Marynen, P., 129
Masana, J., 373
Maschio, G., 73
Masci, B., 356
Masegosa, R. M., 238
Mashelkar, R. A., 76, 292, 387
Mashutina, G. G., 151
Masi, P., 85
Masiulanis, B., 267
Mason, R. G., 375
Masson, P., 38, 43
Masubuchi, T., 153, 319
Masuda, T., 174
Masuhara, H., 402
Masure, M., 30
Mataga, N., 402

Matheson, N. K., 99
Matheson, R. R., 255
Mathew, M. K., 109
Mathias, L. J., 37, 313
Mathieu, D., 383
Mathieu, J., 152
Mathis, A., 199
Mathis, C., 180
Mathur, G. N., 50
Mathur, N. K., 304
Matisova-Rychla, L., 333
Matsubara, A., 403
Matsubara, T., 319
Matsubara, Y., 209
Matsuda, M., 369
Matsui, Y., 353, 355
Matsumoto, A., 54, 179
Matsumoto, D. S., 84, 289
Matsumoto, I., 88, 111
Matsumoto, K., 295, 367
Matsumoto, M., 368
Matsumoto, T., 69, 81, 378, 381
Matsumura, K., 158
Matsunaga, T., 376
Matsuo, C., 371
Matsuo, M., 89, 221, 222
Matsuzaki, K., 30, 33, 35, 36, 38, 39, 45, 47, 52, 176, 178, 180, 222, 394
Matsuzawa, T., 406
Matteucci, M. D., 133
Mattheiem, W., 378
Mattice, W. L., 56, 164, 166, 182, 394, 395, 396
Matusevich, P. A., 347
Matusevich, Y. I., 347
Matusinovic, T., 364
Matuszak, M. L., 339
Matuura, R., 370
Matveev, Y. I., 339
Matvienko, L. G., 362
Matyjaszewski, K., 36
Matynia, T., 78
Matzner, M., 263
Maubert, C., 51
Mauritz, K. A., 377
Maxfield, F. R., 393
Mayne-Banton, V., 409
Mayr, A. J., 356
Mayre, S. H., 289
Mayo, F. R., 333
Mayor, J. M., 148
Mazet, J., 119, 246
Mazon-Arecjederra, J. M., 346
Mazumber, T., 111
Mazur, B., 151
Mazur, J., 160, 262, 397
Mazur, S., 364
Meacock, P. A., 133
Mead, J. W., 336
Mead, K. E., 336
Meagher, E. P., 150
Mebkhout, A., 51
Mechin, R., 41
Mediola, A. J. M., 335

Meek, R. R., 335
Meesiri, W., 82
Megerman, J., 375, 380
Mehendru, P. C., 84, 405
Mehnert, R., 32
Mehrotra, A., 374
Mehta, M., 266
Meier, H., 218, 404
Meijer, H. E. H., 283
Meirovitch, H., 392
Meissner, M., 210
Mejzlik, J., 25, 27
Melby, L. R., 47
Melenevskaya, E. Yu., 40
Mellor, M. T. J., 336
Mellottee, H., 345
Mel'nik, O. A., 54
Mel'nikov, A. B., 88
Melnikov, M. Y., 337, 338
Melnikov, V. P., 349
Mel'nikova, N. E., 30
Melton, L. D., 106
Melveger, A., 381
Menczel, J., 82
Mendelson, M. A., 283
Menezes, E. V., 174, 291
Meng, D. Y., 243
Menger, F. M., 351, 362, 368, 371, 372
Menges, G., 295
Mengoli, G., 73
Menke, K., 187
Mercier, J. P., 342
Merienne, C., 79
Merk, A., 27
Merkireva, A. A., 395
Merle, L., 184
Merle, M., 408
Merle, Y., 184, 336
Merle-Aubry, L., 336, 408
Merlin, A., 50, 71
Merlyn, M., 130
Merrill, E. W., 149
Mertens, J. J. R., 342
Merz, A., 359
Messing, J., 134
Metcalfe, J. C., 351
Metkin, I. A., 150, 151
Metras, F., 182
Metzger, J., 100
Metzger, S. H., 96
Metzner, A. B., 292, 295
Meunier, A. M., 123
Meunier, G., 53
Mewis, J., 280, 291
Meyer, B., 185
Meyer, H., 19
Meyer, J. E., 256
Meyer, K., 26
Meyer, T. J., 74, 319, 366
Meyerhoff, G., 56, 262, 265, 388
Meyers, K. O., 149
Meyland, I., 106
Mhala, M. M., 368

Miano, J. D., 96
Miasa, K., 263
Michailov, G. P., 78
Michal, J., 346
Michel, A., 343, 344
Michel, J., 251
Michel, V., 122
Micheron, F., 209
Middleman, S., 280
Migharesi, C., 381, 383
Mihailov, M., 37
Mijovic, M. V., 408
Mikadza, L. A., 349
Mikawa, H., 52, 400, 403
Mikes, J., 388
Mikhailik, D. M., 338
Mikhailik, V. F., 73
Mikhailova, N. V., 92
Mikheev, Yu. A., 84
Miki, M., 32
Mikitaev, A. K., 347
Mikov, V. L., 348
Mikulásová, D., 164, 226, 245
Milas, M., 120
Mildner, D. F. R., 189
Miles, M. J., 120
Milevskaya, I. S., 91, 92
Milkovich, R., 53
Millard, P. C., 384
Miller, A., 125
Miller, D. C., 410
Miller, D. M., 140
Miller, J., 382
Miller, J. M., 129
Miller, J. W., 171
Miller, K. J., 392
Miller, L. J., 410
Miller, R. L., 111, 390
Miller, R. S., 33
Mills, N. J., 287
Mills, P. J., 217
Milner, R., 45
Milnera, S. M., 80
Milnes, G. J., 348
Milova, L. A., 392
Milstein, O., 105
Minami, T., 358, 379
Minamino, Y., 63
Minato, T., 163
Mineev, S. I., 37
Minett, T. W., 378
Mingshi, S., 242
Minh, L. Y., 341
Minoda, M., 236
Minoura, Y., 53
Minsker, K. S., 30, 343
Mir, L., 235
Mirold, P., 249
Miron, M., 65
Mironov, V. S., 72
Mironova, N. M., 65
Mirsa, S., 79
Mishra, G. C., 30
Mishra, G. V., 34
Mishra, M. K., 51, 79, 87

Mitani, T., 405
Mitchell, G. R., 204, 394
Mitchell, M. J., 360
Mitera, J., 346
Mitomo, H., 89
Mitra, C. K., 392
Mitra, S., 405
Mitsui, H., 363
Mitterer, A., 127
Miura, M., 87, 337
Miura, Y., 87, 364
Miwa, E., 155
Miyaji, H., 215
Miyake, H., 383
Miyaki, Y., 162, 164, 237
Miyamoto, E., 369
Miyamoto, S., 226
Miyamoto, Y., 215
Miyasaka, K., 213
Miyashita, K., 245
Miyata, N., 366
Miyatake, T., 183
Miyazaki, H., 353
Miyazaki, Y., 226
Miyoshi, Y., 30
Mizerovskii, L. N., 37
Mizukami, W., 346
Mizuno, K., 413
Mizunuma, K., 183
Mizutani, T., 404
Mládek, M., 226
Moad, G., 50, 58, 181
Mobbs, R. H., 205
Mochida, K., 353
Mochizuik, T., 383
Modelli, A., 365
Moens, L., 262
Moffatt, J. R., 369, 370
Mohan, S., 115
Mohr, P., 386
Molaire, M. F., 399, 408
Molenaar, I., 380
Molina, P., 314
Molinari, H., 373
Molkin, A. Ya., 85
Moll, M., 121
Molnar, T., 251
Momany, F. A., 391
Monaco, A., 86
Monakov, Y. B., 343
Mones, E. T., 289
Moniz, W. B., 184
Monkenbusch, M., 207, 299, 393
Monnerie, L., 202, 206, 262
Monroe, C. M., 152
Montanari, F., 313, 372, 373
Montaudo, G., 96, 157, 346, 347, 348, 350
Montawari, F., 356
Monte, S. J., 96
Montreuil, J., 122
Moon, K.-Y., 69
Moore, J. C., 264
Moore, W. S., 384

Moran, J. M., 411, 412, 413
Morcellet, J., 251
Morcellet, M., 262, 264, 266
Morcellet-Sauvage, J., 264
Moreau, M., 33, 177
Moreau, E. M., 410
Moreno, R., 114
Morese-Segela, B., 266
Morffew, A. J., 390
Morgan, R. J., 289
Mori, S., 54, 247
Mori, Y., 377, 383, 386
Morii, H., 35, 394
Morikawa, M., 255
Morimoto, Y., 385
Morishima, Y., 403
Morita, S., 412, 413
Moritz, H. U., 56, 388
Morkind, L. A., 72
Moroi, Y., 370
Moroz, E. M., 12
Morozova, L. V., 35
Morozova, T. P., 348
Morrell, S. H., 294
Morris, D. B., 79
Morris, E. R., 101, 104, 105, 106, 107, 108, 118
Morris, V. J., 107, 108, 120
Morrison, S. L., 130
Morrow, W. J., 152
Morshed, M., 77
Morsi, S. E., 333
Mort, J., 405
Mortimer, D. A., 72
Mosbach, K., 328, 364
Moschonas, N., 144
Moscicki, J. K., 255
Mosher, D. F., 124, 377
Moskowitz, J. W., 150
Moss, D. S., 390
Moss, E. K., 96
Moss, G., 393
Moss, R. A., 366, 367, 370
Mostovaya, E. M., 332
Motov, S. A., 347
Motozato, Y., 243
Mottweiler, R., 27
Moucha, A., 85
Mount, E. M., 284
Mount, S. M., 143, 144, 145
Moxley, T. T., 317
Mozzhukhin, D. D., 409
Mrkvickova, L., 226, 233, 245
Much, H., 245
Müller, H., 226
Mueller, M. H., 189
Mueller, R., 34
Mueller, W. M., 355
Münstedt, H., 280
Muinov, T. M., 348
Muir, H., 125
Mukherji, A. K., 244, 245, 247
Mukhopadhyay, G., 52
Mulderije, J. J. H., 234
Muller, A. H. E., 46

Author Index

Muller, H., 87
Mulligan, R., 143
Mulvihill, E. R., 140
Munari, A., 77
Munch, J. P., 150
Munir, A., 37, 38
Munk, P., 235, 252, 259, 262
Munaoz-Escalona, A., 19, 20
Munro, H. S., 338
Munro, M. S., 376
Munstedt, H., 294
Murachev, V. B., 30, 34
Murachi, T., 385
Murakami, K., 56
Murakami, Y., 315, 351, 359, 360, 366, 367, 369, 371
Muraki, M., 87, 364
Murani, M. T., 73
Muraoka, Y., 237
Murata, H., 208
Murata, K., 340
Murayama, T., 150
Murphy, L. A., 112
Murphy, R., 245
Murray, E. J., 105
Murray, J., 380
Murray, P. A., 116
Murray, R. W., 74
Murray, S. G., 319
Musindi, G. M., 213
Mustafa, B. H., 58
Muthukumar, M., 163, 173, 196
Mutschler, W., 383
Muzzarelli, R. A. A., 328
Myers, G. E., 187
Myers, R. T., 359

Naaktgeboren, A. J., 365
Nabeshima, T., 351, 354
Nabizadeh, H., 154
Nagahara, S., 23
Nagai, T., 351
Nagakubo, K., 87, 337
Nagamatsu, T., 323
Nagaoka, S., 383, 386
Nagarajan, M. R., 97, 207
Nagasaka, K., 166
Nagasawa, M., 167, 226, 237, 246
Nagase, Y., 153, 154, 350
Nagaska, S., 377
Nagaya, T., 246, 346
Nagbhusham, T., 198
Nagiev, Z. M., 90
Nagui, M. K., 307
Nagura, S., 234
Nagy, E., 204
Nagy, T. T., 343
Nahas, L. F., 383
Naidenov, V. P., 404
Naik, C. D., 85, 387
Naik, D., 85
Naik, N. R., 86
Naito, I., 409

Naito, Y., 183
Nakae, A., 65
Nakafatumi, Y., 246
Nakagawa, H., 364
Nakahama, S., 42, 325, 330
Nakahara, H., 245
Nakahira, T., 400
Nakai, R., 321
Nakajima, A., 360
Nakajima, M., 383
Nakajima, S., 196
Nakamae, K., 81
Nakamura, A., 179
Nakamura, I., 320
Nakamura, K., 243, 367
Nakamura, M., 100
Nakamura, N., 363
Nakamura, S., 234
Nakamura, T., 154, 350
Nakane, H., 410, 412, 413
Nakano, A., 359, 366, 367, 369, 371
Nakano, H., 362
Nakano, S., 247
Nakata, M., 254
Nakaya, H., 352
Nakayama, T., 410
Nakazawa, N., 385
Nakazumi, H., 356
Nakomura, Y., 88
Nalchadzhyan, S. O., 61
Nallet, F., 197
Namba, S., 409
Nametkin, N. S., 148
Nango, M., 361, 367
Nanjan, M. J., 86
Napolitano, R., 207, 393
Napper, D. H., 61, 62, 68, 389, 395
Nara, S., 101
Narang, C. K., 304
Narasimhan, V., 242, 244, 389
Narducci, P., 30
Narkis, M., 84
Naruchi, K., 352
Nasirov, F. M., 30
Nasonova, T. P., 56
Nassar, R., 389
Nasser, B. E., 29
Nasybullin, S. A., 341
Nate, K., 407
Nathan, P., 383
Nathans, D., 135
Natta, G., 22, 25
Nauflett, G. W., 242
Naumann, D., 118
Naus, M. D., 311
Navard, P., 255
Navare, R., 96
Navas, A. A., 244
Navratil, M., 338
Nawrot, S., 102
Nayak, M. C., 51
Nayak, P. L., 51, 79, 87
Nazran, A. S., 52

Ndong-Nkoume, M., 385
Neamtu, G., 90
Nechvoloclova, E. M., 349
Neckers, D. C., 52, 53, 326, 327, 328, 407
Nedelcheva, M., 101
Nee, G., 314, 365
Neel, B. G., 137, 138
Neenan, T. X., 386
Nefedov, V. I., 362
Neilson, R. H., 155
Nelkin, B. D., 141
Nelsestuen, G. L., 128
Nelson, D., 140
Nelson, R., 377
Nelson, S. D., 383
Nelson, S. E., 369
Nemes, J., 379
Nemethy, G., 392, 393
Nemoto, N., 171
Nemoto, S., 319
Nemoto, T., 131
Nenkov, G., 334
Nesmelov, A. I., 30
Nesterov, V. V., 242
Nestrov, A. E., 259
Nesyaeva, E. I., 151
Netopilik, M., 226, 242
Neuenschwander, P., 176, 394
Neuman, A. W., 376
Neumann, D. L., 184
Neurath, A. R., 385
Neuray, D., 87, 226
Neurohr, K. J., 111
Neverov, A. N., 337
Nevin, A., 191, 192, 206
Neurohr, K. J., 114
Neustadt, P. M., 126
Newkome, G. R., 358
Newlin, D. D., 392
Newman, B. A., 215, 296
Newman, G. R., 385
Newman, S., 267
Newmark, P., 137
Newton, C. R., 133
Newton, M. D., 150
Newton, L. E., 111
Newton, R. A., 94
Newton, R. F., 359
Ng, D., 399
Ng, K. Y., 284
Ngo, N. K., 153
N'Guini, J. B., 319
Nguyen, H. A., 31, 33, 177
Nguyen, T., 312
N'Guyen, T. D., 306, 310
Ni, Z., 54
Niazi, G., 144
Nicholais, L., 381, 383
Nichols, K. H., 73
Nichols, M. F., 386
Nichols, W. K., 378
Nicholson, L. K., 202, 203
Nicholson, P. N., 319
Nicodemo, L., 287

Nicolaides, C. P., 365
Nicolais, L., 84, 287
Nicolaisen, F. M., 106
Nicolini, C. A., 378
Nicolson, G. L., 126
Nieduszynski, I. A., 126
Nielsen, B. L., 351
Nielsen, O. F., 103
Nienhuis, A. W., 140
Nierlich, M., 165, 190, 193, 197
Nieto, J. L., 185
Nigam, A., 313
Niikuni, S., 340
Niino, S., 363
Nikaido, H., 118
Nikitina, L. S., 335
Niknam, M. K., 33, 177, 188
Nikolaev, N., 44
Nikolaev, N. I., 40
Nikolaeva, T. Yu., 148
Nikolayev, A. F., 54
Nikolayeva, S. N., 90
Ninomiya, S.-I., 49
Nishida, K., 371
Nishida, S., 68
Nishide, H., 320
Nishii, K., 88
Nishijima, Y., 401
Nishikido, N., 370
Nishikubo, T., 37, 305, 406, 408
Nishimoto, S., 336
Nishimura, J., 34
Nishimura, K., 369
Nishimura, T., 53
Nishinari, K., 108
Nishio, I., 165
Nishioka, A., 176
Nishioka, N., 167
Nishioka, T., 289
Nishiyama, S., 376
Nishiyama, T., 86, 87
Nishizawa, K., 174
Nissan, R. A., 156
Nitadori, Y., 53
Nitta, M., 37, 408
Niu, S., 362
Niume, K., 91
Niwa, M., 222
Njuguna, H., 346
Noda, I., 167, 226, 237, 246, 254
Noel, H. P., 380
Noggle, J. H., 177
Nogues, P., 54
Noishiki, Y., 377
Nojima, K., 404
Nolte, R. J. M., 362, 365
Nomura, A., 295
Nomura, M., 61, 63, 68, 69, 389
Nomura, Y., 83
Nonnenmacher, D., 111
Nonogaki, S., 406
Noolandi, J., 250
Nordin, J. H., 120
Norisuye, T., 121, 164, 167, 240

North, A. M., 149, 334, 402
Northrup, S. H., 398
Nose, T., 171, 226, 254, 257
Nose, Y., 380
Noskova, M. P., 348
Noufi, R., 74
Novakovsky, V. B., 226
Novekov, D. D., 226
Novikova, Y. I., 7
Novozhilov, B. V., 332
Nowakowska, M., 334
Nozakura, S., 49, 50, 403
Nucci, L., 75
Nuino, K., 413
Nunlist, R., 185
Nuyken, O., 51
Nwachuka, N. E., 251
Nyari, T., 379
Nymberg, D. P., 312
Nys, P. S., 328
Nysenko, Z. N., 36
Nyström, B., 172, 225, 235, 236

Oatley, S. J., 391
Obata, F., 113
Oberthür, R. C., 193
Obrecht, W., 33
O'Brien, J. P., 156
O'Brien, V., 283
Obukhova, S. V., 54
Obuschak, N. D., 75
Ochoa, J. L., 109
Ochsenfeld, W., 328
Ockman, N., 113
O'Connell, C. M., 319
O'Connor, J. M., 96
O'Connor, K. M., 287
O'Connor, M. N., 177
Oda, T., 58
Oda, Y., 112
Odagiri, N., 266
Odashima, K., 360
Odijk, T., 167, 172
O'Donnell, J. H., 56, 183, 184, 346
O'Driscoll, K. F., 182, 319, 366, 388
Oelert, H. H., 348
Oesgh, U., 359
Ofek, I., 116
O'Gara, J. F., 83
Ogata, N., 86, 407
Ogata, T., 255
Ogawa, E., 313
Ogawa, H., 385
Ogawa, K., 104
Ogawa, T., 342
Ogilvie, A. L., 380
Ogino, H., 354
Ogino, K., 257
Ogita, T., 89
Ogiwara, Y., 316, 369
Ogordnikov, I. A., 362
Ogura, K., 54

Ohara, M., 211
Ohbayshi, K., 236
Ohkubo, K., 368
Ohkura, K., 384
Ohlson, J. L., 65
Ohman, P., 384
Ohnishi, Y., 409, 413
Ohno, H., 301, 376
Ohnuma, H., 226
Ohsaku, M., 208
Ohta, K., 46, 180
Ohta, T., 162
Ohta, Y., 174
Ohtani, H., 346
Ohtomo, T., 184
Ohtsuka, Y., 67
Ohyabu, M., 402, 403
Oikawa, S., 412, 413
Oishi, T., 253
Oiwa, M., 54
Okada, A., 52
Okada, M., 36, 37, 179, 180
Okada, T., 183
Okada, Y., 379
Okahata, S., 180
Okamoto, H., 360
Okamoto, Y., 47
Okamura, K., 104
Okamura, S., 32
Okamura, T., 362
Okano, K., 196
Okano, T., 376
Okatova, O. V., 88
Okatova, O. Y., 226
Okawaki, I., 208
Okawara, M., 54, 305, 321, 322, 360
Okieimen, E. F., 56, 344
Okos, M. R., 241, 246
Oktani, N., 313
Oku, J., 363
Okubo, M., 67, 68, 69
Okumara, M., 385
Olabisi, O., 267, 290, 295
Olaj, O. F., 56
Olauemi, J. Y., 262
Olbricht, W. L., 292
Oldberg, A., 126
Oldham, W. G., 409
Olea, F. A., 403
Olivé, S., 13
Ol'khova, O. M., 245
Ol'khovskaya, E. G., 75
Olley, R. H., 22
Olsen, D. B., 378, 383
Olsher, U., 374
Olson, K. W., 392
Olson, S. T., 129
Olson, W. K., 392
Olssen, P., 377
O'Malley, B. W., 139, 140, 144
Omorodian, S. N. E., 246
Omoto, M., 237
Onclin, M. H., 252
Onda, N., 266

Author Index

Onda, Y., 234
O'Neill, H. C., 129
Ong, R., 382
Onishi, Y., 51
Ono, S., 366
Ono, T., 383
Ono, Y., 403
Onogaki, T., 101
Onogi, S., 174
Onopchenko, A., 30
Ooki, H., 79, 261, 261, 402
Oono, Y., 162
Oosta, G. M., 128
Opel, M., 253
Opferman, J., 295
Opitz, G., 346
Oppenheimer, C., 51
Orena, M., 320
Oreste, P., 128
Orgill, D., 383
O'Riordan, S., 379
Orkin, S. H., 144
Orlov, G. I., 404
Orlova, A. P., 34
Oro, L., 384
Ørskov, D., 118
Ørskov, I., 118
Orwoll, R. A., 265
Osa, T., 351, 355
Osada, Y., 57, 386
Osaki, K., 174, 280, 291
Osaki, Y., 207
Osawa, S., 362
Osawa, T., 111, 114, 130
Osawa, Z., 332, 335, 337
Osborn, M. J., 118
Osborne, D., 102
O'Shea, G., 384
Osipchik, V. S., 350
Ostrovidova, G. U., 73
Ostrovskii, V. V., 349
Ostrowski, M. C., 143
Ostrowsky, N., 390
O'Sullivan, D., 410
O'Sullivan, P. W., 184
Otagiri, M., 353
Ototani, N., 127
Otsu, T., 54, 56
Ottino, J. M., 284
Ou, J. J., 170
Ouano, A. C., 241, 406
Ovchinnikov, Y. A., 374
Overbeek, J. T. G., 60
Overberger, C. G., 86, 87, 330, 361, 362
Overeem, T., 57
Owen, A. J., 89, 213
Owen, D. R., 380
Owen, E. D., 335
Oyama, T., 164, 233
Ozaki, F., 89
Ozaki, M., 405
Ozerkovskii, B. V., 55
Ozin, G. A., 319
Ozoe, Y., 353

Pabiot, J., 331
Pacansky, J., 413
Paci, M., 36, 185, 187
Pacitti, A., 384
Packard, R. B., 382
Pacovska, M., 31
Padaki, S. M., 226
Padma, S., 86
Padwa, A. R., 62
Pae, K. D., 296
Page, C. J., 288
Page, M. I., 353, 368, 369
Pagelot, A., 359
Painter, G. R., 374
Painter, P. C., 156, 392, 393
Pak, W., 382
Pakhanov, P. P. M., 87
Pakula, T., 218, 223
Pal, P. K., 387
Palenik, G. J., 54
Paleos, C. M., 371
Pallas, J., 289
Palm, G., 133
Palmiter, R. D., 140, 141
Pamakis, I., 94
Panayotov, I., 31
Panayotov, I. M., 31, 40
Pancheshnikova, R. B., 343
Panda, G., 51
Panda, S. P., 408
Pankova, M., 356
Pankratov, V. A., 349
Pannell, K. H., 356
Panov, Yu. N., 85
Panshin, Yu. A., 55
Pantin, V. I., 371
Papageorge, A. G., 138
Papas, T. S., 137
Papathomas, K. I., 362
Pappas, S. P., 31, 52
Papsidero, L. D., 131
Papulov, Y. G., 396
Paradossi, G., 120
Paradowska, B., 334
Paralikar, K. M., 208
Paramathans, S. V., 99
Paredes, E., 84
Paraedes, P. I., 286
Parfenov, V., 80
Pariiskii, G. B., 348
Parish, C. R., 129
Parizenberg, M. D., 150
Park, C. N., 390
Park, G. S., 307, 344
Park, J. H., 362
Park, L. S., 246
Parker, J. M., 208
Parker, K. J., 213
Parker, P., 390
Parker, R. G., 186
Parlides, C., 381
Parmon, V. N., 362
Parnaby, J., 283
Parsons, G. L., 365
Parsons, I. W., 18, 19, 83

Parsons, J. R., 381
Pascault, J. P., 97, 180, 245
Pasch, H., 245
Pascual, M., 364
Pask, S. D., 31, 35
Pasquali, F., 125
Pasquini, M. A., 356
Pasquon, I., 22, 25
Pass, G., 108
Passaglia, E., 287
Passalacqua, V., 77
Paster, M. D., 62
Pasynkiewicz, S., 9
Patchornik, A., 320, 363
Patel, G. N., 218
Patel, K. L., 370, 371
Patel, R., 96, 388
Patel, R. C., 238
Patel, S. R., 86
Patel, V. V., 73
Pathak, J., 259
Pati, N. C., 79, 87
Patil, A. O., 407
Patil, N. B., 103
Patin, H., 319
Patrick, J., 114
Patsevich, I. V., 362
Patsiga, R. A., 154
Patterson, D. B., 33, 155
Paul, A. M., 56
Paul, D. R., 82, 84, 85, 221, 263, 267, 277, 278
Paul, S., 308
Paul, T. K., 62
Paulides, C. A., 378
Pautov, P. G., 34
Pavis, A. A., 122
Pavlicok, V., 383
Pavlinec, J., 49
Pavlisko, J. A., 361, 362
Pavlov, G. M., 235
Pavlov, P., 86
Pavlova, S. A., 88, 91, 154, 347, 348, 349
Pavlyuchenko, V. N., 60, 64, 69, 71
Pavone, V., 117
Payne, D. R., 208
Paynter, R. W., 376
Pazur, J. H., 129, 130
Pear, M. R., 398
Pearce, E. M., 83, 86, 156, 293, 346, 348
Pearman, G. T., 296
Pearson, D. S., 160
Pease, P. F. W., 409
Pechere, J. C., 380
Pechhold, E., 36
Pechova, L., 109
Pecora, R., 170, 236
Pedemonte, B., 85
Pedemonte, E., 85
Pedone, C., 117
Peeling, J., 332, 338
Peeters, F. A. H., 235, 238

Pegny, A., 255
Peguy, A., 102
Pei, R.X., 408
Peil, A., 360
Pelham, H. R. B., 142
Pelman, P., 385
Pelrus, J., 78
Pembleton, R. G., 206
Penati, A., 94
Penczek, S., 36, 42, 43
Penhale, D. W. H., 385
Penice, R., 96
Penner, J. A., 375, 377
Pennings, A. J., 220
Penot, G., 332
Peo, M., 187
Peppas, N. A., 218, 385
Perahia, D., 398
Pereira, M. C., 108
Pereña, J., 103, 225
Perenyi, K., 332
Perera, J. M., 90
Perez, E., 258
Perez, H. D., 382
Peringa, A. J., 88
Perkins, S. J., 125
Perkins, W. G., 221
Perlin, A. S., 186
Perlman, M., 381
Perrey, H., 65
Pescia, A., 75
Peterlin, A., 213, 221, 360
Petermann, J., 210, 219
Peters, A., 149
Peterson, P. A., 124
Pethrick, R. A., 149, 185, 334, 402
Petiaud, R., 176
Petit, M. A., 307
Petitou, M., 128
Petraccone, V., 207, 393
Petrak, K. L., 408
Petrini, G., 319
Petropoulos, C. C., 410
Petrova, N. A., 154
Petrova, O. M., 349
Petrova, T. L., 184
Petrus, V., 225
Pettman, R. B., 351
Pew, J., 379
Pezzia, G., 350
Pezzin, G., 156, 157
Pfannemüller, B., 100, 255
Pfeiffer, U., 315
Pfeuty, P., 253
Pfister, G., 405
Pham, M. C., 74
Pham, Q. B., 61, 63, 388
Pham, Q. T., 176, 180, 181, 186
Pham, T. M., 102
Phan-Thien, N., 291
Philips, R. L., 385
Phillips, D., 401, 408
Phukan, U. K., 387
Phung, K. V., 309

Piazza, R., 218
Picard, J., 125
Pichot, C., 69
Pickelman, D. M., 65
Picot, C., 149, 192, 195, 197, 199
Piejko, K. E., 47
Piejza, J., 74
Pier, L., 371
Pierre, J. L., 356
Pierschbacher, M., 99, 123, 124
Piet, P., 319
Pietrasanta, Y., 34, 51, 386, 406
Piglowski, J., 258, 267
Piguta, I. K., 335
Piirma, I., I., 60
Pijol, D., 307
Pike, E. R., 390
Pike, G. E., 152
Pilati, F., 77, 87
Pileni, M. P., 371
Pillai, P. K. C., 91
Pincus, F., 82, 206
Pincus, M. R., 117, 391, 392
Pincus, P., 166
Pino, P., 176
Piorkowska, E., 214
Piret, W., 334
Pirotta, C. M., 328
Pirozhnaya, L. N., 55
Pirozzi, B., 207, 393
Pirumyan, G. P., 63
Pistole, T. G., 109
Pitha, J., 351, 386
Pitt, C. G., 80
Pittman, C. U., jun., 53, 54, 55, 178, 318, 365, 410
Pittman, J. F. T., 284
Pitts, E., 408
Pivcová, H., 53
Pizzio, B., 393
Pizzirani, G., 30
Plamthottam, S. S., 39
Plank, E., 383
Platchkova, S., 245
Platzer, N., 34
Plavljonić, B., 71
Playtis, A. J., 47
Plesch, P. H., 31, 33, 35
Pleshivtsev, A. S., 409
Pleskachevskii, Yu. M., 72
Plessi, L., 365
Pletcher, C. H., 128
Pletneva, S., 396
Plimley, S., 179
Plitz, I. M., 182
Plochocki, A. P., 294
Plotnikov, V. D., 55
Plumere, P., 359
Plyashechnik, N. I., 54
Pocci, M., 53
Pochan, D. L. F., 84
Pochan, J. M., 84, 299, 404
Pochinok, V. Ya., 404
Poddubnaya, D. M., 73

Poddubnyi, I. Ya., 152
Podlesskaya, N. K., 344
Podolski, A. F., 41
Podosenova, N. G., 52, 56
Poehlein, G. W., 60, 70
Pogodina, N. V., 87
Pokorny, S., 243, 245
Pokorski, Z., 335
Polakowski, D., 385
Polanska, H., 83
Polikarpoc, I. S., 336
Polina, T. V., 347
Politi, M. J., 372
Polizzolti, G., 222
Polk, H. C., jun., 378
Poll, H. G., 326
Pollack, D., 378
Pollock, E., 380
Pollock, M., 389
Poloczek, J., 86
Poltev, V. I., 392
Polyansky, V. I., 51, 181
Pomakis, T., 76
Pommerening, K., 386
Pongs, O., 143
Ponnusamy, E., 77, 347
Ponnuswamy, P. K., 390
Ponomarenko, V. A., 36, 344
Ponomarev, G. V., 55
Ponomeva, M. A., 154
Ponratham, S., 78
Pons, A. B., 383
Popkowka, J., 78
Popli, R., 287
Porsch, B., 225
Porth, A., 72
Porter, G. T., 245
Porter, R. S., 192, 221, 222, 226, 246
Portetelle, D., 129
Portoukalian, J., 130
Porzi, G., 320
Pospisil, J., 331, 333
Postnikov, L. M., 88
Potier, A., 102
Potier, J., 102
Pottenger, L. A., 126
Potter, H., 132
Pottle, M. S., 392
Potuis, S. P., 79
Pouchly, J., 251
Pouyet, G., 254
Powell, D. A., 105, 106
Power, A. J., 346
Poxton, I. R., 117
Praahan, A. K., 79
Prabhakaran, M., 390
Pradhou, A. K., 87
Prager, S., 168
Pragnell, R. J., 185
Prasad, B. V., 392
Prasadarao, M., 293
Pratt, G. W., 101
Pratten, M. K., 384
Prausnitz, J. M., 252, 253

Author Index

Pravednikov, A. N., 30, 34, 347
Prentis, J. J., 162
Pressman, B. C., 374
Preston, F. J., 96
Preston, J., 167
Price, D., 348
Price, J., 119
Price, J. B., 289
Price, P. B., 410
Price, T. M., 124
Priester, R. D., jun., 53
Prieto, N. E., 372
Prikhodko, P. L., 147
Pringle, O. A., 189, 386
Prinke, G., 411
Priola, A., 34
Prisyazhnyi, V. M., 75
Pritchard, G., 80
Pritchard, M. J., 165, 236
Prohofsky, E. W., 392
Prokharova, L. K., 92
Prokhorov, A. M., 412
Prokof'ev, Y. N., 34
Prokop, Z., 316
Prolongo, M. G., 238
Pron, A., 75
Protasov, V. G., 332
Proudfoot, N. J., 144
Provencher, S. W., 171
Provenzano, F., 53
Prpic, I., 378
Pruckmayr, G., 36
Prud'homme, J., 266
Prud'homme, R. E., 77, 82, 217, 263, 278
Prut, E. V., 349
Przyluski, J., 75, 412
Pshisukha, A. M., 338
Pucci, B., 386
Puchalik, A., 78
Puchin, V. A., 53
Pudov, V. S., 341
Pueppke, S. G., 111
Pukanszky, B., 343
Pukhanskii, M. D., 344
Pundsack, A. L., 207
Purbrick, M. D., 407
Pusztai, A., 114
Putley, J. H., 353
Putnam, B. F., 392, 393
Pyriadi, T., 54

Qian, R., 207
Quack, G., 44
Quadrat, O., 226, 233
Quan, C., 369, 373
Quanten, E., 401
Quarma, J. K., 91
Quattrone, A. J., 376
Queguiner, G., 321
Quici, S., 313
Quigley, G., 392
Quina, F. H., 366, 368
Quinlan, G. L., 30
Quintana, J. R., 233

Quinty, W. C., jun., 383
Quirk, R. P., 42, 43

Raabe, D., 187
Raanby, B., 335
Rabek, J. F., 80, 331, 333, 334, 335, 338
Rabolt, J. F., 211, 300
Rackovsky, S., 391
Raczek, J., 226, 237
Radhakrishna, T. S., 341
Radhakrishnan, S., 405
Radic, D., 234
Rafler, G., 90
Ragab, Y. A., 184
Ragazzi, M., 394
Raghavachari, K., 343
Ragni, R., 384
Rahalker, R. R., 218
Rahman, A., 60
Rajogapalan, K. S., 72
Rajora, P., 387
Rakowsky, T. F., 157
Ralapati, S., 122
Ralien, G. J. L., 375
Rallison, J. M., 292
Ramachandran, K., 156
Raman, A. R., 387
Ramanathan, G. V., 256
Ramchandran, N., 152
Ramirez, A., 289
Ramirez, F., 144
Ramiro Vera, C., 232
Ramos, A. R., 268
Ramsay, J., 389
Ramwell, P. W., 377
Ranby, B., 80, 308, 331
Rand, W. G., 244, 245
Randall, J. C., 182, 183
Ranz, W. E., 96, 284
Rao, M. R., 341
Rapak, A., 306
Rappenecker, G., 100
Rappoport, D. C., 391
Raskus, H. J., 146
Rasmussen, J. K., 51, 372
Rasnby, B., 338
Ratner, B. D., 376
Ratner, M. A., 150
Ratzsch, M., 253
Ratzsch, M. T., 239
Raubach, H., 348
Rauner, F. J., 410
Rausch, M. D., 53
Ravanat, G., 105
Ravindran, K., 51
Ravindranath, K., 76, 387
Ray, A., 90
Ray, G. J., 183
Ray, J. A., 381
Raymond, J., 312
Raynal, B., 53
Raynal, S., 34, 53, 180
Razuvaev, G. A., 350
Razzeca, K. J., 383

Ready, B. W., 259
Rebel, J., jun., 356
Rebenfeld, L., 81, 219
Recca, A., 349
Record, M. T., 256
Reddy, E. P., 137, 138
Reddy, I. A. K., 369
Reddy, M. S., 116
Reddy, R., 144, 145
Reddy, V., 160
Redmond, J., 116
Redpath, A. E. C., 161
Reed, A. M., 80, 241
Reed, P. E., 288
Reed, W., 372
Rees, D. A., 104, 105, 106, 107, 119, 378
Reese, C. B., 96, 347
Refojo, M. F., 152, 386
Regel, W., 179
Regelson, W., 384
Regen, S. L., 312, 313, 314, 373, 374
Regula, D., 381
Rehák, A., 58
Reich, S., 249
Reich, T., 377
Reichert, K. H., 19, 26, 56, 388
Reichert, W. M., 377
Reichmanis, E., 405, 406, 407
Reihanian, H., 172
Reinhold, J., 33
Reinisch, G., 90, 226
Reinohl, V., 338
Reinsborough, V. C., 371
Reiser, A., 408
Rémillard, B., 81, 208
Rempel, G. L., 319, 366
Rempp, P., 38, 43, 45, 149, 153
Renaud, J. M., 266
Reneker, D. H., 210
Rengl, R., 51
Rennard, S. I., 125
Rens, M., 334
Retuert, J., 35
Reuter, P., 313
Revillon, A., 247
Revol, J. F., 103, 120
Revuelta, L. M., 226
Reynaers, H., 80
Reynard, C., 72
Reynold, P. A., 201
Reynolds, R. K., 138
Rha, C., 266
Rhine, W., 360
Rhi-Sausi, J., 281
Rhoades, J. W., 54
Rhodes, C., 140
Rialdi, G., 113
Riande, E., 31, 36, 37, 79, 233, 394, 395
Richards, D. H., 47, 52
Richards, G. D., 283
Richards, R. W., 192, 193, 195, 198, 199

Richardson, G. C., 289
Richardson, J. D., 378
Richter, D., 172, 202, 203, 236
Richters, P., 333
Richters, V. E. M., 362
Rick, P. D., 118
Ricco, M., 395
Riddell, S. Z., 88
Riedeberger, J., 383
Rieder, R., 411
Riesenfeld, J., 127
Riess, G., 258, 341
Rietsch, F., 81, 222, 251
Riffle, J. S., 82, 153
Rigal, G., 34
Rigby, D., 169
Riggs, A. D., 133
Righetti, P. G., 109
Rinaudo, M., 105, 107, 108, 119, 120, 246
Ringo, W. M., jun., 84, 186
Ringold, G., 143
Ringsdorf, H., 384
Rinstead, R. A., 366
Rios, L., 69
Rios Guerrero, L., 69
Ritchey, N. M., 89
Ritte, J., 152
Ritz, E., 384
Riva, F., 205
Rizzardo, E., 50, 58
Robb, E. C., 383
Robb, J. C., 345
Robbins, K. C., 137
Robbins, P. W., 114
Roberson, M. M., 111
Roberts, A. J., 401, 408
Roberts, C. N., 80
Roberts, D. D., 114
Robertson, F. C., 47
Robeson, L. M., 263, 267, 290
Robila, G., 181, 343
Robinowitz, M., 377
Robinson, G., 104, 107
Robinson, I. M., 36
Robinson, R. A., 126
Robinson, S. I., 141
Robinson, W., 346
Rocas, J., 373
Rochas, C., 107, 108
Roche, E., 103
Rockenbauer, A., 333
Rocko, J. M., 380
Rodehearer, G. T., 381
Rodenas, E., 370
Rodin, E. L., 382
Rodionov, A. G., 20
Rodkey, L. S., 131
Rodriguez, F., 53
Rodriguez-Sanches, D., 266
Roe, B. H., 134
Roerdink, E., 263
Rogers, C. E., 332
Rogers, J., 141, 144
Rogovina, L. Z., 149

Rohde, M., 77, 301
Rojas, A. J., 285
Rolla, F., 356, 372
Rolland, M., 296
Rollings, J. E., 241, 246
Roman, I., 289
Román, J. S., 50, 395
Romano, G., 219
Romsted, L. S., 369, 370, 373
Rondelez, F., 172, 235, 236, 254
Roof, L. B., 245
Rooney, J. J., 179
Rooney, J. M., 31, 345
Rooney, M. L., 331
Roots, J., 172, 235, 236
Roovers, J., 41, 160, 174, 226
Roper, B. A., 379
Rose, E., 383
Rose, J., 333
Rose, R. M., 382
Roseman, T. J., 385
Rosen, J. J., 376
Rosenbaum, J., 383
Rosenberg, H., 153
Rosenberg, L. C., 126
Rosenberg, R. D., 128
Rosenvasser, D., 79
Roshchupkin, V. P., 55
Rosin, M. L., 96
Ross, G., 381
Ross, L. R., 39
Rosana, D. M., 354
Ross-Murphy, S. B., 104, 119, 159
Rossmy, G., 97
Rotenberg, E. B., 56
Roth, S., 187
Rothblum, L., 144
Rothenhaeusser, B., 67, 68
Rotne, J., 168
Rottler, R., 26, 180
Roucoux, C., 408
Rourrillon, R., 111
Rowe, J. R., 368
Rowland, R. N., 384
Rowley, J. D., 137
Roy, A., 90, 104, 130, 132
Roy, K. L., 136
Royer, G. P., 351
Roylance, D., 287
Rozhkova, D. A., 64
Rubessa, F., 366
Rubio, S., 51
Rudee, M. L., 124
Rudenko, A. P., 345
Rudin, A., 69, 171, 182, 239, 242, 243, 244, 256, 262, 267, 388, 389
Rueda, D. R., 212
Rüdiger, H., 111
Ruffin, C., 321
Ruland, J., 205, 221
Rullman, J. A. C., 398
Rumack, M. S., 182, 388
Rumpler, M., 381

Rumyantsev, B. M., 403
Runt, J., 210
Runt, J. P., 205
Ruoslahti, E., 123, 124, 126
Ruoslahti, E. H., 99
Rupert, L. A. M., 369
Rupprecht, M. C., 251
Rusanov, A. L., 349
Ruscher, C., 88
Russell, J. C., 366, 370
Russell, T. R., 200
Russo, P. J., 35
Russo, R., 219, 221
Russo, S., 85
Ryabov, Ye. A., 85
Ryabov, M. S., 63
Ryan, M. E., 295
Ryan, T. G., 218
Rybak, W. K., 366
Rybnikar, F., 215
Rychly, J., 333
Rymian, B., 88
Ryssel, H., 411

Saab, H. H., 173, 174
Sabbadin, J., 227
Sabanayagam, P., 380
Sabet, A., 51
Sacchi, M. C., 181
Sacher, R., 411
Sadafule, D. S., 408
Sadanobu, T., 240
Sadhir, R. K., 386
Sadler, D. M., 194, 211, 212
Sadrmohaghegh, C., 332, 333
Saegusa, T., 31, 37, 38, 179
Saenger, W., 351, 374
Sagrario Casas, A., 79
Sahu, C., 79
Said, Z. F. M., 62, 71
Saigo, K., 359
St. Clair, A. K., 91
St. Jacques, M., 266
St. John Manley, R., 221
Saint-Onge, H., 152
St-Pierre, L. E., 151
Saito, A., 383
Saito, K., 245, 342
Saito, M., 226, 237, 240
Saito, S., 66
Saito, Y., 174
Saitoh, M., 160
Saiz, E., 394, 395
Saka, R., 355
Sakai, T., 113
Sakamoto, S., 102
Sakata, Y., 355
Sakharov, A. M., 36
Sakharova, A. A., 54
Sakonaka, A., 358
Sakuma, M., 36, 178
Sakurada, I., 102
Sakurai, Y., 376
Salamone, J., 198
Salamone, J. C., 67, 362

Author Index

Salem, M. A., 333
Sallee, C., 409
Sallet, D., 88
Sallot, P., 258
Salloum, R. J., 97
Salmon, M., 74
Salyer, I. O., 75
Salyn, Y. V., 362
Samal, R. K., 51
Samant, S., 90
Samoson, A., 150, 188
Samson, D., 382
Samuels, R. J., 200
Samui, A. B., 56
Sanchez, I. C., 256
Sander, R., 80
Sanders, E. M., 80
Sandler, S. R., 96
Sandner, M. R., 95
Sandri, S., 320
Sandrick, J. L., 151
Sanford, T. J., 213
Sang, M.-M., 242
Sang, R. D., 79
Sangalov, T. A., 30
Sangen, O., 362
Sanger, F., 134
Sangster, B., 383
Santiago, E. J., 380
Sannikov, S. G., 344
Sano, H., 371
Sano, T., 22
Sansone, G., 144
Santappa, M., 86
Santora, J., 395
Santos, E., 138
Sanui, K., 86, 407
Sarasola, C., 216
Sarasvathy, S., 51, 56
Saratani, Y., 368
Sargant, J. G., 99
Sarhan, A., 328, 364
Sariban, A. A., 397
Saringer, J. H., 281
Sarkar, B., 363
Sarkisyan, V. A., 36
Sarko, A., 104
Sarma, M. H., 392
Sarma, R. H., 392
Sartori, G., 245
Sasagawa, T., 42
Sasaoka, T., 400
Sasson, Y., 365
Sastre, R., 334, 403
Sasuoka, T., 148
Satgurunathan, R., 69
Sathpathy, U., 62
Sato, F., 298
Sato, H., 86, 176, 245, 246
Sato, K., 79, 184, 261, 340, 402
Sato, M., 79, 90, 358, 371
Sato, N., 311, 379
Sato, S., 337
Sato, T., 121, 410
Sato, T. U. S., 87

Sato, Y., 320
Satrusallya, S. C., 51
Sauer, J. A., 289
Saunders, D. W., 281, 294
Saunders, K., 178
Saunders, K. G., 178
Sauvet, G., 30
Sauviat, M., 226
Savehuck, T. M., 337
Savelli, G., 366, 369
Saville, B. P., 89
Savinov, V. M., 86
Savitskaya, E. M., 328
Sawaguchi, T., 340
Sawamoto, M., 33, 35
Sawatari, C., 222
Sayler, D. F., 384
Sazanov, Yu. N., 90, 348
Sazhin, B. I., 332
Sbrana, G., 318, 360, 365
Scafer, L., 254
Scagnolari, F., 365
Scaiano, J. S., 401
Scamporrino, E., 346
Scanlong, D. B., 133
Schaap, A. P., 327
Schachet, E., 308
Schaefer, J., 176
Schäfer, L., 237
Schafer, L., 162
Schaff, J., 91
Schaper, A., 88
Scharling, M., 381
Schaumburg, K., 106
Scheer, D., 378
Scheffler, I. E., 256
Scheidereit, C., 143
Scheinbeim, J. I., 209, 296
Schellenberg, R. K., 89
Schelten, J., 191, 192, 194, 206, 212
Schenkel-Brunner, H., 130
Scheraga, H. A., 117, 390, 391, 392, 393
Scherer, J. R., 210
Scheutz, H., 34
Schibler, V., 143
Schilling, F. C., 79, 84, 175, 177, 179, 181, 182, 183, 186, 394
Schimpfle, H. U., 94
Schindler, A., 80
Schindler, M., 118
Schlick, S., 410
Schlothauer, K., 183
Schmehl, R. H., 366, 370
Schmerr, M. J. F., 129
Schmidt, J. R., 238, 257, 262
Schmidt, L., 295
Schmidt, M., 169, 170, 171, 224, 237
Schmidt, M. A., 410
Schmidt, W., 204
Schnabel, W., 88, 334, 399, 409
Schneider, B., 79

Schneider, C., 179
Schneider, H., 356, 359
Schneider, H. A., 341
Schneider, H. J., 60
Schneider, M., 314
Schneider, N. S., 259
Schneider, P. M., 123
Schnell, D., 26
Schnitz, F. P., 247
Schofield, E., 54
Scholtens, B. J. R., 244
Schon, J., 328
Schopov, I., 348
Schork, F. J., 70
Schowalter, W. R., 292
Schrag, J. L., 171
Schreiber, H. P., 89
Schreiner, H., 144
Schressl, O., 245
Schroeder, A. H., 73
Schröder, E., 237
Schröder, U. K. O., 342
Schroll, A. L., 339
Schroll, W. K., 164
Schroter, B., 187
Schuback, H. R., 204
Schuch, W., 133, 252
Schue, F., 44, 51, 52, 178, 180, 296
Schultz, A., 31
Schutz, H., 31
Schultz, J. M., 213, 219, 294
Schultz, J. S., 375, 377
Schulz, E., 88
Schulz, G., 226
Schulz, G. O., 53
Schulz, G. V., 261
Schulz, R. C., 226
Schumacher, R., 331
Schuppiser, J. L., 71
Schurr, J. M., 167
Schurz, J., 127
Schuster, H., 51
Schutten, J. H., 319
Schwabacher, A. W., 31
Schwartz, A. B., 380
Schwartz, L. J., 183
Schwartz, L. M., 355
Schwartz, T., 227
Schwarz, C., 78
Schway, M. B., 376
Schweizer, R. J., 187
Schwier, I., 101
Sciarratta, G. V., 144
Scola, D. A., 90
Scola, L., 91
Scolnick, E. M., 138
Scopelianos, A. G., 156, 319
Scotney, A., 342, 345
Scott, A. J., 281, 294
Scott, G., 331, 332, 333, 334, 335, 336
Scott, J. C., 187
Scott, R. W., 141
Scott, W. G., 331

Seanor, D. A., 299
Searby, G. M., 236
Searle, R., 106
Sebenik, A., 35, 178
Secui, Y., 148
Sedlacek, B., 165
Sedlar, J., 333, 338
Seeburg, P. H., 134
Seeger, K., 299
Seel, K., 96
Seela, F., 373
Sefcik, M. D., 176
Seferis, J. C., 293
Seferis, J. S., 214
Segal, E., 345
Seganov, I., 89
Segura, C., 364
Seidewand, R. J., 69
Seidl, H., 51
Seiffert, G., 384
Seino, Y., 89
Seitz, W. A., 163
Sek, D., 78
Sekeris, C. E., 145
Seki, K., 86, 404
Sekiguchi, H., 185
Sekiguchi, Y., 342
Sekine, Y., 153, 350
Selden, A. C., 245
Selivanov, G. K., 409
Sell, C. S., 323
Selli, E., 79, 87
Selser, J. C., 236
Selwyn, J. C., 401
Semenov, A. N., 255
Semenova, A. A., 19
Semlyen, J. A., 150, 161, 241
Sen, A., 111
Sen, U. K., 225
Sendt, J., 150
Senear, A. W., 140
Senear, D. F., 112
Senet, J.-P., 53
Senga, M., 54
Sengupta, P. K., 52
Senkevich, S. I., 154
Seno, M., 313
Sentman, R. C., 54
Serafini, S., 176
Serelis, A. K., 50
Serenkova, I. A., 349
Sergeev, G. B., 35
Sergeev, S. A., 16
Sergeeva, L. M., 268
Sergot, Ph., 206
Seropegina, E. N., 338
Serrano, J. J., 386
Serre, B., 51
Serres, C., 183
Servet, A., A., 209
Sessions, W. J., 96
Sessler, G. M., 296
Setinek, K., 316
Setoudeh, E., 333
Seurin-Vellutini, M. J., 236

Sévcénko, V. V., 78
Seyden-Penne, J., 314, 365
Sgonnik, V., 44
Sgonnik, V. N., 40
Shablygin, M. V., 87
Shacklette, L. W., 300
Shadrina, N. E., 344
Shadrina, N. Ye., 55
Shafer, J. A., 112
Shaikh, A. S., 52
Shakhtakhtinskii, T. N., 30
Shamanin, V. V., 40, 41
Shamkina, N. A., 349
Shamoo, A. E., 374
Shander, M., 144
Shannon, T. G., 314, 374
Shapiro, Yu. Ye., 65
Shapoval, G. S., 75
Sharaby, Z., 34, 48, 178
Sharkey, W. H., 47
Sharma, A. K., 386
Sharma, B. L., 91
Sharma, C. P., 381
Sharon, J., 130
Sharon, N., 109, 111, 116
Sharp, D. S., 389
Shator, H., 97
Shaw, B. R., 73
Shaw, M. T., 260, 267, 290, 292
Shaw, T. J., 408
Shcherbakova, L. M., 348
Sheehan, J. K., 126
Sheetz, M. P., 123
Sheffed, H., 90
Sheikh, K., 380
Shelgaev, V. N., 344, 347
Shelgaeva, V. G., 347
Shelykhmanov, E. F., 413
Shen, C. J., 141
Shen, C. S., 95
Shen, M., 57, 290
Shepelev, S. N., 16
Shepherd, L., 187
Shepherd, T. M., 402
Sher, W., 378
Sherman, E. S., 221
Sherratt, D., 135
Sherrington, D. C., 31, 304, 312, 313, 314, 315, 317, 324, 325, 330, 360
Shete, P., 294
Shevchuk, E. S., 75
Shevchuk, L. M., 66
Shibaev, V. A., 348
Shibata, S., 111
Shibayama, M., 266
Shigematsu, K., 358
Shih, Y. J., 71
Shiibashi, T., 240
Shiina, S., 150
Shikawa, H., 113
Shilov, A. E., 27
Shilov, V. V., 53
Shimada, H., 91
Shimada, J., 166

Shimada, S., 210
Shimidzu, T., 319
Shimizu, C., 346
Shimizu, I., 405
Shimizu, Y., 385
Shimoda, T., 222
Shimokohara, S., 403
Shimomura, M., 210, 362
Shimomura, Y., 295
Shimono, T., 342
Shin, J. S., 362
Shinkai, S., 351, 358, 362, 364, 366, 371, 374
Shinohara, I., 155, 266, 328, 376
Shinohara, Y., 45
Shinya, S., 295
Shiokawa, K., 164, 233
Shirai, M., 402, 403
Shirakawa, H., 299
Shire, J. D., 129
Shirinyan, V. T., 344
Shirmsky, V. F., 85
Shirota, Y., 52, 400
Shkurina, G. P., 12
Shlyapintokh, V. Ya., 88, 331, 334
Shlyapnikov, Y. A., 346, 349
Shoji, H., 366
Shoko, N. R., 344
Sholsby, S. E., 79
Shono, T., 342
Shore, G. R., 13
Shors, E., 377
Short, J. N., 29
Shortall, J. B., 95
Shortle, D., 135
Shosenji, H., 366
Showronski, T., 258
Shteiselbein, B. I., 335
Shu, C., 383
Shufan, T., 86
Shumnyi, L. V., 51
Shupik, A. N., 30
Shustov, A. L., 148
Shevchuk, L. M., 69
Shvetsov, O. K., 69
Siadat, B., 69
Siano, D. B., 225
Sibilo, J. P., 81
Sichel, E. K., 91
Sideridou-Karayannidou, I., 86
Sidorova, L. P., 32, 39
Sidorovich, A. V., 92
Siedentop, K. H., 382
Sieger, H., 355
Siekmann, G., 82
Sierra, P., 51
Sierra-Vargas, J., 38
Sietsma, J. H., 121
Signorini, R., 25
Sigwalt, P., 30, 40, 45, 147
Sikorski, A., 160, 397
Silberberg, A., 166
Silverblatt, F. J., 116

Author Index

Silina, N. A., 150, 151
Silinskaya, I. G., 226
Silvestro, G. D., 220
Simanka, A. J., 382
Simek, L., 226
Simeonov, N., 86
Simeonska, T., 86
Simionescu, C., 52, 72, 78
Simionescu, C. F., 75
Simitzis, I., 76, 94
Simkovich, N. N., 336
Simon, A., 5, 7
Simon, I., 141
Simon, M. I., 138
Simon, W., 359
Simonson, R., 104
Sinay, P., 128
Sindorf, D. W., 188
Singh, A., 374
Singh, B. P., 337
Singh, M., 133
Singh, P., 256, 373
Singh, R. P., 334
Singler, R. E., 154, 259
Sinitsina, G. V., 40
Sinkai, S., 370
Sinn, H., 27, 28
Sinn, J. H., 27
Sircar, A. K., 298
Sirkar, K. K., 152
Siroky, M. B., 379
Sirota, A. G., 344
Sisido, M., 361, 398
Sitek, F., 331
Sivanamakrishnan, K. N., 78
Sivaram, S., 183
Sivy, G. T., 345
Sixma, J. J., 383
Sixou, P., 236, 255, 265
Sjoholm, I., 385
Sjoerdsma, S. D., 223
Sket, B., 320, 362
Sklizkova, V. P., 226
Skobochkina, S. V., 407
Skolnick, J., 167, 256, 395, 396
Skondras, P., 249
Skotheim, T., 74
Skoulios, A., 199
Skowronski, T., 267
Skrabut, E. M., 383
Skurat, V. E., 409
Skvortsov, A. M., 397
Sladek, T., 383
Slagowski, E., 165
Slama, Z., 337, 347
Sledz, J., 44, 51, 178, 180
Sloane, N. J. A., 84, 186
Slobodetskaya, E. M., 332
Slobodyanik, V. V., 400, 404
Slominskii, G. L., 149, 150, 153, 339
Slonim, I. Ya., 185
Smeby, L. C., 384
Smedley, J. E., 89
Smela, N., 152

Smetanina, I. Ye., 284
Smets, G., 53
Smid, J., 328
Smidsrød, O., 106
Smirnov, A. L., 184
Smirnov, B. R., 55
Smirnova, G. A., 54
Smirnova, O. V., 349
Smith, A. J. H., 134
Smith, C. Z., 353
Smith, D. A., 104, 347
Smith, D. E., 124
Smith, H. K., 51, 372
Smith, I. C. P., 111, 114
Smith, J., 87
Smith, J. H., 145
Smith, J. S., 346
Smith, P., 102, 221, 290
Smith, P. B., 358
Smith, R. A., 33
Smith, S. A., 83, 152
Smith, T. E., 148, 178
Smith, T. L., 215
Smith, W. A., 84
Smits, H. J. E., 235
Smolders, C. A., 226, 232, 252
Snape, C. E., 185
Snider, M. D., 114
Snuparek, J., 70
Snyder, R. G., 394
Snyder, R. W., 393
Sörvik, E., 56
Sörvik, E. M., 49
Soga, K., 14, 21, 22
Soga, T., 360
Sogah, G. D. Y., 357
Soh, S. K., 56, 61, 388
Sohrab, M. H., 339
Sojka, S. A., 185
Sokolik, I. A., 403
Sokolov, L. B., 86
Sokolov, V. N., 19
Solaro, R., 53, 513
Solc, K., 161, 163, 239, 250, 256
Solomon, D. H., 49, 50, 58, 181
Solov'ev, Yu. V., 70
Soloveva, L. V., 347
Soltes, L., 64, 226, 245
Somani, R. H., 260
Somogyrari, K., 379
Sonderhof, D., 344
Sone, T., 371
Sonnenberg, A. S. M., 121
Sonoda, T., 372
Soong, D. S., 290, 291
Sophianopoulos, A., 113
Sophianopoulos, J. A., 113
Soricelli, R. R., 380
Sornette, D., 390
Sorvik, E., 181, 182
Sotakova, E., 383
Sotalova, O., 383
Sotobayashi, H., 225
Souel, T., 52
Soum, A., 47, 180

Soutar, I., 47, 401
Souter, W. A., 379
Soutif, J.-C., 41
South, M., 37
Southick, G. J., 120
Southwick, J. G., 172
Sowden, L. C., 103
Soy, B. J., 213
Sozzani, P., 178
Spadaro, G., 83
Spange, S., 31, 33
Spanggaard, H., 36, 179, 182
Spanswick, J., 183
Sparrow, D. J., 93
Spassky, N., 217
Spencer, C. P., 186
Spencer, S., 53
Sperling, L. H., 251, 268, 290
Spiegel, J., 343
Spiers, R. P., 281
Spik, G., 122
Spiro, T. C., 74
Spitler, K. G., 94
Spitsyn, V. I., 39
Spitzer, G., 384
Spooner, S. J., 189
Spragg, S. P., 351
Sprenger, R., 382
Springer, H., 80, 81, 217
Spruiell, J. E., 89, 295
Spurr, N. K., 138
Spychaj, T., 245
Squire, D. R., 32
Squire, P. C., 247
Sreenivasan, S., 103
Sridharan, N. S., 289
Srimal, S., 115
Srinivasan, A., 137
Srinivasan, K. R., 129
Srinivasan, M., 86, 90
Srinivasan, R., 80, 409
Srivastava, A. K., 50
Srivastava, S. K., 370
Sri-Widada, J., 145
Staal, W., 138
Stabl, H. G., 64
Stacey, K. A., 363
Stachurski, Z. H., 213
Stackhouse, J. A., 151
Staden, R., 144
Stadler-Szoka, A., 351
Stafford, R. J., 377
Stahl, D., 81
Stahl, G. A., 53
Staikos, G., 249
Stalder, B., 288
Stamatoglou, S. C., 125
Stamberg, J., 234
Stamm, M., 193, 194, 195, 212
Stankevic, I. V., 85
Stanley, J. C., 377
Stannett, V. T., 31, 32, 39, 386
Starch, M. S., 152
Starchenko, L. V., 87
Stark, B. P., 405

Stark, W. J., 382
Starkweather, H. W., jun., 211
Starnes, E. H., jun., 181
Starnes, W. H., jun., 175, 182, 335, 343
Starowieyski, K. B., 9
Stasikelis, P., 383
Stassinakis, C. A., 289
Steele, W. J., 289
Steffan, R., 349
Steffers, F., 67, 68
Stefoni, S., 383
Steger, J. J., 10
Steger, T. R., 176
Stehling, F. C., 56, 182, 394
Stein, R., 192
Stein, R. S., 190, 200, 201, 260
Steinmann, D. K., 201, 210
Steitz, J. A., 144, 145, 146
Stejskal, E. O., 176
Stejskal, J., 236, 261, 270
Stengl, G., 411
Stenius, P., 67
Stepanek, P., 165
Stepanov, Ye. M., 54
Stephenson, J. R., 138
Stephenson, P. J., 186
Steplewski, Z., 130
Steptoe, R. F. T., 150, 161, 169, 192, 241
Sternbach, D. D., 354
Stevens, E. S., 104, 105, 106
Stevens, M. P., 90
Stewart, C. L., 141
Stickler, M., 56
Stickley, M., 388
Stigter, D., 167, 273
Stilbs, P., 119, 227
Stille, J. K., 226, 349, 366
Stille, J. R., 366
Stipanovic, A. J., 104
Stirling, G. C., 202
Stivala, S. S., 119
Stix, W., 212
Stockmayer, W. H., 163, 166, 169, 170, 224
Stoddart, J. F., 351, 355, 374
Stoeckel, J., 70
Stokes, G. M., 91
Štokr, J., 55, 79
Stol, M., 381
Stone-Elander, S. A., 54
Storey, R. F., 38
Strartaas, T. M., 384
Stratta, P., 384
Straube, E., 238
Strauss, H. W., 375
Strecker, G., 122
Street, G. B., 74, 74, 299
Strezelecki, L., 78
Strick, N., 385
Stringer, B., 385
Strivens, T. A., 260
Strizhko, G. D., 338
Strobl, G. R., 206, 218

Strohmeier, W., 79, 81
Strong, A. B., 376
Strosberg, A. D., 111
Struck, E. C., 314, 374
Struik, L. C. E., 287
Struszczyk, H., 157
Strzelbicki, J., 358
Strzelecki, L., 79
Stuhlman, H., 141
Stuhrmann, H., 191
Stumph, W. E., 140
Sturm, J., 391
Sturgeon, P. Z., 80
Stynes, D. V., 360
Subramanian, R. V., 73
Sudakova, S. V., 12
Sudres, P., 34
Sugai, Y., 101
Sugano, H., 391
Suganuma, A., 85
Sugarman, G., 96
Sugawara, S., 408, 409
Suggalt, J. R., 345
Sugi, Y., 67
Sugibayashi, K., 385
Sugimoto, M., 389
Sugimoto, T., 379
Sugimoto, Y., 102
Sugimura, Y., 246, 346
Sugiyama, K., 58, 65
Sukhinin, V. S., 37
Sullivan, B. P., 74
Sullivan, M. J., 188
Sumi, A., 36, 179, 180
Sumitomo, H., 36, 37, 179, 180, 330
Summerfield, G. C., 190
Sun, A., 384
Sun, C., 362
Sun, S. T., 165, 242
Sunamoto, J., 360, 366
Sunberg, D. C., 56, 388
Sundardi, F., 63
Sundelin, J., 124
Sundelof, L.-O., 225
Sung, H. N., 177
Suno, T., 370
Supakorn, O., 36
Suparno, S., 180
Supino, N., 185
Suprunchuk, T., 80
Surolia, A., 109, 111, 115
Surolia, N., 109
Suryanarayan, G. V., 51
Sussman, J. L., 392
Suszko, P. R., 155
Suter, U. W., 176, 394
Suthar, B. P., 86
Sutherland, I. O., 355, 359
Sutin, N., 319
Sutina, O. D., 350
Sutter, D., 141
Suwa, T., 67
Suzuki, A., 81, 222
Suzuki, E., 12

Suzuki, H., 237, 260
Suzuki, K., 119, 121, 340, 352, 367
Suzuki, M., 69, 376
Suzuki, N., 119, 121
Suzuki, T., 48, 80, 177, 247
Suzuki, Y., 266
Svec, F., 243, 309, 316
Svejdova, E., 347
Svensson, J., 381
Svetlov, Ju. E., 226
Swan, K. G., 380
Swanson, D. K., 150
Swanson, S. A. V., 379
Swartz, B. L., 383
Swislow, G., 165
Syatkovskii, A. I., 179
Synder, R. G., 210, 213
Szabo, G., 374
Szejtli, J., 351
Székely-Pécsi, Zs., 81
Szele, I., 356
Szesztay, M., 49, 185, 242
Szeto, R., 148
Szewczyk, P., 241, 389
Szinicz, G., 380
Szwarc, M., 22, 40
Szymarski, W., 88

Tabak, L. A., 116
Tabata, Y., 334
Tabin, C. J., 138
Tabner, B. J., 51
Tabuse, I., 360
Tabushi, I., 351, 354, 355, 359, 373
Tachiya, M., 371
Tada, T., 413
Tada, Y., 90
Tadashi, K., 60
Tadokoro, H., 207, 209, 393
Tadros, Th. F., 199
Taganov, N. G., 56
Tagashina, K., 206
Tagashira, Y., 391
Tagawa, S., 334
Tagawa, T., 85, 315, 387
Tagle, L. H., 86, 90
Taguchi, T., 367
Taha, M., 34, 51
Tai, K., 85, 387
Tait, P. J. T., 13, 15, 16
Tajima, I., 37
Takacs, E., 7, 57
Takada, M., 24
Takada, T., 321
Takadate, A., 353
Takahama, T., 298
Takahashi, M., 55, 161
Takahashi, N., 104
Takahashi, T., 207
Takahashi, Y., 209
Takai, M., 103
Takai, Y., 404, 405
Takaki, M., 38

Author Index

Takamiya, N., 366
Takahashi, M., 83
Takano, M., 366
Takarabe, K., 31, 33
Takasawa, R., 310
Takawashi, K., 355
Takayanagi, M., 222, 255
Takegami, Y., 48
Takeishi, M., 306, 363
Takemori, M. T., 84, 289
Takemura, K., 52
Takemura, T., 206
Takeshita, M., 371
Takeuchi, H., 202, 393
Takeuchi, K. J., 357
Takeuchi, T., 102
Takeuchi, Y., 118, 386
Taki, W., 376
Takido, N., 383
Takigami, S., 88
Talreja, R., 289
Talwar, S. S., 407
Talykov, V. A., 151
Tamada, T., 222
Tamagaki, S., 327
Tamai, N., 402
Tamaki, M., 181
Tamamura, T., 409
Tamano, J., 412, 413
Tameel, H., 81
Tamura, M., 12
Tamura, S., 23
Tan, Z. C. H., 410
Tanabe, T., 184
Tanaka, G., 160, 161, 169, 238, 239, 256
Tanaka, H., 47, 86, 212, 294, 407
Tanaka, M., 331, 342, 402, 403, 407
Tanaka, N., 221
Tanaka, S., 130, 266, 352, 407
Tanaka, T., 165, 224, 237, 245, 251
Tanaka, Y., 32, 176, 246
Tánczos, I., 58
Tandon, P., 85, 387
Tanfani, F., 328
Tang, L., 88
Tang, R., 309
Tang, S. H. K., 153
Tang, Y., 4
Tangari, C., 190
Tanielian, C., 41
Tanigami, T., 213
Taniguchi, H., 100
Taniguchi, M., 376
Taniguchi, Y., 352, 367
Tanner, R. I., 291
Tant, M. R., 153
Tanzawa, H., 386
Tarakanov, O. G., 152
Taran, A. A., 41
Tartof, D., 144
Tarutina, L. I., 55

Tashiro, K., 393
Tasumi, M., 393
Tatarova, L. A., 226
Taut, M. R., 81
Taylor, D. E., 380
Taylor, G. N., 411, 412, 413
Taylor, G. W., 339
Taylor, H. C. R., 358
Taylor, I. E. P., 105
Taylor, J. W., 409
Taylor, L. J., 332
Taylor, L. T., 91
Taylor, P. A., 283
Taylor, P. L., 209, 393
Taylor, R., 185
Tazaki, T., 56
Tazuke, S., 36, 38, 79, 261, 401, 402, 405
Tchir, M. F., 242
Tchir, W. J., 244, 389
Tealdi, A., 255
Teichberg, V. L., 111
Teichner, S. J., 364
Teisseire, B., 385
Teixeira-Barreira, S. R., 41
Tejima, S., 130
Telfer, A. R., 244, 389
Telford, D., 172
Teller, D. C., 112
Temple, G. F., 136
Tench, D., 74
Tengbald, A., 125
Teo, W. H., 205
Tepfer, M., 105
Terado, T., 222
Teramatsu, T., 385
Teramoto, A., 167, 252
Teranishi, H., 85
Terano, M., 21
Teraoka, N., 266
Terashima, M., 130
Terawaki, Y., 181
Terdhman, S., 381
Terlemestan, E., 86
Terlemezyan, L., 37
Terman, L. M., 350
Terrell, D. R., 36, 52, 177
Terui, K., 39
Teshima, H., 340
Tessier, T. G., 410
Tetta, C., 384
Teyssié, P., 78, 95
Tezuka, Y., 38
Thacker, J. G., 381
Thambidorai, D., 115
Thangavel Vijayakumar, C., 80
Tharanathan, R. N., 99
Thayer, A. L., 327
Theobald, W., 384
Theocaris, P. S., 289, 293
Thijs, L., 53
Thøgersen, H., 129
Thom, D., 106
Thomalla, M., 370
Thomas, D. A., 268

Thomas, E. L., 190, 221
Thomas, J., 333
Thomas, J. K., 327
Thomas, L. J., 384
Thomason, J. L., 198, 199
Thompson, L., 66
Thompson, L. F., 346, 411
Thompson, P. E., 145
Thompson, R. W., 62, 389
Thomson, J. W., 381
Thonar, E. J. M. A., 125
Thornton, P. R., 406
Thorpe, F. G., 364
Thorpe, M. F., 164
Thu, C. T., 36
Thubrikar, M., 377
Thulin, B., 364
Thunberg, L., 127
Ti, S. S., 393
Ticha, M., 109
Tidswell, , B. M., 72
Tiedeman, G. T., 185
Tieke, B., 301
Tighe, B. J., 378, 382
Tighzert, H.-L., 186
Tikhonov, K. I., 72
Tikhonova, L. S., 72
Tilghman, S. M., 141
Till, R. J., 384
Tindale, W. B., 376
Tingsvik, K., 104
Tirrell, D. A., 37, 331
Tirrell, M., 56, 57, 260, 388
Titmas, R. C., 133
Titomanlio, G., 291
Titov, A. P., 65
Titova, Z. P., 86
Tkachuk, B. U., 148
Tobias, J. W., 332
Tobias, R. II., 149
Tobin, F. L., 33
Tobin, G. R., 378
Tobochnik, J., 397
Toda, F., 91, 351, 353, 354
Todd, G. P., 389
Togna, D., 376
Toh, H. K., 342
Toirov, A., 348
Tokarev, V. S., 53
Tokiwa, Y., 80
Tokue, I., 150
Tokuzawa, T., 37
Tollefson, N. M., 319
Tolusso, F., 87
Tomahogh, R., 359
Tomas, A., 373
Tomas, M., 245
Tomescu, M., 345
Tominaga, Y., 129
Tomioka, H., 406
Tomita, B., 176
Tomita, M., 342
Tomoi, M., 313, 318, 364
Tompkins, T., 185, 406
Tonellato, V., 367, 368

Tonelli, A. E., 175, 177, 179, 183, 208, 394, 395
Tong, J. P. K., 102
Toniolo, C., 117
Too, J. R., 389
Topiol, S., 150
Toporowski, P. M., 41, 160, 226
Toppet, S., 261
Toptygin, D. Y., 84, 348
Torii, M., 130
Torikai, A., 335, 336
Torkington, J. A., 253
Torp, J., 99
Torri, G., 128
Torrie, G. M., 165
Torroni, S., 365
Tosaka, M., 368
Toselli, M., 113
Toshima, N., 365
Toshio, K., 60
Touloupis, C., 54, 226
Toulouse, G., 253
Tourillon, G., 74
Tovborg, J. P., 334
Townsend, F. M., 385
Townsend, P., 283
Toyoda, K., 409
Toyoda, M., 368
Toyoda, N., 54
Toyoki, C., 346
Toyoshima, N., 33
Trainer, G. L., 353
Tramontini, M., 318
Traylor, T. G., 360
Trecarten, M., 325, 326
Treisman, R., 144
Trekoval, J., 43, 245
Trezov, V. V., 348
Triboulet, J. P., 379
Tricoli, M., 33, 71, 75
Tripathy, A. K., 79, 87
Tripathy, S. K., 216, 393
Tritto, I., 181
Trivedi, P. D., 34
Trofimov, B. A., 35
Trofimova, N. F., 340
Tronick, S. R., 137
Troostwijk, C. B., 357
Trudel, G., 151
Truji, T., 114
Trumbo, D. L., 177
Truss, R. W., 288
Trzebiatowski, T., 206
Tsai, B., 165
Tsai, M.-J., 139
Tsai, T.-T., 153
Tsamantakis, A., 87
Tsang, W. K. W., 291
Tsao, G. T., 241, 246
Tsay, F. D., 336
Tschirwitz, U., 404
Tsendrovskii, V. A., 148
Tseng, S., 215
Tsirgiladze, M. W., 349

Tsubokawa, N., 322, 363
Tsuchida, E., 304, 318, 319, 320, 326, 376
Tsuchiya, K., 408
Tsuchiya, S., 360
Tsuda, K., 52, 54, 169
Tsuda, M., 412, 413
Tsuge, S., 246, 346
Tsui, M. J., 140
Tsuji, A., 369
Tsuji, K., 65
Tsuji, M., 211
Tsuji, S., 385
Tsuji, T., 111
Tsuji, Y., 48
Tsukino, M., 179
Tsukruk, V. V., 53
Tsukube, H., 374
Tsunashima, Y., 171
Tsunoda, T., 407
Tsunooka, M., 331, 407
Tsurami, Y., 184
Tsuruta, H., 335
Tsuruta, T., 53, 376
Tsutsumi, Y., 387
Tsvetanov, Ch. B., 40
Tsvetkov, N. S., 75
Tsvetkov, V. N., 87, 226
Tuan, Y. L., 344
Tuberville, A. W., 382
Tüdös, F., 49, 58, 185, 242, 246, 333, 340, 343
Tuke, M. A., 379
Tulig, T. J., 56, 57, 388
Tulliez, M., 385
Tumangan, V. G., 392
Tundo, P., 313, 372, 373, 374
Tung, L. H., 56, 264
Tunuli, M. S., 371
Turcsanyi, B., 343
Turickova, O., 245
Turner, C., 120
Turner, H. A., 385
Turner, P., 140
Tuzar, Z., 226, 266
Tuzar, Z. K., 85
Tuzi, A., 207, 393
Twieg, R. J., 406
Twigg, A. J., 135
Twisleton, J. F., 201

Uchiyamo, T., 337
Uden, P. C., 346
Udovenko, V. F., 338
Ueda, M., 53, 54, 55, 86, 178, 310, 311, 312, 410
Ueda, T., 407
Uehara, A., 412, 413
Uekama, K., 351, 353
Ueki, S., 22, 23
Uematsu, I., 207, 213
Ueno, A., 355
Ueoka, R., 368
Ui, N., 123
Ulanski, J., 299

Ullman, R., 169, 197
Ulrich, H., 96
Umemura, T., 151
Umeta, N., 306
Ungurenascu, C., 319
Uniyal, S., 376
Unsworth, J., 150
Urakawa, K., 47
Uranek, C. A., 71
Urban, G., 218
Urbanek, E., 383
Urbanek, P., 383
Urman, Ya. G., 90, 185
Urushido, K., 54
Ury, D. N., 391, 392
Uryu, T., 33, 35, 38, 39, 176
Usami, T., 246
Usmani, A. M., 75
Utiyama, H., 236
Utracki, L. A., 294
Uvarov, A. V., 337
Uyttenhoeven, H., 45

Vairon, J. P., 33, 177
Valcarcel, M. V., 314
Valcheva, E., 101
Valenti, B., 222
Valenti, P. C., 327
Valentini, G., 318, 360, 365
Valenzuela, M. I., 35
Valiev, K. A., 409, 412
Vallance, M. A., 97
Valles, E. M., 388
Valltars, B., 381
Van, B. T., 61, 63, 388
Vanarsdale, W. E., 292
Van Bogart, J. W. C., 219
van Brandwijk, R. A. M., 238
Vancea, L., 180
Vancsó-Szmeresányi, I., 81
van den Berg, J. W. A., 226, 232, 234, 262
van den Berghe, H., 129
Vandenberghe, L., 37
Vandenreissche, J., 261
van Der Auweraer, M., 262, 401
van der Eijk, J. M., 362
Vander Hart, D. L., 187
Vanderhoff, J. W., 69
van de Ridder, G., 226, 232
van der Maaten, M. J. 129
van der Pleog, L., 141
Van der Velden, G., 183
van der Weij, F. W., 95
Vaneso, G., 343
Van Gunsteren, W. F., 398
van Heijst, A. N., 383
van Kessel, A., 138
Van Laeken, A., 187
van Landschoot, A., 113
van Leuven, F., 129
Van Luyen, D., 78, 79
Van Noort, R., 151, 376
Vannucchi, S., 125

ature Index 449

Vantelon, J. P., 347
van Wenden, K., 206
van Wijngaarden, H., 295
Van Zandt, L. L., 392, 393
Varadarajan, K., 84
Varga, J., 81
Varhegyi, G., 342
Varkevisser, F. A., 261
Varnell, D. F., 82
Varnell, E. D., 382
Varnell, W. D., 210, 212, 213
Varshavsky, A., 140
Vasile, C., 343, 346
Vasilev, V. G., 149
Vasku, J., 383
Vasquez Vallejo, J. M., 74
Vass, F., 38
Vasuderan, P., 385
Vatuleu, V. N., 206
Vatvars, A., 245
Vaughan, M. H., 247
Vavra, M., 385
Veda, T., 335
Vekatasubban, K. S., 368
Velichkova, R., 31
Velikov, L. V., 409, 412
Velikova, M., 19
Vellutini, M. J., 265
Venczel, G., 351
Venkatachalam, C. M., 391, 392
Venkatachalam, R. S., 385
Venkatarao, K., 51, 56
Venkatesan, T., 413
Venkotesh, G. M., 81
Venturello, P., 373, 374
Vercellone, A., 384
Verdu, J., 331, 332, 335, 343
Vered, Y., 105
Verlaan, J. P. J., 366
Verlová, H., 85
Vermande, P., 342
Vermel, Y. V., 12
Vernaleken, H., 82
Vernon, F., 328
Veron, J., 243
Vert, M., 77, 88
Vértes, E., 49
Vezey, P. N., 318
Vichutinskaya, Ye. V., 88
Vidakovic, P., 172, 235, 254
Vidal, J. L., 245
Vidali, M., 4
Vidotti, G. J., 370
Vieira, J., 134
Vieira, R. C., 368
Vietmeier, J., 326
Vigneron, C., 385
Vijayakumar, C. T., 77, 347
Vijayendran, B. R., 66
Vilasagar, S., 78
Vilenchik, L. Z., 242, 244
Villacorta, G. M., 181
Villiger, B., 123
Vincendon, M., 107

Vincent, B., 199
Vinogradov, G. V., 279, 284
Vinogradova, L. V., 40
Vinogradova, S. V., 90, 91, 349
Vinter, D. W., 377
Viout, P., 309, 356, 365
Viouy, J. L., 202
Viriyayuthakorn, M., 282
Visger, R. L., 96
Vishnevskii, G. E., 340
Vitovskaya, M. G., 226
Vitterio, V., 219, 221
Vitus, D. M., 178
Vlachogiannis, G. J., 119
Vlachopoulos, J., 295
Vladimir, G. M., 75
Vlasova, I. V., 91, 348
Vlegels, M. A., 38
Voegtle, F., 351, 355
Voelzke, W., 67
Vofsi, D., 48, 96, 212
Vogel, K. G., 126
Vogel, M., 69
Vogelstein, B., 141
Vogl, O., 36, 82, 87, 176, 308
Vogt, H. G., 385
Vogt, W., 386
Vogtle, F., 358, 364
Vohlidal, J., 31
Voigt-Martin, I. G., 212
Voliotis, S., 371
Volkov, V. A., 63
Volkova, T. V., 85
Vollmer, H.-J., 27, 28
Volosatov, V. N., 90
Volynskii, A. L., 81
von Hagens, G., 385
Voorhees, K. J., 347
Vorma, D. S., 384
Voronov, S. A., 53, 64, 71
Vorreux, G., 264
Vos, J. G., 319
Vose, J. R., 99, 100
Votavova, E., 243
Voyles, C. R., 378
Vozka, S., 243
Vrentas, J. S., 294
Vygodskii, Ya. S., 90, 91
Vymalalova, Z., 335
Vymazal, Z., 335
Vyshinskaya, L. I., 7

Wada, A., 119, 121
Wada, E., 196
Waddell, W. H., 172
Wadsworth, T. G., 379
Waegall, B., 328
Wagner, A., 411, 413
Wagner, H. M., 407
Wagner, L. J., 156
Wagner, M. H., 290
Wahl, B., 72
Wai, G. K. C., 54
Wai, M. P., 190
Wakuda, T., 353

Waldherr, R., 384
Waldman, J., 219
Waldmon, T., 81
Wales, M., 235
Walker, B. W., 145
Walker, J. E., 138
Walker, N., 84
Walker, P. S., 379
Wall, R., 141, 144
Wallace, D. G., 123
Wallteg, B., 298
Walsh, D. J., 200, 250, 251, 259
Walsh, E. K., 255
Walters, K., 281, 295
Walton, I. G., 93
Wandelt, B., 334, 337, 342, 402
Wandrey, Ch., 226
Wang, A., 392
Wang, C. H., 206
Wang, C. S., 79, 204
Wang, C. T., 151
Wang, F., 207
Wang, G. H., 81
Wang, H. C., 40
Wang, J.-I., 212, 213
Wang, J. L., 266
Wang, K. W., 246
Wang, L. H., 192
Wang, S., 352
Wang, T. T., 251
Ward, I. M., 205, 213, 220, 221, 222, 287, 288
Ward, R. S., 259
Warkentin, J., 52
Warnell, W. D., 205
Warner, F. D., 217
Warner, M., 203
Warner, R. J., 77
Warpehoski, M., 383
Warren, C. D., 122
Warren, L. F., 74
Warrier, L. A. V. R., 84
Warshawsky, A., 327
Warshell, A., 391
Washio, M., 334
Wasiak, A., 81
Wasniowska, K., 115
Wassermann, G. D., 364
Wasylishen, R. E., 319
Watanabe, H., 320, 363
Watanabe, J., 207, 213
Watanabe, M., 33, 301
Watanabe, N., 362
Watanabe, S., 385, 406, 407
Watanabe, T., 89, 363
Watanabe, Y., 48
Watase, M., 108
Waterhouse, J., 306, 309
Watnick, P. I., 187
Watson, J. G., 284
Watson, W. R., 148
Wattel, F., 383
Wattley, R. V., 356
Watts, R. K., 409
Weatherhead, R. H., 363

Weathersby, P. K., 376
Webber, G., 212
Webber, S. E., 157, 401
Weber, E., 355
Weber, J. V., 314
Weber, W. P., 304
Webman, I., 397
Webster, M. F., 281
Weddigen, G., 301
Weeks, J. J., 208
Weghe, T. V. D., 294
Wegner, G., 207, 226, 299, 393
Weill, G., 162, 168, 197, 336
Weinberg, R. A., 137, 138
Weiner, P., 359, 392
Weiner, P. K., 391
Weingartner, B., 146
Weintraub, H., 139
Weintraub, H. M., 140
Weirauch, K., 82
Weisbrod, S., 140
Weise, C., 152
Weiss, A. B., 381
Weissman, S. M., 146
Weissman, S. R., 92
Wells, J. K., 289
Wells, J. L., 298
Welsh, E. J., 101
Welzen, T. L., 244
Wendisch, D., 183
Wendroff, J. H., 89, 261
Wendorff, W., 205
Wenig, W., 82
Wenz, G., 226
Werkman, R. T., 29
Werner, T., 337
Wernicke, R., 243
Wertheimer, M. R., 89
Wesseling, P., 295
Wessels, J. G. H., 121
Wessling, R. A., 65
Westin, E. H., 137
Westmoreland, T. D., 319, 366
Westphal, H. M., 143
Wetton, R. E., 217
Whang, B. C. Y., 61
Whitcomb, G., 409
White, D., 118
White, J. L., 283, 293, 294, 295
White, J. W., 201, 210
White, R., 377
White, S. A., 92
Whitfield, D. M., 351
Whitman, C. B., 152
Whitten, D. G., 366, 370
Whittington, S. G., 162, 165
Whittle, R. R., 319
Whicken, A., 116
Wideroe, T. E., 384
Widmaier, J. M., 251
Wiecko, J., 382
Wignall, C. D., 206
Wignall, G. D., 190, 192, 194, 195, 199, 200, 210
Wiklander, B., 384

Wilczek, L., 37
Wildevuur, C. R., 380
Wilding, M. A., 213, 222
Wilemon, G. M., 365
Wilemski, G., 169
Wiles, D. M., 80, 331, 333
Wilfong, D. E., 266
Wilke, W., 205
Wilkens, B., 413
Wilkes, G. L., 81, 153, 295
Wilkie, C. A., 313
Wilkins, C. W., jun., 405, 406, 407
Wilkins, R. G., 114
Wilkinson, M. C., 60, 69, 71
Wilks, A. F., 142
William, K. W., 74
Williams, A., 363
Williams, C., 165
Williams, D. F., 379, 381
Williams, E. A., 83, 153
Williams, E. C., 124
Williams, E. H., 47
Williams, F., 32
Williams, F. A., 339
Williams, G., 255
Williams, H. L., 95
Williams, J. A., 316, 363, 381
Williams, J. C., 310
Williams, J. G., 284, 288, 289
Williams, J. L. R., 399
Williams, J. W., 152
Williams, M. C., 157, 290
Williams, R., 384
Williams, R. E., 304
Williams, R. F., 310
Williams, R. J. J., 285, 387
Williams, T. J., 112
Williams, W. N., 378
Williamson, A. G., 336
Williamson, F. B., 128
Williamson, F. S., 189
Willing, R. I., 50, 179, 181
Willis, H. A., 79, 218
Willis, M. R., 380
Willson, C. G., 406, 410
Willson, A. M., 91
Wilson, C. G., 330
Wilson, D. P., 266
Wilson, J. F., 377
Wilson, K. G., 253
Wilson, M. C., 381
Wimberley, J. W., 343
Winchester, J. F., 383
Windle, A. H., 204, 394
Windsor, C., 189
Windwer, S., 167
Winhold, S., 209
Winke, W., 205
Winkeler, H. D., 373
Winnan, H. W., 152
Winnik, M. A., 161, 398
Winter, H., 315
Winter, H. H., 281, 290
Winter, H. J., 313

Wipff, G., 359
Wise, E. M., 263
Wisian-Neilson, P., 155
Wissbrun, K. F., 291, 293, 295
Wittekoek, S., 405
Witten, T. A., 162, 170
Wittmann, J. C., 216
Woessner, G., 337
Wohlfarth, Ch., 253
Wohlford, T. L., 91
Wojcik, A., 78
Woldt, R., 28
Wolf, B. A., 237, 238, 249, 252, 257, 259, 262
Wolf, C. J., 337
Wolf, E. D., 411
Wolf, P., 411
Wolf, T. M., 412
Wolfe, J. F., 86
Wolfe, R. A., 185
Wolfer, D., 152
Wolff, R. L., 371
Wolff, T., 371
Wolin, S. L., 144
Wolstenholme, J. B., 351
Wondraczek, R., 31, 34, 38
Wong, A. C., 89
Wong, E., 242
Wong, N.-C., 261
Wong, R. B. K., 102
Wong, T. C., 129
Wong-Staal, F., 137
Woo, M. L., 129
Woo, P. W. K., 114
Woo, S., 139
Woo, S. L., 144
Wood, D. G. M., 344
Wood, L. E., 361
Wood, P. J., 104
Woodbury, C. P., 256
Woodhouse, M. E., 32
Woodley, M. F., 384
Woods, G., 97
Woodward, A. E., 215
Wool, R. P., 287
Wooley, J. C., 144
Worsfold, D. J., 40, 44
Worster, P. M., 324
Wouters, G., 53
Wright, C. B., 380
Wright, J. D., 283
Wright, P. V., 208
Wright, W. W., 83, 344, 348
Wrobel, A. M., 350
Wtochowicz, A., 218
Wu, B. E., 395
Wu, D., 83
Wu, G., 86
Wu, J.-C., 7, 9
Wu, S. K., 338
Wu, S.-R., 54
Wu, W., 194
Wulff, G., 326, 328, 364
Wunderlich, B., 82, 259
Wynford-Thomas, D., 385

Author Index

Wyn-Jones, E., 108
Wynne, K. J., 299
Wypych, J., 335
Xie, Y., 4
Xu, B. Z., 386
Xu, J., 246
Xu, Y., 36
Xu, Z., 226, 242
Yablonskii, O. P., 66
Yaffe, M. B., 287
Yagi, M., 141
Yagi, T., 362
Yahiro, N., 363
Yajima, H., 100
Yakhontova, L. F., 328
Yakimenko, M. N., 412
Yakubi, Y., 342
Yamada, A., 319, 326
Yamada, F., 155
Yamada, K., 102, 222, 352, 366
Yamada, M., 412, 413
Yamada, T., 101, 208
Yamagato, S., 376
Yamaguchi, H., 53
Yamaguchi, K., 42, 330
Yamaguchi, N., 266
Yamaji, I., 33
Yamakawa, H., 159, 166
Yamamoto, K., 101, 114, 240, 336, 389
Yamamoto, M., 352, 401
Yamamoto, Y., 32, 208, 226, 362
Yamamura, K., 354
Yamamura, Z., 359
Yamana, T., 369
Yamanaka, M., 65
Yamanda, Y., 386
Yamaoka, T., 407
Yamashita, S., 34
Yamashita, T., 37, 363
Yamashita, Y., 33
Yamato, H., 323
Yamauchi, T., 337
Yamawaki, J., 372
Yamazaki, N., 42, 54, 330
Yamazaki, S., 67
Yamazaki, T., 208
Yanai, S., 96
Yanazawa, H., 406
Yang, H., 201
Yang, J., 386
Yang, V. W., 145, 146
Yang, Y., 36
Yannas, I. V., 383
Yannoni, C. S., 176, 187
Yano, S., 336
Yanogiliova, N., 84
Yanovskii, D. M., 343
Yansura, D. G., 133
Yarosh, A. A., 344
Yashchuk, V. N., 400, 404
Yashina, V. Z., 185

Yasina, L. L., 341
Yasuda, H., 148, 386
Yasuda, H. K., 386
Yasuda, K., 174
Yasufuku, S., 151
Yasui, T., 155
Yasukawa, T., 56
Yatsenko, I. V., 346
Ye, M.-L., 242
Yeadon, G., 243
Yechevskaya, L. G., 12
Yee, A. F., 83
Yeh, G. S. Y., 79, 204
Yeh, M. Y., 130
Yeh, P.-W., 283
Yemelyanov, D. N., 284
Yemel'yanov, V. N., 226
Yen, N. M., 84
Yenikolopyan, N. S., 55
Yeo, J. K., 268
Yermakov, V. I., 19
Yermakov, Y. I., 16, 18
Yildirim, A. E., 25
Ying, Q.-C., 242
Yokota, A., 410, 412, 413
Yokota, K., 354
Yokouchi, M., 84
Yokoyama, M., 79, 90, 319, 403
Yokoyama, Y., 36
Yoneda, F., 323
Yoneda, K., 413
Yonekawa, T., 376
Yoo, H. J., 281, 295
Yoon, D., 191
York, D. H., 386
Yoshida, J., 317, 321
Yoshida, M., 56, 385
Yoshida, N., 354
Yoshida, Y., 83
Yoshikawa, M., 81
Yoshimatsu, A., 366, 367
Yoshimura, M., 372
Yoshinaga, A., 37, 408
Yoshinaga, H., 368
Yoshino, K., 404
Yoshitake, J., 16
Yoshizaki, T., 166
Yosizawa, Z., 127
You, A., 86
Young, B. R., 377
Young, D. C., 385
Young, G., 107
Young, N. M., 111
Young, N. S., 140
Young, P. R., 141
Young, R. A., 143
Young, R. J., 210
Young, R. N., 40, 45
Young, W. W., 130
Yu, H., 84
Yu, T. L., 172
Yu, Y. C., 71
Yuan, L. C., 355
Yuasa, S., 383
Yuki, H., 46, 47, 180

Yuran, V. S., 332
Zabeau, F., 332
Zaboristov, V. N., 34
Zachariades, A. E., 82, 222, 295
Zadorina, E. N., 340
Zagorska, M., 75, 412
Zaikov, G. E., 343
Zainullina, A. Sh., 89
Zakharov, V. A., 12, 16, 18, 19
Zaki, A. B., 333
Zaklika, K. A., 327
Zakour, R. A., 135
Zamaraev, K. I., 362
Zambelli, A., 22, 25, 181
Zamora, F., 233
Zamotaev, P. V., 332
Zander, A. R., 384
Zanderighi, L., 319
Zane, R. M., 380
Zanette, D., 366
Zang, Q. Z., 78
Zannetti, R., 4, 87
Zannoni, G., 181
Zarian, J., 156
Zaripov, I. N., 341
Zarkhin, L. S., 349
Zarkhina, T. S., 349
Zavada, J., 356
Zavodchikova, N. N., 343
Zawadzki, S., 386
Zeldin, M., 156
Zelei, B., 154, 350
Zelenetskii, A. N., 349
Zelenev, Y. V., 340
Zelenka, T., 79
Zelenkova, T. N., 332
Zerhi, G., 218
Zero, K. M., 236
Zeronian, S. H., 80, 88, 89
Zgonnik, V. N., 40
Zhang, J., 36
Zhang, K.-C., 245
Zhang, S., 86
Zhang, S.-Q., 86
Zhang, S.-Y., 246
Zhang, X., 71, 147
Zhang, W. G., 386
Zhang, Z. M., 149
Zhao, X., 207
Zhdanov, A. A., 150, 153
Zheng, Y. C., 341
Zhikuan, C., 250
Zhamikna, T. P., 244
Zhorina, L. A., 349
Zhorov, B. S., 392
Zhou, Q., 408
Zhou, Q. S., 320
Zhou, Z. L., 290
Zhoung, C.-S., 238
Zhubanov, B. A., 89
Zhukova, T. I., 90
Zhuravleva, I. V., 339
Ziabicki, A., 219
Ziegast, G., 100

Ziegeldorf, R., 221
Ziemiecki, H., 371
Zifferev, G., 56
Zilkha, A., 77
Zimm, B. H., 168, 169, 235
Zimmerhackl, E., 404
Zimmermann, H., 346, 347
Zingg, Q., 376
Ziolkowski, J. J., 366
Zipp, A. P., 319
Ziska, P., 111
Zislina, S. S., 350

Ziv, M. H., 97
Zivny, A., 167, 240, 251
Zolczer, L., 379
Zoller, W., 26
Zollinger, H., 356
Zoppetti, G., 128
Zoran, A., 365
Zotikov, E. G., 52, 56
Zrinyi, M., 251
Zubov, V. P., 57, 184
Zugenmaier, P., 100
Zugrăvescu, I., 90

Zuikov, A. V., 70
Zupan, M., 320
Zupan, Y. J. M., 362
Zupancic, B. G., 314
Zupko, H. M., 251
Zurakowska-Orszagh, J., 91
Zussman, M. P., 37
Zuznetsova, S. V., 334
Zvara, I., 53
Zvonkova, Ye. M., 84
Zwanziger, H., 33
Zytner, Ya. D., 72, 73

RAYMOND H. FOGLER LIBRARY
DATE DUE

BOOKS ARE SUBJECT TO RECALL AFTER TWO WEEKS

JAN 21

DEC 18 1987